FOUNDATIONS OF PARASITOLOGY

"They do certainly give strange and new-fangled names to diseases."
PLATO

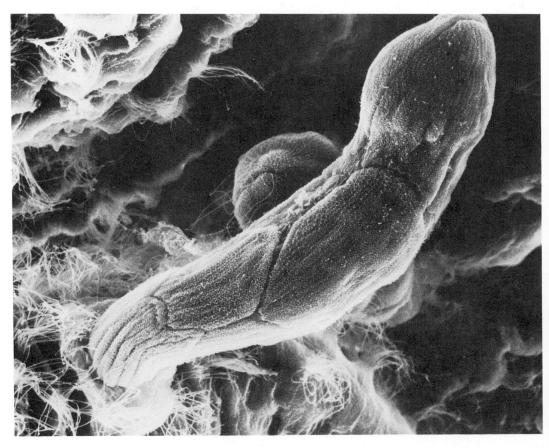

Miracidia of *Fascioloides magna* penetrating the epithelium of a snail. One of the organisms is partially hidden by the miracidium in the central foreground. The outlines of the ciliated epithelial cells, which are about to be shed, are clearly distinguishable. Note also the longer cilia of the snail's epithelium (left foreground). (Courtesy William H. Coil.)

FOUNDATIONS OF
PARASITOLOGY

GERALD D. SCHMIDT

Professor of Zoology/Parasitology
University of Northern Colorado
Greeley, Colorado

LARRY S. ROBERTS

Professor of Zoology
University of Massachusetts
Amherst, Massachusetts

with 889 illustrations

THE C. V. MOSBY COMPANY

Saint Louis   1977

The C. V. Mosby Company
11830 Westline Industrial Drive, St. Louis, Missouri 63141

Library of Congress Cataloging in Publication Data

Schmidt, Gerald D   1934-
   Foundations of parasitology.

   Bibliography: p.
   Includes index.
   1. Parasitology. I. Roberts, Larry S.,
1935-   joint author. II. Title. [DNLM:
1. Parasites. 2. Parasitology. QX4 S351f]
QL757.S35   591.5′24   76-30335
ISBN 0-8016-4345-7

TS/CB/CB 9 8 7 6 5 4 3   02/C/246

PREFACE

This book is intended for use in introductory courses in parasitology. Although each teacher will address the subject differently, we believe that the foundation material for any approach can be found here. The format is basically traditional, with up-to-date contributions from electron microscopy, biochemistry, immunology, and other disciplines woven into the text. We believe that a sound knowledge of the biology, morphology, and ecology of parasites is essential to understanding and appreciating the more subtle aspects of the host-parasite relationship, pathogenesis, and epidemiology.

In the chapters that follow, each group of parasites is described and differentiated from all others, and any peculiarities of the group are discussed. For illustrations of parasitic types and principles, we have emphasized species of medical or veterinary importance, for we all are interested in diseases that may infect our own species or affect our pocketbooks. Many parasites that do not infect humans or domestic animals are still of considerable scientific interest, however, and we have included numerous examples of these. Many of the organisms discussed are available from commercial suppliers, which should aid in planning laboratory exercises. Because of space limitations and because courses in medical entomology are available at most universities and colleges, we have not included chapters on medically important or other parasitic arthropods. An exception is the parasitic Crustacea, which provide fantastic examples and insights into parasitism, but which often are neglected in both medical entomology and invertebrate zoology courses. We feel that coverage of parasitic Crustacea is appropriate in parasitology, and we hope that students and teachers will enjoy our introduction to these organisms.

Physiological adaptations of parasites are essential requirements for the exploitation of their modes of life. Therefore, where possible we have integrated physiological information within each section, organizing a complete foundation upon which the student can build. Some topics, such as energy metabolism and principles of immunology, are offered as separate units, which can be included or deleted at the pleasure of the instructor. Though necessarily brief, we hope that these sections will be understandable to students who have no prior training in the areas.

Any advanced undergraduate or graduate student in biology should have the following prerequisite background to use this textbook: a year of biology, including basic introductory invertebrate zoology, and chemistry through organic, preferably with a first course in biochemistry.

In our years of teaching we have found that students want to know what treatment, if any, is available for the diseases they study. Accordingly, we have included mention of chemotherapy at appropriate places in the book. However, we have tried to name the classes of drugs, or in some cases the drugs of choice, rather than suggesting clinical procedures. We want the reader to have a feeling for what might be done for those who suffer from parasitic diseases.

The world literature in parasitology is already so vast and is expanding so rapidly that it is impossible to review it all. Almost any chapter, or even many parts of chapters, could themselves be expanded to book length, but we hope that we have omitted nothing essential to the purpose of this basic text. We have tried to rely extensively on recently published information to be as timely as possible, but some material doubtless will be dated even as the manuscript goes to press. This is an inevitable consequence in a vital and active field. In our effort to reach a high level of accuracy, we have had each chapter read and corrected by authorities on the various topics. We sincerely hope that any remaining errors are minimal; they are, of course, our responsibility.

No one could hope to write a book such as this without the assistance of many people. We have drawn shamelessly on the expertise of numerous colleagues and friends, whose experiences have exceeded ours in certain specialized fields. Experts who have contributed their time and special knowledge to this book are too numerous to list in entirety, but we especially wish to thank Lawrence R. Ash, W. S. Bailey, Wilbur Bullock, Edelberto J. Cabrera, Donald W. Duszynski, Reino S. Freeman, Arthur G. Humes, Z. Kabata, Richard Komuniecki, Delane C. Kritsky, Robert E. Kuntz, John S. Mackiewicz, Ralph Muller, George S. Nelson, Brent B. Nickol, Robert L. Rausch, J. Teague Self, and Robert B. Short. B. M. Honigberg read all chapters on protozoa and made his personal library available for our use. Many persons and publications gave freely of published and unpublished illustrations. These are greatly appreciated and are credited along with the relevant pictures.

Andra Schmidt transformed our grammar and syntax, making sense out of nonsense, and typed the first two drafts of this book. Joann Sharp typed the final draft, bringing order to the chaotic cut-and-paste copy that we presented her. We offer grateful thanks to them.

Gerald D. Schmidt
Larry S. Roberts

CONTENTS

FOUNDATIONS OF PARASITOLOGY

Chapter 1

INTRODUCTION TO PARASITOLOGY

Big fleas have little fleas
upon their backs to bite 'em,
Little fleas have lesser fleas
and so, ad infinitum.

SWIFT

Few people realize that there are far more kinds of parasitic than nonparasitic organisms in the world. Even if we exclude the viruses and rickettsias, which are all parasitic, and the many kinds of parasitic bacteria and fungi, the parasites are still in the majority. The parasitic way of life, generally speaking, is a highly successful one, for it evolved independently in nearly every phylum of animals, from protozoans to arthropods and chordates, as well as in many plant groups. Organisms that are not parasites are usually hosts. Humans, for example, are hosts to over 100 kinds of parasites, again not counting viruses, bacteria, and fungi. It is unusual to examine a domestic or wild animal without finding at least one species of parasite on or within it. Even animals reared under strict laboratory conditions are commonly infected with protozoans or other parasites. Often the parasites themselves are the hosts of other parasites. It is no wonder then that the science of parasitology has been developed out of efforts to understand the parasites and their relationships with their hosts.

THE RELATIONSHIP OF PARASITOLOGY TO OTHER SCIENCES

Parasitology has passed through a series of stages in its history, each of which, today, is an active discipline in its own right. The first and most obvious stage is the discovery of the parasites themselves. This undoubtedly began in the shadowy eons of prehistory, but the ancient Persians, Egyptians, and Greeks recorded their observations in such a way that later generations could, and often still can, recognize the animals they were writing about. The discovery and naming of unknown parasites, their study, and their arrangement into a classification is an exciting and popular branch of parasitology today.

When people became aware that parasites were troublesome and even serious agents of disease, they began a continuing campaign to heal the infected and eliminate the parasites. The later discovery that other animals could be the **vectors,** or means of dissemination, of parasites opened the door to other approaches for control of parasitic diseases.

Development of better lenses led to basic discoveries in cytology and genetics of parasites that are applicable to all of biology. And, in the twentieth century, refined techniques in physics and chemistry have contributed much to our knowledge of host-parasite relationships. Some of these have added to our understanding of basic biological principles and mechanisms, such as the discovery of cytochrome and the electron transport system by David Keilin in 1925[3] during his investigations on parasitic worms and insects. Today, biochemical techniques are widely used in studies of parasite metabolism, immunology, serology, and chemotherapy. The advent of the electron microscope has resulted in many new discoveries at the subcellular level. Thus, parasitologists employ the tools and concepts of many scientific disciplines in their research.

Parasitology today usually does not include virology, bacteriology, and mycology, because these sciences have developed into disciplines in their own right. Exceptions do occur, however, for it is not uncommon for parasitological research to overlap these areas. Medical entomology, too, has branched off as a separate discipline, but it still is a subject of paramount importance to the parasitologist, who must understand the relationships between arthropods and the parasites they harbor and disperse.

PARASITOLOGY AND HUMAN WELFARE

Human welfare has suffered mightily through the centuries because of parasites. Fleas and bacteria conspired to destroy one third of the population of Europe in the seventeenth century, and malaria, schistosomiasis, and African sleeping sickness have sent untold millions to their graves. Even today, after successful campaigns against yellow fever, malaria, and hookworm in many parts of the world, parasitic diseases in association with nutritional deficiencies are the primary killers of humanity. A recent summary of the worldwide prevalence of selected parasitic diseases shows there are more than enough existing infections for every living person to have one, were they evenly distributed.[4]

Disease category	Number of human infections
All Helminths	3.5 billion
Hookworms	700 million
Schistosomiasis	200 million
Onchocerciasis	40 million
Malaria	25 million

These, of course, are only a few of the many kinds of parasites that infect humans. This points out that parasitic diseases are an important fact of life for many people. The majority of the more serious infections are in the so-called tropic zones of the earth, so most dwellers within the temperate regions are unaware of the magnitude of the problem. For instance, of the approximately 60 million total annual worldwide deaths from all causes, 30 million are children under 5 years. Half of these, 15 million, are attributed to the combination of malnutrition and intestinal infection.[2]

However, the notion held by the average person that we are free of worms in humans in the United States is largely an illusion—an illusion created by the fact that the topic is rarely discussed. It is rarely discussed because of our attitudes that worms are not the sort of thing that refined people talk about, the apparent reluctance of the media to disseminate such information, and the fact that poor people are the ones most seriously affected. Some estimates place the number of children in the United States infected with worms at about 1 million, though this is certainly a gross underestimation if one includes such parasites as pinworms (*Enterobius vermicularis*). Only occasionally is the situation accurately reflected in the popular press: "If I brought in a jar of some child's roundworms, a great many people would be thoroughly nauseated. It is the sort of thing that is left unsaid, undiscussed and unreported throughout the U.S. A good note to close on! Let's not disturb folks. The thought of that jar upsets refined people. Things should be kept in their place, in the . . . well, let's skip it. Sleep well, good people—only a few million kids are affected."[6]

But even though there are many "native-born" parasite infections in the United States, many "tropical" diseases are imported within infected humans coming from endemic areas. After all, one can travel halfway around the world in a day or two. Many thousands of immigrants who are infected with schistosomes, malaria, hookworms, and other parasites—some of which are communicable—currently live in the United States. It is estimated that there are about 100,000 cases of *Schistosoma mansoni* (see Chapter 17) in the continental United States that originated in Puerto Rico. Servicemen returning from abroad often bring parasite infections with them. In 1971 the Center for Disease Control reported 3,047 cases of malaria in the United States, about three fourths of which were acquired in Vietnam. There are still viable infections of filariasis in ex-servicemen who contracted the disease in the South Pacific 25 years ago! A traveler may become infected during a short layover in an airport, and many pathogens find their way into the United States as stowaways on or in imported products. Small wonder, then, that "exotic" diseases confront the general practitioner with more and more frequency. The family physician of one of the authors claims to have treated virtually every major parasitic disease of humans during the years of his practice in Amherst, Massachusetts.

There are other, much less obvious, ways in which parasites affect all of us, even those in comparatively parasite-free areas. Primary among these is malnutrition, as the result of inefficient use of arable land and the inefficient use of food energy. Only 3.4 billion of the 7.8 billion acres of total potentially arable land in the world is now under cultivation.[2] Much of the remaining 4.4 billion acres cannot be developed because of malaria, trypanosomiasis, schistosomiasis, and onchocerciasis. In Africa alone, there is an area of land equal to the size of the United States where people cannot live and grow livestock due to trypanosomes. How many starving people could be fed if this land were cultivated? As many as half of the world's population today is undernourished. The population will double again in 35 years. It is impossible to ignore the potentially devastating effects that worldwide famine will have on all humankind.

Even where food is being produced it is not always used efficiently. Considerable caloric energy is wasted by fevers caused by parasitic infections. Heat production of the human body increases about 7.2% for each degree rise in Fahrenheit. A single, acute day of fever due to malaria requires approximately 5,000 calories, or an energy demand equivalent to 2 days of hard manual labor. To extrapolate, in a population with a 2,200 caloric average diet per day, if 33% had malaria, 90% had a worm burden, and 8% had active tuberculosis (conditions that are repeatedly observed), there would be an energy demand equivalent to 7,500 tons of rice per month per million people over and above normal requirements. That is a waste of 25% to 30% of the total energy yield from grain production in many societies.[5]

Another cause of energy loss is malabsorption of digested food. This is a common occurrence in parasitic infections. It is difficult to quantify this loss, but it undoubtedly is highly significant, especially in those who are undernourished to begin with.

People create much of their own disease

conditions because of high population density and subsequent environmental pollution. Despite great progress in extending water supplies and sewage disposal programs in developing countries, not more than 10% or 15% of the world population is thus served. Usually an adequate water supply has first priority, with sewage disposal running a poor second. When one recalls that most parasite infections are caused by ingesting food or water contaminated with human feces, it is easy to understand why 15 million children die of intestinal infections every year.

At first glance it seems incongruous that the nations that suffer the most from disease are also the nations whose populations are undergoing the most rapid growth. The world's population has doubled three times in the past 200 years, and it will double again in the next 35 years, from 3.5 billion to 7 billion. During this time Latin America will add 400 million, and Asia will double its 1.6 billion to 3.2 billion, which together equal the total current world population. The efforts at family planning are beginning to be felt in several countries, especially those where disease is at a minimum. But what can we say to a mother who wants to have seven children so that three can survive? A Johns Hopkins University study on family planning motivation confirms the importance of child survival to the sustained practice and acceptance of family planning.[2] The parasitologists' role then, together with other medical disciplines, is to help achieve a lower death rate. But it is imperative that this be matched with a concurrent lower birth rate. If not, we are faced with the "parasitologist's dilemma," that of sharply increasing a population that cannot be supported by the resources of the country. For instance, malaria costs the government of India about $21 million a year in death, treatment, and loss of manpower. It has been estimated that malaria can be eradicated from India within 2 years at a cost of about $14 million. But it is a cheerless prospect to contemplate the effect of a sudden increase in population under current levels of birth rate and standards of nutrition in that country. Dr. George Harrar, President of the Rockefeller Foundation, observed: "It would be a melancholy paradox if all the extraordinary social and technical advances that have been made were to bring us to the point where society's sole preoccupation would of necessity become survival rather than fulfillment." Harrar's paradox is already a fact for half of the world. Parasitologists have a unique opportunity to break the deadly cycle by contributing to the global eradication of communicable diseases, while, at the same time, making possible more efficient use of the earth's resources.

PARASITES OF DOMESTIC AND WILD ANIMALS

Both domestic and wild animals are subject to a wide variety of parasites that demand the attention of the parasitologist. Although wild animals are usually infected with several species of parasites, they seldom suffer massive deaths, or **epizootics,** because of the normal dispersal and territorialism of most species. But domesticated animals are usually confined to pastures or pens year after year and often in great numbers, so that the parasite eggs, larvae, and cysts become very dense in the soil, and the burden of adult parasites within each host becomes devastating.[1] For example, the protozoans known as the coccidia thrive under crowded conditions; they may cause up to 100% mortality in poultry flocks, 28% reduction in wool in sheep, and 15% reduction in weight of lambs.[5] In 1965 the U.S. Department of Agriculture estimated the annual loss in the United States as the result of coccidiosis of poultry alone at about $45 million. Many other examples can be given and some are discussed later in this book. Agriculturists, then, are forced to expend much money and energy in combating the phalanx of parasites that attack their animals. Thanks to the continuing efforts of parasitologists all over the world the identifications and life cycles of most parasites of domestic animals are well known. This knowledge, in turn, exposes weaknesses in the biology of these pests and suggests possible methods of control. Similarly, studies on the biochemistry of the organisms continue to suggest modes of action for chemotherapeutic agents.

Less can be done to control parasites of wild animals. While it is true that most wild animals tolerate their parasite burdens fairly well, the animals will succumb when crowded and suffering from malnutrition —just as will domestic animals and humans. For example, the range of the big horn sheep in Colorado has been reduced to a few small areas in the high mountains. They are unable to stray from these areas because of human pressure. Consequently, lungworms, which probably have always been present in big horn sheep, have so increased in numbers that in some herds no lambs survive the first year of life. These herds seem destined for quick extinction unless a means for control of the parasites can be found in the near future.

A curious and tragic circumstance has resulted in the destruction of large game animals in Africa in recent years. These animals are heavily infected with species of *Trypanosoma,* a flagellate protozoan of the blood. The game animals tolerate infection quite well but function as **reservoirs** of infection for domestic animals, which quickly succumb to trypanosomiasis. One means of control employed is the complete destruction of the wild animal reservoirs themselves. Hence, their parasites are the indirect cause of their death. It is hoped that this parasitological quandry will be solved in time to save the magnificent wild animals.

Still another important aspect of animal parasitology is the transmission to humans of parasites normally found in wild and domestic animals. The resultant disease is called a **zoonosis.** Many zoonoses are rare and cause little harm, but some are more common and of prime importance to public health. An example is trichinosis, a serious disease caused by a minute nematode, *Trichinella spiralis.* This worm exists in a **sylvatic cycle** that involves rodents and carnivores and in an **urban cycle** chiefly among rats and swine. People become infected when they enter either cycle, such as by eating undercooked bear or pork. Another zoonosis is echinococcosis or hydatid disease. Here, humans accidentally become infected with larval tapeworms when they ingest eggs from dog feces. *Toxoplasma,* which is normally a parasite of felines and rodents, is now known to cause many human birth defects.

New zoonoses are being recognized

from time to time. It is the obligation of the parasitologist to identify, understand, and suggest means of control of such diseases. The first step is always the proper identification and description of existing parasites, so that other workers can recognize them and refer to them by name in their own work. Thousands of species of parasites of wild animals are still unknown and will occupy the energies of taxonomists for many years to come.

CAREERS IN PARASITOLOGY

It can truly be said that there is an area within parasitology to interest every biologist. The field is large and has so many approaches and subdivisions that anyone who is interested in biological research can find a lifetime's career in parasitology. It is a satisfying career, for one knows that each bit of progress made, however small, contributes to our knowledge of life and to the eventual conquering of disease. As in all scientific endeavor, every major breakthrough depends on many small contributions made, usually independently, by individuals around the world.

The training required to prepare a parasitologist is rigorous. Modern researchers in parasitology are well grounded in physics, chemistry, and mathematics, as well as biology from the subcellular through the organismal and populational levels. Certainly they must be firmly grounded in medical entomology, histology, and basic pathology. Depending on their interests, they may require advanced work in physical chemistry, immunology and serology, genetics, and systematics. Most parasitologists hold a Ph.D. or other doctoral degree, but significant contributions have been made by persons with a master's or

bachelor's degree. Such intense training is understandable, for parasitologists must be familiar with the principles and practices that apply to over a million species of animals; in addition they need thorough knowledge of their fields of specialty. Once they have received their basic training, parasitologists continue to learn during the rest of their lives. Even after retirement, many continue to be active in research for the sheer joy of it. Parasitology does indeed have something for everyone.

REFERENCES

1. Ershov, V. S. 1956. Parasitology and parasitic diseases of livestock. State Publishing House for Agricultural Literature, Moscow.
2. Howard, L. M. 1971. The relevance of parasitology to the growth of nations. J. Parasitol. 57 (Sect. 2, Part 5):143-147.
3. Keilin, D. 1925. On cytochrome, a respiratory pigment, common to animals, yeast and higher plants. Proc. R. Soc. B 98:312-339.
4. Le Riche, W. H. 1967. World incidence and prevalence of the major communicable diseases. In Health of mankind. Little, Brown and Co., Boston. pp. 1-42.
5. Pollack, H. 1968. Disease as a factor in the world food problem. Institute for Defense Analysis, Arlington, Va.
6. TRB. 1969. Sleep well. The New Republic 160(Mar. 6):6.

SUGGESTED READING

Anonymous. 1970. Careers in parasitology, medical zoology, tropical medicine. Am. Soc. Parasitol.

Baer, J. G. 1971. Parasitology in the world today. J. Parasitol. 57(Sect. 2, part 5):136-138.

Cheng, T. C. 1973. The future of parasitology: one person's view. Bios 44:163-171.

Mueller, J. F. 1961. From rags to riches, or the perils of a parasitologist. The New Physician 272:46-50. (Contemporary thoughts by three men who have made great impacts on parasitology.)

Chapter 2

BASIC PRINCIPLES AND CONCEPTS

The host is an island invaded by strangers with different needs, different food requirements, different localities in which to raise their progeny.

TALIAFERRO

DEFINITIONS

The science of parasitology is largely a study of **symbiosis,** especially the form known as parasitism. Although some authors, especially in Europe, restrict the term "symbiosis" to relationships wherein both partners benefit, we prefer to use the term in a wider sense, as originally proposed by de Bary in 1879.[6] We consider any interaction between two organisms to be symbiotic, in keeping with the meaning of the word, "living together" as contrasted with "free-living." The relationship may be of long or short duration, and it may benefit one or both **symbionts** or neither. Usually the symbionts are of different species, but not necessarily. The study of all aspects of symbiosis is called **symbiology.**

For the sake of convenience we can subdivide symbiosis into several categories, based on the amount of interdependence of the symbionts. It should be recognized that not all relationships fit obviously into one category or another, for they often overlap each other, and further, the exact relationship cannot be determined in some cases. Not all authors agree on the definitions of these categories, and some subdivide them further. We have selected definitions that seem to us to be concise and meaningful.

Predation. Predation is obviously a short-term relationship in which one symbiont, the **predator,** benefits at the expense of the other symbiont, the **prey.** Two lives have crossed and interacted to the betterment of one and the detriment of the other. The point here is that the predator kills the prey outright and does not subsist on it while it is alive.

Phoresis. Phoresy exists when two symbionts are merely "traveling together." Neither is physiologically dependent on the other. Usually, one **phoront** is smaller than the other and is mechanically carried about by its larger companion (Fig. 2-1). Examples would be bacteria on the legs of a fly or fungous spores on the feet of a beetle.

Mutualism. In this relationship the partners are called **mutuals** because both members benefit from the association. Mutualism is usually obligatory, for in most cases the mutuals have evolved physiological dependence on one another to such a degree that one cannot survive without the other. A good example is the termite and its intestinal protozoan fauna. Termites cannot digest cellulose fibers because they do not secrete the enzyme cellulase. However, a myriad of flagellate protozoa, which dwell within the termite's gut, synthesize cellulase freely and are able to utilize as nutrient the wood eaten by the termites. The termite is nourished by the fermentation products excreted by the protozoa. That the protozoa are necessary to the termite can be shown by defaunating the insects (killing the protozoa by subjecting their hosts to elevated temperature or oxygen tension); the termites then die, even with plenty of choice wood to eat. The protozoa benefit by living in a stable, secure environment, constantly supplied with food, and by being provided with a low oxygen environment, since they are obligate anaerobes. The termite-flagellate association is but the most often cited example of insect-microbe mutualism. A wide variety of insects have bacteria or yeast-like organisms in their gut or other

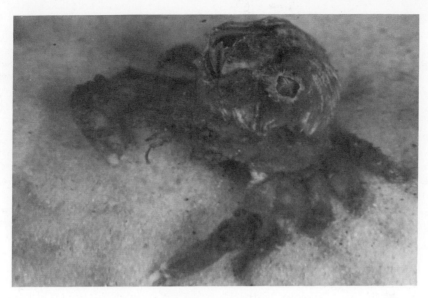

Fig. 2-1. Barnacles attached to the carapace of a crab. Sponges and coelenterates are attached to both, providing several examples of phoresis. (Photograph by Warren Buss.)

organs, and these are physiologically necessary for the insects in almost all cases studied, apparently furnishing vitamins or other micronutrients. Some insects even have specialized organs (**mycetomes**) where they "keep" their microbes, and the symbionts are passed to the progeny transovarially. Other examples of symbiotic mutualism are the alga and fungus that together form a lichen, and the relationship between a cow and the flora and fauna of its rumen.

One form of mutualism that is not obligatory is usually called **cleaning symbiosis.** In this instance certain animals, called cleaners, remove ectoparasites, injured tissues, fungi, and other organisms from a cooperating host. For example, often one or several cleaners establish cleaning stations at a particular location, and fish to be cleaned visit that station repeatedly to enjoy the services of the cleaners. The fish to be cleaned may remain immobile at the cleaning station while the cleaners graze its external surface and enter its mouth and branchial cavity with impunity. There is some evidence that such associations may be in fact obligatory; when all cleaners are carefully removed from a particular area of reef, for example, all the other fish leave too.[13] Some terrestrial cleaning associations are known; two examples are the cleaning of a crocodile's mouth by the Egyptian plover and the cleaning of the rhinoceros by tick birds. Some excellent accounts of a wide variety of mutualistic and related associations are found in the volumes edited by Henry[14] and Cheng.[4]

Commensalism. When one symbiont, the **commensal,** benefits from its relationship with the **host,** but the host neither benefits nor is harmed, the condition is known as commensalism. When the commensal is on the surface of its host, it is an **ectocommensal;** when internal, it is an **endocommensal.** The term means "eating at the same table," and most, but certainly not all, examples of commensalism involve the commensal feeding on unwanted or unusable food captured by the host.

Pilot fishes *(Naucrates)* and remoras (Echeneidae) are often cited as examples of commensals. A remora is a slender fish whose dorsal fin is modified into an adhesive organ, with which it attaches to large fish, turtles, and even submarines! It gets free rides this way and perhaps some crumbs left over when its host makes its kill, but in no way does it harm the host or rob it of food. In fact, it has now been found that some species of remoras perform important cleaning functions for their hosts, feeding at least partially on the host's parasitic copepods.[8] Thus, remoras, the "classic" commensals, are often mutuals. Depending on the species of echeneid and its degree of specialization, the association may be more or less **facultative,** the remora being able to leave its host at will. Other species are more specialized; for example, *Remoropsis pallida,* at certain stages in its life, lives entirely in the branchial chambers of marlin.

An example of an obligatory commensal is *Entamoeba gingivalis,* an ameba that lives in the mouth of humans. Here it feeds on bacteria, food particles, and dead epithelial cells, but never harms the tissues of its host. It cannot live anywhere else, and in fact has no cyst stage to withstand life outside the buccal cavity; it is transmitted from person to person by direct contact. Other, often cited, examples of commensalism are the clownfish that live among the tentacles of sea anemones (although this association too has its mutualistic aspects—if the clownfish lures other fish into the anemone's waiting tentacles); the pearlfish that hides within the respiratory tree of sea cucumbers and the coelomic cavity of certain starfish; and some of the bacteria that live in the intestines of humans. Humans harbor several species of commensal protozoans, and for this reason parasitologists must be able to distinguish between commensal and parasitic species. It is not always easy to determine whether or not a symbiont is harming its host; tapeworms universally are referred to as parasites, yet in at least some cases they have no known ill effect and might possibly be regarded as commensals.[15]

Parasitism. When a symbiont actually does harm its host, it is then a **parasite.** It may harm its host in any of a number of ways: by mechanical injury, such as boring a hole into it; by eating or digesting and absorbing its tissues; by poisoning the host with toxic metabolic products; or simply by robbing the host of nutrition. Most parasites inflict a combination of these conditions on their hosts.

If a parasite lives on the surface of its host, it is called an **ectoparasite;** if internal, it is an **endoparasite.** Most parasites are **obligatory parasites,** that is, they must spend at least a part of their lives as parasites to survive and complete their life cycles. However, many obligatory parasites have free-living stages outside any host, including some periods of time in the external environment within a protective egg shell or cyst. **Facultative parasites** are not normally parasitic but can become so, at least for a time, when they are accidentally eaten or enter a wound or other body orifice. Two examples of facultative parasitism are "vinegar eels"—small roundworms that normally live in decaying fruit but, when eaten, can become true parasites—and the larva of a latrine fly, *Fannia scalaris,* which can live in the ear canal.

When a parasite enters or attaches to the body of a species of host different than its normal one, it is called an **accidental** or **incidental** parasite. For instance, it is common for nematodes, normally parasitic in insects, to live in the intestines of birds or for a rodent flea to bite a dog or human. Accidental parasites usually are unable to stay long on, or live long in, the wrong host. While there, though, they may add a good deal of confusion to the life of a parasite taxonomist!

Some parasites live their entire adult lives within or on their hosts and may be called **permanent** parasites, while a **temporary** or **intermittent** parasite, such as a mosquito or bedbug, only feeds on the host and then leaves. Temporary parasites are often referred to as **micropredators,** in recognition of the fact that they usually "prey" on several different hosts (or the same host individual at several discrete times).

Of course, a parasite sometimes kills its host, but it is clearly not a selective advantage, since the parasite's life also is thereby terminated. To produce little pathogenesis is often regarded as one mark of a well-adapted parasite.

Definitions of hosts

Hosts also are placed into different categories. The host in which a parasite reaches sexual maturity and reproduction is termed the **definitive host.** If there is no sexual reproduction in the life of the parasite, such as an ameba or trypanosome, we arbitrarily call the host we believe is most important the definitive host. An **intermediate host** is one in which some development of the parasite occurs but in which it does not reach maturity. Hence, in the case of the malarial organism, *Plasmodium,* the mosquito is the definitive host, and humans or other vertebrates are the intermediate hosts.

When a parasite enters the body of a host and does not undergo any development but continues to stay alive and be infective to a definitive host, the host is called a **paratenic,** or **transport,** host. Paratenic hosts are often useful, or even necessary, for completion of the life cycle of the parasite, for they may bridge an ecological gap between the intermediate and definitive hosts. An owl may be the definitive host of a thorny-headed worm, while an insect is the intermediate host. The parasite might have little chance of being eaten by the owl while in the insect, but when a shrew eats the insect and the larval worm encysts in the shrew's mesentery, it stands a better chance of a happy life in the alimentary tract of the owl. Further, the shrew might accumulate large numbers of larvae before it becomes an owl's dinner, thereby increasing the chances of both sexes of a dioecious parasite sharing a common host.

Some parasites can live and develop normally in only one or two species of host. These exhibit high **host specificity.** Others have low host specificity or something in between. For example, the pork tapeworm, *Taenia solium,* apparently can mature only in humans, so it has absolute host specificity, while the trichina worm, *Trichinella spiralis,* seems to be able to mature in any warm-blooded vertebrate. If an animal *other* than a human is normally infected with a parasite that can also infect humans, we call that animal a **reservoir host,** even if the animal is a normal host of the parasite. It is a reservoir for a zoonotic infection to people. Examples are the rat with the trichina and dogs, cats, and armadillos with the agent of Chagas' disease, *Trypanosoma cruzi.*

Finally, there are many examples of parasites hosting other parasites, a condition known as **hyperparasitism.** Examples include *Plasmodium* in a mosquito, a tapeworm larva in a flea, an ameba within an opalind protozoan, and a monogenetic trematode *(Udonella)* on a copepod parasite of fish.

In nearly all cases of parasitism the host is a different species than the parasite. Exceptions do occur, however, as in the case of the nematode parasite of rats, *Trichosomoides,* where the male lives its mature life within the uterus of the female worm, obtaining its nourishment from her tissues. This is also the case with *Gyrinicola japonicus,* a nematode parasite of frogs, and in the echiurid worm, *Bonellia viridis,* the female of which is free-living. An even stranger relationship has evolved in some species of anglerfish where the male bites the skin of the female and sucks her blood and tissue fluids for nourishment. Eventually they grow together, and he shares her bloodstream! *Syngamus trachea,* a common hookworm of birds, is known to pierce the body wall and feed on the fluids of other worms of the same species. This might better be called predation, but specimens are found with healed wounds and so were not killed by the attack.

A very obvious example of parasitism within the same species is the **mammalian embryo** and **fetus.** Here we have an organism obtaining its nourishment at the expense of its host, poisoning its host with its metabolic wastes, and physically damaging its host at birth, occasionally even killing it. And for some time after birth, it continues to suck secretions from its mother-host's body, further robbing her of vital calcium, protein, and fat.

The ultimate in intraspecific parasitism would be that of a tumor. Here self (tumor) parasitizes self (host), often very much to the detriment of the host, such as a malignancy, or with only minor inconvenience to the host, such as a wart. The tumor may be stimulated by a virus, but it is still the tumor itself that harms the host.

Modern authors increasingly have realized the utility of de Bary's distinction, but for many people, symbiosis still means mutualism. Nevertheless, symbiosis has evolutionary, morphological, ecological, and physiological implications, whether mutualistic, commensalistic, or parasitic, that a free-living habitus does not.

It must be stressed that definitions are always arbitrary, and when we construct pigeonholes to receive descriptions of situations in the real world, we must not be dismayed to find one or another situation that does not quite fit our assignment. Thus, we may cite symbiotic associations with greater or lesser dependence, of greater or lesser duration, or with some grading over into free-living. Certainly many symbiotic associations cannot be specified with certainty as to what the effects on the hosts are; an apparent case of commensalism may have damaging effects on the host that are subtle, have not been observed, or are only present in certain circumstances. Conversely, a case of assumed parasitism may, on closer inspection, turn out to be commensalism. But definitions are necessary to communication; therefore, we must strive for definitions that are concise, meaningful, and, above all, useful.

EVOLUTION OF PARASITISM

The question of the evolution of parasitism is, as in all speculations regarding evolution, impossible to answer definitely. However, many indirect indications that are impossible to ignore suggest several factors involved in past and continuing adaptations to a parasitic mode of life. Paramount among these is the process of preadaptation, the touchstone of evolutionary philosophy. In a preadapted organism, accumulated mutations and other chance genetic changes give it the potential to live in a different environment. Such potentials may never be realized, but if *by chance* the organism finds itself in an environment for which it is preadapted, it may successfully invade and survive in an entirely new niche.

Such preadaptation may be morphological or physiological and may also allow such organisms to pass from a generalized type of environment to a more specialized one. For example, arthropods of several types may have fed on carrion that was also found and eaten by primitive humans. Any arthropod eaten along with the carrion may well have been digested, but if the colony of pinworm nematodes in its hind gut were preadapted to survive at a higher temperature, they might have then colonized the hind gut or cecum of the human. This isolated population could no longer share the gene pool with its arthropod-dwelling relatives and would thus be free to undergo genetic drift and become different from them. Although the

likelihood of such an event happening seems chancy, at best, it has been such a successful process that nearly every species of living organism has several kinds of parasites peculiar to it. The result is more species of parasites than of nonparasites.

Of course, the immense number of parasites today is the result of other processes as well. Once in the relatively stable environment of a host, selective pressure by the environment is decreased. Both beneficial and nonharmful mutations could accumulate in the gene pool, preadapting the species to invade a new host, organ, or tissue successfully should the opportunity arise. It is also conceivable that a parasite could become nonparasitic, but this probably never occurs, because the organism will have become too specialized, and evolution generally does not progress from the more specialized to the more generalized. The parasite could not compete with well-adjusted organisms already occupying external niches.

The probability that two organisms will establish a host-parasite relationship depends largely on ecological and behavioral factors. In order for a prospective parasite to succeed it must come into frequent contact with a potential host. The behavior of both must favor such contact. Ultimately, the distribution of closely related parasites reflects the opportunities available to them when they became parasitic. This can lead to similar species parasitizing phylogenetically unrelated hosts. For instance, the nematode genus *Molineus* has species in primates, insectivores, rodents, and carnivores, the ancestors of which were available to the ancestors of the worms at the time of their radiation.

Undoubtedly, parasitism arose in many different ways at many different times. Ectoparasites may well have evolved from nonparasitic omnivores or predators or from ancestors that sought out body secretions from nearby animals. A hypothetical example of the latter is seen in certain moths of the family Noctuidae in southeast Asia. Some species feed on plant secretions, while others, such as *Calpe thalictri*, can pierce ripe fruit to suck the sweet juices inside.[2] One species, *Lobocraspis goiseifusa,* gathers around the eyes of a large mammal, such as a cow, and sucks tears

Fig. 2-2. Eight *Lobocraspis goiseifusa* (Lepidoptera, Noctuidae) suck tears from the eye of a banteng, *Bos banteng* in northern Thailand. Note the proboscis of each moth extended to feed at the eye perimeter. (Photograph by Hans Bänziger.)

Fig. 2-3. A noctuid moth, *Calpe eustrigata,* piercing the skin and sucking the blood of a Malayan tapir. This is the only known blood-sucking moth. (Photograph by Hans Bänziger.)

from its orbits (Fig. 2-2). It seems a natural step from such lachryphagous species to the skin-piercing, blood-sucking noctuid, *Calpe eustrigata* (Fig. 2-3).[1] Other ectoparasitic arthropods may have followed a similar road to parasitism, or they may have accomplished the transition in a single step.

Endoparasites probably found entrance into a host easiest through the alimentary tract. Cyst-forming protozoans and nematodes protected by a dense cuticle most readily survived maceration and gastric juices. Species adapted to endocommensalism might have found it possible to go a step further and become parasitic, but once again, it was probably a one-step process in many cases. We have no space here to discuss individual groups of parasites; the interested reader is referred to the useful work edited by Taylor.[26]

Whatever the mode of attack by incipient parasites, they are confronted by numerous barriers that protect the host from such incursions. The temperature of the host places severe restrictions on potential parasites of birds and mammals, for only those preadapted to survive the higher body temperatures could become established. The parasite's size could prevent its effective entrance into a smaller host or one that was a filter feeder. A mechanism that enabled the invader to maintain its position in the host's gut would seem to be a necessity in most cases. This is provided by suckers or hooks in some species or by a strong swimming action, as with many nematodes.

When free-living organisms enter the gut or tissues of a prospective host, they move into a medium of higher osmotic pressure. Only those that can survive this challenge will be successful. The high hydrogen ion concentration of the stomach is a formidable barrier to invasion. Yet, not only do hoards of parasites survive this pitfall, but some also live, mature, and reproduce bathed in the hydrochloric acid medium of the vertebrate stomach. The host's digestive enzymes are sufficient to destroy most would-be colonists, but successful endoparasites secrete substances that neutralize the enzymes, secrete a tegument as fast as it is digested, or have some other specialization that protects them.

Oxygen is in short supply in the digestive tract, but immediately adjacent to the gut mucosa oxygen concentration may approach that of the blood. Oxygen is much lower in the gut lumen, and in deep tissues of large parasites or in areas of heavy bacterial growth, such as the colon, oxygen is consumed faster than it can be diffused, and its concentration approaches zero. A facultative anaerobe, such as a saprophytic protozoan or a nematode, would not find such low oxygen concentration to be a major barrier, but organisms that rely on the Krebs cycle and classical oxidative phosphorylation could not survive. Hence, endoparasites have either been derived from facultatively anaerobic ancestors or have evolved metabolic adaptations to tolerate low oxygen tensions. Interestingly, to a considerable degree this generalization applies also to parasites in the blood and other tissues. Whatever their origin, metabolic adaptations to low oxygen have involved a shortening or modification of oxidative metabolism to the point where glycolysis assumes a major importance in energy derivation. End products of the parasite's energy metabolism, such as lactic, pyruvic, and succinic acids, excreted by many parasites, can then be catabolized by the host in its own energy metabolism.

A major barrier to incursion by parasites is the immune response by the host. This will be discussed in more detail, but it appears that at least two basic "strategies" have evolved. In the first of these, the parasite produces substances that the host interprets as "self"—the parasite masquerades as a host tissue—and thus the host makes no effort to attack. This represents the highest attainable adaptation to para-

sitism and is seldom realized. The second and much more common situation involves a grudging stalemate in which the parasite does not directly cause its host's death, and the host does not react so violently as to destroy the parasite. Such penetration of the host's immunological barriers is a difficult but necessary step in the evolution of a successful endoparasite, and in actuality it also represents the evolution of the host to accommodate the parasite.

The closer the host-parasite relationship is, the more the symbionts present a *common* phenotype that reacts with the environment, with *common* evolutionary pressures. Thus, each member—the host and the parasite—evolves independently, but also the partners evolve together as a unit. In time a given parasite is so adapted to its host species that it cannot mature and survive in any other. Such host specificity may be nearly absolute or may be slight, depending on the degree of adaptation by the symbionts. As each host moves through time and accumulates its infinite series of minute genetic changes, its parasites move with it, adapting to their host's mutations or else perishing. This concept is useful in determining phylogenetic affinities of hosts in some cases. For instance, most shark tapeworms are quite host-specific. The problem of determining affinities between genera and species of shark sometimes can be solved by comparing their tapeworms. In most cases, the more closely related the tapeworms, the more closely related are the sharks. Similar conclusions can be reached when studying the biting lice of birds.

When considering the morphological changes associated with parasitism, we are confronted with two related but opposite phenomena, those of progressive specialization and regressive elimination of specializations. In the former we find structures such as the hooks of tapeworms and Acanthocephala, which cannot be other than adaptations to parasitism. Other examples of favorable morphological adaptations are acetabula, tribocytic organs, and tegumentary spines in digenetic trematodes; posterior clamps and suckers in monogeneans; and teeth and cutting plates in nematodes.

In contrast, structures may be lost or diminished in parasitic animals because they have no selective value. The digestive system is completely absent in tapeworms and thorny-headed worms, although they are thought to have evolved from fully equipped ancestors. Fleas and lice have lost their wings, and in some cases their eyes, while in parasitic nematodes the sensory structures called amphids are much reduced in size. In fact, the loss of sense organs is a common feature of parasitism, although specialized receptors are still found on all metazoan species.

Finally, most successful parasites have overcome the odds against survival by evolving mechanisms that increase their rate of reproduction. For instance, certain large tapeworms and nematodes may produce nearly 1 million embryos a day for months or even years. This adaptation compensates for the massive mortality of offspring while enacting their complex life cycles. Other examples of increased biotic potential are discussed in the following section.

BIOTIC POTENTIAL OF PARASITES

All organisms must reproduce successfully, or else they will join the legions of the extinct. Most parasites, especially endoparasites, are faced with special problems, for they must survive not only the concerted defense efforts of the host, but also the danger-fraught interludes between hosts. All parasites must produce offspring that can infect the next host, and many species have complex life cycles that involve a series of intermediate hosts as well as a definitive host. When one considers that chance governs the successful completion of much of the life cycle of any given parasite, it becomes apparent that the odds against success are nearly overwhelming. Parasites, more than most other groups of organisms, beat the odds by producing enormous numbers of offspring. Most of these perish, but enough have survived to maintain the species that still exist.

Different groups have evolved various methods to solve the problem of mass reproduction. For example, **multiple fission,** or **schizogony,** is found in many parasitic protozoans. Here, instead of a single mitotic division that produces two daughter cells, the nucleus redivides several times before the cytoplasm is divided among them. Then many daughter cells are produced simultaneously, flooding the microenvironment with so many individuals that one or more is likely to infect the next host and initiate the next stage of development. A more detailed discussion of schizogony and similar processes and illustrations of their functions are given in the chapters on Protozoa.

Some metazoans have developed **polyembryony,** in which a single egg develops directly into numerous offspring. Certain species of parasitic wasps, for instance, lay eggs in a host insect, and each egg divides repeatedly to become up to 2,000 wasp larvae. This must surely be one of the most efficient means of reproduction in the animal kingdom, and, as always, it is the host that "pays the bill."

Hermaphroditism solves the problem in many cases. When an individual possesses both male and female reproductive systems, it has solved the problem of finding a mate. Many tapeworms and flukes fertilize their own eggs; this method, while not likely to produce anything original, certainly is efficient in guaranteeing offspring. However, some hermaphrodites must rely on cross-fertilization, while others can swing either way.

Most tapeworms undergo a type of continuous budding, called **strobilization,** in which segments bud at a zone near the hold-fast organ, or scolex. The resulting chain of buds, or **proglottids,** consist of a linear series of zoids, in most cases each with functioning sets of male and female reproductive systems. Thus instead of an animal with a single set of reproductive systems, we find a veritable factory of reproduction. The ultimate must be *Polygonoporus,* a tapeworm of whales, which consists of about 45,000 proglottids, each with 5 to 14 sets of male and female reproductive systems. The reproductive potential of this 100-foot monster is staggering. Yet there are few whales, and the ocean is large; so the chances of survival of even this species remain slim.

Some tapeworms, such as the orders Tetraphyllidea and Trypanorhyncha, break off each proglottid before it matures. The segment then assumes the identity of an independent organism, maturing and copulating with a similar individual. Gravid eggs exit through a uterine pore, and the senile proglottid expires. The curious tapeworm *Haplobothrium,* a parasite of the freshwater fish *Amia calva,* takes this process a step further, for each detached segment undergoes proglottidization and becomes a secondary strobila. Each segment of this chain produces many eggs.

One of the most common means of increasing the biotic potential of species is the production of numerous eggs. Thus the common rat tapeworm, *Hymenolepis diminuta,* produces up to 250,000 eggs per day or 100×10^6 during the life of its host. If all of these could reach maturity in new hosts, they would represent over 20 tons of tapeworm tissue.[20] A female *Ascaris* will produce over 200,000 eggs per day for several months, and the filarial nematode *Wuchereria* produces several million larvae during her lifetime. Many adult parasites appear to have the single function of reproduction, some giving up their lives in the process. Examples include the human pinworm, which leaves the security of the host's rectum to scatter its eggs outside the host, and other nematodes in which the mother's body becomes a dead, parchment-like case for her developing eggs.

Parthenogenesis, that process of egg development without fertilization, occurs in some parasites. Species of the nematode genus *Rhabdias* live as females in the lungs of reptiles and amphibians, giving birth to numerous larvae in the complete absence of males. Parthenogenesis is also known in many parasitic insects and in the nematode genus *Strongyloides.* One peculiar form of reproduction, termed parthenogenesis by some authors and sequential polyembryony by others, is exhibited by the monogenetic flatworm *Gyrodactylus.* When a larva hatches from its eggshell it contains a smaller larva, which in turn contains a still smaller larva. As each matures, it gives birth to its internal twin and then begins producing larvae on its own. This rather elaborate adaptation allows a rapid population build-up from a single, successful larva.

Asexual reproduction is very common in several groups of parasites. Simple mitotic fission is especially important among the protozoans, and indeed is the only form of reproduction in the Sarcomastigophora. Rapid fission often results in millions of offspring in a matter of only a few days.

More complex forms of asexual reproduction are found in the flatworms. The larvae of several tapeworms are capable of external or internal budding of more larvae. The cysticercus larva of *Taenia crassiceps,* for instance, has been found to bud off as many as 100 small bladderworms while in the abdominal cavity of the mouse intermediate host. Each new larva develops a scolex and neck, and when the mouse is eaten by a carnivore, each develops into an adult cestode. The hydatid larva of the tapeworm *Echinococcus granulosus* is capable of budding off hundreds of thousands of larval scolices within itself. When such a larva is eaten by a dog, it produces vast numbers of adult tapeworms.

Perhaps the most bizarre and astonishing asexual reproduction in all zoology is found among the digenetic trematodes, a large, successful group of parasites commonly called the "flukes." These animals undergo a series of reproductive stages, each of which produces the next generation of larvae, rapidly building up a huge population. Although there are many variations, the basic scheme is: egg (miracidium) → sporocyst → redia → cercaria → adult. The microscopic miracidium that hatches from an egg enters the first intermediate host, usually a mollusc, and becomes a sac-like sporocyst. By a form of internal budding (which is still not well understood) the next larval stage, the redia, develops within it. Several rediae are usually produced by a single sporocyst. The rediae break out of the sporocyst and grow, feeding actively on host tissues. Several cercariae form within each redia by a process apparently similar to the formation of the redia. The cercariae pass out of the redia, and the intermediate host, to seek their fortunes elsewhere. There may be more than one generation of sporocysts or rediae, or neither of these may be present. The point is, each egg has the potential of producing hundreds of offspring by these successive asexual stages. When one observes that most flukes give birth to thousands of eggs each day, the biotic potential of these worms is staggering. Parasites are, indeed, masters of reproduction.

PARASITE DISTRIBUTION AND DENSITY

Species of parasites have adapted to virtually every tissue, organ, and space in the body. Parasites that live within tissues are called **histozoic,** while those inhabiting the lumen of the intestine or other hollow organs are said to be **coelozoic.** Most endoparasites live in the digestive system. This cannot be considered a single niche, for many different environments exist between the mouth and the anus. Further, digestive systems vary greatly between species (compare the stomachs of human and ox) and even between different stages in the life of the host, such as a tadpole and a frog. For a review of variations between vertebrate intestinal tracts see Crompton.[9] Even a given level along the length of the alimentary tract cannot be considered a single niche, for there are subtle differences in oxygen and carbon dioxide tension, pH, and other chemical and physical factors between the mucosa and the center of the lumen. Such differences occur even between the tip of a villus and its base, making at least two different niches available for colonization by parasites of suitable sizes. Obviously, then, when two species of parasites are found in the same region of the intestine, one cannot state that they are occupying the same niche; for, while in proximity to one another, they may be in entirely different microenvironments. Thus Schad found eight species of the nematode *Tachygonetria* in the large intestine of the turtle *Testudo graeca,* all apparently occupying different habitats.[21] Obviously, though, one large parasite may occupy several microenvironments simultaneously.

If two species cannot occupy the same niche simultaneously, there must be *competition* between them when they enter a host. As is usually the case in nature, the species that is first established normally will not be displaced by the intruder. How it manages to prevent establishment by the interloper is not well understood, but the phenomenon of premunition may be involved (see p. 26).

The principles outlined previously apply to all areas of the host's body, not just the alimentary tract. Adult parasites are found in blood, skin, nervous tissues, and virtually all other parts of the body, while larval stages often undergo elaborate migrations through distant regions of the body before arriving at their definitive sites. Parasites that enter a host in which they are incapable of maturing often wander about until they die or become dormant for long periods of time within the accidental host's tissues. It has been said that if a host were infected with all the parasites capable of infecting it, and the host tissues were then removed leaving only the parasites, the host could still be recognized! While this may be an exaggeration, it is only slightly so. Consider the human eye, for example, an organ not especially suited to infection by parasites. Yet the retina may be infected by the protozoan *Toxoplasma gondii* and larvae of the nematode *Onchocerca volvulus;* the chamber may harbor the bladderworm larvae of the tapeworms *Taenia solium, T. crassiceps, T. multiceps,* or *Echinococcus granulosus;* the conjunctiva may host a wandering nematode *Loa loa;* while the orbit may be the home of nematodes of the genus *Thelazia.* Parasitologically speaking, the vertebrate body can be considered as a great mass of ecological niches, which have been colonized by a great variety of parasite species.

The **density** of infection varies widely depending on the species involved, age and condition of the host, inter- and intraspecific competition of parasites, and other parameters too numerous to mention. Generally, the larger and older the host, the more parasites it has. Exceptions do occur. Young animals are often susceptible to infections to which their elders are refractory; conversely, young animals may have a short-term immunity due to antibodies received from their mothers. Further, a low grade infection in young or old hosts may result in premunition, that is, concomitant immunity to further infection by that species of parasite.

Still, very heavy parasite burdens sometimes occur. In terminal infections the number of trypanosomes or plasmodia in the blood sometimes rivals the number of erythrocytes. The intestinal protozoan *Giardia* is sometimes so abundant that nearly every cell of intestinal epithelium is covered by one, and over 20 million cysts may be passed in a single stool. We have recovered over 800 tapeworms from a single snipe, a small bird weighing only a few ounces, and more than 1,000 nematodes from a small turtle. Humans are not exempt from massive infections, for in poorly developed communities any individual may have simultaneous and heavy infections of hookworm, whipworm, *Ascaris,* pinworm, *Entamoeba,* and several other parasites, as well as bacterial, fungal, and insect infections. When this total burden is added to a low level of nutrition, the victim is lucky to survive, let alone contribute to civilization.

ECOLOGY OF PARASITIC INFECTIONS

Basically it can be stated that the general principles of ecology apply equally well to parasitic organisms. Therefore these general principles will not be considered in detail here. We will explain the special aspects of parasite ecology later in conjunction with discussion of the parasites themselves, but we will mention a few topics at this point.

Between World War I and World War II, it was noted by the Russian school of Pavlovsky that certain parasitic diseases occur in certain ecosystems and not in others, and that factors making up these ecosystems can be categorized such that they can be recognized wherever they are encountered. Thus, each disease has a natural focus or **nidus.** Discovery of this *natural nidality of infection* was a landmark in the history of parasitology, for it enabled the epidemiologist to recognize "landscapes" where certain diseases could be expected to exist or, equally, where the possibility of their presence could be eliminated. Such **landscape epidemiology** requires thorough knowledge of all parameters that bear on the infection, such as climate, plant and animal populations and their densities, geology, and human activities within the nidus. This holistic approach is best applied to parasitic diseases of wild and domestic animals and to the **zoonoses,** those diseases of animals transmissable to humans. However, the principles of landscape epidemiology can be applied equally to whipworm infections in

mental institutions or *Giardia* outbreaks in swank vacation communities.

The ecology of any disease is profoundly influenced by the **vector** of the parasite. A vector is the means of transmission of a disease organism from one host to another. Thus, water may be the vector for *Entamoeba,* aerosol sputum droplets the vector for tuberculosis, and wind the vector for the fungal disease coccidioidomycosis. As such, they require consideration in landscape epidemiology. Mosquitoes, flies, ticks, and other arthropods are vectors for many parasitic diseases; their biology and their positions in the nidus must be well understood before prevention and eradication of the diseases are possible. Before we can properly understand the factors governing parasitic diseases, we must study not only the parasite and the host, but also all other physical and biological factors making up the community. Once the complex interactions between populations are known, it is possible to outline a scheme to interrupt the interactions and thereby eradicate the diseases.

Many parasites of humans are shared by other vertebrate hosts. These hosts are **reservoirs** that maintain the disease in nature and provide a source for infection of humans. For example, rodents are reservoirs of Oriental sore and kala-azar, and the house mouse is a reservoir of the dwarf tapeworm, *Hymenolepis nana.* The same principle applies to animals other than humans; for wild deer are the reservoir of filarial disease of sheep (*Elaeophora),* and rodents are reservoirs of trichinosis in bears and pigs.

Parasitology, then, is actually a study of ecology, not only of the interactions between the parasite and its host, but also of the factors influencing their life cycles. The level of nutrition among the people in a community is just as important in a parasitic disease as the availability of an intermediate host. The traditional defecation and bathing habits of people may influence a parasitic infection as much as the presence of suitable reservoirs or the mean annual temperature of the community. The parasitologist is faced with the task, sometimes joyful, sometimes not, of unraveling the tangled thread of ecology that binds together parasitic infections.

FROM THE POINT OF VIEW OF THE HOST: DISEASE AND DEFENSE

Progressive specialization of a symbiont increasingly limits its potential host species; in other words, it increases its host specificity. However, in symbiosis we must consider a vital additional factor, that the habitat (host) is a dynamic, living component of the system, reacting to the presence of the symbiont. In this section we will briefly examine some of the host's reactions, showing that the host reactions are equal in importance to parasite specializations in determination of host specificity.

Balance of symbiosis

In any real case of a symbiotic relationship, some kind of balance must be achieved between the symbiont's tendency to damage its host and the host's mechanisms to defend against foreign bodies. This is true whether the relationship fits our definitions of commensalism, mutualism, or parasitism. The host must survive long enough for its symbiont to reproduce and remove its progeny, for death to the host means death to the symbiont. If the balance is tipped too far either way, the host or its symbiont population may be wiped out, conceivably driving either or both to extinction. This probably has occurred many times in the history of life.

Nonpathogenicity, or lack of disease production, is often considered a mark of an evolutionarily long symbiotic relationship. This generalization is open to challenge, however, because some organisms that appear to have been parasites for a long time are demonstrably pathogenic, for example, many trematodes. Certainly, purely from considerations of natural selection, the less pathogenesis the better for the parasite, since it seldom outlives its host; the longer the host lives, the more progeny can be produced by the parasite. Important exceptions to this principle are cases where pathogenesis *facilitates* transmission. In some larval cestodes pathogenesis in the intermediate host makes it easier for that host to be caught and eaten by the definitive host.

Susceptibility and resistance

Before discussing host defense mechanisms, we will distinguish certain terms that often are used with other than their strict biological meanings. **Susceptibility** and **resistance** were well defined by Cheng: "A host is said to be susceptible if it is theoretically capable of being infected by a specific parasite. This implies that the physiological state of the host is such that it will not eliminate the parasite before the parasite has an opportunity to become established in the host. A host is said to be resistant if its physiological state prevents the establishment and survival of a parasite, be it during the initial or a subsequent contact."[3] A corresponding term, but from the point of view of the parasite, would be **infectivity.**

Note that these terms deal only with the ease or difficulty of infection, not with the mechanisms producing the result. The mechanisms that increase resistance (and correspondingly reduce susceptibility and infectivity) may involve either nonspecific attributes of the host, independent of prior contact with the parasite, or specific conditions stimulated as a consequence of previous physiological experience with the infective agent (immune response, see p. 18). Furthermore, the terms are relative, not absolute; for example, an individual organism may be more or less resistant.

The term **immunity** often has been used in the past with a meaning synonymous with "resistance," although it would have been less confusing if "immunity" had always been reserved for conditions arising from the immune response. Many authors distinguish various sorts of "immunity" on the basis of whether or not the immune response is involved. Thus, Sprent (1963) defined **innate immunity** as "mechanisms with which the host is endowed as a species, i.e. those anatomical and physiological features which make the host to a greater or lesser extent unsuitable for a certain range of parasites."[25] **Natural immunity** comprises "those mechanisms which are possessed to some degree by all animals and are physiological processes evolved for the purpose of defence." **Acquired immunity** is immunity arising from a specific immune response, which is stimulated by the presence of a foreign material (antigen) in the host's body. It follows that innate and natural immunity in this sense will play a large role in the determination of host specificity, while acquired immunity may play some role, as in the case of an abnormal host that expels the parasite after a short infection, or no role at all, as in the case of the normal host whose immune response either is absent or does not confer protection.

The immune response

The following treatment deals primarily with acquired immunity in vertebrate hosts; specific immune responses are very poorly known in invertebrates, though they may occur in some cases. There are several readable accounts that supplement our presentation.*

An immune response is stimulated by an **antigen,** and circularly, an antigen is any substance that will stimulate an immune response, or, is **immunogenic.** An immune response is elicitation of specific substances produced by certain types of white blood cells, the **lymphocytes,** and the configuration of these substances depends on the complementary molecular configuration of the antigen. A wide variety of substances can be antigens, but most antigens are proteins, polysaccharides, or nucleic acids with a molecular weight of over 3,000; generally speaking, the substance must be foreign to the individual animal to be antigenic in that animal. When an antigen is introduced into the body of an immunocompetent organism, genetic "triggers" are released in certain lymphocytes, causing repeated cell divisions. Depending on characteristics of the antigen, the lymphocytes stimulated to divide are either those called **T cells** or others referred to as **B cells.** T cells sensitized by a particular antigen have receptor sites on their surfaces that bind with the particular antigen, and B cells secrete proteins called **antibodies** that bind specifically with the antigen. Antigens interact with the T cell surfaces or bind with the antibodies on only one or on several areas known as the **antigenic determinants.** The specificity of the antigenic determinant is the result of its configuration, which binds with the receptor site in a "lock-and-key" manner. Actually, rather simple organic molecules can be immunogenic when they are bound to

*See references 7, 16-19, 23.

a carrier molecule, usually a protein, thus modifying or becoming the determinants of the antigenic complex; such simple molecules are called **haptens.**

The two distinct types of immune responses, **cellular** and **humoral,** are based on different populations of lymphocytes—the T cells and B cells, respectively. T cells are so called because their differentiation has been somehow modified by a sojourn in the **thymus,** an organ of formerly obscure function below the sternum that slowly regresses with age. The appellation B cells arises from the discovery that in birds these cells are processed through the **bursa of Fabricius,** a lymphoid organ attached to the intestine near the cloaca. The equivalent processing center for B cells in mammals appears to be the liver, or possibly the liver and spleen. Both cell types, originate ultimately from stem cells in the embryonic yolk sac in both birds and mammals.

Although cellular and humoral immune responses seem to interact, the distinction has biological significance: humoral immunity seems to be more important in a variety of bacterial infections, and the cellular response—often called **cell-mediated immunity (CMI)**—is of particular importance in tissue rejection reactions and a variety of virus, fungal, and parasitic infections. Within both types of lymphocyte population, there is an extremely large number of subtypes (about 50,000), each so differentiated that it can respond to a particular antigen. Although there are far more than 50,000 possible antigens, apparently there are enough cross-reactions possible that the "lock-and-key" fit is good enough to furnish sufficient diversity for virtually all possible antigens. (**Cross-reaction** is the binding of an antibody or a cell receptor site with an antigen other than the one that would provide an exact "fit.")

The nature of the receptor sites on T cells is unclear, but in the case of B cells, it appears that small amounts of antibody are constantly being produced and incorporated into the cell membrane, even in the absence of prior exposure to the antigen. When an antigen is introduced into an animal, the appropriate one of the 50,000 lymphocytes is sensitized, probably by the antigen binding only to the cell membranes containing the complementary antibody. Most antigens are first processed by T cells before stimulating the B cells to antibody production. The sensitized B lymphocyte differentiates into a "blast" cell that undergoes numerous mitoses, progressively differentiating and maturing. Finally, a mature **plasma cell** results that produces large amounts of antibody—the same antibody its ancestral lymphocyte could produce in only very small amounts, but that was specifically selected by a particular antigen. The plasma cell secretes 2,000 to 3,000 molecules of antibody per second and then dies within a few days.

Following a period of some days after antigen introduction, during which the antibody is very low or undetectable, there is an exponential rise in antibody **titer** (units of antibody per milliliter of serum). Then a plateau is achieved, followed by a decline, though usually not to the previous very low level. The decline occurs because of the short life of the plasma cells and the decay of antibody by serum enzymes and by excretion. If a second dose of antigen **(challenge)** is then given, a very interesting phenomenon occurs. With little or no lag period the antibody titer shoots up to a much higher level than the first plateau, often as much as 10 to 100 times the previous level. This is the **secondary** or **anamnestic** response, as the result of the fact that some of the lymphocytes originally stimulated do not differentiate fully into plasma cells, but rather are transformed into small lymphocytes called **memory cells.** The memory cells may have a very long life, perhaps years, and since they are considerably larger in number than the "virgin" B cells that could make that particular antibody, a much more rapid immune response is mobilized after challenge. Although the antibody-building sites in cell-mediated immunity apparently are on the cells and are not usually regarded as antibody, the dynamics of the primary and secondary responses generally parallel those of the humoral type. For example, in the case of a tissue graft, a primary implantation will even heal into place until the primary response causes its rejection, but if a challenge graft is attempted, the rejection will take place much more rapidly.

Research in the last 10 to 15 years has revealed much of the structure and function of antibodies. Antibodies belong to the family of proteins dissolved in blood serum, which are called **immunoglobulins (Ig).** All types are constructed basically of two shorter chains of amino acids (light chains) and two longer chains (heavy chains) (Fig. 2-4). The chains are joined by covalent (disulfide) and hydrogen bonds. The light chains may be one of two types: kappa or lambda; the heavy chains may be any of five types: mu, gamma, alpha, delta, or epsilon. The latter determines the **class** of the antibodies, referred to as IgM, IgG (now familiar to many people as "gamma globulin"), IgA, IgD, and IgE, respectively. IgM and IgA form oligomers, associations of the two-pair chain groups with one or a few other chain groups forming larger molecules. Part of each heavy and light chain is constant in composition, that is, the amino acid sequence does not vary, and part of each chain varies with the specificity of the antibody. The variable portions of the heavy and light chains lie alongside each other and bear the antigen-binding sites. Generally, a given cell can only produce antibody molecules that are alike in all ways, including class and specificity. However, it does appear that during B cell differentiation the lymphocytes begin producing IgM, with some switching over to IgG production, and some of these then changing to IgA. Whether such a developmental sequence may account for IgD and IgE production is not clear.

The site of antibody production in the body is of interest, since plasma cells normally do not circulate in the blood. It was noted that the embryonic origin of the stem cells for both T and B lymphocytes is the yolk sac. Cells that later become lymphocytes are derived from the bone marrow hemopoietic factory, which also produces other white cells, red blood cells, and platelets. Some of the yolk sac-derived cells apparently may enter the liver or spleen directly in mammals or the bursa of Fabricius in birds—there to be influenced into becoming B cells—while others are processed through the thymus to be differentiated into T cells. Lymphocytes circulate with the blood, and enormous numbers of

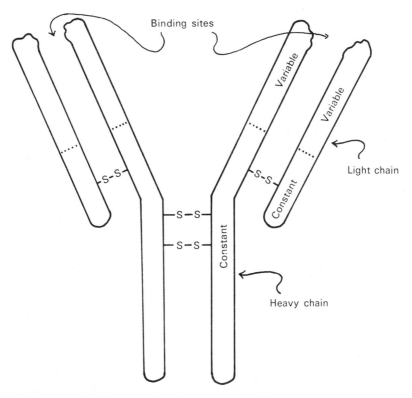

Fig. 2-4. Diagram of an immunoglobulin molecule. The bars represent the polypeptide chains bound together by disulfide and hydrogen bonds. The light chains may be either of two types, kappa or lambda. The class of antibody is determined by the type of heavy chain: mu (IgM), gamma (IgG), alpha (IgA), delta (IgD), or epsilon (IgE). The constant portion of each chain does not vary for a given type or class, and the variable portion varies with the specificity of the antibody. Antigen-binding sites are in clefts formed in the variable portions of the heavy and light chains. IgM normally occurs as a pentamer, five of the structures illustrated being bound together by another chain. IgA may occur as a monomer, dimer, or trimer.

B cells concentrate in the lymph nodes, spleen, and Peyer's patches of the intestine. The initial antigenic stimulus for the B lymphocyte to form a "blast" cell apparently occurs in one of these organs, though a plasma cell-precursor can migrate to other organs, and antibody production by the plasma cells may take place in nodes distant from the original stimulation site. Antibody production in the spleen seems particularly important in blood-borne antigen responses.

In terms of antibody function, we are particularly concerned with the mechanisms by which it enhances the host's defenses against invading organisms. One is by binding to the surface of the invader, thus in some way making it easier for macrophages to engulf the invader by phagocytosis **(opsonization);** another way is by binding to some part of the invader's structure, so that some vital process is restricted; a third is by acting in concert with other blood components, collectively known as **complement,** to lyse the cells of the infective agent. Phagocytosis and complement will be discussed further.

The type of immune response caused by T cells, cell mediated immunity (CMI), is an area being very actively researched in parasitology because it appears that CMI is very important in parasitic infections. A manifestation of CMI, the **delayed hypersensitivity** (DH) reaction, once thought to produce only tissue damage, is now understood to have a basically protective role, and many researchers incorrectly use the terms CMI and DH interchangeably. The initial distinction between this type of immune response and humoral immunity is that contact with the antigen is via sensitized lymphocytes (derived from T cells) rather than circulating antibody. CMI and humoral responses have a number of similarities. Examples are an initial lag phase in the primary response, a secondary response upon challenge, and transformation to blast cells and multiplication in the lymph nodes (though in a different part of the nodes). Another similarity is that **passive** immunization can be accomplished—though in humoral immunity this is done by transfer of serum from one animal to another, and in DH the lymphocytes themselves must be transferred **(adoptive immunity).** The term delayed hypersensitivity is derived from the fact that a period of 24 hours or more elapses between the time of antigen injection and the response to it in an immunized subject. This is in contrast to **immediate hypersensitivity** reactions, mediated by antibodies, in which maximal response is reached within a few minutes or hours. An example of DH that is familiar to most people is the tuberculin diagnostic test. When the nonimmunogenic tuberculin antigen is injected into a person who has had no contact with tuberculosis, there is no reaction. However, if the person has had even a slight tuberculosis infection in the past, after about 4 hours a small swelling appears at the site of injection, increasing to a maximum size and redness after 24 to 48 hours, then gradually subsiding. The mechanism of this reaction, as in other CMI reactions, seems to involve comparatively small numbers of specifically sensitized lymphocytes that are circulating in the blood and arrive randomly at the injection site. This is a *specific* interaction of antigen with lymphocyte, the presence of the sensitized lymphocyte being the result of a prior immunizing dose of antigen, just as in the humoral antibody response.

The effect produced on the affected lymphocyte, however, is a nonspecific one: the cell is stimulated to synthesize and release a series of effector molecules called **lymphokines.** The lymphokines play an important role in inflammation. One lymphokine, the **macrophage migration inhibitory factor (MIF),** inhibits the migration of macrophages from the vicinity of the local reaction, that is, any that arrive there are inhibited from migrating; thus they are present to perform their phagocytic activity. Another lymphokine stimulates phagocytic activity in macrophages, and still others have a chemotactic effect, attracting macrophages and other white cells to the vicinity. In short, the specific interaction of antigen and lymphocyte in the CMI greatly influences subsequent events in a nonspecific response, inflammation. The immune response elicited by at least several parasitic worms is not directly responsible for their expulsion from the host. Rather, the expulsion is caused by the inhospitable chemical conditions in the ensuing inflammation that has been enhanced by the specific CMI.[17]

Nonspecific mechanisms of resistance

Under the heading nonspecific mechanisms of resistance one may place the "accidents" of a host's structural and physiological characteristics that reduce its susceptibility to certain parasites (Sprent's "innate" immunity). This includes the physiological adaptations possessed by a wide range of hosts, evolved as adaptations for defense and effected on first encounter with a wide range of invading organisms (Sprent's "natural" immunity). Examples of the first category include physical barriers, such as a thick, cornified epidermis or other protective external covering, the ability to repair damaged tissue rapidly, and high acidity in the stomach. Such mechanisms are not always easy to distinguish from some "natural" defense adaptations, and the distinction may not be a useful one. In any case, a variety of parasiticidal substances are known to be present in such body secretions of animals as tears, mucus, saliva, and urine. In fact, at least one of what were previously thought to be nonspecific parasiticidal substances is now known to be a class of antibody, IgA. IgA can cross cellular barriers easily. It seems to be an important protective agent in the mucus of the intestinal epithelium, and it is present in mucus in the respiratory tract, in tears, in saliva, and in sweat. And, as indicated, in higher vertebrates the specific defense mechanisms, humoral and cellular immune responses, greatly enhance or modify the nonspecific processes, phagocytosis, inflammation, and the action of complement.

Phagocytosis is one of the most fundamental mechanisms of resistance, occurring in almost all metazoans, as well as being a feeding mechanism in protozoans. Among invertebrates, phagocytosis seems to be one of the most widely prevalent mechanisms by which to deal with foreign particles or invading microorganisms. Indeed, it was from studies on phagocytosis in crustaceans *(Daphnia)* and echinoderms (starfish larvae) that Elie Metchnikoff realized the significance of the process as a defense mechanism. A cell that has this ability is known as a **phagocyte,** and the process involves engulfment of the invading particle within an invagination of the phagocyte's cell membrane. The invagination becomes pinched off, and the particle is thereby enclosed in an intracellular vacuole. Small vesicles in the cell, the **lysosomes,** contain digestive enzymes. The lysosomes fuse with the phagocytic vacuole, and the enzymes attack the particle or organism it contains. The various cells with phagocytic ability are divided into two types: **mobile** and **fixed** phagocytes. The most numerous of the circulating or mobile phagocytes are the **polymorphonuclear leukocytes,** or **granulocytes.** The first name refers to the fact that the nucleus is highly variable in shape, and the second name refers to the many small granules that can be seen in their cytoplasm, especially when stained with a Romanovsky-type stain. According to their staining properties, granulocytes are further subdivided into **neutrophils, eosinophils,** and **basophils.** Neutrophils are the most abundant of the white cells in the blood, and they provide the first line of phagocytic defense in an infection. Eosinophils in normal blood account for about 2% to 5% of the total leukocytes, and basophils are the least numerous at about 0.5%. A high **eosinophilia** (eosinophil count in the blood) is often associated with allergic diseases and, of particular interest for our purposes, with parasitic infections. Eosinophils, however, rarely if ever display phagocytosis, and their function is unknown.

They may play a role in the inflammatory process in conjunction with the mast cells.

Another circulating white cell important in phagocytosis is the **monocyte.** This cell is not considered a granulocyte, though its cytoplasm may contain fine granules. During inflammation, monocytes move into the tissue, there they are called **macrophages,** which are active phagocytes. Macrophages seem to have a role in the humoral immune response as well. Macrophages do not themselves form antibodies, but may "process" the antigen before it really becomes immunogenic for the lymphocytes.

The **fixed** phagocytes taken together form what is known as the **reticuloendothelial system (RE system),** which is spread through a variety of tissues. In the liver sinusoids, the **Kupffer cells** form a network through which blood flows. Also included are the phagocytic reticular cells of the lymphatic tissue, myeloid tissue, and spleen; the lining cells of the sinusoids in the adrenal and hypophysis; certain perivascular cells; and the "dust cells" of the lungs. The phagocytic cells in the liver, the spleen, and the endothelium that lines the blood vessels ingest foreign particles from the blood, as well as senescent or damaged erythrocytes. Macrophages in the lymph nodes help remove foreign particles from the lymph.

Inflammation is a vital process in the mobilization of the body defenses against an invading organism and in the repair of damages thereafter. Although the process itself is nonspecific, its mechanisms are considerably modified by immune processes, that is, the inflammation may be much enhanced in an animal with prior physiological experience with the antigen. If the invading substance or organism is poorly immunogenic but otherwise causes little damage or disturbance to the surrounding tissue cells, the object will be surrounded gradually by phagocytes, immobilized by collagen fibers, and, if possible, ultimately disintegrated. If the invading substance is somewhat more noxious, or the host has been immunized, a more vigorous inflammatory reaction will ensue, initiated by the release of pharmacologically active substances from damaged cells and especially from **mast cells.** Mast cells are related to basophils and are found in the loose connective tissue under the skin and in many mucous surfaces. They have many granules containing histamine, heparin, and serotonin. When these substances are released, small blood vessels in the immediate vicinity dilate and become more engorged. This causes an accumulation of blood and, therefore, redness and warmth in the area (hence the term "inflammation"). The capillary walls become more permeable, allowing blood proteins and plasma to escape into the local interstitial spaces, causing swelling **(edema).** Passing leukocytes tend to adhere to vessel walls in the area, then to migrate through to attack the invader and clean up dead host cells. The first, and initially most numerous, phagocytes are the neutrophils, which may live only a few days. Next the macrophages, either fixed or differentiated from monocytes, will become predominant. Depending on the site, the persistence of the antigen, and other circumstances, the leukocytic infiltration in the area may be prominent for long periods, a condition apparent in numerous parasitic infections. Eventually, the host tends to remove the irritant by destroying it or by walling it off. Repair of the damaged tissue or the "wall" around the irritant often takes the form of a scar; fibroblasts infiltrate the area to form fibrous connective tissue. Some of the ways in which prior immunizing experience enhances the inflammatory response will be apparent from what has been said, for example, opsonization by antibodies, release of macrophage activating factor (MIF), and leukocyte attractants. In addition, the formation of the antigen-antibody complex seems to facilitate release of histamines from the mast cells—a prominent feature in the "immediate hypersensitivity" reaction.

Particularly interesting as an example of a nonspecific defense mechanism interacting with a specific immune phenomenon is the cooperation of antibody with **complement.** After the discovery of antibodies, the presence of another constituent was found to be necessary for certain microorganisms to be killed in immune serum; this substance was called "complement." It is now known that complement is not one

substance, but a complex of 11 proteins, designated by the letters C1 through C9, and the subunits of C1 are called C1q, C1r, and C1s. The complement acts in a complex sequence of enzyme activations. The series of activations eventually damages the microorganism's cell membrane and results in its death, apparently by impairing the integrity of the cell membrane —literally punching holes in it. The first step in the series occurs after an antibody binds to the invading cell, and this must be either IgM or IgG, which are the only immunoglobulins that activate complement. The complement component that recognizes bound antibody is C1q, which in turn binds to the fixed antibody and in doing so becomes an active enzyme. This initiates activation reactions in a definite sequence involving the other complement components. Some of the complement fractions (C3, C5) are cleaved, with one cleavage product binding to the cell surface and the other being released into the fluid phase to mediate the inflammation reaction.

Complement participates in a variety of antigen-antibody reactions, not all of which produce such visible results as lysis of cells. This has led to the development of a means of diagnosis called the **complement fixation test.** The complement fixation test generally can be adapted to diagnose any infection that stimulates antibodies that bind to complement, including many parasitic infections. The most widely known complement fixation test is the Wassermann test for syphilis. In preparation of reagents for such a test, *rabbits* are immunized against *sheep* erythrocytes. Antigen is derived from the parasite to be diagnosed (or a related species), and standard complement is prepared, usually from guinea pig blood. To perform the test, a small amount of serum from the patient is mixed with the known antigen and a quantity of complement. Then the rabbit antibody (against sheep erythrocytes) and the sheep red blood cells are added to the mixture. If antibodies to the specific antigen were present in the patient's serum, then complement will have been fixed before the rabbit antibody and sheep cells are added, and *no lysis* results; therefore, the test is *positive.* In contrast, if antibodies to the infection were not present in the patient's serum, complement will still be available, and the sheep cells will be disrupted; therefore, *lysis* of the sheep cells indicates a *negative* result.

PATHOGENESIS OF PARASITIC INFECTIONS

The pathogenic effects of a parasitic infection may be so subtle as to be unrecognizable, or they may be strikingly obvious. An apparently healthy animal may be host to hundreds of parasitic worms and yet show no signs of distress, at least none that is detectable. On the other hand, another host may be so anemic, unthrifty, and stunted that parasites are undoubtedly the reason for its sad state. The pathogenic effects of parasites are many and varied, but for the sake of convenience can be discussed under the headings of trauma, nutrition-robbing, and poisoning.

Physical **trauma,** or destruction of cells, tissues, or organs by mechanical or chemical means, is common in parasite infections. When an *Ascaris* or hookworm larva penetrates a lung capillary to enter an air space, it damages the blood vessel and causes hemorrhage and possible infection by bacteria that may have been inhaled. The hookworm, after completing its migration to the small intestine, feeds by biting deeply into the mucosa and sucking blood and tissue fluids. The dysentery ameba, *Entamoeba histolytica,* digests away the mucosa of the large intestine, forming ulcers and abscessed pockets that can cause severe disease. Similarly, the protozoa that cause Oriental sore and espundia enter and kill epithelial and connective tissue cells, destroying circumscribed areas of the skin or nasal septum. When the female nematode *Capillaria hepatica* wanders through the liver of its host, she destroys parenchyma as she goes and also prevents regeneration by filling her path with eggs. The rare, proliferating, sparganum-type tapeworm larva tunnels through the dermis of its unfortunate host, destroying tissues as it goes and riddling the skin with thousands of cavities. These are but a few examples of known physical traumas caused by parasites. Many are discussed in later chapters in conjunction with the particular parasites involved.

A less obvious but often pernicious pathogenic situation is diversion of the host's nutritive substances. Tapeworms and Acanthocephala, for instance, lack digestive systems and rely on the host's daily intake for their own food. While most tapeworms absorb so little food in proportion to the amount eaten by the host that the host still manages very well, when the level of subsistance of the host is low, one or two large tapeworms may absorb enough to cause serious deficiencies in the host's diet. The broad fish tapeworm, *Diphyllobothrium latum,* has such a strong affinity for vitamin B_{12} that it absorbs large amounts from the intestinal wall of its host. As B_{12} is necessary for erythrocyte production, a severe anemia may result. The giant nematode *Ascaris lumbricoides* inhabits the small intestine—often in large numbers—and consumes a good deal of food the host intends for itself. The tiny protozoan *Giardia* robs its host in a different way, for it is concave on its ventral surface and applies this suction cup to the surface of an intestinal epithelial cell. When many of these parasites are present, they cover so much intestinal absorptive surface that they interfere with the absorption of nutrients by the host. The unused nutrients then pass uselessly through the intestine and are wasted.

Several kinds of parasites are known to produce **toxins** that damage their hosts in a variety of ways. The protozoan *Trypanosoma cruzi* develops clusters of cells in the smooth and cardiac muscles of its host, and when the parasites degenerate many years later, they release a neurotoxic substance that attacks the autonomic ganglion cells, ruining nervous control of peristalsis and heart contraction. Another trypanosome, *T. brucei gambiense,* which causes African sleeping sickness, produces a neurotoxin that causes severe brain damage to the patient. The malaria organism *Plasmodium* lives for a time in a red blood cell, eventually killing it. When the red cell ruptures, it releases waste products elaborated by the parasite; and, when billions of such cells rupture simultaneously, the system is flooded with poisons, producing the characteristic onset of malarial fever. When female filarial nematodes *Onchocerca volvulus* are located in the skin of the head or neck, many of the larvae they produce are likely to wander into the retinas of the eyes. Here they die and elicit a powerful immune reaction from the host. The combination of poisoning by foreign protein and invasion of the area by host defense cells often destroys the retina, causing permanent blindness. Today there are villages in Africa and Central America where the majority of adults are blind because of this parasite. A similar series of reactions is produced by a different filarial nematode, *Wuchereria bancrofti,* which lives in the lymphatic system. It, too, produces numerous larvae, many of which "get lost" in surrounding tissues. The toxins produced when they die, together with host reactions, result in swelling and thickening of the affected area, sometimes achieving the horrible dimensions of elephantiasis. The waste products of tapeworms occasionally cause a nauseous, dizzy condition called **verminous intoxication.** The common whipworm, *Trichuris trichiura,* is known to cause prolapse, or eversion, of the rectum, presumably by producing a toxic substance that affects the nervous control of the intestinal muscles. These and many other examples demonstrate how parasites poison their hosts, although most of the toxic substances responsible have not been isolated.

One other aspect of parasite pathogenesis should be mentioned, the fact that some parasites serve as vectors for other disease agents. An example is the salmon-poisoning fluke, *Nanophyetus salmincola,* which carries a rickettsia, *Neorickettsia helminthoeca.* When a dog, racoon, fox, or other fish-eating mammal eats a salmonid fish containing larval flukes, it not only becomes infected with flukes but also may acquire the salmon-poisoning disease, which has a high mortality in dogs. The nematode *Heterakis gallinae* transmits the protozoan parasite *Histomonas meleagridis* within its eggs, and when a turkey becomes infected with the relatively harmless nematode, it also receives the highly pathogenic protozoan.

ACCOMMODATION AND TOLERANCE IN THE HOST-PARASITE RELATIONSHIP

Successful parasites of vertebrates have had to evolve one or more tactics to avoid a protective immunity in a given host. Otherwise that host is simply not susceptible. Current thought on the interaction of the immune response with the evolution of the host-parasite relationship follows two lines[12]: (1) the development of some means whereby the parasite becomes immunologically inert, or lacks the capacity for eliciting a protective response and (2) the possibility that the parasite may mask itself in some way with host components so that it fails to stimulate a protective response or else avoids its consequences.

Sprent (1959) suggested two routes by which the first of these might be achieved.[24] In both of these routes, the underlying assumption is that the parasite and host populations select toward mutual tolerance (adaptation tolerance), related to the concept that nonpathogenesis has adaptive value. By one route, both populations would select toward identity of antigenic determinants. Over many generations important antigenic determinants of the parasite, those which might otherwise elicit protective immunity, would become more like host determinants so that the host immune system would recognize the parasite as "self." These would be **common** or **shared** antigens.[10,11] Presumably, parasite antigens that did not elicit protective host responses would be neutral selective characteristics; therefore, though the host might produce antibodies against them, they would remain distinct in evolution. Assuming that the shared antigen was genetically of parasite origin, it would be referred to as an **eclipsed** antigen; if of host origin, it would be a **contaminative** antigen. Antigens may be shared by parasite and host, though whether they are eclipsed antigens is less clear. The second route of immunologic inertness could result from immune "blind spots" in the host response capacity: antigens in the parasite to which the host could not mount a response.

The possibility also exists that the parasite may manage to mask itself with host components so that the host does not recognize it as foreign. This might occur by either of two means. One, which seems to occur in a variety of nematodes, involves adsorption of host antibody to the worm's surface by **heterophile** reactions, or binding of an antigen on the worm's surface with a host antibody not specifically elicited by that antigen. This coats the worm's body with an innocuous layer of host antibody, which prevents attack by immune mechanisms, either because worm antigens are shielded and fail to stimulate a protective response or because protective antibodies are stereochemically inhibited from binding with the antigen by the harmless, bound antibody already present. The globulins IgM, IgG, and IgA from normal human serum adsorb to the cuticle of a variety of infective ascaridoid larvae, to *Nippostrongylus brasiliensis* (a nematode of rats), and even to the free-living nematode *Turbatrix aceti*. Infective larvae of *Ascaris suum* adsorb *human blood group* antibodies, indicating that these nematodes have antigens on their cuticles that correspond to human blood groups such as A, B, and Rh.[22a]

Another means by which parasites can mask themselves from the host immune system is by actually adsorbing *host antigen* so that the host immune system "sees" only "self," not recognizing the parasite as foreign; a fascinating series of experiments with the blood fluke *Schistosoma* suggest that such is the case in this worm.[5] The disease caused by infection with these flukes, schistosomiasis, is one of the most important human diseases in the world. The three species of the worm responsible for the disease in humans differ from each other in a variety of ways, one of which is host specificity. Though all three can infect several mammals other than humans, *S. japonicum* is least host specific, and *S. haematobium* is most. The degree of resistance of a given host is determined partly by innate-natural immunity and partly by acquired immunity. The proportionate importance of either varies. Humans may acquire resistance to reinfection, though our knowledge of acquired immunity to schistosomes in humans is lamentably meager. The response of rhesus monkeys, however, is much better known. These animals show a solid resistance to reinfection after a single previous infection. Initially, the monkeys are highly susceptible, and interestingly, the adult worms from the primary infection persist and are apparently unaffected by the resistance to reinfection that they engender. This has been called

concomitant immunity, or premunition. The eggs released by the adults, the juvenile worms (schistosomula), and the invasive stages (cercariae) are all antigenic to some degree, but antigens released from the adults seem to be most important in stimulating the protective response, that is, the response that facilitates destruction of the schistosomula in a secondary infection. Thus, the adults somehow must protect themselves against the response they stimulate, a response that kills juvenile stages not so protected. Evidence indicates that the adults adsorb host antigens onto their surface, and, by this means, they masquerade as host tissue. For example, if adult worms are removed from mice and transferred surgically to monkeys, the worms stop producing eggs for a time, but then recover and resume normal egg production. However, if the worms from mice are transferred to a monkey that has been previously immunized against mouse red blood cells, the worms are promptly destroyed.

Finally, we point out the survival value of concomitant immunity to the parasite. In the case of the schistosome, the severity of the disease increases with the number of worms present, so if repeated exposure to infection led to a steadily increasing worm load, the host would soon be killed, and with it, the worms it carried.

THE SPECIES PROBLEM AND CLASSIFICATION

The delineation of any species of organism is a difficult task. This applies also to parasitic forms, which in many cases are even more difficult to recognize than free-living species. Many factors contribute to the confusion. Probably the single most important factor is the paucity of known specimens. Many species of parasites, even some of those that infect humans, are known from only a few individuals. Obviously nothing can be known about infraspecific variation of morphological and physiological characteristics of such animals.

But even when parasites are common and easily obtained, their uniqueness is often in question. For example, in well-known forms like *Trypanosoma brucei, Trichinella spiralis,* and *Entamoeba histo-*

lytica, the more we learn about each species the more we become aware that it is not simply a kind of animal different and reproductively isolated from all other types, but rather it is a complex of strains, or races, having slightly different characteristics from each other. Even more sophisticated techniques are then required to separate members of species-flocks—those true species that are morphologically indistinguishable.

Most definitions of species include some considerations of reproductive isolation. Generally speaking, this is a valid concept, but there are so many exceptions that it is useful only as a general principle. For example, sexuality and biparental reproduction are completely unknown for many species. How can the concept of genetic incompatability apply to organisms that reproduce only by mitotic fission? Where would parthenogenesis fit into such a scheme? The series of asexual reproductions exhibited by digenetic trematodes have nothing in common with the usual concept of gamete exchange with resultant diploid recombinations.

Another pitfall in our attempt to recognize species is the tendency among many parasites to alter their form according to age, host, or nutrition. Juvenile parasites often bear no superficial resemblance to their parents. Who would guess that a cysticercus is a tapeworm, that a miracidium is a fluke, or for that matter, that which hatches from a flea's egg is a flea? Some parasites have altered forms, according to the host in which they find themselves. Digenetic trematodes, for example, are notorious for assuming different sizes and shapes when in different host species. This has accounted for many redundant species names, some of which may never be reconciled. The nutrition available to the parasite, the size and age of the host, and the effectiveness of the parasite's and host's defense mechanisms also tend to alter the morphology of the specimen in question. Even as adults some species have alternating parasitic and free-living phases, which are quite unlike one another. The parasite taxonomist must be able to recognize all of these variations.

Surprisingly, the taxonomist usually does recognize the variations. Over 200 years of taxonomy, based mainly on adult morphology, has established our present concept of species delineation. As new information on life cycles, genetics, and other cryptic aspects of the species has become available, it has, in the majority of instances, borne out the conclusions previously reached by an experienced taxonomist, sometimes hundreds of years previously. Recent advances in numerical taxonomy, chemotaxonomy, and serotaxonomy provide useful tools for taxonomic research in some groups. The taxonomist welcomes these additional tools to apply to the perplexing problem of defining a species population.[22]

The names given to parasites reflect the vagueness, eruditeness, and imagination of their discoverers. To the beginning student of parasitology such tongue twisters as *Macracanthorhynchus hirudinaceus* and *Leucochloridium macrostomum* may seem to be insurmountable barriers to true knowledge. Yet when these names are understood as being symbolic of discrete populations of animals, and as being used by scientists in every nation in the world regardless of their native tongue, their usefulness becomes apparent. Every described species must have a name so that we can refer to it and retrieve published knowledge about it in an efficient way. True, the names could just as easily be numerals or other symbols, but the Latin alphabet has been accepted for hundreds of years and has the advantage of possessing a wry charm. Consequently, it is much easier to remember names like *Trichinella spiralis* and *Entamoeba histolytica* than coded groups of numbers. Our system of classification has enabled us to catalog 1.5 million species of organisms in such a way that we can retrieve all published information about any one of them in a very short time, regardless of the language we speak. Scientific names, then, should not be dreaded or avoided, but should be welcomed as a simple scheme that avoids unimaginable chaos.

In this book we have included brief summaries of the classification of each major group. These can be referred to whenever the reader wishes to know where a given taxon is placed in the taxonomic hierarchy.

REFERENCES

1. Bänziger, H. 1968. Preliminary observations on a skin-piercing blood sucking moth (*Calyptra eustrigata* Hmps.) (Lep., Noctuidae) in Malaya. Bull. Ent. Res. 58:154-163.
2. Bänziger, H. 1970. The piercing mechanism of the fruit-piercing moth *Calpe (Calyptra) thalicteri* Bkh. (Noctuidae) with reference to the skin-piercing blood sucking moth *C. eustrigata* Hmps. Acta Trop. 27:54-88.
3. Cheng, T. C. 1964. The biology of animal parasites. W. B. Saunders Co., Philadelphia.
4. Cheng, T. C. 1971. Aspects of the biology of symbiosis. University Park Press, Baltimore.
5. Clegg, J. A. 1972. The schistosome surface in relation to parasitism. In Taylor, A. E. R., and R. Muller, editors. Functional aspects of parasite surfaces. Blackwell Sci. Publ. Oxford. pp. 19-40.
6. Committee on Terminology, Report of the. 1937. The terms *symbiosis, symbiont* and *symbiote.* J. Parasitol. 23:326-329.
7. Cooper, M. D., and A. R. Lawton III. 1974. The development of the immune system. Sci. Am. 231(5):59-72.
8. Cressey, R. F., and E. A. Lachner. 1970. The parasitic copepod diet and life history of diskfishes (Echeneidae). Copeia. No. 2:310-318.
9. Crompton, D. W. T. 1973. The sites occupied by some parasitic helminths in the alimentary tract of vertebrates. Biol. Rev. 48:27-83.
10. Damian, R. T. 1964. Molecular mimicry: antigen sharing by parasite and host and its consequences. Am. Naturalist 98:129-149.
11. Damian, R. T. 1967. Common antigens between adult *Schistosoma mansoni* and the laboratory mouse. J. Parasitol. 53:60-64.
12. Dobson, C. 1972. Immune response to gastrointestinal helminths. In Soulsby, E. J. L. editor. Immunity to animal parasites. Academic Press, New York. pp. 191-222.
13. Feder, H. M. 1966. Cleaning symbiosis in the marine environment. In Henry, S. M., editor. Symbiosis, Vol. 1. Academic Press, Inc., New York. pp. 327-380.
14. Henry, S. M., editor. 1966, 1967. Symbiosis, vols. 1 and 2. Academic Press, Inc., New York.
15. Insler, G. D., and L. S. Roberts. 1976. *Hymenolepis diminuta:* lack of pathogenicity in the healthy rat host. Exp. Parasitol. 39:351-357.
16. Jackson, G. J., R. Herman, and I. Singer. 1969, 1970. Immunity to parasitic animals, vols. 1 and 2. Appleton-Century-Crofts, New York.
17. Larsh, J. E., Jr., and N. F. Weatherly. 1975. Cell-mediated immunity against certain parasitic worms. In Dawes, B., editor. Advances in parasitology. Academic Press, Inc., New York. pp. 183-222.
18. Mayer, M. M. 1973. The complement system. Sci. Am. 229(5):54-66.
19. Nossal, G. J. V. 1969. Antibodies and immunity. Basic Books, Inc., Publishers, New York.
20. Read, C. P. 1972. Animal parasitism. Prentice-Hall, Inc., Englewood Cliffs, N.J.
21. Schad, G. A. 1963. Niche diversification in a parasitic species flock. Nature 198:404-406.
22. Schmidt, G. D., editor. 1969. Problems in systematics of parasites. University Park Press, Baltimore.
22a. Soulsby, E. J. L. 1971. Host reaction and nonreaction to parasitic organisms. In Gaafar, S. M., editor. Pathology of parasitic diseases. Purdue University Press, Lafayette, Ind. pp. 243-257.
23. Soulsby, E. J. L., editor. 1972. Immunity to animal parasites, Academic Press, Inc., New York.
24. Sprent, J. F. A. 1959. Parasitism immunity and evolution, In Leeper, G. S., editor. The evolution of living organisms, Melbourne University Press, Victoria, Australia. pp. 149-165.
25. Sprent, J. F. A. 1963. Parasitism. Williams and Wilkins, Baltimore.
26. Taylor, A. E. R., editor. 1965. Evolution of parasites, Blackwell Sci. Publ., Oxford.

SUGGESTED READING

Baer, J. G. 1952. Ecology of animal parasites. University of Illinois Press, Urbana, Ill.
Cameron, T. W. M. 1964. Host specificity and the evolution of helminthic parasites. In Dawes, B., editor. Advances in parasitology, vol. 2. Academic Press, New York.
Pavlovsky, E. N. 1966. Natural nidality of transmissible diseases, with special reference to the landscape epidemiology of zooanthroponoses (English translation). University of Illinois Press, Urbana, Ill.
Read, C. P. 1970. Parasitism and symbiology. Ronald, New York.

Chapter 3

PARASITIC PROTOZOA: FORM, FUNCTION, AND CLASSIFICATION

Because of their small size, protozoans were not detected until Leeuwenhoek invented his magnifying lenses in the seventeenth century. He recounted his discoveries to the Royal Society of London in a series of letters covering a period between 1674 and 1716. Among his observations were oocysts of a parasite of rabbit livers, the species known today as *Eimeria stiedai.* It was another 154 years before the second sporozoan was found, when in 1828 Delfour described gregarines from the intestines of beetles. Leeuwenhoek also found *Giardia lamblia* in his own diarrheic stools, and he found *Opalina* and *Nyctotherus* in the intestines of frogs. By the middle of the eighteenth century other parasitic protozoans were being reported at a rapid rate, and such discoveries have continued unabated to the present. At least 45,000 species of protozoa have been described to date, many of which are parasitic. Parasitic protozoa still kill, mutilate, and debilitate more people in the world than any other group of disease organisms. Because of this, studies on protozoa occupy a prominent place in parasitology and are covered in some detail in this book. We will begin their study with a review of protozoan form and function.

FORM AND FUNCTION

Every protozoan consists of a single cell, although many species contain more than one nucleus during all or portions of their life cycles. Phenomenal adaptations to wide varieties of ecological niches have evolved, a great many of which resulted in parasitic or other symbiotic associations. The success of protozoans is, to a large extent, the result of their remarkable development of organelles, which perform the same functions as do organs in higher life forms.

The phylum Protozoa consists of a large, heterogeneous assemblage of animals that are almost certainly not monophyletic. The most widely accepted system of classification for Protozoa[6] (see p. 46), recognizes four subphyla, which, were they metazoan groups, would probably be considered separate phyla. Nevertheless, studies of the fine structure of protozoans have shown that most of their organelles do not differ in any basic way in any of these subphyla or from those in metazoan cells. Indeed, Pitelka (1963) concluded "that the fine structure of protozoa is directly and inescapably comparable with that of cells of multicellular organisms," and the "morphologist has to start out by admitting that protozoa are, at the least, cells."[8] In the following discussion we shall attempt to emphasize the basic similarities of organelles in protozoans and those in other kinds of cells, and to use terminology that is consistent with modern knowledge of ultrastructure.

Nucleus and cytoplasm

Together with metazoans and most plants, protozoans are described as **eukaryotes,** that is, their genetic material (**deoxyribonucleic acid** or **DNA**) is carried on well-defined **chromosomes** combined with a protein (**histone),** and the chromosomes are contained within a membrane-bound **nucleus.** In contrast, bacteria and blue-green algae—**prokaryotes**—have their DNA as a long, coiled, single molecule, lying free in the cytoplasm. In addition to their simple chromosome, prokaryotes do not have the elaborate differentiation of membranous organelles characteristic of eukaryotes.

Like all cells, the bodies of protozoans are covered by a trilaminar **unit membrane.** The membrane appears three-layered in electron micrographs because the central lipid portion looks light or clear (electron lucent), and this is enclosed by the darker (electron dense) protein layers. As in other eukaryotes, the nuclei of protozoa are bound by a double unit membrane with pores. The several other membranous organelles characteristic of eukaryotes, such as endoplasmic reticulum, mitochondria, various membrane-bound vesicles, and Golgi bodies, are usually found in protozoa. **Mitochondria,** the organelles that bear the enzymes of oxidative phosphorylation and the tricarboxylic acid cycle, often have tubular rather than lamellar cristae in protozoans, although they may be absent altogether. The cytoplasmic matrix consists of very small granules and filaments suspended in a low-density medium with the physical properties of a colloid. Central and peripheral zones of cytoplasm can often be distinguished as the **endoplasm** and the **ectoplasm.** The endoplasm is in the **sol state** of the colloid, and it bears the nucleus, mitochondria, Golgi bodies and so on. The ectoplasm is often in the **gel state;** it appears more transparent under the light microscope, and it helps give structural rigidity to the protozoan's body. The bases of the flagella or cilia and their associated fibrillar structures, which may be very complex, are embedded in the ectoplasm. The outer membrane and structures immediately beneath the outer membrane often are referred to as the **pellicle.** Pellicular microtubules or fibrils may course just beneath the unit membrane, presumably to contribute structural integrity.

Nuclei of protozoa exhibit a wide variety of appearances, particularly under the light microscope. The most common type of nucleus, in protozoans other than ciliates, is described as **vesicular.** These nuclei are characterized by such an irregular distribution of chromatin material that "clear" areas are apparent in the nuclear sap. Condensations of chromatin within the nucleus may be peripheral or internal. One or more nucleoli may be present. **Endosomes,** conspicuous internal bodies, are thought to be analogous to nucleoli, although they do not disappear during mitosis. Parasitic amebae, trypanosomes, and phytoflagellates have endosomes. **Compact** or **condensed nuclei** are exemplified in the Ciliophora. Ciliates have two types

of nuclei, **macronuclei** and **micronuclei.** Micronuclei, as the name implies, are much smaller than macronuclei. Their major function seems to be the sequestration of genetic material for exchange during **conjugation**—that unique process of sexual reproduction in ciliates. During conjugation the micronuclei undergo meiosis, and haploid micronuclei are exchanged between two fused individuals. During conjugation the macronuclei have been resorbed and are subsequently reformed by division from the micronuclei. Macronuclei take a variety of forms, according to species, but their common function is the genetic direction of the phenotypic expression of the organism (feeding, digestion, locomotion, excretion, and so on). Macronuclei divide amitotically and are hyperpolyploid. They appear "compact" by light microscopy because clear areas of nucleoplasm are not observable, though present. On the electron microscope level, one can distinguish a large number of granules, apparently chromatin, randomly scattered throughout a fine fibrogranular reticulum. Nucleoli are large and sometimes quite numerous in ciliates.

Locomotory organelles

Protozoans move by four basic types of organelles: flagella, cilia, pseudopodia, and undulating ridges. **Flagella** are slender, whip-like structures composed of a central **axoneme** and an outer sheath that is a continuation of the cell membrane. The axoneme consists of nine peripheral and two central microtubules that are enclosed in an inner sheath. The two microtubules are bilateral, and the plane of the flagellar beat is associated with their orientation. The entire unit, the shaft and its basal fibrils and organelles, is called the **kinetid** or **mastigont.** The flagellum may be buried in the cell membrane along much of its length, forming a fin-like **undulating membrane.** A flagellum is capable of a variety of movements, which may be fast or slow, forward, backward, lateral, or spiral, The stroke may originate at the base, thereby propelling the rest of the cell ahead of it, or it may begin at its tip, effecting a force that pulls the cell behind it.

The base of the axoneme terminates in a complicated **root system,** which varies greatly in complexity in different flagellates. The entire flagellar base is sunken into an elongate, blind pouch, the **flagellar reservoir.** The axoneme arises from a small centriole, called a **basal body, kinetosome,** or **blepharoplast** in the cytoplasm. (A second, nonfunctional basal body may be found nearby.) In a simple system, such as in the trypanosomes (see Chapter 4), a large, dark-staining body called a **kinetoplast** is found near the kinetosome. It varies in structure in different species but consists of a double membrane enclosing DNA that has different genetic properties from the nucleus. The kinetoplast is part of a mitochondrion that runs most of the length of the animal's body. It appears to give rise to other mitochondria and divides by binary fission at mitosis. Under light microscopy the kinetosome and kinetoplast are often too close together to be differentiated.

The structure of the kinetosome varies in detail over the wide range of flagellated or ciliated cells, but the basic structure is always similar. Fig. 3-1, which illustrates the structure of a *Paramecium* kinetosome, can be taken as typical.[9] A short cylinder, with a constant diameter throughout its length, is formed by nine groups of three fibrils, and the fibrils in each triplet are so close together that they share a common wall where they touch. If one views a kinetosome from the base distally (Fig. 3-1, *E*), it is clear that the triplets are skewed inward in a clockwise direction, with fine filaments projecting in a cartwheel-like fashion toward a central hub from each triplet. The "lumen" of the kinetosome is open to and appears continuous with the rest of the cytoplasm in the cell. More distally in the kinetosome (Fig. 3-1, *D*), the skewing becomes less, and the cartwheel disappears. Finally, at the distal end *one* fibril in each triplet tapers to an end, and the kinetosome is closed by a somewhat more electron-dense, discoid structure called the **terminal plate** (Fig. 3-1, *C*). The position of the terminal plate is often very close to the cell surface where the shaft of the flagellum begins its protrusion. Distal to the terminal plate is a disc-shaped **axosome,** from which two central fibrils originate and continue throughout the flagellar shaft (Fig. 3-1, *B*). Thus, the flagella of

protozoa, as those of other eukaryotes, have the familiar 9 + 2 structure: a circle of nine pairs of fibrils and two central fibrils. The only motile flagellum without the 9 + 2 pattern yet discovered in eucellular organisms are the 9 + 1 flagella of the sperm tails of flatworms (see Chapter 14). The central and peripheral fibrils, plus the cytoplasm within the cylinder, constitute

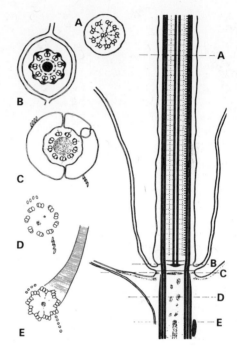

Fig. 3-1. Idealized diagrams of a *Paramecium* kinetid. At the right is shown a longitudinal section through the base of the cilium and the kinetosome, with adjoining parts of the cell membranes and membranes of the pellicular alveoli. At the left are cross sections at the levels indicated by the letters and broken lines. **A,** Free cilium. **B,** Cilium at level of axosome. **C,** Kinetosome at level of terminal plate, surrounded by a pair of alveoli; parasomal sac adjacent to kinetosome, at right; anterolateral and posterior tubular fibrils shown. **D,** Kinetosome at about its middle level, with anterolateral and posterior tubular fibrils. **E,** Base of kinetosome with cartwheel structure, kinetodesmal fibril attachment, and tubular fibrils. (From Pitelka, D. R., and F. M. Child. 1964. In Hutner, S. H., editor. Biochemistry and physiology of Protozoa, vol. 3. Academic Press, Inc., New York. pp. 131-198.)

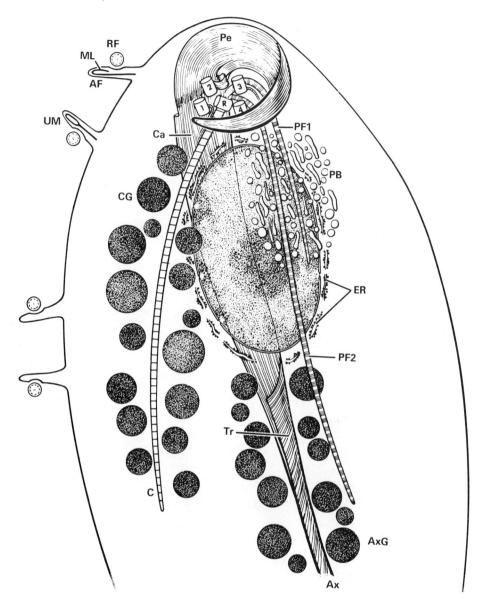

Fig. 3-2. A composite schematic diagram of a flagellate seen from a dorsal and slightly right view. *AF,* Accessory filament; *Ax,* axostyle; *AxG,* paraxostylar granules; *C,* costa; *Ca,* capitulum of the axostyle; *CG,* paracostal granules; *ER,* endoplasmic reticulum; *ML,* marginal lamella; *Pe,* pelta; *PB,* parabasal body; *PF,* parabasal filament; *R,* kinetosome of the recurrent flagellum; *RF,* the recurrent flagellum; *Tr,* trunk of the axostyle; *UM,* undulating membrane; *1* to *4,* kinetosomes of the anterior flagella.

The diagram is based on light and electron microscopic observations. The relationships among the axostyle, pelta, costa, parabasal apparatus, including the body and the one or two filaments, and nucleus were derived largely from light microscopy. In keeping with the system of numbering kinetosomes adopted by Joyon for a *Tritrichomonas augusta*-type organism, the kinetosomes of *T. gallinae* have been numbered clockwise (from the dorsal view) in a manner that would ensure spatial correspondence from *1, 2,* and *3* in *Tritrichomonas* and *Trichomonas.* Kinetosome *4* of the latter genus is missing in the former. (From Mattern, C. F. T., B. M. Honigberg, and W. A. Daniel. 1967. J. Protozool. 14:321. Reprinted with permission of The Society of Protozoologists.)

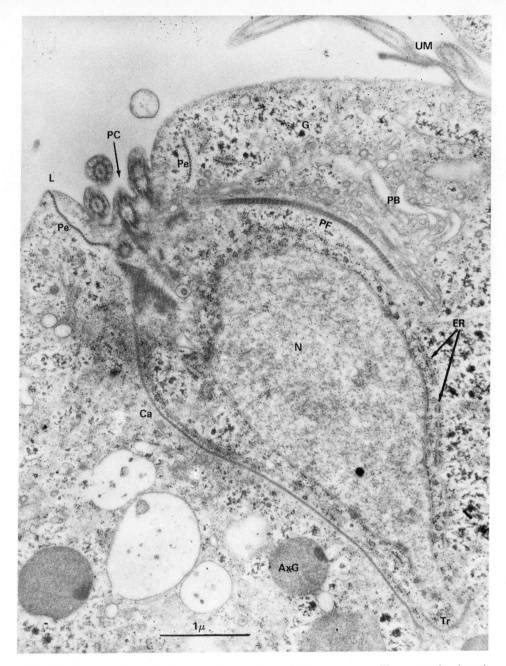

Fig. 3-3. Section through the anterior portion of *Trichomonas gallinae* seen in dorsal and slightly left view. The capitulum *(Ca)* of the axostyle and a few paraxostylar granules *(AxG)* are located ventral to the nucleus *(N)*; the parabasal body *(PB)* and its accompanying filament *(PF)* are dorsal and to the right of the nucleus. The pelta *(Pe)* extends to the extreme anterior end of the organism, terminating at the cell membrane in the area of the periflagellar lip *(L)*. A portion of the edge of the pelta is sectioned in its dorsal right aspect. The proximal segments of the flagella are seen within the periflagellar canal *(PC)*. The undulating membrane *(UM)* with its recurrent flagellum is located dorsally. Near the lower right corner note the proximal part of the axostylar trunk *(Tr)*. Two additional noteworthy features of this micrograph are the absence of structures identifiable as mitochondria and the presence of large numbers of dense granules presumed to be glycogen *(G)*. (×32,600.) (From Mattern, C. F. T., B. M. Honigberg, and W. A. Daniel. 1967. J. Protozool. 14:322. Reprinted with permission of The Society of Protozoologists.)

the **axoneme** of the flagellum. Regularly spaced, short arms often extend from one of the fibrils in each pair in the direction of the next pair (Fig. 3-1, *A*); poorly defined filaments may extend from each peripheral pair of fibrils toward the central two.

The kinetosome is of critical importance in the formation and operation of the individual flagellum. Not only are the nine peripheral pairs of fibrils in the axoneme direct extensions of two fibrils in each triplet in the kinetosome, but the kinetosome is responsible for actual formation of the axoneme. Further, when the cells divide, the kinetosomes replicate themselves (or serve as organizing centers for such replication) prior to forming the new flagella. Thus, the kinetosome closely resembles a structure found in many other animal cells, the **centriole,** which is so important in organizing fibrillar structures (for example, the spindle) during cell division. A more complex flagellar root system is found in a number of species with multiple flagella (Figs. 3-2 and 3-3). Some species have a prominent, striated rod, the **costa,** which courses from one of the kinetosomes along the margin of the organism just beneath the recurrent flagellum and undulating membrane. A tube-like **axostyle,** formed by a sheet of microtubules, may run from the area of the kinetosomes to the posterior end, where it may protrude. A Golgi body may be present, and if a periodic fibril, the **parabasal filament,** runs from the Golgi body to contact a kinetosome, the Golgi body is referred to as a **parabasal body.** The function of the parabasal body is probably similar to that of the Golgi body in other cells.

Some parasites, such as the flagellates found in cockroaches and ruminants, have dense coats of flagella covering their entire bodies. The root systems are correspondingly complex, but will not be discussed here.

One feature that differentiates flagellates with numerous flagella from the ciliates is that division of the body at fission occurs between the rows of flagella (**symmetrogenic**) but across the rows of cilia (**homothetogenic**).

Cilia are miniature flagella, also being composed of a surrounding sheath and an axoneme with two central and nine peripheral microtubules. Because cilia are

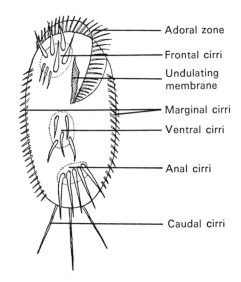

Adoral zone

Frontal cirri

Undulating membrane

Marginal cirri

Ventral cirri

Anal cirri

Caudal cirri

Fig. 3-4. *Stylonychia,* a ciliate protozoan, showing ciliary organelles. (From Kudo, R. 1966. Protozoology, ed. 5. Charles C Thomas, Publisher, Springfield, Ill.)

usually very numerous on ciliate protozoans, their root systems are complex. Each cilium has a basal body (a kinetosome) at the end of its axoneme. A fiber, the kinetodesma (plural: kinetodesmata), arises from each basal body and joins a similar fiber from the adjoining cilium in the same row. The resulting compound fiber of kinetodesmata is called a **kinetodesmose.** The row of kinetosomes and their kinetodesmose is a **kinety,** and all the kineties and associated fibrils constitute the **infraciliature.** The kinetodesmose is thought to coordinate the ciliary beat along each row. The mechanism of coordination between kineties is not known.

The mechanism by which flagella and cilia move is still obscure. The fibrils in the axoneme seem to slide past one another by a mechanism similar to the Hanson-Huxley model of muscle contraction. This hypothesis is supported by the finding that the short arms of each pair of fibrils are the sites of ATPase activity. It seems clear that movement requires ATP. So-called models can be prepared in which the cells are killed and their membrane permeability destroyed; rhythmic ciliary beats then can be restored artificially by the addition of ATP to their medium.

Several varieties of ciliary specialization

have evolved within the ciliate protozoans. One specialization involves the reduction of somatic ciliature across parts of the body. In many cases, cilia fuse to form **ciliary organelles** (Fig. 3-4). Fused somatic cilia form tuft-like brushes of cilia, called **cirri,** in some species. These function like tiny legs, enabling their owner to walk about. Most ciliary specializations occur in the oral region. If a longitudinal row of cilia fuses along its base, it forms an **undulating membrane** (which must not be confused with the undulating membrane of the flagellates). This functions in moving food particles into the oral groove. Short, transverse rows of cilia, fused at their bases to form a triangular flap, are called **membranelles** (Fig. 3-5). These also are organelles that serve to move food particles toward the cytostome.

    Pseudopodia are temporary organelles found in the Sarcodina (and other organisms) that cause the organism to move and aid it in capturing food. They do not occur in all sarcodines, for some amebas flow along with no definite body extensions (**limax forms,** named after the slug *Limax*). Four types of pseudopodia are found among the rest of the Sarcodina (Fig. 3-6). Most of the amebas have **lobopodia,** which are finger-shaped, round-tipped pseudopodia that usually contain both ectoplasm and endoplasm. All parasitic and commensal amebas of humans have this kind of pseudopodia. **Filopodia** are slender, sharp-pointed organelles, composed only of ectoplasm. They are not branched like **rhizopodia,** which branch extensively and fuse together to form netlike meshes. **Axopodia** are like filopodia but contain a slender axial filament composed of microtubules that extend into the interior of the cell.

    Movement by means of pseudopodia is a complex form of protoplasmic streaming. The mechanisms of pseudopodial flow probably vary somewhat in the different groups, and they are not well understood in any of them. The basic pattern is as follows: the ectoplasm is a plasmagel and the endoplasm is a plasmasol. At any point the plasmagel is capable of becoming a plasmasol, at which time the endoplasm streams through the weakened area to become a new pseudopodium. The plasmagel appears to contract somewhat, lending

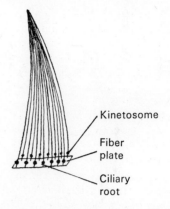

Fig. 3-5. A ciliary membranelle. (From Kudo, R. 1966. Protozoology, ed. 5. Charles C Thomas, Publisher, Springfield, Ill.)

an internal pressure favoring the flow of protoplasm. What was once part of the endoplasmic plasmasol converts quickly to an ectoplasmic plasmagel, stabilizing the boundaries of the cell. While a pseudopodium thus advances, the rear of the cell must become motile in order to keep up with the progress of its front. How it can compensate for the movements of the pseudopodia, how the pseudopodia can form and act when prey is near, how they can go in two directions simultaneously, and why pseudopodia form to remove the animal in one direction when threatened from the opposite side are entirely unknown.

    The surface of an ameba is bounded by a membrane so thin it is often called a protoplasmic surface. As the plasmasol surges forward its surface instantly becomes a gel, forming a tube through which more sol flows. The gel is somewhat contractile, forcing more sol forward in thin waves and filaments that are capable of detecting environmental conditions. This process, which also is found in human white blood cells, is not understood, although several interesting theories have been proposed.[2,3]

    In most sporozoan parasites locomotion is accomplished by **undulating ridges.** The merozoites, ookinetes, and sporozoites appear to glide through fluids with no subcellular motion whatever, but electron microscope studies reveal tiny undulatory waves that form in the cell membrane and pass posteriad. This effectively propels the cell forward, albeit at a slow rate. Subpellicular

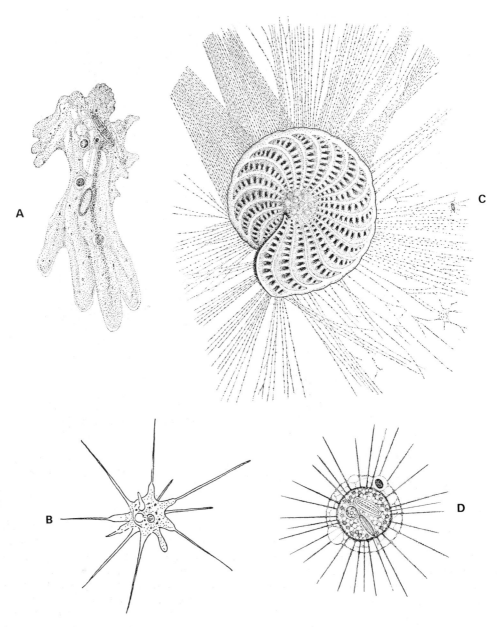

Fig. 3-6. Types of pseudopodia. **A,** Lobopodia; **B,** Filopodia; **C,** Rhizopodia; **D,** Axopodia. (From Kudo, R. 1966. Protozoology, ed. 5. Charles C Thomas, Publisher, Springfield, Ill.)

microtubules probably aid this action in some species. The mechanism is unknown.

Nutrition

Several types of nutrition are found in Protozoa. In **holophytic nutrition** (also known as **photoautotrophic nutrition**) carbohydrates are synthesized by chloroplasts, the organelle of the "typical" plant.

None of the holophytic protozoa is of importance in parasitology. Likewise those protozoa sustained by **autotrophic nutrition** are medically unimportant, for they obtain their energy by oxidation of inorganic substances, such as iron or sulfur.

Holozoic nutrition is typical of many parasitic protozoans, which feed by ingesting entire organisms or particles thereof.

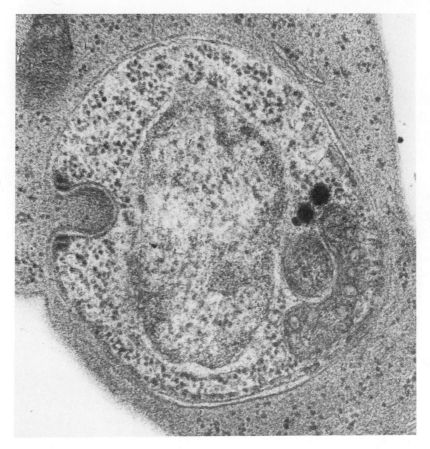

Fig. 3-7. A uninucleate trophozoite of *Plasmodium cathemerium* ingesting host cell cytoplasm through a cytostome. (From Aikawa, M., P. K. Hepier, C. G. Huff, and H. Sprinz. 1966. J. Cell Biol. 41:362.)

Their mouth openings may be temporary, as in amebas, or permanent **cytostomes,** as in most ciliates. Particulate food passes into a food vacuole, which is a digestive organelle that forms around any food thus ingested. Indigestable material is voided either through a temporary opening or through a permanent **cytopyge,** which is found in many ciliates. **Pinocytosis** is an important activity in many protozoa, as is **phagocytosis.** Pinocytosis may also contribute to food intake, but to what extent is not known.[1,5] A submicroscopic **micropyle** is present in *Eimeria* and *Plasmodium* and, in certain stages, is involved in taking in nutrients (Fig. 3-7).

In **saprozoic nutrition,** nutrients are assimilated by diffusion through the cell membrane.

Permeation, the passing of molecules directly through the outer cell membrane, may be one of three different types. **Diffusion** is possible when the cell membrane is permeable to a particular molecule and when the concentration of that molecule is lower inside the cell than outside. Few molecules satisfy these requirements; the diffusion component of nutrition for most protozoans must be negligible, though it may be more important for some parasites. A carrier molecule that has binding sites for the nutrient may be present in the cell membrane. The membrane itself might not be permeable to the nutrient but might be permeable to the carrier, which picks up the nutrient molecule at the outer membrane surface, then releases it into the cytoplasm. This mechanism is known as **facilitated diffusion.** But like free diffusion, facilitated diffusion cannot operate against a concentration gradient. The accumulation of molecules against a concentration

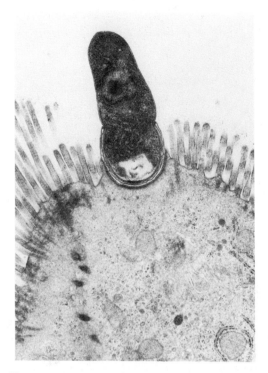

Fig. 3-8. Merozoite of *Cryptosporidium wrairi* penetrating mucosal surface of a guinea pig mucosal cell. (From Vetterling, J. M., and others. 1971. J. Protozool. 18:255. Reprinted with permission of The Society of Protozoologists.)

gradient requires the expenditure of energy and is called **active transport.** It appears that active transport operates via a carrier in the cell membrane, as in facilitated diffusion, but the action of the carrier must be coupled to an energy-yielding metabolic reaction. Obviously, the ultimate value of the molecule or the energy derived from it must be greater than the energy expended in acquiring it. Some important food molecules such as glucose are brought to the cell by active transport.

The nutrition of intracellular parasitic protozoa is so intimately bound to the metabolic activities of the host cell that it almost appears as if the parasitized cell willingly contributes to the welfare of its guest. In some, entry into the host cell is by **endocytosis,** that is, the host cell phagocytizes the parasite. An example is *Leishmania donovani,* which is eaten by reticuloendothelial cells. The host cell forms a membrane-bound vacuole around the par-

asite, but instead of killing the parasite with digestive enzymes, as might be expected, the host cell provides it with nutrients. The host is controlling the flow of materials to the parasite that will kill it.

A different mode of entry into a host cell is utilized by members of the class Microspora (see Chapter 9). The cyst stage of these parasites contains a coiled, hollow filament that apparently is under great pressure. When eaten by the host, which is usually an arthropod, the tubule is forcibly extruded from the cyst and penetrates an adjoining host cell. The organism within the spore (**sporoplasm**) crawls through the tube and enters its host. In this case, the membrane of the parasite is in direct contact with the cytoplasm of the host, with no vacuole being formed around it. The protozoan is then free to assimilate the nutrients it needs.

Active invasion of host cells by motile infective stages of protozoans is another means by which these tiny animals become intracellular parasites. Several of the Coccidia, such as *Toxoplasma, Eimeria,* and *Cryptosporidium* (Fig. 3-8), have been found to penetrate host cells actively, not by injection as in the Microspora, but by a boring action, probably aided by digestive secretions.[10] Once within the cell, the parasite is surrounded by layers of host endoplasmic reticulum, forming a **parasitophorous vacuole** (Fig. 3-9). The host cell then proceeds to provide nourishment for the parasite, as in the case of *Leishmania.*

Whether an intracellular parasite is bound by one membrane or by two, it mainly obtains its nourishment by endocytosis. When a parasitophorous vacuole surrounds the parasite, the host appears to extrude material into it, and the protozoan then takes it up by phagocytosis.[5] How the parasite thus manipulates its host to provide room and board is unknown.

Excretion and osmoregulation

Most protozoans appear to be **ammonotelic,** that is, they excrete most of their nitrogen as ammonia. Most of this readily diffuses directly through the cell membrane into the surrounding medium. Other, sometimes unidentified waste products are also produced, at least by intracellular parasites. These substances are secreted and accumulated within the host

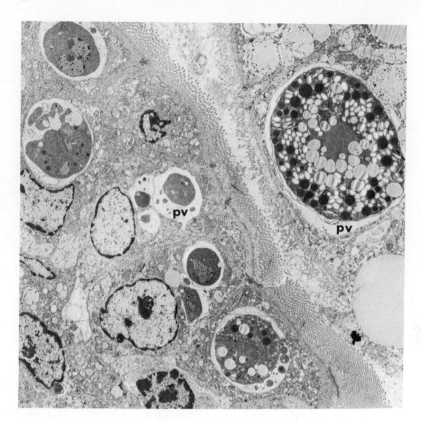

Fig. 3-9. Electron micrograph of several sexual stages of *Eimeria magna* in epithelial cells of the lower two-thirds of the small intestine of a domestic rabbit. Each parasite is surrounded by a parasitophorous vacuole *(pv)*, two of which are labeled, $5^1/_2$ days after experimental inoculation of rabbit with 200,000 oocysts. (×4,100.) (Photograph by Clarence Speer.)

cell and, on the death of the infected cell, have toxic effects on the host. Carbon dioxide, lactate, pyruvate, and short chain fatty acids are also common waste products.

Contractile vacuoles are probably more involved with osmoregulation than with excretion per se. Because free-living, freshwater protozoa are hypertonic to their environment, they imbibe water continuously by osmosis. This is effectively pumped out by the action of contractile vacuoles. Marine species and most parasites do not form these vacuoles, probably because they are more isotonic to their environment. Trypanosomes and *Balantidium,* however, do contain them.

Reproduction

Reproduction in protozoa may be either asexual or sexual, although many species alternate types in their life cycles. Most often, **asexual reproduction** is by **binary fission,** in which the individual divides into two. The plane of fission is random in Sarcodina, longitudinal in flagellates (between kineties, symmetrogenic), and transverse in ciliates (across kineties, homothetogenic). The sequence of division is kinetosome(s), kinetoplast (if present), nucleus, then cytokinesis. With the possible exception of the macronucleus of ciliates, nuclear division during asexual reproduction of protozoa is by mitosis. However, patterns of mitosis are much more diverse among the Protozoa than among the Metazoa. Details of the patterns of diversity are beyond the scope of this book but include the fact that the nuclear membrane often retains its identity through mitosis, that spindle fibers may form within the nuclear membrane, that centrioles may not be present, or that the chromosomes may not

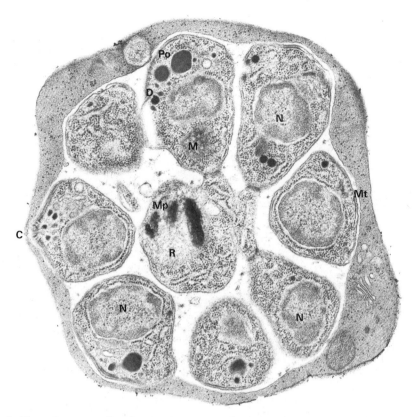

Fig. 3-10. A late stage in the development of *P. cathemerium.* The segmentation has been almost completed and a conoid *(C)*, paired organelles *(Po)*, dense bodies *(D)*, nucleus *(N)*, mitochondrion *(M)*, pellicular complex with microtubules *(Mt)*, and ribosomes are observed in the new merozoites. A residual body *(R)* surrounded by a rim of cytoplasm of the mother schizont contains a cluster of malarial pigment *(Mp)* granules. (×30,000.) (From Aikawa, M. 1966. Am. J. Trop. Med. Hyg. 15:467.)

go through a well-defined cycle of condensation and decondensation. Nevertheless, the essential features of mitosis—replication of the chromosomes and regular distribution of the daughter chromosomes to the daughter nuclei—are always present.

Multiple fission, or **schizogony,** occurs in the Sarcodina and Sporozoa. In this type of division the nucleus and other essential organelles divide repeatedly before cytokinesis; thus, a large number of daughter cells is produced almost simultaneously. During schizogony the cell is called a **schizont** or **segmenter.** The daughter nuclei in the schizont arrange themselves peripherally, and the membranes of the daughter cells form beneath the cell surface of the mother cell, bulging outward (Fig. 3-10). The daughter cells are **merozoites,** and they finally break away from a small residual mass of protoplasm

remaining from the mother cell to initiate another phase of schizogony or begin gametogony. Another type of multiple fission often recognized is **sporogony,** which is multiple fission after the union of gametes (see sexual reproduction).

Several forms of **budding** can be distinguished. **Plasmotomy,** sometimes regarded as budding, is a phenomenon in which a multinucleated individual divides into two or more smaller, but still multinucleated, daughter cells. Plasmotomy itself is not accompanied by mitosis. **External budding** is found among some complex ciliates, such as the Suctoria. Here, nuclear division is followed by unequal cytokinesis, resulting in a smaller daughter cell, which then grows to its adult size.

Internal budding, or **endopolyogeny,** differs from schizogony only in the location of the formation of daughter cells. In

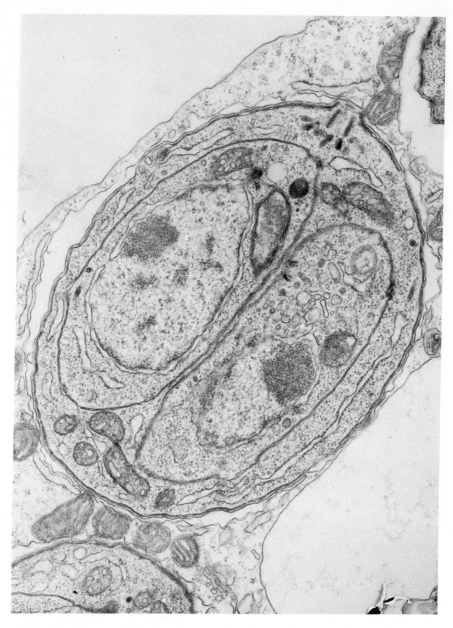

Fig. 3-11. *Toxoplasma gondii* showing two daughter cells in a mother cell, formed by endodyogeny. (From Vivier, E., and A. Petitprez. 1968. J. Cell Biol. 43:337.)

this process the daughter cells begin forming within their cell membranes, distributed throughout the cytoplasm of the mother cell rather than at the periphery. The process occurs in some stages of the schizonts of the Eimeriina. **Endodyogeny** is endopolyogeny in which only two daughter cells are formed (Fig. 3-11).

Sexual reproduction is of two basic types in parasitic Protozoa. Sexual reproduction involves reductional division in meiosis, causing a change from diploidy to haploidy, with a subsequent union of two cells to restore diploidy by amphimixis. Reproduction may be **amphimictic,** involving the union of gametes from two parents, or **automictic,** in which one parent gives rise to both gametes. Uniting gametes may

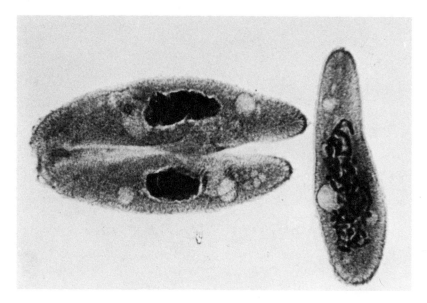

Fig. 3-12. Conjugation of *Paramecium*. (Courtesy Turtox/Cambosco.)

be entire cells or only nuclei. When they are whole cells, the union is called **syngamy.** When only nuclei unite, the process is termed **conjugation.** Conjugation is found only among the ciliates, while syngamy occurs in all other groups where sexual reproduction is found. Meiosis is known in both types of sexual reproduction.

In the majority of the protozoa, including all of the Sporozoa, meiosis occurs in the first division of the zygote (**zygotic meiosis**),[4] and all other stages are haploid. **Intermediary meiosis,** which occurs only in the Foraminifera among the Protozoa but which is widespread in plants, exhibits a regular alternation of haploid and diploid generations.

In syngamy the gametes may be outwardly similar (**isogametes**) or dissimilar (**anisogametes**). Isogamy is most common in the more primitive groups and, therefore, is considered more primitive than anisogamy. Even though isogametes look similar, they will only fuse with isogametes of another "mating type," thus avoiding inbreeding. Differences between these "sexes" in anisogametes vary from slight size differences to marked dimorphism. The larger, more quiescent of the two is the **macrogamete,** which corresponds to the ovum of metazoans. The smaller, more active gamete is the **microgamete,** which corresponds to the spermatozoan, although it is debatable whether "male" and "female" sexes can be distinguished in protozoa, or whether such a distinction is even useful. Fusion of the microgamete and macrogamete produces the **zygote,** which is often a resting stage that overwinters or forms spores that enable survival while between hosts.

Conjugation occurs only in ciliates and varies somewhat in details between species. While most complex in multinucleated protozoans, the process is basically the same as that shown by simple, two-nucleated species. Two individuals ready for conjugation unite, fusing their pellicles at the point of contact (Fig. 3-12). The macronucleus in each disintegrates. The micronucleus of each divides twice to form four haploid pronuclei, three of which disintegrate. The remaining pronucleus divides to form a stationary pronucleus and a wandering pronucleus that passes into the other conjugate. The ciliates then separate, and the pronuclei of each fuse into a zygote nucleus that divides several times to form several sets of primordia of the macronucleus and micronucleus. Each of these then grows to its definitive size, and the sets of nuclei pass into daughter cells at the first two postconjugant fissions. The resultant gene recombination lends renewed vigor to the exconjugants, which then actively reproduce by fission.

Encystment

Many protozoans can secrete a resistant covering and go into a resting stage called a **cyst.** Cyst formation is particularly common among free-living protozoans found in temporary bodies of water that are subject to drying or other harsh conditions, and among parasitic forms that must survive transferral to new hosts.[11] In addition to protection against unfavorable conditions, cysts may serve as sites for reorganization and nuclear division, followed by multiplication after excystment. In a few forms, such as *Ichthyophthirius,* a ciliate parasite of fish, the cyst falls from the host to the substrate and sticks there until excystment occurs.

The conditions favoring encystment are not fully understood, but they are thought in most cases to involve some adverse change in the environment, such as food deficiency, desiccation, increased tonicity of the environment, decreased oxygen concentration, or pH or temperature change. In parasitic species, the normal feeding form (**trophozoite,** also sometimes referred to as the **vegetative** stage) often cannot infect a new host or is too fragile to survive the transfer. Human amebiasis, caused by *Entamoeba histolytica,* is spread by persons who often have no clinical symptoms but who pass cysts in their feces (see Chapter 8). Therefore, understanding the elusive factors that induce cyst formation within the host is important.

During encystation the cyst wall is secreted, and some food reserves such as starch or glycogen are stored. Projecting portions of locomotor organelles are partially or wholly resorbed, and certain other structures, such as contractile vacuoles, may be dedifferentiated. During the process or following soon thereafter, one or more nuclear divisions give the cyst more nuclei than the trophozoite. In the flagellates and amebae cytokinesis occurs in a characteristic division pattern after excystation. In Sporozoa the cystic form is the **oocyst,** which is formed after gamete union and in which multiple fission (**sporogony**) occurs with cytokinesis to produce

sporozoites. In the coccidians the oocyst containing the sporozoites serves as the resistant stage for transferral to a new host, while in the Haemosporina (containing the causative agent of malaria, *Plasmodium*) the oocyst merely serves as a developmental capsule for the sporozoites within the insect host.

In those species in which the cyst is a resistant stage, a return of favorable conditions stimulates excystation. In parasitic forms some degree of specificity in the requisite stimuli provides that excystation will not take place except in the presence of conditions found in the host gut. Mechanisms for excystation may include absorption of water with consequent swelling of the cyst, secretion of lytic enzymes by the protozoan, and action of host digestive enzymes on the cyst wall. Excystation must include reactivation of enzyme pathways that were "turned off" during the resting stage, internal reorganization, and redifferentiation of cytoplasmic and motor organelles.

Metabolism

Because the phylum Protozoa is so large and heterogeneous, and because metabolic studies have been concentrated on so few parasitic species, few generalizations are possible or practical. Therefore, the following considerations will be rather simplified and limited, and we will comment appropriately on specific groups in subsequent chapters.

The main energy in protozoa, as in other cells, is in the form of **high energy phosphate bonds,** primarily in **adenosine triphosphate (ATP).**[7] Energy is released in the step by step, enzymatic oxidation of food molecules, and part of the energy so released is conserved by coupling these oxidations to the phosphorylation of **adenosine diphosphate (ADP)** to ATP. Subsequent hydrolysis of the high energy phosphate bond yields energy to drive other endergonic reactions in the cell. Conceptually, oxidation of the main-energy-source molecule, **glucose,** can be divided into three phases: **glycolysis,** the **Krebs** or **tricarboxylic acid cycle,** and **electron transport.** Glycolysis, sometimes called the Embden-Meyerhof pathway, is the degradation of the six-carbon compound, glu-

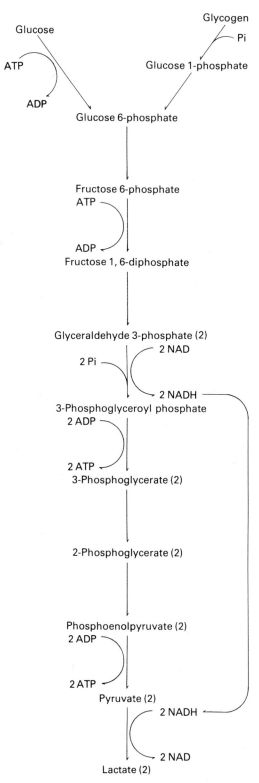

Fig. 3-13. The glycolytic pathway.

cose, to the three-carbon pyruvate (one mole glucose → two moles pyruvate) (Fig. 3-13). The pyruvate then may be reduced to lactate, or it may enter a mitochondrion and be routed into the tricarboxylic acid cycle. To enter the tricarboxylic acid cycle, pyruvate is decarboxylated to a two-carbon group and joined to the acyl carrier, **coenzyme A (CoA),** a low molecular weight compound containing the vitamin pantothenic acid. The resulting compound, **acetyl-CoA,** is condensed with oxaloacetate to form citrate. Through the series of reactions constituting the tricarboxylic acid cycle, the citrate is decarboxylated twice, finally producing another molecule of oxaloacetate to condense with another molecule of acetyl-CoA (Fig. 3-14). Since one mole of glucose produces two pyruvate, the tricarboxylic acid cycle goes through two cycles per mole of glucose.

In overview the reactions of glycolysis and the tricarboxylic acid cycle may be considered the oxidation of the carbons in glucose to carbon dioxide. Of course, in any given oxidation reaction, one compound is always **oxidized (the electron donor),** and another is always **reduced (the electron acceptor).** The tendency of a compound to give up or gain electrons determines whether it is an oxidizing or reducing agent and also the final distribution of electrons at equilibrium. In the oxidation-reduction reactions of glycolysis and then in the tricarboxylic acid cycle, the principal electron acceptor is **nicotinamide-adenine dinucleotide (NAD).** The reduced NAD is reoxidized by transferring its electrons to acceptors in the **electron transport chain.** This is a series of carriers that are alternately reduced and oxidized as they accept electrons and then donate them to the next compound in the chain. Several components in the chain are heme-containing proteins called **cytochromes,** hence the chain is sometimes referred to as the **cytochrome system.** The last electron acceptor to be reduced is molecular oxygen, producing water.

As noted, part of the energy released in the oxidation of glucose is conserved by the synthesis of high energy phosphate bonds in ATP. In glycolysis four moles of ATP are produced per mole of glucose, but two ATPs are used in phosphorylation

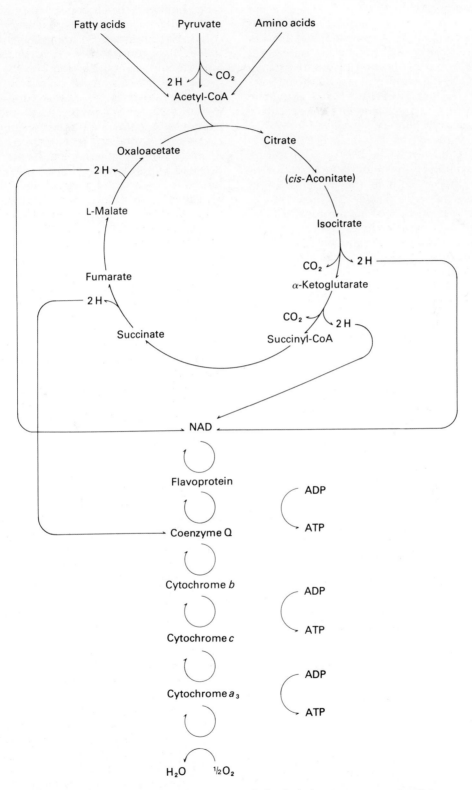

Fig. 3-14. The tricarboxylic acid cycle and classical electron transport system.

"priming" reactions; hence, there is a net gain of two ATPs. Another direct, **substrate level** phosphorylation reaction takes place in the tricarboxylic acid cycle, but a much larger proportion of the energy released by the cycle results from the passage of the reducing power it forms along the electron transport sequence by **oxidative phosphorylation.** Specifically, four ATPs are produced by substrate level (two in glycolysis and two in the Krebs cycle) and 32 by oxidative phosphorylation. Glycolysis can proceed both in the presence or in the absence of molecular oxygen, that is, either **aerobically** or **anaerobically.** The tricarboxylic acid cycle and electron transport are essentially aerobic processes, oxygen being the necessary, ultimate acceptor of all the electrons initially accepted by NAD. When glycolysis is proceeding under aerobic conditions, the two reduced NAD produced in that path may enter the mitochondria and the electron transport system, but under anaerobic conditions the reduced NAD must be reoxidized in another manner. In vertebrate muscle and some other tissues the NAD is reoxidized by reduction of the pyruvate from glycolysis to lactate.

Most of our knowledge of the preceding classical system of energy metabolism has been derived from a few species of mammals and microorganisms. Much of it has evolved from discovery to dogma in the space of a few years. It is now becoming clear, however, that parasites, both protozoan and metazoan, are unexpectedly variable in their energy metabolism, particularly in the portions after glycolysis. Some important biological factors to consider are that many parasites must survive in locations where the oxygen supply is quite limited, and that even in many cases in which oxygen is not limited neither is glucose; therefore, there is no advantage in completely oxidizing glucose. If glucose is in plentiful supply, the organism can live on little more than the energy derived from glycolysis, simply by consuming more glucose, and the partially oxidized products can be excreted as waste. The complete Krebs cycle and cytochrome system then become so much excess metabolic machinery, at least in terms of energy production. However, the problem of reoxidation of the net accumulation of reduced NAD remains, for even without the need for the energy obtainable in subsequent electron transfer, the oxidized compounds must be available for continuous functioning of glycolysis. In some parasites the electrons are transferred to the pyruvate, and lactate is excreted, as in the classical model; but many organisms excrete such compounds as succinate, acetate, and short chain fatty acids as end products of glycolysis. Some metabolic solutions to the problem of NAD oxidation will be mentioned in subsequent chapters.

Classification of Phylum Protozoa*

Subphylum SARCOMASTIGOPHORA

Flagella, pseudopodia, or both types of locomotory organelles; single type of nucleus except in developmental stages of certain Foraminiferida; typically no spore formation; sexuality, when present, essentially syngamy.

Superclass MASTIGOPHORA

One or more flagella typically present in trophozoites; solitary or colonial; asexual reproduction basically by symmetrogenic binary fission; sexual reproduction unknown in many groups; nutrition phototrophic, heterotrophic, or both.

*Modified from the scheme adopted by the Society of Protozoologists in 1964.[6] Groups without important parasitic species are omitted.

Class PHYTOMASTIGOPHOREA

Typically with chromatophores; if chromatophores lost secondarily, relationship to pigmented forms clearly evident; commonly only one or two emergent flagella; ameboid forms frequent in some groups; sexual reproduction known with certainty in few orders; mostly free-living.

Order DINOFLAGELLIDA

Two flagella, typically one transverse and one trailing; body usually grooved transversely and longitudinally, forming girdle and sulcus, each containing a flagellum; chromatophores usually yellow or dark brown, occasionally green or blue-green; many species thecate; starch and lipids as food reserves.

Class ZOOMASTIGOPHOREA

Chromatophores absent; one to many flagella; additional organelles may be present in mastigonts; ameboid forms with or without flagella in some groups; sexuality known in a few groups; species predominantly symbiotic.

Order RHIZOMASTIGIDA

Pseudopodia and one to four flagella.

Order KINETOPLASTIDA

One to four flagella; kinetoplast argentophobic and Feulgen-positive, present as self-replicating organelle with mitochondrial affinities; most species parasitic.

Suborder BODONINA

Typically two unequal flagella, one directed anteriorly, other posteriorly; no undulating membrane; kinetoplast absent secondarily in some species; free-living or parasitic.

Suborder TRYPANOSOMATINA

One flagellum either free or attached to body by means of undulating membrane; all species parasitic.

Order RETORTAMONADIDA

Two to four flagella, one turned posteriorly and associated with ventrally located cytostomal area; cytostome bordered by fibril; parasitic.

Order DIPLOMONADIDA

Body bilaterally symmetrical, with two karyomastigonts, each with four flagella and a set of accessory organelles; most species parasitic.

Order OXYMONADIDA

One or more karyomastigonts, each with four flagella, typically in two pairs, in motile stages; one or more flagella may be turned posteriorly, adhering for some distance to body surface; one to many axostyles; spindle intranuclear division; no Janicki-type parabasal apparatus; sexuality in some species; all parasitic.

Order TRICHOMONADIDA

Typically four to six flagella, of which one is recurrent per mastigont system; undulating membrane, if present, associated with recurrent flagellum; axostyle and nondividing argentophilic Janicki-type parabasal apparatus (an organelle of the Golgi type) in each mastigont; division spindle extra-nuclear; sexuality in some species; true cysts unknown; all parasitic.

Order HYPERMASTIGIDA

Mastigont system with numerous flagella and multiple parabasal apparatus; basal bodies (kinetosomes) distributed in complete or partial circle, in plate or plates, or in longitudinal or spiral rows meetinq anteriorly in centralized structure; nucleus single; division spindle extranuclear; sexuality in some species; all symbiotic.

Suborder LOPHOMONADINA

Extranuclear organelles arranged in one system; typically resorption of all old structures in division, with formation de novo of daughter organelles.

Suborder TRICHONYMPHINA

Organization basically bilateral, with two or occasionally four mastigont systems; typically equal separation of systems in division, with total or partial retention of old structures when new systems are formed.

Superclass OPALINATA

Numerous cilia-like organelles in oblique rows over entire body surface; cytosome absent; two to many nuclei of one type; nuclear division acentric, binary fission generally interkinetal, thus usually symmetrogenic: known life cycles involve syngamy with anisogamous flagellated gametes; all symbiotic.

Order OPALINIDA

With characters of the superclass.

Superclass SARCODINA

Pseudopodia typically present; flagella, when present, restricted to developmental stages; cortical zone of cytoplasm relatively undifferentiated in comparison with other major taxa; body naked or with external or internal tests or skeletons of various types and chemical composition; asexual reproduction by fission; sexual reproduction, if present, with flagellate or, more rarely, ameboid gametes; most species free-living.

Class RHIZOPODEA

Locomotion associated with formation of characteristic lobopodia, filopodia, or reticulopodia; nutrition, phagotrophic.

Subclass LOBOSIA

Pseudopodia typically lobose, rarely filiform or anastomosing.

Order AMOEBIDA

Naked; typically uninucleate; majority freeliving, many parasitic.

Subclass LABYRINTHULIA

Groupings of spindle-shaped individuals that glide along filamentous tracks, forming a slime net; occurrence of ameboid stage in life cycle not clearly established; on marine plants and in soil.

Order LABYRINTHULIDA

With characters of subclass.

Class ACTINOPODEA

Spherical, typically floating forms, some attached secondarily; pseudopodia typically delicate and radiose, either axopodia or with filose or reticulate patterns; naked or with test; test membranous chitinoid, or of silica or strontium sulfate; reproduction asexual and sexual; gametes usually flagellated.

Subclass PROTEOMYXIDIA

Without test; filopodia and reticulopodia formed in some species; flagellated swarmers and cysts present in some species; most species parasitic.

Order PROTEOMYXIDA

With the characters of the subclass.

Subphylum APICOMPLEXA

Apical complex generally consisting of polar ring, micronemes, rhoptries, subpellicular tubules, and conoid present at some stage; micropore(s) generally present; single type of nucleus; cilia and flagella absent except for flagellated microgametes in some groups; cysts often present; all species parasitic.

Class SPOROZOA

Spores typically present; spores simple, without polar filaments and with one to many sporozoites; single type of nucleus; cilia and flagella absent except for flagellated microgametes in some groups; sexuality, when present, syngamy; all species parasitic.

Subclass GREGARINIA

Mature trophozoites extracellular, large; parasites of digestive tract and body cavity of invertebrates.

Order ARCHIGREGARINIDA

Life cycle apparently primitive, characteristically with three schizogonies; parasites of annelids, sipunculids, enteropneustids, and ascidians.

Order EUGREGARINIDA

Schizogony absent; parasites of annelids and arthropods.

Suborder ACEPHALINA

Trophozoite composed of single compartment.

Suborder CEPHALINA

Trophozoite septate, composed of more than one compartment.

Order NEOGREGARINIDA

Schizogony present, presumably reacquired secondarily; parasites of insects.

Subclass COCCIDIA

Mature trophozoites small, typically intracellular.

Order PROTOCOCCIDA

Schizogony absent; two species known in marine annelids.

Order EUCOCCIDA

Schizogony present; asexual and sexual phases in life cycle; in epithelium and blood cells of invertebrates and vertebrates.

Suborder ADELEINA

Macrogametocyte and microgametocyte associated in syzygy during development; microgametocyte usually produces few microgametes; sporozoites enclosed in envelope; monoxenous and heteroxenous.

Suborder EIMERIINA

Macrogametocyte and microgametocyte develop independently; syzygy absent; microgametocyte typically produces many microgametes; zygote nonmotile; oocyst does not increase in size during sporogony; sporozoites typically enclosed in sporocyst; monoxenous or heteroxenous.

Suborder HAEMOSPORINA

Macrogametocyte and microgametocyte develop independently; syzygy absent; microgametocyte produces moderate number of microgametes; zygote motile in some forms; oocysts increase in size during sporogony, sporozoites naked; heteroxenous; schizogony in vertebrate and sporogony in invertebrate host; pigment ordinarily formed from host cell hemoglobin.

Class PIROPLASMEA

Small, piriform, round, rod-shaped or ameboid; spores absent; no flagella or cilia; locomotion by body flexion or gliding; asexual reproduction by binary fission or schizogony; pigment not formed from host cell hemoglobin; heteroxenous; parasitic in vertebrate erythrocytes; known vectors ticks.

Subphylum MYXOSPORA

Spores of multicellular origin; one or more sporoplasms; with two or three (rarely one) valves.

Order MYXOSPORIDA

Spore with one or two sporoplasms and one to six (typically two) polar capsules; each capsule with coiled polar filament; filament probably with anchoring function; spore membrane generally with two, occasionally up to six, valves; coelozoic or histozoic in cold-blooded vertebrates.

Suborder UNIPOLARINA

One to six polar capsules at or near anterior end of spore; capsules sometimes widely separated or located in central area, but polar filaments attached near anterior end.

Suborder BIPOLARINA

Two widely separated polar capsules, one located and opening at or near each end of spore.

Order ACTINOMYXIDA

Spore with three polar capsules, each enclosing polar filament; membrane with three valves; several to many sporoplasms; in invertebrates, especially annelids.

Order HELICOSPORIDA

Spore with three sporoplasms surrounded by spirally coiled, thick filament; spore membrane with one valve; histozoic in insects.

Subphylum MICROSPORA

Spore of unicellular origin; single sporoplasm; single valve; one long, tubular polar filament through which sporoplasm emerges; cytozoic in invertebrates, especially arthropods, and lower (rarely higher) vertebrates.

Order MICROSPORIDA

With characters of the class.

Suborder MONOCNIDINA

Spores independent.

Suborder DICNIDINA

Two spores each with one polar filament, fused at base.

Subphylum CILIOPHORA

Simple cilia or compound ciliary organelles in at least one stage of life cycle; subpellicular infraciliature universally present even when cilia absent; two types of nucleus, except in a few homokaryotic forms; binary fission basically homothetogenic and generally perikinetal; sexuality involving conjugation, autogamy, and cytogamy; nutrition heterotrophic; most species free-living.

Class CILIATA

With the characters of the subphylum.

Subclass HOLOTRICHIA

Somatic ciliature often simple and uniform; buccal ciliature present in only two orders, basically tetrahymenal and generally inconspicuous.

Order TRICHOSTOMATIDA

Somatic ciliature typically uniform, but highly asymmetrical in some forms; vestibular but no buccal ciliature in oral area; some species parasitic.

Order CHONOTRICHIDA

Somatic ciliature absent in mature individuals; vestibular ciliature in apical "funnel" derived from field of ventral cilia present on migratory larval forms; adults vase-shaped, attached to crustaceans by noncontractile, nonscopula-produced stalk; reproduction by budding.

Order APOSTOMATIDA

Somatic ciliature of mature forms spirally arranged; typically with unique rosette near inconspicuous cytosome; polymorphic life cycles with marine crustaceans usually involved as hosts.

Order ASTOMATIDA

Somatic ciliature typically uniform; cytostome absent; often large; some species with endoskeletons and holdfast organelles; catenoid "colonies" typical of some groups; mostly parasitic in oligochaetes.

Order HYMENOSTOMATIDA

Somatic ciliature typically uniform; buccal cavity ventral, with ciliature fundamentally composed of one undulating membrane on the right and with adoral zone of three membranelles on the left; often small.

Suborder TETRAHYMENINA

Oral ciliature usually inconspicuous, composed of undulating membrane and three membranelles; vestibulum seldom present; body typically small.

Suborder PENICULINA

Oral ciliature dominated by presence of "peniculi" deep in buccal cavity; outer vestibulum with uniform vestibular ciliature often present; usually large.

Suborder PLEURONEMATINA

Somatic ciliature usually sparse, but prominent caudal cilium common; oral ciliature dominated by conspicuous external, undulating membrane; adoral zone of membranelles hardly recognizable; vestibulum absent; cytostome typically subequatorial.

Order THIGMOTRICHIDA

Tuft of thigmotactic somatic ciliature typically present near anterior end of body; buccal ciliature, if present, located subequatorially on ventral surface or at posterior end; usually parasitic in or on bivalve molluscs.

Suborder ARHYNCHODINA

Somatic ciliature usually uniform; oral ciliature and cytostome present.

Suborder RHYNCHODINA

Somatic ciliature often reduced, absent in some forms; cytostome replaced by anterior sucker, occasionally on end of tentacle.

Subclass PERITRICHIA

Somatic ciliature essentially absent in mature form; oral ciliature conspicuous, winding around apical pole counterclockwise to cytostome; body often attached to substrate by contractile, scopula-produced stalk or by prominent adhesive basal disc; colonial organization common; migratory larval form with aborally located ciliary girdle.

Order PERITRICHIDA

With characters of the subclass.

Suborder SESSILINA

Predominantly sessile; with contractile or noncontractile stalk; some loricate; solitary or colonial.

Suborder MOBILINA

Motile; without stalk; oral-aboral axis shortened; commonly with prominent adhesive basal disc at aboral end of body; often parasitic on or in aquatic hosts.

Subclass SUCTORIA

Mature stage without external ciliature of any kind; typically sessile forms, attached to substrate by noncontractile, scopula-produced stalk; ingestion through few to many suctorial tentacles; astomatous migratory larval stage produced by budding, with some somatic cilia.

Order SUCTORIDA

With characters of the subclass.

Subclass SPIROTRICHIA

Somatic ciliature sparse in all but one order; cirri dominant feature of one order; buccal ciliature conspicuous, with adoral zone, typically composed of many membranelles, winding clockwise to cytostome; body often large.

Order HETEROTRICHIDA

Somatic ciliature, when present, usually uniform; body frequently large; some species pigmented; a few species loricate with migratory larval forms.

Suborder HETEROTRICHINA

With characters of the order, in the strictest sense.

Suborder LICNOPHORINA

Somatic ciliature absent; conspicuous buccal membranelles on prominent oral disc; elaborate basal disc used for attachment; almost exclusively marine; ectoparasitic.

Order OLIGOTRICHIDA

Somatic ciliature sparse or absent; buccal membranelles conspicuous, often extending around apical end of body; typically small; mostly marine, some commensal in ruminants.

Order ENTODINIOMORPHIDA

Simple somatic ciliature absent; oral membranelles functional in feeding, restricted to small area; other membranellar tufts or zones present in many species; pellicle firm, often drawn out posteriorly into spines; commensal in herbivores.

REFERENCES

1. Aikawa, M. 1971. Parasitological review: *Plasmodium:* the fine structure of malarial parasites. Exp. Parasitol. 30:284-320.
2. Allen, R. D. 1961. A new theory of ameboid movement and protoplasmic streaming. Exp. Cell Res. 8:17-31.
3. Allen, R. D., D. Francis, and R. Zeh. 1971. Direct test of the positive pressure gradient theory of pseudopod extension and retraction in amoebae. Science 174:1237-1240.
4. Grell, K. G. 1973. Protozoology. Springer-Verlag New York Inc., New York.
5. Hammond, D. M., E. Scholtyseck, and B. Chobotar. 1967. Fine structure associated with nutrition of the intracellular parasite *Eimeria auburnensis.* J. Protozool. 14:678-683.
6. Honigberg, B. M., W. Balamuth, E. C. Bovee, and others. 1964. A revised classification of the phylum Protozoa. J. Protozool. 11:7-20.
7. Lehninger, A. L. 1975. Biochemistry, ed. 2. Worth Publishers, Inc., New York.
8. Pitelka, D. R. 1963. Electron-microscopic structure of Protozoa. Pergamon Press, Inc., Elmsford, New York.
9. Pitelka, D. R., and F. M. Child. 1964. The locomotor apparatus of ciliates and flagellates: Relations between structure and function. In Hutner, S. H., editor. Biochemistry and physiology of Protozoa, vol. 3. Academic Press, Inc., New York. pp. 131-198.
10. Sheffield, H. J., and M. L. Melton. 1968. The fine structure and reproduction of *Toxoplasma gondii.* J. Parasitol. 54:209-226.
11. van Wagtendonk, W. J. 1955. Encystment and excystment of Protozoa. In Hutner, S. H., and A. Lwoff, editors. Biochemistry and physiology of Protozoa, vol. 2. Academic Press, Inc., New York.

SUGGESTED READING

Cosgrove, W. B. 1973. Why Kinetoplasts? J. Protozool. 20:191-194.

Hyman, L. H. 1940. The invertebrates: I. Protozoa through Ctenophora. McGraw-Hill Book Co., New York. (An excellent reference to general aspects of the Protozoa.)

Jahn, T. L., and F. F. Jahn. 1949. How to know the Protozoa. William C. Brown Co., Publishers, Dubuque, Iowa. (Identification keys to the common Protozoa.)

Kudo, R. R. 1966. Protozoology, ed. 5. Charles C Thomas, Publisher, Springfield, Ill. (A classic reference to classification of the Protozoa.)

Levine, N. D. 1973. Protozoan parasites of domestic animals and man, ed. 2. Burgess Publishing Co., Minneapolis. (This is a very useful reference to the parasites indicated by the title.)

Smyth, J. D. 1973. Some interface phenomena in parasitic Protozoa and platyhelminths. Can. J. Zool. 51:367-377.

Thompson, P. L. 1971. Perspectives and important needs in parasite chemotherapy. J. Parasitol. 57:3-8. (This article enumerates areas where parasite chemotherapy is currently inadequate.)

Trager, W. 1974. Some aspects of intracellular parasitism. Science 183:269-273. (An interesting short account of intracellular parasitism, with emphasis on nutrition.)

Chapter 4

ORDER KINETOPLASTIDA: TRYPANOSOMES AND THEIR KIN

heteroxenous
single nucleus
kinetoplast
one flagellum

? Haemoflagellates

Family Trypanosomatidae

Most members of the family Trypanosomatidae are **heteroxenous**—during one stage of their lives, they live in the blood and/or fixed tissues of all classes of vertebrates, and during other stages, they live in the intestines of blood-sucking invertebrates. Thus they usually are called hemoflagellates. All species are either elongate with a single flagellum or rounded with a very short, nonprotruding flagellum. All forms have a single nucleus.

single flagellum
single nucleus

The flagellum arises from the kinetosome and, when well developed, propels the organism (Fig. 4-1). The flagellum may also attach the organism to an insect host's gut wall or salivary gland epithelium.[30] Closely associated with, and usually posterior to, the kinetosome is a structure unique to the order, the **kinetoplast.** The sausage or disc-shaped kinetosome contains the mitochondrial DNA, and the mitochondria arise from it. Electron micrographs reveal the DNA fibers running in an anterior-posterior direction within the kinetosome. The kinetoplast and the kinetosome are always closely associated, sometimes so close that they appear as a single body under the light microscope. Most electron micrographs show no physical connection between the kinetosome and kinetoplast, but a few have shown electron-dense "junctions" between the structures in some species.[18]

The family originally parasitized the digestive tract of insects and, possibly, annelids. Many species are still only parasitic within a single arthropod host **(monoxenous).** Some species pass through different morphological stages, depending on the phase of their life cycle and the host they are parasitizing. In the past, these stages were named after the genera they most resembled, such as crithidium or lep-

different morphological stages

tomonad, but currently a nomenclature referring to the flagellum prevails.

Before proceeding to the general characteristics of these stages, we will describe the morphology of one of them, the **trypomastigote** stage, more fully. In this form the kinetoplast and kinetosome are near the posterior end of the body, while the flagellum runs along the surface, usually continuing as a free whip anterior to the body. Running along the axoneme within the flagellar membrane is a **paraxial rod** (Fig. 4-2). The flagellar membrane is closely apposed to the body surface, and when the flagellum beats, that area of the pellicle is pulled up into a fold; the fold plus the flagellum constitute the **undulating membrane.** A second, "barren" kinetosome, without a flagellum, is usually found near the flagellar kinetosome. In the typical, bloodstream form of trypomastigote, a simple mitochondrion with tubular cristae, or without cristae, runs anteriorly from the kinetoplast, and a very short mitochondrion may run posteriorly. Both the anterior and the posterior mitochondria are much larger and more complex, with lamellar cristae, in the insect stage of the organism. At the base of the flagellum and surrounding the kinetosome is a **flagellar pocket** or reservoir. A system of **pellicular microtubules** spirals around the body just beneath the cell membrane (Fig. 4-2). These give supportive resistance to the deformation of the body caused by the beating flagellum.[32] Rough endoplasmic reticulum is well developed, and a Golgi body is found between the nucleus and kinetosome.

The trypomastigote is the definitive stage of the genus *Trypanosoma.* The other forms differ in body shape, position of the kinetosome and kinetoplast, or development of the flagellum (Fig. 4-3). A

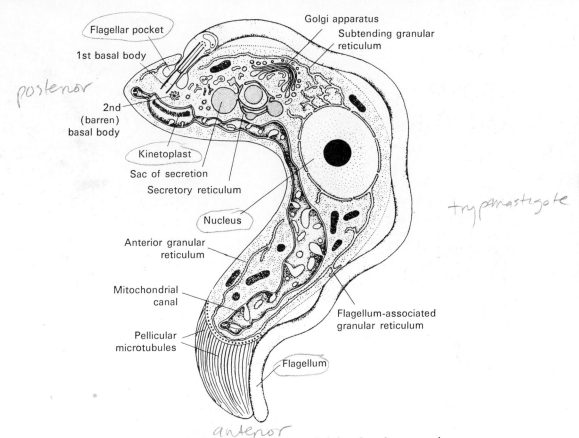

posterior

trypomastigote

anterior

- Flagellar pocket
- 1st basal body
- Golgi apparatus
- Subtending granular reticulum
- 2nd (barren) basal body
- Kinetoplast
- Sac of secretion
- Secretory reticulum
- Nucleus
- Anterior granular reticulum
- Mitochondrial canal
- Pellicular microtubules
- Flagellum-associated granular reticulum
- Flagellum

Fig. 4-1. Diagram to show principal structures revealed by the electron microscope in the bloodsteam trypomastigote form of the salivarian trypanosome, *Trypanosoma congolense*. It is seen cut in sagittal sections, except for most of the shaft of the flagellum and the anterior extremity of the body. (From Vickerman, K. 1969. J. Protozool. 16:54-69.)

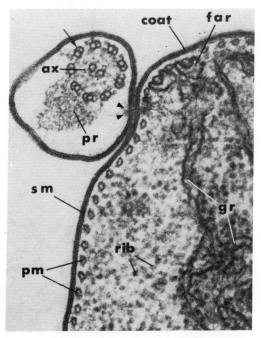

coat   far
ax
pr
s m
gr
pm
rib

Fig. 4-2. *Trypanosoma congolense.* Transverse section of shaft of flagellum and adjacent pellicle in region of attachment. Both flagellum and body surface have a limiting unit membrane *(sm)* covered by a thick coating *(coat)* of dense material. The axoneme *(ax)* of the flagellum shows the partition (arrowed) dividing one of the tubules of each doublet; alongside the axoneme lies the paraxial rod *(pr)*. Pellicular microtubules *(pm)* underlie the surface membrane of the body and a diverticulum *(far)* of the granular reticulum *(gr)* is always found embracing three or four of these microtubules close to the flagellum. Note the fibrous condensations (at heads of arrows) on either side of the opposed surface membranes, apparently "riveting" the flagellum to the body. A row of these "rivets" replace a microtubule along the line of adherence. (×66,000.) (From Vickerman, K. 1969. J. Protozool. 16:54-69.)

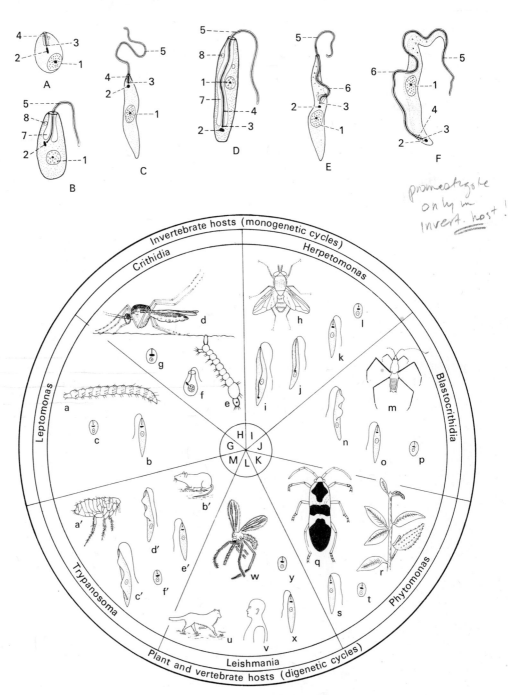

Fig. 4-3. For legend see opposite page.

Leishmania — amastigote

spheroid **amastigote** occurs in the life cycle of some species and is definitive in the genus *Leishmania*. The flagellum is very short, projecting only slightly beyond the flagellar pocket.

In the **promastigote** stage the elongate body has the flagellum extending forward as a functional organelle. The kinetosome and kinetoplast are located in front of the nucleus, near the anterior end of the body. The promastigote form is found in the life cycles of several species while they are in their insect hosts. It is the mature form in the genus *Leptomonas*. If the flagellum emerges through a wide, collar-like process, the type is termed a **choanomastigote**, which is found in some species of *Crithidia* (parasitic in insects).

The **epimastigote** form is encountered in some life cycles. Here, the kinetoplast and kinetosome are still located between the nucleus and the anterior end, but a short undulating membrane lies along the proximal part of the flagellum. The genera *Crithidia* and *Blastocrithidia*, both parasites of insects, exhibit this form during their life spans. Finally, the **opisthomastigote** form is found in *Herpetomonas*, a widespread group of insect parasites. The kinetosome and kinetoplast are located between the nucleus and posterior end, but there is no undulating membrane. The flagellum pierces a long reservoir that passes through the entire length of the body and opens at the anterior end.

GENUS TRYPANOSOMA

All trypanosomes (except *T. equiperdum*) are **heteroxenous**, or at least are transmitted by an animal vector. Various species pass through amastigote, promastigote, epimastigote, and/or trypomastigote stages, with the other forms developing in the invertebrate hosts. Much research has been conducted on this genus because of its extreme importance to the health of humans and domestic animals. Relatively recent reviews are available, dealing with various aspects of the group: physiology and morphology,[8,29,30] chemotherapy,[6,35] taxonomy,[10,23] immunology,[26,31] and evolution.[2,12] Wallace outlined the trypanosomatids of arachnids and insects.[33] It is possible that some species now known only from invertebrates may also be parasites of vertebrates.[17]

Members of the genus *Trypanosoma* are parasites of all classes of vertebrates. Most live in the blood and tissue fluids, but some important ones, such as *T. cruzi*, occupy

T. cruzi — bloodstream AND intracellular

Fig. 4-3. Genera of Trypanosomatidae and their basic life cycles. **A,** *Leishmania* (amastigote form); **B,** *Crithidia* (choanomastigote); **C,** *Leptomonas* (promastigote); **D,** *Herpetomonas* (opisthomastigote); **E,** *Blastocrithidia* (epimastigote); **F,** *Trypanosoma* (trypomastigote); **G,** life cycle of *Leptomonas ctenocephali* with amastigote and promastigote in dog flea *(Ctencephalides canis)*; **H,** life cycle of *Crithidia fasciculata* with amastigote and choanomastigote in mosquitoes; **I,** life cycle of *Herpetomonas muscarum* with amastigote, promastigote, and opisthomastigote in houseflies; **J,** life cycle of *Blastocrithidia gerridis* with amastigote, promastigote, and epimastigote in water striders; **K,** life cycle of *Phytomonas elmassiani* with amastigote and promastigote from milkweed bug *(Oncopeltus fasciatus)* and milkweeds *(Asclepias servus)*; **L,** life cycle of *Leishmania donovani* with promastigotes in sand flies *(Phlebotomus)* and amastigotes in reticuloendothelial cells of humans, dogs, and other mammals; **M,** life cycle of *Trypanosoma lewisi* with trypomastigotes in blood of rats and epimastigotes, promastigotes, and amastigotes in the gut of rat fleas *(Nosopsyllus fasciatus)*. *1,* Nucleus; *2,* kinetoplast; *3,* kinetosome; *4, 5,* axoneme and flagellum; *6,* undulating membrane; *7,* flagellar pocket; *8,* contractile vacuole. *a,* Larval flea; *b,* promastigote; *c,* amastigote; *d,* adult mosquito; *e,* larval mosquito (wriggler); *f,* choanomastigote; *g,* amastigote; *h,* housefly; *i,* fully developed opisthomastigote; *j,* developing opisthomastigote; *k,* promastigote; *l,* amastigote; *m,* water strider; *n,* epimastigote; *o,* promastigote; *p,* amastigote; *q,* common milkweed bug; *r,* milkweed; *s,* promastigote; *t,* amastigote; *u, v,* mammalian hosts (dogs, humans); *w,* sand fly invertebrate host; *x,* promastigote; *y,* amastigote; *a′,* flea invertebrate host; *b′,* rat vertebrate host; *c′,* trypomastigote; *d′,* epimastigote; *e′,* promastigote; *f′,* amastigote. (From Olsen O. W. 1974. Animal parasites, their biology and life cycles. University Park Press, Baltimore.)

intracellular habitats as well. Although other means of transmission exist, the majority are transmitted by blood-feeding invertebrates. Most trypanosomes parasitize animals of no particular importance to humans, but a few species are responsible for misery and privation of enormous proportions. In Africa alone, 4.5 million square miles, an area larger than the United States, cannot support agriculture—not because the land is poor, for it is much like the grasslands of the American west, but because domestic livestock are killed by trypanosomes that are nonfatal parasites of native grazing animals. Thus, semiarid lands that otherwise could support agronomy are denied to millions of persons who most need the protein affordable by the rich soil. More directly affected by trypanosomes are the millions of people in South America who have never known a day of good health because of trypanosome infections.

Trypanosomes are divided into two broad groups, or "sections," based on characteristics of their development in their invertebrate hosts. If a species develops in the front portions of the digestive tract, it is said to develop **anterior station** and is relegated to the section *Salivaria,* which contains several subgenera. When a species develops in the hind gut of its invertebrate host, it is said to develop **posterior station,** and is placed in the section *Stercoraria.* Other developmental and morphological criteria separate the two sections and further aid in placement in the proper section of species that do not require development in an intermediate host (*T. equiperdum, T. equinum*).[10] Classification of the various species into subgenera is based on their physiology, morphology, and biology.

Section Salivaria
Trypanosoma (Trypanozoon) brucei brucei

Three subspecies of *Trypanosoma* are morphologically indistinguishable and traditionally have been considered separate species. These are *T. b. brucei, T. b. gambiense,* and *T. b. rhodesiense.* They vary in infectivity for different species of hosts and produce somewhat different pathological syndromes. It seems clear that the ancestral form in the complex is represented by

T. b. brucei, and the other two should be considered subspecies of *Trypanosoma brucei.*[12,23] These trypanosomes are widely distributed in tropical Africa between 15° N and 25° S latitude, corresponding in distribution with their vectors, tsetse flies (*Glossina* spp.) (Fig. 4-4).

T. b. brucei is fundamentally a parasite of the bloodstream of native antelopes and other African ruminants, causing a disease called **nagana.** Unfortunately, the parasite also infects introduced livestock, including sheep, goat, ox, horse, camel, pig, dog, donkey, and mule. It is quite pathogenic to these animals, as well as to several native species. Humans, however, are not susceptible.

Morphology and life history. In recent years correlated findings from the fields of cytology, biochemistry, and immunology have combined to make the life history of these trypanosomes one of the most fascinating stories of development in parasitology. We will trace the sequence of events in the life history itself and then briefly consider the physiological and ultrastructural changes associated with these events.

Trypanosoma brucei in natural infections tends to be quite **pleomorphic** (polymorphic) in its vertebrate host, ranging from long, slender trypomastigotes with a long free flagellum, through intermediate forms, to short, stumpy individuals with no free flagellum. The small kinetoplast is usually very near the pointed posterior end, and the undulating membrane is conspicuous.

The insect vectors are *Glossina morsitans, G. pallidipes,* and *G. swynnertoni*—common tsetse flies throughout much of Africa. At least 90% of the flies are retractive to infection. When eaten by a susceptible fly, along with a blood meal, *T. brucei* locates in the posterior section of the midgut of the insect, where it multiplies in the trypomastigote form for about 10 days. At the end of this time, the slender individuals produced migrate forward into the foregut, where they are found on the twelfth to twentieth days. They then migrate farther forward into the esophagus, pharynx, and hypopharynx and enter the salivary glands. Once in the salivary glands they transform into the epimastigote form and attach to host cells or lie free in the lumen.

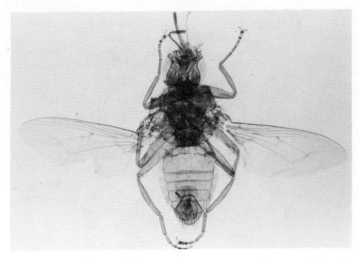

Fig. 4-4. *Glossina,* a tsetse fly. Species in this genus are vectors of some types of *Trypanosoma.* (Photograph by Warren Buss.)

After several generations they transform into the **metacyclic trypomastigote** form, which is small, stumpy, and lacks a free flagellum. The metacyclics are the only stage infective to the vertebrate host. When feeding, the tsetse fly may inoculate a host with up to several thousand protozoans with a single bite. The entire cycle within the fly can be completed in 15 to 35 days.

Once within a vertebrate, the trypanosomes multiply as trypomastigotes in the blood and lymph. Amastigote forms have been reported from the liver and spleen of experimentally infected mice and from the myocardium of monkeys.[22,36] After a variable period of time, many of the trypanosomes invade the central nervous system, multiply, and enter the intercellular spaces within the brain.

Biochemical, ultrastructural, and immunological studies have added greatly to our understanding of trypanosomes.[29] Of considerable value is the fact that essentially pure preparations of certain morphological stages can be obtained. When *T. brucei* is passed by syringe from one vertebrate host to another, the strain tends to become monomorphic after a period of time, consisting only of slender trypomastigotes that are no longer infective to tsetse flies. Their morphology and metabolism correspond to the slender trypomastigote in natural infections. In contrast, when *T.*

brucei is placed in in vitro culture, its morphology and metabolism revert to that found in the fly midgut, with the kinetoplast further from the posterior end and closer to the nucleus. It has been found that the monomorphic, syringe-passed strain depends entirely on glycolysis for its energy production, degrading glucose only as far as pyruvate and having no tricarboxylic acid cycle or oxidative phosphorylation via the classical cytochrome system. The reduced NAD produced in glycolysis is reoxidized by a nonphosphorylating glycerophosphate oxidase system, which, though it requires oxygen, is not sensitive to cyanide. The long, slender trypomastigote is very active, and it consumes substantial quantities of both glucose and oxygen in its inefficient energy production. However, since the blood and lymph have a plentiful supply of both, no selective value is attached to the conservation of either. The situation is quite different when the trypanosome finds itself in a blood clot in its vector's midgut; for, interestingly enough, in this case the form completely degrades glucose via glycolysis, the tricarboxylic acid cycle, and the cyanide-sensitive cytochrome system. The oxygen and glucose consumption of the midgut (or culture) form is only one-tenth that of the bloodstream form.

Ultrastructural observations on the mitochondria in the respective forms corre-

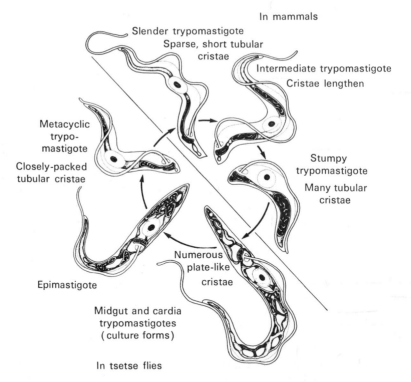

Fig. 4-5. Diagram to show changes in form and structure of the mitochondrion of *T. brucei* throughout its life cycle. The slender bloodstream form lacks a functional Krebs cycle and cytochrome chain. Stumpy forms have a partially functional Krebs cycle but still lack cytochromes. The glycerophosphate oxidase system functions in terminal respiration of bloodstream forms. The fly gut forms have a fully functional mitochondrion with active Krebs cycle and cytochrome chain. Cytochrome oxidase may be associated with the distinctive plate-like cristae of these forms. Reversion to tubular cristae in the salivary gland stages may therefore indicate loss of this electron transfer system. All forms other than slender bloodstream forms gave a positive reaction for NADH-tetrazolium reductase activity. (From Vickerman, K. 1971. In Fallis, A. M., editor. Ecology and physiology of parasites. University of Toronto Press, Toronto.)

late beautifully with the biochemical findings. The long, slender trypomastigote has a single, simple mitochondrion extending anteriorly from its kinetoplast, and the cristae are few, short, and tubular. The midgut stage has an elaborate mitochondrion extending both posteriorly and anteriorly from the kinetoplast, and the cristae are numerous and plate-like. The curious movement of the kinetoplast away from the posterior end in the midgut trypomastigote and anterior to the nucleus in the epimastigote can now be understood as reflecting the elaboration of the posterior section of mitochondrion; it "pushes" the kinetosome forward. Furthermore, present evidence indicates that the short, stumpy form is the only one infective to the tsetse fly and that the intermediate form is transitional from the long, slender noninfective form (Fig. 4-5). Correspondingly, electron microscopy has shown that this transition is marked by increasing elaboration of the mitochondrion; synthesis of mitochondrial enzymes has been shown by cytochemical means. Similarly, the metacyclic form, as though "gearing down" for infection of the vertebrate, has a mitochondrion much like the bloodstream form.

An important remaining question is: what triggers the bloodstream trypomastigotes to become short, stumpy forms infective to flies? A possibility is suggested by some interesting immunological observations. The clinical course characteristic of the infection varies according to the host infected, but in certain hosts (guinea pig, dog, and rabbit) repeated remissions alternate with very high parasitemias. That is, periods with few trypanosomes (and disease symptoms) evident are followed by a large increase in parasite population. This cycle tends to repeat itself until the host dies. The mechanism of the phenomenon lies in the sequential development of variant antigenic types by the trypanosomes over time.[31] The remissions seem to result from the elaboration by the host's protective antibodies that attack the trypanosomes. But the parasites have evolved an amazing subterfuge to escape obliteration by the host's defenses: each time the host's antigens are almost successful, the trypanosomes elude destruction by development of a new antigenic type! Furthermore, the only apparent limit to the number of antigens that can develop in a clone strain of trypanosomes during an untreated infection is the host's life span.[7]

The means by which the parasites achieve this succession of antigenic types could be either by a remarkable mutational rate, whereby a "resistant" antigenic strain is always available as soon as the host produces antibodies, or by a *set genetic program* of successive antigenic types that the trypanosome always goes through during the course of an infection. Present evidence indicates that the latter is the case. Each time a trypanosome strain is passed through a tsetse fly, it loses whatever antigenic type it had and reacquires a common, "basic antigenic type," when it reaches the salivary glands of the fly and develops to the metacyclic stage. Then, upon reinfection of the vertebrate, a predictable sequence of antigenic types begins all over again (A → B → C → D and so on). It appears that the antigen is present in the surface coat of the trypanosome and that it is produced by the Golgi body and released through the flagellar pocket. How are these phenomena related to the production of the short, stumpy forms infective to flies? The remissions, or decline in parasitemias, are accompanied by increases in intermediate and stumpy forms. It appears that stumpy forms are produced in response to unfavorable environmental conditions, and such conditions are provided by the cyclical and repeated elicitation of protective antibodies during the natural course of the infection.[29] Finally, it should be noted that though the remissions are rather variable in human infections, this clinical picture has been known since 1910.[25] However, in some hosts, mice for example, certain strains of trypanosomes are so virulent that the disease is rapidly fatal without intermittent remissions.

Pathogenesis. The clinical course depends on the susceptibility of the host species. Horses, mules, donkeys, some ruminants, and dogs suffer acutely from *T. b. brucei* infections, and they usually survive only 15 days to 4 months. Symptoms include anemia, edema, watery eyes and nose, and fever. Within a few days the animals become emaciated, uncoordinated, and paralyzed, and die shortly afterwards. Blindness as the result of the infection is common in dogs. Cattle are somewhat more refractory to the disease, often surviving for several months after onset of symptoms. Swine usually recover from the infection.

The mechanism of pathogenesis is quite unclear. In the acute infection of small mammals, where death occurs rapidly at a time of high parasitemia, mortality probably is a result of overall disruption of normal physiological processes.[20] In the case of mammals, including humans, present evidence suggests that pathogenesis may be caused in part by the antigenic "performance" of the trypanosomes and the host's immune reactions. Because of the repeated changes in the surface antigens of the parasites, and the fact that these surface antigens are being released into the blood almost constantly (exoantigens), the immune system of the host is greatly stimulated, and huge amounts of immunoglobulins are produced. It has been shown that some of the trypanosome antigens can adsorb to the surface of some host cells, and binding of the specific antibody to the adsorbed antigen, in conjunction with complement, leads to lysis of the *host's* own cells. Lysis of red blood cells by this mechanism may account for the anemia of trypansomiasis.[20] Lysis of mast cells releases pharmacologically active kinins, with resultant reactions in susceptible tissues.

T. vivax - mechanical transmission

Epidemiology and control. Tsetse flies occupy 4.5 million square miles of Africa, making much of that area impractical for human habitation. Trypanosomes of the brucei group do not occur throughout the entire range of tsetse flies, and not all species of *Glossina* are vectors for them. Therefore, transmission varies locally, depending on coincidence of the trypanosome and the proper fly species. *Glossina morsitans* is the most important vector of *T. b. brucei,* and that species occurs in open country, pupating in dry, friable earth.

Control of trypanosomiasis brucei is conducted along several lines, most of which involve the vectors. Tsetse flies are larviparous, and they deposit their young on the soil under brush. Because of this, and because the adults rest in bushes at certain heights above the ground and no higher, brush removal and trimming is a very successful means of control. When wide belts of land are thus cleared, the flies seldom cross them and can be more easily controlled. However, this method is quite expensive and must be followed up every year to remove new growth.

Elimination of the wild game reservoirs has been proposed and practiced in some regions, stimulating an outcry among conservation-minded people all over the world.

Programs have been established in which people simply sit and catch flies that try to bite them. Because the flies only feed during the day, some farmers graze their livestock at night, moving them into enclosures during the day and protecting them from flies with switches.

The most satisfactory means of control is by spraying insecticides by aircraft. DDT and benzene hexachloride are inexpensive and highly effective for this purpose. *Glossina pallidipes* was eradicated from Zululand in this manner at a cost of about 40¢ per acre. The possibilities of harmful side effects of DDT must be carefully weighed against the benefits gained by its use.

Nagana is also caused by *Trypanosoma (Nannomonas) congolense,* which is similar to *T. brucei* but lacks a free flagellum.[28] It occurs in South Africa, where it is the most common trypanosome of large mammals. The life cycle, pathogenesis, and treatment are as for *T. brucei.*

Trypanosoma (Duttonella) vivax is also found in the tsetse fly belt of Africa and has spread to the Western Hemisphere and Mauritius. Very similar to *T. brucei,* it causes a similar disease in similar hosts. In the New World, transmission is mechanical and involves tabanid flies. Pathogenesis and control are as for *T. brucei.* Also, the changes in the mitochondrion through the life cycle are similar in *T. vivax* and *T. congolense* to those in *T. brucei,* described previously. However, these two species appear to retain some mitochondrial function in the bloodstream form.[29]

Trypanosoma (Trypanozoon) brucei gambiense and Trypanosoma (Trypanozoon) brucei rhodesiense

The two subspecies, *T. b. gambiense* and *T. b. rhodesiense,* are the etiological agents of African sleeping sickness. They are morphologically indistinguishable from each other and from *T. b. brucei,* from which they probably evolved. There are physiological differences between them, as they differ in pathogenesis, growth rate, and biology. *Trypanosoma b. rhodesiense* can be cultivated in vitro in the presence of human blood, while *T. b. brucei* cannot. As early as 1917 a German researcher named Taute showed that the nagana trypanosome would not infect humans. He repeatedly inoculated himself and native "volunteers" with nagana-ridden blood: none of them acquired sleeping sickness. His work was largely discounted by British experts[5] for reasons that can only be guessed.

These trypanosomes live in the heart of Africa, from 15° N to 15° S latitude. *Trypanosoma gambiense* causes a chronic form of the disease and is found in west central and central Africa, while *T. rhodesiense* occurs in central and east central Africa and causes a more acute type of infection. Native game animals are thought to serve as reservoirs for Rhodesian trypanosomiasis, but not for Gambian trypanosomiasis.

The biology of these two subspecies parallels that of *T. b. brucei,* with minor differences. The principal vectors of *T. b. gambiense* are *Glossina palpalis* and *G. tachinoides,* while those of *T. b. rhodesiense* are *G. morsitans, G. pallidipes,* and *G. swynnertoni.*

not intracellular T. b. rhodesiense
[T. b. gambiense
Order Kinetoplastida: trypanosomes and their kin **61**

In their vertebrate hosts, these trypanosomes live in the blood, lymph nodes and spleen, and cerebrospinal fluid, just like *T. b. brucei.* They do not invade or live within cells but inhabit connective tissue spaces within various organs and the reticular tissue spaces of the spleen and lymph nodes. They are particularly abundant in the lymph vessels and the intercellular spaces in the brain. Reproduction and distribution in, and transmission by, *Glossina* spp. are identical to that of *T. b. brucei.*

Pathogenesis. A small sore develops at the site of inoculation of metacyclic trypanosomes. This disappears after 1 or 2 weeks, while the protozoa gain entrance to the blood and lymph channels. Reproducing rapidly, they produce a parasitemia and invade nearly all organs of the body. *Trypanosoma b. rhodesiense* rarely invades the nervous system like *T. b. gambiense* but usually causes a more rapid course toward death. The lymph nodes become swollen and congested, especially in the neck, groin, and legs. Swollen nodes at the base of the skull were recognized by slave traders as signs of certain death, and slaves who developed them were routinely thrown overboard by slavers bound for the Caribbean markets. Today, such swollen lymph nodes are called **Winterbottom's sign,** named after the British officer who first described the symptom. The symptoms of illness usually are more marked in nonnative than in native peoples. Intermittent periods of fever accompany the early stages of the disease, and the number of trypanosomes in the circulating blood greatly increases at these times. As previously noted, the successive parasite populations represent sequential antigenic types. With fever there is an increase in swelling of lymph nodes, generalized pain, headache, weakness, and cramps. Infections of *T. b. rhodesiense* cause rapid weight loss and heart involvement. Death may occur within a few months after infection, but *T. b. rhodesiense* causes no somnambulism or other protracted nervous disorders found with *T. b. gambiense* because the host usually dies before these can develop.

When the trypanosomes of *T. b. gambiense* invade the central nervous system, they initiate the chronic, sleeping-sickness stage of infection. Increasing apathy, a disinclination to work, and mental dullness accompany disturbances of coordination. Tremor of the tongue, hands, and trunk is common, and paralysis or convulsions usually follow. Sleepiness increases, with the patient falling asleep even while eating or standing. Finally coma and death ensue. Actually, death may result from any one of a number of causes, including malnutrition, pneumonia, heart failure, other parasitic infections, or the result of a severe fall.

Diagnosis and treatment. Demonstration of the parasite in the blood, bone marrow, or cerebrospinal fluid establishes diagnosis. In native populations where early infections are usually symptomless, a serological test is available.[1]

Arsenical drugs exclusively have been used in the treatment of African trypanosomiases, but these drugs have severe drawbacks. They cause eye damage and are best administered intravenously; furthermore, trypanosomes rapidly become tolerant to them. Other drugs (suramin and pentamidine) have been developed in recent years and have proved to be quite satisfactory in most early cases. Prognosis, however, is poor if the nervous system has become involved.

Control, as with *T. b. brucei,* depends on tsetse fly eradication and, in some areas, game destruction programs. The political situation in Africa, often with decreased cooperation between adjacent nations and tribes, may contribute to new epidemics. At the same time, increased mobility of the human population contributes to wider and faster spread of the disease.

Trypanosoma (Trypanozoon) evansi

T. evansi causes a widespread disease of camels, horses, elephants, deer, and many other mammals. The disease goes by many different names in different languages and countries but is most often called **surra.** It probably was originally a parasite of camels.[9] Today it is distributed throughout the northern half of Africa, Asia Minor, southern Russia, India, southwestern Asia, Indonesia, Philippines, and Central and South America. The disease was introduced by the Spaniards into the western hemisphere in infected horses in the sixteenth century.

This trypanosome is morphologically indistinguishable from *T. brucei.* Typically it is 15 to 34 μm long. Most are slender in shape, but stumpy forms occasionally appear. However, the biology of *T. evansi* is quite different from that of *T. brucei.* The life cycle does not involve *Glossina* or development within an arthropod vector. In most areas, contaminated mouthparts of horseflies *(Tabanus)* mechanically transmit the disease, but *Stomoxys, Lyparosia,* and *Haematopota* can also transmit it. In South America, vampire bats are common vectors of the disease, known there as **murrina**.[11]

The disease is most severe in horses, elephants, and dogs, with nearly 100% fatalities in untreated cases. It is less pathogenic to cattle and buffalo, which may be asymptomatic for months. In camels, surra is serious, but it tends to remain chronic. Pathogenesis, symptoms, and treatment are the same as for *T. brucei.*

T. evansi probably originated from *T. brucei,* when camels were brought into the tsetse fly belt. Subsequently, the organism apparently lost its requirement for development within an insect.

Trypanosoma (Trypanozoon) equinum occurs in South America where it causes a disease in horses similar to surra. The condition is known as "mal de Caderas." *Trypanosoma equinum* is similar to *T. evansi* except that it appears to lack a kinetoplast. Actually, a vestigial kinetoplast can be seen in electron micrographs, but it does not function in activation of the mitochondrion. The condition is known as *dyskinetoplasty. T. brucei* and *T. evansi* can be rendered dyskinetoplastic with certain drugs, and the character is inherited as a mutation. Such organisms can survive as bloodstream parasites but no longer can infect flies. *T. equinum* most likely evolved from *T. evansi.* It also is transmitted mechanically by tabanid flies. Pathogenesis, symptoms, and treatment are as for *T. evansi.*

Trypanosoma (Trypanozoon) equiperdum

Another trypanosome, *T. equiperdum,* also morphologically indistinguishable from *T. brucei,* causes a venereal disease called **dourine** in horses and donkeys. The organisms are transmitted during coitus, and no arthropod vector is known. No doubt this trypanosome also originated from *T. brucei.*

The disease is found in Africa, Asia, southern and eastern Europe, Russia, and Mexico. It was once common in western Europe and North America but has been eradicated from those areas.

Dourine exhibits three stages. In the first, the genitalia become edematous, with a discharge from the urethra and vagina. Areas of the penis or vulva may become depigmented. In the second stage a prominent rash appears on the sides of the body, remaining for 3 or 4 days. The third stage produces paralysis, first of the neck and nostrils, then the hind body; the paralysis finally becomes general. Dourine is usually fatal unless treated.

Diagnosis depends on finding trypanosomes in the blood, genital secretions, or fluids from the large urticarious patches of the skin during the second stage. A complement fixation test is very reliable and was used by USDA personnel to ferret out infective horses during their successful campaign to eradicate the disease in the United States. All horses now entering the United States must be tested for dourine before being admitted.

Section Stercoraria
Trypanosoma (Schizotrypanum) cruzi

T. cruzi carries the unusual distinction of having been discovered and studied several years before it was found to cause a disease. In 1910 a 40-year-old Brazilian, Carlos Chagas, dissected a number of cone-nosed bugs (Hemiptera, family Reduviidae, subfamily Triatominae) and found their hindguts swarming with trypanosomes of the epimastigote type. The cone-nosed bugs are notorious household pests throughout much of South America. They have "bedbug habits," lurking in crevices and crannies during the day, and emerging at night to seek a blood meal from sleeping victims. Their propensity to bite areas of the face where the skin is thin, such as around the eye or edge of a lip, has earned them the common name of "kissing bug."

Chagas sent a number of the bugs to the Oswaldo Cruz Institute, where they were allowed to feed on marmosets and guinea pigs. Trypanosomes appeared in the blood of the animals within a month. Chagas thought the parasites went through a type of schizogony in the lungs, so he named

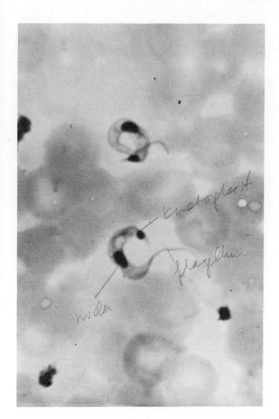

Fig. 4-6. *Trypanosoma cruzi*: trypomastigote form in a blood film. (Courtesy Ann Arbor Biological Center.)

them *Schizotrypanum cruzi*. The name *Schizotrypanum* still is employed by some workers, although most prefer to use it as a subgenus of *Trypanosoma*. By 1916 Chagas demonstrated that an acute, febrile disease, common in children throughout the range of cone-nosed bugs, was always accompanied by the trypanosome. Unfortunately, he thought that goiter and cretinism also were caused by this parasite. When that was disproved, suspicion was cast on the rest of his work. Also, Chagas maintained to near the end of his life that transmission of the disease (that now bears his name) was through the bite of the insect. It was not until the early 1930s that Chagas' disease was proved to be transmitted via the feces of the cone-nosed bug.

Trypanosoma cruzi is distributed throughout most of South and Central America, where it infects 12 million persons. Another 35 million are exposed to infection.[38] In some surveys in Brazil, it has been reported that 30% of all adults die

of *T. cruzi* infection.[14] Many kinds of wild and domestic mammals serve as reservoirs. Animals that live in proximity to humans, such as dogs, cats, opossums, armadillos, and wood rats, are particularly important in the epidemiology of Chagas' disease.

In the United States, *T. cruzi* has been found in Maryland, Georgia, Florida, Texas, Arizona, New Mexico, California, Alabama, and Louisiana. Fourteen species of infected mammals have been found in the United States,[16] but only one indigenous infection in humans is known.[37] Several North American strains have been isolated. They are morphologically indistinguishable from any other *T. cruzi*, but they seem to be much less pathogenic.

Morphology. The trypomastigote form is found in the circulating blood. It is slender, 16 to 20 μm long, and its posterior end is pointed. The free flagellum is moderately long, and the undulating membrane is narrow, with only two or three undulations at a time along its length. The kinetoplast is subterminal and is the largest of any trypanosome; it sometimes causes the body to bulge around it. The protozoan commonly dies in a question mark shape and therefore has that appearance in stained smears (Fig. 4-6).

Amastigotes develop in muscles and other tissues. They are spheroid, 1.5 to 4.0 μm wide, and occur in clusters comprised of many organisms. Intermediate forms are easily found in smears of infected tissues.

Biology (Fig. 4-7). When reduviid bugs feed, they often defecate on the skin of their host. Their feces contain metacyclic trypanosomes, which gain entry into the body of the vertebrate host through the bite, through scratched skin, or, most often, through mucous membranes that are rubbed with fingers contaminated with the insect's feces. Also, reservoir mammals can become infected by eating infected insects.[39] While trypomastigotes are abundant in the blood in early infections, they do not reproduce until they have entered a cell and have transformed into amastigotes. The cells most frequently invaded are reticuloendothelial cells of the spleen, liver, and lymphatics and cells in cardiac, smooth, and skeletal muscles. The nervous system, skin, gonads, intestinal mucosa,

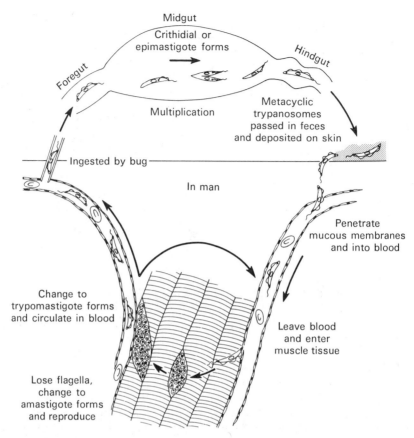

Fig. 4-7. The life cycle of *Trypanosoma cruzi.* (From Adam, K. M. G., J. Paul, and V. Zaman. 1971. Medical and veterinary protozoology. An illustrated guide. Churchill Livingstone, Edinburgh.)

bone marrow, and placenta also are infected in some cases (Fig. 4-8).

The undulating membrane and flagellum disappear soon after the parasite enters a host cell. Repeated binary fission produces so many amastigotes that the host cell soon is killed and lyses. When released, the protozoans attack other cells. Cyst-like pockets of parasites, called **pseudocysts,** form in muscle cells (Fig. 4-9). Intermediate forms (promastigotes and epimastigotes) can be seen in the interstitial spaces. Some of these complete the metamorphosis into trypomastigotes and find their way into the blood.

Trypomastigotes that are ingested by triatomid bugs pass through to the posterior portion of the insect's midgut, where they become short epimastigotes. These multiply by longitudinal fission to become long, slender epimastigotes. Eight to 10 days after infection, short metacyclic trypomastigotes appear in the insect's rectum. These are passed with the feces and can infect a mammal, if rubbed into a mucous membrane or a wound in the skin. It has been observed that the first generation amastigotes in the insect's stomach group together to form aggregated masses.[3] These fuse and may represent a primitive form of sexual reproduction.

Pathogenesis. Entrance of the metacyclic trypanosomes produces an acute local inflammatory reaction. Within 1 to 2 weeks after infection they spread to the regional lymph nodes and begin to multiply in the cells that phagocytose them. The intracellular amastigote undergoes repeated divisions to form large numbers of parasites, producing the so-called **pseudocyst.** After

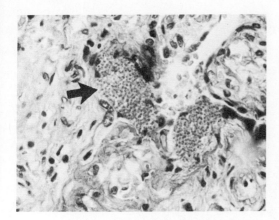

Fig. 4-8. Clusters of *Trypanosoma cruzi* amastigotes (arrow) in placenta. (AFIP photograph neg. no. 63-5589.)

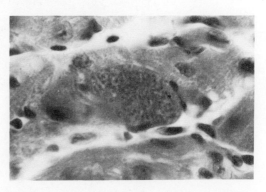

Fig. 4-9. *Trypanosoma cruzi* pseudocyst in cardiac muscle. (Photograph by Warren Buss.)

a few days, some of the organisms retransform into trypomastigotes and burst out of the pseudocyst, destroying the cell that contains them. A generalized parasitemia occurs then, and almost every type of tissue in the body can be invaded, though the parasites show a particular preference for muscle and nerve cells (Fig. 4-10). The reversion to amastigote, pseudocyst formation, retransformation to trypomastigote, and pseudocyst rupture are repeated in the newly invaded cells, and then the process begins again. Rupture of the pseudocyst is accompanied by an acute, local inflammatory response, with degeneration and necrosis (cell or tissue death) of nerve cells in the vicinity, especially ganglion cells. This is the most important pathological change in Chagas' disease, and though the mechanism remains unknown, it has been attributed to the release of toxins upon rupture of the pseudocyst and destruction of the remaining amastigotes.[14]

Chagas' disease manifests *acute* and *chronic* phases. The acute phase is initiated by inoculation into the wound of the trypanosomes from the bug's feces. The local inflammation produces a small red nodule, known as a **chagoma,** with swelling of the regional lymph nodes. In about 50% of the cases, the trypanosomes enter through the conjunctiva of the eye, causing edema of the eyelid and conjunctiva and swelling of the preauricular lymph node. This symptom is known as **Romaña's**

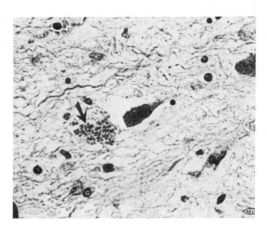

Fig. 4-10. A pseudocyst (arrow) of *Trypanosoma cruzi* in brain tissue. (AFIP photograph neg. no. 67-5313.)

sign. As the acute phase progresses, pseudocysts may be found in almost any organ of the body, although the intensity of attack varies from one patient to another. The heart muscle usually is invaded, with up to 80% of the cardiac ganglion cells being lost. Symptoms of the acute phase include anemia, loss of strength, nervous disorders, chills, muscle and bone pain, and varying degrees of heart failure. Death may ensue 3 to 4 weeks after infection. The acute stage is most common and severe among children less than 5 years old.

The chronic stage is most often seen in adults. Its spectrum of symptoms is pri-

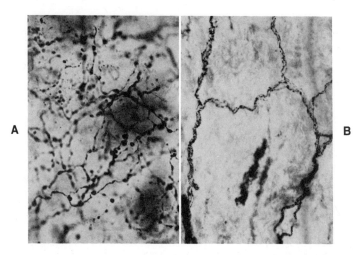

Fig. 4-11. Diaphanised tricuspid valves with zinc-osmium impregnation of nerve fibers. **A,** Normal heart; **B,** Chagas' cardiopathy with marked reduction of nerve fibers. (From Hutt, M. S. R., F. Köberle, and K. Salfelder. 1973. In Spencer, H., editor. Tropical pathology. Springer-Verlag, Berlin.)

marily the result of central and peripheral nervous dysfunction, which may last for many years. Some patients may be virtually asymptomatic and then suddenly succumb to heart failure. Chagas' disease accounts for about 70% of cardiac deaths in young adults in endemic areas. Part of the inefficiency in heart function is caused by loss of muscle tone resulting from the destroyed nerve ganglia (Fig. 4-11). The heart itself becomes greatly enlarged and flabby.

In some regions of South America it is common for the autonomic ganglia of the esophagus or colon to be destroyed. This ruins the tonus of the muscularis, resulting in deranged peristalsis and gradual flabbiness of the organ, which may become huge in diameter and unable to pass materials within it. This advanced condition is called **megaesophagus** or **megacolon,** depending on the organ involved (Fig. 4-12). Advanced megaesophagus may be fatal when the patient can no longer swallow. It has been experimentally demonstrated that testis tubules and epididymis atrophy in chronic cases.[4]

As a point of interest, Charles Darwin was afflicted by a mysterious malady throughout his adult life, following his famous trip to South America. The cause of the bouts of invalidism that plagued Darwin for weeks at a time was quite unknown to attending physicians, but the symptoms conform exactly to those of chronic Chagas' disease. In his *Diary of the Voyage of the H.M.S. Beagle,* Darwin complained of attacks of triatomids, the "great black bug of the Pampas," in native huts. No one can say what further contributions he would have made had he been healthy during his creative life!

Epidemiology. The principal vectors of *T. cruzi* in Brazil are *Panstrongylus megistus, Triatoma sordida,* and *T. brasiliensis;* in Uruguay, Chile, and Argentina, *Triatoma infestans* is the primary culprit. *Rhodnius prolixus* is the main vector in northern South America and in Central America, while species of the *Triatoma protracta* group serve as vectors in Mexico. Several other species of triatomids have been found naturally infected throughout this range. Natural infections in *Triatoma sanguisuga* have been found in the United States. The insects can become infected as nymphs or adults. Triatomids can infect themselves when they feed on each other, presumably by sucking the contents of the intestine. Ticks, sheep keds, and bedbugs have been experimentally infected, but no evidence that they serve as natural vectors has been found. Natural mammalian reservoirs of infection have been mentioned,

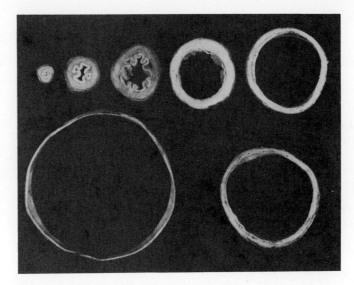

Fig. 4-12. Different stages of chagasic esophagopathy beginning with a normal organ, passing through hypertrophy and dilatation to the final megaesophagus. (From Hutt, M. S. R., F. Köberle, and K. Salfelder. 1973. In Spencer, H., editor. Tropical pathology. Springer-Verlag, Berlin.)

but domestic dogs and cats probably are the most important to human health.

Transmission from human to human during coitus or through mother's milk may be possible, although this has yet to be documented. It has been shown conclusively that *T. cruzi* can and does cross the placental barrier from mother to fetus (Fig. 4-8). Newborn infants with advanced cases of Chagas' disease, including megaesophagus, have been described in Chile.

Finally, the hazard of transmission by blood transfusion from donors with cryptic infection should not be underestimated. The frequency of this mode of transmission now ranks second only to natural vector transmission in endemic areas.[24]

The age of the victim is important in the epidemiology of Chagas' disease. Most new infections are in children from a few weeks to 2 years of age. It is in this age group that the acute phase is most often fatal.

Because the bugs hide by day, primitive or poor quality housing favors their presence. Thatched roofs, cracked walls, and trash-filled rooms are ideal for the breeding and survival of the insects. Misery compounds itself.

Diagnosis and treatment. Diagnosis usually is by demonstration of trypano-somes in blood, cerebrospinal fluid, fixed tissues, or lymph. Trypomastigotes are most abundant in peripheral blood during periods of fever but may be difficult to find at other times or in cases of chronic infection. In these cases, blood can be inoculated into guinea pigs, mice, or other suitable hosts, and the animals in turn can be examined by heart smear or spleen impression. Another method that is widely employed is **xenodiagnosis.** Laboratory-reared triatomids are allowed to feed on the patient, then, after a suitable period of time (10 to 30 days), they are examined for intestinal flagellates. This technique can detect cases when trypanosomes in the blood are too few to be found by ordinary examination of blood films.

Complement fixation or other immunodiagnostic tests are extremely effective in demonstrating chronic cases, although they may give false positive reactions if the patient is infected with *Leishmania* or another species of trypanosome.

Unlike the other trypanosomes of humans, *T. cruzi* does not respond well to chemotherapy. The best drugs kill only the extracellular protozoans, but the intracellular forms defy our best efforts. This seems to be because the reproductive stages, inside living host cells, are shielded from the drugs. The lives and strength of millions of Latin American people depend on the discovery of a drug that is effective against *Trypanosoma cruzi.*

Trypanosoma (Herpetosoma) rangeli

T. rangeli first was found, as was *T. cruzi,* in a triatomid bug in South America. *Rhodnius prolixus* is the most common vector, but *Triatoma dimidiata* and other species will also serve. Development is in the hindgut, and the epimastigote stages that result are from 32 to over 100 μm long. The kinetoplast is minute, and the species can thereby be differentiated from *T. cruzi,* with which it often coexists.

Trypanosma rangeli is common in dogs, cats, and humans in Venezuela, Guatemala, Chile, El Salvador, and Colombia. It has been found in monkeys, anteaters, opossums, and humans in Colombia and Panama. Trypomastigotes are larger than those of *T. cruzi,* being 26 to 36 μm long. The undulating membrane is large and has many curves. The nucleus is pre-equatorial, and the kinetoplast is subterminal.

The method of transmission is unclear. Although development is by posterior station, transmissions by both fecal contamination and by feeding inoculation have been reported.[27] *T. rangeli* multiplies by binary fission in the mammalian host's blood. No intracellular stage is known, and the organism is apparently not pathogenic in humans.

Trypanosoma (Herpetosoma) lewisi
(Fig. 4-13)

T. lewisi is a cosmopolitan parasite of *Rattus* spp. Other rodents apparently are not susceptible, not even mice. The vector is the northern rat flea, *Nosopsyllus fasciatus,* in which the parasite develops inside cells of the posterior midgut. Metacyclic trypomastigotes appear in large numbers in the rectum of the insect, infecting rats that have eaten a flea or its feces. The parasite seems to be nonpathogenic in most cases, perhaps even promoting the growth of its rat host.[19] However, infection may contribute to abortion and arthritis.

Much research has been conducted on this species because of the ease of maintaining it in the laboratory rat. One fascinating subject of this research is the "ablastin" phenomenon.[26,29] **Ablastin** is an antibody or antibody-like substance that arises during the course of an infection. After a rat is infected by the metacyclic trypo-

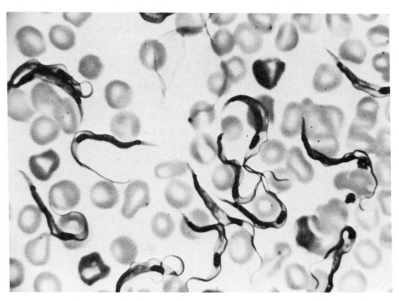

Fig. 4-13. *Trypanosoma lewisi* trypomastigotes in the blood of a rat. (Courtesy Turtox/Cambosco.)

mastigotes, the parasite begins reproducing in the epimastigote form in the visceral blood capillaries. After about 5 days, the trypanosome appears in the peripheral blood as a rather "fat" form, and shortly thereafter a crisis occurs in which most of the trypanosomes are killed by a trypanocidal antibody. A small population of slender trypomastigotes remains, which are infective for the flea but do not reproduce further while in the rat. After a few weeks, the host produces another trypanocidal antibody, which clears the remaining trypanosomes, and the infection is cured. The slender trypomastigotes are sometimes known as "adults," and it has been shown that their reproduction is inhibited by the ablastin. Ablastin is a globulin with many characteristics of a typical antibody, but its action inhibits reproduction. Nucleic acid and protein synthesis by the trypanosome are inhibited, as is uptake of nucleic acid precursors. The protozoan manages to escape the first trypanocidal antibody by a surface antigen change, similar to that described for *T. brucei,* but only one such change occurs. This seems to be an interesting adaptive accommodation of parasite and host: the parasite eludes the first onslaught of its host's immune defenses, but the host "allows" the parasite to remain available for the intermediate host for a longer period, meanwhile holding its reproduction in check. A somewhat less effective ablastin is produced by mice, when infected by the mouse trypanosome *T. musculi.* Interestingly, the two ablastins are cross-reactive, that is, anti-*lewisi* ablastin from a rat inhibits *T. musculi* when injected into a mouse, and anti-*musculi* ablastin inhibits *T. lewisi;* neither trypanosome is infective for the alternate host.

Trypanosoma (Megatrypanum) theileri

T. theileri is a cosmopolitan parasite of cattle. The vectors are horseflies of the genera *Tabanus* and *Haematopota.* The trypanosome reproduces in the fly gut as an epimastigote.

The size of *Trypanosoma theileri* varies with the strain—from 12 to 46 μm, 60 to 70 μm, and even 120 μm in length. The posterior end is pointed, and the kinetoplast is considerably anterior to it. Both trypomastigote and epimastigote forms can be found in the blood. Reproduction in the vertebrate host is in the epimastigote form and apparently occurs extracellularly in the lymphatics.

Trypanosoma theileri is usually nonpathogenic, but, under conditions of stress, it may become quite virulent. When cattle are stressed by immunization against another disease, undergo physical trauma, or become pregnant, the parasite may cause serious disease.

This parasite is rarely found in routine blood films. Detection usually depends on in vitro cultivation from blood samples. In fact, during tissue culture of bovine blood or cells, *T. theileri* is the most commonly found contaminant. Strong evidence points to transplacental transmission. In the United States, a similar trypanosome is also common in deer and elk.

• • •

Other species of *Trypanosoma* are common in other classes of vertebrates. These include, for example, *T. percae* of perch, *T. granulosum* of eels, *T. rotatorium* of frogs, *T. avium* of birds, and incompletely known species in turtles and crocodiles. Trypanosomes are frequently found in a variety of marine fishes of both classes.

GENUS LEISHMANIA

Like the trypanosomes, the leishmanias are heteroxenous. Part of their life cycle is spent in the gut of a fly, where they assume the form of a promastigote, and the remainder of their life cycle is completed in vertebrate tissues, where only the amastigote form is found. Traditionally the amastigote also is known as a **Leishman-Donovan body (L-D body).** Species in humans are widely distributed (Fig. 4-14).

The vertebrate hosts of *Leishmania* spp. are primarily mammals, although nearly a dozen species have been reported from lizards. The mammals most commonly infected are humans, dogs, and several species of rodents. Sandflies of the genera *Phlebotomus* (Fig. 4-15) and *Lutzomyia* are the invertebrate hosts and thereby are the primary vectors of **leishmaniasis.** When they suck the blood of an infected animal, they ingest amastigote forms. These pass to the midgut, where they transform into promastigotes and multiply by binary fis-

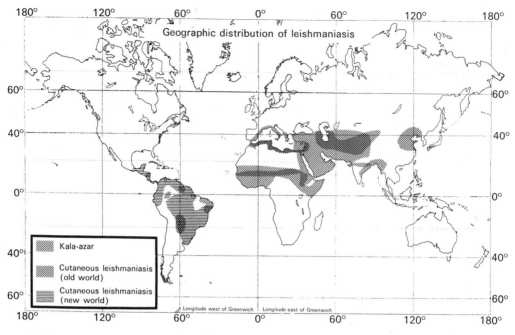

Fig. 4-14. Geographic distribution of leishmaniasis. (AFIP photograph neg. no. 68-1805-2.)

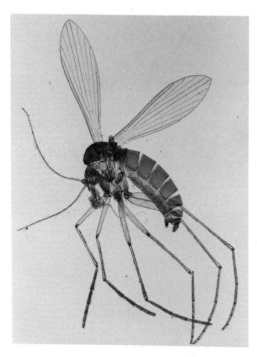

Fig. 4-15. The sandfly *Phlebotomus*, a vector of *Leishmania* spp. (Photograph by Jay Georgi.)

sion. The parasites may attach to the walls of the fly's gut or remain free in the lumen. Masses of promastigotes also may be found in the foregut and hindgut. By the fourth or fifth day postfeeding, promastigotes develop in the esophagus and pharynx. When leishmanias begin to clog up the esophagus, the feeding sandfly pumps its esophageal contents in and out to clear the obstruction, thereby inoculating promastigotes into the skin of a luckless victim. Transmission also can occur when infected sandflies are crushed into the skin or mucous membrane. *Leishmania* promastigotes have been found in the hindgut of *Phlebotomus*,[15] which suggests that infection also may be possible by fecal contamination, as in the case of *Trypanosoma cruzi*.

All amastigotes in vertebrate tissues look alike (Fig. 4-16). They are spheroid to ovoid, usually 2.5 to 5.0 μm wide, although smaller ones are known. In stained preparations, only the nucleus and a very large kinetoplast can be seen, and the cytoplasm appears vacuolated. Exceptionally, a short axoneme is visible within the cytoplasm under the light microscope.

Although all *Leishmania* spp. exhibit the same morphology, they differ clinically, bi-

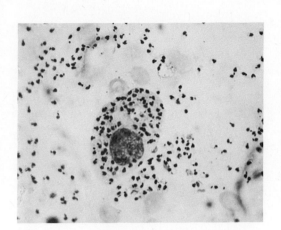

Fig. 4-16. A spleen smear showing numerous intra- and extracellular amastigotes of *Leishmania donovani.* (AFIP photograph neg. no. 55-17580.)

ologically, and serologically. Even so, these characteristics often overlap, so distinctions between species are not clear-cut. Leishmaniases that normally are visceral may become dermal; dermal forms can become mucocutaneous; and an immunodiagnostic test derived from the antigens of one species may give positive reactions in the presence of other species of *Leishmania,* or even *Trypanosoma.* These populations are closely related and are in the process of speciation as the result of recent geographic isolation. It is likely that the transport of slaves to the western world from Africa through the Middle East and Asia spread *Leishmania* into previously uncontaminated areas, where they are now rapidly speciating.

The result of the difficulty in species definition within the genus is that several schemes of classification have been proposed, nearly all of them having some acceptance. The published literature is confusing, since some researchers refer to several species and others consider the same organisms as a single, widespread species with slightly different clinical manifestations but similar or identical immunological properties. While this is rather a nuisance to the student, it is a good demonstration of evolution. The more similarities two or more populations exhibit, the more recently they have diverged from one another.

For the sake of simplicity we recognize three species of *Leishmania* parasites of humans, each distinguished by its medical symptoms. Other probable species exist in reptiles and mammals but will not be considered here. An exception is *L. enriettii,* a parasite of guinea pigs that is easily maintained in the laboratory and is widely used in basic research into *Leishmania* physiology and in applied investigation into chemotherapy and cell-mediated immunology against this "group" of organisms.

Leishmania tropica

L. tropica produces a cutaneous ulcer variously known as Oriental sore, cutaneous leishmaniasis, Jericho boil, Aleppo boil, or Delhi boil. It is found in west-central Africa, the Mediterranean area, the Middle East, and Asia Minor into India.

Morphology and life cycle. The appearance of *L. tropica* is identical to that of the other leishmanias of humans (Fig. 4-16). Sandflies of the genus *Phlebotomus* are the intermediate hosts and vectors. When the fly takes a blood meal containing amastigotes of *L. tropica,* the parasites multiply in the midgut, move to the pharynx, and are inoculated into the next mammalian victim. There, they multiply in the reticuloendothelial system and lymphoid cells of the skin. Few amastigotes are found except in the immediate vicinity of the site of infection, so the sandflies must feed there to become infected.

Pathogenesis. The incubation period lasts from a few days to several months. The first symptom of infection is a small, red papule at the site of the bite. This may disappear in a few weeks, but usually it develops a thin crust that hides a spreading ulcer underneath. Two or more ulcers may coalesce to form a large sore (Fig. 4-17). In uncomplicated cases the ulcer will heal in 2 months to a year, leaving a depressed, unpigmented scar. It is common, however, for secondary infection to occur, including, for example, yaws (a disfiguring disease caused by a spirochete) and myiasis (infection with fly maggots).

Russian scientists in Turkestan recognize two varieties of *L. tropica,* which they distinguish on pathological and epidemiological grounds. The urban type they call *L. tropica* var. *minor,* and the rural type they designate as *L. tropica* var. *major.* The

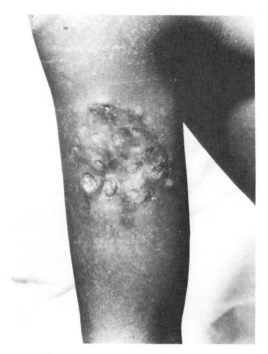

Fig. 4-17. Oriental sore. A complicated case with several lesions. (AFIP photograph neg. no. A-43418-1.)

urban type is found in more densely populated areas. Its papule is dry, persists for months before ulcerating, and has numerous amastigotes within it. By contrast, the rural type is found in sparsely inhabited regions. Its papule ulcerates quickly, is of short duration, and has few amastigotes within it. The two types appear to have different reservoir hosts and, being thus somewhat isolated from one another, would appear to be diverging into new species.

A variant clinical form of the disease is known as **diffuse cutaneous leishmaniasis.** It occurs in South and Central America and Ethiopia. Several months or years after the initial infection by the sandfly, the organisms disseminate widely and cause nodules and plaque-like lesions under the skin. The face, arms, and legs are most commonly affected, and the appearance is similar to lepromatous leprosy. Patients who contract this form are believed to have a specific type of immune deficiency involving the cell-mediated mechanisms.[14]

Treatment. Diagnosis of the infection de-

pends on finding amastigotes. Scrapings from the side or edge of the ulcer, smeared on a slide and stained with Wright's or Giemsa's stain, will show the parasites in endothelial cells and monocytes, even though they cannot be found in the circulating blood. Cultures should be made, in case amastigotes go undetected in the smear. An immunological test is available, but it will show positive in cases that were cured years before.

Diagnosis and control. Antimony compounds are used successfully to treat Oriental sore. Protective immunity following medical treatment seems to be absolute, and immunity following the natural course of the disease is 97% to 98% effective. Recognizing this, some native peoples deliberately inoculate their children on a part of their body normally hidden by clothes. This prevents their later developing a disfiguring scar on an exposed part of the body. In Lebanon and Russia attempts at mass vaccination with promastigotes show promising results. Control, however, ultimately depends on eliminating the sandflies and reducing the population of rodent reservoirs.

Leishmania donovani

In 1900, Sir William Leishman discovered *L. donovani* in spleen smears of a soldier who died of a fever at Dum-Dum, India. The disease was known locally as Dum-Dum fever, or **kala-azar.** Leishman published his observations in 1903, the same year that Charles Donovan found the same parasite in a spleen biopsy. The scientific name is in honor of these men, as is the common name of the amastigote forms, Leishman-Donovan bodies, or L-D bodies for short. The Indian Kala-azar Commission (1931 to 1934) demonstrated the transmission of *L. donovani* by *Phlebotomus.*

Leishmania donovani has a wide distribution, and several varieties are distinguished on epidemiological and clinical grounds. The Mediterranean–Middle Asian form is found throughout the Mediterranean basin and extends through southern Russia to China. Its common vectors are *Phlebotomus major, P. perniciosus, P. chinensis,* and *P. longicuspis;* it has dogs, jackals, and foxes as reservoirs; and it

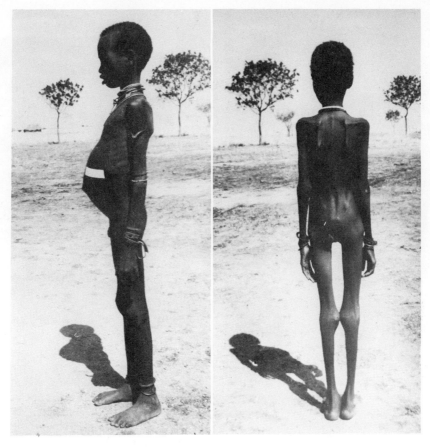

Fig. 4-18. Advanced kala-azar. Boy, about 6 years old, from Sudan, showing extreme hepatosplenomegaly and emaciation typical of advanced kala-azar. (From Hoogstraal, H., and D. Heyneman. 1969. Am. J. Trop. Med. Hyg. 18:1091-1210.)

mainly infects infants. Another variety is found in northeast India and Bangladesh. Its most important vector is *P. argentipes;* it apparently has no natural reservoir, and it mainly infects adults and adolescents. A clinically similar, but more virulent, variety occurs in East Africa. Its vectors are *P. martini* and *P. orientalis,* and the organism may use wild rodents as reservoirs. The New World variety is widespread in Central and South America. It seems to be mainly a zoonotic infection among dogs and foxes, and its vector is *P. longipalpis.* However, to further complicate matters, it is known that the patterns and distribution of the disease are constantly changing.[14]

Morphology. *L. donovani* cannot be differentiated from other species of *Leishmania.* Its rounded or ovoid body measures 2 or 3 μm, with a large nucleus and kinetoplast. It lives within cells of the reticuloendothelial system of the viscera, including spleen, liver, mesenteric lymph nodes, intestine, and bone marrow. Amastigotes have been found in nearly every tissue and fluid of the body.

Life cycle. The life cycle parallels that of *L. tropica.* When a sandfly, *Phlebotomus* spp., ingests amastigote forms along with its blood meal, the parasites lodge in the midgut and begin to multiply. They transform into slender promastigotes and quickly block up the gut of the insect. Soon they can be seen in the esophagus, pharynx, and buccal cavity, where they are injected into a new host with the fly's bite. Not all strains of *L. donovani* are adapted to all species and strains of *Phlebotomus.* Once in the mammalian host the parasite is immediately engulfed by a macrophage,

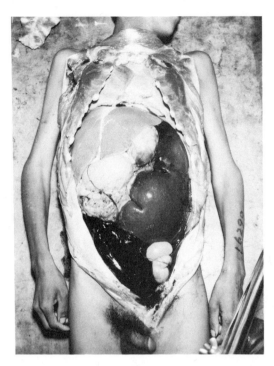

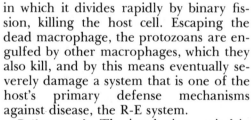

Fig. 4-19. A patient with kala-azar who died of hemorrhage after a spleen biopsy. Note the greatly enlarged spleen. (The dark matter in the lower abdominal cavity is blood.) (AFIP photograph neg. no. A-45364.)

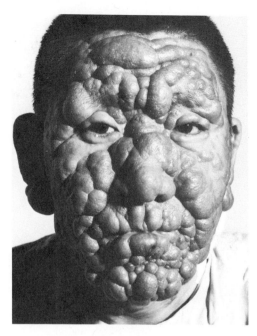

Fig. 4-20. Post–kala-azar dermal leishmanoid. This patient responded very well to treatment, regaining a nearly normal appearance. (Photograph by Robert E. Kuntz.)

in which it divides rapidly by binary fission, killing the host cell. Escaping the dead macrophage, the protozoans are engulfed by other macrophages, which they also kill, and by this means eventually severely damage a system that is one of the host's primary defense mechanisms against disease, the R-E system.

Pathogenesis. The incubation period in humans may be as short as 10 days or as long as a year, but usually is 2 to 4 months. The disease usually begins slowly with low-grade fever and malaise and is followed by progressive wasting and anemia, protrusion of the abdomen from enlarged liver and spleen (Fig. 4-18), and finally by death (in untreated cases) in 2 to 3 years. In some cases the symptoms may be more acute in onset, with chills, fever up to 104° F, and vomiting; death may occur within 6 to 12 months. Accompanying symptoms are edema, especially of the face, bleeding of the mucous membranes, breathing difficulty, and diarrhea. The im-

mediate cause of death often is the invasion of secondary pathogens that the body is unable to combat. A certain proportion of cases, especially in India, recover spontaneously, and post–kala-azar dermal leishmaniasis also is more common in India.

Visceral leishmaniasis may be viewed essentially as a disease of the reticuloendothelial system. The phagocytic cells, which are so important in defending the host against invasion, are themselves the habitat of the parasites. Blood-forming organs, such as spleen and bone marrow, undergo compensatory production of macrophages and other phagocytes (hyperplasia) to the detriment of red cell production. Thus, the spleen and the liver become greatly enlarged (hepatosplenomegaly) (Fig. 4-19), while the patient becomes severely anemic and emaciated.

A condition known as **post–kala-azar dermal leishmanoid** develops in some cases.[21] It is rare in the Mediterranean and

Latin American areas but develops in 5% to 10% of cases in India. The condition usually becomes apparent about 1 to 2 years following inadequate treatment for kala-azar. It is marked by reddish, depigmented nodules in the skin (Fig. 4-20).

Epidemiology. Transmission of visceral leishmaniasis is related to the activities of humans and the biology of sandflies. *Phlebotomus* spp. exist mainly at altitudes under 2,000 feet, most commonly in flat plains areas. Even in desert areas such as in the Sudan, the flies rest in and are protected by cracks in the parched earth and under rocks. In such conditions the flies are active only during certain hours of the day. For humans to become infected, they must be in sandfly areas at those times.

Age of the victim is a factor in the course of the disease, and fatal outcome is most frequent in infants and small children. Males are more often infected than females, most likely as the result of more exposure to sandflies. Poor nutrition, concomitant infection with other pathogens, and other stress factors predispose the patient to lethal consequences.

A wide variety of animals can be infected experimentally, though dogs are the only important reservoir in most areas. Canine infection is less common in India, where it is believed that a fly-to-human relationship is maintained.

Diagnosis. As in *L. tropica,* diagnosis depends on finding L-D bodies in tissues or secretions. Spleen punctures, blood or nasal smears, bone marrow, and others should be examined for the characteristic parasites, and cultures from these and other organs should be attempted. Immunodiagnostic tests are sensitive but cannot differentiate between species of *Leishmania* or *T. cruzi,* or between current and cured cases. Other diseases that might have symptoms similar to kala-azar are typhoid and paratyphoid fevers, malaria, syphilis, tuberculosis, dysentery, and relapsing fevers. Each must be eliminated in the diagnosis of kala-azar.

Control and treatment. Treatment consists of injections of various antimony compounds, together with good nursing care. An adequate diet with high protein and vitamins is essential to combat nutritional deficiency. These antimonial drugs are highly toxic and must be used with great care. Further, relapses and post–kala-azar dermal leishmaniasis may follow insufficient treatment.

Control of sandflies and reservoirs is urgently required in endemic areas.

Leishmania braziliensis

L. braziliensis produces a disease in humans variously known as espundia, uta, chiclero ulcer, or mucocutaneous leishmaniasis. It is found throughout the vast area between central Mexico and northern Argentina, although its range does not extend into the high mountains. The clinical manifestations of the disease vary along its range, which has led to confusion regarding the identity of the organisms responsible. Several species names have been proposed for different clinical and serological types. Once again, it appears that the parasite is rapidly evolving into groups that are adapting to local populations of humans and flies. We believe that it is reasonable to consider the American cutaneous-type leishmaniasis as a single species that causes variable pathologic conditions. Morphologically, *L. braziliensis* cannot be differentiated from *L. tropica* or *L. donovani.* An interesting historical account of this disease, with evidence of its pre-Colombian existence in South America, is given by Hoeppli.[13]

Life cycle and pathogenesis. The life cycle and methods of reproduction are identical to those of *L. donovani* and *L. tropica,* with several species of *Lutzomyia* serving as vectors. Inoculation of promastigotes by the bite of a sandfly causes a small, red papule on the skin. This becomes an itchy, ulcerated vesicle in 1 to 4 weeks and is quite similar at this stage to Oriental sore. This primary lesion heals within 6 to 15 months. The parasite never causes a visceral disease but often develops a secondary lesion on some region of the body. In Mexico and Central America the secondary lesion is often on the ear, resulting in considerable destruction to that organ. It is found commonly infecting "chicleros," those forest-dwelling peoples who glean a living by harvesting the gum of chicle trees. **"Chiclero ulcer"** is the name often given to this clinical manifestation. Several mammalian reservoirs are

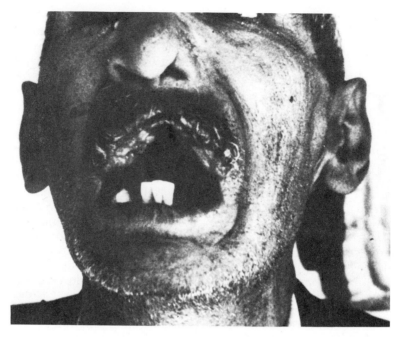

Fig. 4-21. Espundia of 2 years development after 24 years delay in onset. The upper lip, gum, and palate are destroyed. (From Walton, B. C., L. V. Chinel, and O. Eguia y E. 1973. Am. J. Trop. Med. Hyg. 22:696-698.)

known in this area, including spiny rats and other forest rodents, kinkajou, dogs, and cats.

In Venezuela and Paraguay the lesions more often appear as flat, ulcerated plaques that remain open and oozing. The disease is called **pian bois** in that area.

In the more southerly range of *L. braziliensis,* the parasites have a tendency to metastasize or spread directly from the primary lesion to mucocutaneous zones. The secondary lesion may appear before the primary has healed, or it may be many years (up to 24) before secondary symptoms appear.[34]

The secondary lesion often involves the nasal system and buccal mucosa, causing degeneration of the cartilaginous and soft tissues (Fig. 4-21). Necrosis and secondary bacterial infection are common. **Espundia** and **uta** are the names applied to these conditions. The ulceration may involve the lips, palate, and pharynx, leading to great deformity. Invasion of the infection into the larynx and trachea destroys the voice. Rarely, the genitalia may become infected. The condition may last for many years,

and death may result from secondary infection or respiratory complications.

Diagnosis and treatment. Diagnosis is established by finding L-D bodies in affected tissues. Espundia-like conditions are also caused by tuberculosis, leprosy, syphilis, and various fungal and viral diseases, and these must be differentiated in diagnosis. A skin test is available for diagnosis of occult infections, but it cross-reacts with *Trypanosoma cruzi.* Culturing the parasite in vitro is also a valuable technique when L-D bodies cannot be demonstrated in routine microscope preparation.

Treatment is similar to that for kala-azar and tropical sore: antimonial compounds applied on the lesions or injected intravenously. Secondary bacterial infections should be treated with antibiotics. Mucocutaneous lesions are particularly refractory to treatment and require extensive chemotherapy. Relapse is common, but once cured a person usually has lifelong immunity. However, if the infection is not cured but merely becomes occult, there may be a relapse with onset of espundia many years later. Because this is primarily a sylvatic disease, there is little opportunity for its control.

GENUS LEPTOMONAS

Leptomonas, like the other trypanosomatid genera, is parasitic in invertebrates and is of no medical importance. It is variously a promastigote and an intracellular amastigote throughout its monoxenous life cycle. Species are found in molluscs, nematodes, insects, and other protozoa.

GENUS HERPETOMONAS

Members of *Herpetomonas* also are characteristically monoxenous in insects. They pass through amastigote, promastigote, opisthomastigote, and possibly epimastigote stages in their life cycles. In the opisthomastigote, the flagellum arises from a reservoir that runs the entire length of the body.

GENUS CRITHIDIA

Crithidia spp. are tiny (4 to 10 μm) choanoflagellates of insects. They are often clustered together against the intestine of their host. They can assume the amastigote form and are monoxenous.

GENUS BLASTOCRITHIDIA

Blastocrithidia are monoxenous insect parasites, usually found as epimastigotes and amastigotes in the intestines of their hosts. Species are common in water striders (family Gerridae).

GENUS PHYTOMONAS

Phytomonas is a parasite of milkweeds and related plants. It passes through promastigote and amastigote phases in the intestines of certain beetles and appears as promastigotes in the sap (latex) of its plant hosts.

REFERENCES

1. Bailey, N. M. 1967. Recent development in the screening of populations for human trypanosomiasis and their possible application in other immunizing diseases. East Afr. Med. J. 44:475-481.
2. Baker, J. R. 1965. The evolution of parasitic Protozoa. In Taylor, A. E. R., editor. Evolution of parasites. Blackwell Scientific Publication Ltd., Oxford. pp. 1-27.
3. Brener, Z. 1972. A new aspect of *Trypanosoma cruzi* life-cycle in the intermediate host. J. Protozool. 19:23-27.
4. Ferreira, A. L., and M. A. Rossi. 1973. Pathology of the testis and epididymis in the late phase of experimental Chagas' disease. Am. J. Trop. Med. 22:699-704.
5. Foster, W. D. 1965. A history of parasitology. E. & S. Livingston, Edinburgh.
6. Goodwin, L. G. 1964. The chemotherapy of trypanosomiasis. In Hutner, S. M., editor. The biochemistry and physiology of Protozoa, vol. 3. Academic Press, Inc., New York. pp. 495-524.
7. Gray, A. R. 1965. Antigenic variation in clones of *Trypanosoma brucei* I.—immunological relationships of the clones. Ann. Trop. Med. Parasitol. 59:27-36.
8. Guttman, H. N., and F. G. Wallace. 1964. Nutrition and physiology of the Trypanosomatidae. In Hutner, S. M., editor. The biochemistry and physiology of protozoa, vol. 3. Academic Press, Inc., New York. pp. 459-494.
9. Hoare, C. A. 1956. Morphological and taxonomic studies on the mammalian trypanosomes. VIII. Revision of *Trypanosoma evansi.* Parasitology 46:130-172.
10. Hoare, C. A. 1964. Morphological and taxonomic studies on mammalian trypanosomes. X. Revision of the systematics. J. Protozool. 11:200-207.
11. Hoare, C. A. 1965. Vampire bats as vectors and hosts of equine and bovine trypanosomes. Acta Trop. 22:204-216.
12. Hoare, C. A. 1967. Evolutionary trends in mammalian trypanosomes. In Dawes, B., editor. Advances in parasitology, vol. 5. Academic Press, Inc., New York. pp. 47-91.
13. Hoeppli, R. 1969. Parasitic Diseases in Africa and the Western Hemisphere. Early Documentation and Transmission by the Slave Trade. Verlag für Recht und Gesellschaft AG, Basel.
14. Hutt, M. S. R., F. Köberle, and K. Salfelder. 1973. Leishmaniasis and trypanosomiasis. In Spencer, H., editor. Tropical pathology. Springer Verlag, Berlin. pp. 351-398.
15. Johnson, P. T., E. McConnell, and M. Hertig. 1963. Natural infections of leptomonad flagellates in Panamanian *Phlebotomus* sandflies. Exp. Parasitol. 14:107-122.
16. Kagan, I., L. Norman, and D. S. Allain. 1966. Studies on *Trypanosoma cruzi* isolated in the United States: a review. Rev. Biol. Trop. 14:55-73.
17. Levine, N. D. 1973. Protozoan Parasites of Domestic Animals and of Man, ed. 3. Burgess Publishing Co., Minneapolis.
18. Lewis, D. H. 1975. Ultrastructural study of promastigotes of *Leishmania* from reptiles. J. Protozool. 22:344-352.
19. Lincicome, D. R., R. N. Rossan, and W. C. Jones. 1963. Growth of rats infected with *Trypanosoma lewisi.* Exp. Parasitol. 14:54-65.
20. Lumsden, W. H. R. 1971. Pathobiology of

trypanosomiasis. In Gaafar, S. M., editor. Pathology of parasitic diseases. Purdue Research Foundation, Lafayette, Ind. pp. 1-14.

21. Morgan. F. M., R. H. Watten, and R. E. Kuntz. 1962. Post–kala-azar dermal leishmaniasis. A case report from Taiwan (Formosa). J. Formosa Med. Assoc. 61:282-291.

22. Noble, E. R. 1955. The morphology and life cycles of trypanosomes. Rev. Biol. 30:1-28.

23. Ormerod, W. E. 1967. Taxonomy of the sleeping sickness trypanosomes. J. Parasitol. 53:824-830.

24. Rohwedder, R. 1965. Chagas' infection in blood donors and the possibilities of its transmission by means of transfusion. Bull. Chil. Parasitol. 24:88-93.

25. Ross, R., and D. Thomson. 1910. A case of sleeping sickness studied by precise enumerative methods: regular periodical increase of the parasites disclosed. Proc. Roy. Soc. Lond. (Biol.) 82:411-415.

26. Taliaferro, W. H., and L. A. Stauber. 1969. Immunology of protozoan infections. In Chen, T., editor. Research in protozoology, vol. 3. Pergamon Press, Ltd., Oxford. pp. 505-564.

27. Tobie, E. J. 1965. Biological factors influencing transmission of *Trypanosoma rangeli* by *Rhodnius prolixus*. J. Parasitol. 51:837-841.

28. Vickerman, K. 1969. The fine structure of *Trypanosoma congolense* in its bloodstream phase. J. Protozool. 16:54-69.

29. Vickerman, K. 1971. Morphological and physiological considerations of extracellular blood protozoa. In Fallis, A. M., editor. Ecology and physiology of parasites. University of Toronto Press, Toronto. pp. 58-91.

30. Vickerman, K. 1972. The host-parasite interface of parasitic Protozoa. Some problems posed by ultrastructural studies. In Taylor, A. E. R., and R. Muller, editors. Functional aspects of parasite surfaces. Blackwell Scientific Publications Ltd., Oxford. pp. 71-91.

31. Vickerman, K. 1974. Antigenic variation in African trypanosomes. In Parasites in the immunized host: mechanisms of survival (Ciba Foundation Symposium 25, new series). Elsevier, Amsterdam. pp. 53-80.

32. Vickerman, K., and F. E. G. Cox. 1967. The protozoa. Houghton Mifflin Co., Boston.

33. Wallace, F. G. 1966. The trypanosomatid parasites of insects and arachnids. Exp. Parasitol. 18:124-193.

34. Walton, B. C., L. V. Chinel, and O. Eguia y E. 1973. Onset of espundia after many years of occult infection with *Leishmania braziliensis*. Am. J. Trop. Med. Hyg. 22:696-698.

35. Williamson, J. 1962. Chemotherapy and chemoprophylaxis in African trypanosomiasis. Exp. Parasitol. 12:274-367.

36. Woo, P. T. K., and M. A. Soltys. 1970. Animals as reservoir hosts of human trypanosomes. J. Wildl. Dis. 6:313-322.

37. Woody, N. C., and H. B. Woody. 1955. American trypanosomiasis (Chagas' disease). First indigenous case in the United States. J.A.M.A. 159:476-477.

38. World Health Organization. 1960. Chagas' disease. Report of a study group. Technical Report Series no. 202. Geneva.

39. Yaeger, R. G. 1971. Transmission of *Trypanosoma cruzi* infection to opossums via the oral route. J. Parasitol. 57:1375-1376.

SUGGESTED READING

Adler, S. 1964. Leishmania. In Dawes, B., editor. Advances in parasitology, vol. 2. Academic Press, Inc., New York. pp 1-34. (An advanced treatise on the subject. Recommended reading for all who are interested in the Trypanosomatidae.)

Bardsley, J. E., and R. Harmsen. 1973. The trypanosomes of Anura. In Dawes, B., editor. Advances in parasitology, vol. 11. Academic Press, Inc., New York. pp. 1-73.

Foster, W. D. 1965. A history of parasitology. E. & S. Livingstone, Edinburgh. (Chapter X, "The Trypanosomes," is a very interesting account of the history of knowledge about this group.)

Hoogstraal, H., and D. Heyneman. 1969. Leishmaniasis in the Sudan Republic. 30. Final epidemiological report. Am. J. Trop. Med. Hyg. 18:1089-1210. (This is an extensive account of all aspects of leishmaniasis by two men who have an unashamed love for humanity. It should be required reading for all students of parasitology, and it will stand by itself as an example of what scientific writing should be.)

Lumsden, W. H. R. 1965. Biological aspects of trypanosomiasis research. In Dawes, B., editor. Advances in parasitology, vol. 3. Academic Press, Inc., New York. pp. 1-57. (An advanced treatment of the subject.)

Olivier, M. C., L. J. Olivier, and D. B. Segal. 1972. A bibliography of Chagas' disease. Index-Cat. Med. Vet. Zool. Special Publ. no. 2.

Chapter 5

OTHER FLAGELLATE PROTOZOANS

Several groups of flagellate protozoans have members that are parasitic in or on invertebrate and vertebrate animals. Certain dinoflagellates, for example, parasitize copepods, diatoms, and pelagic invertebrates. Although such forms are interesting, limited space allows us to consider only a few examples. Representative species are drawn from four orders.

ORDER RETORTAMONADIDA
Family Retortamonadidae

Two species in the family Retortamonadidae are commonly found in humans. Though they are apparently harmless commensals, they are worthy of note because they can easily be mistaken for highly pathogenic species.

Chilomastix mesnili (Fig. 5-1)

Chilomastix mesnili infects about 3.5% of the population of the United States and 6% of the world population.[3] It lives in the cecum and colon of humans, chimpanzees, orangutans, monkeys, and pigs. Other species are known in other mammals, birds, reptiles, amphibians, fishes, leeches, and insects.

The living trophozoite is piriform, with the posterior end drawn out into a blunt point, and is 6 to 24 μm by 3 to 10 μm. There is a longitudinal, **spiral groove** in the surface of the middle of the body, but this is usually visible only on living specimens. A sunken **cytostomal groove** is prominent near the anterior end. Along each side of the cytostome runs a cytoplasmic **cytostomal fibril,** presumably strengthening the lips of the cytostome. The cytostome leads into the cytopharynx, where endocytosis takes place. Four flagella, one longer than the others, emerge from kinetosomes on the anterior end of the body, and the kinetosomes are interconnected by microfibrillar material.[4] One of the flagella is very short and delicate, curving back into the cytostome, where it

undulates. The large nucleus is near the front end.

A cyst stage occurs, especially in formed stools (Fig. 5-2). A typical cyst is thick-walled, 6.5 to 10 μm long, and pear or lemon-shaped. It has a single nucleus and retains all the cytoplasmic organelles, including cytostomal fibrils, kinetosomes, and flagella.

Transmission is by ingestion of cysts, for trophozoites cannot survive stomach acid. Fecal contamination of drinking water is the most important means of transmission.

Chilomastix mesnili usually is considered to be nonpathogenic, although Mueller (1959) suggested that it might cause a watery stool in some instances.[19]

Retortamonas intestinalis (Fig. 5-3)

Retortamonas intestinalis is a tiny protozoan that is basically similar to *Chilomastix mesnili,* but the trophozoite is only 4 μm to 9 μm long. Further, it has only two flagella, one of which extends anteriad, while the other emerges from the cytostomal groove and trails posteriad. The living trophozoite usually extends into a blunt point at its posterior end, but it bends to round up in fixed specimens. The ovoid to pear-shaped cysts contain a single nucleus.

Like *C. mesnili,* this species is probably a harmless commensal. It lives in the cecum and large intestine of monkeys and chimpanzees, as well as humans, and apparently is not a common symbiont anywhere in the world.

ORDER DIPLOMONADIDA
Family Hexamitidae

Members of the Hexamitidae are easily recognized because they have two equal nuclei lying side by side. There are several species in five genera, most of which are parasitic in vertebrates or invertebrates. One species is a parasite of humans. It will serve to illustrate the genus *Giardia,* while

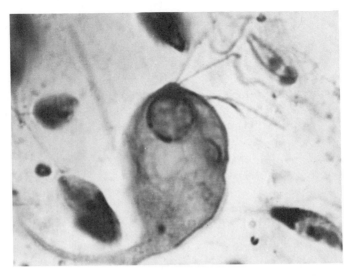

Fig. 5-1. *Chilomastix caulleryi*, trophozoite, which is quite similar morphologically to *C. mesnili.* Note the four flagella and the cytostomal fibrils.

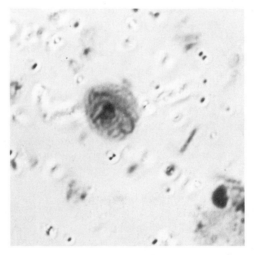

Fig. 5-2. A cyst of *Chilomastix mesnili* from a human stool, showing the characteristic lemon or pear shape. The large, irregular karyosome and the cytostomal fibrils are visible.

Hexamita meleagridis is an example of a related species in domestic animals.

Giardia lamblia

Giardia lamblia was first discovered in 1681 by Leeuwenhoek, who found it in his own stools. The taxonomy of the species was confused in the nineteenth century, and even today it is often called *Giardia intestinalis* (which is probably the correct name). The species is cosmopolitan in distribution, but is most common in warm climates, and children are especially susceptible. It is the most common flagellate of the human digestive tract.

Morphology. The trophozoite (Fig. 5-4) is rounded at the anterior end, while its posterior end is pointed. The organism is dorsoventrally flattened and is convex on the dorsal surface. The flattened ventral

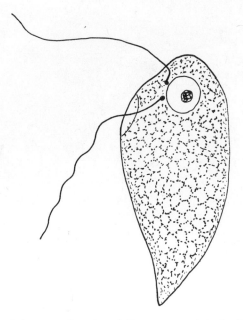

Fig. 5-3. A trophozoite of *Retortamonas intestinalis.*

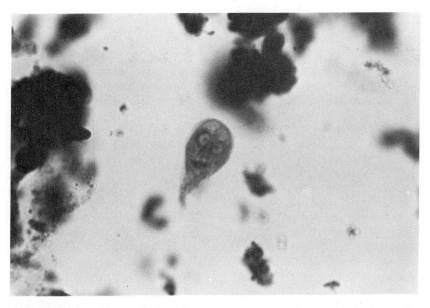

Fig. 5-4. *Giardia lamblia:* trophozoite in a human stool. (Photograph by James Jensen.)

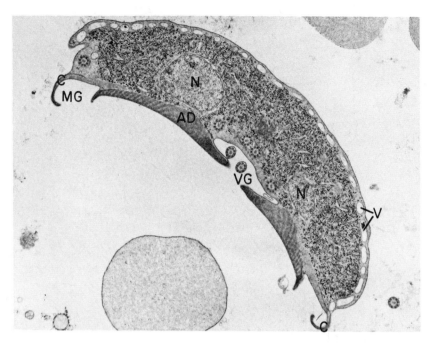

Fig. 5-5. A transverse section through *Giardia muris,* showing its discoid shape, general topography, and cytoplasmic structures. The marginal groove *(MG)* is the space between the striated rim of cytoplasm *(C)* and the lateral ridge of the adhesive disc *(AD)*. The ventral groove *(VG)* is the space between the medial lips of the two lobes of the adhesive disc. One nucleus *(N)* is over each lobe of the disc. The central area of cytoplasm contains granules and clefts, whereas the dorsal portion of the cytoplasm is agranular and is occupied by vacuoles *(V)*. The flagella are cut in cross section. (×15,500.) (From Friend, D. S. 1966. J. Cell Biol. 29:317-332.)

surface bears a concave, bilobed **adhesive disc** (Fig. 5-5). The so-called adhesive disc actually is a rigid structure, reinforced by microtubules and fibrous ribbons, surrounded by a flexible, apparently contractile, striated rim of cytoplasm. Application of this flexible rim to the host's intestinal cell, working in conjunction with the **ventral flagella,** found in the **ventral groove,** is responsible for the organism's remarkable ability to adhere to the host cell (Fig. 5-6). The pair of ventral flagella, as well as three more pairs of flagella—the **anterior, posterior,** and **caudal flagella**—all arise from kinetosomes located between the anterior portions of the two nuclei (Fig. 5-7). The axonemes of all flagella course through the cytoplasm for some distance before emerging from the cytoplasm; those of the anterior flagella actually cross and emerge laterally from the adhesive disc area on the side opposite

from their respective kinetosomes. A pair of large, curved, transverse, dark-staining **median bodies** lie behind the adhesive disc. These bodies are unique to *Giardia.* Various authors have regarded them as parabasal bodies, kinetoplasts, or chromatoid bodies, but ultrastructural studies have shown that they are none of these.[6,9] Their function is obscure, though it has been suggested that they may help support the posterior end of the organism, or they may be involved in its energy metabolism. There is no true axostyle; the structure so described by previous authors is formed by the intracytoplasmic axonemes of the ventral flagella and associated groups of microtubules. Interestingly, no mitochondria, smooth endoplasmic reticulum, Golgi bodies, or lysosomes have been found.[9]

The overall effect of the two nuclei behind the lobes of the adhesive disc and the median bodies is that of a wry little face

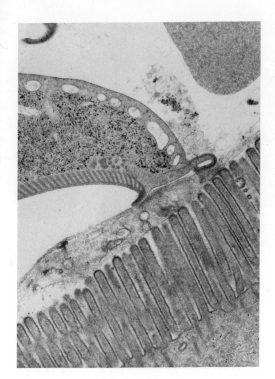

Fig. 5-6. The periphery of *Giardia muris* in contact with the mucous stream covering the microvilli of a duodenal epithelial cell. It appears that the peripheral flange of striated cytoplasm is the grasping organelle of the ventral surface. (×33,000.) (From Friend, D. S. 1966. J. Cell Biol. 29:317-332.)

that seems to be peering back at the observer.

Life cycle. *Giardia lamblia* lives in the duodenum, jejunum, and upper ileum of humans, with the adhesive disc fitting over the surface of an epithelial cell. In severe infections nearly every cell has its free surface covered by a parasite. The protozoa can swim rapidly using their flagella.

Trophozoites divide by binary fission. First the nuclei divide, then the locomotor apparatus and the sucking disc, and finally the cytoplasm. Enormous numbers can build up rapidly in this way. It has been calculated that a single diarrheic stool can contain 14 billion parasites, while a stool in a moderate infection may contain 300 million cysts.[5]

In the small intestine, and in watery stools, only the trophic stage can be found. But as the feces enter the colon and begin to dehydrate, the parasites become encysted. First the flagella shorten and no longer project. The cytoplasm condenses and secretes a thick, hyaline cyst wall. The ovoid cysts (Fig. 5-8) are 8 to 12 μm by 7 to 10 μm in size. Newly formed cysts have two nuclei, but older ones have four. Soon, the sucking disc and the locomotor apparatus are doubled, and the Siamese-twinned flagellates are ready to emerge. When swallowed by the host, they pass safely through the stomach and excyst in the duodenum, immediately completing the division of the cytoplasm. The flagella grow out and the parasites are once again at home.

Pathogenesis. Many cases of infection show no evidence of disease. Apparently, some persons are more sensitive to the presence of *G. lamblia* than others, and considerable evidence suggests that some protective immunity can be acquired. In other cases there is a marked increase of mucous production, diarrhea, dehydration, intestinal pain, flatulence, and weight loss. The stool is fatty but never contains blood. The protozoan does not lyse host cells but appears to feed on mucous secretions. A dense coating of flagellates on the intestinal epithelium interferes with the absorption of fats and other nutrients, which probably triggers the onset of disease. The gallbladder may become infected, which can cause jaundice and colic. The disease is not fatal but can be intensely discomforting.

Epidemiology. Giardiasis is highly contagious. If one member of a family catches it, others will usually become infected. Transmission depends on mature cysts being swallowed; prevention, therefore, depends on a high level of sanitation.

A summary of surveys of 134,966 people throughout the world showed that the prevalence of the infection was from 2.4% to 67.5%.[3] It was found in 7.4% of 35,299 persons tested in the United States.

Outbreaks continue to flare up in the United States, often without regard for the affluence of the people involved. For instance, an epidemic occurred in Aspen, Colorado, during the 1965 to 1966 ski season, with at least 11% of 1,094 skiers infected.[18] Five percent of the permanent population remained infected after the epidemic.[10] In addition to the frequency of infection in the United States, giardiasis

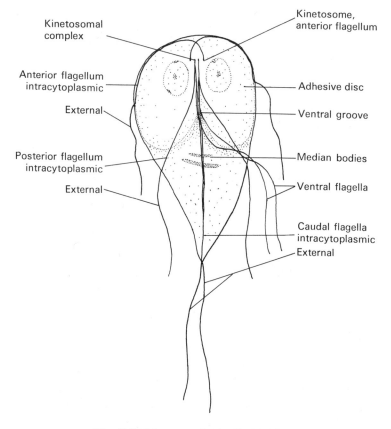

Kinetosomal complex

Kinetosome, anterior flagellum

Anterior flagellum intracytoplasmic

Adhesive disc

External

Ventral groove

Posterior flagellum intracytoplasmic

Median bodies

External

Ventral flagella

Caudal flagella intracytoplasmic

External

Fig. 5-7. Diagram of *Giardia lamblia.*

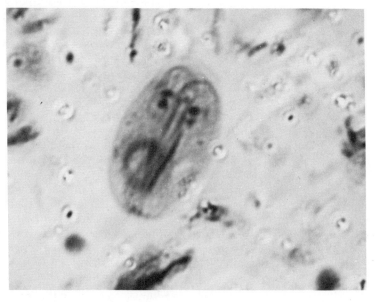

Fig. 5-8. A cyst of *Giardia lamblia* in a human stool. The karyosomes of all four cyst nuclei are visible in this photograph, as well as several intracytoplasmic axonemes.

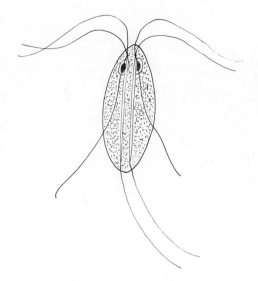

Fig. 5-9. *Hexamita meleagridis:* diagram of a trophozoite.

has become a "traveler's disease" for persons visiting the Soviet Union. Since the increase in travel to Russia began in about 1970, hundreds of people have brought the infection back, and the disease has become "serious and has reached epidemic proportions."[1,2]

Diagnosis and treatment. Recognition of trophozoites or cysts in stained fecal smears is adequate for diagnosis. However, an otherwise benign infection with the flagellates may coexist with a peptic ulcer, enteritis, tumor, or strongyloidiasis, any of which could actually be causing the symptoms. Treatment with quinacrine or metronidazole usually effects complete cure within a few days. All members of a family should be treated simultaneously to avoid reinfection of the others.

Other *Giardia* spp. occur in a wide variety of mammals, including dogs, cats, rodents, cattle, and others. It has not been established whether any of these can infect humans. They all may be varieties of the same species.[8]

Hexamita meleagridis (Fig. 5-9)

Hexamita meleagridis is a parasite of the small intestine of young galliform birds, including turkey, quail, pheasant, partridge, and peafowl. It is known from the United States, Great Britain, and South America, although it probably is common

elsewhere. In the United States, at least, it causes millions of dollars in loss to the turkey industry every year.

Morphologicaly, *Hexamita* is quite similar to *Giardia,* being elongate, with two nuclei and four pairs of flagella. However, it is smaller, has no sucking disc, has karyosomes two-thirds the size of the nuclei, and has no median bodies. Like *Giardia,* the kinetosomes are grouped anterior to and between the nuclei, but three pairs of axonemes emerge anteriorly, while one pair courses intracytoplasmically. The intracytoplasmic axonemes run posteriorly along granular lines and emerge to become the posterior flagella.

The life cycle is essentially the same as for *Giardia,* except that birds are the normal hosts rather than mammals.

Like *Giardia lamblia,* hexamitosis is mainly a disease of young animals. Adults may be infected but are symptomless, thereby becoming reservoirs of infection.

Hexamitosis also exhibits a pathogenesis similar to that of giardiasis. Infected chicks have a foamy or watery diarrhea, a ruffled, unkempt appearance, and soon become weak and listless. Mortality in a flock may range from 7% to 80% in very young birds. Those that survive are somewhat immune to the disease but commonly are stunted in size. They become the most ready source of infection for new broods of birds.

Diagnosis is possible by recognizing trophic or cystic stages in feces or intestinal scrapings. No completely satisfactory treatment is available, but prevention in domestic flocks is possible by proper management and sanitation. Separation of chicks from adults birds is mandatory.

ORDER TRICHOMONADIDA
Family Trichomonadidae

The many members of this family are rather similar in structure. They are easily recognized because they have an anterior tuft of flagella, a stout, median rod (the **axostyle**), and an **undulating membrane** along the recurrent flagellum. They are found in intestinal or reproductive tracts of vertebrates and invertebrates, with one group occurring exclusively in the gut of termites. Three species are common in humans, and one is of extreme importance

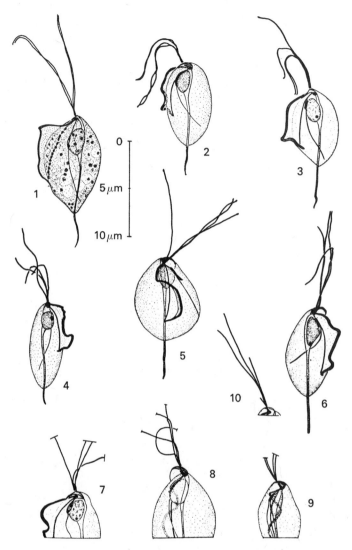

Fig. 5-10. *Trichomonas tenax,* typical trophozoites. (From Honigberg, B. M., and J. J. Lee. 1959. Am. J. Hyg. 69:183.)

in domestic ruminants. These will serve to illustrate the order.

The three trichomonads of humans, *T. tenax, T. vaginalis,* and *Pentatrichomonas hominis,* are similar enough morphologically to have been considered conspecific by many taxonomists. More recently there has been a wide recognition of the differences between *P. hominis* and the other two. As currently defined, the genus *Trichomonas* contains only three species, *T. tenax, T. vaginalis,* and a species found in birds, *T. gallinae,* which is more like *T. tenax* than is *T. vaginalis.*[12]

Trichomonas tenax (Fig. 5-10)

Trichomonas tenax was first discovered by O. F. Müller in 1773, when he examined an aqueous culture of tartar from teeth. *T. tenax* is now known to have worldwide distribution.

Morphology. Like all species of *Trichomonas, T. tenax* has only a trophic stage. It is an oblong cell 5 to 16 μm long by 2 to 15 μm wide, with size varying according to strain. There are four anterior, free flagella, with a fifth flagellum curving back along the margin of an undulating membrane and ending posterior to the middle of the body[17] (Fig. 3-3). The recurrent flagellum is not enclosed by the undulating membrane but is closely associated with it in a shallow groove. A densely staining lamellar structure (**accessory filament**) courses within the undulating membrane along its length. A **costa** arises in the kinetosome complex and runs superficially beneath and generally parallel to the serpentine path of the undulating membrane. The costa distinguishes the Trichomonadidae from other families in its order. It is a rod-like structure with complex cross-striations, which probably serves as a strong, flexible support in the region of the undulating membrane.

A parabasal body (Golgi body) lies near the nucleus with the parabasal filament running from the kinetosome complex, through or very near the parabasal body, and ending in the posterior portion of the body. A small, "minor" parabasal filament, which is inconspicuous in light microscope preparations, has been shown in other trichomonads, and it probably is present in *T. tenax* as well. The tube-like axostyle extends from the area of the kinetosomes posteriorly to protrude from the end of the body (covered by cell membrane). The axostylar tube is formed by a sheet of microtubules, and its anterior, middle, and posterior parts are known as **capitulum, trunk,** and **caudal tip,** respectively. Toward the capitulum, the tubular trunk opens out to curve around the nucleus, and the microtubules of the capitulum slightly overlap the curving, collar-like **pelta.** The pelta is also comprised of a sheet of microtubules and appears to function in supporting the "periflagellar canal," a shallow depression in the anterior end from which all the flagella emerge (Fig. 3-2). A cytostome is not present. Concentrations of **paracostal granules** often can be found along the costa. Other species of *Trichomonas* have **paraxostylar granules** along the axostyle.

Biology. *Trichomonas tenax* can live only in the mouth and, apparently, cannot survive passage through the digestive tract. Transmission, then, is direct, usually by kissing or common use of eating or drinking utensils. Trophozoites divide by binary fission. They are harmless commensals, feeding on microorganisms and cellular debris. They are most abundant between the teeth and gums, in pus pockets, tooth cavities, and crypts of the tonsils, but they also have been found in the lungs and trachea. *T. tenax* is resistant to changes in temperature and will live for several hours in drinking water. Thus, the "communal dipper" may be a route of infection in some situations. Good oral hygiene seems to decrease or even eliminate the infection.

Trichomonas vaginalis (Fig. 5-11)

Trichomonas vaginalis was first found by Donné in 1836 in purulent vaginal secretions. In 1837 he named it *Trichomonas vaginalis*, thereby creating the genus. It is a cosmopolitan species, found in the reproductive tracts of both men and women the world over. Donné thought the organism had three anterior hairs, which is what prompted the generic name. Today, species of trichomonads with three anterior flagella are placed in the genus *Tritrichomonas*.

Morphology. *Trichomonas vaginalis* is very similar to *T. tenax* but differs in the following ways. It is somewhat larger, 7 to 32 μm long by 5 to 12 μm wide. Its undulating membrane is relatively shorter, and there are more granules along the axostyle and costa. In living and appropriately fixed and stained specimens, the constancy in presence and arrangement of the paraxostylar granules is the best criterion for distinguishing *T. vaginalis* from other *Trichomonas*.[14] *T. vaginalis* frequently produces pseudopodia.

Biology. *T. vaginalis* lives in the vagina and urethra of women and in the prostate, seminal vesicles, and urethra of men. It is transmitted primarily by sexual intercourse,[15] although it has been found in newborn infants. Its occasional presence in very young children, including virginal females, suggests that the infection can be contracted from soiled washcloths, towels, and clothing. Viable cultures of the organism have been obtained from damp cloth as long as 24 hours after inoculation. The acidity of the normal vagina (pH 4.0 to 4.5) ordinarily discourages infection, but once established, the organism itself causes a shift toward alkalinity (pH 5.0 to 6.0), which further encourages its growth.

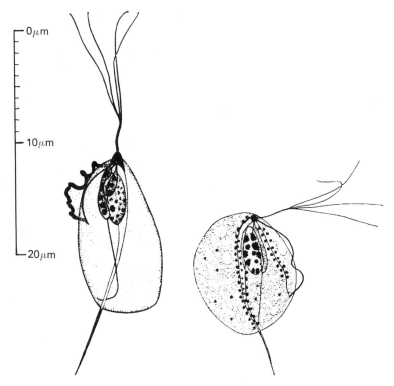

Fig. 5-11. Typical trophozoites of *Trichomonas vaginalis*. (From Honigberg, B. M., and V. M. King. 1964. J. Parasitol. 50:345-364.)

Pathogenesis. Most strains are of such low pathogenicity that the infected person is virtually asymptomatic. However, other strains cause an intense inflammation, with itching and a copious white discharge that is swarming with trichomonads. They feed on bacteria, leukocytes, and cell exudates and are themselves ingested by monocytes. Like all mastigophorans, *T. vaginalis* divides by longitudinal fission, and, like all trichomonads, it does not form cysts.

A few days after infection there is a degeneration of the vaginal epithelium followed by leukocytic infiltration. The vaginal secretions become abundant, white or greenish, and the tissues become intensely inflamed. An acute infection will usually become chronic, with a lessening of symptoms, but will occasionally flare up again. In men the infection is usually asymptomatic, although there may be an irritating urethritis or prostatitis.

Diagnosis depends on recognizing the trichomonad in a secretion. There is also a promising indirect hemagglutination test.[11] Oral drugs, such as metronidazole (Flagyl), usually cure infection in about 5 days. Some apparently recalcitrant cases may be caused by reinfection by the sexual partner. Suppositories and douches are useful in promoting an acid pH of the vagina. Sexual partners should be treated simultaneously to avoid reinfection.

Pentatrichomonas hominis (Fig. 5-12)

The third trichomonad of humans is a harmless commensal of the intestinal tract. It was first found by Davaine, who named it *Cercomonas hominis* in 1860. Traditionally, it has been called *Trichomonas hominis*, but since most specimens actually bear five anterior flagella, the organism has been assigned to the genus *Pentatrichomonas*. Next to *Giardia lamblia* this is the most common intestinal flagellate of humans. It is also known in other primates and in various domestic animals. The prevalence among 13,517 persons examined in the United States was 0.6%.[3]

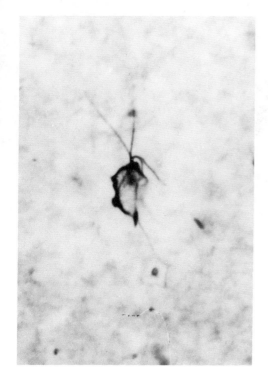

Fig. 5-12. *Pentatrichomonas hominis,* trophozoite. (Photograph by Peter Diffley.)

Morphology. This species is superficially similar to *T. tenax* and *T. vaginalis* but differs in several respects. Its size is 8 μm to 20 μm by 3 μm to 14 μm. Five anterior flagella are present in most specimens, although individuals with fewer are sometimes found. The arrangement is referred to as "four-plus-one" since the fifth flagellum originates and beats independently of the others.[12] A recurrent (sixth) flagellum is aligned alongside the undulating membrane, as in *T. tenax* and *T. vaginalis,* but in contrast to those two species, the recurrent flagellum in *P. hominis* continues as a long, free flagellum past the posterior end of the body. Axostyle, pelta, parabasal body, "major" and "minor" parabasal filaments, costa, and paracostal granules are present. Paraxostylar granules are absent.

Biology. *Pentatrichomonas hominis* lives in the large intestine and cecum, where it divides by binary fission, often building up incredible numbers. It feeds on bacteria and debris, probably taking them in with active pseudopodia. The organism often is present in routine examinations of diar-

rheic stools, but there is no indication that it contributes to this or other disease conditions. In formed stools the flagellates are rounded and dormant, but not encysted. They are difficult to identify at this stage, for they do not move, and the structures normally characteristic for the species cannot be distinguished.

The organism apparently can survive acidic conditions of the stomach, and transmission occurs by contamination. Filth flies can serve as mechanical vectors. Higher prevalence is correlated with unsanitary conditions.

Diagnosis depends on identification of the animal in fecal preparations, and prevention depends on personal and community sanitation. The organism cannot establish in the mouth or urogenital tract.

Tritrichomonas foetus (Fig. 5-13)

Tritrichomonas foetus is responsible for a serious genital infection in cattle, zebu, and possibly other large mammals. It is probably the third leading cause of abortion in cattle, after brucellosis and leptospirosis. *T. foetus* is especially common in Europe and the United States. The Department of Agriculture estimated losses as the result of *T. foetus* in the United States between 1951 and 1960 at $8.04 million.

Morphology. The cell is spindle to pear-shaped, 10 to 25 μm long by 3 to 15 μm wide. There are three anterior flagella, and the fourth, the recurrent flagellum, extends free from the posterior of the body about the length of the anterior flagella. The mastigont system is generally similar in organization to those trichomonads described previously, but it is even more complex and will not be detailed here.[13] The costa is prominent and, though similar in position and function to those of the other trichomonads, differs in ultrastructural detail, resembling a parabasal filament in this respect. The structure of the undulating membrane is curious, consisting of two parts. The proximal part is a fold-like differentiation of the dorsal body surface, and the distal part, which contains the axoneme of the recurrent flagellum, courses along the rim of the proximal part with no obvious physical connection to it. The thick axostyle pro-

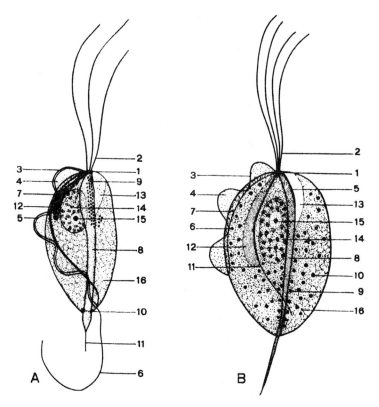

Fig. 5-13. A, Diagrammatic drawing of *Tritrichomonas foetus* (Riedmuller): *1*, kineto-somes; *2*, anterior flagella; *3*, posterior flagellum as margin of the *4*, undulating membrane; *5*, accessory filament in the undulating membrane; *6*, posterior free flagellum; *7*, costa; *8*, axostyle; *9*, endoaxostylar granules in the capitulum of the axostyle; *10*, chromatic ring about the axostyle at point of emergence from the body; *11*, terminal spine of axostyle; *12*, parabasal body; *13*, ventral part of capitulum of axostyle; *14*, nucleus; *15*, karyosome; *16*, undifferentiated cytoplasm. **B,** Diagrammatic drawing of *Trichomonas vaginalis* from human vagina to compare with *Tritrichomonas foetus.* Index numbers refer to corresponding parts in figure **A,** except as follows: *5*, parablepharoplast bar; *6*, paracostal granules (not always in a definite row); *9*, row of granules along the axostyle; *10*, chromatic guanules commonly present in the cytoplasm; *11*, parabasal fibril. (From Wenrich, D. H., and M. A. Emmerson. 1933. J. Morphol. 55:195.)

trudes from the posterior end of the body. Numerous paraxostylar granules are present in the posterior part of the organism, just anterior to the point of the axostyle, and these are apparent in the light microscope preparations as the "chromatic ring."

Biology. These trichomonads live in the preputial cavity of the bull, although the testes, epididymis, and seminal vesicles also may be infected. In the cow, the flagellates first infect the vagina, causing a vaginitis, then move into the uterus. After es-

tablishing in the uterus, they may disappear from the vagina or remain there as a low-grade infection. Bovine genital trichomoniasis is a venereal disease transmitted by coitus, although transmission by artificial insemination is possible. Trichomonads multiply by longitudinal fission and form no cyst.

Pathogenesis. The most characteristic sign of bovine trichomoniasis is early abortion, which usually happens 1 to 16 weeks after insemination. The owner may not notice that the cow has aborted, because of the small size of the fetus, and therefore believes she did not conceive. The cow may recover spontaneously if all of the fetal membranes are passed after abortion; but, if they remain, she usually develops chronic endometritis, which may cause permanent sterility. Normal gestation and delivery occasionally happen with an infected animal. Cows that recover from trichomoniasis are usually immune to further infection.

Pathogenesis is not observable in bulls, but an infected bull is worthless as a breeding animal; unless treated, it usually remains infected permanently. Treatment is expensive, difficult, and not always effective. Considering the immense prices paid for top quality bulls, the loss of a single animal may bankrupt the breeder.

Epidemiology. It has not been determined definitely whether *T. foetus* can infect animals other than cattle and closely related species. Experimental infections have been established in rabbits, guinea pigs, hamsters, dogs, goats, sheep, and pigs. Trichonomads similar to *T. foetus* have been found as natural infections in pigs and horses. Whether these can be transmitted to cattle by contamination is not known, but experiments demonstrate that it is likely.

Trichomonads can survive freezing in semen ampules, although some media are more detrimental than others. This precludes use, by artificial methods, of semen from infected bulls.

Diagnosis, treatment, and control. Direct identification of protozoans from smears or culture remains the only sure means of diagnosis, although a mucus agglutination test is available. In light infections a direct smear of mucus or exudate is sufficient. They can be obtained from amniotic or allantoic fluid, vaginal or uterine exudates, placenta, fetal tissues or fluids, or preputial washings from bulls. Flagellates fluctuate in numbers in bulls; in cows they are most numerous in the vagina 2 or 3 weeks after infection.

No satisfactory treatment is known for cows, but the infection is usually self-limiting in them, with subsequent, partial immunity. Bulls can be treated if the condition has not spread to the inner genital tubes and testes. Treatment is usually attempted only on exceptionally valuable animals, as it is a tedious, expensive task. Preputial infection is treated by massaging antitrichomonal salves or ointments into the penis, after it has been let down by nerve block or by injection of a tranquilizer into the penis retractor muscles. Repeated treatment is usually necessary. Systemic drugs show promise of becoming the standard method of treatment.

Control of bovine genital trichomoniasis depends on proper herd management. Cows that have been infected should be bred only by artificial insemination to avoid infecting new bulls. Bulls should be examined before purchase, with a wary eye for infection in the resident herd. Unless they are quite valuable, infected bulls should be killed. Like any venereal disease, this one can be controlled and eventually eliminated with proper treatment and reporting, but the disease is likely to remain a problem for some time.

Family Monocercomonadidae

The Monocercomonadidae show strong affinities with the Sarcodina, for in some species pseudopodia are well developed, an undulating membrane is absent, and flagella tend to be reduced. Most species are parasites of insects, but three genera infect domestic animals. One of these is economically important and has evolved a unique mode of transmission: in the egg of a nematode.

Histomonas meleagridis (Fig. 5-14)

Histomonas meleagridis is a cosmopolitan parasite of gallinaceous fowl, including chickens, turkeys, peafowl, and pheasant, causing a severe disease, known variously as blackhead, infectious enterohepatitis, and histomoniasis. The disease is more virulent in some species of host than others: chickens show disease less often than do turkeys, for example. The U.S. Department of Agriculture estimated that the loss in the United States as the result of histomoniasis in chickens and turkeys during 1951 to 1960 amounted to $9.3 million.

The taxonomic history of *H. meleagridis* has been very confused, because of its polymorphism in different situations. At various times it has been confused with amebas, coccidia, fungi, and *Trichomonas*.

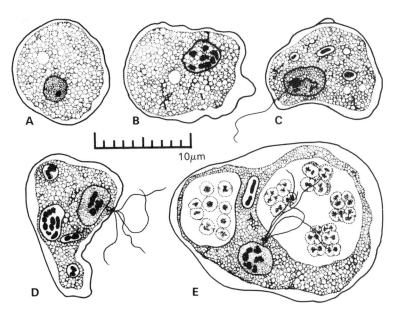

Fig. 5-14. Examples of *Histomonas meleagridis* **(A, B, C)** compared with *Parahistomonas wenrichi* (**D** and **E**), showing for each species variations associated with environmental conditions. **A,** Tissue type *H. meleagridis* in fresh preparation from liver lesion; viewed with phase contrast. **B,** *H. meleagridis* in transitional stage in lumen of the cecum. Pseudopodia have been formed, and the distribution of chromatin suggests that binary fission is approaching. However, the flagellum has not yet appeared. **C,** An organism in same cecal preparation as **B,** but this one completely adapted as a lumen dweller. **D,** Small *P. wenrichi*, structurally distinguishable from *H. meleagridis* by presence of four flagella. **E,** *P. wenrichi* as viewed in stained smear from cecum in which packets of *Sarcina* were abundant. All figures from camera lucida tracings. (From Lund, E. E. 1969. In Brandly, C. A., and C. E. Cornelius, editors. Advances in veterinary science and comparative medicine. Academic Press, Inc., New York.)

Even the disease that it causes has been attributed to different organisms, from amebas to viruses. Today, much is known about the organism, and its biology and pathogenesis are less mysterious.

Morphology. *Histomonas meleagridis* is pleomorphic, its stages changing size and shape in response to environmental factors. There is no cyst in the life cycle, only various trophic stages. When they are found in the lumen of the cecum (which is rare) or in culture, they are ameboid, 5 to 30 μm in diameter, and almost always with only one flagellum. However, there are usually four kinetosomes, the basic number for trichomonads, though this condition has been attributed to duplication of the kinetic apparatus in preparation for mitosis.[20] More likely, they are vestiges of an ancestral type with four flagella. The nucleus is vesicular and often

has a distinct endosome. One can usually discern an outer, clear ectoplasm and an inner, granular endoplasm. Food vacuoles may contain host blood cells, bacteria, or starch granules. Electron microscope studies have revealed a pelta, a V-shaped parabasal body, a parabasal filament, and a structure resembling an axostyle (Fig. 5-15). These cannot be seen with light microscopy, but their presence supports placement of *Histomonas* in the order Trichomonadida. No mitochondria have been observed.

The forms within the tissues (Fig. 5-16) have no flagella, although the kinetosomes are present near the nucleus. Three extracellular, intertissue stages have been described from the vertebrate host. The **invasive form** is 8 to 17 μm in diameter and resembles a tiny ameba. It has very active pseudopodia and a finely granular

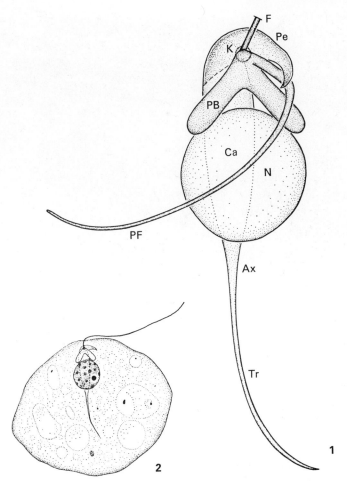

Fig. 5-15. *Histomonas meleagridis.* **1,** Composite, schematic diagram of the mastigont system and nucleus as seen from a dorsal and somewhat right view. **2,** Composite diagram of an organism, with the mastigont system seen in the same view as in part **1.** The flagellum arises from the kinetosomal complex just anterior to the V-shaped parabasal body. The cytoplasm appears highly vacuolated and contains ingested bacteria and rice starch. The nucleus has uniformly dispersed accumulations of chromatin and a nucleolus surrounded by a clear halo. (×4,270.) (From Honigberg, B. M., and C. J. Bennett. 1971. J. Protozool. 18:688.)

endoplasm. It feeds by phagocytosis; food vacuoles contain particles but not bacteria. The parasite is found in new lesions in the host's cecal wall and liver.

The **vegetative stage** is in older lesions and is somewhat larger, measuring 12 to 21 μm wide. These are less active than the invasive forms and are often packed tightly together in liver or cecal lesions. The food vacuoles seldom contain large particles; the protozoans appear to feed on host tissues, which they first dissolve with proteolytic enzymes.

The **resistant stage,** ironically, is no more resistant than either of the others. It is compact, 4 to 11 μm in diameter, and the ectoplasm forms a dense layer around it. These may be found scattered singly or packed together in hepatic lesions. They too absorb predigested cells.

Biology and epidemiology. Like all flagellate protozoa, *Histomonas meleagridis* divides by binary fission. This enables them to build up large numbers in a short time. No cysts or sexual stages are found in the life cycle.

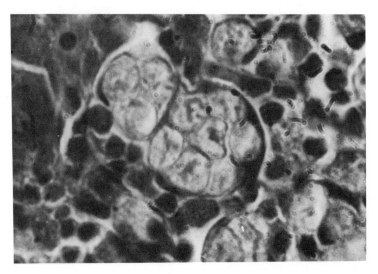

Fig. 5-16. Histomonads in liver of young turkey. In addition to organisms seen in two clusters, at least two appear singly. (×2,000.) (Photograph by Everett E. Lund).

The trophozoite can live only a short time in expelled feces. Further, if ingested shortly after being expelled, it stands little chance of surviving the host's stomach acids. Certain factors can, and sometimes do, conspire to allow infection by trophozoites. If a great many are eaten, a few may make it through the upper digestive tract. The pH of the stomach is critical to the parasites, with acidity being deadly. If trophozoites are eaten with certain foods that raise the stomach pH, they may survive to initiate a new infection. This can be the means of an epizootic in a dense flock of birds.

The most important, and by far the most interesting, mode of transmission is within the egg of the cecal nematode, *Heterakis gallinarum*. Since the protozoan undergoes development and multiplication in the nematode, the worm can be considered a true intermediate host.[16] After being ingested by the worm, the flagellates enter the nematode's intestinal cells, multiply, then break out into the pseudocoel and invade the germinative area of the nematode's ovary. There they feed and multiply extracellularly and move down the ovary with the developing oogonia, then penetrate the oocytes (Fig. 5-17). Feeding and multiplication continue in the oocytes and newly formed eggs. Passing out of the mother worm and out of the bird with its feces, the protozoan divides rapidly, invading the tissues of the larval nematode, especially those of the digestive and reproductive systems. Interestingly, *Histomonas* also parasitizes the reproductive system of the male nematodes.[16] Presumably, it could be transmitted to the female during copulation, thus constituting a venereal infection of nematodes!

Infected eggs can survive for at least 2 years in the soil. If the worm eggs are eaten by an appropriate bird, they hatch in the intestine, and the larval *Heterakis* passes down into the cecum, where *Histomonas* is free to leave its temporary host to begin residence in a more permanent one.

Earthworms are important paratenic hosts of both *Heterakis* and its contained *Histomonas*. When eaten by an earthworm, the nematode eggs will hatch, releasing second stage larvae that become dormant in the earthworm's tissues. When the earthworm is eaten by a gallinaceous fowl, the *Heterakis* larvae are released, and the bird becomes infected by two kinds of parasites at once. Earthworms can serve to maintain the parasites in the soil for long periods of time. Chickens are the most important reservoirs of infection because they are less often affected by *Histomonas*

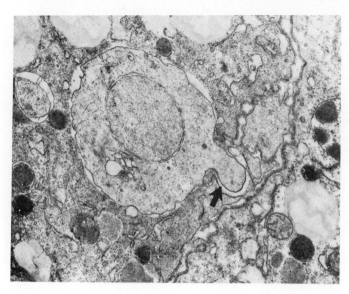

Fig. 5-17. Electron micrograph of section through the growth zone of the ovary of *Heterakis gallinarum* to show *Histomonas meleagridis* in the process of entering an oocyte (arrow). (From Lee, D. L. 1971. In Fallis, A. M. editor. Ecology and physiology of parasites. University of Toronto Press, Toronto.)

than are turkeys. Because *Heterakis* eggs and infected earthworms can survive for so long in the soil, it is nearly impossible to raise uninfected turkeys in the same yards where chickens have lived.

Pathogenesis (Fig. 5-18). Turkeys are most susceptible during the ages of 3 and 12 weeks, although they can become infected as adults. In very young poults, losses may approach 100% of the flock. Chickens are less prone to the diseases, but outbreaks among young birds have been reported. Quails and partridges show varying degrees of susceptibility.

The principal lesions of histomoniasis are found in the cecum and liver. At first, pinpoint ulcers are formed in the cecum. These may enlarge until nearly the entire mucosa is involved. The ceca often become filled with cheesy, foul-smelling plugs that adhere to the cecal walls. Complete perforation of the cecum, with peritonitis and adhesions, can occur. The ceca are usually enlarged and inflamed. Liver lesions are rounded, with whitish or greenish areas of necrosis. Their size varies, and they penetrate deep into the parenchyma.

Infected birds show signs of droopiness, ruffled feathers, and hanging wings and tail. The skin of the head turns black in some cases, giving the disease the name **blackhead.** Other diseases can cause this symptom, however. There is usually yellowish diarrhea.

It has been shown that *H. meleagridis* by itself is incapable of causing blackhead, but does so only in the presence of intestinal bacteria of several species, especially *Escherichia coli* and *Clostridium perfringens.* Birds that survive are immune for life. A related histomonad, *Parahistomonas wenrichi,* also is transmitted by *Heterakis* but is not pathogenic.

Diagnosis, treatment, and control. Cecal and liver lesions are diagnostic. Scrapings of these organs will reveal histomonads, thereby distinguishing the disease from coccidiosis.

Several types of drugs are used in prevention and treatment, including nitrofurans, thiazole derivatives, and phenylarsonic acid derivatives. These successfully inhibit, suppress, or cure the disease, but some have undesirable side effects, such as delaying sexual maturity of the bird. Treatment of birds with phenothiazine, a nematocide, to eliminate *Heterakis* is effective in preventing future outbreaks, as *H.*

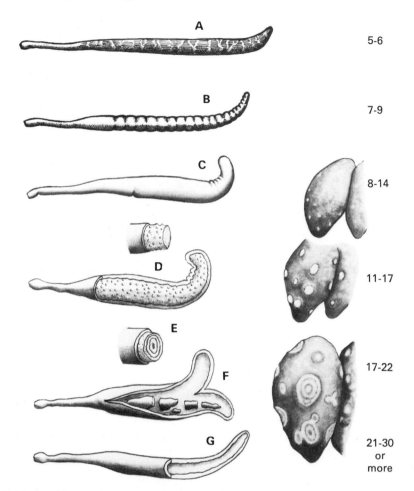

Fig. 5-18. Semidiagrammatic representation of the progress of histomoniasis in young turkeys. **A,** A normal cecum. **B,** Hyperemic and annulated. **C,** Cecum distended with mucus or containing a soft core (first indications of liver lesions may appear). **D,** The core becomes firm and the liver lesions grow. **E,** Later, the firm core may be seen to be composed of concentric cylinders of caseous material, corresponding roughly to the days of their deposition. However, the innermost may have liquefied. If the bird lives long enough, the core may liquefy or fragment, **F,** or both and be voided, leaving, **G,** the thick-walled cecum. It may require 2 to 3 weeks to return to normal. Meanwhile, the liver lesions have grown larger, often have elevated margins and depressed centers, and may show concentric rings. If the bird lives, the liver may bear scars 3 to 4 weeks after the histomonads have ceased to multiply. (From Lund, E. E. 1969. In Brandly, C. A., and C. E. Cornelius, editors. Advances in veterinary science and comparative medicine. Academic Press, Inc., New York.)

meleagridis cannot survive in the soil by itself.

Control depends on effective management techniques, such as rearing young birds on hardware cloth above the ground, keeping young birds on dry ground, and controlling *Heterakis*. Pasture rotation of *Heterakis*-free flocks is also successful.

Superclass Opalinata
ORDER OPALINIDA
Family Opalinidae

There are about 150 species of opalinids, most of which live in the intestines of amphibians. They are of no economic or medical importance but are of zoologi-

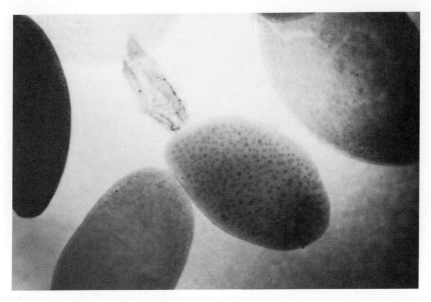

Fig. 5-19. Opalinids from the rectum of a frog. The dark spots are nuclei. (Photograph by Peter Diffley.)

cal interest because of their peculiar morphology and the fact that their reproductive cycles apparently are controlled by host hormones.[7] Further, they are commonly encountered in routine dissections of frogs in teaching laboratories.

Numerous oblique rows of cilia occur over the entire body surface of opalinids, giving them a strong resemblance to ciliates (Fig. 5-19), and they traditionally have been classified with the Ciliophora. However, they have several important differences from ciliates and are now placed as a separate superclass in the subphylum Sarcomastigophora. For example, opalinids have two to many nuclei of similar structure and reproduce sexually by anisogamous syngamy. Asexually, they undergo binary fission between kineties.

Adult opalinids reproduce asexually by binary fission in the rectum of frogs and toads during the summer, fall, and winter. In the spring, which is their host's breeding season, they accelerate divisions and produce small, precystic forms. These then form cysts and pass out with the feces of the host. When the cysts are eaten by tadpoles, male and female gametes excyst and fuse to form the zygote, which resumes asexual reproduction. The exact chemical identity of the compound(s) that stimulates encystment is not known, but present evidence indicates that it is one or more breakdown products of steroid hormones excreted in the frog's urine. This is an interesting example of a physiological adaptation to ensure the production of infective stages at the time and place of new host availability. The effectiveness of the adaptation is attested to by the prevalence of opalinids in frogs and toads.

A curious symbiosis is found in *Zelleriella opisthocarya,* a parasite of toads, and *Entamoeba* sp., in which over 200 cysts of the ameba were found in one opalinid.[21]

REFERENCES

1. Altman, L. K. 1970. Doctors say a parasitic disease infects U.S. visitors to Russia. New York Times, Dec. 6.
2. Altman, L. K. 1974. Serious infection from drinking water afflicts visitors to Soviet. New York Times, Mar. 10.
3. Belding, D. L. 1965. Textbook of clinical parasitology, ed. 3. Appleton-Century-Crofts, New York.
4. Brugerolle, G. 1973. Etude ultrastructurale du trophozoite et du kyste chez le genre *Chilomastix* Alexeieff, 1910 (Zoomastigophorea, Retortamonadida Grassé, 1952). J. Protozool. 20:574-585.
5. Chandler, A. C., and C. P. Read. 1961. Introduction to parasitology, ed. 10. John Wiley & Sons, Inc., New York.

6. Cheissin, E. M. 1964. Ultrastructure of *Lamblia duodenalis*. I. Body surface, sucking disc, and median bodies. J. Protozool. 11:91-98.

7. El Mofty, M. M., and I. A. Sadek. 1973. The mechanism of action of adrenaline in the induction of sexual reproduction (encystation) in *Opalina sudafricana* parasitic in *Bufo regularis*. Int. J. Parasitol. 3:425-431.

8. Filice, F. P. 1952. Studies on the cytology and life history of a *Giardia* from the laboratory rat. University of California Press (Zool.) 57:53-143.

9. Friend, D. S. 1966. The fine structure of *Giardia muris*. J. Cell Biol. 29:317-332.

10. Gleason, N. N., M. S. Horwitz, L. H. Newton, and G. T. Moore. 1970. A stool survey for enteric organisms in Aspen, Colorado. Am. J. Trop. Med. Hyg. 19:480-484.

11. Hoffman, B. 1966. An evaluation of the use of the indirect hemagglutination method in the serodiagnostic of trichomonadosis. Wiadomosci Parazytol. 12:392-397.

12. Honigberg, B. M. 1963. Evolutionary and systematic relationships in the flagellate Order Trichomonadida Kirby. J. Protozool. 10:20-63.

13. Honigberg, B. M., C. F. T. Mattern, and W. A. Daniel. 1971. Fine structure of the mastigont system in *Tritrichomonas foetus* (Riedmüller). J. Protozool. 18:183-198.

14. Honigberg, B. M., and V. M. King. 1964. Structure of *Trichomonas vaginalis* Donne. J. Parasitol. 50:345-364.

15. Jírovic, O. 1965. Neuere Forschungen über *Trichomonas vaginalis* und vaginale Trichomonosis. Angew. Parasitol. 6:202-210.

16. Lee, D. L. 1971. Helminths as vectors of microorganisms. In Fallis, A. M., editor. Ecology and physiology of parasites. University of Toronto Press, Toronto. pp. 104-122.

17. Mattern, C. F. T., B. M. Honigberg, and W. A. Daniel. 1967. The mastigont system of *Trichomonas gallinae* (Rivolta) as revealed by electron microscopy. J. Protozool. 14: 320-339.

18. Moore, G. T., W. M. Cross, D. McGuire, and others. 1970. Epidemic giardiasis at a ski resort. N. Engl. J. Med. 281:402-407.

19. Mueller, J. F. 1959. Is *Chilomastix* a pathogen? J. Parasitol. 45:170.

20. Schuster, F. L. 1968. Ultrastructure of *Histomonas meleagridis* (Smith) Tyzzer, a parasitic amebo-flagellate. J. Parasitol. 54:725-737.

21. Stabler, R. M., and T. Chen. 1936. Observations on an *Endomoeba* parasitizing opalinid ciliates. Biol. Bull. 70:56-71.

SUGGESTED READING

Honigberg, B. M. 1970. Trichomonads. In Jackson, G. J., R. Herman, and I. Singer, editors. Immunity to parasitic animals, vol 2, part 6. Appleton-Century-Crofts, New York. pp. 469-550.

Rybicka, R., B. M. Honigberg, and S. C. Holt. 1972. Fine structure of the mastigont system in culture forms of *Histomonas meleagridis* (Smith). J. Protozool. 18:687-697.

Trussell, R. E. 1947. *Trichomonas vaginalis and Trichomoniasis*. Charles C Thomas, Publisher, Springfield, Ill.

Chapter 6

SUPERCLASS SARCODINA: AMEBAS

Students of biology are introduced to amebas (order Amoebida) early in their careers. Most are left with the impression that amebas are harmless, microscopic creatures that spend their lives aimlessly wandering about in mud, water, and soil, occasionally catching a luckless ciliate for food and unemotionally reproducing by binary fission. Actually, this is a pretty fair account of most amebas. However, a few species are parasites of other organisms, and one or two are responsible for much misery and death of humans. Still others are commensals. These must be recognized, however, to differentiate them from the pathogenic species.

The superclass Sarcodina evolved from the Mastigophora, but as traditionally conceived, the group is polyphyletic. One line passes from the flagellate *Tetramitus,* which includes flagellate and ameboid stages (the flagellate stage has a permanent cytostome), through *Naegleria* to *Vahlkampfia.* The life cycle of *Naegleria* also includes flagellate and ameboid stages (see Chapter 3), but no permanent cytostome is found in them. *Vahlkampfia* has no flagellate stage, but its ameboid stage is very like that of *Naegleria.* Another line passes from the ameboid-flagellate *Histomonas* to the related ameba *Dientamoeba,* which itself is considered a flagellate by some parasitologists.[3,10] No doubt other lines of evolution have evolved in the superclass.

Of the many families of amebas, only the Endamoebidae has species of great medical or economic importance. A second family, Dimastigamoebidae, has species that can become facultatively parasitic in humans.

FAMILY ENDAMOEBIDAE

Species in the Endamoebidae family are parasites or commensals of the digestive systems of arthropods and vertebrates. The genera and species are differentiated on the basis of nuclear structure. Four genera contain known parasites or commensals of humans and domestic animals: *Entamoeba, Endolimax, Iodamoeba,* and *Dientamoeba.* We will follow the traditional arrangement, placing *Dientamoeba* in the Endamoebidae, in spite of strong evidence that the genus is a flagellate (without a flagellum) and should be regarded as belonging to the order Trichomonadida (see p. 110).

Genus Entamoeba

Species of *Entamoeba* have a nucleus that is quite vesicular and that has a small endosome at or near the center. Chromatin granules are arranged around the periphery of the nucleus and, in some species, also around the endosome. The cytoplasm contains a variety of food vacuoles, often containing particles of food being digested, usually bacteria or starch grains.[11] On the ultrastructural level, both lysosomes and some endoplasmic reticulum are seen, and ribosomes are abundant. Golgi bodies and mitochondria apparently are absent. Curious, small **helical bodies** can be seen widely distributed in the cytoplasm of some trophozoites. These bodies are 0.3 to 1 μm in length and are ribonucleoprotein, perhaps some form of "packaged" messenger RNA. These become aggregated in a crystalline array in some species[13] and then are visible by light microscopy as **chromatoidal bars.** These bodies stain darkly with basic dyes and have been known by parasitologists for many years, though their RNA character has only recently been discovered. The chromatoidal bars may be blunt rods or splinter-shaped, according to species, and in some species they are noticeable only in young cysts. As the cyst ages, the bars apparently are dissembled and disappear.

Species of *Entamoeba* are found in both vertebrate and invertebrate hosts. The taxonomic difficulties involved in separating them have been summarized by Levine.[10]

Four species are common in humans (*E. histolytica, E. hartmanni, E. coli,* and *E. gingivalis*) and will be considered here in some detail.

Entamoeba histolytica

Dysentery, both bacterial and amebic, has long been known as a handmaiden of war, often inflicting more casualties than bullets and bombs. Accounts of epidemics of dysentery accompany nearly every thorough account of war, from antiquity to the prison-camp horrors of World War II and Viet Nam. Captain James Cook's first voyage met with amebic disaster in Batavia, Java, and modern tourists, too, often find themselves similarly afflicted upon visiting foreign ports.

The history of acquired knowledge of the parasite *Entamoeba histolytica* is rampant with confusion and false conclusions. An interesting account has been prepared by Foster.[7]

The ameba was first discovered by a clinical assistant, Dr. F. Lösch, in St. Petersburg (now Leningrad), Russia, in 1873. The patient, a young peasant with bloody dysentery, was passing large numbers of amebas in his stools. Many of these, Lösch observed, contained erythrocytes in their food vacuoles. He successfully infected a dog by injecting amebas from his patient into the dog's rectum. On dissection Lösch found the dog's colonic mucosa riddled with ulcers that contained amebas. His human patient soon died, and at autopsy Lösch found identical ulcers in the intestinal mucosa. Despite these clear-cut observations, Lösch concluded that the ulcers were caused by some other agent and that the amebas merely interfered with their healing. It was nearly 40 years before it was generally accepted that an intestinal ameba can cause disease.

A major part of the problem was the then unrecognized fact that several species of amebas are found in the human intestine. Once this was established and nonpathogenic species were delineated, only one species-complex remained that appeared to cause disease, and only occasionally at that. In 1903, Schaudinn named this group *Entamoeba histolytica*,[15] although the trivial *"coli"* was already applied to it by Lösch (as *Amoeba coli*). Schaudinn applied

the latter name to a nonpathogenic species that he named *Entamoeba coli.*

Through the years it became obvious that *E. histolytica* occurs in two sizes. The smaller sized amebas have trophozoites 12 to 15 μm in diameter and cysts 5 to 9 μm wide. This form is encountered in about one-third of those who harbor amebas and is not associated with disease. The larger form has trophozoites 20 to 30 μm in diameter and cysts 10 to 20 μm wide. The larger form may actually consist of two races, one sometimes pathogenic and the other always a commensal.

The small, nonpathogenic type is considered here as a separate species called *Entamoeba hartmanni*. Its life cycle, general morphology, and overall appearance, with the exception of size, are identical to those of *E. histolytica*. The burden of proper identification is placed on the diagnostician, whose diagnosis may save the life of the patient or add the burden of unnecessary medication.

A third species, *Entamoeba moshkovskii,* is identical in morphology to *E. histolytica,* but it is not a symbiont. It dwells in sewage and is often mistaken for a parasite of humans. Indeed, it may be a strain' that recently derived from one of the symbionts of humans.

The pathogenic strain of *Entamoeba histolytica* is still one of the most important parasites of humans, and we will consider it further.

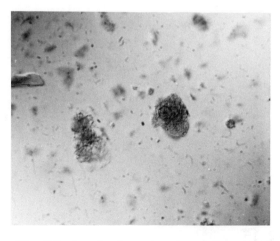

Fig. 6-1. A stained trophozoite of *Entamoeba histolytica* from a human stool.

Morphology and life cycle. There are several successive stages in the life cycle of *Entamoeba histolytica:* the **trophozoite, precyst, cyst, metacyst,** and **metacystic trophozoite.**

While most trophozoites (Fig. 6-1) fall into the range of 20 to 30 μm in diameter, occasional specimens are as small as 10 μm or as large as 60 μm. In the intestine and in freshly passed, unformed stools, the parasites actively crawl about, their short, blunt pseudopodia rapidly extending and withdrawing in a nearly random fashion. The clear ectoplasm is rather thin but is clearly differentiated from the granular endoplasm. It is difficult to see the nucleus in living specimens, but nuclear morphology may be distinguished after fixing and staining with iron-hematoxylin. The nucleus is spherical and is about one-sixth to one-fifth the diameter of the cell. A prominent endosome is located in the center of the nucleus and delicate, achromatic fibrils radiate from it to the inner surface of the nuclear membrane. Chromatin is absent from a wide area surrounding the karyosome but is concentrated in granules or plaques on the inner surface of the nuclear membrane. This gives the appearance of a dark circle with a bull's-eye in the center.

The nuclear membrane itself is quite thin.

Food vacuoles are common in the cytoplasm of active trophozoites and may contain host erythrocytes in samples from diarrheic stools (Fig. 6-2). Granules typical for all amebas are numerous in the endoplasm. Chromatoidal bars are not found in this stage.

In a normal, asymptomatic infection, the amebas are carried out in formed stools. As the fecal matter passes posteriad and becomes dehydrated, the ameba is stimulated to encyst. Cysts are not found in the stools of patients with dysentery, nor are cysts formed by the amebas when they have invaded the tissues of the host. Trophozoites passed in stools are unable to encyst. At the onset of encystment, the trophozoite disgorges any undigested food it may contain and condenses into a sphere. This is called the **precyst.** A precyst is so rich in glycogen that a large glycogen vacuole may occupy most of the cytoplasm in the young cyst. The chromatoidal bars that form typically are rounded at the ends. The bars may be short and thick, thin and curved, spherical, or very irregular in shape, but they do not have the splinter-like appearance found in *E. coli.*

The precyst rapidly secretes a thin,

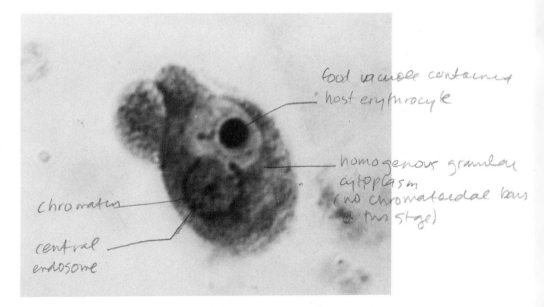

Fig. 6-2. A trophozoite of *Entamoeba histolytica* with a host erythrocyte in a food vacuole. Note the centrally placed karyosome and the peripheral chromatin.

tough hyaline **cyst wall** around itself to form a **cyst.** The cyst may be somewhat ovoid or elongate, but it usually is spheroid. It is commonly 10 to 20 μm wide but may be as small as 5 μm. The young cyst has only a single nucleus, but this

rapidly divides twice to form two- and four-nucleus stages (Fig. 6-3). As the nuclear division proceeds and the cyst matures, the glycogen vacuole and chromatoidal bodies disappear. In semiformed stools one can find precysts and cysts with one to four

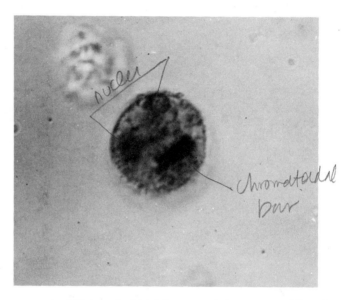

Fig. 6-3. A young cyst of *Entamoeba histolytica,* containing two nuclei and a prominent chromatoid bar.

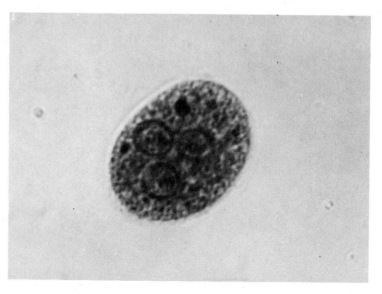

Fig. 6-4. Three of the four nuclei are in focus in this metacyst of *Entamoeba histolytica,* and two small chromatoidal bodies can be seen.

nuclei, but cysts with four nuclei (**meta-cysts**) are most common in formed stools (Fig. 6-4). This stage can survive outside the host and can infect a new one. After excysting in the small intestine, both the cytoplasm and nuclei divide to form eight small amebulae, or **metacystic tropho-zoites.** These are basically similar to mature trophozoites except in size.

Biology. To quote Elsdon-Dew,[6] "Were one tenth, nay, one hundredth, of the alleged carriers of this parasite to suffer even in minor degree, then the amoeba would rank as the major scourge of mankind." Trophozoites may live and multiply indefinitely within the crypts of the mucosa of the large intestine, apparently feeding on starches and mucous secretions and interacting metabolically with enteric bacteria. However, such trophozoites commonly initiate tissue invasion when they hydrolyse mucosal cells and absorb the predigested product. At this stage they no longer require the presence of bacteria to meet their nutritional requirements. It has been shown that under optimum conditions of pH and ionic concentrations, nonvirulent *E. histolytica* becomes virulent when it contains a certain ratio of starch to cholesterol.[16] Both pathogenic and nonpathogenic strains can possess proteolytic enzymes that presumably would make tissue invasion possible. Virulence of a particular strain can be attenuated by in vitro cultivation and sometimes restored by passage through certain experimental hosts. The complex of factors involved in the environmental conditions in the host are even more difficult to untangle because the conditions mutually interact. The oxidation-reduction potential and the pH of the gut contents influence invasiveness, but these conditions are determined largely by the bacterial flora, which is in turn influenced by the host's diet and perhaps even its overall nutritional state. It is believed that one reason newcomers to areas of endemicity suffer more than the local population may be because of differences in their bacterial flora.

Invasive organisms erode ulcers into the intestinal wall, eventually reaching the submucosa and underlying blood vessels. From there, they may travel with the blood to other sites in the body, such as the liver, lungs, or skin. Although these endogenous forms are active, healthy amebas that multiply rapidly, they are on a dead-end course. They cannot leave the host and infect others and so must perish with their luckless benefactor.

Mature cysts in the large intestine, on the other hand, leave the host in great numbers. The host that produces such cysts is usually asymptomatic or only mildly afflicted. Cysts of *E. histolytica* can remain viable and infective in a moist, cool environment for at least 12 days, and in water they can live up to 30 days. They are rapidly killed by putrefaction, desiccation, and temperatures below $-5°$ C and over $40°$ C. They can withstand passage through the intestines of flies and cockroaches. The cysts are resistant to levels of chlorine normally used for water purification.

When swallowed, the cyst passes through the stomach unharmed and shows no activity while in an acidic environment. When it reaches the alkaline medium of the small intestine, the metacyst begins to move within the cyst wall, which rapidly weakens and tears. The quadrinucleate ameba emerges and divides into amebulae that are swept downward into the cecum. This is the first opportunity the organism has to colonize, and its success depends on one or more metacystic trophozoites making contact with the mucosa. Obviously, its chances for survival are improved when large numbers of cysts are swallowed.

Pathogenesis. *Entamoeba histolytica* is unique among the amebas of humans in its ability to hydrolyse host tissues. Once in contact with the mucosa, the amebas secrete proteolytic enzymes, which enable them to penetrate the epithelium and begin moving deeper. The **intestinal lesion** (Fig. 6-5) usually develops initially in the cecum, appendix, or upper colon, then spreads the length of the colon. The number of parasites builds up in the ulcer, increasing the speed of mucosal destruction. The muscularis mucosae is somewhat of a barrier to further progress, and pockets of amebas form, communicating with the lumen of the intestine through a slender, duct-like ulcer. The lesion may stop at the basement membrane or at the muscularis mucosae and then begin erod-

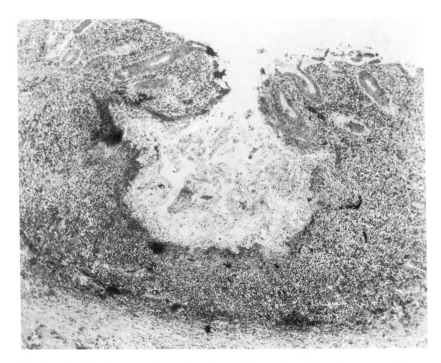

Fig. 6-5. A typical flask-shaped amebic ulcer of the colon. Extensive tissue destruction has resulted from invasion by *Entamoeba histolytica*. (AFIP neg. no. N-44718.)

ing laterally, causing broad, shallow areas of necrosis. The tissues may heal nearly as fast as they are destroyed, or the entire mucosa may become pocked. These early lesions usually are not complicated by bacterial invasion, and there is little cellular response by the host. In older lesions, the amebas, assisted by bacteria, may break through the muscularis mucosae, infiltrate the submucosa, and even penetrate the muscle layers and serosa. This enables trophozoites to be carried by blood and lymph to ectopic sites throughout the body where secondary lesions then form. A high percentage of deaths result from perforated colons with concomitant peritonitis. Surgical repair of perforation is difficult because a heavily ulcerated colon becomes very delicate.

Sometimes a granulomatous mass, called an **ameboma,** forms in the wall of the intestine and may obstruct the bowel. It is the result of cellular responses to a chronic ulcer and often still contains active trophozoites. The condition is rare except in Central and South America.

Secondary lesions have been found in nearly every organ of the body, but the liver is most commonly affected (about 5% of all cases). Regardless of the secondary site, the initial infection was an intestinal abscess, even though it may have gone undetected. **Hepatic amebiasis** results when trophozoites enter the mesenteric venules and travel to the liver via the hepatic portal system. They digest their way through the portal capillaries and enter the sinusoids, where they begin to form abscesses. The lesions thus produced may remain pinpoint size, or they may continue to grow, sometimes reaching the size of grapefruit. The center of the abscess is filled with necrotic fluid; a median zone consists of liver stroma, and the outer zone consists of liver tissue being attacked by amebas, though it is bacteriologically sterile. The abscess may rupture, pouring debris and organisms into the body cavity, where they attack other organs.

Pulmonary amebiasis is the next most common secondary lesion. It usually develops by metastasis from a hepatic lesion but may originate independently. Most cases originate when a liver abscess ruptures through the diaphragm. Other ectopic sites occasionally encountered are the brain, skin, and penis (possibly acquired venereally). Rare ectopic sites are kidneys, adrenals, spleen, male and female genitalia, pericardium, and others. As a rule, all ectopic abscesses are bacterially sterile.

Symptoms of infection vary greatly between cases. The strain of *E. histolytica* present, the host's natural or acquired resistance to that strain, and the host's physical and emotional condition when challenged all affect the course of the disease in any individual. When conditions are appropriate, a highly pathogenic strain can cause a sudden onset of severe disease. This usually is the case with water-borne epidemics. More commonly, the disease develops slowly, with intermittent diarrhea, cramps, vomiting, and general malaise. Infection in the cecal area may mimic the symptoms of appendicitis. Some patients tolerate intestinal amebiasis for years with no sign of colitis (but are passing cysts), then suddenly succumb to an ectopic lesion. Depending on the number and distribution of intestinal lesions, the patient might develop pain in the entire abdomen, fulminating diarrhea, dehydration, and loss of blood. Amebic diarrhea is marked by bouts of abdominal discomfort with four to six loose stools per day but little fever.

Acute amebic dysentery is a less common condition, but the sufferer from this affliction can best be described as miserable. The onset may be sudden after an incubation period of 8 to 10 days or after a long period as an asymptomatic cyst passer. In acute onset there may be headache, fever, severe abdominal cramps, and sometimes prolonged, ineffective straining at stool. There is an average of 15 to 20 stools per day, consisting of liquid feces flecked with bloody mucus. Death may occur from peritonitis, resulting from gut perforation or from cardiac failure and exhaustion. Bacterial involvement may lead to extensive scarring of the intestinal wall, with subsequent loss of peristalsis. Symptoms arising from ectopic lesions are typical for any lesion of the affected organ.

Epidemiology. *Entamoeba histolytica* is found throughout the world. Approximately 400 million persons are infected, of which about 100 million suffer acute or chronic effects of the disease. Although clinical amebiasis is most prevalent in tropical and subtropical areas, the parasite is well established from Alaska to the southern tip of Argentina. The incidence of infection varies widely, depending on local conditions, from less than 1% in Canada and Alaska, to 5% in the contiguous United States, to 40% in many tropical areas.[17] The incidence in the United States may be much higher among particular groups, such as persons in mental hospitals or orphanages. Age influences the incidence of infection, for children under 5 years old have a *lower* rate than other age groups. In the United States the greatest incidence is in the age group 26 to 30. The higher incidence in the tropics results from lower standards of sanitation and the greater longevity of cysts in a favorable environment. The onset of the disease in persons who travel from temperate regions to endemic tropical areas may partly be the result of lessened resistance from the stress of travel and unaccustomed heat, in addition to the change in bacterial flora in the gut, as mentioned previously. All races are equally susceptible.

The manner of disposal of human wastes in a given area is the most important factor in the epidemiology of this organism. Transmission depends heavily on contaminated food and water. Filth flies, particularly *Musca domestica,* and cockroaches also are important mechanical vectors of cysts. Their sticky, bristly appendages can quite easily carry cysts from a fresh stool to the dinner table, and the habit of the housefly to vomit and defecate while it feeds has been shown to be an important means of transmission. Polluted water supplies, such as wells, ditches, and springs, are common sources of infection. Instances of careless plumbing have been known, in which sanitary drains were connected to freshwater pipes with resultant epidemics. Carriers (cyst passers) handling food can infect the rest of their family group or hundreds of people if the carrier works in a restaurant. The use of human feces as fertilizer in Asia, Europe, and South America contributes heavily to transmission.

While humans are the most important reservoir of this disease, dogs, pigs, and monkeys are also implicated.

Diagnosis and treatment. Demonstration of trophozoites or cysts is necessary for the accurate diagnosis of *E. histolytica*. However, a large proportion of patients with extraintestinal amebiasis have no concurrent intestinal infection; therefore, diagnosis in such cases must be primarily by clinical and immunological means. X-rays and other means of scanning the liver may be useful in diagnosing abscesses. Examination of stool samples is the most effective means of diagnosis of gut infection. A direct smear examined either as a wet mount or fixed and stained will usually detect heavy infections. Even so, repeated examinations may be necessary; one of us found abundant trophozoites in the stool of a hospital patient after negative findings on 3 previous days. Lighter infections of cyst passers may be detected by using concentration techniques, such as zinc sulfate flotation.

Immunological diagnosis is promising but has yet to be perfected. Fluorescent antibody technique has some value in diagnosis but cannot differentiate *E. histolytica* from *E. hartmanni*. Serological procedures that have been adopted widely are the hemagglutination test and agar gel diffusion. Many other diseases can easily be confused with amebiasis; on the hospital chart of the patient mentioned, a dozen other possible explanations for his persistent diarrhea had been listed (but not amebiasis). Hence, demonstration of the organism itself is nearly mandatory in diagnosis.

There are several drugs with high efficacy against colonic amebiasis. Most fall into the categories of arsanilic acid derivatives, iodochlorhydroxyquinolines, and other synthetic and natural chemicals. Antibiotics, especially tetracycline, are useful as bactericidal adjuvants. These drugs are not as effective in ectopic infections, where chloroquine phosphate and niridazole show promise of effacacy. Metronidazole (Flagyl) has become the drug of choice in treatment of amebiasis. It is low in toxicity and is effective against both extraintestinal and colonic infections.

Metabolism. The metabolism of *E. histo-lytica* has received some attention, but a detailed picture has yet to emerge.[1,12] It had been believed that the organism was an obligate anaerobe because it requires a low oxidation-reduction potential for optimum growth. It has neither a cytochrome system nor mitochondria, and although it consumes oxygen when it is available, the oxygen consumption is not inhibited by cyanide. It has now been shown that the ameba can multiply in the presence of low oxygen concentrations and that exogenous glucose greatly stimulates oxygen consumption. Anaerobically, the organism ferments glucose to ethanol and acetate in a ratio of 3 to 1 and evolves carbon dioxide and molecular hydrogen. Aerobically, the ratio of ethanol and acetate is reversed, and hydrogen is not produced. It is reasonable to suppose that the ethanol is produced by a decarboxylation and then a reduction of the pyruvate from glycolysis, reoxidizing the NAD, as is the case in many bacteria and yeasts.

Entamoeba coli

Entamoeba coli often coexists with *E. histolytica* and, in the living trophozoite stage, is difficult to differentiate from it. Unlike *E. histolytica*, however, *E. coli* is a commensal that never lyses its host's tissues. It feeds on bacteria, other protozoans, yeasts, and occasional blood cells that may be casually available to it. The diagnostician must identify this species correctly; for if it is incorrectly diagnosed as *E. histolytica*, the patient may be submitted to unnecessary drug therapy.

Entamoeba coli is more common than *E. histolytica*, partly because of its superior ability to survive in putrefaction and partly because it does not kill its host. For example, Hitchcock (1950) reported 51% prevalence of infection with *Entamoeba coli* in Eskimos at Anchorage, Alaska, and 22% in the conterminous United States.[8]

Morphology. The trophozoite of *E. coli* (Fig. 6-6) is 15 to 50 μm (usually 20 to 30) in diameter and is superficially identical to that of *E. histolytica*. However, their nuclei differ: the endosome of *E. coli* is usually eccentrically placed, while that of *E. histolytica* is central. Also, the chromatin lining the nuclear membrane is usually coarser, with larger granules, than that of *E. histo-*

lytica. The food vacuoles of *E. coli* are more likely to contain bacteria and other intestinal symbionts than those of *E. histolytica,* although both may ingest available blood cells.

Encystation follows the same pattern as for *E. histolytica.* A precyst is formed, which rapidly secretes the cyst wall. The young cyst usually has a dense mass of chromatoidal bars that are splinter-shaped, rather than blunt as in *E. histolytica.* As the cyst matures, the nucleus divides repeatedly to form eight nuclei (Fig. 6-7). Rarely, up to 16 nuclei may be produced. The cysts vary in size from 10 to 33 μm in diameter.

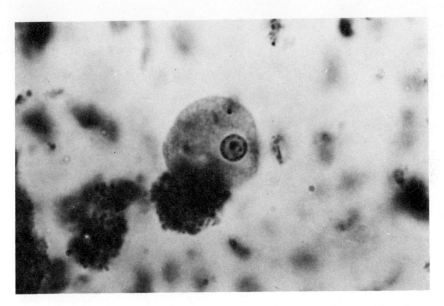

Fig. 6-6. A trophozoite of *Entamoeba coli,* a commensal in the human digestive tract. Note the characteristic, eccentrically located endosome. (Photograph by Peter Diffley.)

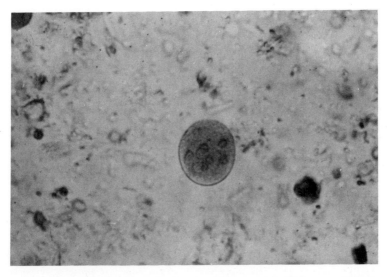

Fig. 6-7. A metacyst of *Entamoeba coli,* showing eight nuclei. (Photograph by David Oetinger.)

Biology. Infection and migration to the large intestine are identical to those of *E. histolytica*. The octonucleate metacyst produces eight to 16 metacystic trophozoites, which first colonize the cecum and then the general colon. Infection is by contamination—in some areas of the world reaching nearly 100% infection. Obviously, this is a reflection of the level of the sanitation and water treatment. Because *E. coli* is a commensal, no treatment is required. However, infection with this protozoan indicates that opportunities exist for ingestion of *E. histolytica*.

Entamoeba gingivalis

Entamoeba gingivalis was the first ameba of humans to be described. It is present in all populations, dwelling only in the mouth. Like *E. coli,* it is a commensal and is of interest to parasitologists as another example of niche location and speciation.

Morphology. Only the trophozoite has been found, and encystment probably does not occur. The trophozoite (Fig. 6-8) is 10 to 20 μm (exceptionally 5 to 35 μm in diameter) and is quite transparent in life. It moves rather quickly, by means of numerous blunt pseudopodia. The spheroid nucleus is 2 to 4 μm in diameter and has a small, nearly central endosome. As in all members of this genus, the chromatin is concentrated on the inner surface of the nuclear membrane. Food vacuoles

are numerous and contain cellular debris, bacteria, and occasional blood cells.

Biology. *Entamoeba gingivalis* lives on the surface of the teeth and gums, in the gingival pockets near the base of the teeth, and sometimes in the crypts of the tonsils. The organisms often are abundant in cases of gum or tonsil disease, but no evidence shows that they cause these conditions. More likely, the protozoans multiply rapidly with the increased abundance of food. They even seem to fare well on dentures, if the devices are not kept clean. The commensal also infects other primates, dogs, and cats.

Because no cyst is formed, transmission must be direct from one individual to another, by kissing, by droplet spray, or by sharing eating utensils. Up to 95% of persons with unhygenic mouths may be infected, and up to 50% of persons with healthy mouths may harbor this ameba.[9]

Genus Endolimax
Endolimax nana

Members of the genus *Endolimax* live in both vertebrates and invertebrates. These amebas are small, with a vesicular nucleus. The endosome is comparatively large and irregular and is attached to the nuclear membrane by achromatic threads. Encystment occurs in the life cycle.

Endolimax nana lives in the large intes-

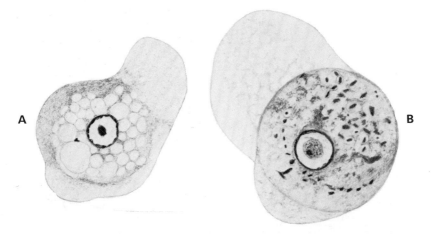

Fig. 6-8. *Entamoeba gingivalis.* **A,** Small trophozoite; **B,** larger specimen with numerous small food vacuoles containing bacteria. (From Kofoid, C. A., and O. Swezy. 1924. U. Calif. Publ. Zool. 26:165-198.)

tine of humans, mainly at the level of the cecum, and feeds on bacteria. Like *Entamoeba coli* it is a commensal.

Morphology. The trophozoite of this tiny ameba (Fig. 6-9) measures 6 to 15 μm in diameter, but it is usually less than 10 μm. The ectoplasm is a thin layer surrounding the granular endoplasm. The pseudopodia are short and blunt, and the ameba moves very slowly, characteristics from which its name is derived: "dwarf internal slug." The nucleus is small and contains a large centrally or eccentrically located endosome. The marginal chromatin is a thin layer. Large glycogen vacuoles are often present, and food vacuoles contain bacteria, plant cells, and debris.

Encystment follows the same pattern as *E. coli* and *E. histolytica.* The precyst secretes a cyst wall, and the young cyst thus formed includes glycogen granules and, occasionally, small curved chromatoidal bars. The mature cyst is 5 to 14 μm in diameter and contains four nuclei.

Biology. As with other cyst-forming amebas that infect humans, the mature cyst must be swallowed. The metacyst excysts in the small intestine and colonization begins in the upper large intestine. Incidence of infection parallels that of *Entamoeba coli* and reflects the degree of sanitation practiced within a community. The cyst is more susceptible to putrefaction and desiccation than is that of *E. coli.* While the protozoan is not a pathogen, its presence indicates that opportunities exist for infection by disease-causing organisms.

Genus Iodamoeba
Iodamoeba buetschlii

The genus *Iodamoeba* has only one species, and it infects humans, other primates and pigs. Its distribution is worldwide. This is the most common ameba of swine, which probably is its original host. The prevalence of *I. buetschlii* in humans is 4% to 8%, considerably lower than *Entamoeba coli* or *Endolimax nana.*

Morphology. The trophozoite (Fig. 6-10) is usually 9 to 14 μm long but may range from 4 to 20 μm. It moves slowly by means of short, blunt pseudopodia. The ectoplasm is not clearly demarcated from the granular endoplasm. The nucleus is relatively large and vesicular, containing a

large endosome that is surrounded by lightly staining granules about midway between it and the nuclear membrane. Achromatic strands extend between the endosome and the nuclear membrane, which has no peripheral granules. Food

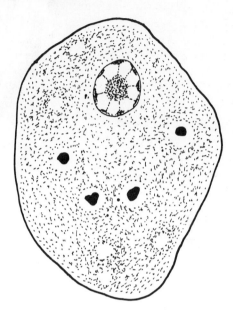

Fig. 6-9. *Endolimax nana,* trophozoite.

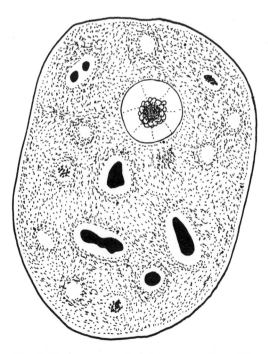

Fig. 6-10. A trophozoite of *Iodamoeba buetschlii.*

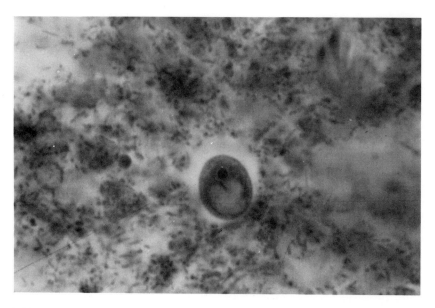

Fig. 6-11. *Iodamoeba buetschlii*, cyst in human feces. Note the large iodinophilous vacuole. (Photograph by James Jensen.)

vacuoles usually contain bacteria and yeasts.

The precyst is usually oblong and contains no undigested food. It secretes the cyst wall that also is usually oblong, measuring 6 to 15 μm long. The mature cyst (Fig. 6-11) nearly always has only one nucleus. A large conspicuous glycogen vacuole stains deeply with iodine, hence the generic name.

Biology. *Iodamoeba buetschlii* lives in the large intestine, mainly in the cecal area, where it feeds on intestinal flora. Infection spreads by contamination, for mature cysts must be swallowed to induce infection. It is possible that humans become infected through pig feces, as well as human feces.

Although *I. buetschlii* usually is harmless to humans and other vertebrates, it has, in a few cases, induced ectopic abscesses like those of *E. histolytica*.

Genus Dientamoeba
Dientamoeba fragilis

Dientamoeba fragilis has long been recognized as being unlike other members of the Endamoebidae for several reasons: a large proportion of individuals have two nuclei, the nuclear structure is rather unlike other Endamoebidae, an extranuclear spindle is present during division, and cysts are not formed. The last is a charac-

teristic shared with a more typical member of the family, *Entamoeba gingivalis*. Over 35 years ago Dobell believed that *D. fragilis* was closely related to the ameboflagellate *Histomonas*.[5] On the basis of ultrastructural and immunological evidence, Honigberg placed *Dientamoeba* in a subfamily of the Monocercomonadidae in the flagellate order Trichomonadida.[3] This placement seems to reflect the phylogenetic relationship of the organism rather than the fact that it moves by pseudopodia instead of flagella, and the genus will probably be placed in the Mastigophora in the future. *Dientamoeba fragilis*, infecting about 4% of the human population, is the only species known in the genus.

Morphology. Only trophozoites are known in this species; cysts are not formed. The trophozoites (Fig. 6-12) are very delicate and disintegrate rapidly in feces or water. They are 6 to 12 μm in diameter, and the ectoplasm is somewhat differentiated from the endoplasm. There is usually a single, broad pseudopodium. The food vacuoles contain bacteria, yeasts, starch granules, and cellular debris. About 60% of the amebas contain two nuclei, which are connected to each other by a filament, observable by light microscopy; the rest have only one nucleus. By electron microscopy one can discern that the filament

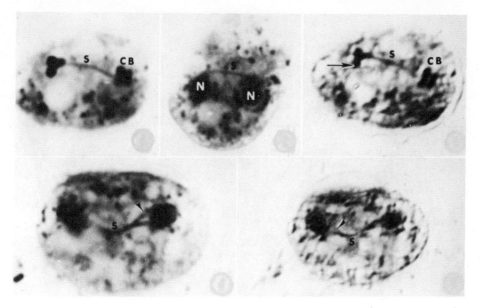

Fig. 6-12. *Dientamoeba fragilis:* photomicrographs of binucleate organisms. Four chroma-tin bodies *(CB)* can be resolved within the telophase nucleus of the organism, shown in the first and third figures. The extranuclear spindle *(S)* extends between the nuclei *(N)* in all figures. Note the branching of the spindle (arrowheads) near the nucleus in the fourth and fifth figures. Bouin's fixative. First, second, and fourth figures: bright field (×4,950); third and fifth figures: Nomarski differential interference (×3,650). (From Camp, R. R., C. F. T. Mattern, and B. M. Honigberg. 1974. J. Protozool. 21:69-82.)

connecting the nuclei is a division spindle composed of microtubles; the binucleate individuals are, in reality, in an arrested telophase. The endosome is eccentric, sometimes fragmented or peripheral in the nucleus, and concentrations of chro-matin are usually not apparent. A filament and Golgi apparatus are present, which are reminiscent of the parabasal fibers and parabasal bodies found in *Histomonas* and trichomonads. There are no kinetosomes or centrioles.

Biology. *Dientamoeba fragilis* lives in the large intestine, especially in the cecal area. It feeds mainly on debris and usually is con-sidered a commensal. However, it may cause episodes of flatulence with mild diarrhea.

The mode of transmission is unknown, as the parasite does not form cysts, and it cannot survive the upper digestive tract. It is possible that the organism survives transmission in the eggs of a parasitic nem-atode, as does its relative, *Histomonas* (see Chapter 5). Small, ameboid organisms re-sembling *D. fragilis* have been found in the eggs of the common human pinworm, *En-terobius vermicularis,* and there is epidemio-logical evidence that the nematode may be the vector of the protozoan.[2]

FAMILY DIMASTIGAMOEBIDAE

Dimastigamoebidae are soil and water inhabitants and are mainly coprophilous. They show affinities with the Mastigoph-ora, for they possess a flagellated stage as well as an ameboid form. Binary fission seems to take place only in the ameboid form; thus, these are diphasic amebas with the ameboid stage predominating over the flagellated stage. Although the several genera and species in this family live in stagnant water, polluted soil, sewage dis-posal systems, and the like, a few are able to become facultative parasites in verte-brates. There have been isolated reports of parasitism by species of *Naegleria, Hart-mannella,* and *Acanthamoeba.* Some of these may have been misidentified, but one dominant genus involved has been *Naeg-leria.*

Naegleria fowleri

The flagellated stage of the species *Naegleria fowleri* bears two long flagella, is rather elongate, and does not form pseudopodia; the ameboid stage has blunt pseudopodia. The nucleus is vesicular and has a large endosome and peripheral granules. A contractile vacuole is conspicuous in free-living forms. Food vacuoles contain bacteria and the cyst has a single nucleus.

Since 1964 over 60 cases of fatal human meningoencephalitis resulting from this, or from closely related forms, have been recorded from widely separated parts of the world: for example, from Czechoslovakia, the United States, and Australia. *Naegleria fowleri* has been isolated and cultured from many fatal cases. These amebas kill a variety of laboratory animals when injected intranasally, intravenously, or intracerebrally.[4,14] The *Naegleria* in the host do not form cysts or have flagella, and the food vacuoles contain host cell debris, rather than bacteria. In contrast, cysts of *Acanthamoeba* can be found in cases of meningoencephalitis caused by that genus.

Most cases were contracted in swimming pools or lakes. It is possible that trophozoites are forced deep into the nasal passages when the victim dives into the water; the amebas then migrate along the olfactory nerves, through the cribiform plate, and into the cranium. Death from brain tissue destruction is rapid, and no cure is known.

REFERENCES

1. Bryant, C. 1970. Electron transport in parasitic helminths and protozoa. In Dawes, B., editor. Advances in parasitology, vol. 8. Academic Press, Inc., New York. pp. 139-172.
2. Burrows, R. B., and M. A. Swerdlow. 1956. *Enterobius vermicularis* as a probable vector of *Dientamoeba fragilis*. Am. J. Trop. Med. Hyg. 5:258-265.
3. Camp, R. R., C. F. T. Mattern, and B. M. Honigberg. 1974. Study of *Dientamoeba fragilis* Jepps and Dobell. I. Electronmicroscopic observations of the binucleate stages. II. Taxonomic position and revision of the genus. J. Protozool. 21:69-82.
4. Chang, S. H. 1974. Etiological, pathological, epidemiological, and diagnostical consideration of primary amoebic meningoencephalitis. CRC Crit. Rev. Microbiol. 3:135-159.
5. Dobell, C. 1940. Researches on the intestinal protozoa of monkeys and man. X. The life history of *Dientamoeba fragilis*—observations, experiments and speculations. Parasitol. 32:417-459.
6. Elsdon-Dew, R. 1964. Amoebiasis. Exp. Parasitol. 15:87-96.
7. Foster, W. D. 1965. A history of parasitology. E. & S. Livingstone, Edinburgh.
8. Hitchcock, D. J. 1950. Parasitological study on the Eskimos in the Bethel area of Alaska. J. Parasitol. 36:232-234.
9. Jaskoski, B. J. 1963. Incidence of oral Protozoa. Trans. Am. Microsc. Soc. 82:418-420.
10. Levine, N. D. 1973. Protozoan parasites of domestic animals and of man, ed. 2. Burgess Publishing Co., Minneapolis.
11. Ludvík, J., and A. C. Shipstone. 1970. The ultrastructure of *Entamoeba histolytica*. Bull. WHO 43:301-308.
12. Montalvo, R. E., R. E. Reeves, and L. G. Warren. 1971. Aerobic and anaerobic metabolism in *Entamoeba histolytica*. Exp. Parasitol. 30:249-256.
13. Morgan, R. S., and B. G. Uzman. 1966. Nature of the packing of ribosomes within chromatoid bodies. Science 152:214-216.
14. Phillips, B. P. 1974. *Naegleria*: another pathogenic ameba. Studies in germfree guinea pigs. Am. J. Trop. Med. Hyg. 23:850-855.
15. Schaudinn, F. 1903. Untersuchungen über die Fortpflanzung einiger Rhizopoden. Arb. Kaiserl. Gesundh.-Amte 19:547-576.
16. Sharma, R. 1959. Effect of cholesterol on the growth and virulence of *Entamoeba histolytica*. Trans. R. Soc. Trop. Med Hyg. 53:278-281.
17. Spencer, H. 1973. Amoebiasis. In Spencer, H., editor. Tropical pathology. Springer-Verlag, Heidelberg. pp. 271-297.

SUGGESTED READING

Chang, S. L. 1971. Small, free-living amebas: cultivation, quantitation, identification, classification, pathogenesis, and resistance. In Cheng, T. C., editor. Current Topics in Comparative Pathobiology, vol. 1. Academic Press, Inc., New York. pp. 202-254. (A review of the facultatively parasitic amebas.)

Elsdon-Dew, R. 1968. The epidemiology of amoebiasis. In Dawes, B., editor. Advances in Parasitology, vol. 6. Academic Press, Inc., New York. pp. 1-62. (An excellent review by the world authority on the subject.)

Hoare, C. A. 1958. The enigma of host-parasite relations in amebiasis. Rice Inst. Pamphlet 45:23-35. (Very interesting reading.)

Lösch, F. A. 1875. Massive development of amebas in the large intestine. (Translated by Kean, B. H., and K. E. Mott, 1975.) Am. J. Trop. Med. Hyg. 24:383-392.

Neal, R. A. 1966. Experimental studies on *Entamoeba* with reference to speciation. In Dawes, B., editor. Advances in Parasitology, vol. 4. Academic Press, Inc., New York. pp. 1-51. (This is a summary of the several modern approaches to definition of species in this genus.)

Chapter 7

SUBPHYLUM APICOMPLEXA: GREGARINES, COCCIDIANS, AND RELATED ORGANISMS

Protozoologists have become increasingly dissatisfied with the grouping of parasitic protozoa traditionally assigned to the class or subphylum Sporozoa, and Levine proposed a new subphylum, the Apicomplexa, to be comprised of organisms that possessed a certain combination of structures, the **apical complex,** distinguishable with the electron microscope.[15] The structures that make up the apical complex will be described further, but they typically include a polar ring, micronemes, rhoptries, subpellicular tubules, micropore(s) (cytostome), and a conoid. Members of the Apicomplexa have a single type of nucleus and no cilia or flagella, except for the flagellated microgametes in some groups. The subphylum contains two classes, the Sporozoa and the Piroplasmea (see Chapter 8). Among other differences, the Sporozoa have a well-developed apical complex, while the apical complex is reduced in the Piroplasmea.

Class Sporozoa

Members of class Sporozoa are, without exception, parasites. All classes of vertebrates and most invertebrates harbor sporozoans. Several species are of exceptional importance to people, for they kill and weaken most of the domestic animal species and thereby cause millions of dollars of loss to agriculture every year. Furthermore, sporozoans are responsible for malaria, the most important disease of humans in the world today.

Most sporozoans produce a resistant spore, or oocyst (containing sporozoites), which survives the elements between hosts (Fig. 7-1). In some, the spore wall has been eliminated, for the development of the sporozoites is completed within an invertebrate vector.

Locomotor organelles are not as obvious as they are in the other subphyla of Protozoa. Pseudopodia are found only in some tiny, intracellular forms; flagella occur only on gametes of a few species, and a very few have cilia-like appendages. Various species have sucker-like depressions, knobs, hooks, myonemes, and/or internal fibrils that aid in limited locomotion. The myonemes and fibrils form tiny waves of contraction across the body surfaces; these can propel the animal slowly through a liquid medium.

Both asexual and sexual reproduction are known in many sporozoans. Asexual reproduction is by either binary or multiple fission or by endopolyogeny. Sexual reproduction is by isogamous or anisogamous fusion, and in many cases, this stage marks the onset of spore formation.

SUBCLASS GREGARINIA

The large group of sporozoans known as Gregarinia parasitizes only invertebrates. Because gregarines are widespread, common, and may be large in size, they are often used as models in elementary and advanced zoology laboratories. Accordingly, we will discuss some examples to outline the characteristics of the group.

Multiple fission, or **schizogony,** occurs in a few families of gregarines (orders Archigregarinida and Neogregarinida). These, however, are not commonly encountered and so will not be discussed. Most gregarines (order Eugregarinida) have no schizogony and do produce spores in their life cycles. They range in size from only a few micrometers in diameter to at least 10 mm long. Some are so large that nineteenth century zoologists placed them among the worms!

In some species, the **acephaline** gregarines, the body consists of a single unit

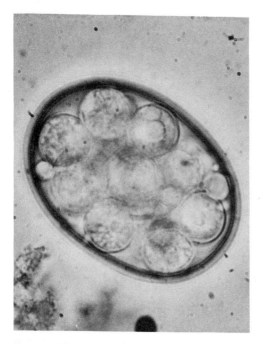

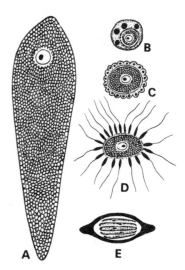

Fig. 7-1. Oocyst of *Adelina* sp. from a common garter snake, *Thamnophis sirtalis.* The oocysts of this genus contain numerous sporocysts. (Photograph by Richard Wacha.)

Fig. 7-2. Stages in the life cycle of *Monocystis lumbrici,* an acephaline gregarine of earthworms. **A,** A mature sporont, from the host's seminal vesicle. **B-D,** Intracellular stages within the sperm morulae (mother sperm cells) of the earthworm. In **D,** a coating of earthworm sperm tails is all that is left of the sperm morulae; the rest has been consumed by the parasite. **E,** Spore, containing eight sporozoites. (From Jahn, T. L. 1949. How to know the Protozoa. William C. Brown Co. Publishers, Dubuque, Iowa.)

that may have an anterior anchoring device, the **mucron.** In the **cephaline** species, the body is divided by a septum into an anterior **protomerite** and a posterior **deutomerite** that contains the nucleus. Sometimes the protomerite bears an anterior anchoring device, the **epimerite.** Mucrons and epimerites are considered modified conoids (see p. 116). In both the cephalines and acephalines, the host becomes infected by swallowing spores. Most parasitize the body cavity, intestine, or reproductive system of their hosts.

Order Eugregarinida
Suborder Acephalina

Monocystis lumbrici (Fig. 7-2). *Monocystis lumbrici* lives in the seminal vesicle of *Lumbricus terrestris* and related earthworms, and it is easily demonstrated in the laboratory.

The worm becomes infected when it ingests a spore containing several **sporozoites.** These hatch in the gizzard, and the released sporozoites penetrate the intestinal wall, enter the dorsal vessel, and move forward to the hearts. Then they leave the circulatory system and penetrate the seminal vesicles, where they enter into the sperm-forming cells (blastophores) in the vesicle wall. After a short period of growth, during which the parasites destroy the developing spermatocytes, they enter the lumen of the vesicle where they become mature trophozoites, or **sporadins (gamonts),** measuring about 200 μm long by 65 μm wide. They attach to cells in the region of the sperm tunnel and undergo a form of union called **syzygy,** in which two or more sporadins connect with one another. The anterior organism is called the **primite** and the posterior, the **satellite.** The differences between the two are unclear, but they do exhibit different staining reactions. After syzygy the gamonts flatten against each other and surround themselves with a common cyst envelope, forming the **gametocyst.** Although "united" in the same cyst, the gamonts are still morphologically distinct, and each undergoes numerous nuclear divisions. The many small nuclei move to the periphery of the cytoplasm and, taking a small portion of the cytoplasm with them, bud off to become gametes. Some of the cytoplasm of each gamont remains, and this fuses to become the **residual body.** The gametes from each gamont are morphologically distinguishable and are thus anisogametes. Many species of gregarines have isogamy, however.[9] The fusion of a pair of gametes to form a zygote is followed by the secretion of the **spore** or **oocyst membrane** around that zygote, and three cell divisions (sporogony) to form eight **sporozoites.** Thus, each gametocyst now contains many oocysts, and the new host may become infected by eating a gametocyst or, if that body ruptures, an oocyst. As in other sporozoa, meiosis is zygotic: only the zygote is diploid, and reduction division in sporogony returns the sporozoites to the haploid condition. The gametocyst or oocyst passes from the host through the sperm duct to be ingested by another worm.

Suborder Cephalina

Gregarina polymorpha. *Gregarina polymorpha* are common parasites of the mealworm, *Tenebrio molitor,* that usually infect colonies of the beetles maintained in the laboratory. The sporadins are cylindrical, up to 350 μm long by 100 μm wide, with a small, globular epimerite that is inserted into a host cell. The entire life cycle takes place within the midgut of the mealworm larva or adult. The sporadins undergo syzygy, having detached from the host intestinal epithelium, leaving the epimerite behind. The protomerite is dome-shaped and the deutomerite is cylindrical, is rounded posteriorly, and has a nucleus with an endosome. A gametocyst is formed that in turn produces gametes and zygotes. The cysts pass out with the feces of the host, to be eaten by the next mealworm.

When the mealworms are well-fed and healthy, infection with *G. polymorpha* causes no obvious ill effects. But when the beetles are stressed by malnutrition, the large gregarines take their toll: the pupae are smaller than normal and fewer of them survive metamorphosis.

SUBCLASS COCCIDIA

Unlike the subclass Gregarinia, members of the subclass Coccidia are quite small, with a prevalent intracellular development and with no epimerite or mucron. Some species are monoxenous, while others require two hosts to survive. Coccidia live in the digestive tract, epithelium, liver, kidney, blood cells, and other tissues of vertebrates and invertebrates.

The typical coccidian life cycle (Fig. 7-7) includes three major phases: schizogony, gametogony, and sporogony. The infective stage is a rod- or banana-shaped sporozoite that enters a host cell and begins to develop. The organism becomes an ameboid trophozoite that fragments by schizogony to form more rod- or banana-shaped **merozoites,** which then escape from the host cell. These enter other cells to initiate further schizogony or transform into a gamont **(gametogony).** Gamonts produce "male" **microgametocytes** or "female" **macrogametocytes.** Most species are thus anisogamous, and the macrogametocyte develops directly into a comparatively large, rounded macrogamete. The macrogamete is an ovoid body filled with globules of a refractile material and has a central nucleus. The microgametocyte undergoes multiple fission to form tiny, biflagellated microgametes. Fertilization produces a zygote. Multiple fission of the zygote (sporogony) produces the sporozoite-filled oocyst. Fig. 7-3 illustrates oocyst variations among genera in the Coccidia. In monoxenous life cycles all of the stages occur in a single host, although the oocyst matures in the oxygen-rich, lower temperature environment outside a host. The sporozoites are then released when the sporulated oocyst is eaten by another host. In heteroxenous life cycles, schizogony and a part of gametogony occur in vertebrate host, while sporogony occurs in an invertebrate, and the sporozoites are transmitted by the bite of the invertebrate.

The ultrastructure of the sporozoites and merozoites in the subclass is typical of the Apicomplexa.[15] These are banana-shaped organisms (Fig. 7-4), somewhat more attenuated at the anterior, apical

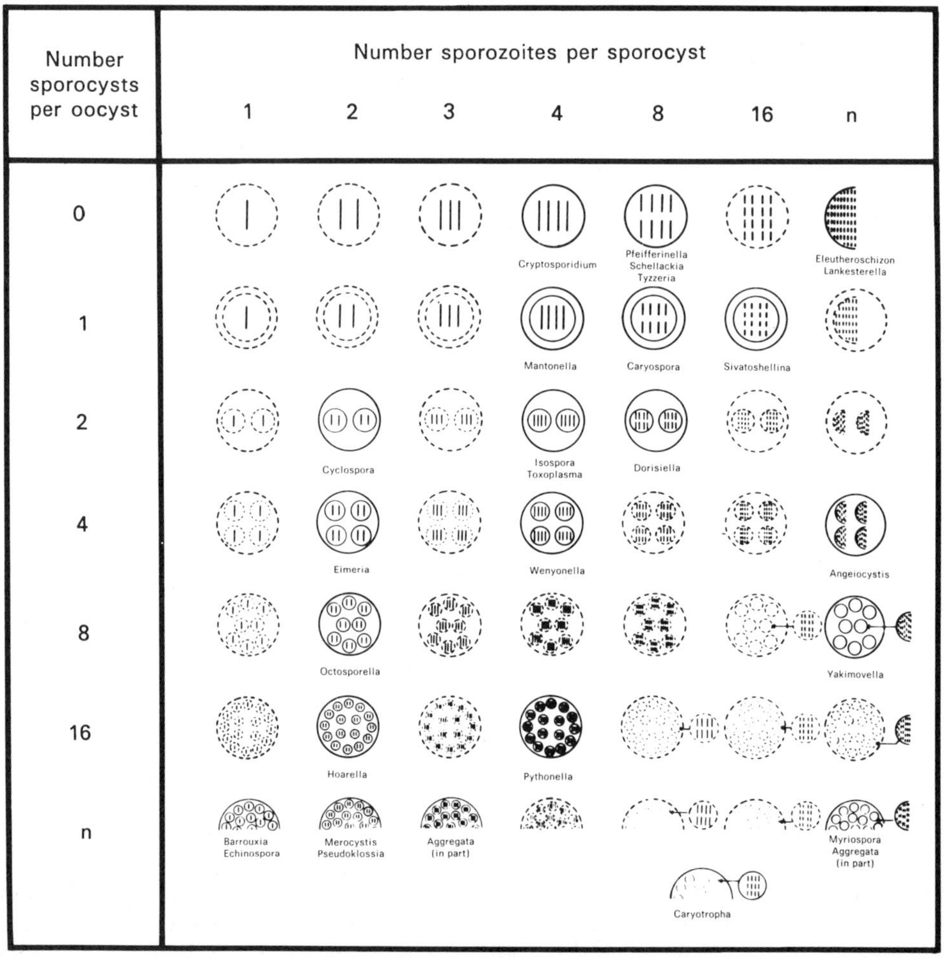

Number sporocysts per oocyst	Number sporozoites per sporocyst						
	1	2	3	4	8	16	n
0				Cryptosporidium	Pfeifferinella Schellackia Tyzzeria		Eleutheroschizon Lankesterella
1				Mantonella	Caryospora	Sivatoshellina	
2		Cyclospora		Isospora Toxoplasma	Dorisiella		
4		Eimeria		Wenyonella			Angeiocystis
8		Octosporella					Yakimovella
16		Hoarella		Pythonella			
n	Barrouxia Echinospora	Merocystis Pseudoklossia	Aggregata (in part)				Myriospora Aggregata (in part)

Caryotropha

Fig. 7-3. Number of sporocysts per oocyst and of sporozoites per sporocyst in genera of the suborder Eimerina. (In the genera without sporocysts the numbers of sporozoites per oocyst are given.) (From Levine, N. D. 1973. In Hammond, D. M., and P. L. Long. The Coccidia. University Park Press, Baltimore.)

complex end. A constant feature of the apical complex is one or two **polar rings,** electron-dense structures just beneath the cell membrane, which encircle the anterior tip. A **conoid** is found in members of the suborder Eimeriina. This structure is a truncated cone of spirally arranged fibrillar structures just within the polar rings. **Subpellicular microtubules** radiate from the polar rings and run posteriorly, parallel to the axis of the body. These organelles probably serve as structural elements and may be involved with locomotive function. From two to several elongate, electron-dense bodies, the **rhoptries,** extend to the cell membrane within the polar rings (and conoid, if present). Smaller, more convoluted elongate bodies, the **micronemes** also extend posteriorly from the apical complex. The ducts of the micronemes apparently run anteriorly into the rhoptries or join a common duct system with the rhoptries to lead to the cell surface of the apex. The contents of the rhoptries and micronemes seems similar in electron micrographs, and it is thought that this material is secreted during penetration of the host cell and is of aid in that process. Along the side of the organism are one or more **cytostomes,** which function in ingestion of

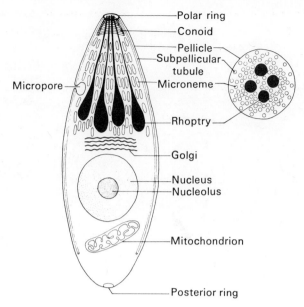

Fig. 7-4. An apicomplexan sporozoite or merozoite, illustrating the apical complex. (From Levine, N. D. 1973. In Hammond, D. M., and P. L. Long. The Coccidia. University Park Press, Baltimore.)

food material during the intracellular life of the parasite. The edges of the cytostome are marked by two concentric, electron-dense rings, located just beneath the cell membrane. As host cytoplasm or other food matter within the parasitophorus vacuole is pulled through the rings, the parasite's cell membrane invaginates accordingly and finally pinches off to form a membrane-bound food vacuole. With the exception of the cytostome, the structures described above dedifferentiate and disappear after the sporozoite or merozoite penetrates the host cell to become a trophozoite. Members of the subphyla Microspora and Myxospora, which formerly were included in the Sporozoa, lack these structures at all stages of their life cycles.

Order Eucoccida
Suborder Adeleina

In the Adeleina the macrogametocyte and microgametocyte are associated in syzygy during development; therefore, this is considered the most primitive suborder in the Eucoccida. The microgametocyte produces only one to four microgametes, the sporozoite is surrounded by a membrane, and endodyogeny is absent. The life cycles are either monoxenous or heteroxenous.

Family Haemogregarinidae

Haemogregarina stepanowi (Fig. 7-5). *Haemogregarina stepanowi* is a parasite of a European turtle, *Emys orbicularis,* and a leech, *Placobdella catenigra.* Similar species are common in turtles and frogs in the United States, and the following description of *H. stepanowi* essentially can be applied to them. Sporogony occurs in the leech, and schizogony occurs in the turtle.

BIOLOGY. Trophozoites live in the circulating erythrocytes of the turtle. They become U-shaped as they grow, and the unequal arms of the U finally fuse together to form an ovoid body. This becomes a macroschizont, and the erythrocyte that bears it lodges in the bone marrow. Schizonts of this generation produce 13 to 24 large merozoites that enter other erythrocytes and become microschizonts. The microschizonts produce only six smaller merozoites. When these enter erythrocytes, they become gametocytes, thus ending the schizogonous cycle. The gametocytes are elongate. The macrogametocyte has a small nucleus, and the microgametocyte has a large nucleus and dark-staining

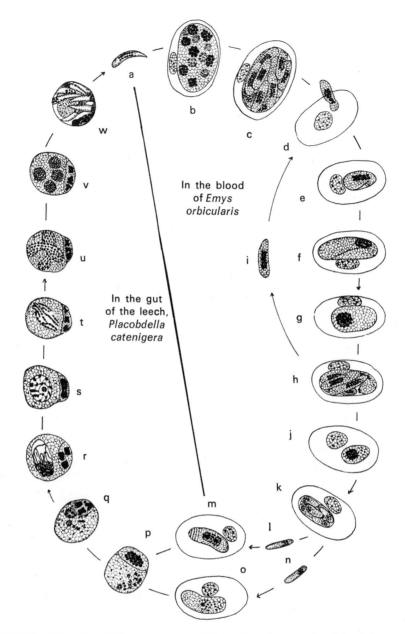

Fig. 7-5. The life cycle of *Haemogregarina stepanowi. a,* Sporozoite; *b-i,* schizogony; *j, k,* gametocyte formation; *l, m,* microgametocytes; *n, o,* macrogametocytes; *p, q,* association of gametocytes; *r,* fertilization; *s-w,* division of the zygote nucleus to form eight sporozoites. (From Kudo, R. R. 1966. Protozoology, ed. 5. Charles C Thomas, Publisher, Springfield, Ill.)

transverse bands at the anterior end. No further development takes place in the turtle.

When a leech ingests infected erythrocytes, the gametocytes are released. A macrogametocyte fuses in syzygy with a microgametocyte, and the pair becomes surrounded by a thin oocyst membrane. The nucleus of the microgametocyte divides, and the cell forms four microgametes, one of which fertilizes the macrogamete. Eight sporozoites develop from the zygote and, when mature, break out of the thin-walled oocyst into the intestinal lumen. The sporozoites then enter the circulatory system and migrate to the salivary glands, where they are injected into a new turtle host when the leech feeds.

Haemogregarines of terrestrial reptiles apparently are transmitted by mites. The means of transmission of haemogregarines of fish is unknown.

Suborder Eimeriina

In this suborder the macrogamete and microgamete develop independently without syzygy. The microgametocyte produces many microgametes, and the sporozoites are enclosed in a sporocyst. This is a very large group with many families and thousands of species. Very few species are found in humans, but many parasitize domestic animals, making this suborder of Eucoccidia one of the most important parasites in agriculture. The history and taxonomy of the Eimeriina have been reviewed by Levine.[16]

Family Eimeriidae. The family Eimeriidae contains 14 genera (Fig. 7-3) of which two, *Eimeria* and *Isospora,* are of great importance. The common name "coccidian" is applied to any member of this family. Development is within the cells of a single host, although the last stages of sporogony usually are exogenous. The oocyst lacks an attachment organelle and contains from none to many sporocysts, each with one or more sporozoites. The microgametes have two or three flagella. The genera are differentiated by the number of sporocysts and/or the number of sporozoites in each sporocyst.

A typical oocyst *(Eimeria)* is diagrammed in Fig. 7-6. The oocyst wall is of two layers, and with the electron microscope a membrane surrounding the outer wall may be distinguished.[20] In many species there is a tiny opening at one end of the oocyst, the **micropyle,** and this may be covered by the **micropylar cap.** A refractile **polar granule** may lie somewhere within the oocyst. The oocyst wall (and probably the sporocyst wall, too) is of a resistant material that helps the organism to survive harsh conditions in the external environment. Wilson and Fairbairn showed that the composition of the wall was of a chitin-

like substance, but not chitin, since it does not contain N-*acetyl*-glucosamine.[23] The chitin-like substance is probably found in the inner layer of the wall because that layer is resistant to sodium hypochlorite.[20]

Most species form sporocysts, which contain the sporozoites, within the oocyst. During the sporogony to form the sporozoites, the cytoplasmic material not incorporated into the sporozoites forms the **oocyst residuum.** In like manner, some material may be left over within the sporocysts to become the **sporocyst residuum.** However, it appears that the sporocyst residuum is more than a depository for waste. It contains a large amount of lipid that seems to be an important source of energy for the sporozoites during their sojourn outside a host.[23] The sporocyst wall consists of a thin outer, granular layer surrounded by two membranes and a thick, fibrous inner layer. At one end of the sporocyst, a small gap in the inner layer is plugged with a homogeneous **Stieda body.** In some species additional plug material underlies the Stieda body, and this is designated the **substiedal body.** When the sporocysts reach the intestine of a new host or are treated in vitro with trypsin and bile salt, the Stieda body is digested, the substiedal body pops out, and the sporozoites wriggle through the small opening thus created.[20] In addition to the apical complex, nucleus, and so on, the sporozoites themselves may contain one or more prominent **refractile bodies** of unknown function.

The size and shape of the oocyst and its contents, the presence or absence of several of the above mentioned structures, and the texture of the outer wall are all useful taxonomic characters. Identification of a coccidian usually can be accomplished by examining its oocyst, which is remarkably constant in its characters in a given species.

Host specificity is probably more rigid in this family, especially in the genus *Eimeria,* than in any other invasive organisms. Not only are they often restricted to a certain host species, but a given species of *Eimeria* may be limited to certain organ systems, narrow zones in that system, specific kinds of cells in that zone, and even specific locations within the cells.[19] One

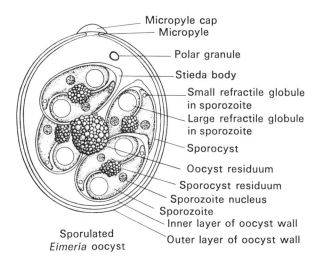

Fig. 7-6. Structure of sporulated *Eimeria* oocyst. (From Levine, N. D. 1961. Protozoan parasites of domestic animals and of man, ed. 2. Burgess Publishing Co., Minneapolis.)

species may be found only at the tips of the intestinal villi, another in the crypts at the bases of the villi, and a third in the interior of the villi, all in the same host. Some species develop below the nucleus of the host cell, others develop above it, and a few develop within it. Most coccidia inhabit the digestive tract, but a few are found in other organs, such as the liver and kidney.

The number of species of coccidia is staggering. Levine and Ivens recognized 204 species of *Eimeria* and ten of *Isospora* in rodents, but they estimated that there must be at least 2,700 species of *Eimeria* in rodents alone.[18] It has been calculated that *Eimeria* has been described from only 1.2% of the chordate species and 5.7% of the mammals in the world.[14] If all chordates were examined, Levine estimated that 34,000 species of *Eimeria* would be found—3,500 of them in mammals—and if all animals were examined, a total of 45,000 species would be found of this single genus.[13] That is nearly equivalent to the number of all living and fossil protozoa described so far.

Eimeria tenella (Fig. 7-7). *Eimeria tenella* lives in the epithelium of the intestinal ceca of chickens, where it produces considerable destruction of tissues, causing high mortality in young birds. This and related species are of such consequence that all commercial feeds for young chickens now contain anticoccidial agents.

BIOLOGY. A chicken becomes infected when it swallows food or water that is contaminated with sporulated oocysts. The oocyst of this species is ovoid, smooth, 14 to 31 μm long by 9 to 25 μm wide. When the oocyst wall ruptures in the bird's gizzard, the activated sporozoites escape the sporocyst in the small intestine. Once in the cecum the sporozoites first enter the cells of the surface epithelium and pass through the basement membrane into the lamina propria. There they are engulfed by macrophages that carry them to the glands of Lieberkühn. They then escape the macrophages and enter into a glandular epithelial cell of the crypt, where they locate between the nucleus and the basement membrane.

Within the epithelial cell the sporozoite becomes a trophozoite, feeding on the host cell and enlarging to become a schizont. During schizogony (sometimes here called **merogony**), the schizont separates into about 900 first-generation merozoites, each about 2 to 4 μm long. They break out into the lumen of the cecum about $2^1/_2$ to 3 days after infection, destroying the host cell. Each first-generation merozoite enters another cecal epithelial cell to initiate the second endogenous generation. The merozoite develops into a schizont that lives between the nucleus and the free border of the host cell. A great many will form schizonts in the lamina propria under the basement membrane.

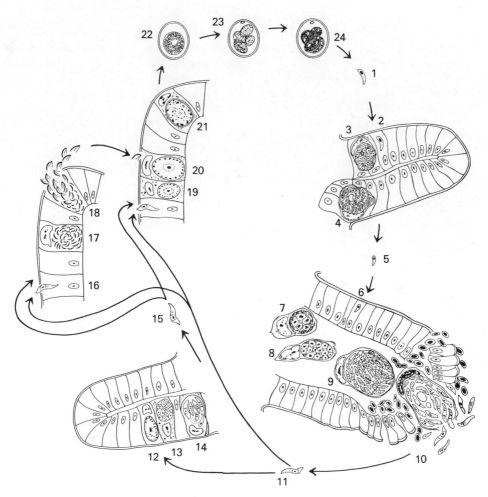

Fig. 7-7. Life cycle of the chicken coccidium, *Eimeria tenella*. A sporozoite *(1)* enters an intestinal endothelial cell *(2)*, rounds up, grows, and becomes a first-generation schizont *(3)*. This produces a large number of first-generation merozoites *(4)*, which break out of the host cell *(5)*, enter new intestinal endothelial cells *(6)*, round up, grow, and become second-generation schizonts *(7, 8)*. These produce a large number of second-generation merozoites *(9, 10)*, which break out of the host cell *(11)*. Some enter new host intestinal endothelial cells and round up to become third-generation schizonts *(12, 13)*, which produce third-generation merozoites *(14)*. The third-generation merozoites *(15)* and the great majority of second-generation merozoites *(11)* enter new host intestinal endothelial cells. Some become microgametocytes *(16, 17)*, which produce a large number of microgametes *(18)*. Others turn into macrogametes *(19, 20)*. The macrogametes are fertilized by the microgametes and become zygotes *(21)*, which lay down a heavy wall around themselves and turn into young oocysts. These break out of the host cell and pass out in the feces *(22)*. The oocysts then sporulate. The sporont throws off a polar body and forms four sporoblasts *(23)*, each of which forms a sporocyst containing two sporozoites *(24)*. When the sporulated oocyst *(24)* is ingested by a chicken, the sporozoites are released *(1)*. (From Levine, N. D. 1961. Protozoan parasites of domestic animals and of man. ed. 2. Burgess Publishing Co., Minneapolis.)

About 200 to 350 second-generation merozoites, each about 16 μm long, are then formed by schizogony. These rupture the host cell and enter the lumen of the cecum about 5 days after infection. Some of those merozoites enter new cells to initiate a third generation of schizogony below the nucleus, producing four to 30 third-generation merozoites, each about 7 μm long. Many merozoites are engulfed and digested by macrophages during these cycles of schizogony.

Some of the second-generation merozoites enter new epithelial cells in the cecum to begin gametogony. Most develop into macrogametocytes. Both male and female gamonts lie between the host cell nucleus and the basement membrane. The microgametocyte buds to form many slender, biflagellated microgametes that leave the host cell and enter cells containing macrogametes; there fertilization takes place.

The macrogamete has many granules of two types. Immediately after fertilization these granules pass peripherally toward the surface of the zygote, flatten out, and coalesce to form first the outer, then the inner layer of the oocyst wall. This coalescence takes place within the cell membrane of the zygote, and that membrane becomes the covering of the outer wall. The oocyst then is released from the host cell and moves with the cecal contents into the large intestine to be passed out of the body with the feces. Oocysts appear in the feces 7 days after infection. Oocysts are passed for several days because not all second-generation merozoites reenter host cells at the same time; further, oocysts often will remain in the lumen of the cecum for some time before moving to the large intestine.

The freshly passed oocyst contains a single cell, the **sporont.** Sporogony (often called "sporulation"), or development of the sporont into sporocysts and sporozoites, is exogenous. The sporont is diploid, and the first division is reductional, a polar body being expelled. The haploid number of chromosomes is two. The sporont divides into four **sporoblasts,** each of which forms a sporocyst containing two sporozoites. Sporulation takes 2 days at summertime temperature, whereupon the oocysts are infective.

Although the organisms can survive anaerobic conditions, as might be found in freshly passed feces, the metabolism of sporulation is an aerobic process and will not proceed in the absence of oxygen.[23] Development also is strongly and reversibly inhibited by cyanide, indicating that the cytochrome system is probably very important in the energetics of sporulation. Oxygen consumption is high at first but falls steadily as sporulation is completed. The organisms have large amounts of glycogen, which is rapidly consumed, and measurements of the respiratory quotient indicate that they depend primarily on carbohydrate oxidation for energy during sporoblast formation, then change over to lipid for energy as sporulation is completed. Thus, the biochemistry suggests an interesting developmental control in metabolism: first a rapid burst of energy fuels sporulation, and then a shift to a low level of maintenence metabolism conserves resources until a new host is reached.

Coccidial infections are self-limiting. That is, asexual reproduction does not continue indefinitely. If the chicken survives through oocyst release, it recovers. It may become reinfected, but a primary infection usually imparts some degree of protective immunity to the host.

The number of oocysts produced in any infection can be astounding. Theoretically, one oocyst of *E. tenella,* containing eight sporozoites, can produce 2.52 million second-generation merozoites, most of which will become macrogametes and thereby oocysts. The actual numbers of oocysts produced are far fewer than their theoretical potential, however. Many merozoites and sporozoites are discharged with the feces before they can penetrate host cells, and many are destroyed by host defenses. There normally is a complete replacement of cecal epithelium about every 2 days, so any merozoite or sporozoite that invades a cell that is about to be sloughed is out of luck. Young chickens are more susceptible to infection and discharge more oocysts than do older birds.

PATHOGENESIS. Cecal coccidiosis is a serious disease that causes a bloody diarrhea, sloughing of patches of epithelium, and, commonly, death of the host. When merozoites emerge, especially from the lamina propria, host tissues are disrupted.

Fig. 7-8. Cecum of chicken, opened to show patches of hemorrhage caused by *Eimeria tenella.* (Photograph by James Jensen.)

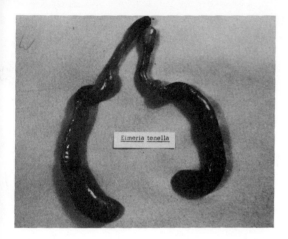

Fig. 7-9. Ceca of chicken infected by *Eimeria tenella.* Note distention caused by clotted blood and debris and dark color from hemorrhage. (Photograph by James Jensen.)

The large schizonts, especially when packed close together, disrupt the delicate capillaries that service the epithelium, further altering normal physiology of the tissues and also causing hemorrhage (Fig. 7-8). A hard core of the clotted blood and cell debris often plugs up the cecum, causing necrosis of that organ (Fig. 7-9). Birds that are not killed outright by the infection become unthrifty, listless, and susceptible to predation and other diseases. The U.S. Department of Agriculture estimated that loss to poultry farmers in the United States alone in 1965 was $34.854 million, not counting the extra cost of medicated feeds and added labor.

Many useful drugs are available as pro-

phylaxis against coccidiosis. However, once infection is established there is no effective chemotherapy. Therefore, a coccidiostat must be administered continuously in food or water to prevent an outbreak of disease. These compounds affect the schizont primarily, so the host can still build up an immunity in response to invading sporozoites.

Other Eimeria species. The number of species of *Eimeria* is so large that it is impossible to discuss more than a representative species, as we have done above with *E. tenella*. Levine has summarized those species that parasitize domestic animals.[17] Some of the most common species are *E. auburnensis* and *E. bovis* in cattle, *E. ovina* in sheep, *E. debliecki* and *E. porci* in pigs, *E. stiedai* in rabbits, *E. necatrix* and *E. acervulina* in chickens, *E. meleagridis* in turkeys, and *E. anatis* in ducks. All have life cycles similar to that of *E. tenella* but differing in details.

Isospora canis (Fig. 7-10). *Isospora* species are basically similar to *Eimeria,* except that the oocyst contains two sporocysts, each with four sporozoites. *Isospora canis* is a common, cosmopolitan parasite of dogs. It occurs in about 11% of dogs in North America. Its development has been studied in cell culture.[6]

Sporulated oocysts are slightly ovoid, 35 to 42 µm by 27 to 33 µm, and have a smooth, thin wall. A micropyle, oocyst residuum, and oocyst polar granule are absent. The endogenous stages develop in the epithelium and lamina propria of the villi of the small intestine and in the epithelium of the large intestine. Oocysts are passed 10 days after infection.

Heavy infections cause inflammation and hemorrhage of the intestine. Puppies with severe infections have diarrhea, are dull, weak, and anemic, and have a slight temperature. They gradually recover after about a week.

Other Isospora species. *Isospora* species are not as numerous as *Eimeria;* neither are they as host-specific. There is ample evidence that many species can infect a broad spectrum of potential hosts. However, some species are limited to certain hosts. It is known that some *Isospora* have extraintestinal stages in intermediate hosts. Frenkel and Dubey found stages of the cat parasites *Isospora felis* and *I. rivolta*

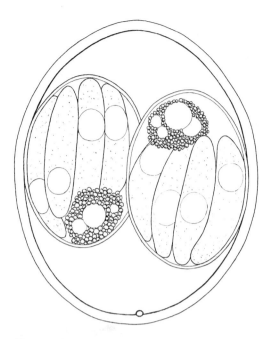

Fig. 7-10. *Isospora canis*, sporulated oocyst. (From Levine, N. D., and V. Ivens. 1965. J. Parasitol. 51:859-864.)

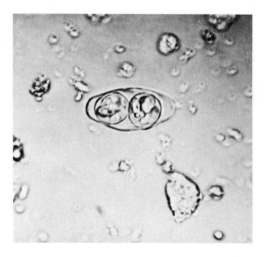

Fig. 7-11. *Isospora belli* oocyst. (From Beck, J. W., and J. E. Davies. 1972. Medical parasitology, ed. 2. The C. V. Mosby Company, St. Louis.)

in the mesenteric lymph nodes of mice, rats, and hamsters 3 to 14 days after they were fed oocysts.[8] Suspensions of lung, spleen, and liver of these mice were infective when fed to kittens. The same researchers also found these coccidia developing in the liver, spleen, mesenteric lymph nodes, and other tissues of cats.[3] The finding of intermediate hosts and extraintestinal development in the life cycles of coccidia previously thought to be monoxenous raised a question: are the life cycles of coccidia as simple as previously thought? Transmission by carnivorism may explain the prevalence of some coccidia in regions where oocysts are rapidly killed by cold, heat, or dryness. The relationships of the isosporans of cats and dogs to *Toxoplasma* and *Sarcocystis* will be discussed.

Isospora bigemina is found in dogs, cats, and other carnivores; *I. felis* is known in cats but cannot infect dogs; *I. suis* infects pigs, and *I. gallinae* is a parasite of chickens. Three rare species of *Isospora* are known from humans: *I. belli* (Fig. 7-11), *I. hominis,* and *I. natalensis.* One or more of these may be zoonoses. The oocyst of *I. hominis,* for example, is essentially indistinguishable from that of *I. bigemina.* Little is known about the biology of any species from humans, but infections appear to be more common than previously thought (see Brandberg, Goldberg, and Briedenbach, 1970).

Family Sarcocystidae. The taxonomic relationships of the members of the family Sarcocystidae have been unclear; sexual forms have been found only recently. We are here following Levine, who placed *Sarcocystis* and *Toxoplasma* in subfamilies of the Sarcocystidae.[16] Members of the family form conspicuous cysts (**zoitocysts**) in tissue other than (or in addition to) the gut of their hosts. These cysts contain **bradyzoites** that multiply by endodyogeny. Sexual reproduction does occur in at least some species.

The subfamily Besnoitiinae contains *Besnoitia* spp., a poorly known genus with thick, laminated, nucleated zoitocyst walls. *Besnoitia besnoiti* forms its cysts in the skin and connective tissue of cattle and wild ruminants. It causes a debilitating and sometimes fatal disease, particularly important in south and central Africa. Some other species of *Besnoitia* are *B. bennetti* in horses, *B. tarandi* in reindeer and caribou, and *B. jellisoni* in rodents.*

---

*For further treatment of *Besnoitia,* see Levine.[17]

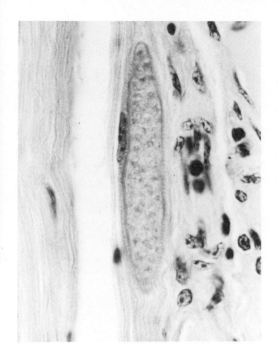

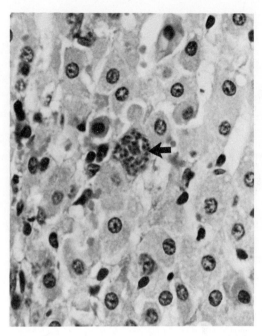

Fig. 7-12. Zoitocyst of *Sarcocystis fusiformis* in muscle of experimentally infected calf. (Photograph by Ronald Fayer.)

Fig. 7-13. *Sarcocystis fusiformis* schizont or young zoitocyst (arrow) in adrenal gland of experimentally infected calf. (From Fayer, R., and A. J. Johnson. 1973. J. Parasitol. 59:1135-1137.)

Subfamily Sarcocystinae. The subfamily Sarcocystinae contains the common genus *Sarcocystis* and the poorly known species *Arthrocystis galli,* found once in a chicken in India. *Sarcocystis* spp. are abundant in cattle, sheep, swine, and horses and are often found in birds. It appears to be rather common in humans but is rare in such carnivores as dogs and cats. More than 50 species have been named, but many of these are probably invalid, for the parasite is not very host specific and tends to be polymorphic in different hosts. For example, when bradyzoites from sheep are fed to guinea pigs, they develop into zoites that are only half their former size, and the zoitocyst lacks alveoli that are present when the parasite grows in sheep. Further, the cyst wall, which is used as a taxonomic character, changes in form and composition as it ages. Various specific names, then, are retained at present because synonymies have yet to be clarified.

MORPHOLOGY. The zoitocysts (Figs. 7-12 and 7-13) are formed in the striated and cardiac muscles, rarely in the brain or glands. They are especially abundant in the wall of the esophagus, masseter mus-

cles, tongue, diaphragm, and heavy muscles of the body. In birds their site preference is the pectoralis muscles. The cysts are also known as **sarcocysts** or **Miescher's tubules** and are large enough to see with the unaided eye. They usually have internal septa and compartments. They are elongate, cylindroid, or spindle-shaped, but they may be irregular. They lie within a muscle fiber, in the same plane as the muscle bundle. The overall size varies, reaching 1 cm in diameter in some cases, but they usually are 1 or 2 mm in diameter and 1 cm or less long. The structure of the cyst wall varies between "species" and in different stages of development of the parasite. In some cases the outer wall is smooth, and in others it has an outer layer of fibers, the **cytophaneres,** which radiate out into the muscle (Fig. 7-12). The origin of the cyst wall is controversial, some authors concluding that it is of host origin, others maintaining that it is of parasite origin. It may well be derived from both sources.

As the cyst grows old, it breaks down,

releasing zoites into the blood, where they reach the intestinal tract and are passed out with the feces. They have also been found in nasal secretions, and it is possible that they enter several types of body secretions.

The bradyzoites are banana-shaped, 5 to 12 μm long, with the anterior end pointed and the posterior end rounded. They have a typical apical complex, and they move by flexion, gliding, or twisting. Generally speaking, they are very similar to merozoites of the Eimeriidae.

BIOLOGY. Much that has been written about the life cycle of *Sarcocystis* now appears inaccurate. Many scattered observations that presumably concerned this genus actually dealt with mixtures of stages of *Toxoplasma* and fungi, as well as of *Sarcocystis*. It was long thought that carnivorism of sarcocysts was the main means of transmission, for infection was maintained through several generations of laboratory mice that had been fed infected mouse flesh. However, there was a high rate of enzootic infection in laboratory mice throughout the world, presumably through fecal contamination. Further, carnivores are rarely infected, while herbivores commonly are. Among natural infections in birds, ducks are the most frequent hosts. Fayer (1972) maintained *Sarcocystis* sp. from a grackle in bovine cell culture, and he observed formation of macrogametes and microgametes with subsequent formation of cyst-like bodies.[4]

When *Sarcocystis tenella* from sheep is fed to cats, the cats begin shedding sporulated sporocysts after a 12-day prepatent period.[21] Dogs are not infected by *S. tenella*, but when fed *S. fusiformis* from cattle, they give rise to sporocysts and occasional oocysts. Cats and humans can also be infected. The oocysts and sporocysts are of the *Isospora* type (Fig. 7-14).

Fayer (1974) traced the endogenous development of the sexual cycle of *Sarcocystis fusiformis* (Fig. 7-15) in the small intestine of dogs that had been fed bovine heart infected with cysts.[5] He found macrogametes, oocysts, and sporocysts in the villi of the dogs and recovered sporocysts from their feces. One dog that was fed sarcocysts passed oocysts of *Isospora bigemina*, thus raising the question that some "*Sarcocystis*" might in fact be a parenteral form of

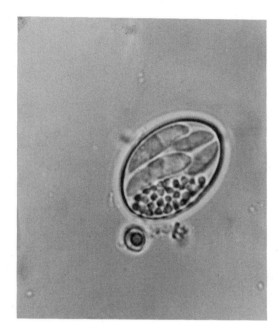

Fig. 7-14. Sporocyst of *Sarcocystis fusiformis* from dog. (Photograph by Ronald Fayer.)

Isospora. Dogs inoculated with sporocysts obtained from other dogs did not become infected. *Sarcocystis miescheriana* from pigs produced oocysts and sporocysts in humans.[10] *Isospora hominis*, then, may be *Sarcocystis fusiformis* and/or *S. miescheriana*, while *Isospora felis* may represent *S. tenella* or *S. fusiformis*.

Thus it appears now that *Sarcocystis* can be transmitted between various hosts through fecal contamination with bradyzoites, by ingestion of sporulated oocysts, and possibly by carnivorism of Miescher's tubules (Fig. 7-16).

PATHOGENESIS. *Sarcocystis* infections usually are not considered very pathogenic. However, very heavy infections can cause emaciation, weakness, paralysis, and death. The cyst destroys the muscle cell it occupies, and as the cyst grows, it causes pressure atrophy of adjacent cells. When the cysts break down, they release a powerful toxin called **sarcocystin,** which affects the central nervous system, heart, adrenal glands, liver, and intestine. Large amounts are lethal. *Sarcocystis fusiformis* is highly significant in bovine abortion. No treatment for sarcocystosis is known.

Subfamily Toxoplasmatinae. The sub-

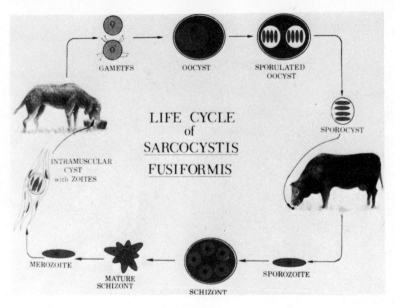

Fig. 7-15. Life cycle of *Sarcocystis fusiformis*. (Courtesy Ronald Fayer. Drawing by R. B. Ewing.)

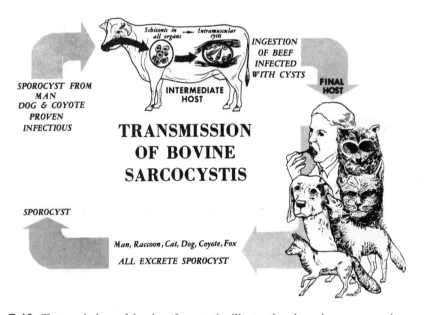

Fig. 7-16. Transmission of bovine *Sarcocystis*, illustrating how humans can become infected. (Courtesy Ronald Fayer. Drawing by R. B. Ewing.)

family Toxoplasmatinae contains the ubiquitous genus *Toxoplasma* and the little-known *Frenkelia* and *Hammondia,* each with a single known species. *Frenkelia microti* forms thin-walled cysts in the brain of voles and muskrats. *Hammondia* is known from mice. Both are of no known medical or economic importance and so will not be considered further. *Toxoplasma gondii,* however, has come to be recognized as an important pathogen in humans and will be discussed in some detail.

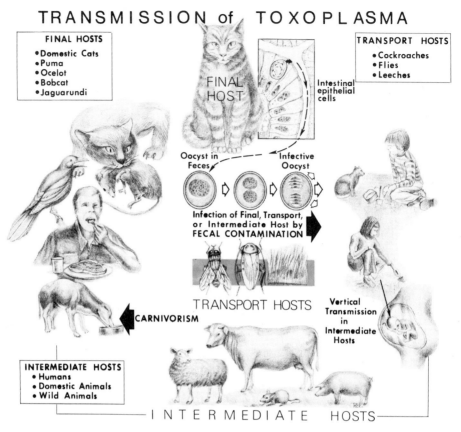

Fig. 7-17. Transmission of *Toxoplasma gondii*. (Courtesy Ronald Fayer. 1976. National Wool Grower 66:22. Drawing by R. B. Ewing.)

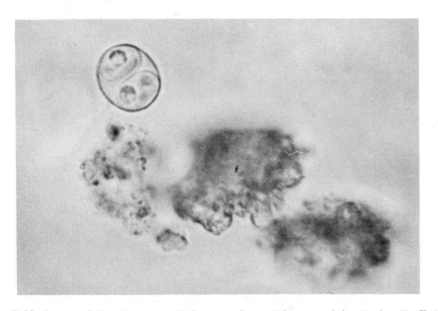

Fig. 7-18. Oocyst of *Toxoplasma gondii* from cat feces. (Photograph by Harley Sheffield.)

Toxoplasma gondii (Fig. 7-17). Like *Trypanosoma cruzi, Toxoplasma gondii* was discovered before it was known to cause disease in humans. It was first discovered in 1908 in a desert rodent, the gondi, in a colony maintained in the Pasteur Institute in Tunis. Since then, the parasite has been found in nearly every country of the world in many species of carnivores, insectivores, rodents, pigs, herbivores, primates, and other mammals, as well as in birds. We now realize that it is cosmopolitan in the human population and can cause disease. The importance of the organism as a human pathogen has stimulated a huge amount of research in recent years. A 1963 bibliography on the subject contained 3,706 references, and Jacobs (1973) reported over 2,000 references for the years 1967 through 1972.[11] Thus, the onetime obscure protozoan parasite of an obscure African rodent has become one of the most exciting subjects in parasitology.

BIOLOGY. The biology of *Toxoplasma* has been reviewed recently by Frenkel, among others, and terminology in the account to follow adopts Frenkel's usage.[7]

Toxoplasma is an intracellular parasite of many kinds of tissues, including muscle, intestinal epithelium, and others. In heavy acute infections, the organism can be found free in the blood and peritoneal exudate. It may inhabit the nucleus of the host cell but usually lives in the cytoplasm. The life cycle includes intestinal-epithelial **(enteroepithelial)** and **extraintestinal** stages in domestic cats and other felines, but extraintestinal stages only in other hosts. Sexual reproduction of *Toxoplasma* occurs while in the cat, and only asexual reproduction is known in other hosts.

Extraintestinal stages begin when a cat or other host ingests a sporulated oocyst or sporocyst. Ingested tachyzoites or bradyzoites also are infective. Intrauterine infection is possible (see pathogenesis). The oocyst (Fig. 7-18) is 10 to 13 μm by 9 to 11 μm and is basically identical in appearance to that of *Isospora bigemina,* a common coccidian of cats. There are four sporozoites in each of two sporocysts per oocyst. There is no oocyst residuum or polar granule, and the sporocysts have a sporocyst residuum but no Stieda body.

The sporozoites escape from the sporocysts and the oocyst in the small intestine. In cats some of the sporozoites enter epithelial cells and remain to initiate the enteroepithelial cycle, while others penetrate through the mucosa to begin development in the lamina propria, mesenteric lymph nodes and other distant organs, and white blood cells. In hosts other than cats, there is no enteroepithelial development; the sporozoite enters a host cell and begins multiplying by endodyogeny. These rapidly dividing cells in acute infections are called **tachyzoites** (Fig. 7-19). Eight to 16 tachyzoites accumulate within the host cell's parasitophorous vacuole before the cell disintegrates, releasing the parasites to infect new cells. These accumulations of tachyzoites in a cell are called **groups.** Tachyzoites apparently are less resistant to stomach secretions; therefore, they are less important sources of infection than other stages.

As infection becomes chronic, the zoites that affect brain, heart, and skeletal muscles, multiply much more slowly than in the acute phase. They are now called **bradyzoites,** and they accumulate in large numbers within a host cell. They become surrounded by a tough wall and are called cysts or **zoitocysts** (Fig. 7-20). Previous authors have referred to the cysts, and also the groups of tachyzoites, as "pseudocysts" on the basis that the cyst wall is of host origin; but since the origin of the cyst wall has not been firmly established, "cyst" is prefered. Cysts may persist for months or even years after infection, particularly in nervous tissue. Cyst formation coincides with the time of development of immunity to new infection, which is usually permanent. If immunity wanes, released bradyzoites can boost the immunity to its prior level. This protection against superinfection by the presence of the infectious agent in the body is called **premunition.** Immunity to *Toxoplasma* is both of the humoral and cell-mediated type; the relative importance of each in protection has not been established. The tough, thin cyst wall, except when the cyst breaks down, effectively separates the parasite from the host, and an inflammatory reaction is not elicited. The cyst wall and its bradyzoites develop intracellularly, but they may eventually be-

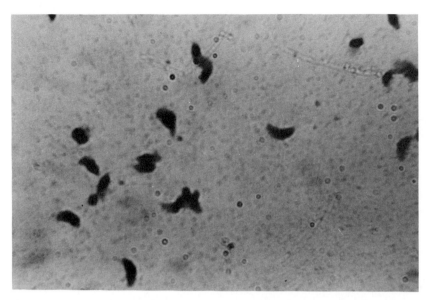

Fig. 7-19. Tachyzoites of *Toxoplasma gondii*. (Photograph by Warren Buss.)

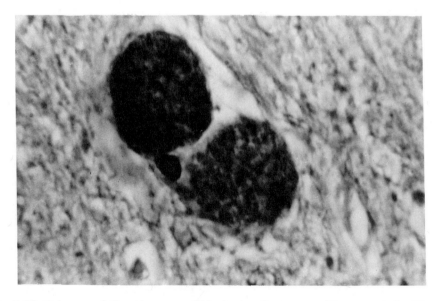

Fig. 7-20. Zoitocyst of *Toxoplasma gondii* in brain of a mouse. (Photograph by Warren Buss.)

come extracellular because of distention and rupture of the host cell. Bradyzoites are resistant to digestion by pepsin and trypsin, and, when eaten, they can infect a new host.

Enteroepithelial stages are initiated when a cat ingests zoitocysts containing bradyzoites, oocysts containing sporozoites, or, occasionally, tachyzoites. Another possible means of epithelial infection is by migration of extraintestinal zoites into the intestinal lining within the cat. Once inside an epithelial cell of the small intestine or colon, the parasite becomes a

Cat: final host   human: intermediate host

trophozoite that grows and prepares for schizogony. At least five different strains have been studied well enough to allow characterization of the enteroepithelial stages.[7] These strains differ in duration of stages, number of merozoites produced, shape, and other details. Basically, from two to 40 merozoites are produced by schizogony, endopolyogeny, or endodyogeny, and these initiate subsequent asexual stages. The number of schizogonous cycles is variable, but gametocytes are produced within 3 to 15 days after cyst-induced infection. Gametocytes develop throughout the small intestine but are more common in the ileum. From 2% to 4% of the gametocytes are male, and each produces about 12 microgametes. Oocysts appear in the cat's feces from 3 to 5 days after infection by cysts, with peak production occurring between 5 to 8 days. Oocysts are the *Isospora*-type, require oxygen for sporulation, and sporulate in 1 to 5 days. The sequence of events after the cat and other hosts ingest tachyzoites and bradyzoites has not been studied in detail at this writing.

Frenkel (1973) pointed out that *Toxoplasma* may be in transition to a heteroxenous life cycle from an ancestral monoxenous life cycle, as found in *Eimeria* and *Isospora*.[7] A similar observation could be made in regard to *Sarcocystis*. At present *Toxoplasma* has a facultatively heteroxenous life cycle, with ready transmissability by carnivorism of the incipient intermediate hosts by cats or by direct infection with oocysts.

One final note of interest is that in the Pasteur Institute in Tunis in 1908, when the gondis were brought in from the field and died, the source of their infection was never established. But it is known that a cat was roaming the laboratory.[11]

PATHOGENESIS. In view of the fact that antibody to *Toxoplasma* is widely prevalent in humans throughout the world, yet clinical toxoplasmosis is rare, it is clear that most infections are asymptomatic. Several factors influence this phenomenon: the virulence of the strain of *Toxoplasma*, the susceptibility of the individual host and of the host species, the age of the host, and the degree of acquired immunity of the host. Pigs are more susceptible than cattle;

white mice are more susceptible than white rats; chickens are more susceptible than carnivores. The reasons for natural resistance or susceptibility to infection are not known.

Tachyzoites proliferate in many tissues and tend to kill host cells at a faster rate than the normal turnover of such cells. Enteroepithelial cells, on the other hand, normally live only a few days, especially at the tips of the villi. Therefore, the extraepithelial stages, particularly in sites such as the retina or brain, tend to cause more serious lesions than those in the intestinal epithelium.

Since there seems to be an age resistance, infections of adults or past-weaning juveniles are asymptomatic, although exceptions do occur. Asymptomatic infections can suddenly become fulminating if immunosuppressive drugs such as corticosteroids are employed for other conditions. Symptomatic infections can be classified as acute, subacute, and chronic.

In most **acute infections** the intestine is the first site of infection. Cats infected by oocysts usually show little pathology beyond loss of individual epithelial cells, and these are rapidly replaced. In massive infections, however, intestinal lesions can kill kittens in 2 or 3 weeks. The first extraintestinal sites to be infected in both cats and other hosts, including humans, are the mesenteric lymph nodes and the parenchyma of the liver. These, too, have rapid regeneration of cells and perform an effective preliminary screening of the parasites. The most common symptom of acute toxoplasmosis is painful, swollen lymph glands in the cervical, supraclavicular, and inguinal regions. This symptom may be associated with fever, headache, muscle pain, anemia, and sometimes lung complications. This syndrome can be mistaken easily for the flu. Acute symptomatic infection can—though rarely does—cause death. If immunity develops slowly, the condition can be prolonged and is then called subacute.

In **subacute infections,** pathogenic conditions are extended. Tachyzoites continue to destroy cells, causing extensive lesions in the lung, liver, heart, brain, and eyes. Damage may be more extensive in the central nervous system than in unrelated

pathogenesis 1) virulence of strain
2) state of acquired immunity
3) age of host
4) susceptibility of host

organs because of lower immunocompetence in these tissues.

Chronic infection results when immunity builds up sufficiently to depress tachyzoite proliferation. This coincides with the formation of cysts. These cysts can remain intact for years and produce no obvious clinical effect. Occasionally, a cyst wall will break down, releasing bradyzoites; most of these are killed by host reactions, although some may form new cysts. Death of the bradyzoites stimulates an intense hypersensitive inflammatory reaction, the area of which, in the brain, is gradually replaced by nodules of glial cells. If many such nodules are formed, the host may develop symptoms of chronic encephalitis, with spastic paralysis in some cases. Repeated infections of retinal cells by tachyzoites can destroy the retina. Cysts and cyst rupture in the retina and choroid can also lead to blindness. Other kinds of extensive pathology that can occur in chronic toxoplasmosis are myocarditis, with permanent heart damage and with pneumonia.

The most tragic form of this disease is **congenital toxoplasmosis.** If a mother contracts acute toxoplasmosis at the time of her child's conception or during pregnancy, the organisms will often infect her developing fetus. Fortunately, most neonatal infections are asymptomatic but a significant number do cause death or disability to newborns. It is generally assumed that *Toxoplasma* crosses the placental barrier from the mother's blood, but, because the uterus itself is commonly heavily infected, direct transmission cannot be ruled out.

Stillbirths and spontaneous abortions may result from fetal infection with *Toxoplasma* in humans and other animals. Sheep seem to be particularly susceptible, and *Toxoplasma*-caused abortions in this host often reach epidemic proportions. This disease is said to account for half of all ovine abortions in England and New Zealand.[1]

In a study of more than 25,000 pregnant women in France, no case of congenital toxoplasmosis was found whenever maternal infection occurred prior to pregnancy.[2] However, of 118 cases of maternal infection near the time of or during pregnancy, there were nine abortions or neonatal deaths without confirmation by examination of the fetus, 39 cases of acute congenital toxoplasmosis with two deaths, and 28 cases of subclinical infection. The remainder of fetuses were free of infection. Maternal infection in the first 3 months of pregnancy results in more extensive pathogenesis, but transmission to the fetus is more frequent if the maternal infection occurs in the third trimester.

Relatively common lesions in congenital toxoplasmosis are hydrocephalus, microcephaly, cerebral calcification, chorioretinitis, and psychomotor disturbances. In cases of twins, one may have severe symptoms while the other shows no overt evidence of infection. In children who survive infection, there is often congenital damage to the brain, manifested as mental retardation and epileptic seizures. Thus it appears that toxoplasmosis is a major cause of human birth defects.

EPIDEMIOLOGY. While toxoplasmosis usually affects only scattered individuals, small epidemics occur from time to time. For example, several medical students were infected simultaneously by wolfing down undercooked hamburgers between classes.[12] In 1969, at a university in São Paulo, Brazil, 110 persons were diagnosed with acute toxoplasmosis in a 3-month period. Most admitted to eating undercooked meat. Therefore, raw meat seems to be an important source of infection. One large steer or pig might conceivably be the source of an epidemic at any time. One looks at the fad of backyard cooking and the American's fondness for rare beef and wonders how many cases of toxoplasmosis are thus acquired every day.

While beef is certainly a potential source of infection, pork and lamb are much more likely to be contaminated. Freezing at $-14°$ C for even a few hours apparently will kill all cysts. To avoid a multitude of parasites, persons who insist on eating undercooked meat would do well to see that it has been hard-frozen.

Feral and domestic cats will continue to be a source of infection to humans. Stray cats lead to problems of several kinds and are reservoirs of several diseases; efforts should be made to keep their numbers down. A more difficult problem to resolve is the household pet, the tabby that spends most of its time in a close, symbiotic relationship with its owners. Any cat, no matter how well fed and protected, may be passing oocysts of *Toxoplasma,* although for only a few days after infection. The possibilities are particularly alarming if someone in the house becomes pregnant. Certainly, a woman who knows she is pregnant should never empty the litterbox or clean up after the cat's occasional indiscretion. (Emptying the box every 2 days should be effective, for the cysts require 3 days to sporulate.) It may be well at that stage to have the cat tested for *Toxoplasma* antibodies and to see if it is passing oocysts. Also, because children's sandboxes become a haven for any cats in the neighborhood, they should have tightly fitting covers. This also will protect children from larva migrans from hookworms and ascaridoid larvae.

Filth flies and cockroaches are capable of carrying *Toxoplasma* oocysts from cat feces to the dinner table.[22] Earthworms may serve to move oocysts from where cats have buried them to the surface of the ground.

Toxoplasma tachyzoites have been isolated in humans from nasal and eye secretions, milk, saliva, urine, and feces. The role of any of these in spreading infection is unknown, but it seems reasonable that any or all may be involved. Whole blood or leukocyte transfusions and organ transplants are also potential sources of infection, made more important because the recipient may be immunodeficient because of disease or treatment.

DIAGNOSIS AND TREATMENT. Specific diagnosis in humans is based on one or more laboratory tests. Demonstration of the organism at necropsy or biopsy is definitive. Intraperitoneal inoculation of a biopsy of lymph node, liver, or spleen into mice is useful and accurate. Demonstration of specific antibody, using the complement fixation technique combined with the Sabin-Feldman dye test or hemagglutination tests in conjunction with the dye test, gives accurate results. The toxoplasmin skin test is widely employed in mass surveys to demonstrate *Toxoplasma* antibody but is not reliable because of cross-reaction with other antigens.

No satisfactory chemotherapy for toxoplasmosis has been developed at the time of this writing. Pyrimethamine with triple sulfonamide gives good results in ocular infections but may cause macrocytic anemia. Prevention would seem to be the best recourse available at this time.

REFERENCES

1. Beverly, J. K. A., W. A. Watson, and J. B. Spence. 1971. The pathology of the foetus in ovine abortion due to toxoplasmosis. Vet. Rec. 88:174-178.
2. Couvreur, J. 1971. Prospective study in pregnant women with a special reference to the outcome of the foetus. In Hentsch, D., editor. Toxoplasmosis. Huber, Bern. pp. 119-135.
3. Dubey, J. P., and J. K. Frenkel. 1972. Extra-intestinal stages of *Isospora felis* and *I. rivolta* (Protozoa: Eimeriidae) in cats. J. Protozool. 19:89-92.
4. Fayer, R. 1972. Gametogony of *Sarcocystis* sp. in cell culture. Science 175:65-67.
5. Fayer, R. 1974. Development of *Sarcocystis fusiformis* in the small intestine of the dog. J. Parasitol. 60:660-665.
6. Fayer, R., and J. L. Mahrt. 1972. Development of *Isospora canis* (Protozoa: Sporozoa) in cell culture. Z. Parasitenk. 38:313-318.
7. Frenkel, J. K. 1973. Toxoplasmosis: parasite life cycles, pathology, and immunology. In Hammond, D. M., editor. The Coccidia. *Eimeria, Isospora, Toxoplasma,* and related genera. University Park Press, Baltimore, pp. 343-410.
8. Frenkel, J. K., and J. P. Dubey. 1972. Rodents as transport hosts for cat coccidia, *Isospora felis* and *I. rivolta.* J. Infect. Dis. 125:69-72.
9. Grell, K. G. 1973. Protozoology. Springer-Verlag, Heidelberg.
10. Heydorn, A. O., and M. Rommel. 1972. Beiträge zum Lebenszyklus der Sarkosporidien. II. Hund und Katze als Überträger der Sarkosporidien des Rindes. Berl. Münch. Tierärztl. Wochenschr. 85:121-123.
11. Jacobs, L. 1973. New knowledge of *Toxoplasma* and toxoplasmosis. In Dawes, B., editor. Advances in parasitology, vol. II. Academic Press, Inc., New York. pp. 631-669.
12. Kean, B. H., A. C. Kimball, and W. N. Christenson. 1969. An epidemic of acute toxoplasmosis. J.A.M.A. 208:1002-1004.
13. Levine, N. D. 1962. Protozoology today. J. Protozool. 9:1-6.
14. Levine, N. D. 1963. Coccidiosis. Ann. Rev. Microbiol. 17:179-198.
15. Levine, N. D. 1970. Taxonomy of the Sporozoa. J. Parasitol. 56(II):208-209.
16. Levine, N. D. 1973a. Introduction, history and taxonomy. In Hammond, D. M., editor. The Coccidia. *Eimeria, Isospora, Toxoplasma,* and re-

lated genera. University Park Press, Baltimore. pp. 1-22.

17. Levine, N. D. 1973b. Protozoan parasites of domestic animals and of man, ed. 2. Burgess Publishing Co., Minneapolis.

18. Levine, N. D., and V. Ivens. 1965 The coccidian parasites (Protozoa, Sporozoa) of rodents. Ill. Biol. Monogr. 33. University of Illinois Press, Urbana, Ill.

19. Marquardt, W. C. 1973. Host and site specificity in the Coccidia. In Hammond, D. M., editor. The Coccidia. *Eimeria, Isospora, Toxoplasma* and related genera. University Park Press, Baltimore. pp. 23-43.

20. Roberts, W. L., C. A. Speer, and D. M. Hammond. 1970. Electron and light microscope studies of the oocyst walls, sporocysts, and encysting sporozoites of *Eimeria callospermophili* and *E. larimerensis.* J. Parasitol. 56:918-926.

21. Rommel, M., A. O. Heydorn, and F. Gruber. 1972. Beiträge zum Lebenzyklus der Sarkosporidien. I. Die Sporozyste von *S. tenella* in der Fäzes der Katze. Berl. Münch. Tierärztl. Wochenschr. 85:101-105.

22. Wallace, G. D. 1971. Experimental transmission of *Toxoplasma gondii* by filth flies. Am. J. Trop. Med. Hyg. 20:411-413.

23. Wilson, P. A. G., and D. Fairbairn. 1961. Biochemistry of sporulation in oocysts of *Eimeria acervulina.* J. Protozool. 8:410-416.

SUGGESTED READING

Brandberg, L. L., S. B. Goldberg, and W. C. Breidenbach. 1970. Human coccidiosis—a possible cause of malabsorption. The life cycle in small-bowel mucosal biopsies as a diagnostic feature. N. Engl. J. Med. 283:1306-1313. (A study of six cases and a review of previous reports. Endogenous stages are illustrated for the first time.)

Desmonts, G., and J. Couveur. 1974. Congenital toxoplasmosis. N. Engl. J. Med. 290:1110-1116. (A study of 378 pregnancies.)

Feldman, H. A. 1974. Congenital toxoplasmosis, at long last . . . . N. Engl. J. Med. 290:1138-1140. (A short summary of the discovery of congenital toxoplasmosis.)

Hammond, D. M. 1973. The Coccidia. *Eimeria, Isospora, Toxoplasma,* and related genera. University Park Press, Baltimore. (This will undoubtedly be the standard reference on the subject for many years.)

Jacobs, L. 1967. *Toxoplasma* and toxoplasmosis. In Dawes, B., editor. Advances in parasitology, vol. 5. Academic Press, Inc., New York. pp. 1-45. (An excellent review of toxoplasmosis.)

Jacobs, L. 1973. New knowledge of *Toxoplasma* and toxoplasmosis. In Dawes, B., editor. Advances in parasitology, vol. II. Academic Press, Inc., New York. pp. 631-669. (This is an excellent reference, especially for details on individual species.)

Levine, N. D. 1973. Protozoan parasites of domestic animals and of man, ed. 2. Burgess Publishing Co., Minneapolis. (This is an excellent reference, especially for details on individual species.)

Chapter 8

SUBPHYLUM APICOMPLEXA: MALARIA AND PIROPLASMS

Suborder Haemosporina

The Haemosporina contains the three families Plasmodiidae, Haemoproteidae, and Leucocytozoidae, which are the malaria and malaria-like organisms. Syzygy is absent in these parasites. The macrogametocyte and microgametocyte develop independently, the microgametocyte producing about eight flagellated gametes. The zygote is motile and is called an ookinete; the sporozoites are within a spore wall. Endodyogeny is absent. These protozoans are heteroxenous, with merozoites produced in the vertebrate host and sporozoites developing in the invertebrate host.

Although most species in the Haemosporina are parasites of wild animals and appear to cause little harm in most cases, a few cause diseases that are among the worst scourges of humankind. Indeed, malaria has played an important part in the rise and fall of nations and has killed untold millions the world over. John F. Kennedy said in 1962:

For centuries, malaria has outranked warfare as a source of human suffering. Over the past generation it has killed millions of human beings and sapped the strength of hundreds of millions more. It continues to be a heavy drag on man's efforts to advance his agriculture and industry.[13]

Despite the efforts of 90 countries joined in a common effort to eradicate malaria, it remains the most important disease in the world today in terms of lives lost and economic burden. Progress has been made, however. Between 1948 and 1965 the number of cases was cut from a worldwide total of 350 million to fewer than 100 million. In some countries, such as the United States, eradication of endemic malaria is complete.

At this writing approximately 1,472 million persons live in malarious areas of the world. This unprotected population lives in countries without the administrative, financial, and manpower resources necessary for eradication.

Family Plasmodiidae

The Plasmodiidae contains the single genus *Plasmodium*. Recent studies with the electron microscope reveal the basic similarity of *Plasmodium* to the coccidia. The merozoites and sporozoites have the typical constituents of the apical complex, supporting the phylogenetic relationship of these two groups. When in a host cell, the parasite usually produces a pigment called **hemozoin** from host hemoglobin. The presence of this pigment separates the species in Plasmodiidae from the closely related Haemoproteidae and Leucocytozoidae. It is possible that these parasites evolved from the coccidia of vertebrates rather than invertebrates, with mites or other blood-suckers initiating the cycle in arthropods.

History

History is, after all, a review of past experiences which influence present events.

ELVIO H. SADUN

Because malaria is still the most important disease of humankind, we think it is of value to relate the history of its conquest in considerable detail.

Malaria has been known since antiquity, with recognizable descriptions of the disease recorded in various Egyptian papyri. The Ebers papyrus (1550 B.C.) mentions fevers, splenomegaly, and the use of oil of the Balamites tree as a mosquito repellent. Hieroglyphs on the walls of the ancient Temple of Denderah in Egypt describe an intermittent fever following the flooding of the Nile.[8] Hippocrates studied medicine in Egypt and clearly described quotidian, tertian, and quartan fevers with splenomegaly. He believed that bile was the cause of the fevers. Greek states built beautiful cities in the lowlands only to see them devastated by the disease, and wealthy Greeks and Romans traditionally summered in the highlands to escape the heat, mosquitoes, and mysterious fevers. Herodotus (c. 500-424 B.C.) states that Egyptian fishermen slept with their nets arranged around their beds so that mosquitoes could not reach them. Homer also noted that malaria is most prevalent in late summer for in the *Iliad* (XXII, 31) we read ". . . like that star which comes on in the autumn . . . , the star they give the name of Orian's dog which is brightest among the stars, and yet is wrought as a sign of evil and brings on the great fever for unfortunate mortals." Medieval England saw crusaders falter and fail as they encountered malaria. As had happened before and has happened since, malaria killed more warriors than did warfare. When Europeans imported slaves and returned their colonial armies to their continent, they brought malaria with them, increasing the concentration of the disease with devasting results.

Throughout history, a connection between swamps and fevers has been recognized. It was commonly concluded that the disease was contracted by breathing "bad air," or "malaria." This theory flourished until near the end of the nineteenth century. Another name for the disease, "paludism" (marsh disease), is still in common use in the world.

There has been much speculation as to whether malaria existed in the Western Hemisphere before the Spanish conquest. It seems inconceivable that the great Olmec and Mayan civilizations could have developed in regions that are now highly malarious. The Spanish Conquistadores made no mention of fevers during the early years of the conquest, and in fact they holidayed in Guayaquil and the coastal area near Veracruz, regions that soon after became very unhealthy because of malaria. Even Balboa, while traversing the Isthmus of Panama, did not mention any encounters with malaria. It therefore seems likely that malaria was introduced into the New World by the Spaniards and

their African slaves. Yet, nagging evidence that Africans reached South America during pre-Columbian times suggests that, while improbable, it is not impossible that malaria existed in localized areas of the continent prior to the Spanish conquest,[10] could have been brought from Oceania or from Asia via the Bering Strait, or could have been introduced by the Vikings.

No progress was made in the etiology of malaria until 1847 when Meckel observed black pigment granules in the blood and spleen of a patient who died of the disease. He even stated that the granules lay within protoplasmic masses. Was he the first to actually see the parasite? In 1879, Afanasiev suggested that the granules caused the disease.

During the next 30 years physicians and scientists of high stature searched diligently for the cause of the disease and its means of transmission to people. It remained for two obscure army medical officers, working in their spare time, under primitive and difficult circumstances, to make these cardinal discoveries.

Most research was directed at finding an infective organism in water or in the air. Many false hopes were generated when a previously unknown ameba or fungus was discovered, and when *Bacillus malariae* was declared to be the causative organism by Klebs (German) and equally prestigious Tomasi-Crudelli (Italian), few doubted the truth of their momentous discovery.

Meanwhile, in North Africa, far from academic circles, a young French Army doctor named Louis Alphonse Laveran decided that the mysterious pigment in his malarious patients would be a good starting point for further research. He observed the pigment not only free in the plasma but also within leukocytes, and he saw clear bodies within erythrocytes. As the hyaline bodies of irregular shape grew he saw the erythocytes grow pale and pigment form within them. He little doubted the parasitic nature of the organisms he saw. Then, on November 6, 1880, he witnessed one of the most dramatic events in protozoology: the formation of male gametes by the process of exflagellation. He quickly wrote of his discovery, reporting it on November 23, 1880, to the Academy of Medicine of Paris, where much skepticism was offered his report. Most scientists were loath to abandon the Klebs–Tomasi-Crudelli bacillus in favor of a protozoan that an army doctor claimed to have discovered in Algeria. His "organisms" were assumed to be degenerating blood cells. Laveran was able to ignore his detractors after 1884, when he had the satisfaction of demonstrating exflagellation to no less a personage than Louis Pasteur. With Pasteur's verification of his discovery, the rest of the world slowly acknowledged the truth. In 1885 Camillo Golgi differentiated between species of *Plasmodium* and demonstrated the synchronism of the parasite in relation to paroxysm, thus describing for the first time a "biological clock."

Laveran accurately described the male and female gametes, the trophozoite, and the schizont while working with a poor, low-power microscope and with unstained preparations. By 1890 several scientists in different parts of the world verified his findings. In 1891 Romanovski, in Russia, developed a new method of staining blood smears based on methylene blue and eosin. Modifications of his stain are still in wide use.

The mode of transmission of malaria was, however, still unknown. Although theories were rampant ("bad night air" was still a popular candidate), few were as well thought out as that of Patrick Manson, who favored the idea of the possibility of transmission by mosquitoes. True, he was conditioned by the proof of mosquitoes as vectors of filariasis, which gave him some insight. Surgeon-Major Ronald Ross was 38 years old when he met Manson for the first time, while on leave from the Indian Medical Service. Finding in Ross a man who was interested in malaria and who could test his theories for him, Manson lost no time in convincing Ross that malaria was caused by a protozoan parasite. For the next several years, in India, Ronald Ross worked during every spare minute, searching for the mosquito stages of malaria that he was certain existed. Dissecting mosquitoes at random and also after allowing them to feed on malarious patients, he found many parasites but none of them proved to be what he searched for. During this time he had a steady correspondence with Manson, who encouraged him and brought his discourses to the learned societies of England. Ross left a wonderful record of his moods of excitement, frustra-

tion, disappointment, and triumph; in addition he was a sensitive poet. His journals also contain long quotations from Manson's letters written to him at that time.

Ross's first significant observation was that exflagellation normally occurs in the stomach of a mosquito, rather than in the blood as was currently thought. At this time he was posted to Bangalore to help fight a cholera epidemic, the first in a series of frustrating interruptions by superiors who had no concept of the importance of the work Ross was doing in his spare time. Returning from Bangalore, he continued the search for further development of the parasite within the mosquito. Failing this he concluded that he had been working with the wrong kinds of mosquitoes (*Culex* and *Stegomyia*). He tried other kinds and was led astray time after time by gregarines and other mosquito parasites, each of which had to be eliminated as possible malaria organisms by laborious experimentation. After 2 years of work, which his superior officers ignored as harmless lunacy, he seemed to have reached an impasse. He was eligible for retirement soon and was determined to try "one more desperate effort to solve the Great Problem." He toiled far into the nights, dissecting mosquitoes, in a hot little office. He could not use the overhead fan lest it blow his mosquitoes away, and swarms of gnats and mosquitoes avenged themselves "for the death of their friends." At last, late in the night of August 16, 1897, he dissected some "dappled-winged" mosquitoes *(Anopheles)* that had fed on a malaria patient, and he found some pigmented, spherical bodies in the walls of the insects' stomachs. The next day he dissected his last remaining specimen and found the spheroid cells had grown. They were most certainly the malaria parasites! That night he penned in a notebook:

> This day designing God
> Hath put into my hand
> A wondrous thing. And God
> Be praised. At his command
> I have found thy secret deeds
> Oh million-murdering Death.
>
> I know that this little thing
> A myriad men will save—
> Oh death where is thy sting?
> Thy victory oh Grave?

He reported his discovery to Manson and immediately set about breeding the correct kind of mosquito in preparation for the first step of transmitting the disease from the insect to humans. He was immediately posted to Bombay, where he could do no further research on human malaria, but found similar organisms *(Plasmodium relictum)* in birds. He repeated his feeding experiments with mosquitoes and found similar parasites, when they fed on infected birds. He also found that the spheroid bodies ruptured, releasing thousands of tiny bodies that dispersed throughout the insect's body, including into the salivary glands. Through Manson he reported to the world how malaria is transmitted by mosquitoes. It remained only for a single experiment to prove the transmission to humans. Ross never did it. The authorities were so impressed they ordered Ross to work out the biology of kala-azar in another part of India. This seems to have broken his spirit, for he never really tried again to finish the study. The concentration had made him ill, his eyes were bothering him, and his microscope had rusted tight from his sweat. Anyway, he was a physician, not a zoologist, and was only interested in learning how to prevent the disease, not in the finer points of the parasite's biology. This he considered to be done, and he retired from the Army. He was awarded the Nobel Prize in Medicine in 1902 and was knighted in 1911. He died in 1932 after a distinguished postarmy career in education and research.

Unfortunately, the history of malariology is tarnished by strife and bitterness. Several persons who were working on the life cycle of the parasite claimed credit for the discovery that pointed to the means of control for malaria. Italian, German, and American scientists all made important contributions to the solution of the problem of malaria transmission. Several of these, including Ross himself, spent a good portion of their lives quibbling about priorities in the discoveries. Manson-Bahr (1963) gives a fascinating account of the personalities of the men who conquered the malaria life cycle.[20] Credit for completing the life cycle of malaria should go to Amigo Bignami and Giovanni Grassi, who

experimentally transmitted it from mosquito to human in 1898.

Although the life cycle of malaria was thought to be known after Ross' work, it remained to be found that there were stages in the liver.

In the early twentieth century, the cycle was thought to go from the blood to the mosquito back to the blood. This concept gained support from the published work of Fritz Schaudin, who claimed to have seen sporozoites penetrating red blood cells and transforming into trophozoites.

Schaudin's work remained unchallenged until World War I, when a fact began to emerge that could not be explained by the direct cycle between mosquito and blood. Quinine is a well-known antimalarial drug. Its effect is only on the erythrocytic forms. However, it was found that soldiers treated with the drug were cured, that is, there were no blood parasites, but when the treatment was stopped and the patients moved to a nonmalarious area, parasites would return to the blood at certain time intervals.

In 1920, Wagner and Jauregg discovered that the high fevers of malaria could be used to cure neurosyphilis. From this work two additional facts emerged. When a patient was infected with parasitized blood, the incubation period could be shortened or lengthened by changing the number of parasites injected. However, the bite of one or 200 infected mosquitoes did not alter the incubation period.

In 1938, James and Tate discovered the exoerythrocytic stages of *P. gallinaceum.* After this discovery, large-scale work began in order to find the exoerythocytic stages of human malaria. Finally in 1948 Shortt and Garnham demonstrated the exoerythrocytic stages of *Plasmodium cynomolgi* in monkey and *P. vivax* in humans.

These historical notes cannot be concluded without mention of a man who managed to apply these early discoveries for the immense benefit of his country and humankind: William C. Gorgas. Gorgas was the medical officer put in charge of the Sanitation Department of the Canal Zone, when the United States undertook to build the Panama Canal; were it not for his mosquito control measures, malaria and yellow fever would have defeated

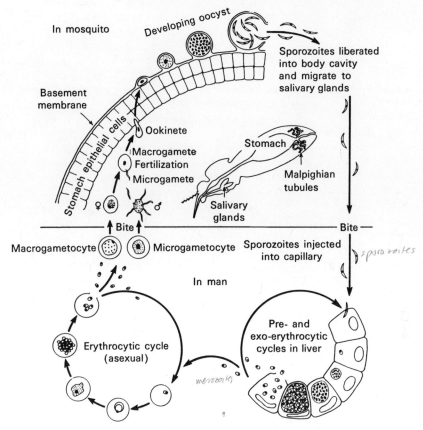

Fig. 8-1. The life cycle of *Plasmodium vivax*. (From Adam, K., J. Paul, and V. Zaman. 1971. Medical and veterinary protozoology. An illustrated guide. Churchill Livingston, Edinburgh.)

American attempts to build the Canal, just as they had the French. In July, 1906, the malaria rate in the Canal Zone was 1,263 hospital admissions per 1,000 population![24] Gorgas' work reduced the rate to 76 hospital admissions per 1,000 in 1913, saving his country $80 million and the lives of 71,000 fellow humans. Gorgas became a hero in his lifetime: the President made him Surgeon General, Congress promoted him, Oxford gave him an honorary Doctor of Science, and the King of England made him a knight. Sir William Osler stated, "There is nothing to match the work of Gorgas in the history of human achievement." It is a sad commentary on our cultural memory that the name of Gorgas is now known by so few, while we find it easy to remember the names of generals and tyrants who caused great bloodshed.

Life cycle and general morphology
(Fig. 8-1)

We give here a general account of the development and structure of malaria parasites, without reference to particular species. Specific morphological details for each species will be found in the pages to follow. *Plasmodium* spp. require two types of hosts: an invertebrate (mosquito) and a vertebrate (reptile, bird, or mammal). Technically, the invertebrate can be considered the definitive host because sexual reproduction occurs there. Asexual reproduction takes place in the tissues of a vertebrate, which thus can be called the intermediate host. However, it has been pointed out that the gametocytes actually form in the blood of the vertebrate, and fertilization occurs while still in this medium in the stomach of the mosquito. By

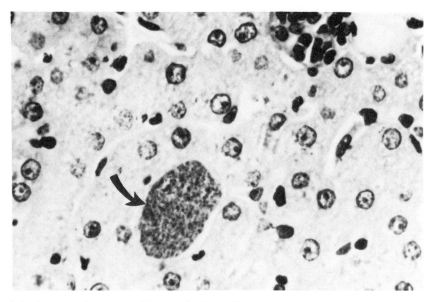

Fig. 8-2. A preerythrocytic schizont of *Plasmodium* (arrow) in liver tissue. (Photograph by Peter Diffley.)

this reasoning the vertebrate is the definitive host.[3] We should also observe that *Plasmodium* was probably derived from an ancestral coccidian whose asexual and sexual reproduction took place in the same (presumably vertebrate) host.

Vertebrate phases. When an infected mosquito takes blood from a vertebrate, she injects saliva containing tiny, elongate sporozoites into the bloodstream. The sporozoite basically is similar in morphology to that of *Eimeria* and other coccidia. It is about 10 to 15 μm long by 1.0 μm in diameter and has a pellicle composed of a thin outer membrane, a doubled inner membrane, and a layer of subpellicular microtubules. There are three polar rings. The rhoptries are long, extending to the midportion of the organism, and much of the rest of the anterior cytoplasm is taken up by the micronemes. An apparently nonfunctional cytostome is present, and there is a mitochondrion in the posterior of the sporozoite.[1]

After being injected into the bloodstream, the sporozoites quickly disappear (within an hour) from the circulating blood. Their immediate fate was a great mystery until the mid-1940s, when it was shown that they enter the parenchyma of

the liver or other internal organ, depending on the species of *Plasmodium*. This initiates a series of asexual reproductions known as the **preerythrocytic cycle,** or **primary exoerythrocytic schizogony,** often abbreviated as the **PE** or **EE** stages. Once within a hepatic cell, the parasite metamorphoses into a feeding trophozoite. The organelles of the apical complex disappear, and the trophozoite feeds on the cytoplasm of the host cell via the cytostome and, in the species in mammals, by pinocytosis.

After about a week, depending on the species, the trophozoite is mature and begins schizogony. Numerous daughter nuclei are first formed, transforming the parasite into a schizont (Fig. 8-2), also known as a **cryptozoite.** During the nuclear divisions, the nuclear membranes persist, and the microtubular spindle fibers are formed within the nucleus. The mitochondrion becomes larger during the growth of the trophozoite, forms buds, then breaks up into many mitochondria. Elements of the apical complex form subjacent to the outer membrane, and schizogony proceeds as previously described. The merozoites thus formed after cytokinesis are referred to in this EE stage as

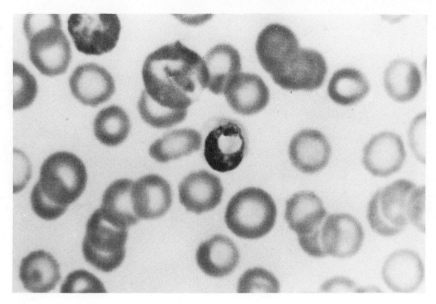

Fig. 8-3. An erythrocyte containing a ring-stage trophozoite of *Plasmodium vivax.* (Courtesy Turtox/Cambosco.)

metacryptozoites. The merozoites are much shorter than sporozoites—2.5 μm long by 1.5 μm diameter—and have small teardrop-shaped rhoptries and small, oval micronemes. The merozoites enter new liver parenchyma cells to form new schizonts and then merozoites. Cycles of EE schizogony may go on indefinitely in some species (for example, in *P. malariae*), or they may be restricted to a single generation in others (for example, in *P. falciparum*).

Eventually, merozoites leave liver cells to penetrate erythrocytes in the blood, initiating the **erythrocytic cycle.** Upon entry into an erythrocyte, the merozoite again transforms into a trophozoite. The host cytoplasm ingested by the trophozoite forms a large food vacuole, giving the young *Plasmodium* the appearance of a ring of cytoplasm with the nucleus conspicuously displayed at one edge (Fig. 8-3). The distinctiveness of the "signet-ring stage" is accentuated by the Romanovsky stains: the parasite cytoplasm is blue and the nucleus is red. As the trophozoite grows (Fig. 8-4), its food vacuoles become less noticeable by light microscopy, but pigment granules of **hemozoin** in the vacuoles may become apparent. Hemozoin is the end product of

the parasite's digestion of the host's hemoglobin. It contains the heme moiety with an attached fragment of the globin, and it is insoluble. The parasite rapidly develops into a schizont (Fig. 8-5). The stage in the erythrocytic schizogony at which the cytoplasm is coalescing around the individual nuclei, prior to cytokinesis, is called the **segmenter.** When development of the merozoites is completed, the host cell ruptures, releasing parasite metabolic wastes and residual body, including hemozoin. The metabolic wastes thus released are one factor responsible for the characteristic symptoms of malaria, though hemozoin itself is nontoxic. A great many of the merozoites are ingested and destroyed by reticuloendothelial cells and leukocytes, but even so, the number of parasitized host cells may become astronomical because erythrocytic schizogony takes only from 1 to 4 days, depending on the species.

After an indeterminate number of asexual generations, some merozoites enter erythrocytes and become **macrogamonts (macrogametocytes)** and **microgamonts (microgametocytes)** (Fig. 8-6). The size and shape of these cells are characteristic for each species; they also contain hemozoin. Gametocytogenesis may occur in the

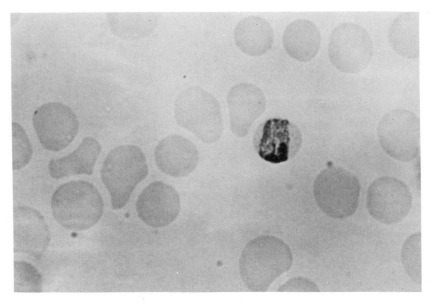

Fig. 8-4. A later trophozoite of *Plasmodium malariae* within an erythrocyte. (Photograph by Peter Diffley.)

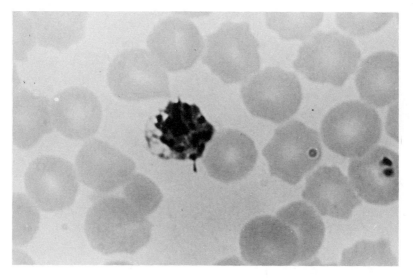

Fig. 8-5. A schizont of *Plasmodium vivax*. The 16 nuclei and some clumped pigment can be distinguished in this photograph.

liver stages as well. Unless they are ingested by a mosquito, gametocytes soon die and are phagocytized by the reticuloendothelial system.

Invertebrate stages. When erythrocytes containing gametocytes are imbibed by an unsuitable mosquito, they are digested along with the blood. However, if a suscep-

tible mosquito is the diner, the gametocytes develop into gametes. Though this development would only be in a female mosquito in nature, since only females feed on vertebrate blood, males of appropriate species can support development after experimental infection with the parasite in the laboratory. Suitable hosts for the

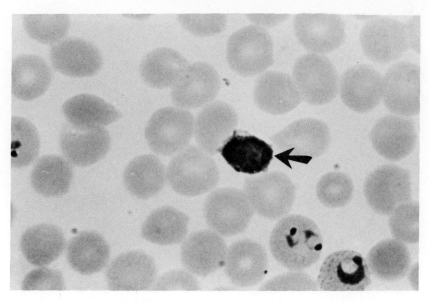

Fig. 8-6. A microgametocyte (arrow) and ring stages of *Plasmodium vivax.* Note the Schüffner's dots on the lower right-hand erythrocyte. (Photograph by Peter Diffley.)

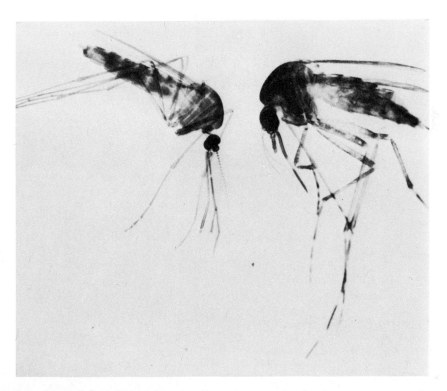

Fig. 8-7. *Anopheles quadrimaculatus,* a common mosquito in the eastern, central, and southern United States; once the most important vector of malaria in that area. (Courtesy Ann Arbor Biological Center.)

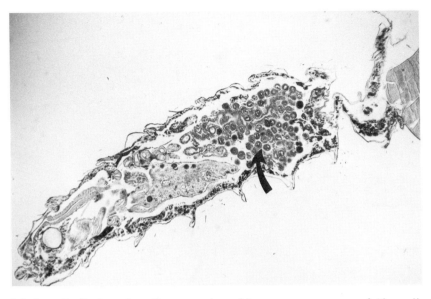

Fig. 8-8. Longitudinal section of a mosquito, with numerous oocysts of *Plasmodium* in the hemocoel (arrow). (Photograph by Warren Buss.)

Plasmodium spp. of humans are a wide variety of *Anopheles* spp. (Fig. 8-7). After release from its enclosing erythrocyte, maturation of the macrogametocyte to the macrogamete involves little obvious change other than a shift of the nucleus toward the periphery. In contrast, the microgametocyte displays a rather astonishing transformation, **exflagellation.** As the microgametocyte becomes extracellular, within 10 to 12 minutes its nucleus divides repeatedly to form six to eight daughter nuclei, each of which is associated with the elements of a developing axoneme. The doubled outer membrane of the microgametocyte becomes interrupted, the flagellar buds with their associated nuclei move peripherally between the interruptions, then continue outward covered by the outer membrane of the gametocyte. These break free and are the microgametes. Their life span is short, since they contain little more than the nuclear chromatin and the flagellum covered by a membrane. The microgamete swims about until it finds a macrogamete, which it penetrates and fertilizes. The resultant diploid zygote quickly elongates to become a motile **ookinete.** The ookinete is reminiscent of a sporozoite and merozoite in morphology. It is 10 to 12 μm in length and has polar rings and subpellicular microtubules, but it has no rhoptries or micronemes.

The ookinete penetrates the peritrophic membrane in the mosquito's gut, migrates to the hemocoel side of the gut, and begins its transformation into an oocyst. The oocyst (Fig. 8-8) is covered by an electron-dense capsule and soon extends out into the insect's hemocoel. The initial division(s) of its nucleus is reductional, meiosis taking place immediately after zygote formation as in other Sporozoa. The oocyst reorganizes internally into a number of haploid nucleated masses called **sporoblasts,** and the cytoplasm contains many ribosomes, endoplasmic reticulum, mitochondria, and other inclusions. The sporoblasts in turn divide repeatedly to form thousands of sporozoites (Fig. 8-9). These break out of the oocyst into the hemocoel and migrate throughout the mosquito's body. On contacting the salivary gland, sporozoites enter its channels and can be injected into a new host at the next feeding.

Sporozoite development takes from 10 days to 2 weeks, depending on the species of *Plasmodium* and the temperature. Once infected, a mosquito remains infective for life, capable of transmitting malaria to every susceptible vertebrate it bites.

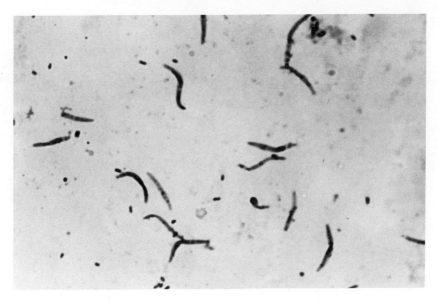

Fig. 8-9. *Plasmodium* sporozoites. (Photograph by Peter Diffley.)

Plasmodium is sometimes transmitted by means other than the bite of a mosquito. The blood cycle may be initiated by blood transfusion, by malaria therapy of certain paralytic diseases, by syringe-passed infection among drug addicts, or (rarely) by congenital infection.

Classification of Plasmodium

The genus *Plasmodium* can conveniently be divided into nine subgenera, of which three occur in mammals, four in birds, and two in lizards.

Plasmodium (Plasmodium), type species *P. (P.) malariae*, with round gametocytes and large erythrocytic schizonts. Found in primates.

Plasmodium (Laverania), type species *P. (L.) falciparum*, with elongate, crescentic gametocytes and large erythrocytic schizonts. Found in primates.

Plasmodium (Vinckeia), type species *P. (V.) bubalis*, with round gametocytes and small erythrocytic schizonts. Found in rodents, ruminants, and other mammals except primates.

Plasmodium (Giovannolaia), type species *P. (G.) circumflexum*, with elongate gametocytes and large erythrocytic schizonts. Found in birds.

Plasmodium (Huffia), type species *P. (H.) elongatum*, with elongate gametocytes and large erythrocytic

schizonts. Development mainly in the immature hemopoietic system of birds.

Plasmodium (Haemamoeba), type species *P. (H.) relictum*, with round gametocytes and large erythrocytic schizonts. Parasites of birds.

Plasmodium (Novyella), type species *P. (N.) vaughani*, with elongate or oval gametocytes and small erythrocytic schizonts. Parasites of birds.

Plasmodium (Carinia), type species *P. (C.) minasense*, with small erythrocytic schizonts. Parasites of lizards.

Plasmodium (Sauramoeba), type species *P. (S.) agamae*, with large erythrocytic schizonts. Parasites of lizards.

One hundred twenty-seven species of *Plasmodium* were discussed by Garnham.[6] A few of these are of doubtful status, but several can be separated into well-defined subspecies. Species of *Plasmodium* are not difficult to distinguish after training. Most are parasites of birds, others are in rodents, primates, reptiles, and others. Some species are very useful in laboratory studies of immunity, physiology, and so forth, such as the rodent parasite *Plasmodium berghei* and the chicken parasite *P. gallinaceum*. Still other species, normally parasitic in nonhuman primates, occasionally infect humans as zoonoses, or can be acquired by humans when infected experimentally.

Such are *P. schwetzi* of chimpanzees and gorilla, *P. eylesi* of Malayan gibbons, *P. cynomolgi*, *P. knowlesi* and *P. inui* of Oriental monkeys, *P. simium* and *P. brasilianum* of New World monkeys, and *P. shortii* of Indian and Ceylonese monkeys. The importance of these species to human medicine is for the most part unassessed; surely they are potential disease agents, at least to the individual who may be exposed under unusual circumstances. In fact, some of the foregoing (*P. simium* and *P. brasilianum*) may be conspecific with some species usually considered parasites of humans.[14] Humans are normal hosts for four species of *Plasmodium,* and these will be treated in more detail.

Plasmodium parasitic in humans

Plasmodium (Plasmodium) vivax. *P. vivax* is the cause of **benign tertian malaria,** also known as **vivax malaria** or **tertian ague.** When early Italian investigators noted the actively motile trophozoites of the organism within host corpuscles, they nicknamed it "vivace," foreshadowing the Latin name "vivax" which later was accepted as its trivial name. The designation "tertian" is based on the fact that fever paroxysms typically recur every 48 hours, and the name is derived from the ancient Roman custom of calling the day of an event the first day, 48 hours later hence being the third. The species flourishes best in temperate zones, rarely as far north as Manchuria, Siberia, Norway, and Sweden and as far south as Argentina and South Africa. Because malaria eradication campaigns have been so successful in many of the temperate areas of the world, however, the disease has practically disappeared from them. Most vivax malaria today is found in Asia; about 40% of malaria among U.S. military personnel in Vietnam resulted from *P. vivax.*[2] It is common in North Africa but drops off in tropical Africa to very low levels, partly because of a natural resistance of black people to infection with this species. About 43% of malaria in the world is caused by vivax.

Sporozoites that are 10 to 14 μm long invade cells of the liver parenchyma within a few hours after injection with the mosquito's saliva. By the seventh day the exoerythrocytic schizont is an oval body about 40 μm long, has blue-staining cytoplasm, a few large vacuoles, and lightly staining nuclei. Upon maturity, the vacuoles disappear, and about 10,000 merozoites are produced. Some of the merozoites may invade erythrocytes to begin the erythrocytic cycles, while others invade new parenchymal cells of the liver. The secondary exoerythrocytic schizonts form on the fifteenth day after infection. They resemble seventh or eighth day schizonts but are not numerous and do not produce as many merozoites. Exoerythrocytic schizogony may persist as long as 8 years, giving rise to periodic **relapses,** caused by new invasions of the red cells. Such relapses are characteristic of vivax malaria, and the patient is in normal health during the intervening periods of latency. It is believed that the relapses result from genetic differences in the original sporozoites, that is, some give rise to tissue schizonts that take much longer to mature.[2a] However, occurrence of relapses may also be related to the immune state of the host (see immunity, p. 153).

Plasmodium vivax merozoites invade only young erythrocytes, the reticulocytes, and apparently are unable to penetrate mature red cells. Soon after invasion of the erythrocyte and formation of the ring stage, the cytoplasm becomes actively ameboid, throwing out pseudopodia in all directions and fully justifying the name "vivax." Infection of the same erythrocyte with more than one trophozoite may occur, but not commonly. As the trophozoite grows, the red cell enlarges, loses its pink color, and develops a peculiar stippling known as **Schüffner's dots.** These dots are visible only after Romanovsky staining, and they may obscure details of the parasite within. Ring stages occupy about one third to one half the erythrocyte, and the trophozoite occupies about two thirds of the red cell after 24 hours. The vacuole disappears, the organism becomes more sluggish, and hemozoin granules accumulate as the trophozoite grows. By 36 to 42 hours after infection, nuclear division begins and is repeated four times, yielding 16 nuclei in the mature schizont. Fewer nuclei may be produced, especially in older infections or those interfered with by host immunity or chemotherapy. Once schizogony has begun, the pigment granules accumulate into two or three masses in the parasite, ultimately to be left in the residual body and engulfed by the host's reticuloendothelial system. The rounded merozoites, about 1.5 μm in diameter, immediately attack new erythrocytes. Erythrocytic schizogony takes about 48 hours, although early in the disease there are usually two populations, each maturing on alternate days, resulting in a daily, or **quotidian,** periodicity (see pathology, p. 151).

Some merozoites develop into gametocytes rather than into schizonts. The factors determining the fate of a given merozoite are not known, but since gametocytes have been found as early as the first day of parasitemia in rare instances, it is possible for exoerythrocytic merozoites to produce gametocytes. The stained macrogametocyte has bright blue, rounded cytoplasm and a nucleus that is compact and dark-staining by Romanovsky methods. Dark brown hemozoin granules are abundant throughout the cytoplasm. The mature macrogametocyte fills most of the enlarged erythrocyte and measures about 10 μm wide. The rounded microgametocyte is more gray than blue with Romanovsky stains. Its nucleus is more diffuse and is much larger, sometimes half the diameter of the entire parasite, and the pigment granules are coarser and more unevenly distributed than in the macrogametocyte. The mature microgametocyte is smaller than the macrogametocyte and usually does not fill the erythrocyte.

Gametocytes take 4 days to mature, twice that of schizonts. Macrogametocytes often outnumber microgametocytes by two to one. A single host cell may contain both a gametocyte and a schizont.

Formation of zygote, ookinete, and oocyst are as described above. The oocyst may reach a size of 50 μm and produce up to 10,000 sporozoites. The mature oocyst ruptures after 9 days at 25° C. If the ambient temperature is too high or too low, the oocyst blackens with pigment and degenerates, a phenomenon noted by Ross. Too many developing oocysts kill the mosquito before the sporozoites are developed.

Plasmodium (Laverania) falciparum. Malaria known as **malignant tertian, subtertian,** or **estivoautumnal (E-A)** is caused by *P. falciparum,* the most virulent of the *Plasmodium* species in humans. It was nearly cosmopolitan at one time, with a concentration in the tropics and subtropics. It still extends into the temperate zone in some areas, although it has been eradicated in the United States, the Balkans, and around the Mediterranean. Nevertheless, falciparum malaria reigns supreme as the greatest killer of humanity in the tropical zones of the world today, accounting for about 50% of all malaria.

Among the many cases studied by Alfonse Laveran, those people suffering from "malignant tertian malaria" interested him the most. He had long noticed a distinct darkening of the gray matter of the brain and abundant pigment in other tissues of his deceased patients. When, in 1880, he saw crescent-shaped bodies in the blood and watched them exflagellate, he knew he had found living parasites. The confusion that surrounded the correct name for this species was great, until 1954 when the International Commission of Zoological Nomenclature validated the trivial name *falciparum.*

Malignant tertian malaria is usually blamed for the decline of the ancient Greek civilization, the halting of Alexander the Great's progress to the East, and the destruction of some of the Crusades. In more modern times, the Macedonian campaign of World War I was destroyed by falciparum malaria, and the disease caused more mortality than battles in some theaters of World War II.

As in other species, the exoerythrocytic schizont grows in liver cells. It is more irregularly shaped than that of *P. vivax,* with projections extending in all directions by the fifth day. The schizont ruptures in about 5½ days, releasing about 30,000 merozoites. There seems to be no second exoerythrocytic cycle, and true relapses do not occur. It should be noted, however, that recrudescences of the disease may follow remissions up to 1 year, occasionally 2 or 3, after initial infection, apparently because of small populations of the parasites remaining in the red cells.

Merozoites can invade erythrocytes of any age, including reticulocytes; therefore, falciparum malaria is characterized by much higher parasitemias than the other

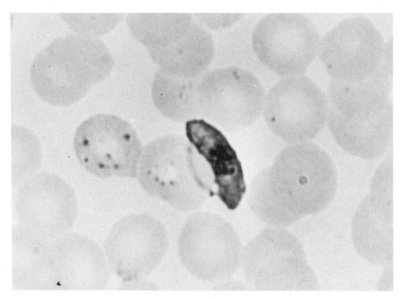

Fig. 8-10. A typical, crescent-shaped macrogametocyte of *Plasmodium falciparum.*

types. For reasons not yet understood, parasitized red cells tend to remain in the deep tissues, particularly spleen and bone marrow, during schizogony. When gametocytes develop, the erythrocytes containing them return again to the peripheral circulation; hence, one usually observes only ring stages and/or gametocytes in blood smears from patients with falciparum malaria. If schizogony is well synchronized, parasites may be practically absent from peripheral blood toward the end of the 48-hour cycle.

The ring stage trophozoite is the smallest of any *Plasmodium* of humans, about 1.2 μm. The following are observed more frequently in *P. falciparum* than in other species: (1) **accole** or **appliqué** forms, (2) ring stages with two chromatin dots (nuclei), and (3) infection of a single erythrocyte with more than one parasite. "Appliqué" or "accole" forms are ring stages that lie very close to the surface of the red cell, appearing to be "applied" to the cell. The frequency of multiple infections in the same cell has led some parasitologists to believe that the ring stages divided and that the binucleate rings are division stages. As it grows, the protozoan extends wispy pseudopodia, but it is never as active as *P. vivax*. The infected erythrocyte de-

velops irregular blotches on its surface known as **Maurer's clefts.** These are much larger than the fine Schüffner's dots found in *P. vivax* infections. The exact cause of Maurer's clefts is unknown, but they appear to attend degenerative changes within the host cell.

The mature schizont is less symmetrical than those of the other species infecting humans. It develops eight to 32 merozoites, with 16 being the usual number. In contrast to the normal situation, schizonts may be fairly common in peripheral blood in some geographic areas. This may reflect strain differences. The erythrocytic cycle takes 48 hours, but the periodicity is not as marked as in *P. vivax,* and it may vary considerably with the strain of parasite. Very heavy parasitemias may occur, with over 65% of the erythrocytes containing parasites; a density of 25% is usually fatal. Two or three parasites per milliliter of blood may be sufficient to cause disease symptoms.

In *P. vivax* the gametocytes may appear in the peripheral blood almost at the same time as the trophozoites, but in *P. falciparum* the sexual stages take nearly 10 days to develop and then appear in large numbers. They develop in the blood spaces of the spleen and bone marrow, first assuming bizarre, irregular shapes, then becoming round, and finally changing into the crescent shape so distinctive of the species (Fig. 8-10). The mature microgametocyte is 9 to 11 μm long, has blunt ends, and a diffuse nucleus extending over half the length of the organism. Hemozoin granules cluster in the nuclear zone. The cell stains light blue to pinkish with Romanovski stains. The macrogametocyte is more slender and has slightly pointed ends. It is 12 to 14 μm long, stains a darker blue, and has a more compact nucleus. Pigment granules also cluster around the nucleus of the female cell. This differs from *P. vivax,* where pigment is diffuse throughout the cytoplasm.

Plasmodium (Plasmodium) malariae. Quartan malaria, with paroxysms every 72 hours, is caused by *Plasmodium malariae.* It was recognized by the early Greeks because the timing of the fevers differed from the tertian malarias. Although Laveran saw and even illustrated the characteristic schizonts of this parasite, he refused to believe it was different from *P. falciparum.* In 1885 Golgi differentiated the tertian and quartan fevers and gave a very accurate description of what is now known as *P. malariae.*

Plasmodium malariae is a cosmopolitan parasite but does not have a continuous distribution anywhere. It is common in many regions of tropical Africa, Burma, parts of India, Sri Lanka (Ceylon), Malaya, Java, New Guinea, and Europe. It is also distributed in the New World, including Jamaica, Guadeloupe, British Guiana, Brazil, Panama, and at one time the United States. The peculiar distribution of this parasite has never been satisfactorily explained. Two likely, but opposite, theories are that either it was recently a parasite of simian primates, and with the decline of simian populations it too is in the decline; or that it was originally a parasite of ancient human populations and is declining with the improvement and migration of peoples. It may be the only species of human malaria that also regularly lives in wild animals. Chimpanzees are infected at about the same rate as humans, but are unimportant as reservoirs, for they do not live side-by-side with people. Some workers believe that *P. brasilianum* is really

P. malariae in New World monkeys.[14] This species accounts for about 7% of malaria in the world.

Exoerythrocytic schizogony is completed in 13 to 16 days. Nothing is known of secondary exoerythrocytic stages in *P. malariae,* but since relapses can occur for up to 53 years, successive liver cycles are probable.[7]

Erythrocytic forms build up slowly in the blood; the characteristic symptoms of the disease may appear before it is possible to find the parasites in blood smears. The ring forms are less ameboid than those of *P. vivax,* and the cytoplasm is somewhat thicker. Rings often retain their shape for up to 48 hours, finally transforming into an elongate "band form," which begins to collect pigment along one edge (Fig. 8-4). The nucleus divides into six to 12 merozoites at 72 hours. The segmenter is strikingly symmetrical and is called a "rosette," or "daisy-head." Parasitemia is characteristically low, with one parasite per 20,000 red cells representing a high figure for this species. This low density is accounted for by the fact that the merozoites apparently can invade only aging erythrocytes, which are soon to be removed from circulation by the normal process of blood destruction.

Gametocytes probably develop in the internal organs, for immature forms are rare in peripheral blood. They are slow to develop in sporozoite-induced infections. The microgametocyte fills the entire host cell. It has a nucleus that occupies at least half the volume of the parasite. The remaining cytoplasm stains a grayish green color, mainly because of the diffuse hemozoin granules within it. Macrogametocytes are nearly impossible to identify, appearing identical to large, uninuclear asexual stages.

Plasmodium (Plasmodium) ovale. This species causes **ovale** or **mild tertian malaria** and is the rarest of the four malaria parasites of humans. It is confined mainly to the tropics, although it has been reported from Europe and the United States. While common on the west coast of Africa, which may be its original home, the species is scarce in central Africa and present but not abundant in east Africa. It is known also in India, the Philippine Islands, New

Guinea, and Vietnam. *P. ovale* is difficult to diagnose because of its similarity to *P. vivax.*

The youngest ring stages have a large, round nucleus and a rather small vacuole that disappears early. The mature schizont is oval, or spheroid, and is about half the size of the host cell. Eight merozoites are usually formed, with a range of four to 16. Schüffner's dots appear early in the infected blood cells. They are very numerous, larger than those in *P. vivax* infections, and stain a brighter red color.

Gametocytes of *P. ovale* take longer to appear in the blood than do those of other species. Three weeks after infection they are numerous enough to infect mosquitoes regularly. Gametocytes of both sexes are about 9 μm in diameter and contain pigment granules in concentric rings or irregular nodes. The macrogametocyte has purplish cytoplasm and a small nucleus, usually on one side. The stained microgametocyte has bluer cytoplasm and a large nucleus nearly half the size of the parasite.

Malaria: the disease

Certain disease aspects of *Plasmodium* have been mentioned in the preceding pages; the following is a brief consideration of the subject, particularly in relation to pathogenesis and public health. We urge the reader to consult further references for more complete treatment.[17-19,24]

Diagnosis. Diagnosis depends to some extent on the clinical manifestations of the disease, but most important is demonstration of the parasites in stained smears of peripheral blood. Technical details can be found in many texts and laboratory manuals of medical parasitology. Characteristic morphology of the respective species has been noted, but the most useful criteria for differential diagnosis are summarized in Table 1.

Pathology. The major clinical manifestations of malaria may be attributed to two general factors; (1) the host inflammatory response, which produces the characteristic chills and fever, as well as other related phenomena, and (2) anemia, arising from the enormous destruction of red blood cells. Severity of the disease is correlated with the species producing it—falciparum malaria being most serious and vivax and ovale being least dangerous.

The main causes of the anemia are hemolysis of the erythrocytes, both parasitized and nonparasitized, inability of the body to recycle the iron bound in the insoluble hemozoin, and preferential attack of reticulocytes, particularly in vivax malaria. Why such large numbers of nonparasitized red cells are lysed is still not understood, but autoimmune processes are implicated, at least in later stages. Hemolysis of erythrocytes leads to an increase in blood bilirubin, a breakdown product of hemoglobin. When excretion cannot keep up with formation of bilirubin, jaundice yellows the skin. The hemozoin is taken up by circulating leukocytes and is deposited

Table 1. Criteria for differential diagnosis of *Plasmodium* spp. in humans*

P. vivax	P. falciparum	P. ovale	P. malariae
Trophozoites ameboid	Larger trophozoites and schizonts not usually in peripheral blood	Trophozoites not ameboid	Trophozoites often band form
Segmenters form about 16 merozoites		Segmenters usually form eight merozoites	Segmenters usually form eight merozoites
Host cell enlarged, decolorized, frequently with Schüffner's dots	Ring stages small, often with two chromatin dots	Host cells somewhat enlarged, sometimes decolorized with oval distortion, Schüffner's dots heavy	Host cells not enlarged, decolorized, or stippled
Parasites relatively large	Appliqué forms frequent		Hemozoin granules large, abundant
	Multiple infections frequent		
	Gametocytes crescent-shaped		

*Modified from Russell, P. F., and others. 1946. Practical malariology. W. B. Saunders Co., Philadelphia.

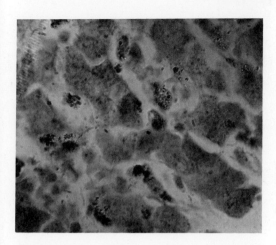

Fig. 8-11. Liver tissue with numerous deposits of malarial pigment.

in the reticuloendothelial system. In severe cases the viscera, especially the liver, spleen, and brain, become blackish or slaty in color as the result of pigment deposition (Fig. 8-11).

Maegraith and Fletcher have held that "The pathophysiological pattern of malaria is essentially inflammatory and nonspecific in nature. . . ."[19] Certainly, fever is a common, nonspecific reaction of the body to infection, functioning at least in part to increase the rate of metabolic reactions important in host defenses. Fever in malaria is correlated with the maturation of a generation of merozoites and the rupture of the red blood cells that contain them. It is widely believed that fever is stimulated by the excretory products of the parasites, released when the erythrocytes lyse, but the exact nature of such substances is not known. There is evidence of production of cytotoxic factors by the parasites; oxidative phosphorylation and respiration are inhibited in mitochondria from infected animals, and damage to liver cells can be observed on the ultrastructural level.[19]

A few days before the first paroxysm, the patient may feel malaise, muscle pain, headache, loss of appetite, and slight fever; or the first paroxysm may occur abruptly, without any prior symptoms. A typical attack of benign tertian or quartan malaria begins with a feeling of intense cold as the hypothalamus, the body's ther-

mostat, is activated, and the temperature then rises rapidly to 104° to 106° F. The teeth chatter, and the bed may rattle from the victim's shivering. Nausea and vomiting are usual. The hot stage begins after $^{1}/_{2}$ to 1 hour, with intense headache and feeling of intense heat. Often a mild delirium stage lasts for several hours. As copious perspiration signals the end of the hot stage, the temperature drops back to normal within 2 to 3 hours, and the entire paroxysm is over within 8 to 12 hours. The person may sleep for a while after an episode and feel fairly well until the next paroxysm. The foregoing time periods for the stages are usually somewhat shorter in quartan malaria, and the paroxysms recur every 72 hours. In vivax malaria the periodicity is often quotidian early in the infection, since two populations of merozoites usually mature on alternate days. "Double" and "triple" quartan infections also are known. Only after one or more groups drops out does the fever become tertian or quartan, and the patient experiences the classical good and bad days.

Because the synchrony in falciparum malaria is much less marked, the onset is often more gradual, and the hot stage is extended. The fever episodes may be continuous or fluctuating, but the patient does not feel well between paroxysms, as in vivax and quartan malaria. In cases in which some synchrony develops, each episode lasts 20 to 36 hours, rather than 8 to 12, accompanied by much nausea, vomiting, and delirium. Concurrent infections with *P. vivax* and *P. falciparum* are not uncommon. Falciparum malaria is always serious, and sometimes it may abruptly produce a **pernicious** or **malignant** form of the disease that may be rapidly fatal. Several clinical types are recognized: bilious remittent fever, cerebral malaria, and algid malaria.

Bilious remittent fever is the most common, and least dangerous, pernicious malaria. It is marked by severe nausea, profuse and continuous vomiting, and sometimes hemorrhage from the stomach. Jaundice usually appears about the second day, which is later than in blackwater fever. The urine contains bile pigment, and the fever tends to be high and remittent. **Cerebral malaria** may be gradual in onset, but it is frequently sudden; a progressive headache may be followed by coma. An uncontrollable rise in temperature to above 108° F may occur. Sudden onset may be marked by mania and psychotic symptoms or convulsions, especially in children. Death may ensue within a matter of hours. Initial stages of cerebral malaria have sometimes been mistaken for acute alcoholism, usually with disastrous consequences. **Algid malaria** is a condition analogous to cerebral malaria, except that the gut and other abdominal viscera are involved. The skin is cold and clammy, but internal temperature is high. Extreme exhaustion and loss of consciousness usually occur. Two types of algid malaria are **gastric,** with persistent vomiting, and **dysenteric,** with bloody, diarrheic stools, and the blood in the stools containing enormous numbers of parasites. The direct cause of the pernicious malarias, especially cerebral and algid, has traditionally been cited as a "plugging" of the capillaries in the affected organs by clots (Fig. 8-12). However, it seems more likely that the conditions are caused by a manifestation of the inflammatory response, leading to circulatory stasis in the affected areas.[19] It appears that vascular permeability increases, perhaps as the result of the release of kinins, that protein and accompanying water are lost from the blood to the tissues, and that blood ceases to flow because of the local concentration of plasma and erythrocytes. Consequently, the area served by that capillary bed becomes anoxic. This theory is supported by the fact that administration of chloroquine or quinine to the comatose falciparum patient is followed in a few hours by release from coma (if such release is to occur at all)—much more rapidly than any possible effect of the drug on the parasite itself. Chloroquine and quinine, in addition to their anti-*Plasmodium* properties, are anti-inflammatory agents; a similar effect can be achieved by administration of cortisone.

General mortality of the various types of pernicious malaria is 25% to 50%.

Blackwater fever is another grave condition associated with falciparum malaria, but its clinical picture is distinct from the foregoing. It is an acute, massive lysis of erythrocytes, marked by high levels of free

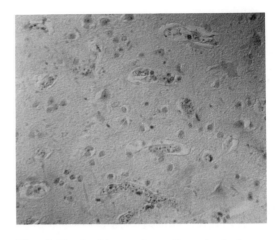

Fig. 8-12. A thin section of cerebral tissue, demonstrating capillaries plugged with erythrocytes infected by *Plasmodium falciparum.*

hemoglobin and its breakdown products in the blood and urine and by renal insufficiency. Because of the presence of hemoglobin and its products in the urine, the fluid is quite dark, hence the name of the condition. There is a prostrating fever, jaundice, and persistent vomiting. Renal failure is usually the immediate cause of death. Damage to the kidney is now thought to result from renal anoxia, reducing efficiency of the glomerular filtration and tubular resorption. The massive hemolysis is not directly attributable to the parasites; the organisms frequently cannot be demonstrated. However, the condition is almost always associated with areas of *P. falciparum* hyperendemicity, found in persons with prior falciparum malaria, and very frequently with irregular or inadequate treatment for the infection. Inadequate suppressive or therapeutic doses of quinine most often have been implicated, but many cases have been reported following treatment with quinacrine and pamaquine and can occur in persons who have not been treated at all. It is now believed that blackwater fever is an autoimmune phenomenon and is triggered by some stimulus that results in release of large amounts of antibodies, which act as hemolysins into the circulation. Mortality is 20% to 50%.

Therapeutic malaria. Intentionally induced malaria can be of therapeutic value in certain paralytic diseases, especially neurosyphilis. Infections can be induced either by sporozoites from the bite of an infected mosquito or by merozoites in blood from a malaria patient. If the infection is merozoite induced, no exoerythrocytic cycle precedes the first paroxysm; instead the parasites continue with erythrocytic schizogony. The mechanism of the therapeutic effect is not known but may be related to the immune response, the antibodies synthesized against the *Plasmodium* having some cross-reaction against the spirochete. It is noteworthy that malaria is commonly associated with positive Wassermann and Kahn reactions, serological tests used in diagnosis of syphilis.[17]

Immunity. Despite the fact that much of the disease results from the inflammatory and immune responses of the host, host defenses are vital in limiting the infection. One vivax segmenter producing 24 merozoites every 48 hours would give rise to 4.59 billion parasites within 14 days, and the host would soon be destroyed if the organisms continued reproducing unchecked.[14] The development of some protective immunity is evident in malaria, and we will consider only briefly some of the practical effects. Relapses and recrudescences may be associated with lowered antibody titers or increased ability of the parasite to deal with the antibody, but they may depend on genetic differences in schizont populations. Symptoms in a relapse are usually less severe than in the primary attack, but the parasitemia is higher. After the primary attack and between relapses, the patient may have a **tolerance** to the effects of the organisms and in fact may have as high a circulating parasitemia as during the primary attack, though remaining asymptomatic. Such tolerant carriers are very important in the epidemiology of the disease. The protective immunity is primarily a premunition, that is, resistance to superinfection. It is effective only as long as a small, residual population of parasites are present; if the person is completely cured, susceptibility returns. Thus, in highly endemic areas, infants are protected by maternal antibodies, and young children are at greatest risk after weaning. If the child survives its first attack, then its immunity will be continuously stimu-

lated by the bites of infected mosquitoes as long as it lives in the malarious area. Nonimmune adults are highly susceptible. Immunity is species-specific and to some degree strain-specific, so that a person may risk a new infection by migrating from one malarious area to another. Falciparum malaria is unmitigated in its severity in a person who is immune to vivax malaria.

Black people are much less susceptible to vivax malaria than are white people, and falciparum malaria in blacks is somewhat less severe. That this is a genetic resistance distinct from acquired immunity is shown by the fact that black people living in the United States still have some degree of resistance. Among U.S. Army troops in Vietnam, black soldiers were infected by falciparum less often than white soldiers, and their attacks were less severe.[9] Other factors that can contribute to genetic resistance are certain heritable anemias: sickle cell, favism, and thalassemia. Though these conditions are of negative selective value in themselves, they have been selected for in certain populations since they confer resistance to falciparum malaria. The most famous of these is **sickle cell anemia.** In persons homozygous for this trait, a glutamic acid residue in the amino acid sequence of hemoglobin is replaced by a valine, interfering with the conformation of the hemoglobin and oxygen-carrying capacity of the erythrocytes. These people usually die before the age of 30. In heterozygotes some of the hemoglobin is normal and they can live relatively normal lives, but the presence of the abnormal hemoglobin inhibits growth and development of *P. falciparum* in their erythrocytes. The selective pressure of malaria in Africa has led to maintenance of this otherwise undesirable gene in the population. This legacy has very unfortunate consequences when the people are no longer threatened by malaria, as in the United States, where one in ten Americans of African ancestry is heterozygous for the sickle cell gene, and one in 400 is homozygous.

Epidemiology, control, and treatment. In light of the prevalence and seriousness of the disease, epidemiology and control are extremely important, and thorough consideration is far beyond the scope of this book. Some aspects of these have been

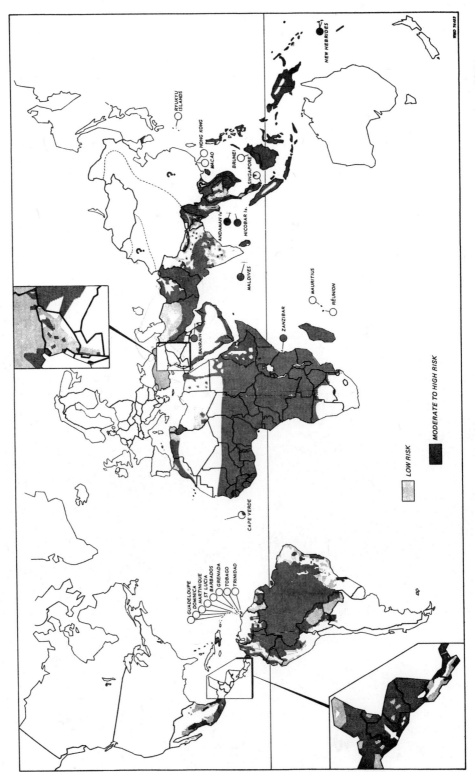

Fig. 8-13. Areas of risk for malaria transmission—December, 1974. (WHO Weekly Epidemiological Record no. 45, 1975.)

touched on in the preceding pages, and the following will give the reader additional insight into the problem involved.

A variety of interrelated factors contribute to the level of transmission of the disease in a given area (Fig. 8-13). The following are the most important (modified from Mackie, Hunter, and Worth[17]):

1. The reservoir—the prevalence of the infection in humans, and in some cases other primates, with high enough parasitemias to infect mosquitos. This would include persons with symptomatic disease and tolerant individuals.
2. The vector—suitability of the local anophelines as hosts, their breeding, flight, and resting behavior, feeding preferences, and abundance.
3. The new host—availability of nonimmune hosts.
4. Local climatic conditions.
5. Local geographic and hydrographic conditions and human activities that determine availability of mosquito breeding areas.

All of these factors must be thoroughly studied and understood before undertaking a malaria control program with any hope of success.

Of the approximately 200 species of *Anopheles*, some are more suitable hosts for *Plasmodium* than others. Of those that are good hosts, some prefer animal blood other than human; therefore, transmission may be influenced by the proximity with which humans live to other animals. The preferred breeding and resting places are very important. Some species breed only in fresh water, others in brackish; some like standing water around human habitations, such as puddles, or trash that collects water, such as bottles and broken coconut shells. Water, vegetation, and amount of shade are important, as are whether the species enters dwellings and rests there after feeding and whether the species flies some distance from breeding areas. *Anopheles* spp. exhibit an astonishing variety of such preferences; two specific examples can be cited for illustration. *Anopheles darlingi* is the most dangerous vector in South America, extending from Venezuela to southern Brazil, breeding in shady, fresh water among debris and vegetation. It in-

vades houses and prefers human blood. *Anopheles bellator* is an important vector in cocoa-growing areas of Trinidad and coastal states of southern Brazil, breeding in partial shade in the "vases" of epiphytic bromeliads (those plants that grow attached to trees and collect water in the center of their leaf rosettes). It prefers humans but enters dwellings only occasionally and returns to the forest. The importance of thorough investigation of such factors is demonstrated by cases in which swamps have been flooded with sea water to destroy the breeding habitat of the species, only to create extensive breeding areas for a brackish water species that turned out to be just as effective a vector.

Valuable actions in mosquito control include destruction of breeding places when possible or practical, introduction of mosquito predators such as the mosquito-eating fish, *Gambusia*, and judicious use of insecticides. The efficacy and economy of DDT have been a boon to such efforts in underdeveloped countries. Though we now seem to be more aware of the supposed environmental dangers of DDT, we consider these dangers preferable and minor compared to the miseries of malaria. Unfortunately, reports of DDT-resistant strains of *Anopheles* are increasing, and this phase of the battle will become more difficult in coming years. For exterminating susceptible *Anopheles* that enter dwellings and rest there after feeding, spraying the insides of houses with residual insecticides can be effective and cheap, without incurring any environmental penalty.

Appropriate drug treatment of persons with the disease, as well as prophylactic drug treatment of newcomers to malarious areas, are integral parts of malaria control. The Chinese found the first antimalarial drug, Ch'ang shan, an herb containing an active alkaloid, but the Europeans were almost medically powerless until **quinine** was discovered in the nineteenth century. Extracts of bark from Peruvian trees had been used with varying success to treat malaria, but alkaloids from the bark of the tree *Cinchona ledgeriana* proved to be dependable and effective. The most widely used of these alkaloids has been quinine. Its mode of action is still not known, but it does disrupt erythrocytic schizogony; it has no effect on sporozoites, exoerythrocytic forms, or falciparum gametocytes and only little effect on gametocytes of the other species. Only two synthetic antimalarials, pamaquine and quinacrine, were discovered prior to World War II. But Japanese capture of cinchona plantations early in the war created a severe quinine shortage in the United States, stimulating a burst of investigation that produced a number of important drugs. The most successful were **chloroquine** and a related compound, **amodiaquin**. These remain the drugs of choice in treatment of vivax, malariae, and ovale malarias, for they destroy gametocytes as well as erythrocytic schizonts. Chloroquine and amodiaquin do not affect falciparum gametocytes. These drugs in combination with others, such as primaquin (to kill the schizonts), seemed to be effective in malaria control until chloroquine-resistant falciparum arose in the 1960s. Chlorquine resistance proved to be a major problem of the armed forces in Vietnam, particularly when some additional quinine resistance was shown. Recalcitrant infections had to be treated with intravenous quinine and a combination of other drugs.[2,21] In light of the multidrug resistance in various strains of *P. falciparum*, it is clear that the search for satisfactory malaria treatments must continue; perhaps the answer lies in the development of vaccines.

In the 1950s it was widely thought that because technical knowledge was adequate, a modicum of effort and money could achieve the eradication of malaria from large areas of the globe. Its scourge would be only history. Such views were naively optimistic. Not only did we not anticipate insecticide-resistant *Anopheles*, drug-resistant *Plasmodium*, and animal reservoirs, insufficient account was taken of the enormous logistical problems of control in wilderness areas and of dealing with primitive peoples, nor were the disruptive effects of wars and political upheavals on control programs considered. Malaria will be with us for a long time, probably as long as there are people.

Metabolism of Plasmodium

Metabolic studies on malaria parasites are of great importance and interest because of their practical yield in terms of drug discovery, design, and use. In spite of much effort and a large literature, many gaps still frustrate our knowledge because the metabolism of these intracellular parasites is so difficult to study. One must use infected blood, parasitized erythrocytes, or parasites freed from their host cells, each of which has serious drawbacks. We cannot assume that any abnormality in the metabolism of infected blood results from metabolism of *Plasmodium*—noninfected cells in the infected blood may have altered normal metabolic parameters.[4] Methods used to separate infected from noninfected cells or to release the parasites from the cells may render the material unsuitable for certain metabolic studies.

Selected features of *Plasmodium* metabolism follow, and we direct the reader to reviews by Moulder,[22] Fulton,[5] Fletcher and Maegraith,[4] Thompson and Werbel,[27] and Warhurst[28] for more information.

Energy metabolism. The presence and importance of glycolysis in the degradation of glucose by *Plasmodium* are well established, although subsequent steps are quite unclear. This is complicated by the fact that malaria species from birds have recognizable mitochondria, while unequivocal mitochondria have been demonstrated in very few species from mammals.[26] The bird plasmodia apparently do have a functional tricarboxylic acid cycle, but the existence of the complete cycle in the erythrocytic stages of the mammalian parasites is doubtful. Membranous structures in some of the mammalian species may represent mitochondria because of certain mitochondrial enzymes demonstrated in them cytochemically (NADH- and NADPH-dehydrogenases and cytochrome oxidase). Interestingly, the spororgonic stages of these organisms in the mosquito possess prominent, cristate mitochondria, reflecting perhaps a developmental change in metabolic pattern analogous to that observed in trypanosomes.[11]

The erythrocytic forms of *Plasmodium* appear to be facultative anaerobes, consuming oxygen when it is available, although the bird plasmodia depend heavily on glycolysis for energy. They convert four to six molecules of glucose to lactate for every one they oxidize completely. A limiting factor may be the parasite's inability to synthesize coenzyme A, which it must obtain from its host, this cofactor being necessary to introduce the two-carbon fragment into the tricarboxylic acid cycle (see p. 45). Supplies of CoA in the mammalian erythrocyte may be even more limited and may impose restrictions on any CoA-dependent reaction. The fate of the oxygen consumed has yet to be resolved; a classical electron transport system has not been demonstrated. The organisms can be cultivated in vitro in erythrocytes at very low oxygen concentrations, though cultivation longer than one or two schizogonic cycles has not been accomplished at this writing. Neither are all of the breakdown products of glucose in glycolysis known; only one mole of lactate, rather than two as would be expected (see Fig. 3-13), is produced from one mole of glucose. Both bird and mammal plasmodia "fix" CO_2 into phosphoenolpyruvate as do numerous other parasites (see discussion of energy metabolism, Fig. 20-35). In plasmodia the CO_2-fixation reaction can be catalyzed by either phosphoenolpyruvate carboxykinase or phosphoenolpyruvate carboxylase. Chloroquine and quinine inhibit both of the enzymes, possibly accounting for the antimalarial activity of these drugs. The significance of the CO_2-fixation is not clearly understood; it may be to reoxidise NADH produced in glycolysis, or its reactions may function to maintain levels of intermediates for use in other cycles.

The **pentose phosphate pathway** or **hexose monophosphate shunt** is an important and interesting metabolic pathway in *Plasmodium*. This path has several known functions in various systems, and its importance to plasmodia is probably twofold: to furnish pentoses from hexoses for use in synthesis of nucleic acids (but note that *Plasmodium* apparently lacks a full complement of enzymes for nucleic acid synthesis, to be discussed further), and to provide reducing power in the form of NADPH. The first step in the path is the dehydrogenation of glucose-6-phosphate to 6-phosphogluconate by the enzyme glucose-6-phosphate dehydrogenase (G6PDH), and the next reaction is oxidation and decarboxylation of the 6-phosphogluconate to D-ribose-5-phosphate (a pentose) by 6-phosphogluconate dehydrogenase (6PGDH). Present evidence indicates that the plasmodia are entirely dependent on G6PDH and possibly 6PGDH and the entire pathway from the host cell.[4] This dependency becomes even more interesting when it is observed that persons with a genetic deficiency in erythrocytic G6PDH, or **favism,** are more resistant to malaria. Favism is a sex-linked trait in which ingestion of various substances such as aspirin, the antimalarial drug primaquine, sulfonamides, or the broad bean *Vicia favia* brings on a hemolytic crisis in the female homozygote or male hemizygote. The gene is relatively frequent in black people and some Mediterranean white people.[15] As the trait is expressed as a mosaic, even heterozygotes have some red cells deficient in G6PDH. Therefore, all conditions, heterozygous, homozygous, and hemizygous, are protected to some extent against *P. falciparum.*[16]

Digestive metabolism. That the parasites digest host hemoglobin, leaving the iron-containing residue with a peptide attached (hemozoin), deserves further comment. The plasmodia depend heavily on this protein source; the trophozoites substantially reduce the hemoglobin content of the erythrocyte. The mechanism of ingestion is by the cytostome, by phagocytosis, or by pinocytosis. The cytostome is the primary means of ingestion by the bird plasmodia, and it functions in at least some mammalian malaria parasites.[26] Digestion occurs within food vacuoles, and the action of quinine and chloroquine seems to interfere with digestion in the vacuole or at the vacuole membrane or perhaps, to affect the way the parasites deal with the pigment.[28]

Sickle-cell hemoglobin (HbS) is difficult for the parasite to ingest. Substitution of the charged glutamic acid for the neutral valine makes HbS less soluble in water, and solutions of it become quite viscous. The increase in viscosity could interfere with phagotrophy or pinocytosis, making the individual with HbS more resistant to infection.[22] An alternative hypothesis is that the reticuloendothelial system "recognizes" the cells with HbS and clears them from the circulation more readily, thus destroying and deleting from the parasite population all those organisms in HbS cells (see Honigberg, 1967).

Synthetic metabolism. As a specialized parasite, *Plasmodium* appears to depend on its host cell for a variety of molecules other than the strictly nutritional ones. Specific requirements for maintenance of the parasites free of host cells are pyruvate, malate, NAD, ATP, coenzyme A, and folinic acid (see Honigberg, 1967). The inability of the organisms to synthesize CoA has been mentioned. They are unable to synthesize the purine ring de novo, thus requiring an exogenous source of purines for DNA and RNA synthesis. The purine source seems to be hypoxanthine "salvaged" from the normal purine catabolism of the host cell (see Manandhar and van Dyke, 1975).

Several aspects of synthetic metabolism in *Plasmodium* have offered opportunites for attack with antimalarial drugs. Though plasmodia have cytoplasmic ribosomes of the eukaryotic type, several antibiotics that specifically inhibit prokaryotic (and mitochondrial) protein synthesis, for example, tetracycline and tetracycline derivatives, have a considerable (but still unexplained) antimalarial potency. Antibiotics have only recently been used extensively in malaria therapy because they are effective less rapidly than convential antimalarials and because of apprehensions relative to development of resistant bacteria. However, antibiotics in conjunction with quinine are being used on chloroquine-resistant falciparum.

Tetrahydrofolate is a cofactor that is very important in the transfer of one-carbon groups in various biosynthetic pathways in both prokaryotes and eukaryotes. Mammals require a precursor form, **folic acid**, as a vitamin, and dietary deficiency in this vitamin inhibits growth and produces various forms of anemia, particularly because of impaired synthesis of purines and the pyrimidine thymine. In contrast, *Plasmodium* (in common with bacteria) synthesizes tetrahydrofolate from simpler precursors, including *p*-aminobenzoic acid, glutamic acid, and a pteridine; the organisms are apparently unable to assimilate folic acid. Classes of compounds known as **sulfones** and **sulfonamides** block incorporation of *p*-aminobenzoic acid, and some of these (for example, dapsone and sulfadiazine) are effective antimalarials. In both the mammalian pathway and the plasmodial-bacterial pathway, an intermediate product is dihydrofolate, which must be reduced to tetrahydrofolate by the enzyme **dihydrofolate reductase**. Also, this enzyme is necessary for tetrahydrofolate regeneration from dihydrofolate, which is produced in a vital reaction for which tetrahydrofolate is a cofactor: thymidylic acid synthesis. Thus, the enzyme is vital to both parasite and host, but fortunately, the dihydrofolate reductases from the two sources are quite different in several respects. These differences include pH optima, molecular weight, and most importantly, affinity for certain inhibitors.[12] Over a 1,000 times higher concentration of the antimetabolites pyrimethamine and trimethoprim are required to produce 50% inhibition of the mammalian enzyme as contrasted to the plasmodial one. Therefore, these drugs have been used as potent antimalarials.

Unfortunately, drug resistance to any of the foregoing compounds is fairly easy to produce in strains of *Plasmodium*. However, they have considerable value when used in combination with other drugs (for example, with chloroquine or quinine) to completely cure an infection.

Family Haemoproteidae

Protozoans belonging to the Haemoproteidae are primarily parasites of birds and reptiles and have their sexual phases in insects other than mosquitoes. Exoerythrocytic schizogony occurs in endothelial cells, the merozoites produced enter erythrocytes to become pigmented gametocytes in the circulating blood (Fig. 8-14). There are several genera: *Haemoproteus, Parahaemoproteus, Haemocystidium,* and *Simondia* in birds and reptiles, and *Hepatocystis, Nycteria,* and *Polychromophilus* in mammals.

Haemoproteus columbae is a cosmopolitan parasite of pigeons. The definitive hosts and vectors of this parasite are several species of ectoparasitic flies in the family Hippoboscidae (Fig. 8-15). Sporozoites are injected with their bite. Exoerythrocytic schizogony is completed in about 25 days in the capillary endothelium of the lungs, with thousands of merozoites produced from each schizont. Merozoites presum-

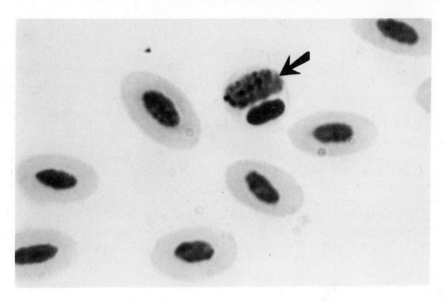

Fig. 8-14. A gametocyte of *Haemoproteus* (arrow) in a magpie's erythrocyte. (Photograph by Warren Buss.)

ably can develop directly from a schizont, or the schizont can break into numerous multinuclear "cytomeres." In this case, the host endothelial cell breaks down, releasing the cytomeres, which usually lodge in the capillary lumen, where they grow, become branched, and rupture, producing many thousands of merozoites. A few of these may attack other endothelial cells, but most enter erythrocytes and develop into gametocytes. At first they resemble ring stages of *Plasmodium*, but they grow into mature microgametocytes or macrogametocytes in 5 or 6 days. Multiple infections of young forms in a single red blood cell are common, but one rarely finds more than one mature parasite per cell.

The mature macrogametocyte is 14 μm long and grows in a curve around the nucleus. Its granular cytoplasm stains a deep blue color and contains about 14 small, dark brown pigment granules. The nucleus is small. The microgametocyte is 13 μm long, less curved, has lighter-colored cytoplasm, and has six to eight pigment granules. The nucleus is diffuse.

Exflagellation occurs in the stomach of the fly, producing four to eight microgametes. The ookinete is like that of *Plasmodium* except there is a mass of pigment at its posterior end. It penetrates the intesti-

Fig. 8-15. A wingless, parasitic fly of the family Hippoboscidae. Several species in this family are vectors of *Haemoproteus*. (Photograph by Jay Georgi.)

nal epithelium and encysts between the muscle layers. The oocyst grows to maturity by 9 days, measuring 40 μm in diameter. Myriads of sporozoites are released when the oocyst ruptures. Many of the sporozoites reach the salivary glands by the following day. Flies remain infected throughout the winter and can transmit infection to young squabs the following spring.

The pathogenesis in pigeons is slight, and infected birds usually show no signs of disease. Exceptionally, birds appear restless and lose their appetite. The air spaces of the lungs may become congested and some anemia may result from loss of functioning erythrocytes. The spleen and liver may be enlarged and dark with pigment.

Over 80 species of *Haemoproteus* have been named from birds, mainly Columbiformes. The actual number may be much less than that, for life cycles of most of them are unknown.

Of the other genera in Haemoproteidae, *Hepatocystis* spp. parasitize African and Oriental monkeys, lemurs, bats, squirrels, and chevrotains; *Nycteria* and *Polychromophilus* are in bats; *Simondia* is in turtles; *Haemocystidium* lives in lizards; and *Parahaemoproteus* is common in a wide variety of birds.

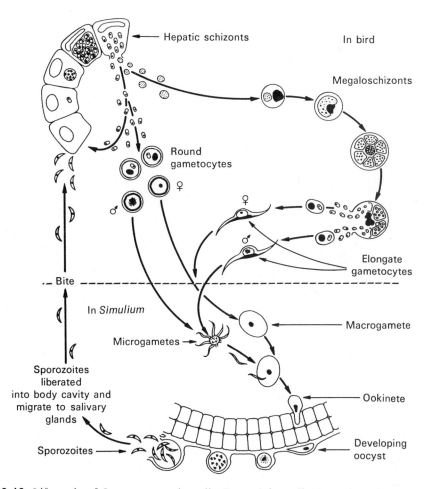

Fig. 8-16. Life cycle of *Leucocytozoon simondi.* (From Adam, K., J. Paul, and V. Zaman. 1971. Medical and veterinary protozoology. An illustrated guide. Churchill Livingston, Edinburgh.)

Family Leucocytozoidae (Fig. 8-16)

Species in the family Leucocytozoidae are parasites of birds. Schizogony is in fixed tissues, gametogony is in both leukocytes and immature erythrocytes of the vertebrate, and sporogony occurs in insects other than mosquitoes. Pigment is absent from all phases of the life cycles. There are two genera: *Akiba*, with only one species, in chickens; and *Leucocytozoon*, with about 60 species, in various birds. The Leucocytozoidae are the most important blood protozoa of birds, for they are very pathogenic in both domestic and wild hosts.

Leucocytozoon simondi is a circumboreal parasite of ducks, geese, and swans. The definitive hosts and vectors are blackflies, family Simuliidae (Fig. 8-17). The sporozoite, which is about 9 μm long, is injected into the avian host when the blackfly feeds. If the sporozoite enters a cell of the liver parenchyma, it becomes a small schizont, 11 to 18 μm in diameter, which produces merozoites in 4 to 6 days. If, however, the sporozoite is ingested by a macrophage in the brain, heart, liver, kidney, lymphoid

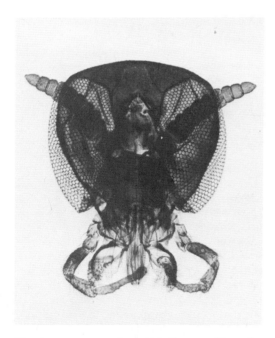

Fig. 8-17. The head of *Simulium* sp. Note the short, cutting-type mandibles. (Photograph by Jay Georgi.)

tissues, or other organ, it develops into a huge megaloschizont up to 165 μm in diameter. It is possible that merozoites from small, hepatic schizonts also become megaloschizonts when ingested by macrophages. The large form is more common than the small hepatic schizont.

The megaloschizont divides internally into **primary cytomeres,** which in turn multiply in the same manner. Successive cytomeres become smaller and finally multiply by schizogony into merozoites. Millions of merozoites may be released from a single megaloschizont.

Merozoites penetrate leukocytes or developing erythrocytes to become gametocytes (Fig. 8-18). Gametocytes of both sexes are 12 to 14 μm in diameter in fixed smears and may reach 22 μm in living cells. The macrogametocyte has a discrete, red-staining nucleus. The male cell is pale-staining and has a diffuse nucleus that takes up most of the space within the cell. As the gametocytes mature, they cause their host cells to become very elongate and spindle-shaped.

Exflagellation produces eight microgametes and begins only 3 minutes after being eaten by the fly. A typical ookinete entering an intestinal cell becomes a mature oocyst within 5 days. Only 30 to 50 sporozoites form and slowly leave the oocyst. Rather than entering the salivary glands of the vector, they enter the proboscis directly and are transmitted by contamination or are washed in by saliva.

Leucocytozoon simondi is highly pathogenic for ducks and geese, especially young birds. The death rate in ducklings may reach 85%; older ducks are more resistant, and the disease runs a slower course in them, but they still may succumb. Anemia is a prominent symptom of leukocytozoonosis, as are elevated numbers of leukocytes. The liver enlarges and becomes necrotic, while the spleen may increase to as much as twenty times the normal size. *Leucocytozoon simondi* probably kills the host by destroying vital tissues, such as brain and heart. An outstanding feature of an outbreak of leukocytozoonosis is the suddeness of its onset. A flock of ducklings may appear normal in the morning, become ill in the afternoon, and be dead by the next morning. Birds that

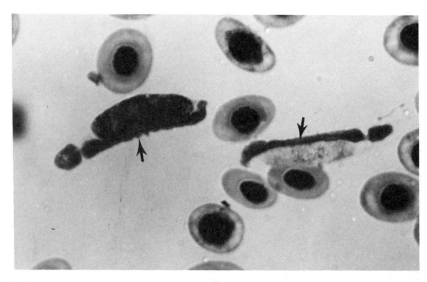

Fig. 8-18. An avian leukocyte infected with gametocytes of *Leucocytozoon simondi*. The host cell nuclei are indicated by the arrows. A macrogametocyte is on the left, and the pale-staining microgametocyte is on the right.

survive are prone to relapses but, as the result of premunition, are generally immune to reinfection.

Another species of importance is *L. smithi*, which can devastate domestic and wild turkey flocks. Its life cycle is similar to that of *L. simondi*.

CLASS PIROPLASMEA

Members of the class Piroplasmea are small parasites of ticks and mammals. They do not produce spores, flagella, cilia, or true pseudopodia; their locomotion, when necessary, is accomplished by body flexion or gliding. No stages produce intracellular pigment. Asexual reproduction is in the erythrocytes or other blood cells of mammals, by binary fission or schizogony. Sexual reproduction is unknown. The components of the apical complex are reduced but have enough similarity to sporozoans to warrant placement in the subphylum Apicomplexa.

The single order, Piroplasmida, contains the two families Babesiidae and Theileriidae, both of which are of considerable veterinary importance.

Family Babesiidae

Babesiids are usually described from their stages in the red blood cells of verte-brates. They are piriform, round, or oval parasites of erythrocytes, lymphocytes, histiocytes, erythroblasts, or other blood cells of mammals and of various tissues of ticks. The apical complex is reduced to a polar ring, rhoptries, microneme, and subpellicular microtubles. A cytostome is present in at least some species. Schizogony occurs in ticks. By far the most important species in America is *Babesia bigemina*, the causative agent of **babesiosis**, or **Texas redwater fever**, in cattle.

Babesia bigemina

By 1890 the entire southeastern United States was plagued by a disease of cattle, variously called Texas cattle fever, redwater fever, or haemoglobinurea. Infected cattle usually had bloody urine resulting from massive destruction of erythrocytes, and they often died within a week after symptoms first appeared. The death rate was much lower in cattle that had been reared in an enzootic area than in northern animals that were brought south. Also, it was noticed that when southern herds were driven or shipped north and penned with northern animals the latter rapidly succumbed to the disease. The cause of redwater fever and its mode of dissemination were a mystery when Theobald

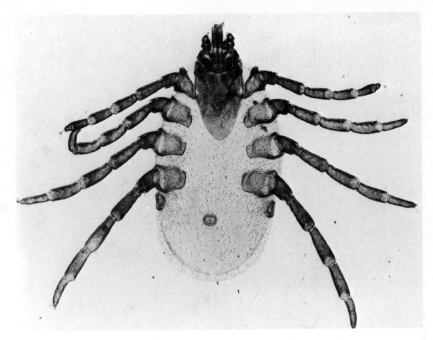

Fig. 8-19. *Boophilus annulatus*, the vector of *Babesia bigemina*. (Photograph by Jay Georgi.)

Smith and Frank Kilbourne began their investigations of this disease in the early 1880s. In a series of intelligent painstaking experiments, they showed that the tick *Boophilus annulatus* (Fig. 8-19) was the vector and alternate host of a tiny protozoan parasite that inhabited the red blood cells of cattle and killed these relatively immense animals.[25] This not only pointed the way to an effective means of control but was also the first proof that a protozoan parasite could develop in and be transmitted by an arthropod. This book is replete with other examples of this phenomenon.

Babesia bigemina infects a wide variety of ruminants, such as deer, water buffalo, and zebu, in addition to cattle. When in the erythrocyte of the vertebrate host, the parasite is pear-shaped or round, or occasionally irregularly shaped, and is 4 μm long by 1.5 μm wide. They usually are seen in pairs within the erythrocyte (hence the name *bigemina*, the twins) and are often united at the pointed tips (Fig. 8-20). At the light microscope level, they appear to be undergoing binary fission, but the electron microscope has revealed that the process is a kind of binary schizogony, a

budding analogous to that occurring in the Haemosporina, with redifferentiation of the apical complex and merozoite formation.[1]

Biology. The infective stage of *Babesia* in the tick is called a **vermicle**. It is about 2 μm long and is piriform, spherical, or ovoid in shape. After completing development, the vermicles reach the salivary glands of the tick and are injected with its bite. There is no exoerythrocytic schizogony in the vertebrate. The parasites immediately enter erythrocytes, where they become trophozoites, undergo binary schizogony, and ultimately kill their host cell. The merozoites attack other red blood cells, building up an immense population in a short time. Trophozoites lie in parasitophorous vacuoles that appear as clear areas in the host cell by light microscopy. This asexual cycle continues indefinitely, or until the host succumbs. Erythrocytic phases are reduced or apparently absent in resistant hosts.

Babesia bigemina is transmitted by ticks of the genus *Boophilus*, and the distribution of babesiosis is limited by the distribution of the tick. *Boophilus annulatus* is the

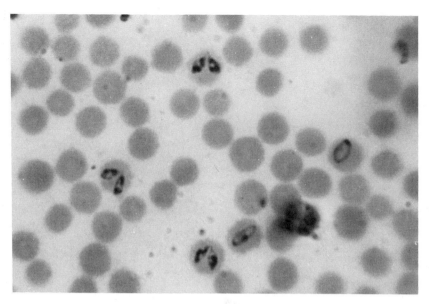

Fig. 8-20. *Babesia bigemina* trophozoites in the erythrocytes of a cow. (Photograph by Warren Buss.)

vector in the United States. It is a one-host tick, feeding, maturing, and mating on a single host. After engorging and mating, the female tick drops to the ground, lays her eggs, and dies. The larval, six-legged ticks that hatch from the eggs climb onto vegetation and attach to animals that brush by the plants.

One would think that a one-host tick would be a poor vector, for if they do not feed on successive hosts, how can they transmit pathogens from one animal to another? This question was answered when it was discovered that the protozoan infects the developing eggs in the ovary of the tick, a phenomenon called **transovarian transmission.**

When ingested by a feeding tick, the parasites are freed from their dead host cells by digestion. They become polymorphic at this stage, assuming a variety of shapes. It is possible, but not certain, that fertilization of one type of cell by another occurs at this time. At any rate, after 24 hours, the parasites are cigar-shaped bodies 8 to 10 μm long. They penetrate into the tick's intestinal epithelium and transform into spheroid bodies up to 16 μm in diameter within 2 days. At this time the nucleus undergoes schizogony, and the resultant vermicles, or merozoites, which

are 9 to 13 μm long, migrate into the hemocoel of their host. There, they attack the cells of the Malpighian tubules, where they once again undergo multiple fission. The merozoites of this generation enter the ovaries and the eggs they contain. At first they are scattered randomly throughout the egg, but after the egg is laid and begins developing, the parasites migrate to the intestinal epithelium of the larva, where schizogony takes place. Merozoites enter the hemocoels of the embryonic ticks, migrate to the salivary glands, and again undergo multiple fission in host cells. This generation of vermicles consists of enormous numbers of minute, piriform bodies about 2 or 3 μm long by 1 or 2 μm wide. They enter the channels of the salivary glands and are injected into the vertebrate host by the feeding tick.[23]

Although this is the life cycle as it occurs in a one-host tick, two- and three-host ticks serve as hosts and vectors of *B. bigemina* in other parts of the world. In these cases, transovarian transmission is not required and does not occur. All instars of such ticks can transmit the disease.

Pathology. *Babesia bigemina* is unusual in that the disease it causes is more severe in adult cattle than in calves. Calves less than a year old are seldom seriously affected, but the mortality in acute, untreated adult cattle is as high as 50% to 90%. The incubation period is 8 to 15 days, but an acutely ill animal may die only 4 to 8 days after infection. The first symptom is a sudden rise in temperature to 106° to 108° F; this may persist for a week or more. Infected animals rapidly become dull and listless and lose their appetite. Up to 75% of the erythrocytes may be destroyed in fatal cases, but even in milder infections so many erythrocytes are destroyed that a severe anemia results. Mechanisms for clearance of hemoglobin and its breakdown products are overloaded, so jaundice results and much excess hemoglobin is excreted by the kidneys, giving the urine the red color mentioned earlier. Chronically infected animals remain thin, weak, and out of condition for several weeks before recovering. Damage to internal organs is described by Levine.[14]

Cattle that recover are usually immune for life because of sterile immunity or, more commonly, premuniton. There are strain differences in the degree of immunity obtained, and, further, there is little cross-reaction between *B. bigemina* and other species of *Babesia*.

For unknown reasons drugs that are effective against trypanosomes are also effective against *Babesia*. A number of chemotherapeutic agents are available, some allowing recovery but leaving latent infection, others effecting a complete cure. It should be remembered that elimination of all parasites also eliminates premunition.

Infection can be prevented by tick control—the means by which redwater fever was eliminated from the United States. Regular dipping of cattle with a tickicide effectively eliminates the tick vector, most especially if it is a one-host species. Another method that has been used is artificial premunizing of young animals with a mild strain of *Babesia* before shipping them to enzootic areas.

Other species of Babesiidae

Cattle seem particularly suitable as hosts to piroplasms. Other species of *Babesia* in cattle are *B. bovis*, in Europe, Russia, and Africa; *B. berbera*, in Russia, north Africa, and the Middle East; *B. divergens*, in western and central Europe; *B. argentina*, in South America, Central America, and Australia; and *B. major*, in north Africa, Europe, and Russia. Several other species are known from deer, sheep, goats, dogs, cats, and other mammals, as well as birds. Their biology, pathogenesis, and control are generally the same as for *B. bigemina*.

At least ten cases of *Babesia* in humans are known: two were caused by *B. divergens*, a cattle parasite, and the others were caused by species normally parasitic in rodents. In three of the cases (two of which were fatal) the patients had been splenectomized some time before infection. The remainder of the cases were in nonsplenectomized patients, all on Nantucket Island off the coast of Massachusetts. They were caused by *B. microti*, a parasite of meadow voles and other rodents; one case was reported in 1969, one in 1973, and a small epidemic of five in the summer of 1975.[9a] Significance of the occurrence of so many cases of such a rare infection in such a small area can only be speculated on at present, but the observation is epidemiologically startling, at least.

Family Theileriidae

In members of this family the apical complex is much reduced, including only rhoptries. The conoid, subpellicular tubules, micronemes, and polar ring are absent. The Theileriidae parasitize blood cells of mammals, where they undergo schizogony in erythrocytes. The vectors are hard ticks of the family Ixodidae. Sexual reproduction has been said to occur in the tick but this is now doubted. Schizogony, or at least repeated binary fission, does occur in the tick's tissues. Several members of this family infect cattle, sheep, and goats, causing a disease called **theileriosis,** which results in heavy losses in Africa, Asia, and Southern Europe.

Theileria parva

Theileria parva causes a disease called **East Coast fever** in cattle, zebu, and Cape buffalo. It has been one of the most important diseases of cattle in south, east, and central Africa, although it has been eliminated from most of south Africa. The forms within erythrocytes have blue cytoplasm and a red nucleus in one end, after Romanovsky staining. At least 80% of them are rod-shaped, about 1.5 to 2.0 μm by 0.5 to 1.0 μm in size. Oval and ring- or comma-shaped forms are also found.

Biology. East Coast fever, like redwater fever, is a disease of ticks and cattle, flourishing in both. The principle vector is the brown cattle tick, *Rhipicephalus appendiculatus*, a three-host species. Other ticks, including one- and two-host ticks, can also serve as hosts for this parasite.

When the tick feeds, it injects all of the piroplasms present in its salivary glands into the next host. There they enter lymphocytes within lymphoid tissue, grow, and undergo schizogony. Schizonts, called **Koch's blue bodies**, can be seen in circulating lymphocytes within 3 days postinfection. Two types of schizonts are recognized. The first generation in lymph cells are **macroschizonts** and produce about 90 macromerozoites, each 2.0 to 2.5 μm in diameter. Some of these enter other lymph cells, especially in fixed tissues, and initiate further generations of macroschizonts. Others enter lymphocytes and become **microschizonts,** producing 80 to 90 micromerozoites, each 0.7 to 1.0 μm wide. If microschizonts rupture while in lymphoid tissues, the micromerozoites enter new lymph cells, maintaining the lymphatic infection. But if they rupture in the circulating blood, the micromerozoites enter erythrocytes to become the "piroplasms," typical of the disease. Apparently, the parasites do not multiply in erythrocytes.

Ticks of all instars can acquire infection when they feed on blood containing piroplasms. However, because three-host ticks drop off the host to ecdyse immediately after feeding, only the nymph and adult are infective to cattle. Transovarian transmission does not occur, as it does in *Babesia*.

Ingested erythrocytes are digested, releasing the piroplasms that enter the tissues of the tick. Details of early biology in the tick are unknown, but the parasites appear within cells of the salivary glands in 24 to 48 hours. Sexual reproduction in the tick has been neither proved nor disproved. If it does occur, subsequent development in the salivary gland is sporogony; if not, it is schizogony. At any rate, rapid multiple division of the parasite's nucleus produces over 30,000 tiny parasites, which

distend and finally rupture the host cell. The merozoites (sporozoites?) enter the lumen of the salivary gland to be injected when the tick next feeds. Development in the nymphal tick requires 3 days, while it takes 4 1/2 days in the adult.

Pathogenesis. As in babesiosis, calves are more resistant to *Theileria parva* than are adult cattle. Still, *T. parva* is highly pathogenic: strains with low pathogenicity kill around 23% of infected cattle, while highly pathogenic strains kill 90% to 100%. Symptoms such as high fever first appear 8 to 15 days after infection. Other signs are nasal discharge, runny eyes, swollen lymph nodes, weakness, emaciation, and diarrhea. Hematuria and anemia are unusual, although blood is often present in feces.

Animals that recover from theileriosis are immune from further infection, without premunition. Diagnosis depends on finding the parasites in blood or lymph smears. No drug is known to be effective once symptoms appear; however some of the tetracyclines prevent clinical disease, if given during the incubation period. Control depends on tick control and quarantine rules.

Other species of *Theileria* are *T. annulata, T. mutans, T. hirei, T. ovis,* and *T. camelensis,* all parasites of ruminants. Other genera in the family are *Haematoxenus,* in cattle and zebu, and *Cytauxzoon* in antelope, both in Africa.

REFERENCES

1. Aikawa, M., and C. R. Sterling. 1974. Intracellular parasitic Protozoa. Academic Press, Inc., New York.
2. Canfield, C. J. 1972. Malaria in U.S. military personnel 1965—1971. In Sadun, E. H., editor. Basic research in malaria. Special issue, Proc. Helm. Soc. Wash. 39:15-18.
2a. Coatney, G. R. 1976. Relapse in malaria—an enigma. J. Parasitol. 62:3-9.
3. Corradetti, A. 1950. Ospite definitive e ospite intermedio di parassiti della malaria. Riv. Parasitol. 11:89.
4. Fletcher, A., and B. Maegraith. 1972. The metabolism of the malaria parasite and its host. In Dawes, B., editor. Advances in parasitology, vol. 10. Academic Press, Inc., New York. pp. 31-48.
5. Fulton, J. D. 1969. Metabolism and pathogenic mechanisms of parasitic protozoa. In Chen, T., editor, Research in protozoology, vol. 3. Pergamon Press, Oxford. pp. 389-505.

6. Garnham, P. C. C. 1966 Malaria parasites and other *Haemosporidia*. Blackwell Scientific Publications, Oxford.

7. Guazzi, M., and S. Grazi. 1963. Consideratione sa un caso di malaria quartana recidivante dopo se anni di latenza. Riv. Malar. 42:55-59.

8. Halawani, A., and A. A. Shawarby. 1957. Malaria in Egypt. J. Egypt. Med. Assoc. 40:753-792.

9. Hall, A. P., and C. J. Canfield. 1972. Resistant falciparum malaria in Vietnam: its rarity in Negro soldiers. In Sadun, E. H., editor. Basic research in malaria. Special issue, Proc. Helm. Soc. Wash. 39:66-70.

9a. Healy, G. R., A. Spielman, and N. Gleason. 1976. Human babesiosis: reservoir of infection on Nantucket Island. Science 192:479-480.

10. Hoeppli, R. 1969. Parasitic diseases in Africa and the western hemisphere. Early documentation and transmission by the slave trade. Verlag für Recht und Gesellschaft Ag, Basel.

11. Howells, R. E. 1970. Mitochondrial changes during the life cycle of *Plasmodium berghei*. Ann. Trop. Med. Parasitol. 64:181-187.

12. Jaffe, J. J. 1972. Dihydrofolate reductases in parasitic protozoa and helminths. In Van den Bossche, H., editor. Comparative biochemistry of parasites. Academic Press, Inc., New York. pp. 219-233.

13. Kennedy, J. F. 1962. Message on first day of issue of U.S. malaria eradication stamp.

14. Levine, N. D. 1973. Protozoan parasites of domestic animals and of man. ed. 2. Burgess Publishing Co., Minneapolis.

15. Levitan, M., and A. Montagu. 1971. Textbook of human genetics. Oxford University Press, New York.

16. Luzzatto, L., E. A. Usanga, and S. Reddy. 1969. Glucose-6-phosphate dehydrogenase deficient red cells: resistance to infection by malarial parasites. Science 164:839-842.

17. Mackie, T. T., G. W. Hunter, III, and C. B. Worth. 1954. A manual of tropical medicine, ed. 2. W. B. Saunders Co., Philadelphia.

18. Maegraith, B. G. 1973. Malaria. In Spencer, H., Tropical pathology. Springer-Verlag, Heidelberg. pp. 319-349.

19. Maegraith, B. G., and A. Fletcher. 1972. The pathogenesis of mammalian malaria. In Dawes, B., editor. Advances in parasitology, vol. 10. Academic Press, Inc., New York. pp. 49-75.

20. Manson-Bahr, P. 1963. The story of malaria: the drama and the actors. Int. Rev. Trop. Med. 2:329-390.

21. Modell W. 1968. Malaria and victory in Vietnam. Science 162:1346-1352.

22. Moulder, J. W. 1962. The biochemistry of intracellular parasitism. University of Chicago Press, Chicago.

23. Rick, R. F. 1964. The life cycle of *Babesia bigemina* (Smith and Kilbourne, 1893) in the tick vector *Boophilus microplus* (Canastrini). Aust. J. Agr. Res. 15:802-821.

24. Russell, P. F., L. S. West, and R. D. Manwell. 1946. Practical Malariology. W. B. Saunders Co., Philadelphia.

25. Smith, T., and F. L. Kilbourne. 1893. Investigations into the nature, causation, and prevention of Texas or southern cattle fever. U.S. Dept. Agr. Bur. Anim. Indust. Bull. 1.

26. Sterling, C. R., M. Aikawa, and R. S. Nussenzweig. 1972. Morphological divergence in a mammalian malarial parasite: the fine structure of *Plasmodium brasilianum*. In Sadun, E. H., editor. Basic research in malaria. Special issue, Proc. Helm. Soc. Wash. 39:109-128.

27. Thompson, P. E., and L. M. Werbel. 1972. Antimalarial agents, chemistry and pharmacology. Academic Press, Inc., New York.

28. Warhurst, D. C. 1973. Chemotherapeutic agents and malaria research. In Taylor, A. E. R., and R. Muller, editors. Chemotherapeutic agents in the study of parasites. Symposia of British Society for Parasitology, vol. 11. Blackwell Scientific Publication, Oxford. pp. 1-28.

SUGGESTED READING

Honigberg, B. M. 1967. Chemistry of parasitism among some protozoa. In Florkin, M., and B. T. Scheer, editors. Chemical zoology, vol. I. Academic Press, Inc., New York. pp. 695-814.

Manandhar, M. S. P., and K. van Dyke. 1975. Detailed purine salvage metabolism in and outside the free malarial parasite. Exp. Parasitol. 37:138-146.

Chapter 9

SUBPHYLA MYXOSPORA AND MICROSPORA: THE POLAR CAPSULE PROTOZOA

Protozoa of the subphyla Myxospora and Microspora have a spore stage that is easily recognized by the presence of one or more **polar capsules,** organelles containing a coiled, filament-like tubule. When eaten, the polar capsules explode, piercing the intestinal epithelium of the host with the extending filament, thus making a wound through which the ameba-like **sporoplasm** can enter. All spores are enclosed in a dense covering consisting of one, two, or three valves. Their life cycles are direct; most parasitize invertebrates, but some attack lower vertebrates and, rarely, humans. They are probably a major natural control of insect populations.

SUBPHYLUM MYXOSPORA

In the subphylum Myxospora, the spores are of multicellular origin and are surrounded by two or three valves of various shapes (Fig. 9-1). They have two or more polar capsules and are parasites of lower vertebrates, especially fishes. A few are reported from amphibians and reptiles, but none is known from birds or mammals. From one to four polar capsules can be found at one end of the spore, except in the family Myxidiidae where one capsule is located at each end of the spore. Next to the polar capsules is an ameboid sporoplasm that is infective to the host. Some species have large vacuoles in the sporoplasm that stain readily with iodine, and are therefore called **iodinophilous vacuoles.** The valves join at a **sutural plane** that is either twisted or straight. The valves may bear various markings and often are extended as pointed processes at the "posterior" end. Over 700 species are described in this subphylum. They are very host and tissue specific.

Family Myxosomatidae

Of the many families of myxosporideans, few are more striking in appearance and importance than the Myxosomatidae. Fish parasites, they have two or four polar capsules in the spore stage, and their sporoplasm lacks iodinophilous vacuoles. One species is of circumboreal importance to salmonid fishes, including trout.

Myxosoma cerebralis. *Myxosoma cerebralis* causes **whirling disease** in salmonids, so called because fish with the disease swim in circles when disturbed or feeding. The parasite appears to have been endemic in brown trout, *Salmo trutta,* in central Europe to southeast Asia, and it causes no disease symptoms in that host. The disease was first noticed in 1900 after the introduction of the rainbow trout, *Salmo gairdneri,* to Europe. Since then it has spread to other localities in Europe, including Sweden and Scotland, to the United States, to South Africa, and to New Zealand.[1] Whirling disease results in high mortality in very young fish and causes corresponding economic loss. If a fish survives, damage to the cranium and vertebrae causes crippling and malformation.

Morphology. The mature spore of *M. cerebralis* (Fig. 9-1, *D*) is broadly oval in shape, with thick sutural ridges on the edges of the valves. It measures 7.4 to 9.7 μm long by 7 to 10 μm wide. The entire spore is covered with a mucoid-like envelope. Two polar capsules are found at the anterior end, each with a filament twisted into five or six coils. Each polar capsule lies within a polar cell that also contains a nucleus. The sporoplasm contains two nuclei—presumably haploid—numerous ribosomes, mitochondria, and other typical organelles.[9] The nuclei of the two val-

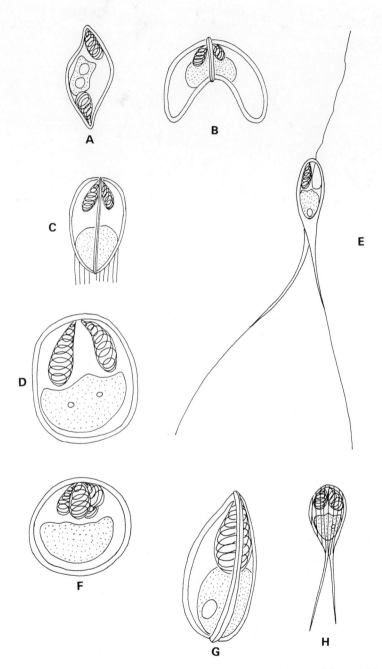

Fig. 9-1. Spores of representative genera of myxosporans. **A,** *Myxidium,* with a polar capsule at each end; **B,** *Ceratomyxa;* **C,** *Spherospora;* **D,** *Myxosoma;* **E,** *Henneguya;* **F,** *Chloromyxum,* with four polar capsules; **G,** *Thelohanellus,* with single polar capsule; **H,** *Myxobilatus.*

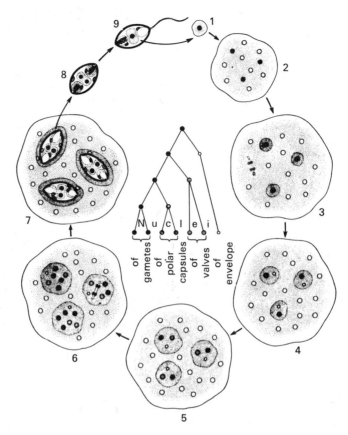

Fig. 9-2. Diagram of the development of a myxosporidian. *1,* Uninucleate amebula. *2,* Multinucleate plasmodium: differentiation into generative (dark) and somatic (light) nuclei. *3,* Segregation of the sporoblasts. *4-6,* Different stages of nuclear multiplication in the sporoblasts, corresponding to the divisional sequence shown in the middle of the diagram (envelope nucleus, white; valve nuclei, cross-hatched; nuclei of the polar capsule, dotted; gamete nuclei [germ line!], black). *7,* Plasmodium with spores. *8,* Single spore with binucleate amebula. *9,* Single spore with uninucleate amebula and a polar capsule with discharged polar filament. (From Grell, K. G. 1973. Protozoology. Springer-Verlag, New York.)

vogenic cells may be seen lying adjacent to the inner surface of each valve.

Biology. The life history of *Myxosoma cerebralis* has not been fully described, but the outlines are now known, and the rest may be inferred from knowledge of related species. Infection probably occurs when very young fish eat spores from the bottom of ponds or streams. The valves open and the polar capsules shoot out their filaments in a manner analogous to eversion of the finger of a glove. The polar filament creates a wound in the intestinal epithelium through which the sporoplasm can enter. Either before infection or soon thereafter, the two nuclei of the sporoplasm fuse; it is believed that the nuclei are haploid, and the fusion is autogamous. The sporoplasm makes its way to its preferred site, the cartilage of the head and spine. There it begins to grow, the nuclei dividing repeatedly. There is no corresponding cytokinesis, and by 4 months the multinucleated trophozoite will have reached a diameter of 1 mm (some species can reach a size of several millimeters). It apparently feeds by digesting the surrounding cartilage, thus creating the cavity within which it lies. During the course of the nuclear divisions, two types of nuclei can be distinguished, **generative** and **somatic** (Fig. 9-2). As development proceeds, a certain amount of cytoplasm becomes segregated around each generative nucleus to form a separate cell within the trophozoite. These cells will produce the spores, hence are called **sporoblasts.** Because in most species each will give rise to more than one spore, they are called **pansporoblasts.** Each pansporoblast in *M. cerebralis* will produce two spores. The generative nucleus for each spore will divide four times, one of the daughter nuclei of each division remaining generative, the other becoming somatic. The first somatic daughter nucleus will form the outer envelope of the spore, the second will divide again to give rise to the valvogenic cells, and the third nucleus will divide to produce the nuclei of the polar cells. Thus, the spore of the Myxopora is of multicellular origin. The fourth division of the generative nucleus produces the two nuclei of the sporoplasm, and this (or one of the preceding divisions) is reductional so that the nuclei of the sporoplasm are haploid.

The cavities within the cartilage become packed with spores by 8 months postinfection. Spores may live in the fish for 3 or more years. How they escape into the water is speculative, but it seems reasonable to assume that when the host is devoured by a larger fish or other piscivorous predator, such as a kingfisher or heron, the spores are released by digestion of their former home. The crippling effect of the parasite would make the host especially vulnerable to predation. Passing through the predator's intestinal tract unscathed, the parasite would settle to the bottom of the pond to become infective to new fry. The spores require an "aging" period of at least 4 months before they become infective to young trout.[4]

Pathogenesis. The main pathogenic effects of this disease can be attributed to damage to the cartilage in the axial skeleton of young fish, consequent interference with function of adjacent neural structures, and subsequent granuloma formation in healing of the lesions. Invasion of the cartilaginous capsule of the auditory-equilibrium organ behind the eye interferes with coordinated swimming; thus, when the fish is disturbed or tries to feed,

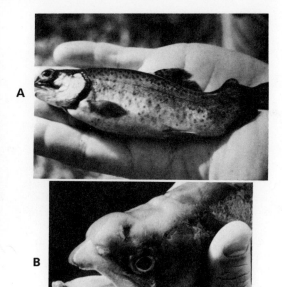

Fig. 9-3. Axial skeleton deformities in rainbow trout that have recovered from whirling disease *(Myxosoma cerebralis).* **A,** Note bulging eyes, shortened operculum, and both dorsoventral and lateral curvature of the spinal column (lordosis and scoliosis). **B,** Note gaping, underslung jaw and grotesque cranial granuloma. (Photographs are of living fish.)

it begins to whirl frantically, as if chasing its tail. It may become so exhausted by this futile activity that it sinks to the bottom and lies on its side until it regains its strength. Predation most likely occurs at this stage.[4] Often the cartilage of the spine is invaded, especially posterior the twenty-sixth vertebra. Function of the sympathetic nerves controlling the melanocytes is interfered with, and the posterior part of the fish becomes very dark, producing the "black tail." If the fish survives, granulomatous tissue infiltration of the skeleton may produce permanent deformities: misshapen head, permanently open or twisted lower jaw, or severe spinal curvature (scoliosis) (Fig. 9-3).

Epidemiology and prevention. In ponds where infected fish are held, it seems clear that spores can accumulate, whether by release from dead and decomposing fish, passage through predators, or some kind of escape from the tissue of infected living fish. Severity of an outbreak will depend on the degree of contamination of a pond, and light infections will cause little or no overt disease. Spores are resistant to drying and freezing, surviving for a long period of time, up to 18 days at $-20°$ C.[5]

No effective treatment for infected fish is known, and such fish should be destroyed by burial or incineration. Great care should be exercised to avoid transferring spores to uncontaminated hatcheries or streams, either by live fish that might be carriers or by feeding possibly contaminated food materials to hatchery fish. Earthen and concrete ponds where infected fish have been held can be disinfected by draining and treating with calcium cyanamide or quicklime. Application of ultraviolet irradiation can disinfect contaminated water.[3]

Other families, genera, and species of myxosporidians are common, widespread parasites of fish, amphibians, and reptiles. For general reviews and keys see Noble,[10] Kudo,[7] Hoffman and others,[4,6] and Hoffman.[2]

SUBPHYLUM MICROSPORA

The subphylum Microspora includes some 200 species of intracellular parasites of invertebrates and lower (rarely higher) vertebrates. They have been found in protozoa, platyhelminths, nematodes, bryozoans, rotifers, annelids, all classes of arthropods, fishes, amphibians, reptiles, and a few mammals. One species is known to infect humans. Numerous species are quite pathogenic, and several are of economic importance. The spores are unicellular in origin and have a single sporoplasm; the spore walls are not divided into separate valves. The subphylum now contains two classes, the Microsporea (with a polar filament) and the Haplosporea (without a polar filament).[8] The Haplosporea are parasites of molluscs and other invertebrates and will not be considered further.

Class Microsporea

The spore is the most conspicuous and morphologically distinctive stage in the life cycle of Microsporea. Spores are ovoid, spheroid, or cylindroid in shape. The

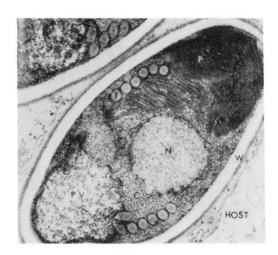

Fig. 9-4. *Nosema lophii* spore displaying polaroplast *(P)*, nucleus *(N)*, ribosome-rich cytoplasm *(c)*, polar tube *(T)*, posterior vacuole *(PV)*, and wall *(W)*. (From Weidner, E. 1970. Zeitschrift für Parasitenkunde 40:230-234.)

spore wall consists of two layers, the outer being proteinaceous and the inner chitinous. The wall is dense and refractile; its resistant properties contribute greatly to the survival of the spores. The spores are only 3 to 6 μm in length, and little structure can be discerned under the light microscope other than an apparent vacuole at one or both ends. It has been long known that they possessed a **polar filament,** since the structure could be stimulated to extrude artificially.

Possession of this structure has been the basis for uniting the microsporidians with the myxosporideans in the Cnidospora. However, use of the electron microscope has shown that the polar filaments of the two groups are basically dissimilar.[11] There is no polar capsule in the microsporidians, nor is the polar filament formed by a separate capsulogenic cell. At the ultrastructural level one can see a small **polar cap** or **sac** covering the attached end of the filament and just overlying the **polaroplast** (the apparent anterior vacuole) (Fig. 9-4). The ameboid **sporoplasm** surrounds the extrusion apparatus, with its nucleus and most of its cytoplasm lying within the coils of the filament. A **posterior vacuole** may be found at the end opposite the polaroplast. The cytoplasm of the sporoplasm has many free ribosomes but no mitochon-

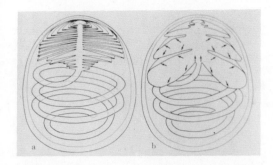

Fig. 9-5. *Nosema lophii:* diagrammatic interpretation of polaroplast membranes before *(a)* and during *(b)* collapse of polar sac and polar aperture before extrusion. (From Weidner, E. 1970. Zeitschrift für Parasitenkunde 40:230-234.)

dria or endoplasmic reticulum. The membrane and matrix of the polar cap are continuous with a highly pleated membrane comprising the polaroplast. This in turn is continuous with the polar end of the polar filament.[14] When polar filament extrusion is stimulated while in the host, a permeability change in the polar cap apparently allows water to enter the spore, and the filament is expelled explosively, simultaneously turning "inside-out." The stacked membrane in the polaroplast is unfolded as the filament is discharged and contributes to the expelled filament, so that it is much longer than when it is coiled within the spore (Fig. 9-5). The force with which the filament is extruded causes it to penetrate any cell in its path, and the sporoplasm flows through the tubular filament, thereby gaining access to its host cell. The end of the filament within the host cell expands to enclose the sporoplasm and becomes the parasite's new outer membrane.

The nuclei of the intracellular trophozoite divide repeatedly, and the organism becomes a large, multinucleate plasmodium. Finally cytokinesis takes place, and the process may then be repeated. In some species the nuclei may be associated in pairs (**diplokarya**), but whether such association has anything to do with sexual reproduction, or autogamy, is unknown. The multiple fission of the trophozoites is usually regarded as schizogony, but the process may not be strictly analogous to the schizogony found in the Apicomplexa. Sporogenesis occurs when the nuclear di-

visions of the monokaryotic or dikaryotic trophozoites give rise to nuclei destined to become spore nuclei. The multinucleate organism is now known as a **pansporoblast** or **sporoblast mother cell.** Each nucleus in the pansporoblast will become a spore nucleus; none is somatic, as they are in the myxosporidians. Cytokinesis of the pansporoblast produces the sporoblasts, which further differentiate into spores. The extensive endoplasmic reticulum in the sporoblast contributes to the polar filament and polaroplast. Mitochondria are not present at any stage.

Family Nosematidae

The Nosematidae is the largest family of microsporidians, with seven genera. The genera are separated on the basis of the number of spores produced by each sporoblast mother cell during the life cycle (from one to 16).

Nosema apis. *Nosema apis* is a common parasite of honeybees in many parts of the world, causing much loss annually to beekeepers. It infects the epithelial cells in the midgut of the insect. Infected bees lose strength, become listless, and die. Although the ovaries of the queen are not directly infected, they degenerate when her intestinal epithelium is damaged, an example of parasitic castration. The disease is variously known as nosema-disease, spring dwindling, bee dysentery, bee sickness, and May sickness.

The spore of *N. apis* is oval, measuring 4 to 6 μm long by 2 to 4 μm wide. The extended filament is 250 to 400 μm long. Infected bees defecate spores that are infective to other bees when the spores are eaten. Swallowed spores enter the midgut and lodge on the peritrophic membrane. Extruded filaments pierce the peritrophic membrane and intestinal epithelium, and the sporoplasm enters the epithelial cell. The entire process is accomplished within 30 minutes. Sporogony takes place in the second multiple fission generation, and the spores rupture the host cell to be passed with the feces. The entire life history in the bee is completed in 4 to 7 days. Destruction of the intestinal epithelium kills the host.

Other Nosema species. Besides *N. apis,* a few other of the many species in this genus are of known direct importance to humans, though many additional ones may be important biological controls of insect populations. *Nosema bombycis* is a parasite of silk moth larvae, flourishing in the unnaturally crowded conditions of silkworm culture. The parasite affects nearly all tissues of the insect's body, including the intestinal epithelium. Parasitized larvae show brown or black spots on their bodies, giving them a peppered appearance. There is a high rate of mortality. Pasteur, in 1870, devoted considerable effort to understanding and controlling this disease and is credited with saving the silk industry in the French colonies. The life cycle of *N. bombycis* is basically similar to that of *N. apis* and can be completed in 4 days.

Nosema cuniculi is common in laboratory mice and rabbits, and it is also known in dogs, rats, and guinea pigs, usually in the brain. It may be transmitted in body exudates or transplacentally. Although damage is usually minimal, the infection can be fatal.

Nosema connori was described from a fatal infection of a human infant.[12] The only detail of the internal structure of the organism presented was that the sporoblast has two nuclei. Little else is known about the parasite, and it has only been provisionally relegated to the genus *Nosema*.

Nosema michaelis is a parasite of the economically important blue crab, *Callinectes sapidus,* which it kills 15 to 29 days after infection.[13] *N. michaelis* undergoes schizogony in the intestinal epithelium, and then undergoes sporogony in the striated muscle, which is extensively damaged.

Species of *Glugea, Plistophora,* and *Nosema* parasitize fish, including several economically important groups, and serious epizootics have been reported.

REFERENCES

1. Hewitt, G. C., and R. W. Little. 1972. Whirling disease in New Zealand trout caused by *Myxosoma cerebralis* (Hofer, 1903) (Protozoa: Myxosporida). N. Zeal. J. Mar. Freshwater Res. 6:1-10.
2. Hoffman, G. L. 1967. Parasites of North American freshwater fishes. University of California Press, Berkeley.
3. Hoffman, G. L. 1975. Whirling disease *(Myxosoma cerebralis):* control with ultraviolet irradiation and effect on fish. J. Wildlife Dis. 11:505-507.
4. Hoffman, G. L., C. E. Dunbar, and A. Bradford. 1969. Whirling disease of trouts caused by *Myxosoma cerebralis* in the United States. U.S. Department of Interior, Fish and Wildlife Service, Special Scientific Report, Fisheries no. 427 (1962 report issued with addendum, 1969).
5. Hoffman, G. L., and R. E. Putz. 1969. Host susceptibility and the effect of aging, freezing, heat, and chemicals on spores of *Myxosoma cerebralis.* Progressive Fish-Culturist 31:35-37.
6. Hoffman, G. L., R. E. Putz, and C. E. Dunbar. 1965. Studies on *Myxosoma cartilaginis* n. sp. (Protozoa: Myxosporidea) of centrarchid fish and a synopsis of the *Myxosoma* of North American freshwater fishes. J. Protozool. 12:319-332.
7. Kudo, R. R. 1966. Protozoology, ed. 5. Charles C Thomas, Publisher, Springfield, Ill.
8. Levine, N. D. 1970. Taxonomy of the sporozoa. J. Parasitol. 56 (sec. II, part 1):208-209.
9. Lom, J., and P. dePuytorac. 1965. Studies on the myxosporidean ultrastructure and polar capsule development. Protistologica 1:53-65.
10. Noble, E. R. 1944. Life cycles in the Myxosporidea. Q. Rev. Biol. 19:213-235.
11. Sprague, V. 1966. Suggested changes in "A revised classification of the phylum Protozoa," with particular reference to the position of the haplosporidans. Syst. Zool. 15:345-349.
12. Sprague, V. 1974. *Nosema connori* n. sp., a microsporidian parasite of man. Trans. Am. Microsc. Soc. 93:400-403.
13. Weidner, E. 1970. Ultrastructural study of microsporidian development. 1. *Nosema* sp. Sprague, 1965, in *Callinectes sapidus* Rathbun. Z. Zellforsch. 105:33-54.
14. Weidner, E. 1972. Ultrastructural study of microsporidian invasion into cells. Z. Parasitenkd. 40:227-242.

Chapter 10

SUBPHYLUM CILIOPHORA: CILIATED PROTOZOANS

CLASS CILIATA

The possession of simple cilia or compound ciliary organelles in at least one stage of their life cycle is the most conspicuous feature of the Ciliata. A compound subpellicular infraciliature is universally present, even when cilia are absent. Most species have one or more macronuclei and micronuclei, and fission is homothetogenic. Some species exhibit sexual reproduction involving conjugation, autogamy, and cytogamy. Though each cilium has a kinetosome, centrioles functioning as such are absent. Most ciliates are free-living, but many are commensals of vertebrates and invertebrates, and a few are parasitic. The following examples will illustrate the class.

Order Trichostomatida

In the Trichostomatida the somatic ciliature is typically distributed uniformly over the body, although exceptions do occur. There is no buccal ciliature in the oral areas, but cilia are found deeper in the cytopharynx.

Family Balantidiidae

The family Balantidiidae has the single genus *Balantidium,* species of which are found in the intestines of crustaceans, insects, fishes, amphibians, and mammals. The cytostome is found at the anterior end and a cytopyge is present at the posterior tip.

Balantidium coli. *Balantidium coli* is the largest protozoan parasite of humans. It is most common in tropical zones, but it is present throughout the temperate climes as well. The epidemiology and effects on the host are somewhat similar to those of *Entamoeba histolytica.* The organism appears to be basically a parasite of pigs, with strains adapted to various other hosts.

Morphology. Trophozoites (Fig. 10-1) of *B. coli* are oblong, spheroid, or more slender, 30 to 150 μm long by 25 to 120 μm wide. Encysted stages (Fig. 10-2), which are most commonly found in stools, are spheroid or ovoid, measuring 40 to 60 μm in diameter. The macronucleus is a large, sausage-shaped structure. The single micronucleus is much smaller and is often hidden from view by the macronucleus. There are two contractile vacuoles, one near the middle of the body and the other near the posterior end. Food vacuoles contain erythrocytes, cell fragments, starch granules, and fecal and other debris. Living trophozoites and cysts are yellowish or greenish.

Biology. The ciliate *B. coli* lives in the cecum and colon of humans, pigs, guinea pigs, rats and many other mammals. It is not readily transmissible from one species of host to another, however, for it seems to require a period of time to adjust to the symbiotic flora of a new host. But when adapted to a host species, the protozoan flourishes and can become a serious pathogen, particularly in humans. In animals other than primates, the organism is unable to initiate a lesion by itself, but it can become a secondary invader if the mucosa is breached by another means.

The trophozoite multiplies by transverse fission. Conjugation has been observed in culture but may occur only rarely, if at all, in nature. The cyst stage, which is basically a dormant time in the life of the parasite, has no sexual or asexual reproduction. Encystment is instigated by dehydration of feces as they pass posteriad in the rectum. They can encyst after being passed in stools—an important factor in the epidemiology of the disease. Infection occurs when the cyst is ingested, usually in contaminated food or water. Unencysted tro-

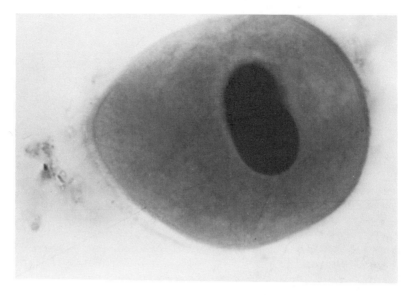

Fig. 10-1. A trophozoite of *Balantidium coli.* (Photograph by James Jensen.)

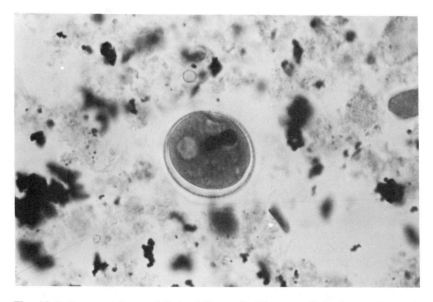

Fig. 10-2. Encysted form of *Balantidium coli.* (Photograph by James Jensen.)

phozoites may live up to 10 days and may possibly be infective if eaten, although this is unlikely under normal circumstances. Since *B. coli* is destroyed by a pH lower than 5.0, infection is most likely to occur in malnourished persons with low stomach acidity.

Pathogenesis. Under ordinary conditions, the trophozoite feeds much like a *Paramecium,* ingesting particles with the cy-topharynx and cytostome. But sometimes it appears that the organisms can produce proteolytic enzymes that digest away the intestinal epithelium of the host. Production of hyaluronidase has been detected, and this enzyme could help enlarge the ulcer. The ulcer usually is flask-shaped, like an amebic ulcer, with a narrow neck leading into an undermining sac-like cavity in the submucosa. The colonic ulceration produces lymphocytic infiltration with few polymorphonuclear leukocytes, and hemorrhage and secondary bacterial invasion may follow. Fulminating cases may produce necrosis and sloughing of the overlying mucosa and occasionally perforation of the large intestine, as in amebic dysentery. Death often follows at this stage. Secondary foci, such as the liver, may become infected. Urogenital organs are sometimes attacked following contamination, and vaginal, uterine, and bladder infections have been discovered.

Epidemiology. Balantidiasis in humans is most common in the Philippines but can be found nearly anywhere in the world, especially among those who are in close contact with swine. Generally, the disease is considered rare and is found in less than 1% of the human population. Higher infection rates have been reported among institutionalized persons. However, in pigs the infection rate may be 20% to 100%. Primates other than humans sometimes are infected and may represent a reservoir of infection to humans, although the reverse is probably more likely. The cysts can remain alive for weeks in pig feces, if the feces do not dry out. The pig is probably the usual source of infection for humans, but the relationship is not clear. The protozoans in swine are essentially nonpathogenic and are considered by some a separate species, *B. suis.* There may be differing strains of *B. coli* that vary in their adaptability to humans.

Treatment and control. Several drugs are used to combat infections of *B. coli,* including carbarsone, diiodohydroxyquin, and tetracycline. The infection often disappears spontaneously in healthy individuals, or it can become symptomless, making the person a carrier. Prevention and control duplicate those of *Entamoeba histolytica,* except that particular care should be taken by those who work with pigs.

Other species of *Balantidium* are *B. praenucleatum,* common in the intestines of American and Oriental cockroaches; *B. duodeni* in frogs; and *B. caviae* in guinea pigs.

Order Hymenostomatida

The ciliature is typically uniform in the Hymenostomatida. The cytostome and pharynx are ventral, with one undulating membranelle on the right side and three membranelles on the left. Most species are very small in size, but *Ichthyophthirius multifiliis* is a very large one.

Family Ophryoglenidae

The family Ophryoglenidae contains one genus of parasites, most of which are unimportant to humans. One species, however, is a very common pest in fresh-water aquaria, causing much loss of exotic fishes.

Ichthyophthirius multifiliis (Fig. 10-3). *Ichthyophthirius multifiliis* causes a common disease in aquarium and wild fresh-water fish, known as **"ick"** to many fish culturists. It attacks the epidermis, cornea, and gill filaments.

Morphology. Adult trophozoites are up to 1 mm in diameter. The macronucleus is a large, horseshoe-shaped body that encircles the tiny micronucleus. Each of several contractile vacuoles has its own micropore in the pellicle. A permanent cytopyge is located at the posterior end of the animal.

Biology. Mature trophozoites form pustules in the skin of their fish hosts (Fig. 10-4). They are set free and swim feebly about when the pustules rupture, finally settling on the bottom of their environment or on vegetation. Within an hour the ciliate secretes a thick, gelatinous cyst about itself and begins a series of transverse fissions. The daughter trophozoites, or **swarmers,** number up to 100 in a single cyst, depending on the original size of the mother trophozoite. Swarmers are piriform, ciliated cells, 30 to 50 μm long with a conspicuous, spherical nucleus and a single contractile vacuole. About 7 or 8 hours after leaving a fish host, the swarmers have completed development and pierce the cyst wall to become infective to a new host. The pointed anterior end is unciliated and is used to bore actively into the epidermis of its host. Unattached individuals do not live more than 2 days.

When in contact with a fish, the swarmer burrows into the epidermis and forms galleries in it that are soon occupied by many other individuals. However, reproduction does not occur in the host. Within 3 days

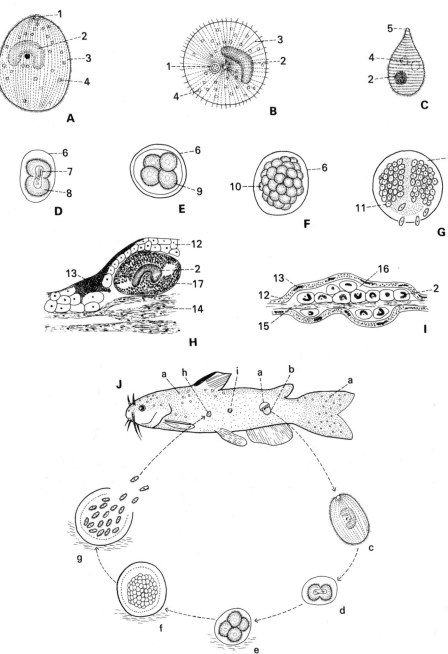

Fig. 10-3. For legend see opposite page.

Fig. 10-3. The life cycle of *Ichthyophthirius multifiliis*. **A,** Fully developed trophozoite of *Ichthyophthirius multifiliis* from pustule. **B,** Anterior end of fully developed trophozoite. **C,** Swarmer from cyst. **D, E,** First and second divisions of encysted trophozoite. **F,** Later stage of cystic multiplication. **G,** Cyst filled with swarmers, some of which are escaping into water. **H,** Section of skin of fish, showing full-grown trophozoite embedded in it. **I,** Section of tail of carp, showing ciliates developing in pustule. **J,** Infected bullhead *(Ameiurus melas)*.

1, Cytostome; *2,* macronucleus with nearby micronucleus; *3,* longitudinal rows of cilia; *4,* contractile vacuoles; *5,* boring or penetrating apparatus; *6,* cyst; *7,* dividing of macronucleus; *8,* two daughter cells formed by first division; *9,* four daughter cells formed by second division in cyst; *10,* numerous daughter cells; *11,* swarmers; *12,* epidermis of fish skin; *13,* pigment cell in epidermis; *14,* dermis; *15,* cartilaginous skeleton of tail of carp; *16,* pustule containing trophozoites; *17,* trophozoite under skin.

a, Pustules; *b,* trophozoite escaping from pustule into water; *c,* trophozoite free in water; *d,* encysted trophozoite on bottom of pond in first division, showing two daughter cells; *e,* cyst in second division with four daughter cells; *f,* cyst with many daughter cells; *g,* ruptured cyst liberating swarmers; *h,* swarmer attached to skin; *i,* swarmer partially embedded in skin. (From Olsen, O. W. 1974. Animal parasites, their life cycles and ecology. University Park Press, Baltimore.)

Fig. 10-4. Sunfish infected with *Ichthyophthirius multifiliis*. Note the light-colored pustules in the skin. (From Hoffman, G. 1977. In Kreier, J. Protozoa of medical and veterinary interest. Academic Press, Inc., New York.)

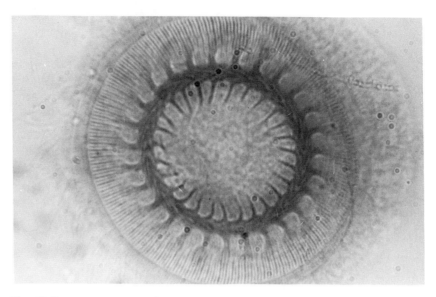

Fig. 10-5. *Trichodina* sp. from the gill of a fish. (Photograph by Warren Buss.)

the organism has enlarged and differentiated into the adult.

Pathogenesis. Grayish pustules form wherever the parasites colonize in the skin. Epidermal cells combat the irritation by producing much mucus, but many die and are sloughed. When many parasites attack the gill filaments, they so interfere with gas exchange that the fish may die.

Aquarium fish can be treated successfully with very dilute concentrations of formaldehyde, malachite green, or methylene blue.

Order Peritrichida
Family Urceolariidae

Somatic ciliature is essentially absent in mature Peritrichida. The oral ciliature is conspicuous, winding around the apical pole counterclockwise to the cytostome. The organisms often attach to the substrate by a contractile stalk (as in *Vorticella*) or by a prominent adhesive basal disc. Most species are free-living, but one family has numerous parasitic forms.

Species in the family Urceolariidae lack stalks and are mobile. The oral-aboral axis is shortened, with a prominent basal disc usually at the aboral pole. A protoplasmic fringe, or velum, lies on the margin of the basal disc, and a circle of strong cilia lies underneath. A second circle of cilia, above

the disc, cannot always be found. The biology of the group is poorly known. The family contains five genera, with *Trichodina* being a typical example.

Trichodina spp. Members of this genus parasitize a wide variety of aquatic invertebrates, fish, and amphibians. The basal disc contains a corona of hard, pointed "teeth" that aid the parasite in attaching to its host (Fig. 10-5). The number, arrangement, and shapes of these teeth are useful taxonomic characters. The buccal ciliary spiral makes more than one, but fewer than two, complete turns. Species of *Trichodina* may cause some damage to the gills of fish, but most produce little pathogenic effect and are of interest only as beautiful examples of highly evolved protozoans with incredibly specialized organelles. Typical examples are *T. californica* on the gills of salmon, *T. pediculus* on *Hydra*, and *T. urinicola* in the urinary bladder of amphibians.

Order Heterotrichida
Family Plagiotomidae

Somatic ciliature is sparse in most species in this order, but, when present, it is usually uniform. Buccal ciliature is conspicuous, with the aboral zone typically composed of one to many membranelles or undulating membranes that wind clock-

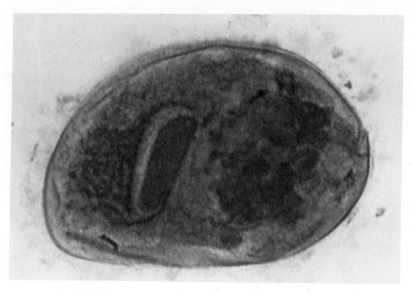

Fig. 10-6. *Nyctotherus cordiformis,* trophozoite from the colon of a frog. (Photograph by Warren Buss.)

wise to the cytostome. Most species are quite large.

Plagiotomidae are robust parasites of the intestine of vertebrates and invertebrates. The entire body has tiny cilia arranged in longitudinal rows. A single undulating membrane extends from the anterior end to deep within the cytopharynx.

The most common genus is *Nyctotherus,* which is easily obtained for laboratory use. These ciliates (Fig. 10-6) are ovoid to kidney-shaped, with the cytostome on one side. The anterior half contains a massive macronucleus, with a small micronucleus nearby. The genus has numerous species, some of which are useful in routine laboratory exercises. Common species are *N.*

ovalis in cockroaches and *N. cordiformis* in the colon of frogs and toads.

SUGGESTED READING

Bykhovskaya-Pavlovskaya, I. E. 1962. Key to the parasites of freshwater fish. Academy of Science, Moscow. English translation, Israel Program for Scientific Translations, Jerusalem (1964). (An outstanding reference to ciliate parasites.)

Hoffman, G. L. 1967. Parasites of North American freshwater fishes. University of California Press, Berkeley. (Ciliates of North American fishes are listed in this useful reference work.)

Kudo, R. 1966. Protozoology, ed. 5. Charles C. Thomas, Publisher, Springfield, Ill. (A standard reference to ciliate parasites.)

Levine, N. D. 1973. Protozoan parasites of domestic animals and of man, ed. 2. Burgess Publishing Co., Minneapolis.

Chapter 11

PHYLUM MESOZOA: PIONEERS OR DEGENERATES?

Mesozoans are tiny, ciliated animals that parasitize marine invertebrates. Their affinities with other phyla are obscure, chiefly because of the simplicity of their structure and their unusual biology. Digestive, circulatory, nervous, and excretory systems are lacking. Basically, a mesozoan's body is made of two layers of cells, but these are not homologous with the endoderm and ectoderm of diploblastic animals.

There are two distinct groups that traditionally are placed in the phylum Mesozoa; the class Dicyemida and class Orthonectida. However, these two classes are so different in morphology and life cycles that they probably should be placed in separate phyla.[1] But since no one has so far formally proposed the separation, we will include both within the phylum Mesozoa, while recognizing the artificiality of the scheme.

CLASS DICYEMIDA

Dicyemids are parasites of the renal organs of cephalopods, either lying free in the kidney sac or attached to the renal appendages of the vena cava. Partial life cycles are known for a few species, but certain details are lacking in all cases. Interesting histories of the group were presented by Stunkard.[7,8]

Morphology and biology. The earliest known stage in the cephalopod is a ciliated larva, the **larval stem nematogen,** which swims freely in the kidney of its host. Its body is composed of a **polar cap,** or **calotte,** and a **trunk.** The calotte is made up of two tiers of cells, usually with four or five cells in each. The anterior tier is called the **propolar,** and the posterior is the **metapolar.** The cells in the two tiers may be arranged opposite or alternate to each other, depending on the genus.

The trunk is comprised of relatively large **axial cells** that are surrounded by a single layer of ciliated, external cells. This outer, ciliated somatoderm is formed by **parapolar cells** immediately behind the calotte, several **diapolar cells** covering most of the trunk, and one or two **uropolar cells** at the posterior end. The inner layer of the trunk, which may penetrate the calotte, is made up of two or three axial cells. Thus, the larval stem nematogen of *Pseudicyema truncatum,* from the European cuttlefish, is composed of four propolars, four metapolars, three parapolars, 16 diapolars, one uropolar, and three axial cells. The number and arrangement of cells are of taxonomic importance. The outer surface of the exterior cells (except in stem nematogens) have slender projections called **ruffles,** which fuse occasionally to form endocytotic vesicles.[4]

The axial cells of the larval stem nematogen each contains a **vegetative nucleus** and a **germinative nucleus.** The germinative nucleus becomes an **agamete.** The stem nematogen grows larger while agametes continue to divide, becoming aggregates of cells, in a process much like the asexual, internal reproduction (germ balls) found in miracidia, sporocysts, and rediae of digenetic trematodes (see Chapter 16). The animal is now called an **adult stem nematogen.**

Within the axial cell, agametes develop into vermiform embryos (Fig. 11-1) of **primary nematogens** that escape the body of the stem nematogen and attach to the kidney tissues of the host. Primary nematogens of *P. truncatum* consist of four propolars, four metapolars, two parapolars, 10 to 15 diapolars and uropolars, and a single axial cell. Agametes within the axial cell of the primary nematogen produce many generations of identical vermiform embryos that develop into primary nematogens, building up a massive infection in the

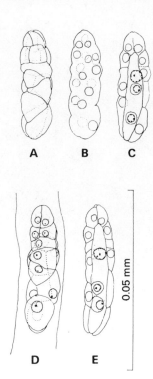

A   B   C

D   E

0.05 mm

Fig. 11-1. Primary nematogens developing within adult stem nematogens of *Dicyema typoides.* (**A-C** immature) **A,** Outline of somatic cells; **B,** somatic cell nuclei; **C,** optical section. (**D** and **E** mature) **D,** Somatic cell outlines and nuclei; **E,** optical section. (From Short, R. B. 1964. J. Parasitol. 50:646-651.)

cephalopod. When the host becomes sexually mature, the production of primary nematogens ceases. Instead, the vermiform embryos form stages that become primary **rhombogens,** a form similar to nematogens in cell number and distribution but with a different method of reproduction and with lipoprotein- and glycogen-filled somatic cells. These cells may become so engorged that they swell out, and the animal appears lumpy. Some primary nematogens metamorphose directly into rhombogens, and rhombogens derived by this route are referred to as **secondary rhombogens.**

Both primary and secondary rhombogens produce agametes in the axial cell that divide to become nonciliated **infusorigens.** An infusorigen is a mass of reproductive cells that represents either a hermaphroditic, sexual stage or a hermaphroditic gonad.[5] It remains within the axial

cell and produces male and female gametes, which fuse in fertilization. The zygotes detach from the infusorigen, and each then divides to become a hollow, ciliated ovoid stage called **infusoriform larva,** which is the most complex stage in the life cycle.[6] This microscopic larva consists of a fixed number of cells: two large, **apical cells** with short cilia and several large ciliated cells that cover most of its surface. In the center of the larva is the **urn,** made up of four **urn cells** surrounded almost completely by two **capsule cells.** Anterior to the urn cells is a small **urn cavity,** bound anteriorly by two cells with cilia extending into the urn cavity. The infusoriform larva escapes from the axial cell and parent rhombogen and leaves the host. Its immediate fate is unknown, for attempts to infect new hosts with it have failed. It is possible that an alternate or intermediate host exists in the life cycle.

The occurrence of two types of embryos, vermiform and infusoriform, gave rise to the name "dicyemides" (from the Greek *dis* meaning two and *kyema* meaning embryo).

CLASS ORTHONECTIDA

The Orthonectida are quite different from the Dicyemida in their biology and morphology. The 17 known species parasitize marine invertebrates, including brittle stars, nemerteans, annelids, turbellarians, and molluscs. Complete life cycles are known for some.

Morphology and biology. The best known orthonectid is *Rhopalura ophiocomae,* a parasite of brittle stars along the coast of Europe. Both sexual and asexual stages exist in the life cycle.

A **plasmodium stage** lives in the tissues and spaces of the gonads and genitorespiratory bursae of the ophiuroid *Amphipholis squamata* and may spread into the aboral side of the central disc, around the digestive system, and into the arms. Developing host ova degenerate, with ultimate castration, but male gonads usually are unaffected.[2] The multinucleate plasmodia are usually male or female but are sometimes hermaphroditic. Some of the nuclei are vegetative, while others are agametes that divide to form balls of cells called **morulas.** Each morula differentiates into an adult male or female, with a ciliated somatoderm

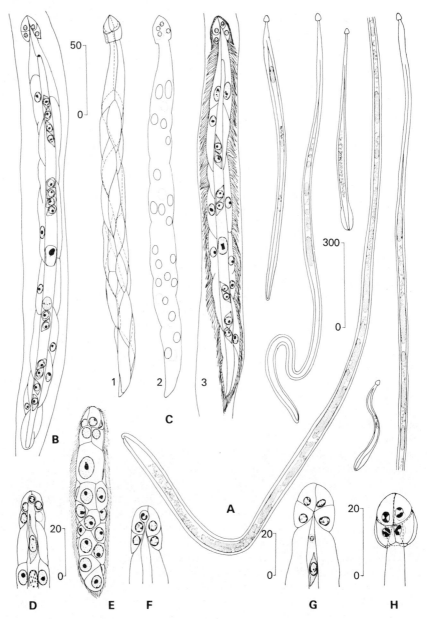

Fig. 11-2. *Dicyemennea antarcticensis,* representing developing and adult stages of a dicye-mid mesozoan. (Scale between **B** and **C** applies to both. Scale between **D** and **E** applies to both in μm. Scale to left of **G** also applies to **F.**) **A,** Entire nematogens. **B-G,** Vermiform embryos within axial cells of nematogens. **B,** Mature vermiform embryo, optical section. **C,** Mature vermiform embryo: *1,* peripheral cell outline; *2* position of peripheral cell nuclei; *3,* optical section. **D,** Anterior end of young vermiform embryo, optical section, showing anterior abortive axial cell. **E,** Young vermiform embryo, surface view, showing large parapolar cell. **F,** Anterior end of young vermiform embryo, optical section, showing anterior abortive axial cell, apparently being pinched off and squeezed out between propolar cells. **G,** Anterior end of young vermiform embryo, optical section, showing anterior abortive axial cell, apparently degenerating within functional axial cell. **H,** Young nematogen, anterior end. (From Short, R. B., and F. G. Hochberg, Jr. 1970. J. Parasitol. 56:517-522.)

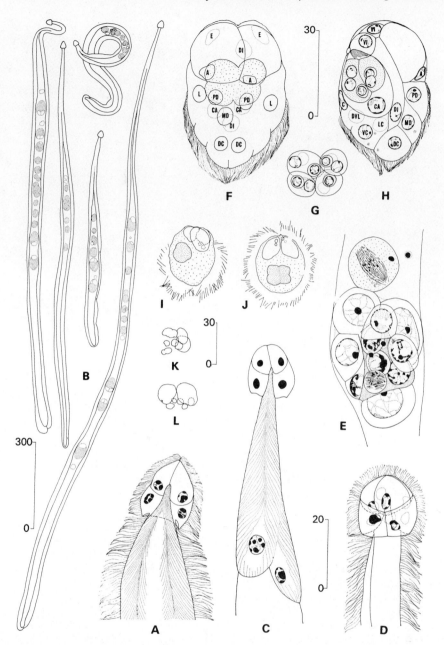

Fig. 11-3. *D. antarcticensis* life stages (continued from Fig. 11-2). (Scale between **C** and **D** also applies to **A**. Scale between **F** and **H** also applies to **E** and **G**. Scale to right of **K** also applies to **I, J,** and **L.**) **I-L** were drawn from temporary seawater-formalin preparations. **A,** Nematogen, anterior end; **B,** entire rhombogens; **C** and **D,** rhombogens, anterior ends; **E,** infusorigen; **F-J,** infusoriform larvae. Abbreviations denoting cells (in **F,** only nuclei of cells are shown): *A,* apical; *CA,* capsule; *C,* couvercle; *DC,* dorsal caudal; *DI,* dorsal internal; *E,* enveloping; *L,* lateral; *LC,* lateral caudal; *MD,* median dorsal; *PD,* paired dorsal; *VI,* ventral internal; *V1,* first ventral. **F,** Dorsal view, position of urn cells stippled. **G,** Urn cells; urn cell formula, 4 (2 + 1). **H,** Side view, optical section. **I** and **J,** Views to show relative sizes of refringent bodies and urn cells. **K** and **L,** Refringent bodies. (From Short, R. B., and F. G. Hochberg, Jr. 1970. J. Parasitol. 56:517-522.)

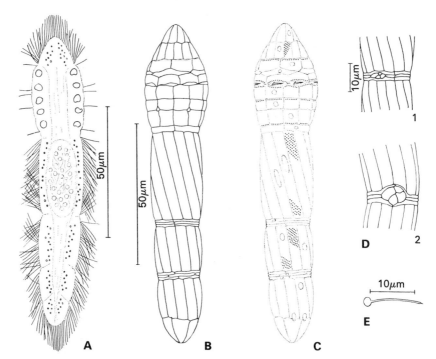

Fig. 11-4. *Rhopalura ophiocomae*, representing adult stages of an orthonectid mesozoan. Male: **A,** Living individual, as seen in optical section, showing distribution of cilia, lipid inclusions, crystal-like inclusions of the second superficial division of the body, and testis. **B,** Boundaries of jacket cells, at the surface; silver nitrate impregnation. **C,** Distribution of kinetosomes and approximate proportions of nuclei in representative jacket cells; diagrammatic, combining cell boundaries as seen in silver nitrate preparations with kinetosomes and nuclei demonstrated by impregnation with Protargol. **D,** Genital pore and adjacent cells in mature *(1)* and nearly mature *(2)* individuals; silver nitrate impregnation. **E,** Living sperm. (From Kozloff, E. N. 1969. J. Parasitol. 55:171-195.)

of **jacket cells** and numerous internal cells that become gametes. Monoecious plasmodia that produce both male and female offspring may represent the fusion of two separate, younger plasmodia. Male ciliated forms are elongate and 90 to 130 μm long. Constrictions around the body divide it into a conical cap, a middle, and a terminal portion. A genital pore, through which sperm escape, is located in one of the constrictions. Jacket cells are arranged in rings around the body; the number of rings and their arrangement is of taxonomic importance.

There are two types of females in this species. One type is elongated, 235 to 260 μm long and 65 to 80 μm wide, while the other is ovoid, 125 to 140 μm long by 65 to 70 μm wide. Otherwise, the two forms are similar to each other and differ from

the male in lacking constrictions that divide the body into zones. The female genital pore is located at about midbody. The oocytes are tightly packed in the center of the body.

Males and females emerge from the plasmodia and escape from the ophiuroid into the sea. There, tailed sperms are somehow transferred into females, where they fertilize the ova. Within 24 hours after fertilization the zygote has developed into a multicellular, ciliated larva that is born through the genital pore of its mother and enters the genital opening of a new host.

It is not known whether a plasmodium is derived from an entire ciliated larva or from certain of its cells or whether one larva can propagate more than one plasmodium.

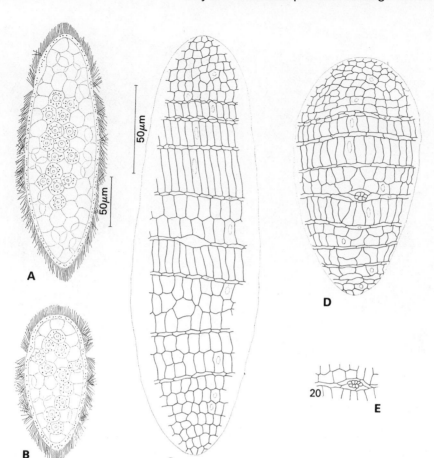

Fig. 11-5. Adult stages of *R. ophiocomae* (continued from Fig. 11-4). Female: **A,** Living specimen of elongated type, as seen in optical section. **B,** Living specimen of ovoid type, as seen in optical section. **C,** Boundaries of jacket cells of elongated type; silver nitrate impregnation. (The cells surrounding the genital pore have been omitted because they were not distinct; approximate proportions of nuclei of representative cells are based on specimens impregnated with Protargol.) **D,** Cell boundaries of ovoid type; silver nitrate impregnation. **E,** Genital pore of ovoid type; silver nitrate impregnation. (From Kozloff, E. N. 1969. J. Parasitol. 55:171-195.)

The **phylogenetic position** of the Mesozoa is most obscure. Early taxonomists placed them between protozoans and sponges because of their cilia, small size, and simple cellularity. Certainly their structure and life cycles are no more complex than those of some Protozoa. A good argument has been made for considering dicyemids to be primitive or degenerate Platyhelminthes. The ciliated larva is similar to a miracidium in some ways, and the internal reproduction by agametes in nematogens and rhombogens parallels similar processes in germinal sacs of digenetic trematodes.

Another possibility is that mesozoans represent one or two independent lines of evolution that have proceeded no further, so they cannot be aligned with higher forms. The question cannot be answered with conviction at this time.

PHYSIOLOGY AND HOST-PARASITE RELATIONSHIPS

What little is known of the physiology of the Mesozoa was reviewed by McCon-

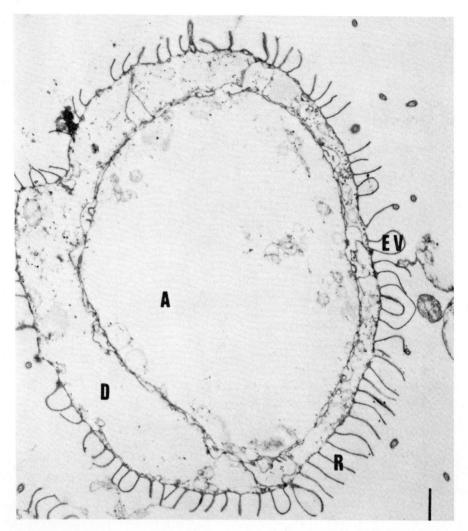

Fig. 11-6. Nematogen of *Dicyema aegira,* transverse section through diapolar cells and axial cells. Note scarcity of organelles, imparting hyaline appearance to cells. Ruffles on dipolar cells fused distally at several locations around periphery, forming large endocytotic vesicles. (× 8,500.) *A,* Axial cell; *D,* dipolar cell; *EV,* endocytotic vesicle; *R,* ruffle membrane. (From Ridley, R. K. 1968. J. Parasitol. 54:975-998.)

naughey.[3] Most of what is known concerns the dicyemids and is based on the early observations of Nouvel.[4] Good ultrastructural studies of both dicyemids and orthonectids are available.[2,5,6]

Most dicyemids attach themselves loosely to the lining of the cephalopod kidney by their anterior cilia. They are easily dislodged and can swim about freely in their host's urine. The relationship appears to be entirely commensalistic; no pathogenic consequences of the infection can be dis-

cerned. However, a few species have morphological adaptations for gripping the renal cell surface more firmly, and upon dislodgement, the renal tissue shows an eroded appearance.

The ruffle membrane surface of the nematogens and rhombogens (Fig. 11-6) is evidently an elaboration to facilitate uptake of nutrients. Ridley (1968) showed evidence that the membranes could fuse at various points and form endocytotic vesicles, and "transmembranosis" was demonstrated by uptake of ferritin.[5] The peripheral cells of infusoriform larvae do not have ruffle membranes but do have microvilli.[6] Clearly, the nutritive substances of nematogens and rhombogens must be derived largely or entirely from the host's urine, while the infusoriform must live for a period on stored food molecules. Oxygen is very low or absent in the cephalopod's urine, and the nematogens and rhombogens apparently are obligate anaerobes. The organisms live longer in vitro with nitrogen, or even in the presence of cyanide, than those maintained in urine under air or in the absence of cyanide. The infusoriform can live anaerobically only until its glycogen supply is consumed. Adult orthonectids, on the other hand, require aerobic conditions.

A plethora of questions remains. Especially in light of modern knowledge of the terminal reactions of glycolysis in parasitic platyhelminthes and nematodes, it would be fascinating to compare the analogous metabolic pathways of mesozoans. Numerous other intriguing questions were enumerated by McConnaughey, including the following:

1. What are the actual nutritional requirements of the mesozoa, and to what extent are these met by the host's urine?
2. What are the factors responsible for differentiation of the nematogens to rhombogens, and do these include host hormones?
3. What is responsible for the host specificity, and how do the invasive stages find and invade young specimens of the correct host species?
4. Why is cephalopod urine usually sterile except for the mesozoa, in spite of the fact that it is a good culture medium?

Classification of Phylum Mesozoa

Class Dicyemida
  Order Dicyemida
    Family Dicyemidae
      Genera: *Dicyema, Pseudicyema, Pleodicyema, Dicyemennea*
  Order Heterocyemida
    Family Conocyemidae
      Genera: *Conocyema, Microcyema*
Class Orthonectida
  Order Orthonectida
    Family Rhopaluridae
      Genera: *Rhopalura, Stoecharthrum*
    Family Pelmatosphaeridae
      Genus: *Pelmatosphaera*

REFERENCES

1. Dodson, E. O. 1956. A note on the systematic position of the Mesozoa. Syst. Zool. 5:37-40.
2. Kozloff, E. N. 1969. Morphology of the orthonectid *Rhopalura ophiocomae.* J. Parasitol. 55:171-195.
3. McConnaughey, B. H. 1968. The Mesozoa. In Florkin, M., and B. T. Scheer, editors. Chemical zoology, vol. 2. Porifera, Coelenterata, and Platyhelminthes. Academic Press, Inc., New York. pp. 537-570.
4. Nouvel, H. 1933. Recherches sur la cytologie, la physiologie et la biologie des dicyemides. Ann. Inst. Oceanogr. 13:163-255.
5. Ridley, R. K. 1968. Electron microscopic studies on dicyemid Mesozoa. I. Vermiform stages. J. Parasitol. 54:975-998.
6. Ridley, R. K. 1969. Electron microscopic studies on dicyemid Mesozoa. II. Infusorigen and infusoriform stages. J. Parasitol. 55:779-793.
7. Stunkard, H. W.. 1954. The life history and systematic relations of the Mesozoa. Q. Rev. Biol. 29:230-244.
8. Stunkard, H. W. 1972. Clarification of taxonomy in the Mesozoa. Syst. Zool. 21:210-214.

SUGGESTED READING

Grassé, P. P., and M. Caullery. 1961. Embranchement des mésozoaires. In Grassé, P., editor. Traité de zoologie: anatomie, systématique, biologie, vol. 4, Plathelminthes. Mésozoaires, Acanthocéphales, Némertiens. Masson et Cie. Paris. pp. 693-729. (A modern summary of the group.)
McConnaughey, B. H. 1968. The Mesozoa. In Florkin, M., and B. T. Scheer, editors. Chemical zoology, vol, 2. Porifera, coelenterata, and platyhelminthes. Academic Press, Inc. New York. pp. 537-570. (This is an excellent review of the physiology of Mesozoa.)

Chapter 12

INTRODUCTION TO PHYLUM PLATYHELMINTHES AND CLASS TURBELLARIA

The Platyhelminthes display several phylogenetic advances over what may be considered more primitive phyla, such as Porifera and Coelenterata. They are bilaterally symmetrical and have a definite "head end," with associated sensory and motor nerve elements. This increase in nervous function enabled them to invade a wide variety of ecological niches, including the bodies of other kinds of animals. In fact, most platyhelminthes are parasitic. A peculiarity of their physiology is their apparent inability to synthesize fatty acids and sterols de novo,[6] which may explain why flatworms are most often symbiotic with other organisms, either as commensals or parasites. The free-living acoel turbellarians, generally considered the most primitive worms in the phylum, also seem to lack this ability, indicating that the parasites may not have lost it secondarily as a response to parasitism.

Flatworms are so called because most are dorsoventrally flattened. They are usually leaf-shaped or oval, but some are very elongate, such as the tapeworms. They range in size from nearly microscopic to almost 100 feet in length. A coelom has not evolved in this phylum.

The **tegument** varies in structure between classes. Generally speaking, the Turbellaria and some free-living stages of Cestoidea and Trematoda have a ciliated epithelium, which is their primary mode of locomotion. This epithelium is very thin, being formed of a single layer of cells, and contains many glandular cells and ducts from subepithelial glands. Sensory nerve endings are abundant in the epithelium. The Trematoda and Cestoidea have lost the cilia except in certain larval stages. Instead, the tegument is a syncytial layer, the nuclei of which are in cell bodies (perikarya) located beneath a superficial muscle layer. Embedded in the tegument in most free-living turbellarians and in the trematode *Rhabdiopoeus,* are numerous rod-like bodies called **rhabdites.** Their function is not clear, but various authors have attributed lubrication, adhesion, and predator repellent to them; they are generally absent in symbiotic Turbellaria.

Most of the body of a flatworm is made up of **parenchyma,** a loosely arranged mass of fibers and cells of several types. Some of these cells are secretory, others store food or waste products, while still others have huge mitochondria and function in regeneration. The internal organs are so intimately embedded in the parenchyma that it is nearly impossible to dissect them out.

Coursing through the parenchyma are **muscle fibers.** These are rarely striated and are usually arranged in one or two longitudinal layers near the surface of the body. Circular and dorsoventral fibers also occur.

The **nervous system** in the primitive turbellarians consists of a simple nerve plexus under the epithelium with a slight concentration of cell bodies near the anterior end. In the more advanced turbellarians and in the trematodes and cestodes, the nerve system is a "ladder-type," with a complex ganglion near the anterior end and with longitudinal nerve trunks extending from it to near the posterior end of the body (Figs. 20-13 and 20-14). The number of trunks varies, but most are lateral and are connected by transverse commissures. Sensory elements are abundant, especially in the Turbellaria. Tactile cells, chemoreceptors, eye-spots, and, rarely, statocysts have been found.

A **digestive system** is completely absent in cestodes. Primitive turbellarians and a few trematodes *(Anenterotrema, Austromicrophallus)* have only a mouth but no permanent gut, food being digested by individual cells of the parenchyma. Most flatworms have a mouth near the anterior end, and many turbellarians and most trematodes have a muscular **pharynx,** behind the mouth, with which they suck in food. The gut varies from a simple sac to a highly branched tube, but only rarely does the flatworm have an anus. Digestion is mainly extracellular, with some phagocytosis by intestinal epithelium. Undigested wastes are eliminated through the mouth.

The functional unit of the **excretory system** is the **flame cell,** or **protonephridium** (Fig. 20-17). This is a single cell with a tuft of cilia that extends into a delicate tubule. Excess water, which may contain soluble nitrogenous wastes, is forced into the tubule, which joins with other tubules, eventually to be eliminated through one or more excretory pores. Some species have an excretory bladder just inside the pore. Because the excreta is mainly excess water, this is often referred to as an **osmoregulatory system,** with excretion of other wastes considered as a secondary function.

The **reproductive systems** follow a common pattern in all Platyhelminthes. Yet, extreme variations of the common pattern are found between groups. Most species are monoecious but a few are dioecious. Because the reproductive organs are so important in identification of parasites, and therefore are considered in great detail for each group, we will not discuss them here. Most hermaphrodites can fertilize their own eggs, but cross-fertilization is also known for many. Some turbellarians and cestodes practice **hypodermic impregnation,** which is sperm transfer by piercing the body wall with a male organ, the **penis** or **cirrus,** and injecting sperm into the parenchyma of the recipient. How the sperm find their way into the female system is not known. Most worms, however, deposit sperm directly into the female tract. The young are usually born within

egg membranes, but a few species are viviparous or ovoviviparous. Asexual reproduction is also common in trematodes and a few cestodes.

CLASSIFICATION OF PLATYHELMINTHES*

(1) Class TURBELLARIA (p. 191)

Mostly free-living worms in terrestrial, freshwater, and marine environments. Some are commensals or parasites of invertebrates, especially of echinoderms and molluscs.

(2) Class MONOGENEA (p. 198)

All are parasitic, mainly on the skin of fishes. While most are ectoparasites, a few live within the stomadaeum, proctodaeum, or their diverticulae.

(3) Class TREMATODA (p. 219)

All are parasitic, mainly in the digestive tract of all classes of vertebrates. There are three subclasses.

Subclass DIGENEA

At least two hosts in the life cycle, the first almost always a mollusc. Perhaps most diversification in bony marine fishes, although many species in all other groups of vertebrates.

Subclass ASPIDOGASTREA

Most have only one host, a mollusc. A few mature in marine turtles or rays and have a mollusc or lobster intermediate host.

Subclass DIDYMOZOIDEA

Tissue-dwelling parasites of fishes. No complete life cycle is known, but an intermediate host may not be required.

(4) Class CESTOIDEA (p. 321)

All are parasites, being common in all classes of vertebrates. An intermediate host is required for almost all species.

CLASS TURBELLARIA

Most turbellarians are free-living predators, but each of the five orders contains species that maintain varying degrees and types of symbiosis. Of these, most are symbionts of echinoderms, but others are found on or in crustaceans, sipunculids,

*There are several different schemes of classification for the phylum platyhelminthes because agreement has not been reached on which morphological and biological characters are most important in reflecting natural relationships. The classification that we propose here is a middle-of-the road scheme, which we have found to be practical and universally understandable. We accept four classes in the phylum.

arthropods, annelids, molluscs, coelenterates, other turbellarians, and fishes. At least 27 families have symbiotic species. A considerable degree of host specificity is manifested by these worms. Most symbionts are commensals, with few being true parasites. Several degrees of these relationships are known within the class. While it is tempting to array these in a series of ectocommensals, endocommensals, ectoparasites, and so on, to postulate how parasitism evolved in this phylum, it is clear that most individual cases are the end results of their particular situation and have not given rise to succeedingly complex associations. This is a predominately parasitic phylum, however, and a brief study of the commensal and parasitic turbellarians might indicate trends toward parasitism as it is found in the other classes.

Order Acoela

The acoels are entirely marine and are from 1 to several millimeters long. They possess several primitive characteristics, including the absence of an excretory system, pharynx, and permanent gut, and many have no rhabdites. Most are free-living, feeding on algae, protozoa, bacteria, and various other microscopic organisms. A temporary gut with a syncytial lining appears whenever food is ingested, and digestion occurs in vacuoles within it. After digestion is completed the gut disappears.

Few species have adopted a symbiotic existence, and it is difficult to decide which, if any, are true parasites. *Ectocotyla paguri* (Fig. 12-1) is the only ectocommensal known.[3] It lives on hermit crabs, but nothing is known of its biology or feeding habits. Several species of acoels live in the intestines of Echinoidea and Holothuroidea. It is not known if any are parasites, but because no apparent harm comes to the hosts they are usually considered endocommensals. *Meara stichopi*, which lives in the intestine and body cavity of a sea-cucumber, *Stichopus tremulus*, has a cellular rather than syncytial gut, has a thicker epidermis than most free-living acoels, and has fewer mucous glands.

Order Rhabdocoela

Most symbiotic turbellarians belong to this order. Again, most appear to be commensals, but a few are definitely parasitic.

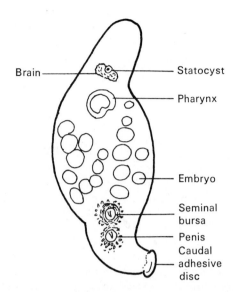

Fig. 12-1. *Ectocotyla paguri,* an ectocommensal on hermit crabs. (From Hyman, L. H. The invertebrates: Platyhelminthes and Rhynchocoela. 1951. McGraw-Hill Book Co., New York.)

Rhabdocoels are small, like acoels, but they have a permanent, saccular gut and a muscular pharynx. Most are predators of small invertebrates. Of the four suborders in the order, only two, Lecithophora and Temnocephalida, have symbiotic species. Within the Lecithophora, one family has parasitic members of considerable interest.

Suborder Lecithophora

Family Fecampidae. *Fecampia erythrocephala* (Fig. 12-2) lives in the hemocoel of decapod crustaceans.[1] During their development in the host, the young worms lose their eyes, mouth, and pharynx, and they absorb nutrients from the blood of their host. When sexually mature, they leave the host by an unknown process and produce several bottle-shaped cocoons, which they cement to the substrate. Each cocoon contains two eggs and several vitelline cells that produce two ciliated, motile juveniles. These swim about until contacting a crustacean. Their mode of entry into the host also is not known. The host is not killed by the parasite but obviously provides nourishment for the worm. *Fecampia* may illustrate a hypothetical stage in the origin of the Digenea.

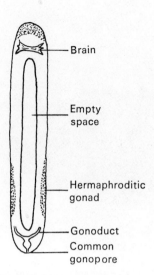

Fig. 12-2. *Fecampia,* from the hemocoel of a crab. Adult form without mouth or pharynx. (Adapted from Caullery, M., and F. Mesnil. 1903. Ann. Fac. Sci. Marseille 13:131-167.)

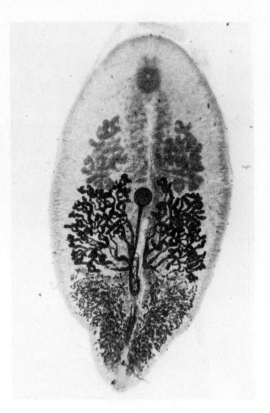

Fig. 12-3. *Syndesmis* sp., a rhabdocoel turbellarian from the intestine of a sea urchin. (Photograph by Warren Buss.)

Kronborgia amphipodicola is very unusual among the Turbellaria because it is dioecious.[2] Further, there is pronounced sexual dimorphism: the males are 4 to 5 mm long, while the females are 20 to 30 mm long and can stretch to 45 mm. Both sexes lack eyes and digestive systems at all stages of their life cycles. They mature in the hemocoel of the tube-dwelling amphipod *Amphiscela macrocephala,* with the male near the anterior end and the female filling the rest of the available space.[2] Upon reaching sexual maturity, the worms burrow out of the posterior end of the host, which becomes paralyzed and quickly dies. As if to add insult to injury, before the host is killed, it is castrated. After emergence from the amphipod, the female worm quickly secretes a cocoon around herself and attaches the cocoon to the wall of the burrow, protruding from it 2 to 3 cm. The male enters the cocoon, crawls down to the female, and inseminates her. He then leaves the cocoon and dies. The female produces thousands of capsules, each with two eggs and some vitelline cells and then also dies. A ciliated larva hatches from each egg and eventually encysts on the cuticle of another amphipod. While in the cyst, the larva bores a hole through the

host's body wall and enters the hemocoel to begin its parasitic existence.

The ultrastructure of *K. amphipodicola* has been studied.[5] The lateral membranes of the epidermal cells break down, and the epidermis thus becomes syncytial. Though short microvilli are not unusual on the outer surface of epithelial cells of free-living Turbellaria, the microvilli of *K. amphipodicola* are quite long and constitute an adaptation for increasing surface area to absorb nutrients. Subepidermal gland cells with long processes extending to the surface are thought to function in the escape of the worm from its host and in construction of the cocoon.

Family Umagillidae. Although no member of the family Umagillidae is unquestionably a parasite, we include it here because several of its species are easily available, making it possible for most interested persons to observe commensal turbellarians.

Most umagillids live in the digestive tract

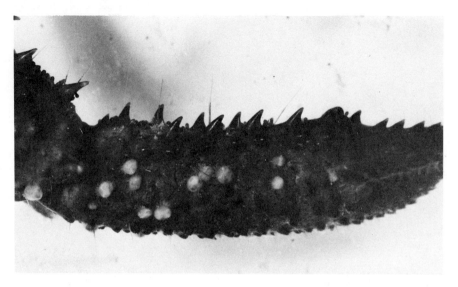

Fig. 12-4. Chela of New Zealand crayfish with several temnocephalans crawling on it. (Photograph by Wallaceville Animal Research Centre, New Zealand. Courtesy William B. Nutting.)

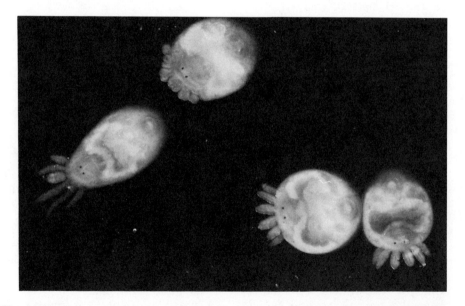

Fig. 12-5. Live temnocephalan turbellarians from a New Zealand crayfish. (Photograph by Wallaceville Animal Research Centre, New Zealand. Courtesy William B. Nutting.)

or coelom of Holothurioidea or Echinoidea. Crinoidea and sipunculids are also infected. No evidence has been found that umagillids harm their hosts in any way, so they are usually considered to be commensals. Very little is known of their biology and physiology, however, so this designation must be tentative. In one case, *Syndesmis antillarum* was observed to ingest host coelomocytes along with commensal ciliate protozoa, suggesting a shift of emphasis in feeding behavior.[4]

Syndesmis spp. (Fig. 12-3) are cosmopolitan in sea urchins and are therefore available to nearly any college laboratory with preserved or living sea urchins in its stock. Very little is known of their biology, but they appear to be excellent subjects for study.

Suborder Temnocephalida

These turbellarians are the only freshwater symbionts in the class. Most are ectocommensals on crustaceans (Fig. 12-4) in South and Central America, Australia, New Zealand (Fig. 12-5), Madagascar, Ceylon, and India, while a few are known from Europe. A few species occur on turtles, molluscs, and freshwater hydromedusae. Probably they are much more widespread but have gone undiscovered or unrecognized as the result of the paucity of trained specialists.

Temnocephalids are small and flattened, with tentacles at the anterior end and a weak, adhesive sucker at the posterior end (Fig. 12-6). They have leech-like movements, alternately attaching with the tentacles and posterior sucker. The tegument is syncytial with no or very few cilia and with a structure like that of trematodes, though adequate electron microscope studies have not been done. Rhabdites are located only at the anterior end, and mucous glands are mainly around the posterior sucker.

The biology of temnocephalids is simple, as far as it is known. Eggs are laid in capsules and attached to the exoskeleton of the host. Each hatches as an immature adult and matures with no further ado. What happens to those that are lost at ecdysis of the host is unknown, and, for that matter, the fate of the adults at that time is also unknown. It is possible that a free-living stage is present in the life cycle of these worms but has yet to be found.

The pattern of nutrition apparently does not differ from those of free-living rhabdocoels, with protozoa, bacteria, rotifers, nematodes, and other microscopic creatures serving as food. The host serves only as a substratum for attachment.

Order Alloeocoela

Alloeocoels are basically intermediate between Acoela and Tricladida and have an irregular gut. Most are marine, but a few inhabit brackish or fresh water, and a few are terrestrial. Several are commensal on snails, clams, and crustacea, but

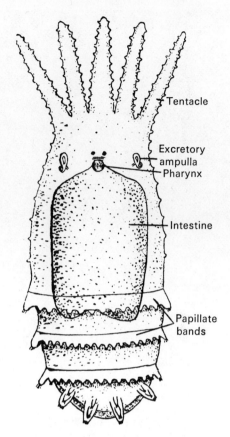

Fig. 12-6. *Craspedella,* a temnocephalid turbellarian. (Adapted from Haswell, O. 1893. A monograph of the Temnocephaleae. Macleay Mem. vol., Linnean Soc. New South Wales, 93-152.)

Ichthyophaga subcutanea is clearly a parasite of marine teleost fishes. It lives in cysts under the skin in the branchial and anal regions of its host and apparently ingests blood. Morphologically, it has nonparasite features, such as eyes and a ciliated epithelium.[8]

Monocelis sp. lives within the valves of tidal barnacles and snails during low tide but returns to the open water when the tide is in. This may illustrate a case of incipient endosymbiosis.

Order Tricladida

Tricladida are large worms, up to 50 cm in length, that occupy marine, freshwater, and terrestrial habitats. They are easily recognized by their tripartite intestine. Nearly all are free-living predators, feed-

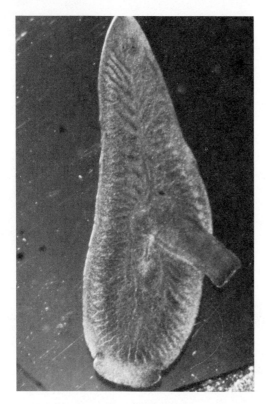

Fig. 12-7. *Bdellura candida,* a triclad turbellarian from the gills of a horseshoe crab. Note the eyespots and the huge midventral pharynx. (Photograph by Warren Buss.)

ing on small invertebrates and sucking the contents out of larger ones by means of their eversible pharynges.

Three genera, *Bdelloura, Syncoelidium,* and *Ectoplana,* live on the book gills of horseshoe crabs, *Limulus polyphemus.* Of these *Bdelloura candida* (Fig. 12-7) is the most common. It has a large adhesive disc at its posterior end and well-developed eyespots. Apparently it feeds on particles of food torn apart by the gnathobases of its host and washed back to the gill area. No evidence of harm to its host has been detected. It lays its eggs in capsules on the book gill lamellae. The triclads may migrate from one horseshoe crab to another during copulation of their hosts, a sort of marine, verminous venereal disease! The biology and physiology of these worms would surely prove to be a rewarding area of research.

Order Polycladida

The polyclads have a complex gut with many radiating branches. Except for one freshwater species they are all marine. No parasites are known in this group, and the few reported "commensals" are suspect of even that high degree of symbiosis. Although some species are found together with hermit crabs, they are also found in empty shells. Others, such as the "oyster leech," *Stylochus frontalis,* live between the valves of oysters and are predators on the original owner, devouring large pieces of it at a time.[7]

The truly parasitic turbellarians show structural changes expected with their specialized way of life: losses of ciliated epidermis, eyes, mucous glands, and rhabdites. The various commensals, however, show few or no specializations over their free-living brethren. The prevalence of rhabdocoels in echinoderms may simply be a result of the rich fauna of ciliate, protozoan commensals in the latter, which offer rich pickings for the former. The origin of trematodes and cestodes from acoel ancestors, which became adapted to endocommensalism within molluscs and crustaceans, is not difficult to visualize.

REFERENCES

1. Caullery, M., and F. Mesnil. 1903. Recherches sur les *Fecampia* Giard. Turbellariés Rhabdocoeles, parasites internes des Crustacés. Ann. Fac. Sci. Marseille 13:131-168.
2. Christiansen, A. P., and B. Kanneworff. 1965. Life history and biology of *Kronborgia amphipodicola* Christiansen and Kanneworff (Turbellaria, Neorhabdocoela). Ophelia 2:237-251.
3. Hyman, L. H. 1951. The invertebrates: 2. Platyhelminthes and Rhynchocoela. McGraw-Hill Book Co., New York.
4. Jennings, J. B., and D. F. Mettrick. 1968. Observations on the ecology, morphology and nutrition of the rhabdocoel turbellarian *Syndesmis franciscana* (Lehman, 1946) in Jamaica. Caribb. J. Sci. 8:57-69.
5. Lee, D. L. 1972. The structure of the helminth cuticle. In Dawes, B., editor. Advances in parasitology, vol. 10. Academic Press, Inc., New York. pp. 347-379.
6. Meyer, F., and H. Meyer. 1972. Loss of fatty acid biosynthesis in flatworms. In Van den Bossche, H., editor. Comparative biochemistry of parasites. Academic Press, Inc., New York. pp. 383-393.
7. Pearse, A. S., and G. W. Wharton. 1938. The oyster "leech" *Stylochus inimicus* Palomi, associated

with oysters on the coasts of Florida. Ecol. Monogr. 8:605-655.

8. Syriamiatnikova, I. P. 1949. A new turbellarian of fish, *Ichthyophaga subcutanea* n. g. n. sp. C. R. Acad. Nauk 68:805-808.

SUGGESTED READING

Baer, J. F. 1961. Class des Temnocéphales. In Grassé, G., editor. Traité de zoologie. Anatomie, systematique, biologie. vol. IV. Plathelminthes, Mesozoaires, Acanthocéphales, Némertiens. Masson et Cie., Paris. pp. 213-241. (A fairly up-to-date account of this group.)

Jennings, J. B. 1971. Parasitism and commensalism in the Turbellaria. In Dawes, B., editor. Advances in parasitology, vol. 9. Academic Press, Inc., New York. pp. 1-32. (The most readable account of the subject. Recommended for all parasitologists.)

Chapter 13

CLASS MONOGENEA

The Monogenea are hermaphroditic flatworms that are mainly external parasites of vertebrates, especially fishes. Some species are found internally, however, in diverticulae of the stomodeum or proctodeum and also in the ureters of fish and the bladders of turtles and frogs. A single species is known from mammals: *Oculotrema hippopotami* from the eye of the hippopotamus.[22] Nevertheless, monogeneans are primarily fish parasites, particularly of the gills and external surfaces. Although a few fish deaths have been attributed to monogeneans in nature, the worms are not usually regarded as hazardous to wild populations. However, like copepods and numerous other fish pathogens, monogeneans become a serious threat when fish are crowded together, as in fish farming.

The group has been somewhat neglected by parasitologists, despite their remarkable morphology and life cycles and their considerable economic importance. Probably fewer than half of the existing species have been described.[21]

These worms were aligned loosely with digenetic trematodes by early parasitologists. The first comprehensive overview of the group was by Braun in 1889 to 1893.[1] Next, Fuhrmann, in 1928, helped to establish the Monogenea as a category separate from the Digenea, although closely allied with it.[4] Bychowsky, in 1937, was apparently the first to propose Monogenea as a separate class, apart from and equal to Digenea.[2] This point of view was not adopted widely until quite recently, and even today it is not universally accepted. We believe that the separation of these worms into two classes is fully justified on morphological and biological grounds.

Monogeneans are often very particular about both the species of host and the site where they live on that host, restricting themselves to very narrow niches in many cases. Thus, one species may live only at the base of a gill filament, while another

is found only at its tip. Further, many species are found on certain gill arches but not on others within the same fish. It is possible that such niche-specificity is influenced by the physical attachment abilities of the highly specialized, posterior attachment organ, the **opisthaptor.** A similar phenomenon is known in the case of the scolex of tetraphyllidean cestodes of elasmobranchs. Some monogeneans remain fixed to the original site of attachment and cannot relocate later. Others, especially those on the skin, move about actively, leech-like, relocating at will. Certain species are found only on young fishes, while others occur only on mature fishes. Nutritional requirements of the parasites may play a role in the determination of such host specificity, but in some cases the free-swimming larvae are particularly attracted by mucus produced by the epidermis of their host species.[9]

The life span of monogeneans varies from a few days to several years. Many are incapable of living more than a short time after the death of the host. Monogeneans can seem to be absent from a fish population, when actually the worms had dropped off after the fish were caught. For this reason, dead fish should not be transported to the laboratory in water; far better to risk them drying so that the worms will remain on the gills. A recommended procedure is to place the fish on ice until examination, or to remove the gills and drop them into 10% formalin while still in the field.

FORM AND FUNCTION

Body form. Monogeneans are basically bilaterally symmetrical, with partial asymmetry superimposed on a few species, particularly involving the opisthaptor. The body can be subdivided roughly into the following regions: **Cephalic region** (anterior to pharynx) **trunk** (body proper), **peduncle** (portion of body tapered pos-

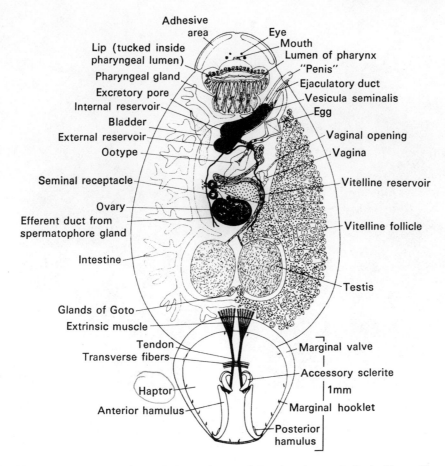

Fig. 13-1. The anatomy of an adult specimen of *E. soleae* (ventral view). (From Kearn, G. C. 1971. In Fallis, M. Ecology and physiology of parasites. University of Toronto Press, Toronto.)

teriorly), and **haptor** or opisthaptor (Fig. 13-1).

Most monogeneans are quite small, but a few are large, ranging from 0.03 to 20 mm long. Marine forms are usually larger than those from fresh water. All are capable of stretching and compressing their bodies, so one must take care that the worms are properly relaxed before fixing, or they may be contracted considerably. The dorsal side of the body is usually convex, while the ventral side is concave. The body is usually colorless or gray, but eggs, internal organs, or ingested food may cause it to be red, pink, brown, yellow, or black.

The anterior end of the body bears various adhesive and feeding organs, collectively called the **prohaptor.** There are two main types of prohaptors; those that

are not connected with the mouth funnel and those that are. The first (Fig. 13-2) is found on the more primitive types, in which the head end usually is truncated, lobated, or broadly rounded. This group usually bears **cephalic** or **head glands,** which are unicellular organs that release sticky substances through individual or groups of ducts. The utility of the substances produced by the head glands for adhesion is clear to anyone who has watched a monogenean with no anterior sucker progress in an inchworm-like fashion, alternately attaching and releasing the anterior and posterior ends, down a fish gill filament. In one species of *Gyrodactylus,* at least three different head gland types have been recognized. Two to eight clusters of such ducts, called **head organs,** are usual. These areas usually bear dense, long

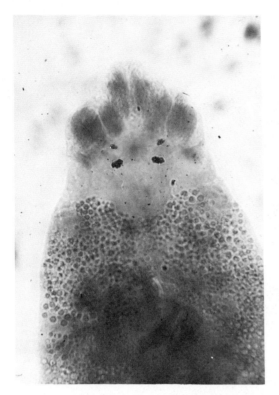

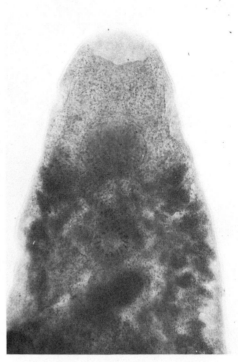

Fig. 13-2. Primitive prohaptor, not connected to mouth funnel. (Photograph by Warren Buss.)

Fig. 13-3. Advanced prohaptor, with an oral sucker. (Photograph by Warren Buss.)

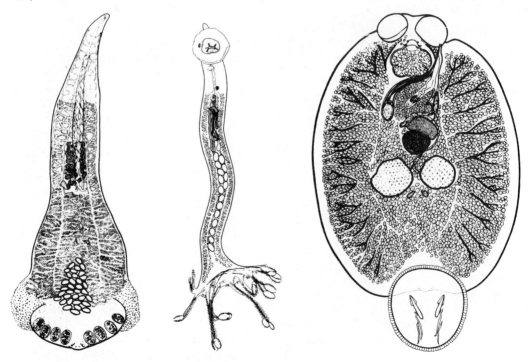

Fig. 13-4. For legend see opposite page.

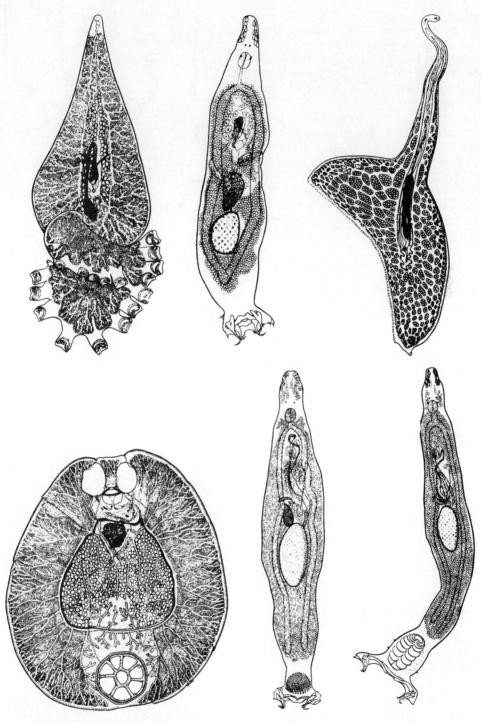

Fig. 13-4. A variety of monogeneans, showing variations of opisthaptors. (From Yamaguti, S. 1971. Monogenetic trematodes of Hawaiian fishes. University Press of Hawaii,
Honolulu.)

microvilli on the tegument, in contrast to the short, scattered microvilli on the remainder of the body. These microvilli may function to spread and mix the secretions of the different types of head glands. Some species in this group have shallow, muscular **bothria,** which serve as suckers, in conjunction with the head gland secretions. Most species have two bothria, but some species have four.

The second, and more advanced, type of prohaptor (Fig. 13-3) involves specializations of the mouth and buccal funnel. The simplest types have an **oral sucker** that surrounds the mouth. This may be a slightly muscular anterior rim of the mouth or a powerful circumoral sucker. A large group of advanced monogeneans have two **muscular suckers** embedded within the walls of the buccal funnel. The oral sucker may be supplied with gland cells that, as indicated by available evidence, also function in adhesion. Some species in this group also have head glands.

The posterior end of all monogeneans also bears a highly characteristic organ, the opisthaptor (Fig. 13-4). It is clear that in a group whose primary habitat is the surface and gills of fish, great adaptive value will accrue to an efficient attachment organ that prevents dislodgment by strong water currents, particularly one that will allow the mouth end to "hang downstream" and graze at will. In the Monogenea, the opisthaptor is such an organ, and it has been, unsurprisingly, subjected to immense adaptive radiation in that group.

The opisthaptor may extend for a considerable distance anteriorly along the trunk of the worm or may be confined to the posterior extremity. It may be sharply delineated from the body by a peduncle or may be merely a broad continuation of it. Opisthaptors develop into one of two basic types during ontogeny. The larva that hatches from the egg always has a tiny opisthaptor armed with sclerotized hooks or spines. This is retained in the adults of most species and either expands into the definitive opisthaptor or remains juvenile, while the adult organ develops from other sources near or surrounding it. In this first, basic type, the muscles expand into a large disc that often has shallow **loculi** or well-

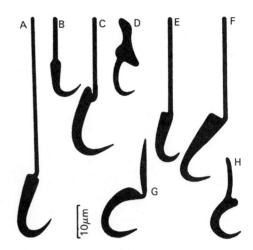

Fig. 13-5. Marginal opisthaptor hooklets from representative families of diclidophoridean Monogenea. (From Llewellyn, J. 1963. In Dawes, B. Advances in parasitology, vol. 1. Academic Press, Inc., New York.)

developed **suckers** as well as large hooks called **anchors,** referred to by some workers as **hamuli.** The tiny hooklets of the larva usually can be found on the margins or ventral surface of the opisthaptor (Fig. 13-5).

The second type loses the larval opisthaptor entirely or retains it as a tiny organ on (or in) the adult opisthaptor or attached at the end of a muscular appendage (**lappet** or **appendix**). In this large group the opisthaptor is the most highly specialized, being profoundly subdivided into individual adhesive organs and/or equipped with complex sclerotized clamps, or muscular valves, commonly with attendant hooks.

Because it has undergone such adaptive radiation and varies considerably between species, while remaining fairly constant within a species population, the opisthaptor is an important taxonomic character. Further, the sclerotized hooks, bars, clamps, and so on which are easily studied, are heavily relied on by specialists studying this group. It is possible that more attention to the anatomy of other organs would lead to a more natural classification of the monogeneans.

Hooks are characterized as being "marginal" or "central." **Marginal hooks,** which are not always strictly marginal, are usually

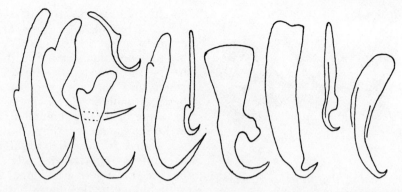

Fig. 13-6. Typical central hooks from opisthaptors of monogeneans. (From Bychowsky, B. E. 1961. Monogenetic trematodes, their systematics and phylogeny. American Institute of Biological Science, Washington, D.C.)

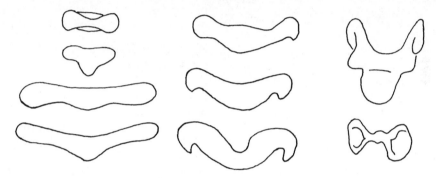

Fig. 13-7. Representative connecting bars, or accessory sclerites, of monogenean opisthaptors. (From Bychowsky, B. E. 1961. Monogenetic trematodes, their systematics and phylogeny. American Institute of Biological Science, Washington, D.C.)

the very tiny hooklets of the larval opisthaptor, some of which may be missing. It takes careful and skillful microscopy to find these. **Central hooks** (Fig. 13-6) are the larger anchors, or hamuli, and occur in one to three pairs, usually in the center of the opisthaptor, although they may be displaced to the side or posterior margin of the disc. Central hooks often have **connecting bars** or **accessory sclerites** supporting them (Fig. 13-7). The homologies of middle hooks and their supporting bars are not always clear. The protein in the anchors and marginal hooks appears to be keratin.

Rarely, **supplementary discs** or **compensating discs** are developed near the base of the opisthaptor (family Diplectanidae). These are accessory to the opisthaptor and are not technically a part of the larval or adult opisthaptor. They consist of a series of sclerotized lamellae or spines. **Suckers** are found on the ventral surface of the opisthaptors of many species. Their number ranges from two to eight.

Complex **clamps** are found on many species of highly evolved monogeneans. On some species the clamp is muscular, while on others it is mainly sclerotized. It functions as a pinching mechanism, aiding in adherence to the host. While many variations of structure occur, all are based on a single, primitive type of clamp (Fig. 13-8). The identity of the material of which the clamps are constructed is enigmatic; it is not keratin, chitin, quinone-tanned protein, or collagen.[13] The number of clamps varies from eight to several hundred, distributed symmetrically in some species and asymmetrically in others. Finally, there are varying combinations of hooks, suckers, and clamps in several families.

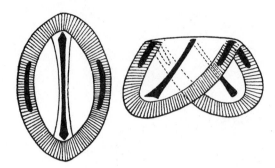

Fig. 13-8. Diagram of the primitive attaching clamp. On the left it is fully open, on the right it is partially closed. The sclerotized parts are black, the musculature is cross-hatched. (From Bychowsky, B. E. 1961. Monogenetic trematodes, their systematics and phylogeny. American Institute of Biological Science, Washington, D.C.)

Tegument

As in the digeneans and cestodes, the tegument of monogeneans traditionally has been referred to as a cuticle because light microscopists could discern little structure within it. However, by use of the electron microscope, the "cuticle" has now been recognized as a living tissue, the **tegument.** Its fundamental structure appears to be similar to those other symbiotic platyhelminths, digeneans, and cestodes, with some noteworthy differences. The surface layer of the tegument is, as in cestodes and digeneans, a syncytial stratum, laden with vesicles of various types and mitochondria, bounded externally by a plasma membrane and glycocalyx and internally by a membrane and basal lamina. This stratum is the **distal cytoplasm,** and it is connected by trabeculae to the "cell bodies" or **perikarya,** located internal to a superficial muscle layer. Such is the case in all species studied so far except *Gyrodactylus,* in which trabeculae and perikarya could not be observed. *Gyrodactylus* is very peculiar in other respects and will be discussed further. Often the outer surface of the tegument is supplied with short, scattered microvilli; in some species these are absent, and shallow pits are observed.

A very curious condition has been reported in certain species. Some areas of the body are without a tegument, and large pieces of the tegument are only loosely connected to the surface. In these areas the basal lamina constitutes the external covering. Rhode (1975) contended that the condition was not artifactual, but that pieces of the tegument were being secreted into the environment, and that these cases might offer a clue to the adaptive value of the tegumental arrangement of the monogeneans, digeneans, and cestodes, that is, syncytial distal cytoplasm with internal perikarya.[19] He suggested that the transfer of an original superficial epithelium into the interior of the body may be a way to prevent permanent damage by the hosts. Since the tegument may be subjected to such damaging influences as host secretions, it then can easily be replaced from the cell machinery that is still intact and can be protected further below.

Muscular and nervous systems

The main musculature, other than that in the opisthaptor, appears to be the **superficial muscles,** just below the distal cytoplasm of the tegument, arranged in circular, diagonal, and longitudinal layers. The muscles of the opisthaptor in the suckers, or inserted on the hooks and accessory sclerites, are clearly important in adhesion. The mechanics of their operation have been explained in several species, and one example is *Entobdella soleae* (Fig. 13-9).[9] This species lives on the skin of the sole, and its opisthaptor is well adapted to anchor the animal firmly in its relatively smooth and exposed site on its host. The disc-shaped opisthaptor forms an effective suction cup. Prominent muscles in the peduncle are inserted on a tendon that passes down to near the ventral surface of the disc, up over a notch in the accessory sclerites, and then to the proximal end of the large anchors. Contraction of the muscles erects the accessory sclerites, so that their distal ends direct down against the fish's skin and their proximal ends serve as a prop toward which the proximal end of the anchors are pulled. This action tends to lift the center area of the opisthaptor, thus reducing pressure and creating suction, at the same time that the distal, pointed ends of the anchors are pushed downward to penetrate the host's epidermis.

Studies on neural and neuromuscular

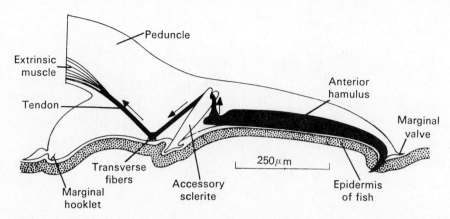

Fig. 13-9. A diagrammatic parasagittal section through the adhesive organ of *E. soleae.* The arrows show the direction of movement of the tendon when the extrinsic muscle contracts. The posterior hamulus (anchor) has been omitted. (From Kearn, G. C. 1971. In Fallis, A. M. Ecology and physiology of parasites. University of Toronto Press, Toronto.)

transmission have not been carried out, but cholinesterase was demonstrated in the nervous system of *Diclidophora merlangi*[7]; therefore, at least some fibers are probably cholinergic.

The general pattern of the nervous system is the ladder type with **cerebral ganglia** in the anterior, with several nerve trunks coursing posteriorly from them. The nerve trunks are connected by the ladder commissures, and additional nerves emanate from the cerebral ganglia to connect with the pharyngeal commissure. As would be expected, the adhesive organs of the opisthaptor are well innervated.

Monogeneans have a fairly wide variety of sense organs. Most have pigmented eyes in the free-swimming larval stage. Oncomiracidia of Monopisthocotylea usually have four eyespots, and these persist in the adult, perhaps somewhat reduced, while the two larval eyespots of Polyopisthocotylea are lost during maturation. These are rhabdomeric eyes similar to those found in Turbellaria and some larval Digenea. In addition, what appears to be a nonpigmented ciliary photoreceptor has been found in the larva of *Entobdella,* with counterparts in structures described in larval Digenea. Several different types of ciliary sense organs in the tegument have been described, including single receptors (one modified cilium in a single nerve ending) and compound receptors (consisting both

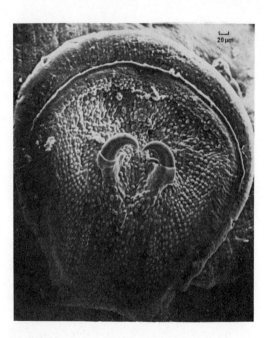

Fig. 13-10. Scanning electron micrograph of the opisthaptor of *Entobdella soleae,* showing the papillae that are thought to be sensory. (From Lyons, K. M. 1973. Z. Zellforsch., vol. 137.)

of several associated nerve endings, each with a single cilium, and of one or a few nerves, each with many cilia).[15]

Finally, a very interesting nonciliated sense organ occurs on the opisthaptor of *Entobdella.* The disc surface of the opisthaptor is covered with over 800 small papil-

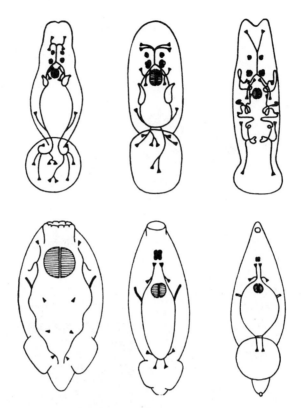

Fig. 13-11. The osmoregulatory system of oncomiracidia, showing patterns of flame cell distribution (From Llewellyn, J. 1963. In Dawes, B., editor. Advances in parasitology, vol. 6. Academic Press, Inc., New York.)

lae (Fig. 13-10), and beneath the tegument of each papilla are packed nerve endings that are doubled over and piled stack-like on top of one another. The function of these peculiar organs is believed to be mechanoreception, perhaps to sense contact with the host or detect local tensions in the opisthaptor. It is not known whether similar organs occur in other monogeneans.[16]

Osmoregulatory system

The excretory system has not been used as a tool for systematics in this group as it has in the Digenea. Typical of the Platyhelminthes, the excretory unit is the **flame cell protonephridium.** Thin-walled capillaries lead from these to fuse with a succession of ducts leading to two lateral **excretory pores** near the anterior end of the worm (Fig. 13-11). The terminal ducts are often each equipped with a contractile bladder at their distal ends.

The fine structure of the excretory system has been studied only in two species of *Polystomoides.*[19] The flame cell is generally similar to that of Digenea and Cestoda with minor differences. The internal surface area of the tubules is increased in a manner differing from that in either of the other groups, that is, by strongly reticulated walls (Fig. 13-12). Lateral or nonterminal flames are frequent.

Acquisition of nutrients

The mouth and buccal funnel often have associated suckers. Behind the buccal funnel a short **prepharynx** is followed by a muscular and glandular **pharynx.** This powerful sucking apparatus draws food into the system. In *Entobdella soleae* the pharynx can be everted and the pharyngeal lips closely applied to the host's skin.[9] The pharyngeal glands secrete a strong protease that erodes the host epidermis, and the lysed products are sucked up into the worm's gut by the pharyngeal muscles. Fortunately for the fish, its epidermis is capable of rapid migration and regener-

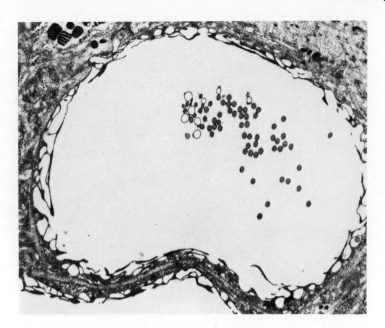

Fig. 13-12. Cross-section of an excretory tubule of *Polystomoides*. Note the reticulate walls and the median cilia. (From Rohde, K. 1975. Int. J. Parasitol. 3:331.)

ation to close the wound left by the parasite's feeding.

Posterior to the pharynx may be an **esophagus,** although it is absent in many species. The esophagus may be simple or have lateral branches and may have unicellular digestive glands opening into it.

In most monogeneans the **intestine** divides into two lateral **crura,** which are often highly branched and may even connect along their length. If the crura join near the posterior end of the body, it is common for a single tube to continue posteriad for some distance. There is no anus. Digestion and ultrastructure of the gut of several species have been studied.[5,8,19] Monopisthocotyleans seem to feed mostly on mucus and epithelial cells of their hosts, although feeding on blood has been demonstrated in *Dactylogyrus* and some other Monopisthocotylea. In contrast, blood appears to be the dominant component of the diet of Polyopisthocotylea. It was believed formerly that the unusable breakdown product of hemoglobin digestion, hematin, was eliminated in the gut of polyopisthocotyleans by sloughing off gut cells containing it, so that parts of the cecal wall were denuded. However, ultrastructural studies on *Polystomoides* and *Diclidophora*

have shown that the cecal epithelium is not discontinuous, but that the hematin-containing cells are interspersed with a different kind of cell called the "connecting cell."[6] In *Diclidophora* both the hematin cell and the connecting cell have their luminal surface increased by long, thin lamellae, but only the hematin cell has such lamellae in *Polystomoides*. It appears that the digestion of hemoglobin in *Diclidophora*, at least, is mostly or entirely intracellular, the protein being taken into the cell by pinocytosis and digested within an extensive, intracellular reticular space, and the hematin being subsequently extruded by temporary connections between the reticular system and the gut lumen. Finally, undigestible particles are eliminated through the mouth in all monogeneans.

Male reproductive system (Fig. 13-13)

Monogeneans are hermaphroditic with cross-fertilization usually taking place. This is epitomized by the genus *Diplozoon*, where individuals completely fuse into pairs with their genital ducts together, the ultimate in "oneness."

Testes usually are rounded or ovoid, but they may be lobated. Most species have only one testis, but the number varies ac-

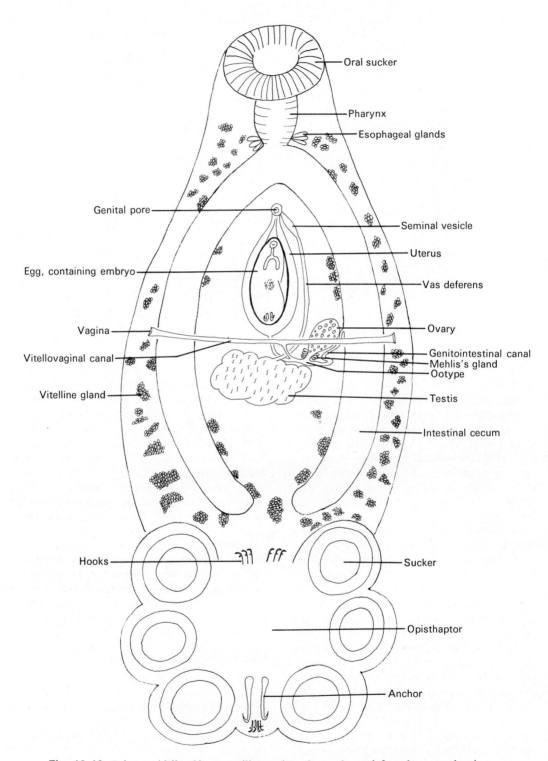

Fig. 13-13. *Polystomoidella oblongum,* illustrating the male and female reproductive systems. (Redrawn from Cable, R. M. 1958. An illustrated laboratory manual for parasitology. Burgess Publishing Co., Minneapolis.)

cording to species, and one species has over 200 per individual. Each testis has a **vas efferens,** which expands or fuses into an **ejaculatory duct.** There is no trace of a cirrus pouch or eversible cirrus, in the sense of those in cestodes or trematodes. In some cases the ejaculatory duct is simple and terminates within a shallow, sometimes sucker-like, **genital atrium,** which propels sperm into the female system at copulation. A higher degree of development is found in many species where the tissues surrounding the terminal ejaculatory duct are thickened and muscular, forming a papilla-like **penis.** Hooks of consistent size and form for each species commonly arm the distal end of the penis. In many, the lining of the distal ejaculatory duct is sclerotized, sometimes for a considerable portion of its length. (We will use the term **sclerotized** although the chemical nature of the stabilized protein is unknown.) A simple, sac-like seminal vesicle is present in some species. Unicellular **prostatic glands** are usually present.

Still another type of copulatory organ exists in several families, where the ejaculatory duct joins with a complex, **sclerotized copulatory apparatus.** These vary widely between species but are similar within a species and, therefore, are important taxonomic characters. The structures are contained in a membranous sac and are controlled by muscles.

Female reproductive system

The single **ovary** of all species of Monogenea is usually anterior to the testes. Between species it varies in shape from round or oval to elongate or lobated. The **oviduct** leaves the ovary and courses toward the ootype, receiving the vitelline, vaginal, and genitointestinal ducts along the way. More specifically, the oviduct extends from the ovary to the confluence with the vitelline duct; the remainder is often referred to as the **female sex duct.** A seminal receptacle is present, either as a simple swelling of the oviduct or as a special sac with a separate duct to the oviduct.

The **vitellaria** are quite abundant, usually extending throughout the parenchyma and often even into the opisthaptor. Despite their many ramifications, the vitellaria consist basically of left and right

groups. Each has its efferent duct; they fuse midventrally near the oviduct, forming a small **vitelline reservoir.** Each vitelline follicle consists of a few cells surrounded by a thin, muscular membrane. The vitelline ducts are lined with ciliated epithelium.

The vagina may be present or not, or when present may be doubled. Vaginal openings are dorsal, ventral, or lateral. The terminal portion is sclerotized in some species, and in others the vaginal pore is multiple or surrounded by spines. In some species, such as *Entobdella,* the vagina may be much smaller than the penis, and sperm transfer is achieved by deposition of a spermatophore adjacent to the vagina of the mating partner, rather than by direct copulation. *Diclidophora merlangi,* which does not have a vagina, practices a kind of hypodermic impregnation.[17] The sucker-like penis of one individual attaches at a ventrolateral position posterior to the genital openings of its partner, draws up a papilla of tegument into the penis, and breaches the tegument with spines in the penis. Sperms enter and make their way between cells of the partner to the seminal receptacle, a distance of 1 to 2 mm.

A very curious structure called the **genitointestinal canal** is present in most Polyopistocotylea. It is a connection between the oviduct and a leg of the intestine or one of its branches. The function, if any, of the duct is unknown. Sometimes yolk granules and sperm are observed in the gut, presumably having arrived there via the genitointestinal canal. One theory is that the canal represents a vestige of a mechanism by which eggs are passed into the intestine to be expelled through the mouth, and another theory is that "surplus" reproductive materials are digested and reabsorbed in the gut.[19] The canal is found in many turbellarians, especially polyclads, and its function there is obscure also.

After being fertilized in the oviduct or ovary itself, the zygote and attendant vitelline cells pass into the ootype, a muscular expansion of the female duct (Fig. 13-14). In those species studied, the **Mehlis's glands** around the ootype are comprised of two cell types, mucous and serous. (The ootype epithelium may also be secretory.)

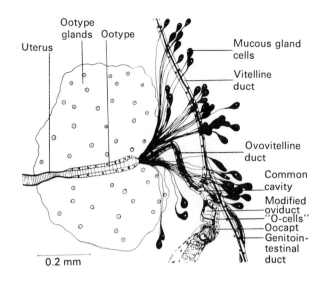

Fig. 13-14. *Polystomoides malayi:* ventral view of the female reproductive tract in the region of the ootype. (From Rohde, K., and A. Ebrahimzadeh. 1969. Zeitschrift fur Pärasitenkunde 33:113.)

The function of Mehlis's gland is not known. It was formerly thought to contribute shell material, but in the Monogenea, as in the Digenea and Cestoda, the shell material seems to come from the vitelline cells. The shape of the egg is apparently determined by the walls of the ootype. In *Entobdella* the tetrahedral egg shape is imparted by four pads in the ootype walls.[9] The eggs of many monogeneans have a filament at one or both ends, also characteristic of a given species. The filament may have an adhesive property and serve to attach the egg to the host or substrate where it falls after release into the open water.

It is generally believed that the protein in the eggshell is stabilized by a process of quinone-tanning to form sclerotin, but recent work indicates that the stabilization may not be by quinone tanning but by means of dityrosine and disulfide links as in resilin and keratin.[18]

Although many eggs may be produced (*Polystoma* produces one to three eggs every 10 to 15 seconds), they are passed out of the worm fairly rapidly; therefore, not many may be found within the parent at one time. Some species may store a few

eggs in the ootype and then pass them to the outside directly through a pore; but in most species, the eggs pass from the ootype into a uterus, which courses anteriad to open into the genital atrium, together with the ejaculatory duct. Hence, the uterus, at least in most cases, does not function as a vagina as in digeneans.

DEVELOPMENT

The life cycles of a few species of Monogenea have been very well studied, while little or nothing is known about most. With the exception of the viviparous Gyrodactylidae, monogeneans usually have a simple, direct life cycle involving an egg, oncomiracidium, and adult. Some evidence suggests that two species of gastrocotylids that parasitize predatory fish do not infect their definitive hosts directly but undergo a period of development on fish preyed on by the parasite's definitive hosts.[11]

Oncomiracidium

The oncomiracidium (Fig. 13-15) hatches from the egg and rather resembles a ciliate protozoan in size and shape. It is elongate and bears three zones of cilia, one in the middle and one at each end. The zones of ciliated epidermal cells are separated by an interciliary, nonnucleated syncytium. It has been shown in *Entobdella* that the nuclei of the interciliary regions are actually extruded during embryogenesis. Subsequently, the perikarya of the "presumptive adult" tegument, which are located within the superficial muscle layer, extend processes out to underlay the ciliated cells and join the syncytial interciliary regions. The animal is thus ready for rapid shedding of the ciliated cells upon attachment to the host; the stimulus for this shedding in *Entobdella* is mucus from the host epidermis, and the shedding takes only 30 seconds.

The oncomiracidium has cephalic glands with efferent ducts opening on the anterior margin and, as previously noted, has one or two pairs of eyes. The digestive tract is well differentiated, and the excretory pores are already formed. The posterior end always is developed into an attachment organ that bears hook sclerites, and these sclerites are retained in the adult. The larvae swim about until they contact

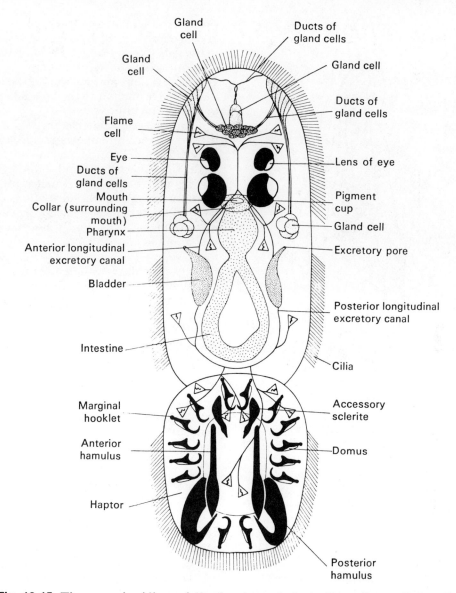

Fig. 13-15. The oncomiracidium of *E. soleae* (ventral view). (From Kearn, G. C. 1971. In Fallis, A. M. Ecology and physiology of parasites. University of Toronto Press, Toronto.)

a host; then they attach, lose their ciliated cells, and develop into adults. Rates of development into adults are largely unknown.

Inasmuch as the oncomiracidium is free-swimming and the primary hosts of Monogenea are fish, with some in other vertebrates that are aquatic or amphibious, it would seem that infection of new hosts would not be a problem. However, the free-swimming life of the oncomiracidium is short, its potential hosts are widely dispersed most of the time and, in any case, can swim much faster than the larva. In addition, potential hosts may not even be present in the aquatic habitat except during breeding season. Thus it is of great selective value for the worm's egg production to be closely related to its host's reproduction; for instance, to coincide with a time when the host will be concentrated in spawning areas. Other features of the

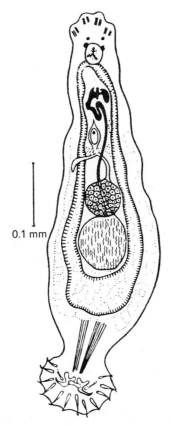

0.1 mm

Fig. 13-16. *Dactylogyrus vastator.* (From Bychowsky. 1933.)

host's habit will also enhance chances for infection. Such correlation has been shown in several species, and similar adaptations are probably more widely prevalent.[12] Some of these adaptations will be illustrated in the discussion of life cycles that follows. Of course, some hosts are quite lethargic or are available over large periods of time in circumscribed areas, and in these cases there may be little or no correlation of monogenean reproduction with host habits.

Dactylogyrus spp. (Fig. 13-16)

A large number of species in this genus have been described, and some of them, such as *D. vastator, D. anchoratus,* and *D. extensus,* are of great economic importance as pathogens of hatchery fishes. *Dactylogyrus* has large anchors on its opisthaptor and lives on the gill filaments of its host. Heavy infections cause loss of blood, ero-

sion of epithelium, and access for secondary bacterial or fungal infections. Irritation to the gills stimulates increased mucus production, which often smothers the fish. Heavy infections may kill the host; massive die-offs are common in the crowded situations of fish culture ponds.

The life cycles and factors influencing the economically important species are reasonably well known.[3] The main features of the life cycle correspond to the preceding general outline. *D. vastator* on carp shows marked seasonal fluctuations correlated with temperature. Each worm deposits four to ten eggs per 24 hours during the summer, and this rate increases with increasing temperature. The eggs require four to five days with temperatures between 20° to 28° C for embryonation, but this rate slows with lower temperatures, down to 4° C, at which point development is completely suppressed. The adult worms are adversely affected by lower temperatures, so that the number of parasites on the fish decreases greatly during the winter. The net effect is that the parasite population builds up over the summer, but the eggs deposited toward the end of the season winter over and result in a mass emergence in the spring to infest the young-of-the-year fish.

Gyrodactylus spp. (Fig. 13-17)

In spite of the (unfortunate) similarity in name and pathological effects with *Dactylogyrus, Gyrodactylus* is a very different organism. It is also very significant economically, being an important pest particularly of trout, bluegills, and goldfish in fish ponds. The family Gyrodactylidae is very unusual among the Monogenea in that it is viviparous. The young are retained in the uterus until they develop into functional subadults. Inside such a developing larva, one can often see a second larva developing, with a third larva inside of it and a fourth inside the third! The exact mechanism of this unique embryogenesis is unknown, but it may be considered a type of sequential polyembryony; up to four individuals usually result from one zygote. After birth, the young worm begins feeding on its host and gives birth to the larva "remaining" inside. Only then can an egg from its own ovary be fertilized and repeat the sequence. As only a day or so is required for a worm to mature after birth and give birth to another worm already developing within it, massive infection can build up quickly.

Certain other peculiarities of *Gyrodactylus* have been described that may be correlated with its unusual embryogenesis. One of these is the possession of an **ovovitellarium,** a fused mass of ova and vitelline cells. Another is the fact that the "distal cytoplasm" of the tegument apparently is not connected to perikarya in the parenchyma. The epidermis of the developing embryo has nuclei that are not present in the adult tegument. The tegument of *Gyrodactylus* may be embryologically equivalent to the ciliated epidermal cells on the oncomiracidium of other monogeneans.[14]

Not having an oncomiracidium, *Gyrodactylus* must depend on transmission of the adult or subadult from one host to another. Since these forms appear unable to swim, it is clear that the prospective host must be quite close to the worm's current host for the transferral to take place, but it is not known how the worm detects the proximity of a new potential host. Random leaps are unlikely; infection can spread through schooled fish in a pond.[12]

Polystoma integerrimum (Fig. 13-18)

This parasite of Old World frogs is of particular interest because it is known that the worm's reproductive cycle is synchronized with that of its host by means of host hormones, a mechanism to provide a ready supply of hosts to the hatching oncomiracidia. Further, two different types of adults develop: normal and neotenic.

Adult worms live in the urinary bladder of their host. They are dormant during the winter, while the frogs hibernate, but become active in the spring along with their hosts. When the frog's gonads begin to swell and produce gametes, the worms begin to copulate and produce eggs that are released into the surrounding urine. In the laboratory, maturation and stimulation of worm gamete production can be elicited by injecting the frogs with pituitary extract, through it is not known whether the effect is direct via the gonadotropins or stimulated by the injection via the gonadal hormones of the host.[20] In nature, the

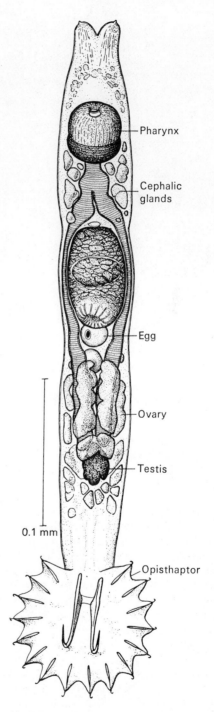

Fig. 13-17. *Gyrodactylus cylindriformis,* ventral view. (From Mueller, J. F., and H. J. Van Cleave. 1932. Roosevelt Wildlife Annals.)

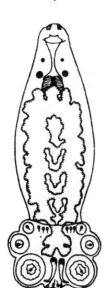

Fig. 13-18. *Polystoma integerrimum,* a parasite of Old World frogs. (From Zeller. 1872.)

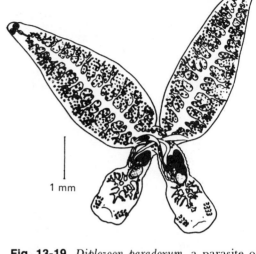

1 mm

Fig. 13-19. *Diplozoon paradoxum,* a parasite of freshwater fishes in Europe and Asia. (From Bychowsky, B. E., and L. F. Nagibina. 1959.)

eggs are voided into the water in the frog's spawning area. Depending on the temperature, the oncomiracidia hatch in 20 to 50 days. By then the frog's eggs have developed into tadpoles that will be the next host generation. Tadpoles first have external, then internal gills, breathing by sucking water into the mouth, over the gills and out the side of the pharynx through slits, much like fish. An oncomiracidium contacting a gill attaches, metamorphoses, and begins producing eggs within 20 to 25 days.

The gill form of *P. integerrimum* is considerably different in morphology from the bladder form. Its body is narrower, and the opisthaptor is not sharply set off from the body. The intestine has fewer lateral branches, and the ovary is a different shape. Further, there is no uterus or genitointestinal canal. Some authors consider the gill stage to be neotenic.[1a]

Eggs from these worms hatch in 15 to 20 days, when the water is somewhat warmer. These larvae also attach to the gills of tadpoles, but by now the tadpoles are older, and worm maturation is delayed. Then the tadpoles begin their metamorphosis, including resorption of the gills, the worms migrate to the bladder. The migration is over the ventral skin of the tadpole at night and only takes about 1 minute.[11]

Oddly, even when the metamorphosing tadpole is exposed to newly hatched larvae, the worms go first to the gills and then to the bladder, and this has been taken as evidence that the migration is controlled endogenously and not by stimuli from the host. It takes much longer, 4 or 5 years, for the bladder form to mature and begin egg production.

A similar species, *P. nearcticum,* occurs in tree frogs in the United States. It also has a neotenic gill form but no slowly developing and then migrating immature gill form. Oncomiracidia enter the cloaca directly when they contact metamorphosing tadpoles. Larvae of *Protopolystoma xenopi* also enter the cloaca of its host, *Xenopus laevis,* directly. Interestingly, *Xenopus* remains in water all year round, and reproduction of *Protopolystoma* continues correspondingly.

Diplozoon paradoxum (Fig. 13-19)

Diplozoon paradoxum is a common parasite of the gills of species of European cyprinid fishes. Like *Dactylogyrus, Diplozoon* exhibits a strong seasonal variation in its reproductive activity. Virtually no gametes are produced during the winter, but gonads begin to function during the spring, reaching a peak during May to June and continuing through the summer. The eggs, which have a long, coiled filament at their ends, can hatch about 10 days after they are deposited, and light intensity and turbulence of the water, as might be caused by host feeding or spawning activity, stimulate hatching. The oncomiracidium bears two clamps on its opisthaptor with which it attaches to a gill filament; it then loses its cilia almost immediately. The worm feeds and begins to grow, adding another pair of clamps to the opisthaptor. A small sucker also appears on the ventral surface and a tiny papilla on the dorsal surface, slightly more posterior than the sucker. When this stage (Fig. 13-20) was first discovered, it was thought to represent a new genus, and it was named *Diporpa.* When *Diporpa* was recognized to be a juvenile stage of *Diplozoon,* the stage was referred to as a **diporpa larva.** Curiously, in contrast to *Dactylogyrus,* young-of-the-year fish are rarely infected with *Diplozoon,* not even diporpae.

A diporpa larva can live for several months, but it cannot develop further until encountering another diporpa, and unless that happens, the diporpa usually perishes by winter. When one diporpa finds another, each attaches its sucker to the dorsal papilla of the other. Thus begins one of the most intimate associations of two individuals in the animal kingdom. The two worms fuse completely, with no trace of partitions separating them. The fusion stimulates maturation. Gonads appear; the male genital duct of one terminates near the female genital duct of the other, permitting cross-fertilization. Two more pairs of clamps develop in the opisthaptor of each. Adults apparently can live in this state for several years.

METABOLISM

Although ultrastructure, behavior, and some aspects of their physiology have received attention in recent years, knowledge of the metabolism of monogeneans is extremely meager. Ectoparasitic species store little glycogen, but endoparasitic forms store about the same amounts as digeneans.[20] In light of the fact that these ectoparasites live in habitats with medium to high oxygen concentrations, it would not be surprising if they depended heavily on

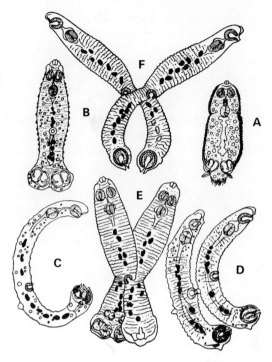

Fig. 13-20. Development of *Diplozoon paradoxum.* **A,** Freshly hatched, free-swimming larva. **B,** Diporpa larva. **C-F,** Diporpa larvae attaching themselves to one another. (From Baer, J. G. 1952. Ecology of animal parasites. University of Illinois Press, Urbana, Ill.)

aerobic energy metabolism. Indeed, certain observations imply that this may be true. *Entobdella soleae,* whose habitat is the lower surface of a bottom-dwelling flatfish, has characteristic, rhythmic, undulating body movements that appear to be clearly ventilatory in function. The frequency and amplitude of the undulations are increased by lower oxygen concentration and increased temperature. Further, when ambient oxygen concentration is high, the parasite is somewhat contracted and wrinkled. At lower oxygen concentrations, the worm stretches out, increasing its surface area and decreasing the distance through which oxygen must diffuse to the interior. Research in monogenean metabolism will surely be rewarding for those who enter the field.

CLASSIFICATION OF CLASS MONOGENEA

Hermaphroditic, dorsoventrally flattened, elongate or oval worms with a syncytial tegument. Conspicuous posterior adhesive organ present (opisthaptor) that is muscular, sometimes divided into loculi, usually with sclerotized anchors, hooks, and/or clamps; often subdivided into individual suckers or clamps without sclerites. Accessory lappets may occur. An anterior adhesive organ (prohaptor) usually present, consisting of one or two suckers, grooves, glands, or expanded ducts from deeper glands. Eyes, when present, usually of two pairs.

Mouth near anterior, pharynx usually present. Gut usually with two simple or branched stems often anastomosing posteriorly; rarely a single, median tube or sac. Gut absent in Gyrocotylidea.

From one to many testes present. Vas deferens commonly convoluted. Seminal vesicle, prostate, and cirrus pouch present or absent. Cirrus present or absent, armed or not, with or without an accessory piece. Male genital pore usually in atrium common with female pore. Atrium present or absent, armed or not. Genital pores ventral or marginal.

Ovary single, variable in shape. Genitointestinal duct present or absent. Seminal receptacle present, uterus short. Reproduction oviparous or viviparous. Vitelline follicles extensive, usually lateral. Vagina present or absent, single or double, with or without sclerotizations, with dorsal, ventral, or lateral pore.

Two lateral osmoregulatory canals present, each with expanded vesicle opening dorsally near anterior end.

Parasites on or in aquatic vertebrates, especially fishes, or rarely on aquatic invertebrates. Cosmopolitan.

Subclass MONOPISTHOCOTYLEA

Opisthaptor a single unit that may be subdivided into shallow loculi, usually developed directly from the larval haptor. One, two, or three pairs of large anchors usually present, commonly with tiny marginal hooks. Prohaptor glandular or with paired suckers or pseudosuckers. Oral sucker absent. Eyes often present. Genitointestinal canal absent. Eggs usually with polar filaments. Seminal receptacle is an enlargement of vagina.

Order ACANTHOCOTYLOIDEA

Opisthaptor developed separately from larval haptor, which remains as a tiny structure with 14 marginal and two central hooklets. Functional haptor muscular, without anchors, but radially arranged spines or septa may be present. Parasites of marine fishes. Cosmopolitan.

Family: Acanthocotylidae.

Order CAPSALOIDEA

Opisthaptor large, circular, muscular, often divided by septa into shallow loculi. Anchors, when present, lacking connecting bars. Marginal hooklets present or absent. Prohaptor, when present, with two glandular areas, two lateral suckers, or single pseudosucker. Cirrus always lacking accessory piece. Parasites of fishes. Cosmopolitan.

Families: Bothitrematidae, Calceostomatidae, Capsalidae, Dactylogyridae, Dioncidae, Diplectanidae, Loimoidae, Microbothriidae, Monocotylidae, Protogyrodactylidae, Tetraoncidae, Tetraoncoididae.

Order GYRODACTYLOIDEA

Opisthaptor rounded or bilobed, with none, one, or two pairs of anchors supported by one, two, or three bars. Marginal larval hooks present. Prohaptor with groups of cephalic glands with ducts opening on margin of anterior end. Parasites of fishes, amphibians, cephalopods, and crustaceans. Cosmopolitan.

Family: Gyrodactylidae.

Order UDONELLOIDEA

Opisthaptor muscular, lacking armature or septa. Prohaptor poorly developed, but with lateral head organs or pseudosuckers. Parasites of marine fishes or of copepods parasitic on marine fishes.

Family: Udonellidae.

Subclass GYROCOTYLIDEA

Body flattened, elongated, with indistinct holdfast mechanism at anterior end. Posterior end forming a crenulated rosette or a long slender cylinder. Alimentary tract absent. Genital pores anterior; male pore ventral, female pore dorsal. Testes anterior in two lateral fields. Ovary posterior. Vitellaria follicular, lateral. Uterine pore anterior, ventral. Intestinal parasite of chimaeroid rishes. (This group usually has been considered a member of the Cestodaria in the Cestoidea, but we agree with Llewellyn [1965], who regards them as an endoparasitic monogenean in which the gut has been lost.[10])

Family: Gyrocotylidae.

Subclass POLYOPISTHOCOTYLEA

Opisthaptor complex, with suckers, clamps, or anchor complexes, commonly subdivided. Larval haptor absent or reduced to pad-supporting terminal anchors. Marginal hooklets usually absent.

Mouth surrounded by a sucker, a striated fringe, or with paired suckers inside buccal cavity. Prohaptor usually without adhesive glands. Eyes usually absent. Genitointestinal canal usually present. Gut usually with two crura, sometimes joined posteriorly. Testes usually numerous. Eggs commonly with polar filaments. Seminal receptacle present or absent.

Order AVIELLOIDEA

Opisthaptor at the end of long stalk, with six suckers on margin and four large anchors in middle. Intestine a simple unbranched sac. Prohaptor with two or three pairs of glandular organs. Four eyes present. Testis single. Eggs lacking polar filaments. Parasites of freshwater teleosts. Lake Baikal, Russia.

Family: Aviellidae.

Order CHIMAERICOLOIDEA

Posterior portion of body elongate, slender, with small opisthaptor present near larval haptor. Opisthaptor with two rows of simple clamps. Prohaptor a simple, weak, circumoral sucker. Eyes and intraoral suckers absent. Reproductive organs in anterior portion of body. Parasites of Holocephali. Atlantic, Mediterranean, Pacific.

Family: Chimaericolidae.

Order DICLIDOPHOROIDEA

Opisthaptor commonly subdivided into two lateral rows of four each, suckers or clamps. Prohaptor usually simple. Two small suckers present in oral cavity. Intestine bifurcate. Eyes absent. Parasites of fishes or crustaceans parasitic on fishes. Cosmopolitan.

Families: Dactylocotylidae, Diclidophoridae, Discocotylidae, Gastrocotylidae, Hexostomatidae, Macrovalvitrematidae, Mazocraeidae, Octolabeidae, Plectanocotylidae, Protomicrocotylidae, Pterinotrematidae.

Order DICLYBOTHRIOIDEA

Opisthaptor with three pairs of clamps or suckers, each surrounding a large anchor. Also there is a posterior appendix on the opisthaptor that bears three pairs of large and one pair of very small hooks, and in some species also a rudimentary pair of suckers on the posterior margin. Prohaptor with two lateral, sucker-shaped depressions. Two pairs of eyes present. Intestine bifurcated, branched, anastomosing near posterior end of body. Parasites of Acipenseriformes, Selachii, and Polyodontidae. Cosmopolitan.

Families: Diclybothriidae, Hexabothriidae.

Order DIPLOZOOIDEA

Adults permanently fused in pairs forming an X shape. Opisthaptor rectangular or bilobed, with four pairs, or numerous, clamps. One pair of posterior anchors also present in some species. Intestine a single tube with many branches. Genital pore in posterior half of body. Parasites of the gills of freshwater fishes. Cosmopolitan.

Family: Diplozoidae.

Order MEGALONCOIDEA

Opisthaptor complex, with three pairs of anchor complexes, each consisting of several sclerotized elements. Marginal hooklets and eyes absent. Two terminal appendices, each with two pairs of hooks, present on posterior margin of opisthaptor of some species. Prohaptor consists of two suckers in the buccal cavity. Parasites of marine teleosts. Japan, China.

Families: Anchorophoridae, Megaloncidae.

Order MICROCOTYLOIDEA

Opisthaptor symmetrical or asymmetrical, with numerous clamps arranged symmetrically or asymmetrically. Anchors may also be present at posterior end. Prohaptor consists of two suckers in the buccal cavity. Intestine bifurcate, not anastomosed posteriorly. Eyes absent. Parasites of marine fishes. Cosmopolitan.

Families: Allopyragraphoridae, Axinidae, Cemocotylidae, Heteromicrocotylidae, Microcotylidae, Pyragraphoridae.

Order POLYSTOMATOIDEA

Opisthaptor with two or six well-developed suckers, with or without anchors, with larval hooklets present. Mouth surrounded by oral sucker. Intestine bifurcate, rejoined or not. Eyes usually absent. Parasites of fishes, amphibians, reptiles, and on eyes of hippopotami. Cosmopolitan.

Families: Polystomatidae, Sphyranuridae.

REFERENCES

1. Braun, M. 1889-1893. Trematodes Rudolphi 1808. In Bronn's Klassen and Ordnugen des Tierreichs 4:306-925.
1a. Baer, J. G., and L. Euzet. Classe des monogènes. Monogenoidea Bychowsky. In Grassé, P., editor. Traité de zoologie. Anatomie, systématique, biologie, vol. iv. Masson et Cie, Paris.
2. Bychowsky, B. E., 1937. Ontogenese und phylogenetische Beziehungen der parasitischen Platyhelminthes. Izvest. Akad. Nauk SSSR Seria Biol. 4:1353-1383.
3. Bychowsky, B. E. 1957. Monogenetic trematodes: their systematics and phylogeny. Akad. Nauk SSSR. (English Translation: Hargis, W. J., editor. American Institute of Biological Sciences, Washington, D.C.

4. Fuhrmann, O. 1928. Trematoda. Zwite Klasse der Cladus Platyhelminthes. In Kukenthal's and Krumbach's Handbuch der Zoologie 2:1-140.

5. Halton, D. W. 1974. Hemoglobin absorption in the gut of a monogenetic trematode, *Diclidophora merlangi*. J. Parasitol. 60:59-66.

6. Halton, D. W., E. Dermott, and G. P. Morris. 1968. Electron microscope studies on *Diclidophora merlangi* (Monogenea: Polyopisthocotylea). I. Ultrastructure of the cecal epithelium. J. Parasitol. 54:909-916.

7. Halton, D. W., and G. P. Morris. 1969. Occurrence of cholinesterase and ciliated sensory structures in fish gill-fluke, *Diclidophora merlangi* (Trematoda: Monogenea). Zeitschr. Parasitenk. 33:21-30.

8. Jennings, J. B. 1968. Nutrition and digestion. In Florkin, M., and B. T. Scheer, editors. Chemical zoology, vol. II, Section III, Platyhelminthes, Mesozoa. Academic Press, Inc., New York. pp. 303-326.

9. Kearn, G. C. 1971. The physiology and behaviour of the monogenean skin parasite *Entobdella soleae* in relation to its host *(Solea solea)*. In Fallis, A. M., editor, Ecology and physiology of parasites. University of Toronto Press, Toronto. pp. 161-187.

10. Llewellyn, J. 1965. The evolution of parasitic platyhelminths. In Taylor, A. E. R., editor. Evolution of parasites, third symposium of the British Society for Parasitology, Blackwell Scientific Publications, Oxford. pp. 47-78.

11. Llewellyn, J. 1968. Larvae and larval development of monogeneans. In Dawes, B., editor. Advances in parasitology, vol. 6. Academic Press, Inc. New York. pp. 373-383.

12. Llewellyn, J. 1972. Behaviour of monogeneans. In Canning, E. U., and C. A. Wright, editors. Behavioural aspects of parasite transmission. Linnaean Society of London. Academic Press, Inc., London. pp. 19-30.

13. Lyons, K. M. 1966. The chemical nature and evolutionary significance of monogenean attachment sclerites. Parasitology 56:63-100.

14. Lyons, K. M. 1970. Fine structure of the outer epidermis of the viviparous monogenean *Gyrodactylus* sp. from the skin of *Gasterosteus aculeatus*. J. Parasitol. 56:1110-1117.

15. Lyons, K. M. 1972. Sense organs of monogeneans. In Canning, E. U., and C. A. Wright, editors. Behavioural aspects of parasite transmission. Linnaean Society of London. Academic Press, Inc., London. pp. 181-199.

16. Lyons, K. M. 1973. The epidermis and sense organs of the Monogenea and some related groups. In Dawes, B., editor. Advances in parasitology, vol. 11. Academic Press, Inc., New York. pp. 193-232.

17. Macdonald, S., and J. Caley. 1975. Sexual reproduction in the monogenean *Diclidophora merlangi*: tissue penetration by sperms. Zeitschr. Parasitenk. 45:323-334.

18. Ramalingam, K. 1973. Chemical nature of the egg shell in helminths: II. Mode of stabilization of egg shells of monogenetic trematodes. Exp. Parasitol. 34:115-122.

19. Rohde, K. 1975. Fine structure of the Monogenea, especially *Polystomoides* Ward. In Dawes, B., editor, Advances in parasitology, vol. 13. Academic Press, Inc., New York. pp. 1-33.

20. Smyth, J. D. 1966. The physiology of trematodes. Oliver and Boyd, Edinburgh.

21. Sproston, N. G. 1946. A synopsis of the monogenetic trematodes. Trans. Zool. Soc. London 25:185-600.

22. Thurston, J. P., and R. M. Laws. 1965. *Oculotrema hippopotami* (Trematoda: Monogenea) in Uganda. Nature (London) 205:1127.

SUGGESTED READING

Hargis, W. J., Jr., A. R. Lawler, R. Morales-Alamo, and D. E. Zwerner. 1969. Bibliography of the monogenetic trematode literature of the world. Virginia Institute of Marine Science, Special Scientific Report, Gloucester Point, Va. 55:1-95.

Hargis, W. J., Jr. 1957. The host specificity of monogenetic trematodes. Exp. Parasitol. 6:620-625.

Hyman, L. H. 1951. The invertebrates: Platyhelminthes and Rhynchocoela. The Acoelomate Bilateria, vol. 2. McGraw-Hill Book Co., New York. (An important source of information on the morphology of the Monogenea.)

Llewellyn, J. 1963. Larvae and larval development of monogeneans. In Dawes, B., editor. Advances in parasitology, vol. 1. Academic Press, Inc., New York. pp. 287-326. (This is an excellent review on the subject, in which the author makes important phylogenetic hypotheses.)

Sproston, N. G. 1946. A synopsis of the monogenetic trematodes. Trans. Zool. Soc. London 25:185-600. (An extremely important monograph on the systematics of the group.)

Yamaguti, S. 1963. Systema Helminthum, vol. IV. Interscience Publishers, New York. (This is an easy-to-use key to all genera of Monogenea, with many illustrations.)

Chapter 14

CLASS TREMATODA: SUBCLASS ASPIDOGASTREA

The Aspidogastrea constitute a small group of Digenea-like worms that seem poorly adapted to parasitism. Though a discrete group, they are much more similar to the Digenea than to the Monogenea and so are given status as a subclass of the Trematoda. They have established a loosely parasitic relationship with molluscs, in most species, but some are facultative or obligate parasites of fishes or turtles.

Two other names have often been used for this group, namely Aspidocotylea and Aspidobothria. While the latter undoubtedly has priority, most of the literature has accumulated under the name Aspidogastrea, which is the title we shall use here.

By any name, this group of organisms has attracted little attention because they are of no medical or known economic importance. Despite this, these innocuous little worms are of considerable biological interest, for they seem to represent a step between free-living and parasitic organisms. A comprehensive review of current knowledge of Aspidogastrea was presented by Rohde.[1]

FORM AND FUNCTION

Body form. Externally, aspidogastreans exhibit three basic types of anatomy, corresponding to the three families that have been established for them. The Aspidogastridae (Fig. 14-1) have a huge **ventral sucker,** extending most of the length of the body. This sucker (also known as an **opisthaptor** or **Baer's disc**) has muscular **septa** in longitudinal and transverse rows, dividing it into shallow depressions called **alveoli** or **loculi.** The number, shape, and arrangement of these loculi are of considerable taxonomic importance. Hooks or other sclerotized structures are never present. Between the marginal loculi there are usually **marginal bodies,** which are se-

cretory organs, or short tentacles, also presumably secretory in nature. Exceptionally, both are absent.

In the Stichocotylidae (Fig. 14-9), a longitudinal series of individual suckers occurs instead of a single complex of loculi; while in Rugogastridae, the ventral holdfast is made up of transverse **rugae** (Fig. 14-2).

Marginal bodies are round to oval organs located between the marginal alveoli of the ventral disc of most aspidogastreans. They are connected to each other by fine ducts and consist of gland cells, storage chambers, and secretory ducts. Although a sensory function has been suggested for the marginal bodies, there is no indication that their function is other than secretory. The tentacles of *Lophotaspis* (Fig. 14-3) are probably modified marginal organs.

The **longitudinal septum** is a very peculiar morphological characteristic of the Aspidogastrea. It is a horizontal layer of connective tissue and muscle in the anterior part of the body, projecting like a shelf and dividing the body into dorsal and ventral compartments. The function of the septum is not known, but it might be correlated with pressures exerted by contraction of the giant ventral sucker.

Tegument. Though only one species (*Multicotyle purvisi*) has been studied adequately, the tegument of this species seems to be basically similar to that of other groups of parasitic flatworms. It is syncytial and has an outer stratum of **distal cytoplasm,** containing numerous vesicles of various types, and mitochondria. The tegumental nuclei are in **perikarya** internal to the superficial muscle layer and connected to the distal cytoplasm by trabeculae. The perikarya are rich in Golgi complexes. A mucoid layer of variable

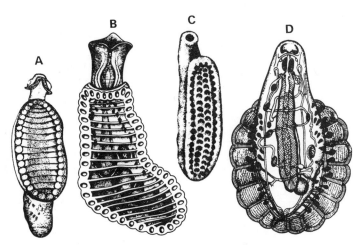

Fig. 14-1. Examples of the family Aspidogastridae. **A,** *Lobatostomum ringens;* **B,** *Cotylogaster michaelis;* **C,** *Lophotaspis vallei;* **D,** *Cotylaspis insignis.* (From various sources.)

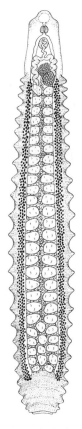

Fig. 14-2. *Rugogaster hydrolagi* from the rectal gland of a ratfish. (From Schell, S. C. 1973. J. Parasitol. 59:803-805.)

thickness is found on the outer surface membrane, and in some areas the surface membrane has rib-like elevations to support the thick mucoid layer.

Digestive system. The digestive tract is simple. The **mouth** is funnel-like in some species, while in others it is surrounded by a muscular sucker or several muscular lobes. At the base of the mouth funnel is a spheroid **pharynx,** a powerful muscular pump. The **intestine** or **cecum** is a single, simple sac that usually extends to near the posterior end of the body. Its epithelial cells bear a complex reticulum of lamellae on their luminal surface, presumably vastly increasing the absorptive surface. A layer of muscles, usually of both circular and longitudinal fibers, surrounds the cecum.

Osmoregulatory system. This system consists of numerous **flame cell protonephridia** connected to capillaries feeding into larger excretory ducts and eventually into an **excretory bladder** near the posterior end of the body. The flame cells are peculiar in that their ciliary membranes continue beyond the tips of the cilia and anchor apically in the cytoplasm of the flame cell. Lateral or nonterminal ciliary flames have been reported in a number of species. The small capillaries have numerous microvilli projecting into their lumina, and the larger capillaries and excretory ducts are abundantly provided with lamellar projections of their surface mem-

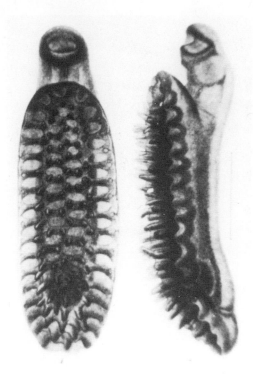

Fig. 14-3. *Lophotaspis interiora* from an alligator snapping turtle. (From Ward, H. B., and S. H. Hopkins. 1932. J. Parasitol. 18:69-78.)

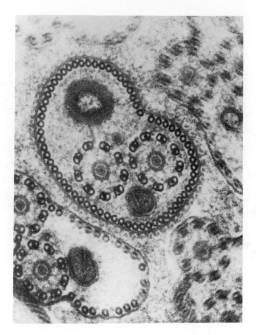

Fig. 14-4. Cross-section of sperm filament of *Aspidogaster conchicola.* (Photograph by Ronald P. Hathaway.)

branes, thus suggesting secretory-absorptive function. The **excretory pore** is dorsosubterminal or terminal and usually single.

Nervous system. The nervous system of aspidogastreans is very complex for a parasitic flatworm, reminiscent of a condition more typical of free-living forms. As in the Turbellaria, there is a complex set of anterior nerves called the **cerebral commissure** and a modified, ladder-type peripheral system. A wide variety of sensory receptors have been observed, mostly around the mouth and on the margins of the ventral disc. In a specimen of *Multicotyle purvisi* 6.1 mm long, Rohde counted 360 dorsal and 260 ventral receptors in the prepharyngeal region, and 140 in the oral cavity, not counting free nerve endings below the tegument.[1] There is a complex system of connectives and commissures in the ventral disc and walls of the alveoli, indicating a high degree of neuromuscular coordination. The septum, intestine, pharynx, prepharynx, cirrus pouch, uterus, and genital and excretory openings are all innervated by plexuses. Some cells in the nervous system are positive for paraldehyde-fuchsin stain, indicating possible neurosecretory function.

Reproductive systems. The male reproductive system of Aspidogastrea is similar to that of the Digenea (Fig. 16-16). One, two, or many **testes** are present, located posterior to the ovary. The **vas deferens** expands to form an **external seminal vesicle** before it enters the **cirrus pouch** to become the **ejaculatory duct.** A cirrus pouch is absent in some species. The **cirrus** is unarmed and opens through the genital pore into a common genital atrium, located on the midventral surface just anterior to the leading margin of the ventral disc. The axial complexes in the spermatozoan filament have the 9 + 1 structure, as is often the case in platyhelminth sperm (Fig. 14-4).

The female reproductive system consists of an ovary, vitelline cells, uterus, and associated ducts. The **ovary** is lobated or smooth and empties its products into an **oviduct.** The oviduct is peculiar among the Platyhelminthes in that its lumen is divided into many tiny chambers by septa,

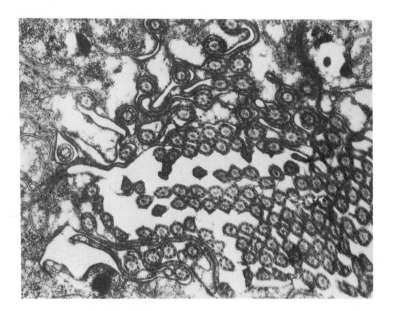

Fig. 14-5. Section of proximal oviduct of *Aspidogaster conchicola,* showing the cilia that line much of its length. (Photograph by Ronald P. Hathaway.)

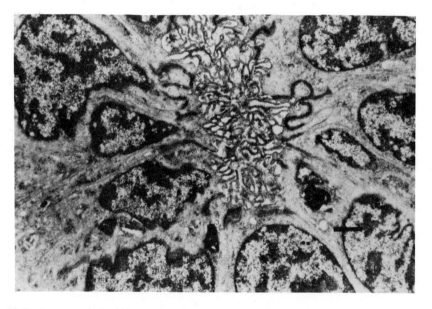

Fig. 14-6. Ootype of *Aspidogaster conchicola,* surrounded by Mehlis's gland cells. (Photograph by Ronald P. Hathaway.)

and the lining along much of its length is ciliated (Fig. 14-5). Each septum has a small hole in it through which the eggs pass. The oviduct empties into the **ootype,** which is surrounded by Mehlis's gland cells (Fig. 14-6). There is a short tube lead-ing from the ootype, which ends blindly in the parenchyma or, in a few cases, con-nects with the excretory canal. This tube is called **Laurer's canal** and probably rep-resents a vestigial vagina.

Vitelline follicles are in two lateral

fields, each of which has a main **vitelline duct** that fuses with that from the other field to form a small **vitelline reservoir,** which in turn opens into the ootype. Finally, a **uterus** extends from the ootype to course toward the genital atrium, usually with a posterior loop and anterior, distal stem. The distal end of the uterus has powerful muscles in its walls and is called the **metraterm.** This propels the eggs out of the system.

Aspidogastreans are apparently self-fertilizing, with the cirrus depositing sperms in the terminal end of the uterus, which serves as a vagina.

DEVELOPMENT

As in other platyhelminths with separate vitellaria, the eggs of aspidobothreans are ectolecithal, that is, most of the yolk supply of the embryo is derived from separate cells packaged with the zygote inside the egg shell. Some species are completely embryonated when they pass from the parent, and they hatch within a matter of hours, while others require 3 to 4 weeks of embryonation in the external environment. The larvae (Fig. 14-7) hatching from the egg in most species have a number of ciliary tufts that are effective in swimming. They have a mouth, pharynx, simple gut, and a prominent posterior-ventral disc without alveoli; there are no hooks. As the worm develops in its host, alveoli begin to form, tier by tier, in the anterior part of the ventral disc. The original cup of the disc remains apparent for some time behind the new ventral sucker, then disappears forever. The larval ultrastructure has been studied only in *Multicotyle purvisi.* The tegument pattern is quite similar to that of the adult—with the distal cytoplasmic, syncytial layer at the surface and with internal perikarya. Between the ciliary tufts, and covering most of the body, the tegument surface bears unique filiform structures called **microfila.** These have one central filament and about nine to 12 peripheral filaments, differing from microvilli in that they do not have a cytoplasmic core. Their function is unknown, but it has been suggested that they help the larva to float.[1]

Most aspidogastreans have a direct life cycle, requiring no intermediate host. Those parasitic in vertebrates do appear

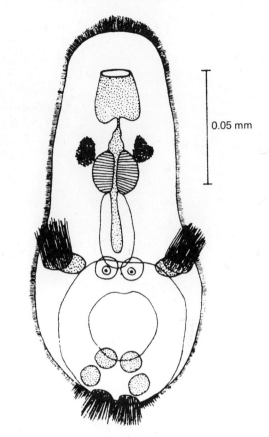

Fig. 14-7. Ciliated larva of *Multicotyle purvisi.* (From Rohde, K. 1968. Zeitschrift für Parasitenkunde 30:78-94.)

to require an intermediate host; no case is known in which the free larva is directly infective to vertebrates. Individuals can be removed from their definitive hosts and are capable of surviving for several days in water or saline, suggesting that they are rather generalized physiologically and not highly specialized for parasitism. Further, if they are eaten by a fish or turtle, they can live for a considerable length of time in this new host. Therefore, it is not uncommon to find an aspidogastrean in the intestine of a fish, though it normally parasitizes a mollusc.

The following life cycles illustrate the biology of two families in the subclass.

Aspidogaster conchicola (Fig. 14-8)

This common representative of the Aspidogastridae is most often found in the pericardial cavity of freshwater clams in

Fig. 14-8. *Aspidogaster conchicola,* a common parasite of freshwater clams. Lateral and ventral views. (Photograph by Warren Buss.)

Europe, Africa, and North America, although it is known from other molluscs, fishes, and turtles. The adult is 2.5 to 3 mm long by 1 mm wide; it is oval, with a long, mobile "neck" with a buccal funnel at its end. The loculi on the ventral sucker are arrayed in four longitudinal rows, totaling 64 to 66.

When the eggs hatch within the host mollusc, the young can develop without further migration. If the egg or larva leaves the mollusc and is drawn into the incurrent siphon of the same or another clam, it can reach the nephridiopore and migrate through the kidney into the pericardium. The larva is 13 to 17 μm long at hatching, lacks external cilia, and bears a simple posterior sucker without loculi. Growth and metamorphosis are rapid.

Lophotaspis vallei, also in Aspidogastridae, may use a marine snail as intermediate host. Mature forms have been found in marine turtles, but it is possible that they normally mature in molluscs.

Stichocotyle nephropsis (Fig. 14-9)

This parasite lives in the bile ducts of rays in the Atlantic Ocean. It has been found in lobsters and other crustaceans and is thought to utilize them as intermediate hosts. The adult is slender, 115 mm long, and has 24 to 30 separate suckers along

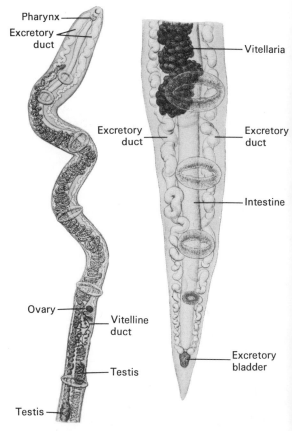

Fig. 14-9. *Stichocotyle nephropsis* from the bile ducts of rays. (From Odhner, T. 1910. Kungl. Svanska Vetern. Handl. 45:3-16.)

its ventral surface. This is the only species in Stichocotylidae and the only aspidogastrean found in crustaceans. It is entirely possible that the crustacean is not a normal host in the life cycle of this parasite, and that its occurence in these animals is accidental.

PHYLOGENETIC CONSIDERATIONS

Superficially, the Aspidogastrea appear to be a link between the Monogenea and Digenea. Their anatomy is rather digenean, while their biology is suggestive of the Monogenea. This tiny group of worms displays sufficient individuality that it becomes apparent that they are distinct and separate from both.

Aspidogastreans differ morphologically from both Digenea and Monogenea in that the ventral sucker develops as a new structure, unrelated to any homologue in larvae or adults of the other two groups. The frontal septum of aspidogastreans is not found in either other group, nor is the septate, ciliated oviduct. Yet, the predominance of mollusc hosts and the presence of Laurer's canal and a highly developed nervous system are suggestive of Digenea. The simple, sac-like cecum and undemanding physiological requirements are primitive, more in keeping with the Turbellaria.

Clearly, then, the Aspidogastrea is a small group of animals, found mainly in molluscs, that appears not to be highly specialized for parasitism. They retain sufficient identity to justify their placement in a group of their own, closest to Digenea, but undoubtedly as a subclass of the Trematoda.

Classification of Aspidogastrea

Subclass ASPIDOGASTREA

Trematodes with single, large, ventral sucker subdivided by septa into numerous, shallow loculi, or with one ventral row of individual suckers. No sclerotized armature on any species. Mouth with or without sucker, sometimes lobated. Pharynx well developed. Intestine with a single or double median sac. Testis single, double, or numerous. Cirrus pouch present or absent. Genital pores median, in front of sucker. Ovary single, pretesticular. Vagina absent, Laurer's canal sometimes present. Vitellaria follicular, usually lateral but occasionally otherwise. Eggs lacking polar prolongations. Excretory pores on or near posterior end. Development direct, without metamorphosis. Parasites of molluscs, fish, and turtles.

Family ASPIDOGASTRIDAE

Body oval or elongate. Ventral sucker with numerous shallow loculi. One or two testes present.

Vitellaria follicular, lateral. Parasites of molluscs, fishes, or turtles. Cosmopolitan.

Family STICHOCOTYLIDAE

Body elongate, slender. Ventral surface with longitudinal row of separate suckers. Two testes present. Vitellaria tubular, unpaired. Parasites of Batoidea.

Family RUGOGASTRIDAE

Body elongate. Most of ventral and lateral body surface has transverse rugae. Musculature of buccal funnel weakly developed. Pharynx, prepharynx, and esophagus present. Two ceca. Testes multiple. Ovary pretesticular. Laurer's canal present, seminal receptacle absent. Vitellaria distributed along ceca. Uterus ventral to testes. Eggs operculate. Parasites of Holocephali.

REFERENCE

1. Rohde, K. 1972. The Aspidogastrea, especially *Multicotyle purvisi* Dawes, 1941. In Dawes, B., editor. Advances in parasitology, vol. 10. Academic Press, Inc., New York. pp. 77-151.

SUGGESTED READING

Baer, J. G., and C. Joyeux. 1961. Classe des trématodes. (Trematoda Rudolphi.) In Grassé, P., editor. Traité de zoologie. Anatomie, systématique, biologie, vol. IV. Masson et Cie., Paris. pp. 561-570.

(This brief discussion gives all the salient knowledge of the Aspidogastrea.)

Dawes, B. 1946. The Trematoda. Cambridge University Press, London. pp. 37-44. (This is a thorough taxonomic review of the Aspidogastrea.)

Dollfus, R. P. 1958. Trematodes. Sous-classe Aspidogastrea. Ann. Parasitol. 33:305-395. (A detailed summary of knowledge of this group to 1958.)

Yamaguti, S. 1963. Systema Helminthum. vol. IV. Interscience Publishers, New York. (The most useful taxonomic treatment of the group.)

Chapter 15

SUBCLASS DIDYMOZOIDEA

This small group of trematodes shows many affinities with the Digenea and in fact traditionally has been placed within that subclass. However, because of certain morphological and biological peculiarities of these worms, it seems best to remove them from the Digenea.

All known species are tissue-swelling parasites of fishes, especially those in marine environments. Most live within gills, often in pairs, but other organs are also infected, such as skin, kidney, and ovary. A review of the group is given by Baer and Joyeux.[1]

Morphology. The body of didymozoids is elongated, usually flattened but occasionally globular or lobated. Most are only a few millimeters long but *Nematobothrium* reaches a length of 2.5 meters, and a worm found in an ocean sunfish (Fig. 15-1) may reach a length of 40 feet![2]

In several genera, such as *Didymozoon* and *Wedlia,* the anterior end of the body is long and thin and inserts at an angle somewhere along the length of the stout or globular posterior body (Fig. 15-2). The body is long and thread-like in *Nematobothrium, Metanematobothrium, Atalostrophion,* and others. In *Diplotrema* and *Phacelotrema,* two or even three individuals become completely fused at their posterior ends.

Many species are dioecious and exhibit striking sexual dimorphism. Most live in encysted pairs.

The mouth is at or near the anterior end of all species. An oral sucker is absent in some but present in varying degrees of development in others. A ventral sucker is apparently absent in most species of didymozoids or is present in young individuals only. A true pharynx is absent or feebly developed. The intestine has two limbs that may reach to near the posterior end of the body or undergo varying degrees of atrophy. The intestine is completely absent in a few species.

The excretory system is of the flame bulb protonephridia type. A voluminous excretory bladder extends to near the pharynx in some species; in some it has no external opening.[3]

The gonads of the Didymozoidea are generally tubular and filiform. One or two thread-like testes occur, but a cirrus pouch is absent. The vas deferens may be expanded into a terminal seminal vesicle. The single ovary is also thread-like and leads to a short oviduct that is surrounded for most of its length by voluminous Mehlis's glands. An ootype and Laurer's canal are absent. The uterus extends to near the anterior end of the body, then descends to near the posterior end, and then winds to near the mouth, where it opens with the male pore. The eggs are very small, operculate, and numerous.

Biology. A complete life cycle is known for no species of Didymozoidea. The fragmentary knowledge available suggests that development is direct, without the intervention of an intermediate host.

Some species are undoubtedly hermaphroditic, probably fertilizing their own eggs. Others are dioecious with sexual dimorphism, but vestigial organs of the opposite sex may be present.[4] Sex determination has not been demonstrated experimentally, but available evidence indicates that the sexuality of one individual may be determined by the presence of another individual. Ishii showed that the first worm in a location develops into a female, while any later arrival has its femininity inhibited and thus becomes a male.[5] A similar phenomenon is known among some molluscs, echiurids, and parasitic crustacea and is suspected in dioecious tapeworms.

The egg contains a larva that appears to be directly infective to the definitive host. On hatching, the unciliated larva has a short gut with an oral sucker that is surrounded by two or more circles of spines. The mode of infection to the definitive

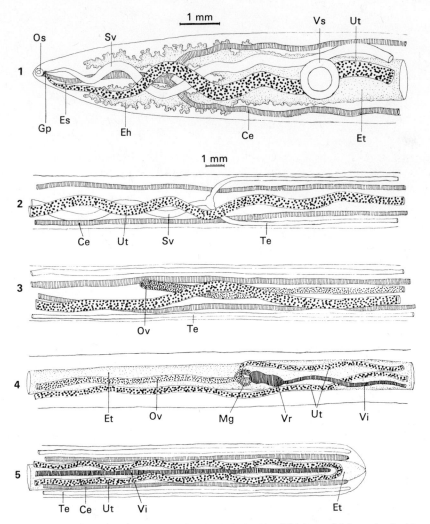

Fig. 15-1. Schematic drawings of *Nematobibothrioides histoidii* from *Mola mola*. **1,** Anterior end of worm. Note minute oral sucker, relatively large ventral sucker, absence of cirrus apparatus, and diverticulated excretory horns. **2,** Union of testes to form seminal vesicle. The excretory tube is omitted to make other organs stand out more clearly. **3,** Beginning of the ovary. All organs are filamentous. Excretory tube omitted for clarity. **4,** Ovary entering female genital complex at Mehlis's gland. Testes and ceca omitted for clarity. **5,** Posterior end of body.

Abbreviations: *Ce,* cecum; *Eh,* excretory horn; *Es,* esophagus; *Et,* excretory tube; *Gp,* genital pore; *Mg,* Mehlis's gland; *Os,* oral sucker; *Ov,* ovary; *Sv,* seminal vesicle; *Te,* testis; *Ut,* uterus; *Vi,* vitellarium; *Vr,* vitelline reservoir; *Vs,* ventral sucker. (From Noble, G. A. 1975. J. Parasitol. 61:224-227.)

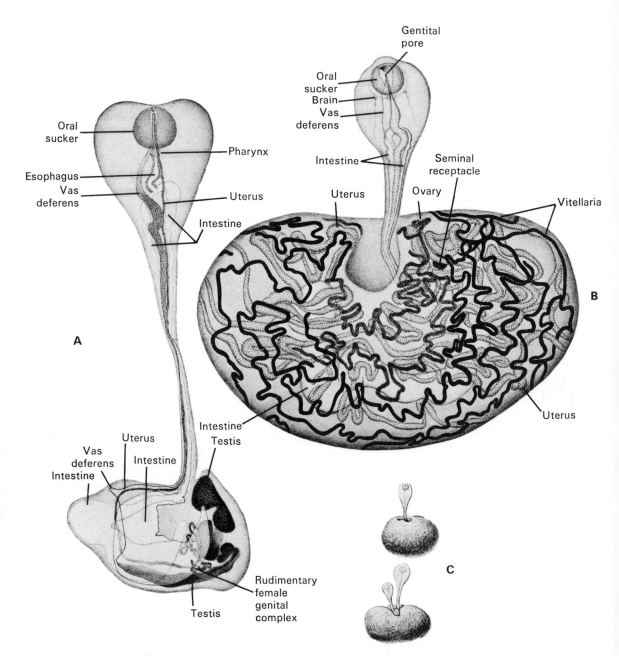

Fig. 15-2. *Koellikeria bipartita,* a common parasite of the intestinal mucosa of various tunas. **A,** Male. **B,** Female. **C,** Female (top) and male inside cavity of female (bottom). (Adapted from Odhner, T. 1910. Zur Anatomie der Didymozoen: ein Getrenntgeschlechtlicher Trematode mit rudimentärem Hermaphroditismus. Almqvist and Wiksells, Uppsala.)

host is unknown; it may be by direct penetration to the site of maturation or by migration to that site through the bloodstream.

Attempts to infect snails and crustaceans with *Ovarionematobothrium texomensis,* a parasite of the ovaries of buffalo fishes in North America, have been unsuccessful.[6] This worm is only found in mature ovaries. Immature worms have never been reported. Presumably the worms are stimulated to mature by the host's hormones, then die, disintegrate, and pass out of the fish at spawning. Many fish do not spawn, in which case the ovaries and worms are resorbed. It has been shown that less than 7% of the parasite's eggs are viable in nonspawning fish.[7] How fish become infected is unknown.

Classification of Didymozoidea

Family DIDYMOZOIDAE

With the previously discussed characters of the subclass.

Subfamilies: Didymozoinae, Adenodidymocystiinae, Annulocystiinae, Colocyntotrematinae, Didymocodiinae, Gonopodasmiinae, Koellikeriinae, Metadidymozoinae, Nemathobothriinae, Neodidymozoinae, Neodiplotrematinae, Nephrodidymotrematinae, Opepherocystiinae, Opepherotrematinae, Osteodidymocodiinae, Patellokoellikeriinae, Phacelotrematinae, Philopinninae, Pseudocolocyntotrematinae, Reniforminae, Sicuotrematinae, Skrjabinozoinae.

REFERENCES

1. Baer, J. G., and C. Joyeux. 1961. Sous-classe des Didymozoides. Didymozoidea Subcl. nov. In Grassé, P. P., editor. Traité de zoologie. Anatomie, Systématique, biologie, tome IV, fascicule I. Masson et Cie., Paris. pp. 678-685.
2. Noble, G. A. 1975. Description of *Nematobibothrioides histoidii* (Noble, 1974) (Trematoda: Didymozoidea) and comparison with other genera. J. Parasitol. 61:224-227.
3. Yamaguti, S. 1951. Studies on the helminth fauna of Japan, part 48. Trematodes of fishes. X. Arb. Med. Fak. Okayama 7:315-334.
4. Williams, H. H. 1959. The anatomy of *Köllikeria filicolis* (Rudolphi, 1819), Cobbold, 1860 (Trematoda: Digenea) showing that the sexes are not entirely separate as hitherto believed. Parasitology 49:39-53.
5. Ishii, N. 1935. Studies on the family Didymozoonidae (Monticelli, 1888). Jpn. J. Zool. 6:279-335.
6. Self, J. T., L. E. Peters, and C. E. Davis. 1963. The egg, miracidium, and adult of *Nematobothrium texomensis* (Trematoda: Digenea). J. Parasitol. 49:731-736.
7. Whittaker, F. H. 1973. Application of histochemistry in studies on the life cycle of *Nematobothrium texomensis* (Trematoda: Didymozoidae). Helminthologia 11:217-220 (dated 1970).

SUGGESTED READING

Yamaguti, S. 1971. Synopsis of digenetic trematodes of vertebrates, vol. I. Keigaku, Tokyo. pp. 248-276. (Diagnostic keys and descriptions of all subfamilies and genera, by the foremost authority on the group.)

Chapter 16

SUBCLASS DIGENEA: FORM, FUNCTION, BIOLOGY, AND CLASSIFICATION

The digenetic trematodes, or flukes, are among the most common and abundant of parasitic worms, second only to nematodes in their distribution. They are parasites of all classes of vertebrates, especially marine fishes, and some species, as adults or juveniles, inhabit nearly every organ of the vertebrate body. Their development occurs in at least two hosts, the first a mollusc or, very rarely, an annelid. Many species include a second and even a third intermediate host in their life cycles. Several species cause economic losses to society through infections of domestic animals, while others are medically important parasites of humans. Because of their importance, the Digenea have stimulated vast amounts of research, and the literature on the group is immense. We will summarize the morphology and biology of the group, illustrating it with some of the more important species.

Trematode development will be considered in detail later, but it is necessary to outline a "typical" life cycle briefly here. A ciliated, free-swimming larva, the **miracidium,** emerges from the egg and penetrates the first intermediate host, a snail. At the time of penetration, or soon after, the ciliated epithelium is discarded, and the miracidium metamorphoses into a rather simple, sac-like form, the **sporocyst.** Within the sporocyst, a number of embryos develop asexually to become **rediae.** The redia is somewhat more differentiated than the sporocyst, possessing, for example, a pharynx and a gut, neither of which were present in the miracidium or the sporocyst. Additional embryos develop within the redia, and these become **cercariae.** The cercaria emerges from the snail and usually has a tail to aid in swimming. Though many species require further development as **metacercariae** before they are infective to the definitive host, the cercariae are properly considered juveniles; they have organs that will develop into the adult digestive tract and suckers, and genital primordia are often present. The fully developed, encysted metacercaria is infective to the definitive host and develops there into the adult trematode.

FORM AND FUNCTION
Body form

Flukes exhibit a great variety of shapes and sizes, as well as variations in internal anatomy. They range in size from the tiny *Levinseniella minuta,* only 0.16 mm long, to the giant *Fascioloides magna* that reaches 5.7 cm in length and 2.5 cm in width. Members of the Didymozoida may reach several meters in length, but we exclude them from the Digenea.

Most flukes are dorsoventrally flattened and oval in shape, but some are as thick as they are wide; some species are filiform, round, or even wider than they are long. Flukes usually possess a powerful oral sucker that surrounds the mouth, and most also have a midventral acetabulum or ventral sucker. The words **distome, monostome,** and **amphistome** are sometimes used as descriptive terms, though they formerly had taxonomic significance, and, of course, they refer to suckers not mouths. If a worm has only an oral sucker, it is called a monostome (Fig. 16-1); with an oral sucker and an acetabulum at the posterior end of the body, it is an amphistome (Fig. 16-2); and if the acetabulum is elsewhere on the ventral surface, the worm is referred to as a distome (Fig. 16-3). The oral sucker may have muscular lappets, as in *Bunodera* (Fig. 16-4), or there may be an anterior adhesive organ with tentacles, as in *Bucephalus* (Fig. 16-5). *Rhopalias,* a parasite of opossums, has a spiny,

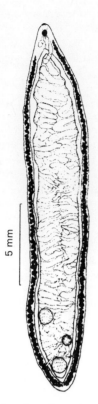

Fig. 16-1. *Cyclocoelum lanceolatum,* a common monostome fluke from the air sacs of shore birds.

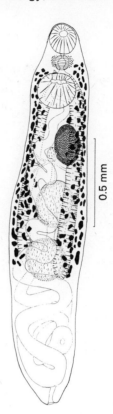

Fig. 16-3. *Alloglossidium hirudicola,* a distome trematode from leeches. (From Schmidt, G. D., and K. Chaloupka. 1969. J. Parasitol. 55:1185-1186.)

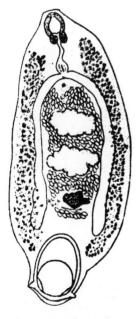

Fig. 16-2. *Zygocotyle lunata,* an amphistome fluke from ducks. (From Travassos, L. 1934. Memorias do Instituto Oswaldo Cruz 29:19-178.)

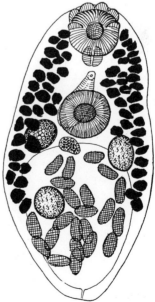

Fig. 16-4. *Bunodera sacculata* from yellow perch. Note the muscular lappets on the oral sucker. (From van Cleave, H. J., and J. F. Mueller. 1932. Roosevelt Wildl. Ann. 3:9-71.)

Fig. 16-5. *Bucephalus polymorphus* from European and Asian fishes. The mouth is midventral and the genital pore is at the posterior end. The front end bears a muscular holdfast with tentacles. (From Yamaguti, S. 1971. Synopsis of digenetic trematodes of vertebrates. Keigaku Publishing Co., Tokyo.)

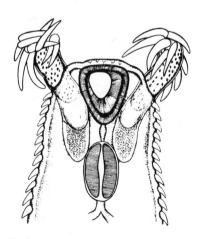

Fig. 16-6. *Rhopalias coronatus*, a parasite of American opossums. Retractable proboscides are located on each side of the mouth. (From Yamaguti, S. 1971. Synopsis of digenetic trematodes of vertebrates. Keigaku Publishing Co., Tokyo.)

retractable proboscis on each side of the oral sucker (Fig. 16-6). In species of Hemiuridae the posterior part of the body telescopes into the anterior portion.

Tegument

The tegument of trematodes, like cestodes, traditionally has been considered a nonliving, secreted cuticle, and likewise, study with the electron microscope has revealed that the body covering of trematodes is a living, complex tissue. In contrast with the cestodes, trematodes have a gut, and it has been assumed that the digestive tract functions in the absorption of nutrients. Perhaps for that reason, the tegument of trematodes has excited somewhat less investigation than that of cestodes; nevertheless, substantial knowledge of tegument structure in several species has accumulated (Lumsden, 1975). In addition, the tegument may be important in nutrient absorption, even though a gut is present (Pappas and Read, 1975).

In common with the Monogenea and the Cestoda, digenetic trematodes have a "sunken" epidermis, that is, there is a distal, anucleate layer. The cell bodies containing the nuclei (perikarya) lie beneath a superficial layer of muscles, connected to the distal cytoplasm via trabeculae (Fig. 16-7). Because the distal cytoplasm is continuous, with no intervening cell membranes, the tegument is syncytial. Although this is the same general organization found in the cestodes, trematode tegument differs in many details, and striking differences in structure may occur in the same individual from one region of the body to another. Ornamentation such as spines are often present in certain areas of the trematode's body, and these may be discernible with the light microscope, but the scanning electron microscope is the instrument of choice for such studies. The oral and ventral suckers of *Schistosoma mansoni* are densely beset with spines, and much of the male's dorsum bears bosses with 50 to 250 spines.[54] Papillae, many with crater-like sensory openings, are interspersed (Fig. 16-8). Bosses are absent from the male's gynecophoral canal and from the female, but the females have many *anteriorly directed* spines on their posterior ends (Fig. 16-9). The spines consist

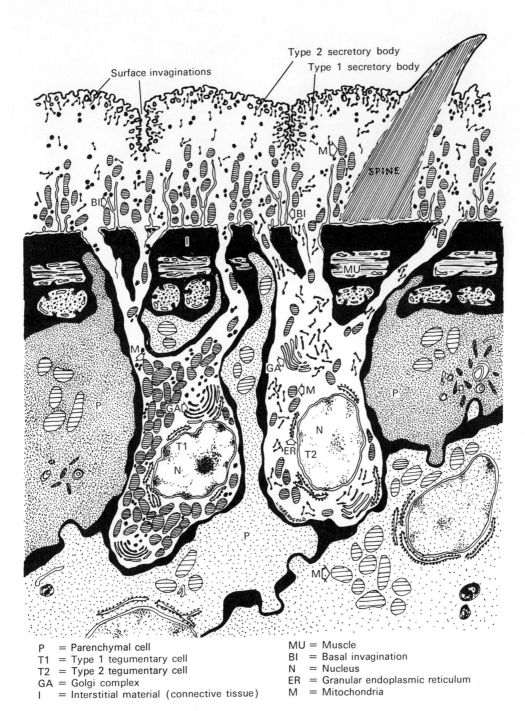

P = Parenchymal cell
T1 = Type 1 tegumentary cell
T2 = Type 2 tegumentary cell
GA = Golgi complex
I = Interstitial material (connective tissue)

MU = Muscle
BI = Basal invagination
N = Nucleus
ER = Granular endoplasmic reticulum
M = Mitochondria

Fig. 16-7. Diagrammatic drawing of the structure of the tegument of *Fasciola hepatica.* (Drawing by L. T. Threadgold.)

Fig. 16-8. Tegument of *Schistosoma mansoni:* dorsal region in distal third of male, showing spines in interbossal spaces and papillae with openings. (From Miller, F. H., G. S. Tulloch, and R. E. Kuntz. 1972. J. Parasitol. 58:693-698.)

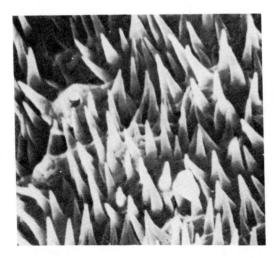

Fig. 16-9. *Schistosoma mansoni:* dense covering of forward-projecting spines at posterior tip of female. (From Miller, F. H., G. S. Tulloch, and R. E. Kuntz. 1972. J. Parasitol. 58:693-698.)

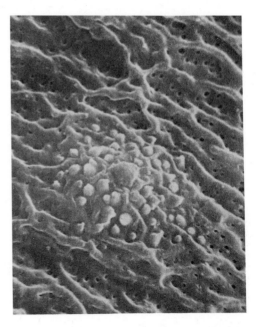

Fig. 16-10. A tubercle and spines on the tegument of a male *Schistosoma mansoni.* Note the many pits in the surface. (From Hockley, D. 1973. In Dawes, B., editor. Advances in parasitology, vol. 11. Academic Press, Inc., New York.)

of a crystalline protein; their bases lie above the basement membrane of the distal cytoplasm, and their apexes project above the surface, though generally are covered by the outer plasma membrane.

The distal cytoplasm usually contains vesicular inclusions, more or less dense, and sometimes several recognizable types in the tegument of the same worm. The function of the vesicles is unclear, though in some cases they contribute to the outer surface. The surface membrane of *S. mansoni* is continuously renewed by multilaminate vesicles moving outward through the distal cytoplasm, perhaps to replace membrane damaged by host antibodies.[39] In *Megalodiscus* the contents of some vesicles seem to be emptied to the outside.[2]

The vesicles of the distal cytoplasm are produced in Golgi bodies found in the perikarya and passed outward via the trabeculae, though Golgi bodies occasionally occur in the distal cytoplasm as well.[3]

Mitochondria are found in the distal cytoplasm in most species examined, though not in *Megalodiscus.*

The outer surface of adult trematodes in general is not modified for absorption by extensive increment in area, though exceptions to this occur. The adhesive organ of *Cyathocotyle,* a strigeoid, is densely beset with long microvilli, strongly suggesting absorptive function.[33] The tegument of *Schistosoma* is penetrated by many deep pits (Fig. 16-10).

Trematodes are particularly appropri-

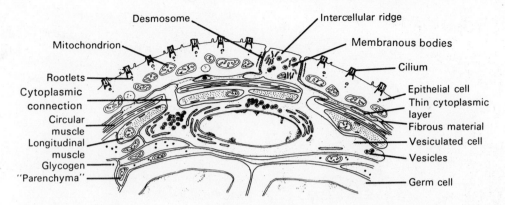

Fig. 16-11. Miracidium of *Fasciola hepatica*. Line drawing reconstruction of transverse segment of body wall in region of the germ cell cavity. (From Wilson, R. A. 1969. J. Parasitol. 55:124-133.)

ate organisms to use for investigation of an old question in zoology: what is the embryonic origin of the body covering of the parasitic flatworms? Hyman (1951), in accord with others working at the time, believed that the "trematodes lack an epidermis and are clothed instead with a resistant cuticle."[40] She enumerated theories as to its homology and origin as follows: (1) the cuticle is an altered and degenerated epidermis; (2) it is the basement membrane of the former epidermis; (3) it is the outer layer of an insunk epidermis, the cells and nuclei of which have sunk beneath the subcuticular musculature; (4) no epidermal cells are involved, mesenchymal (parenchymal) cells secrete the cuticle; and (5) cuticle is secreted by ordinary mesenchyme, not by special cells. Although she regarded the fourth and fifth as the most likely, some modification of the third may be closer to the truth.[49] We shall let readers draw their own conclusions after consideration of the information that follows.

Miracidia of *Fasciola* and *Schistosoma* (at least) are covered by ciliated epithelial cells with nuclei, as is typical for such cells.[85] The epithelial cells are interrupted by "intercellular ridges," extensions of cells whose perikarya lie beneath the superficial muscle layer and that bear no cilia, though some microvilli may be present (Fig. 16-11). Upon loss of the ciliated epithelium, metamorphosis to the sporocyst involves a spreading of the distal cytoplasm over the worm's surface, though whether this comes from the intercellular ridges is unclear. Well-developed microvilli are present on the surface of both the sporocyst and redia. The **luminal surface** of the tegumental cells in the redia may be thrown into a large number of flattened sheets that extend to other cells in the body wall and to the cercarial embryos contained in the lumen. Nutritive molecules such as glucose and molecules up to the size of horseradish peroxidase may be passed through the tegument to the developing cercariae.[29]

The early embryos of the cercariae are covered with a primary epidermis below which a definitive epithelium forms. The nuclei of this secondary epithelium sink into the parenchyma, and the final form of the cercarial tegument results in an organization similar to that of the adult. Cystogenic cells in the parenchyma begin to secrete cyst material, which is passed into the distal cytoplasm of the tegument. The metacercarial cyst is formed when the cercarial tegument is sloughed off and the cyst material it contains undergoes chemical and/or physical changes to envelop the worm in its cyst. The cystogenic cells in the parenchyma then flow toward the surface, their nuclei being retained beneath the superficial muscles, and a thin layer of cytoplasm spreads over the organism to become the definitive adult tegument. A contrasting mode of cercarial tegument formation has been reported for *Schistosoma mansoni*.[37] The definitive tegument, including its nuclei, is formed beneath the primitive tegument and is syncytial. Subsequently, the nuclei of the tegument become pyknotic and are lost, as processes from subtegumental cells grow outward

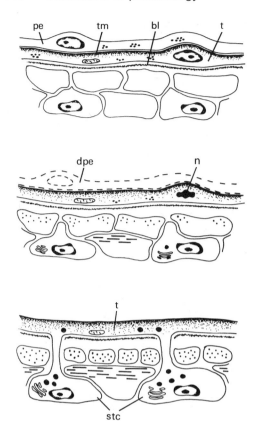

Fig. 16-12. Diagram summarizing three stages in the formation of the tegument during cercarial development. *First,* germ ball covered with a primitive epithelium *(pe)* and the tegument *(t)*, which has a thickened outer membrane *(tm)* and an underlying basal lamina *(bl)*. *Center,* young cercaria with a degenerating primitive epithelium *(dpe)* and a pyknotic tegumental nucleus *(n)*. *Bottom,* cercaria nearly ready to emerge from the sporocyst; the primitive epithelium has been lost and the tegument *(t)* is connected to nucleated subtegumental cells *(stc)*. (From Hockley, D. J. 1973. In Dawes, B., editor. Advances in parasitology, vol. 11. Academic Press, Inc., New York.)

and join the distal cytoplasmic layer (Fig. 16-12). The end result is the same.

The tegument is variously interrupted by cytoplasmic projection of gland cells, by openings of excretory pores, and by nerve endings. Both miracidia and cercariae may have penetration glands that open at the anterior, and the adults of some species have prominent glandular organs opening to the exterior.

Muscular and nervous systems

The muscles that occur most consistently throughout the Digenea are the superficial muscle layers (formerly called "subcuticular"), and these usually are comprised of circular, longitudinal, and diagonal layers enveloping the rest of the body like a sheath below the tegument. The degree of muscularization varies considerably in the group; some species have a rather feeble musculature, some are very robust and strong, and some fit all conditions in between. The deep musculature found in cestodes is generally absent in trematodes. Muscles are often more prominent in the anterior parts of the body, and strands connecting the dorsal to the ventral superficial muscles are usually found in the lateral areas. The fibers are smooth, and the nuclei occur in perikarya called "myoblasts" connected to the fiber bundles and located in various sites around the body, often in syncytial clusters. The suckers and pharynx are supplied with radial fibers, often very strongly developed. A network of fibers may surround the intestinal ceca, helping to fill and empty these structures.

The organization of the nervous system is the typical platyhelminth ladder type. There are a pair of cerebral ganglia connected by a supraesophageal commissure. From these, several nerves issue anteriorly, and three main pairs of trunks—dorsal, lateral, and ventral—supply the posterior parts of the body. The ventral nerves are usually best developed, and a variable number of commissures link these and the other longitudinal nerves. Branches provide motor and sensory endings to muscles and tegument. The anterior end, especially the oral sucker, is well supplied with sensory endings.

Sensory endings in the Digenea are a very interesting array of types, particularly in the miracidia and cercariae, those freeswimming stages that must find a new host within a very short time. The adults require no orientation to such stimuli as light and gravity, and in the few forms in which the ultrastructure has been studied, only one type of sensory ending has been described.[38] This is a bulbous nerve ending in the tegument that has a short, modified cilium projecting from it, and the cilium is enclosed throughout its length by a thin

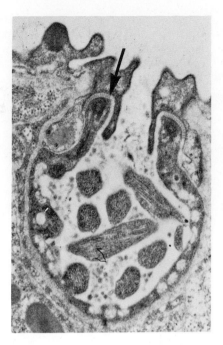

Fig. 16-13. Multiciliated pit in anterior body tegument. Arrow indicates septate desmosome. (×36,000.) (From Morris, G. P. 1971. Zeitschrift für Parasitenkunde 36:20.)

layer of tegument. The general structure is quite similar to sense organs described in cestodes (see Fig. 20-15). These structures have generally been regarded as tangoreceptors in trematodes.

Cercariae and miracidia show more variety in sense organs, doubtlessly related to the adaptive value of finding a host quickly.[4] Uniciliate, bulbous endings are found on the anterior portion of the cercariae of *Schistosoma mansoni,* similar to but smaller than those on the adult. The tegumentary sheath opens at the ciliary apex. In addition, a bulbous type with a long (7 μm), unsheathed cilium is widely distributed over the body of the cercaria, and its lateral areas bear small, flask-shaped endings containing five or six cilia and opening to the outside via a 0.2 μm pore (Fig. 16-13). This latter type is thought to be a chemosensory ending. Most trematode miracidia bear a pair of conspicuous lateral papillae between the first and second series of plates of ciliated epithelium. In *S. mansoni, Fasciola hepatica, Diplostomum spathaceum,* and probably other trematodes as well, these papillae consist of a large, bulbous nerve ending (or two endings) covered with an evagination of the intercellular ridge. The function of these endings has been rather puzzling since there is no obvious organelle for the transduction of stimuli. A uniciliate bulbous ending, as described, is always located just anterior to each papilla, and, assuming some flexibility of the papilla and that its contents are slightly different in specific gravity from the rest of the miracidium, it could impinge on the ciliated ending and give the organism information regarding its orientation with respect to gravity.[4]

Eyespots are present in many species of miracidia and in some cercariae. While also present in some adult trematodes, eyespots do not appear to function in them. The structure of those in several different miracidia have been investigated and are generally similar to such organs found in Turbellaria and some Annelida. They consist of one or two cup-shaped pigment cells surrounding the parallel rhabdomeric microvilli of one or more retinular cells (Fig. 16-14). The mitochondria of the retinular cells are packed in a mass near the rhabdomere. Because the rhabdomeres are the photoreceptors, the cup shape of the pigment cells allows the organism to distinguish light direction. Interestingly, some miracidia do not have eyespots yet can orient with respect to light. Some cells in the miracidia of *D. spathaceum* and *S. mansoni* have large vacuoles, and into these vacuoles project a number of cilia, each of which has a conspicuous membrane evagination. These membranes are stacked in a lamellar fashion, and these might be photoreceptors, thus providing, in the case of *S. mansoni,* a means of light sensitivity for a miracidium without eyespots.[4]

Another apparent chemoreceptor described in the miracidium of *D. spathaceum* consists of two dorsal papillae between the first series of ciliated plates. Each papilla consists of a nerve ending and has radiating from it a number of modified cilia, these being parallel to the surface of the miracidium. These sensory endings are strikingly similar to the olfactory receptors of vertebrate nasal epithelium!

Acetylcholine appears to be an important neurotransmitter in trematodes, and

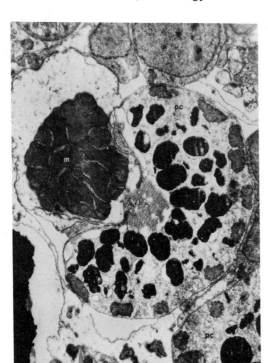

Fig. 16-14. Ultrastructure of eyespot in miracidium of *Diplostomum spathaceum*. Only one retinular cell is visible in this section. Note the closely packed mitochondria and the rhabdomeric microvilli. The position of the cilium-derived photoreceptor (top left) relative to that of the eyespot is also shown. (×15,500.) *m*, Mitochondrion; *pc*, pigment cell; *rb*,, rhabdomere. (From Brooker, B. E. 1972. In Canning, E. U., and C. A. Wright, editors. Behavioural aspects of parasite transmission. Academic Press, Inc., New York.)

5-hydroxytryptamine is also of some importance.[8] Acetylcholine (ACh) and its synthetic and degradative enzymes, acetylcholinesterase and choline acetylase, have been found in *S. mansoni* and *Fasciola hepatica*. Motor activity of *S. mansoni* is stimulated by ACh blocking agents; therefore, it is concluded that the cholinergic endings inhibit muscle activity. The 5-HT is the corresponding excitatory neurotransmitter in *S. mansoni* because incubation in 5-HT or in compounds that cause release or inhibit reuptake of 5-HT by its storage sites causes greatly increased motility. Some properties of the cholinergic endings have been investigated, and several differences between the worm's and the host's receptors have been shown. They respond differently to certain drugs, and the kinetic properties and effect of inhibitors of the worm's acetylcholinesterase and choline acetylase differ from the host's.

Osmoregulatory system

The osmoregulatory system of Digenea is based on the flame bulb **protonephridium,** so-called because it is closed at the inner end and opens to the exterior via a pore. The **flame bulb** or cell is flask-shaped and contains a tuft of fused cilia to provide the motive force for the fluid in the system. The number of flame bulbs, from a few to very many, depends on the species and its size or the extent of its parenchyma. Since the parenchyma is substantially solid, the more massive the worm, the more extensive the flame bulb system required to drain it. Some forms have developed accessory "circulatory" systems. Each flame cell extends processes into the surrounding parenchyma in a "star-like" fashion.[16] The ductules of the flame cells join collecting ducts, those on each side eventually feeding into a bladder in the adult that opens to the outside with a single pore. The pore is almost always located near the posterior end of the worm. The bladder is usually referred to as the **excretory bladder,** and we will use the adjectives osmoregulatory and excretory interchangeably in reference to this system, though there is little evidence to indicate whether either or both is most appropriate. In some trematodes the walls of the collecting ducts are supplied with microvilli, indicating that some transfer of substances, absorption or secretion, is probably occurring.[75] That the system is osmoregulatory may be inferred from the fact that among free-living Platyhelminthes, freshwater Turbellaria have much better developed protonephridial systems than do marine planarians. Trematodes normally have two free-swimming stages in which an efficient water-pumping system is bound to be necessary in those that occur in fresh water.

The embryogenesis of the excretory bladder is of interest and forms the basis for the distinction between the superorders Epitheliocystidia and Anepitheliocystidia. In the developing cercaria the collecting ducts open via a pore on each side of the animal, and in the Anepitheliocystidia the distal portions of the ducts fuse to form a bladder opening to the outside by a single pore. The thin walls of the bladder are derived from the membranous lining of the collecting ducts; some thickness may be added by a subsequent concentration of muscles and other tissues from the body wall, but epithelial cells are not present. In contrast, in the developing cercaria of an epitheliocystidian a number of mesenchymal cells condense around the posterior fusion of the collecting ducts and become an epithelial lining of the bladder. Though this initially seems an insignificant character upon which to base superorders, it correlates well with evidence based on life cycles and comparative morphology and thus reflects phylogenetic relationships.[48] In addition, the shape of the bladder, whether it is Y-, V-, or I- shaped, may have diagnostic value at lower taxonomic levels.

A supplementary "lymphatic system" is found in several families. This is an independent system of irregular, fluid-filled, contractile tubules of uncertain function. Its presence is sometimes used as a taxonomic character.

The primary nitrogenous excretory product of trematodes appears to be ammonia, though excretion of uric acid and urea have been reported. Whether the excretion takes place via the tegument, ceca, or excretory system is not known.

Acquisition of nutrients and digestion

One might assume that the presence of oral sucker, pharynx, gut, and so on in trematodes would make the question of how these worms acquire and digest food a straightforward one. For a variety of reasons, that assumption is not true. Not the least of these is the variety of habitats that different species occupy within the vertebrate host, the difficulty of in vitro cultivation, and the lack of assurance that the worms are behaving "normally" even when in vitro maintenance or cultivation is accomplished. Further, it is becoming clear that trematodes may acquire a substantial quantity of nutrients through their external surface, as well as through their gut (Pappas and Read, 1975).

Halton studied feeding and digestion in a variety of trematodes by histological and histochemical methods.[35] Eight species from six distinct habitats within the vertebrate host were studied, and substantial diversity in digestive mechanisms and morphology was observed. Two lung flukes of frogs, *Haematoloechus medioplexus* and *Haplometra cylindracea,* feed predominately on blood from the capillaries. Both species draw a plug of tissue into their oral sucker, then erode its surface by a pumping action of the strong, muscular pharynx. Other trematode species characteristically found in the intestine, urinary bladder, rectum, and bile ducts feed more or less by the same mechanism, although their food may consist of less blood and more mucus and tissue from the wall of their habitat, and it may even include gut contents (as does *Diplodiscus subclavatus* in the rectum of frogs). In apharyngate species that feed by this mechanism, the walls of the esophagus are quite muscular, and this apparently serves the function of the pharynx. In contrast, *Schistosoma mansoni,* living in the blood vessels of the hepatic portal system and immersed in its semifluid blood food, has no necessity to breach host tissues, and interestingly enough, this species has neither pharynx nor muscular esophagus. Digestion in most species studied is predominately extracellular in the ceca, but in *Fasciola hepatica* it occurs by a combination of intracellular and extracellular processes. One of the frog lung flukes, *H. cylindracea,* has pear-shaped gland cells in its anterior, and a nonspecific esterase is secreted from these cells through the tegument of the oral sucker, beginning the digestive process even before the food is drawn into the ceca. Various degrees of adaptation exist in the blood-feeding flukes, concerning their abilities to eliminate the unwanted iron component of the hemoglobin molecule. In *F. hepatica,* in which the final digestion of hemoglobin is intracellular, the iron is excreted through the excretory system and tegument. The fate of the iron in *H. cylindracea* is unclear, but apparently it is stored, within the worm, tightly bound to protein. The extracellular digestion in *H. medioplexus* and *S. mansoni* produces the end product hematin within the cecal lumen, and this is periodically regurgitated. All species investigated have microvilli on the gastrodermal

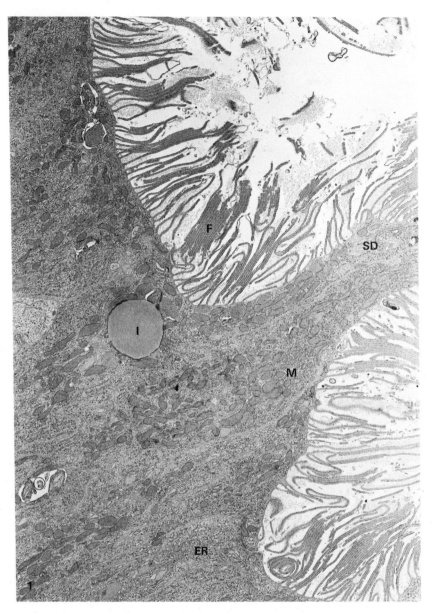

Fig. 16-15 Apical portion of the cecal epithelium of *Paragonimus kellicotti*. The apical surface has numerous folds *(F)* extending into the lumen. The cecal epithelial cells are joined by septate desmosomes *(SD)*. The cytoplasm contains a well-developed granular endoplasmic reticulum *(ER)* and numerous mitochondria *(M)*. An inclusion *(I)* is indicated. (From Dike, S. C. 1969. J. Parasitol. 55:113.)

cells, though these varied from short (1 to 15 μm) and irregular to long (10 to 20 μm) and are organized into a definite brush border, according to species. The ceca of trematodes apparently do not bear any gland cells, but the gastrodermal cells themselves may secrete some digestive enzymes in certain species: proteases, a dipeptidase, an aminopeptidase, lipases, acid and alkaline phosphates, and esterases have been detected.[71,77] A protease from *S. mansoni* has marked substrate specificity for hemoglobin and an optimum pH of 3.9.[79] Interestingly, female *S.*

mansoni ingest far more blood than males, and this is attributed to the requirement of protein for egg production.

The fine structure of the gut cells has been studied for a variety of species.[26] It is clear that the structures interpreted as microvilli in several species at the light microscope level are in fact flattened plate-like or lamelloid processes projecting into the lumen (Fig. 16-15), for example, in *Haematoloechus medioplexus*, *Schistosoma mansoni*, *Fasciola hepatica*, and *Paragonimus kellicotti*. However, digitiform processes more like microvilli are borne by the gut cells of *Gorgodera amplicava*.[25] In either case the absorptive surface area is vastly greater than if the cell surface were flat. Within the gut cells of both *G. amplicava* and *H. medioplexus* are abundant granular endoplasmic reticulum, many mitochondria, and frequent Golgi bodies and membrane-bound, vesicular inclusions. In *P. kellicotti* and *H. medioplexus* is found high activity of acid phosphatase in the vesicles, and, after incubation in ferritin, the material is found within them. No evidence of "transmembranosis" has been found, but the vesicles may be lysosomes that would function in degradation of nutritive materials after phagocytosis.

Although the surface area of the tegument usually is not increased greatly by microvillar-type structures as in cestodes, in light of other similarities, it is not surprising that trematodes can absorb small molecules through the tegument. The uptake of nutrients through the surface of sporocysts and rediae has already been mentioned. Absorption of nutrients through the tegument of adults has been studied by radioautography, by ligature of the anterior end, and by incubation in radioactive substrate. In the few species examined, glucose was absorbed through the tegument and not via the gut, though it has not always been clear whether the worms might not have been able to absorb this hexose by the intestinal route had they been feeding normally.[60] In the case of *Philophthalmus megalurus*, it must be assumed that the trematodes were "feeding" in vitro since they absorbed the amino acids tyrosine and leucine only through the gut, while glucose was absorbed mostly through the tegument.[56] Thymidine was absorbed by *P. megalurus* by *both* routes. *Gorgoderina* can absorb tyrosine, thymidine, adenosine, and glucose through its tegument, while *Haematoloechus* can absorb glucose via its tegument but arginine only by its gut.[58-60] Also, it is known that *Schistosoma mansoni* takes in glucose only through its tegument.[69]

Fasciola and *Fascioloides* absorb several amino acids via their tegument.[41] Therefore, while generalizations cannot be made, it does appear that there is an interesting degree of specificity in absorption sites on the gut and tegument among the various trematodes.

Reproductive systems

Most trematodes are hermaphroditic (important exceptions are the schistosomes), and some are capable of self-fertilization. Others, however, require cross-fertilization to produce viable progeny. When individual *Philophthalmus megalurus* are alone in infected hosts, a large proportion will inseminate themselves, but when other individuals are available, cross-fertilization is the rule.[57]

Male reproductive system (Fig. 16-16). The male reproductive system usually has two testes, although some species have from one to several dozen. Their shape varies from round to highly dendritic, according to species. Each testis has a vas efferens that connects with the other to form a vas deferens. This courses toward the genital pore, which is usually found within a shallow genital atrium. The genital atrium is most often on the midventral surface, anterior to the acetabulum; but it can be found nearly anywhere, including at the posterior end, beside the mouth, or even dorsal to the mouth in some species. Before reaching the genital pore, the vas deferens usually enters a muscular cirrus pouch where it may expand into an **internal seminal vesicle** for sperm storage. Constricting again, the duct forms a thin ejaculatory duct, which extends the rest of length of the cirrus pouch and forms, at its distal end, a muscular cirrus. The cirrus is the male copulatory organ. It can be invaginated into the cirrus pouch and evaginated for transfer of sperm to the female system. The cirrus may be naked or covered with spines of different sizes. The

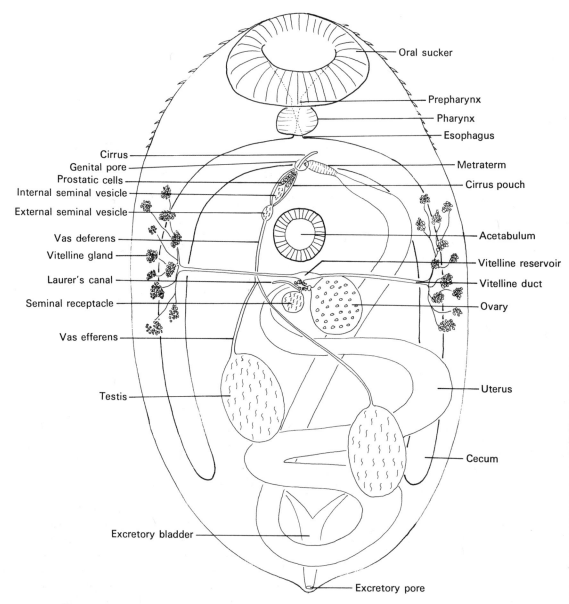

Fig. 16-16. A diagrammatic representation of a digenetic fluke, showing male and female reproductive systems.

ejaculatory duct is usually surrounded by numerous unicellular **prostate gland cells.** At this point a muscular dilation may form a **pars prostatica.**

Much variation in these terminal organs occurs between families, genera, and species. The cirrus pouch and prostate gland may be absent, with the vas deferens expanded into a powerful seminal vesicle that opens through the genital pore, as in *Clonorchis.* The vas deferens may expand into an **external seminal vesicle** before continuing into the cirrus pouch. Other, more specialized modifications are described and illustrated by Yamaguti (1971).[86]

Female reproductive system (Fig. 16-16). There is a single ovary in the female reproductive tract, and the organ is usually round or oval, but it may be lobated or even branched. The short oviduct is provided with a proximal sphincter, the **ovi-**

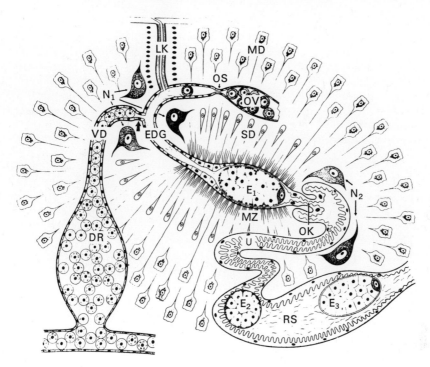

Fig. 16-17. Schematic representation of the oogenotop of *Fasciola hepatica. DR,* vitelline reservoir; E_1, egg in ootype, with beginning shell formation; E_2, egg in the lower uterus, shell granules almost coalesced; E_3, egg in the uterus, shell formation almost completed; *EDG,* ovovitelline duct; *LK,* Laurer's canal; *MD,* mucous cells of the Mehlis's gland; *MZ,* area where Mehlis's glands open into the ootype; N_1, nerve plexus I; N_2, nerve plexus II; *OK,* ootype sphincter; *OS,* oviduct sphincter; *OV,* oviduct; *RS,* seminal receptacle; *SD,* serous cells of the Mehlis's gland; *U,* uterus; *VD,* vitelline duct. (From Gönnert, R. 1962. Zeitschrift für Parasitenkunde 21:477.)

capt, that controls the passage of ova. The oviduct and most of the rest of the female ducts are ciliated. A seminal receptacle forms as an outpocketing of the wall of the oviduct. It may be large or small, but it is nearly always present. At the base of the seminal receptacle there often arises a slender tube, the Laurer's canal, which ends blindly in the parenchyma or opens through the tegument. The Laurer's canal is probably a vestigial vagina that no longer functions as such (with a few possible exceptions), but it may serve to store sperm in some species.

Unlike other animals, but in common with the cestodes and some Turbellaria, the yolk is not stored in the ovum but is contributed by separate cells, called vitelline cells. The vitelline cells are produced in follicular vitelline glands, usually arranged in two lateral fields and connected by ductules to the main right and left vitel-

line ducts. These ducts carry the vitelline cells to a single, median vitelline reservoir, from which extends the common vitelline duct joining the oviduct. The distribution of vitelline glands tends to be constant within a species and so is an important taxonomic character. After the junction with the common vitelline duct, the oviduct expands slightly to form the ootype. Numerous unicellular Mehlis's glands surround the ootype and deposit their products into it by means of tiny ducts.

The structural complex just described (Fig. 16-17), as well as the upper uterus, is called the "egg-forming apparatus," or **oogenotop**.[34] Beyond the ootype, the female duct expands to form the uterus, which extends to the female genital pore. The uterus may be short and fairly straight, or it may be long and coiled or folded. The distal end of the uterus is often quite muscular and is called the metraterm. The metraterm functions as ovijector and as a vagina. The female genital pore opens near the male pore, usually together with it in the genital atrium. In some species, such as in the Heterophyidae, the genital atrium is surrounded by a muscular sucker called a **gonotyl**.

At the time the ova leave the ovary, they may not have completed meiosis and thus are not, strictly speaking, ova at all but oocytes. Meiosis is completed after sperm penetration. The first meiotic division may reach pachytene or diplotene, at which point meiotic activity is arrested, and the chromosomes may return to a diffuse state. After sperm penetration, the chromosomes quickly reappear as bivalents and proceed from the first meiotic metaphase. The two meiotic divisions occur, with extrusion of polar bodies, and the male and female pronuclei fuse.[12,44]

As the oocyte leaves the ovary and proceeds down the oviduct, it becomes associated with several vitelline cells and a sperm emerging from the seminal receptacle. These all come together in the area of the ootype, and there are contributions from the cells of the Mehlis's gland, as well. It was long thought that the Mehlis's gland contributed the shell material, and the organ was often referred to as the "shell gland" in older texts. However, it is now clear that the shell precursors are supplied by the vitelline cells, and the function of the Mehlis's gland is much in doubt. There appear to be two kinds of cells in the Mehlis's gland, serous cells and mucous cells.[34] Evidence suggests that the functions of the secretions of these cells may include the following: (1) formation of a "template" membrane around the ovum-vitelline cell mass on which the shell material is laid down, (2) stimulation of the vitelline cells to release their shell globules, (3) activation of the sperm as it passes from the seminal receptacle, and (4) lubrication of the uterus.[13] The universal presence of Mehlis's gland in trematodes (and cestodes, too) is strong evidence of an important function; it is curious that the function has been so difficult to ascertain.

Curious also is the perplexing problem of the chemical mechanism by which proteins in the "egg" shell are stabilized. "Stabilization" of structural proteins (for example, sclerotin, keratin, and resilin) to impart qualities of physical strength and inertness occurs by cross-linking amino acid moieties in adjacent protein chains. It has been thought that this occurs in trematode eggs by a quinone-tanning process, the enzyme **phenolase** catalyzing hydroxylation of tyrosyl moieties to dihydroxyphenylalanine (dopa), then oxidation to dopa-quinone and cross-linkage with free amino groups in the presence of oxygen to yield sclerotin[77]:

Protein with tyrosyl moiety

Protein with dopa moiety

Dopa-quinone linked to free amino group in adjacent protein chain

Clear demonstration that the above reactions describe the true situation in trematode eggshells has been difficult, and the conclusion that they occur in shell stabilization has been based on observations that (1) freshly formed eggshells darken on exposure to air and bleach after treatment with sodium hypochlorite and (2) vitelline cells contain the precursors of phenolic-tanning—tyrosine-rich protein, phenol, and phenolase.[67] More recently, considerable doubt has been cast on this conclusion by the failure to find phenolase in some species or to find quinones or quinone-tanned proteins in the eggshells themselves. In contrast, Ramalingam[66,67] has shown that the shell proteins of *Fasciola hepatica* and of some monogenetic trematodes are stabilized by disulfide links, as in keratin, and by dityrosine links, as in resilin.[1]

Keratin-type disulfide
cross linkage

Resilin-type dityrosine
cross linkage

DEVELOPMENT
Life cycle

As the name Digenea (two beginnings) suggests, at least two hosts serve in the life cycle of a typical digenetic fluke. One is a vertebrate (with a few exceptions) in which sexual reproduction occurs, and the other is usually a mollusc in which one or more generations are produced by a very unusual type of asexual reproduction that will be discussed further. A few species have asexual generations that develop in annelids.

This alternation of sexual and asexual generations in different hosts is one of the most striking biological phenomena. The variability and complexity of life cycles and ontogeny have stimulated the imaginations of zoologists for over a hundred years, creating a huge literature on the subject. Even so, many mysteries remain, and research on questions of trematode life cycles is still quite active.

As many as six recognizably different body forms may develop during the life cycle of a single species of trematode (see p. 255 for summary). In a given species, certain stages may be repeated during ontogeny, while stages found in other species may be absent. So many variations occur that few generalizations are possible. Therefore, we will first examine each form separately and then illustrate the subject with a few examples.

Embryogenesis. Apart from the fact that the embryo produced by the sexual adult begins with a fertilized egg, the early embryogenesis of progeny produced asexually and sexually is basically similar. The first cleavage produces a **somatic cell** and a **propagatory cell,** and these are cytologically distinguishable. Daughter cells of the somatic cell will contribute to the body tissues of the embryo, whether miracidium, sporocyst, redia, or cercaria. Further divisions of the propagatory cell may each produce another somatic cell and another propagatory cell, but at some point, propagatory cell divisions produce only more propagatory cells. Each of these will become an additional embryo in the miracidium, sporocyst, or redia. In the developing cercaria, the propagatory cells become the gonad primordia. Thus, the propagatory cells are the germinal cells in the asexually reproducing forms, and they give rise to the germ cells in the sexual adult. As noted previously, the miracidium metamorphoses into the sporocyst; but if a sporocyst stage is absent in a particular species, redial embryos develop in the miracidium to be released after penetration of the intermediate host. The youngest embryos developing in a given stage are usually seen in the posterior portions of its body, and these are often referred to as **germ balls.**

The nature of the asexual reproduction has long been controversial, and it has at various times been thought to represent **budding, polyembryony,** or **parthenogenesis.** The view of early zoologists—that it was an example of metagenesis (strictly speaking, an alternation of generations in which the asexual generation reproduces by budding)—was discarded when it was realized that the specific reproductive cells (the propagatory cells) were kept segregated in the germinal sacs. The most widely held opinion has been that the process was one of sequential polyembryony,[22,23] that is, production of multiple embryos from the same zygote with no intervening gamete production as, for example, in monozygotic twins in humans. Parthenogenesis, or reproduction in which there is no fertilization of the female gamete by a male gamete, was not thought to occur because of the failure to confirm any meiotic phenomena in the propagatory cells. However, cytological evidence shows that the process may indeed be parthenogenesis, at least in *Philophthalmus megalurus*.[44] The daughter propagatory cells at a certain point go through a meiotic prophase up to diakinesis in which a haploid number of bivalents are observed. The cells then return to an interphase condition, grow, and become the "typical" germinal cell of sporocysts and rediae, which then cleaves to become the embryo of the next generation. This process is not unlike that seen in parthenogenesis of some other invertebrates. Yet, the evidence for parthogenesis has been questioned, and an argument has been made for regarding the process as budding.[18]

Egg. The structure referred to as an "egg" of trematodes is not an ovum, but the developing (or developed) embryo enclosed by its shell, or capsule. The egg capsule of most flukes has an **operculum** at one end, through which the larva eventually will escape. It is not clear how the operculum is formed, but it appears that the embryo presses pseudopodium-like processes against the inner surface of the shell while it is being formed, thereby forming a circular groove. An operculum is absent in the eggshell of blood flukes. Considerable variation exists in the shapes and sizes of fluke eggs, as well as in the thickness and coloration of the capsules.

In many species the egg contains a fully developed miracidium by the time it leaves the parent, while in others development has advanced to only a few cell divisions by that time. In some species (*Cyclocoelum* and *Heronimus*) the miracidium hatches while still in the uterus. For those eggs that embryonate in the external environment, certain factors are known to influence embryonation. Water is necessary, for the eggs desiccate rapidly in dry conditions. Development is stimulated by high oxygen tension, although eggs can remain viable for long periods under conditions of low oxygen. Eggs of *Fasciola hepatica* will not develop outside a pH range of 4.2 to 9.0.[70] Temperature is critical, as would be expected. Thus, *F. hepatica* takes 23 weeks to develop at 10° C, while it only takes 8 days at 30° C. However, above 30° C development again slows and completely stops at 37° C. Eggs are killed rapidly at freezing. Light may be a factor influencing de-

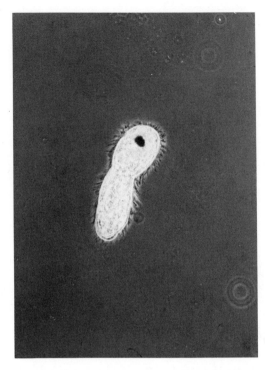

Fig. 16-18. A miracidium of *Alaria* sp. (Photograph by Jay Georgi.)

velopment in some species, but this has been little investigated.

Eggs of many species will hatch freely in water, while others hatch only when eaten by a suitable intermediate host. Factors stimulating hatching have been investigated for several species. Light and osmotic pressure are important in those species that hatch in water, and osmotic pressure, carbon dioxide tension, and, probably, host enzymes initiate hatching in those that must be eaten before they will hatch. The mechanism of hatching has been studied most in *Fasciola hepatica*.[84] The miracidium is surrounded by a thin **vitelline membrane,** which also encloses a pad-like **viscous cushion** between the anterior end of the miracidium and the operculum. Light stimulates hatching activity. Some factor apparently is released by the miracidium that alters the permeability of the membrane enclosing the viscous cushion. The latter structure contains a mucopolysaccharide that becomes hydrated and greatly expands the volume of the viscous cushion. The considerable increase

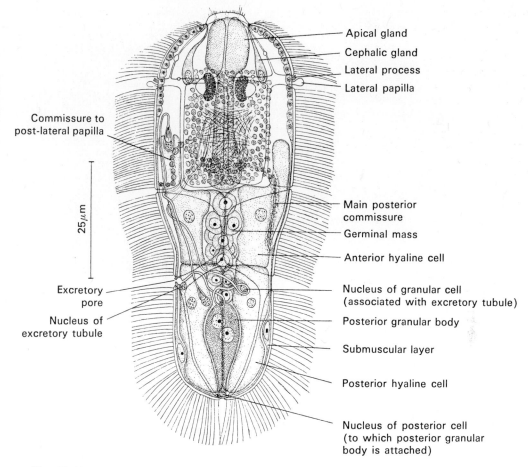

Apical gland
Cephalic gland
Lateral process
Lateral papilla

Commissure to
post-lateral papilla

25 μm

Main posterior
commissure
Germinal mass
Anterior hyaline cell

Excretory
pore

Nucleus of
excretory tubule

Nucleus of granular cell
(associated with excretory tubule)
Posterior granular body

Submuscular layer

Posterior hyaline cell

Nucleus of posterior cell
(to which posterior granular
body is attached)

Fig. 16-19. Miracidium of *Neodiplostomum intermedium,* dorsal view. (From Pearson, J. C. 1961. Parasitology 51: 133-172.)

in pressure within the egg causes the operculum to pop open, remaining attached at one point, and the miracidium rapidly escapes, propelled by its cilia. There is no evidence that a hatching enzyme is produced. Hatching of nonoperculated eggs, such as those of *Schistosoma,* is little understood.

Miracidium. The typical miracidium (Fig. 16-18) is a tiny, ciliated organism that could easily be mistaken for a protozoan by the casual observer. It probably is quite similar to the acoeloid ancestor of the Platyhelminthes. It is piriform, with a retractable **apical papilla** at the anterior end. The apical papilla has no cilia but bears five pairs of duct openings from glands and two pairs of sensory nerve endings (Fig. 16-19). The gland ducts connect with **pen-**etration glands inside the body. A prominent **apical gland** can be seen in the anterior third of the body. This probably also secretes histolytic enzymes. An apical stylet is present on some species, while spines are found on others. The sensory nerve ending connect with nerve cell bodies that in turn communicate with a large ganglion. Miracidia have a variety of sensory organs and endings, including adaptations for photo-, chemo-, tango-, and statoreception.

The outer surface of a miracidium is covered by flat, ciliated epidermal cells, the number and shape of which are constant for a species. Underlying this are longitudinal and circular muscle fibers. Cilia are restricted to protruding **ciliated bars** in the genus *Leucochloridiomorpha* (Brachylaimi-

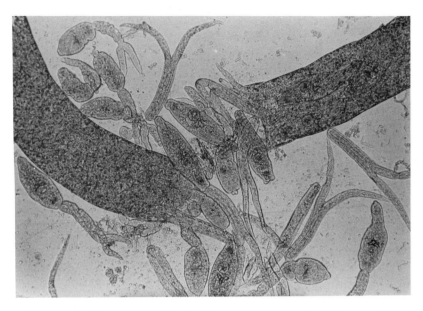

Fig. 16-20. A ruptured sporocyst releasing furcocercous cercariae. (Photograph by James Jensen.)

dae) and family Bucephalidae, and they are absent altogether in the families Azygiidae and Hemiuridae. One or two pairs of protonephridia are connected to a pair of posterolateral excretory pores.

In the posterior half of the miracidium are found propagatory cells or germ balls (embryos), which will be carried into the sporocyst stage to initiate further generations.

Free-swimming miracidia are very active, swimming at a rate of about 2 mm per second, and they must find a suitable molluscan host rapidly, for they can survive as free-living organisms for only a few hours. Snail-finding behavior of miracidia has been reviewed.[80] In many cases the mucus produced by the mollusc is a powerful attractant for miracidia.

On contacting a proper mollusc, the miracidium attaches to it with the apical papilla, which actively contracts and extends, undergoing an auger-like motion. Cytolysis of snail tissues can be seen as the miracidium embeds itself deeper and deeper (see frontispiece). As penetration proceeds, the miracidium loses its ciliated epithelium, although this may be delayed until penetration is complete. It takes about 30 minutes for the miracidium to complete penetration and begin the next phase of its life cycle as the sporocyst.

Sporocyst. Metamorphosis of the miracidium into the sporocyst involves extensive changes. In addition to the loss of the ciliated epithelial cells, the new tegument with its microvilli forms as described previously (see p. 235) The subtegumental muscle layer of the miracidium is retained, as are the protonephridia, but all other miracidial structures generally disappear. The sporocyst has no mouth or digestive system; it absorbs nutrients from the host tissue, with which it is in intimate contact, and the entire structure serves only to nurture the developing embryos. The sporocyst (or other stage with the parthenogenetic embryos developing within it, that is, the miracidium or redia) may be referred to as a **germinal sac.** Often sporocysts grow near the site of penetration, such as foot, antenna, or gill, but they may be found in any tissue, depending on the species, and sometimes they may become very slender and extended, branched, or ramified.

The embryos in the sporocyst may develop into another sporocyst generation **(daughter sporocysts),** or into a different form of germinal sac, the redia, or directly to cercariae (Fig. 16-20).

A very interesting sporocyst adaptation is found in the genus *Leucochloridium.* The highly branched daughter sporocysts contain encysted cercariae, and they extend as swollen, brightly colored brood sacs into the tentacles of their snail host. There they pulsate rapidly: 70 times per minute at summer temperatures.[83] Their color and movement serve to attract the attention (and appetite) of their bird definitive hosts.

Redia. Rediae burst their way out of the sporocyst or leave through a terminal birth pore and usually migrate to the hepatopancreas or gonad of the molluscan host. They are commonly elongate, blunt at the posterior end, and may have one or more stumpy appendages called **procrusculi** (Fig. 16-21). More active than most sporocysts, they crawl about within their host. They have a rudimentary, but functional, digestive system, consisting of a mouth, muscular pharynx, and short, unbranched gut. Rediae pump food into their gut by means of the pharyngeal muscles, as previously described in adults, and the luminal surface of the gut is greatly amplified by flattened, lamelloid or ribbonlike processes.[46] The gut cells are apparently capable of phagocytosis. The outer surface of the tegument also functions in absorption of food, and it is provided with microvilli or lamelloid processes.

The embryos in the redia develop into daughter rediae or into the next stage, the cercaria, which emerges through a birth pore near the pharynx. It appears that rediae must reach a certain population density before they stop producing more rediae and begin producing cercariae, for young rediae have been transplanted from one snail to another through more than 40 generations without cercariae being developed.[32] This is an interesting parallel to certain free-living invertebrates that reproduce parthenogenetically only as long as certain environmental conditions are maintained.[15]

Cercaria. The cercaria represents the juvenile stage of the vertebrate-inhabiting adult. There are many varieties of cercariae, and most have specializations that enable them to survive a brief free-living existence and make themselves available to their definitive hosts (Fig. 16-22). Most have tails which aid them in swimming, but many have rudimentary tails or none at all; these can only creep about, or they may remain within the sporocyst or redia that

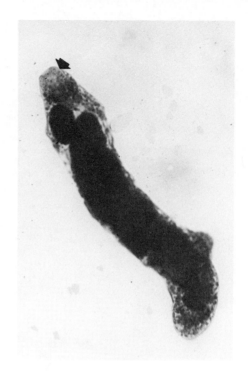

Fig. 16-21. Redia of *Echinostoma revolutum.* Note the large, muscular pharynx (arrow) just inside the mouth. The bulge near the posterior end is a procrusculus. (Photograph by Warren Buss.)

produced them until they are eaten by a definitive host.

The structure of a cercaria is easily studied, and cercarial morphology often has been considered a more reliable indication of phylogenetic relationships between families than the morphology of the adults. Cercariae are widely distributed, abundant, and easily found; hence they have attracted much attention from zoologists. The name *Cercaria* can be used properly in a generic sense for a species in which the adult form is unknown, as is done with the term "microfilaria" among some nematodes.

Most cercariae have a mouth near the anterior end, although it is midventral in the Bucephalidae. The mouth is usually surrounded by an oral sucker, and a prepharynx, muscular pharynx, and a forked intestine are normally present. Each branch of the intestine is simple, even those that are ramified in the adult. Many cercariae have various glands opening

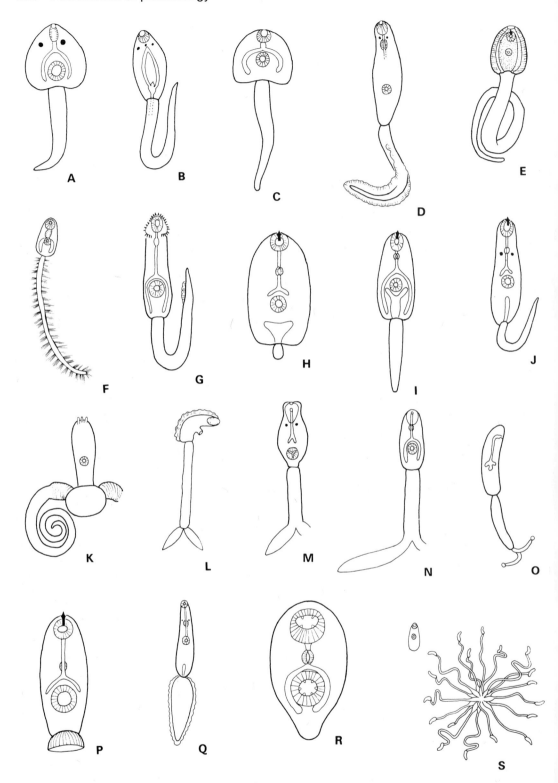

Fig. 16-22. For legend see opposite page.

near the anterior margin, often called "penetration glands" because of their assumed function. The schistosome cercacia, which has been best studied in this regard, has no less than four distinguishable types of such glands[78]:

1. **Escape glands**—so called because their contents are expelled during emergence of the cercaria from the snail, but their function is not known
2. **Head gland**—its secretion is emitted into the matrix of the tegument and is thought to function in the postpenetration adjustment of the schistosomule
3. **Postacetabular glands**—produce mucus, help cercariae adhere to surfaces, and other possible functions
4. **Preacetabular glands**—secretion contains calcium and a variety of enzymes including a protease. The function of these glands seems most important in actual penetration of host skin.

Secretory **cystogenic cells** are particularly prominent in cercariae that will encyst on vegetation or other objects. *Fasciola hepatica* has cystogenic cells of four distinct types, recognized by the different staining characteristics of their contents and by their ultrastructure. Each type contains the precursors for a distinct layer of the cyst wall.[28]

Many morphological variations exist in cercariae that are constant within a species (or larger taxon), and so certain descriptive terms are of value in categorizing the different varieties. Some of the more commonly used terms are as follows:

1. **Xiphidiocercaria**—cercaria with stylet in anterior margin of oral sucker
2. **Ophthalmocercaria**—cercaria with eyespots
3. **Cercariaeum**—cercaria without a tail
4. **Microcercous**—cercaria with small, knob-like tail
5. **Furcocercous**—cercaria with forked tail
6. **Cercariae ornatae**—cercaria tail with a fin

The excretory system is well developed in the cercaria; the embryogenesis of the excretory vesicle has already been described (see p. 239), and it provides the basis for separation of the two superorders, Epitheliocystidia and Anepitheliocystidia. In some cercariae the excretory vesicle empties through one or two pores in the tail. Because the number and arrangement of the protonephridia are constant for a species, these are important taxonomic characters. Each flame cell has a tiny **capillary duct** that joins with others to form the **anterior** or **posterior collecting tubules,** one each of which join to form an **accessory tubule.** Accessory tubules join to form a **common collecting tubule** on each side (Fig. 16-23). When the common collecting tubules extend to the region of the midbody and then fuse with the excretory vesicle, the cercaria is called **mesostomate.** If they extend to near the anterior end and then pass posteriad to join the vesicle, the cercaria is known as **stenostomate.** The number arrangement of flame cells can be expressed conveniently by the *flame cell formula.* For example, $2[(3 + 3) + (3 + 3)]$ means that both sides of the cercaria, *2,* have three flame cells on each of two accessory tubules, *(3 + 3),* on the anterior collecting tubule, plus the same arrangement on the posterior collecting tubule.

Fig. 16-22. A few of the many types of cercariae. **A,** Amphistome cercaria; **B,** monostome cercaria; **C,** gymnocephalus cercaria; **D,** gymnocephalus cercaria of pleurolophocercous type; **E,** cystophorous cercaria; **F,** trichocercous cercaria; **G,** echinostome cercaria; **H,** microcercous cercaria; **I,** xiphidiocercaria; **J,** ophthalmoxiphidiocercaria; **K-O,** furcocercous types of cercariae; **K,** gasterostome cercaria; **L,** lophocercous cercaria; **M,** apharyngeate furcocercous cercaria; **N,** pharyngeate furcocercous cercaria; **O,** apharyngeate monostome furcocercous cercaria without oral sucker; **P,** cotylocercous cercaria; **Q,** rhopalocercous cercaria; **R,** cercariaea; **S,** rattenkönig or rat-king cercariae. (From Olsen, O. W. 1974. Animal parasites, their life cycles and ecology. University Park Press, Baltimore.)

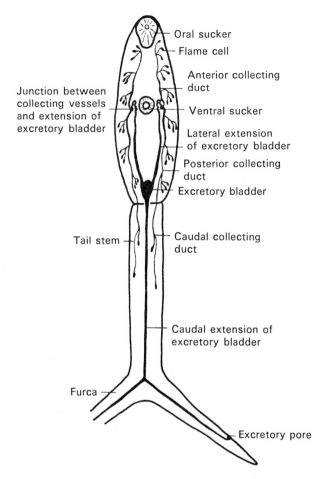

Fig. 16-23. Diagrammatic representation of the excretory system of a fork-tailed cercaria. The caudal flame cells are absent from the tail in nonfurcate forms. (From Erasmus, D. A. 1972. The biology of trematodes. Crane, Russak & Co., New York.)

Mature cercariae emerge from the mollusc and begin to seek their next host. Many remarkable adaptations can be found among cercariae that enable them to do this. Most are active swimmers, of course, and rely on chance to place them in contact with an appropriate organism. Some species are photopositive, dispersing themselves as they swim toward the surface of the water, but then become photonegative and return to the bottom where the next host is. Some opisthorchoid cercariae remain quiescent on the bottom until a fish swims over them; the resulting shadow activates them to swim upward. Some plagiorchoid cercariae cease swimming when in a current; hence, when drawn over the gills of a crustacean host they can attach and penetrate rather than swimming on. Large, pigmented azygiids and bivesiculids are enticing to fish, which eat them and become infected. Some cercariae float; some unite in clusters; some creep on the bottom. In the case of certain cystophorous hemiurid cercariae, the body is withdrawn into the tail, which becomes a complex exploding device. When eaten by a crustacean, the tail explodes, forcing the body into the host's hemocoel! Those and many more adaptations help to ensure that the trematode will reach its next host.

Mesocercaria. Species of the strigeoid genus *Alaria* have a unique larval form, the **mesocercaria,** which is intermediate between the cercaria and metacercaria (Fig. 17-4).

Metacercaria. Between the cercaria and the adult is a quiescent stage, the metacercaria, although this stage is absent in the families of blood flukes. The metacercaria is usually encysted, but this stage in *Brachycoelium, Halipegus,* and *Panopistus* is not. Most metacercariae are found in or on an intermediate host, but some (Fasciolidae, Notocotylidae and Paramphistomidae) encyst on aquatic vegetation, sticks, and rocks or even free in the water.

The cercaria's first step in encysting is to cast off its tail. Cyst formation is most elaborate in metacercariae encysting on inanimate objects or vegetation. The cystogenic cells of *Fasciola hepatica* are of four types, each with the precursors of a different cyst layer. During its development in the redia, the epithelium of the cercaria becomes distended with the products of two of these: protein granules for the outer layer (layer I) and granules of mucoprotein and mucopolysaccharide to form layer II.[30] Layer I is thought to be a tanned protein, though the evidence here is not conclusive. The granules appear to be passed from the cytoplasm of the cystogenic cells to that of the epithelial cells (cytocrine secretion). At the time of cyst formation, the cercarial epithelium is sloughed off and disintegrates, the granules within it forming layers I and II (holocrine secretion). The mode of release of layer III material is unclear. The cystogenic cells of layer IV contain large quantities of **rodlets** or **batonnets.** Ultrastructurally, the batonnets can be seen to consist of lamellae tightly rolled in a scroll-like fashion. After the cercarial epithelium is sloughed off, the batonnet cells move peripherally, and their cytoplasm, with the contained batonnets, moves out to cover the surface of the trematode, while their nuclei (perikarya) remain internal to the superficial muscle layer. Thus, the batonnet cells actually become the definitive tegument of the adult. Membranes of the vesicles containing the batonnets become continuous with the outer cell membranes, and the batonnets are thereby secreted without destroying the integrity of the tegument (eccrine secretion). After secretion, the batonnets unroll and form a highly laminated inner cyst wall; the unrolling

process is aided by movements of the metacercaria within the cyst. The protein of layer IV is stabilized by disulphide links and is considered keratin.[27]

Although the cysts formed by metacercariae in intermediate hosts are thinner and simpler, much of the formation is probably similar to the foregoing. A host response usually involves coating the cyst with fibrous tissue, and melanin is often deposited in the host capsule, such as in "black spot" in the skin of fishes (Fig. 17-5).

Development that occurs in the metacercaria varies widely according to species, from those in which a metacercaria is absent (*Schistosoma*) to those in which the gonads mature and viable eggs are produced (*Proterometra*). Often, some amount of development is necessary in the metacercaria before the organism is infective for the definitive host. We can arrange metacercariae into three broad groups on this basis[31]:

1. Species whose metacercariae encyst in the open, on vegetation and inanimate objects, for example, *Fasciola*. Members of this group can infect the definitive host almost immediately after encystment, in some cases in only a few hours, with no growth occurring.

2. Species that do not grow in the intermediate host, but that require at least several days of physiological development to infect the definitive host, for example, Echinostomatidae.

3. Species whose metacercariae undergo growth and metamorphosis before they enter their resting stage in the second intemediate host; and they usually require a period of weeks for this development, for example, Diplostomatidae.

These developmental groups are correlated with the longevity of the metacercariae: those in group 1 must live on stored food and can survive the shortest time before reaching the definitive host, while those in groups 2 and 3 obtain some nutrient from their intermediate hosts and so can remain viable for the longest periods—in one case up to 7 years. After the required development, the metacercaria goes into its quiescent stage and remains in readiness to excyst upon reaching the definitive host. *Zoogonus lasius,* a typical example of group 2, has a high rate of metabolism for the first few days after infecting its second intermediate host, a nereid polychaete, then drops to a low level, only to return to a high rate upon excystment.[81]

The metacercarial stage has a high selective value for many species of trematodes. It may provide a means for transmission to a definitive host that does not feed on the first intermediate host or is not in the environment of the mollusc, or it may permit survival over unfavorable periods, such as the absence of the definitive host during a particular season.

Development in the definitive host. Once the cercaria or metacercaria has reached its definitive host, it matures in a variety of ways: either by penetration (if a cercaria) or by excystation (if a metacercaria), then by migration, growth, and morphogenesis to gamete production. If the species does not have a metacercaria and the cercaria penetrates the definitive host directly, as in the schistosomes, the most extensive growth, differentiation, and migration will be necessary. At the other extreme, the adult form may have already been attained as a metacercaria, the gonads may almost be mature, or some eggs may even be present in the uterus, and little more than excystation is needed before the production of progeny *(Bucephalopsis, Coitocaecum, Transversotrema).* A very few species *(Proctoeces subtenuis* and *Proterometra dickermani)* reach sexual reproduction in the mollusc and apparently do not have a vertebrate definitive host. For instance *Alloglossidium hirudicola* develops to sexual maturity in a leech. These are sometimes considered examples of progenesis or neoteny.

Normally, development in the definitive host begins with excystation of the metacercaria, and those species with the heaviest, most complex cysts appear to require the most complex stimuli for excystment, like those with cysts on vegetation, for example, *Fasciola hepatica.* The outer cyst of *F. hepatica* is largely removed by digestive enzymes, but escape from the inner cyst requires the presence of a temperature of about 39° C, a low oxidation—reduction potential, carbon dioxide, and bile. It should be noted that this combination of conditions is not likely to be present anywhere but in the intestine of a homoiothermic vertebrate, and, like the conditions required for the hatching or exsheathment of some nematodes, the requirements constitute an adaptation that avoids premature escape from the protective coverings. Such an adaptation is less important to metacercariae that are not subjected to the widely varying physical conditions in the external environment, such as those encysted within a second intermediate host. These have thinner cysts and excyst upon treatment with digestive enzymes.

After excystment in the intestine, a more or less extensive migration is necessary if the final site is in some other tissue. The main sites of the "tissue" parasites are the liver, lungs, and circulatory system. Probably the most common way to reach the liver is via the bile duct *(Dicrocoelium dendriticum),* but *F. hepatica* burrows through the gut wall into the peritoneal cavity and finally, wandering through the tissues, reaches the liver. *Clonorchis sinensis* usually penetrates the gut wall and is carried to the liver by the hepatic portal system. *Paragonimus westermani* penetrates the gut wall, undergoes a developmental phase of about a week in the abdominal wall, then reenters the abdominal cavity and makes its way through the diaphragm to the lungs.

A most remarkable physiological aspect of trematode life cycles is the sequence of totally different habitats in which the various stages must survive, with physiological adjustments that must often be made extremely rapidly. As the egg passes from the vertebrate, it must be able to withstand the rigors of the external environment in fresh or sea water, if only for a period of hours, before it can reach haven in the mollusc. There, conditions are quite different from both the water and the vertebrate. Its physiological capacities must again be readjusted upon escape from the intermediate host and again upon reaching the second intermediate or definitive host. Environmental change may be somewhat less dramatic if the second intermediate host is a vertebrate, but often it is an invertebrate. Although the adjustments must be extensive, the nature of these physiological adjustments made by trematodes during their life cycles has been little investigated, the most studied in this respect being *Schistosoma*.[19]

Penetration of the definitive host is a hazardous phase of the life cycle of schistosomes, and it requires an enormous amount of energy. The hazards include a combination of the dramatic changes in the physical environment, in the physical and chemical nature of the host skin through which it must penetrate, and in host defense mechanisms. Depending on the host species, losses at this barrier may be as high as 50%, and the glycogen content of the newly penetrated schistosomula (**schistosomule** is the name given the young developing worm) is only 6% of that found in cercariae. Among the most severe physical conditions the organism must survive is the sequence of changes in ambient osmotic pressure.The osmotic pressure of fresh water is considerably below that in the snail, and that in the vertebrate is twice as great as the mollusc. Assuming that the osmotic pressure of the cercarial tissues approximates that in the snail, the trematode must avoid taking up water after it leaves the snail and avoid a serious water loss after it penetrates the vertebrate. Aside from the possible role of the osmoregulatory organs (protonephridia and others), there appear to be major changes in the character, and probably permeability, of the cercarial surface. The cercarial surface is coated with a fibrillar layer or glycocalyx, which is lost upon penetration of the vertebrate, and with it is lost the ability to survive in fresh water; 90% of schistosomula recovered from mouse skin 30 minutes after penetration die rapidly if returned to fresh water. That chemical changes have occurred is indicated by the fact that the schistosomule surface is much less easily dissolved by a number of chemical reagents, including 8M urea, than that of the cercaria. The antigenic determinants of the schistosomule as compared to the cercaria are changed also. When cercaria are incubated in immune serum, a thick envelope forms around them called the CHR (for cercarienhüllenreaktion), but schistosomula do not give this reaction.

The most important stimulus for penetration of schistosome cercariae seems to be the skin lipid film. Human skin surface lipid applied to the walls of their glass container will cause cercariae to attempt to penetrate it, lose their tails, evacuate their preacetabular glands, and become intolerant to water. In vivo it has been shown that skin cells are destroyed by the lytic secretions of the schistosomule.

Following penetration, the tegument of the developing schistosomule undergoes a remarkable morphogenesis. Within 30 minutes numerous subtegumental cells have connected to the distal cytoplasm and are passing abundant "laminated bodies" into it. These bodies have *two* trilaminate limiting membranes (therefore, heptalaminate). These bodies move to the surface of the tegument to become the new tegumental outer membrane, the old cercarial outer membrane, along with its remaining glycocalyx, being cast off. Three hours after penetration, the schistosomule outer membrane is almost entirely heptalaminate. During the next 2 weeks the main changes in the tegument are a considerable increase in thickness and the development of many invaginations and deep pits. These pits increase the surface area fourfold between 7 and 14 days postpenetration. It may be assumed that this represents an adaptation for nutrient absorption through the tegument.

Summary of life cycle. The basic pattern of a digenetic trematode life cycle is:

egg → miracidium →
     sporocyst → redia → cercaria →
               metacercaria → adult

The student should learn this pattern well, for it is the theme on which to base the variations, the most common of which are (1) more than one generation of sporocysts or rediae, (2) deletion of either sporocyst or redial generations, and (3) deletion of metacercaria. Much less common are cases in which miracidia are produced by sporocysts, and forms with adult morphology are in the mollusc, producing cercariae (these in turn lose their tails and produce another generation of cercariae!).[14,42] Fig. 16-24 shows some possible life cycles.

Phylogeny of digenetic trematodes

Numerous schemes have been suggested for the origin of the Digenea. Various authors have derived the ancestral form from Monogenea, Aspidogastrea, and even insects.[62,76] However, most authorities today believe that digenetic trematodes

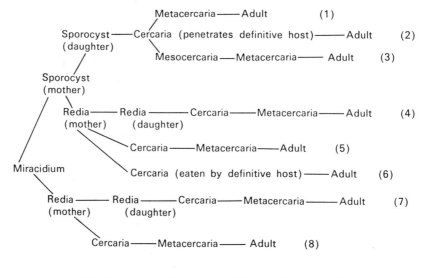

(1) *Diplostomum flexicaudum* (Cort and Brooks, 1928)
(2) *Trichobilharzia physellae* (Talbot, 1936)
(3) *Alaria mustelae* Bosma, 1931
(4) *Fasciola hepatica* Linnaeus, 1758
(5) *Metorchis conjunctus* (Cobbold, 1860)
(6) *Proterometra dickermani* Anderson, 1962
(7) *Stichorchis subtriquetrus* (Rudolphi, 1814)
(8) *Caecincola parvulus* Marshall and Gilbert, 1905

Fig. 16-24. Some possible life cycles of digenetic trematodes. (From Schell, S. C. 1970. How to know the trematodes. William C. Brown Co., Publishers, Dubuque, Iowa.)

evolved from free-living turbellarians, probably rhabdocoels or rhabdocoel-like ancestors.[52,61] Whatever the ancestral digenean, any system of their phylogeny must rationalize the evolution of their complex life cycles in terms of natural selection, a most perplexing task.

Digeneans display much more host specificity to their molluscan hosts than to their vertebrate hosts. This may imply that they established themselves as parasites of molluscs first, then added a vertebrate host as a later adaptation. It is not difficult to imagine a small, rhabdocoel-like worm, which fed on soft-bodied invertebrates, invading the mantle cavity of a mollusc and feeding on its tissues. In fact, the main hosts of known endocommensal rhabdocoels are molluscs and echinoderms. The stage of the protodigenean in the mollusc was probably a developmental one, with a free-living, sexually reproducing adult. There are several reasons for believing this. First, a free reproductive stage would be of selective value in dispersion and

transferral to new hosts. Precisely this "life style" is show by *Fecampia,* a rhabdocoel symbiont of various marine crustaceans. Second, the possession of a cercarial stage is surprisingly ubiquitous among digeneans, and most of these are adapted for swimming. Those without tails show evidence that the structure has been secondarily lost.

If one grants that the present adult represents the ancestral adult, which was free-living, it is clear that additional (parthenogenetic) multiplication in the mollusc would have been advantageous also, and the alternation of the two reproductive generations could have been established. It is likely that such free-living adults would often be eaten by fish, and those individuals in the population that could survive and maintain themselves in the fish's digestive tract for a period of time would have selective advantage in extending their reproductive life. *Fecampia,* for example, dies after depositing its eggs.

Further evidence that the parasite was

originally free-living as an adult is demonstrated by the fluke still having to leave the snail to infect the next host. With few exceptions, the fluke is incapable of infecting a definitive host while still in its first intermediate host, even if eaten. In most cases when a life cycle requires that the fluke be eaten within a mollusc, it has left its first host and penetrated a second to become infective. It is not unlikely that the miracidium represents the larval form of the fluke's ancestor; all digeneans still have them, even though are not all now free-swimming.

Once the basic two-host cycle with two reproducing generations was established in the protodigenean, it is less difficult to visualize how further elaborations of the life cycle could have been selected for. The ecological value of the metacercaria, already mentioned, may have had some protective function as well. The mesocercaria of *Alaria* has a clear value in that the definitive host normally does not feed on the second intermediate host. Many other examples could be cited.

It may be assumed that the digenean adaptation to vertebrate hosts has occurred relatively recently. Digenetic trematodes are very common in members of all classes of vertebrates except Chondrichthyes; extremely few species of digeneans are found in sharks and rays. The urea in the tissue of most elasmobranchs, which plays such an important role in their osmoregulation, is quite toxic to the flukes on which it has been tested. It is supposed that elasmobranchs did not have digenean parasites when that particular osmoregulatory adaptation was evolved and that the urea has since proved a barrier to invasion of the elasmobranch habitat by flukes. The situation is quite the contrary with the cestodes; sharks and rays have a rich tapeworm fauna, and their cestodes either tolerate the urea or degrade it.[68]

METABOLISM
Energy metabolism

With certain reservations, the metabolism of trematodes has received considerable attention.[6,21,77] The first of the reservations, of course, is that compared to certain vertebrates, trematode metabolism (in common with many other invertebrate groups) is meagerly known. Second, the metabolism of larval stages has received scant attention. Third, the number of species investigated has been very small. Like *Hymenolepis diminuta* among the cestodes, *Fasciola hepatica* has been a favorite species for biochemical work because of its size, availability, and added interest of economic importance. In addition, because of their tremendous medical importance and the critical need for more effective chemotherapeutic agents to fight them, the metabolism of schistosomes has been fairly well investigated. Information on genera other than *Fasciola* and *Schistosoma* is sparse.

In terms of energy derivation from nutrient molecules, adult cestodes and trematodes have much in common. Their main sources of energy are from the degradation of carbohydrate from glycogen and glucose. They are facultative anaerobes, and, even in the presence of oxygen, they depend greatly on glycolysis and excrete large amounts of short-chain acid end products[36] (see Fig. 20-35). In other words, the energy potential in the glucose molecule is far from completely realized. The worm has what is, for all practical purposes, an inexhaustible food supply. It may be subjected to very low oxygen concentrations part or all of the time, and its metabolic end products can be further catabolized by its host. Even in the case of the schistosomes, which live in the blood and presumably have an abundant oxygen supply, glycolysis is the main energetic pathway. The worms can survive 5 to 6 days under nitrogen or in 1 mM cyanide, but they do consume oxygen when it is available. At least 25% of the energy production of adult *S. mansoni* is via oxidative phosphorylation. Inhibitions of oxygen uptake by cyanide, comparable to those to be expected in normal aerobic tissues, have been shown in several other trematodes. Cheah and Prichard concluded that *F. hepatica* has a branched electron transport system: one being the classical mammalian type, with cytochrome a_3, and the other with cytochrome o as terminal electron acceptors. Propionate production by *F. hepatica* was reduced by 30% under aerobic as opposed to anaerobic conditions. Cheah and Pritchard believed that the branched chain is an adaptation of large parasites to low environmental oxygen.[17] However, NADH–cytochrome c oxidoreductase, succinate–cytochrome c oxidoreductase, NADH oxidase, and cytochrome c–oxygen oxidoreductase were all present in the trematode.

The ability to catabolize certain of the citric acid cycle intermediates and the presence of various enzymes in that cycle have been shown in several trematode species. In fact, all enzymes necessary for a functional citric acid cycle are in *Fasciola hepatica*[63] but the levels of aconitase and isocitrate dehydrogenase activities are so low that the cycle is considered of minor importance at most. There is evidence for a functional citric acid cycle in schistosomes,[21] but its significance is still not clear. The primary role of some of the enzymes may lie in metabolic paths other than the Krebs cycle (succinic dehydrogenase and malic dehydrogenase) (see Fig. 20-35). Some enzymes may be vestiges of previous phylogenetic or ontogenetic stages, or the cycle may simply function at a lower level than usual in strict aerobes.

Although little is known of the lipid metabolism, we have no evidence that lipids are used as energy sources or energy storage compounds. However, trematodes may contain considerable lipid, and sizable quantities may be excreted. *F. hepatica* excretes about 2% of its wet weight per day as polar and neutral lipids (including cholesterol and its esters), and this is mainly via its excretory system.[11]

Consequently, it is not surprising that digeneans contain large amounts of stored glycogen: 9% to 30% of dry weight, according to species. Amounts in female *S. mansoni* are unusually low, only about 3.5% of dry weight. Although the glycogen content of cestodes may range higher than 30%, it is still surprising that trematodes store so much, even tissue-dwelling species, since the availability of their food should not be subject to the vagaries of their host's feeding schedule, as it is with cestodes. In cases where measurements have been performed, a large proportion of the trematode's glycogen is consumed under starvation conditions in vitro. In fact, the maintenance of a high glycogen concentration in the worms may be of critical importance. The action of niridazole,

an antischistosomal drug, has been attributed to the fact that it causes glycogen depletion in the schistosome, and the mechanism of action is very interesting. The glucose moieties in glycogen are mobilized for glycolysis by the action of glycogen phosphorylase, as in other systems, and the extent of the mobilization is controlled by how much of the enzyme is in the physiologically active *a* form. Niridazole inhibits the conversion of phosphorylase *a* to the inactive *b* form; thus, the phosphorolysis of glycogen is uncontrolled, the glycogen stores of the worm are depleted, and it is finally killed if the niridazole concentration is maintained.[9a] As with any good chemotherapeutic agent, the corresponding host enzyme is not affected.

The dependence of trematodes on glycolysis is illustrated by the action of another chemotherapeutic regime against schistosomes. The organic trivalent antimonials have been the traditional drugs for use in schistosomiasis (and are still the only ones available for use in the United States). The antimonials inhibit the phosphofructokinase (PFK) of the schistosome. PFK catalyzes the phosphorylation of fructose-6-phosphate (F6P) to fructose-1,6-diphosphate, and this reaction is a rate-limiting step in glycolysis. The drugs exert their inhibition by a reversible interaction with an F6P site on the enzyme and displacement of F6P from that site.[9] Unfortunately, although the schistosome PFK is 65 to 80 times more sensitive to the antimonials than in mammalian PFK, considerable toxic effects, such as gastrointestinal disturbance, severe headaches, and possibly hepatitis, may occur in the patient. Therefore, the trivalent antimonials are far from ideal drugs for schistosomiasis.

As in cestodes, the terminal reactions in the glycolytic sequence may be quite different from those to which we are accustomed in mammals. An exception to this statement, however, is found in the schistosomes. These worms are referred to as **homolactic producers;** *S. mansoni* consumes glucose equivalent to up to 20% of its dry weight per hour, and over 80% of this is accounted for as lactate.[7] *Dicrocoelium dendriticum* and *Fasciola hepatica* also consume considerable glucose (though at a much lower rate than *S. mansoni*) but with different glycolytic end products. *Di-*

crocoelium dendriticum produces about 40% lactate, 30% acetate, 30% propionate, and 3% succinate, while *F. hepatica* excretes 8% lactate, 24% acetate, 68% propionate, and only traces of succinate.[45] The small amount of lactate produced by *F. hepatica* is presumably explained by the low activity of its lactic dehydrogenase.[65] All four of the short chain acids seem to be produced via carbon dioxide fixation by phosphoenol pyruvate carboxykinase (PEPCK), since the activity of pyruvate kinase (PK) is very low in *Dicrocoelium* and *Fasciola* (see Fig. 20-35), and pyruvate is apparently produced by decarboxylation of malate by malate dehydrogenase (decarboxylating). Activity of PK in *Schistosoma* is much higher than that of PEPCK. The precise route of propionate production has not been determined, but it is thought to be a decarboxylation of succinate after formation of succinyl CoA.

The phosphofructokinase reaction is rate-limiting in *Fasciola,* as in *Schistosoma,* and 5-hydroxytryptamine (5-HT), which is known to be an excitatory neurotransmitter in *Schistosoma,* has a very interesting effect on PFK in *Fasciola.* The amine does occur in and is synthesized by *F. hepatica,* and the activity of PFK in homogenates of the fluke is increased by 5-HT. Cyclic 3′, 5′-AMP duplicates the effect of the 5-HT, and it appears that the activation of the PFK is a result of 5-HT stimulation of cyclic 3′, 5′-AMP synthesis. Epinephrine, a related catecholamine, which increases formation of cyclic 3′, 5′-AMP in particulate fractions of the mammalian liver, has no effect on the same fractions from the fluke (Mansour, 1967).

A pentose phosphate pathway may function in schistosomes, but critical enzymes for a glyoxylate cycle have not been found. In contrast, the pentose cycle in *Fasciola* appears to be minimal but enzymes necessary for the glyoxylate path are all present.[10,64]

The astonishing ability of trematodes to survive the radical changes in environment requires important adjustments in their energy metabolism. Clearly, the ability to derive every possible ATP from every glucose unit would be of great selective value to the free-swimming miracidium or cercaria that does not feed. In all species investigated so far, the miracidia and cercariae are obligate aerobes, killed by short exposures to anaerobiosis. Pyruvate is used rapidly by cercariae of *S. mansoni;* carbon dioxide is produced from all three of the pyruvate carbons. The miracidium may have a functional citric acid cycle within the egg. Study of the sporocysts is difficult because of the problem of separating host from parasite tissue; but the same drugs that kill sporocysts within the snail affect the adults, and sporocysts (or the cercariae within) produce lactic acid under aerobic conditions. Therefore, the metabolism of the sporocysts is thought to be like that of the adults. Use of inhibitors suggests that a functional citric acid cycle is present in cercariae; they exhibit a Pasteur effect, and cytochromes a/a_3, *b,* and *c* are all present. Immediately after penetration, schistosome energy metabolism seems to undergo a major adjustment. The ability to utilize pyruvate drops dramatically, and the schistosomulae again produce lactate aerobically.[5,20] They still use oxygen when available, as do the adults, and oxygen consumption of both cercariae and schistosomulae is inhibited by more than 80% in the presence of 0.2 mM cyanide. Such developmental studies on the metabolism of other trematodes would be very interesting, but few have been reported. The oxygen consumption of adult *Gynaecotyla adunca,* an intestinal parasite of fish and birds, dropped sharply 24 hours after excystment, then even more after 48 and 72 hours.

Transamination ability appears limited, but the α-ketoglutarate-glutamate transaminase reaction is active.[24,82]

Ammonia and urea are both important end products in degradation of nitrogenous compounds in *Fasciola* and *Schistosoma,* and both worms excrete several amino acids as well.[55,72] A full complement of the enzymes necessary for the ornithine-urea cycle is not present in *Fasciola;* the urea produced must be by other pathways.[43]

Synthetic metabolism

Stimulated by the search for chemotherapeutic agents, purine metabolism in *Schistosoma* has been studied. *Schistosoma mansoni* cannot synthesize nucleotides de novo.[74] Neither glycine nor glucose is incorporated into purine nucleotide bases, but adenine is taken up from the medium and synthesized into the nucleotides. Kurelec showed that *F. hepatica* and *Paramphistomum cervi* could not synthesize carbamyl phosphate and concluded that they were dependent on their hosts for both pyrimidines and arginine.[47] In light of the high arginine requirement of *S. mansoni,* it would seem probable that the situation is the same in that species.

The requirement of schistosomes for arginine is so high, in fact, that they reduce serum arginine to almost zero in mice with severe infections. The worms more rapidly take up arginine than histidine, tryptophan, or methionine. Proline from the host is also rapidly consumed by the male schistosome via both the gut and the tegument, but only a little is absorbed by the tegument of the female.[73] Interestingly, the proline consumed is concentrated in the ventral arms of the gynecophoral canal, the contact region with the female. Glycogen concentrations in male and female schistosomes fluctuate in a parallel manner.[51] This and the foregoing have suggested that the embrace of the male has a nutritive as well as a sexual function.

It has already been pointed out that trematodes excrete lipids, and this seems a rather profligate custom since it appears that they cannot synthesize their own complex lipids. Meyer and others (1970) compared the synthetic capacity of *S. mansoni* to the free-living planarian, *Dugesia dorotocephala,* and they found that neither flatworm could synthesize fatty acids or sterols de novo.[53] Both can synthesize their complex lipids provided that they are supplied with a source of long-chain fatty acids. They noted that of the Platyhelminthes studied so far, including two species of cestodes, all lack the biosynthetic pathways for all three classes of lipids: sterols, saturated fatty acids, and unsaturated fatty acids. If this deficiency is common to the entire phylum, that may be a factor predisposing its members to a symbiotic way of life.

CLASSIFICATION OF THE SUBCLASS DIGENEA

The classification of Digenea is rather unstable, particularly at the higher categories. This is partly because of the im-

mense diversity between species and partly because of the fact that entirely new forms are being discovered every year. Of equal importance, however, is that the biology and ontogeny of most species are unknown, leaving the systematist with only adult morphology on which to base the classification.

The following system is based on La Rue[48] the most widely accepted system of classification of the higher taxa, and modified by various authors. This system is based primarily on characters of larval forms, with emphasis on the excretory system. It is workable, so long as these details are known for at least one species within a family; several families are of unknown relationships because we lack such information about them. The lists of families are fairly complete, but not all specialists will agree with their placement.

Superorder ANEPITHELIOCYSTIDIA
Embryonic excretory bladder retained in the adult, not replaced by mesodermal cells; thin, not epithelial. Cercaria with forked or simple tail; never with oral stylet.

Order STRIGEATA
Cercariae with forked tail, usually with two suckers. Miracidium usually with two pairs of protonephridia.

Superfamily STRIGEOIDEA
Cercaria with thin tail bearing two long rami; pharynx present. Adult body usually divided by construction into anterior and posterior portions. Accessory suckers and/or penetration glands often present on anterior portion of body. Genital pore usually terminal. Parasites of reptiles, birds and mammals.

Families: Bolbocephalidae, Brauninidae, Cyathocotylidae, Diplostomatidae, Proterodiplostomatidae, Strigeidae

Superfamily CLINOSTOMATOIDEA
Cercaria with short rami on tail; oral sucker replaced with protractile penetration organ, ventral sucker rudimentary; pigmented eyespots present. Parasites of reptiles, birds, and mammals.

Family: Clinostomatidae

Superfamily SCHISTOSOMATOIDEA
Cercaria with short caudal rami; pharynx absent, oral sucker replaced with protractile penetration organ. Eyespots pigmented or not. Adults in

vascular system of definitive host. No second intermediate host in life cycle. Parasites of fishes, reptiles, birds, and mammals.

Families: Aporocotylidae, Sanguinicolidae, Schistosomatidae, Spirorchiidae

Superfamily AZYGIOIDEA

Cercaria furcocystocercous; oral sucker present, ventral sucker sometimes absent. Adults often very large. Parasites of fishes.

Families: Aphanhysteridae, Azygiidae, Bivesiculidae

Superfamily TRANSVERSOTREMATOIDEA

Cercaria with short rami; base of tail with two appendages; pharynx absent. Adult transversely elongated, intestine fused at distal ends. Parasites of the dermis of fishes.

Family: Transversotrematidae

Superfamily CYCLOCOELOIDEA

Cercaria with short, bilobed tail, or tail absent. Adults elongate, flat; acetabulum absent or rudimentary. Oral sucker present or absent. Intestine united at posterior end of body. Parasites of the respiratory system of birds.

Family: Cyclocoelidae

Superfamily BRACHYLAEMOIDEA

Cercaria with or without tail, usually developing in terrestrial snails. Genital pore postequatorial. Parasites of amphibians, birds, and mammals.

Families: Brachylaemidae, Harmotrematidae, Leucochloridiidae, Liolopidae, Ovariopteridae, Thapariellidae

Superfamily FELLODISTOMATOIDEA

Cercaria furcocercous, developing in marine bivalves. Adults plump. Ceca sometimes united posteriorly. Parasites of marine fishes.

Families: Fellodistomatidae, Maseniidae, Monodhelminthidae

Superfamily BUCEPHALOIDEA

Mouth midventral in both cercaria and adult. Tail of cercaria very short, with two very long rami. Acetabulum absent. Parasites of fishes and amphibians.

Families: Bucephalidae, Sinicovothylacidae

Order ECHINOSTOMATA

Miracidium with single pair of protonephridia. Cercaria with simple tail.

Superfamily ECHINOSTOMATOIDEA

Cercaria and adult often with circumoral collar usually armed with spines. Parasites of reptiles, birds, and mammals.

Families: Balfouridae, Campulidae, Cathemaisiidae, Echinostomatidae, Fasciolidae, Haplosplanchnidae, Philophthalmidae, Psilotrematidae, Rhopaliasidae, Rhytidodidae

Superfamily PARAMPHISTOMOIDEA

Cercaria without penetration glands; monostomes or amphistomes. Adults always amphistomes. No second intermediate host in life cycle. Adults parasites of fishes, amphibians, reptiles, birds, and mammals.

Families: Angiodictyidae, Gastrodiscidae, Gastrothylacidae, Heronimidae, Paramphistomidae, Mesometridae

Superfamily NOTOCOTYLOIDEA

Cercaria monostomous, without pharynx. Adults monostomous, occasionally with tegumental glands on ventral surface. Parasites of birds and mammals.

Families: Notocotylidae, Pronocephalidae, Rhabdiopoeidae

Superorder EPITHELIOCYSTIDIA

Wall of the embryonic excretory bladder replaced by epithelial cells of mesodermal origin. Cercaria with simple tail. Oral stylet common.

Order PLAGIORCHIATA

Superfamily PLAGIORCHIOIDEA

Cercaria typical distomes, with oral stylet. Metacercaria usually in invertebrates. Parasites of fishes, amphibians, reptiles, birds, and mammals.

Families: Anchitrematidae, Batrachotrematidae, Brachycoelidae, Cephalogonimidae, Dicrocoeliidae, Dolochoperidae, Echinoporidae, Eucotylidae, Haematoloechidae, Haplometridae, Lecithodendriidae, Lissorchiidae, Macroderidae, Macroderoididae, Mesotretidae, Microphallidae, Ochetosomatidae, Omphalometridae, Pachypsolidae, Plagiorchiidae, Plectognathotrematidae, Prosthogonimidae, Stomylotrematidae, Urotrematidae

Superfamily ALLOCREADIOIDEA

Cercariae of many varied types, usually with eyespots. Adults also variable, with or without eye spots; oral sucker usually simple, but occasionally with appendages. Acetabulum in anterior half of body. Testes always in hind body. Ovary nearly always pretesticular. Parasites of fishes, amphibians, reptiles, and mammals.

Families: Acanthocolpidae, Allocreadiidae, Apocreadiidae, Collyriclidae, Gekkonotrematidae, Gorgocephalidae, Homalometridae, Lepocreadiidae, Megaperidae, Monorchiidae, Octotestidae, Opecoelidae, Opistholebetidae, Gorgoderidae, Glyauchenidae, Schistorchiidae, Tetracladiidae, Troglotrematidae, Zoogonidae

Order OPISTHORCHIATA
Cercaria with excretory vessels in the tail; oral stylet never present.

Superfamily ISOPARORCHIOIDEA
Adults parasitic in swim bladder of fishes. Body large, usually plump. Testes postacetabular. Vitellaria posterior.

Families: Aerobiotrematidae, Albulatrematidae, Cylindrorchiidae, Dictysarcidae, Isopharorchiidae, Pelorohelminthidae, Tetrasteridae

Superfamily OPISTHORCHIOIDEA
Cercaria with well-developed penetration glands. Oral sucker protractile, ventral sucker rudimentary. Tail variable. Parasites of fishes, amphibians, reptiles, birds, and mammals.

Families: Acanthostomidae, Cryptogonimidae, Heterophyidae, Opisthorchiidae

Superfamily HEMIUROIDEA
Adults: cuticle usually unspined. Tail appendage sometimes present. Testes usually anterior to ovary. Vitellaria usually postovarian. Parasites of fishes.

Families: Accacoeliidae, Bathycotylidae, Botulidae, Dinuridae, Halipegidae, Hemiceridae, Hirudinellidae, Lampritrematidae, Lecithasteridae, Lecithochiriidae, Mabiaramidae, Ptychogonimidae, Sclerodistomidae, Syncoeliidae

• • •

Families of uncertain relationship
The following families, while justifiably distinct, have unknown life cycles and ontogeny and thus cannot be aligned with major groups at this time.

Families: Acanthocollaritrematidae, Achillurbainiidae, Atractotrematidae, Botulisaccidae, Braunotrematidae, Callodistomatidae, Cortrematidae, Diplangidae, Eumegacetidae, Haploporidae, Jubilariidae, Laterotrematidae, Lobatovitelliovariidae, Mesotretidae, Meristocotylidae, Moreauiidae, Nasitrematidae, Ommatobrephidae, Pholeteridae, Prostogonotrematidae, Sigmaperidae, Stomylotrematidae, Treptodemidae, Waretrematidae.

REFERENCES
1. Andersen, S. O. 1971. Resilin. In Florkin, M., and E. H. Stotz, editors. Comprehensive biochemistry. Elsevier, Amsterdam. 26C:633-657.
2. Bogitsh, B. J. 1968. Cytochemical and ultrastructural observation on the tegument of the trematode *Megalodiscus temperatus*. Trans. Am. Microsc. Soc. 87:477-486.
3. Bogitsh, B. J. 1971. Golgi complexes in the tegument of *Haematoloechus medioplexus*. J. Parasitol. 57:1373-1374.
4. Brooker, B. E. 1972. The sense organs of trematode miracidia. In Canning, E. U., and C. A. Wright, editors. Behavioural aspects of parasite transmission. Linnean Society Academic Press, Inc., London. pp. 171-180.
5. Bruce, J. I., E. Weiss, M. A. Stirewalt, and D. R. Lincicome. 1969. *Schistosoma mansoni:* glycogen content and utilization of glucose, pyruvate, glutamate, and citric acid cycle intermediates by cercariae and schistosomules. Exp. Parasitol. 26:29-40.
6. Bryant, C. 1975. Carbon dioxide utilization, and the regulation of respiratory metabolic pathways in parasitic helminths. In Dawes, B., editor. Advances in parasitology, vol. 13. Academic Press, Inc., New York. pp. 36-69.
7. Bueding, E. 1950. Carbohydrate metabolism of *Schistosoma mansoni*. J. Gen. Physiol. 33:475-495.
8. Bueding, E., and J. Bennett. 1972. Neurotransmitters in trematodes. In Van den Bossche, H., editor. Comparative biochemistry of parasites. Academic Press, Inc., New York. pp. 95-99.
9. Bueding, E., and J. Fisher. 1966. Factors affecting the inhibition of phosphofructokinase activity of *Schistosoma mansoni* by trivalent antimonials. Biochem. Pharmacol. 15:1197-1211.
9a. Bueding, E., and J. Fisher. 1970. Biochemical effects of niridazole on *Schistosoma mansoni*. Molec. Pharmacol. 6:532-539.
10. Buist, R. A., and P. J. Schofield. 1971. Some aspects of the glucose metabolism of *Fasciola hepatica*. Int. J. Biochem. 2:377-383.
11. Burren, C. H., I. Ehrlich, and P. Johnson. 1967. Excretion of lipids by the liver fluke (*Fasciola hepatica* L). Lipids 2:353-356.
12. Burton, P. R. 1960. Gametogenesis and fertilization in the frog lung fluke, *Haematoloechus medioplexus* Stafford (Trematoda: Plagiorchiidae). J. Morphol. 107:92-122.
13. Burton, P.R. 1967. Fine structure of the reproductive system of a frog lung fluke. I. Mehlis' gland and associated ducts. J. Parasitol. 53:540-555.
14. Cable, R. M. 1965. Thereby hangs a tail. J. Parasitol. 51:3-12.
15. Cable R. M. 1971. Parthenogenesis in parasitic helminths. Am. Zoologist 11:267-272.
16. Cardell, R. R., Jr. 1962. Observations on the ultrastructure of the body of the cercaria of

Himasthla quissetensis (Miller and Northup, 1926). Trans. Am. Microsc. Soc. 81:124-131.

17. Cheah, K. S., and R. K. Prichard. 1975. The electron transport systems of *Fasciola hepatica* mitochondria. Int. J. Parasitol. 5:183-186.

18. Clark, W. C. 1974. Interpretation of life history patterns in the Digenea. Int. J. Parasitol. 4:115-123.

19. Clegg, J. A. 1972. The schistosome surface in relation to parasitism. In Taylor, A. E. R. and R. Muller, editors. Functional aspects of parasite surfaces, vol. 10. Blackwell Scientific Publications, Oxford. pp. 23-40.

20. Coles, G. C. 1972. Oxidative phosphorylation in adult *Schistosoma mansoni*. Nature (London) 240:488-489.

21. Coles, G. C. 1973. The metabolish of schistosomes: a review. Int. J. Biochem. 4:319-337.

22. Cort, W. W. 1944. The germ cell cycle in the digenetic trematodes. Q. Rev. Biol. 19:275-284.

23. Cort, W. W., D. J. Ameel, and Anne Van der Woude. 1954. Parasitological Reviews—germinal development in the sporocysts and rediae of the digenetic trematodes. Exp. Parasitol. 3:185-225.

24. Daugherty, J. W. 1952. Intermediary protein metabolism in helminths. I. Transaminase reactions in *Fasciola hepatica*. Exp. Parasitol. 1:331-338.

25. Dike, S. C. 1967. Ultrastructure of the ceca of the digenetic trematodes *Gorgodera amplicava* and *Haematoloechus medioplexus*. J. Parasitol. 53:1173-1185.

26. Dike, S. C. 1969. Acid phosphatase activity and ferritin incorporation in the ceca of digenetic trematodes. J. Parasitol. 55:111-123.

27. Dixon, K. E. 1965. The structure and histochemistry of the cyst wall of the metacercaria of *Fasciola hepatica* L. Parasitology 55:215-226.

28. Dixon, K. E. 1966. A morphological and histochemical study of the cystogenic cells of the cercaria of *Fasciola hepatica* L. Parasitology 56:287-297.

29. Dixon, K. E. 1970. Absorption by developing cercariae of *Cloacitrema narrabeenensis* (Philophthalmidae). J. Parasitol. 56(4, part 2):416-417.

30. Dixon, K. E., and E. H. Mercer. 1967. The formation of the cyst wall in the metacercaria of *Fasciola hepatica*. L. Zeitschr. Zellforsch. 77:345-360.

31. Dönges, J. 1969. Entwicklungs-und Lebensdauer von Metacercarien. Zeitschr. Parasitenk. 31:340-366.

32. Dönges, J. 1970. Transplantation of rediae—a device for solving special problems in trematodology. J. Parasitol. 54:82-83.

33. Erasmus, D. A. 1967. The host-parasite interface of *Cyathocotyle bushiensis* Khan, 1962 (Trematoda: Strigeoidea) II. Electron microscope studies of the tegument. J. Parasitol. 53:703-714.

34. Gönnert, R. 1962. Histologische Untersuchungen über den Feinbau der Eibildungsstatte (Oogeno-top) von *Fasciola hepatica*. Zeitschr. Parasitenk. 21:475-492.

35. Halton, D.W. 1967. Observations on the nutrition of digenetic trematodes. Parasitology 57:639-660.

36. Hochachka, P. W., and T. Mustafa. 1972. Invertebrate facultative anaerobiosis. Science 178:1056-1060.

37. Hockley, D. J. 1972. *Schistosoma mansoni:* the development of the cercarial tegument. Parasitology 64:245-252.

38. Hockley, D. J. 1973. Ultrastructure of the tegument of *Schistosoma*. In Dawes, B., editor, Advances in parasitology, vol. 11. Academic, Press, Inc., New York. pp. 233-305.

39. Hockley, D. J., and D. J. McLaren. 1973. *Schistosoma mansoni:* changes in the outer membrane of the tegument during development from cercaria to adult worm. Int. J. Parasitol. 3:13-25.

40. Hyman, L. H. 1951. The invertebrates: Platyhelminthes and Rhynchocoela. The acoelomate bilateria, vol. 2. McGraw-Hill Book Co., New York.

41. Isseroff, H., and C. P. Read. 1969. Studies on membrane transport. VI. Absorption of amino acids by fascioliid trematodes. Comp. Biochem. Physiol. 30:1153-1159.

42. James, B. L. 1964, The life cycle of *Parvatrema homoeotecnum* sp. nov. (Trematoda: Digenea) and a review of the family Gymnophallidae Morozov, 1955. Parasitology 54:1-41.

43. Janssens, P.A., and C. Bryant. 1969. The ornithine-urea cycle in some parasitic helminths. Comp. Biochem. Physiol. 30:261-272.

44. Khalil, G. M., and R. M. Cable. 1968. Germinal development in *Philophthalmus megalurus* (Cort, 1914) (Trematoda: Digenea). Zeitschr. Parasitenk. 31:211-231.

45. Köhler, P., and D. F. Stahel. 1972. Metabolic end products of anaerobic carbohydrate metabolism of *Dicrocoelium dendriticum* (Trematoda). Comp. Biochem. Physiol. 43B:733-741.

46. Krupa, P. L., A. K. Bal, and G. H. Cousineau. 1967. Ultrastructure of the redia of *Cryptocotyle lingua*. J. Parasitol. 53:725-734.

47. Kurelec, B. 1972. Lack of carbamyl phosphate synthesis in some parasitic platyhelminths. Comp. Biochem. Physiol. 43B:769-780.

48. La Rue, G. R. 1957. The classification of digenetic Trematoda: a review and a new system. Exp. Parasitol. 6:306-349.

49. Lee, D. L. 1966. The structure and composition of the helminth cuticle. In Dawes, B. editor. Advances in parasitology, vol. 4. Academic Press, Inc., New York, pp. 187-254.

50. Lee, D. L. 1972. The structure of the helminth cuticle. In Dawes, B., editor. Advances in parasitology, vol. 10. Academic Press, Inc. New York, pp. 347-379.

51. Lennox, R. W., and E. L. Schiller. 1972. Changes in dry weight and glycogen content as criteria for measuring the postcercarial growth and de-

velopment of *Schistosoma mansoni.* J. Parasitol. 58:489-494.

52. Llewellyn, J. 1965. The evolution of parasitic platyhelminths. In Taylor A., editor. Evolution of parasites. Blackwell Scientific Publications, Oxford. pp. 47-78.

53. Meyer, F., H. Meyer, and E. Bueding. 1970. Lipid metabolism in the parasitic and free-living flatworms, *Schistosoma mansoni* and *Dugesia dorotocephala.* Biochim. Biophys. Acta 210:257-266.

54. Miller, F. H., Jr., G. S. Tulloch, and R. E. Kuntz. 1972. Scanning electron microscopy of integumental surface of *Schistosoma mansoni.* J. Parasitol. 58:693-698.

55. Moss, G. D. 1970. The excretory metabolism of the endoparasitic digenean *Fasciola hepatica* and its relationship to its respiratory metabolism. Parasitology 60:1-19.

56. Nollen, P. M. 1968. Uptake and incorporation of glucose, tyrosine, leucine, and thymidine by adult *Philophthalmus megalurus* (Cort, 1914) (Trematoda), as determined by autoradiography. J. Parasitol. 54:295-304.

57. Nollen, P. M. 1968. Autoradiographic studies on reproduction in *Philophthalmus megalurus* (Cort, 1914) (Trematoda). J. Parasitol. 54:43-48.

58. Nollen, P. M., A. L. Restaino, and R. A. Alberico. 1973. *Gorgoderina attenuata:* uptake and incorporation of tyrosine, thymidine, and adenosine. Exp. Parasitol. 33:468-476.

59. Pappas, P. W. 1971. *Haematoloechus medioplexus:* Uptake, localization, and fate of tritiated arginine. Exp. Parasitol. 30:102-119.

60. Parkening, T. A., and A. D. Johnson. 1969. Glucose uptake in *Haematoloechus medioplexus* and *Gorgoderina* trematodes. Exp. Parasitol. 25:358-367.

61. Pearson, J. C. 1972. A phylogeny of life-cycle patterns of the Digenea. In Dawes, B. editor. Advances in parasitology, vol. 10. Academic Press, Inc. New York. pp. 153-189.

62. Pigulevskii, S. V. 1958. On the question of the phylogeny of flatworms. Rabot. Gelmintol. 80 Let. Skrjabin, 265-270.

63. Prichard, R. K., and P. J. Schofield. 1968. A comparative study of the tricarboxylic acid cycle enzymes in *Fasciola hepatica* and rat liver. Comp. Biochem. Physiol. 25:1005-1019.

64. Prichard, R. K., and P. J. Schofield. 1969. The glyoxylate cycle, fructose-1,6-diphosphatase and glyconeogenesis in *Fasciola hepatica.* Comp. Biochem. Physiol. 29:581-590.

65. Prichard, R. K., and P. J. Schofield. 1968. The glycolytic pathway in adult liver fluke, *Fasciola hepatica.* Comp. Biochem. Physiol. 24:697-710.

66. Ramalingam, K. 1973. The chemical nature of the eggshell of helminths: I. Absence of quinone tanning in the eggshell of the liver fluke, *Fasciola hepatica.* Int. J. Parasitol. 3:67-75.

67. Ramalingam, K. 1973. Chemical nature of the eggshell in helminths: II. Mode of stabilization of eggshells of monogenetic trematodes. Exp. Parasitol. 34:115-122.

68. Read, C. P., L. T. Douglas, and J. E. Simmons, Jr. 1959. Urea and osmotic properties of tapeworms form elasmobranchs. Exp. Parasitol. 8:58-75.

69. Rogers, S. H., and E. Bueding. 1975. Anatomical localization of glucose uptake by *Schistosoma mansoni* adults. Int. J. Parasitol. 5:369-371.

70. Rowcliffe, S. A., and C. B. Ollerenshaw. 1960. Observations on the bionomics of the egg of *Fasciola hepatica.* Ann. Trop. Med. Parasitol. 54:172-181.

71. Sauer, M. C. V., and A. W. Senft. 1972. Properties of a proteolytic enzyme from *Schistosoma mansoni.* Comp. Biochem. Physiol. 42B:205-220.

72. Senft, A. W. 1963. Observations on amino acid metabolism of *Schistosoma mansoni* in a chemically defined medium. Ann. N.Y. Acad. Sci. 113:272-288.

73. Senft, A. W. 1968. Studies in proline metabolism by *Schistosoma mansoni.* I. Radioautography following in vitro exposure to radioproline C^{14}. Comp. Biochem. Physiol. 27:251-261.

74. Senft, A. W., R. P. Miech, P. R. Brown, and D. G. Senft. 1972. Purine metabolism in *Schistosoma mansoni.* Int. J. Parasitol. 2:249-260.

75. Senft, A. W., D. E. Philpott, and A. H. Pelofsky. 1961. Electron microscope observations of the integument, flame cells, and gut of *Schistosoma mansoni.* J. Parasitol. 47:217-229.

76. Sinitsin, D. 1931. Studien über die Phylogenie der Trematoden IV. The life histories of *Plagioporus siliculus* and *Plagioporus virens,* with special reference to the origin of Digenea. Z. Wiss. Zool. 138:409-456.

77. Smyth, J. D. 1966. The physiology of trematodes. W. H. Freeman, San Francisco.

78. Stirewalt, M. A. 1974. *Schistosoma mansoni:* cercaria to schistosomule. In Dawes, B., editor. Advances in parasitology, vol. 12. Academic Press, Inc., New York. pp. 115-182.

79. Timms, A. R., and E. Bueding. 1959. Studies of a proteolytic enzyme from *Schistosoma mansoni.* Br. J. Pharmacol. Chemother. 14:68-73.

80. Ulmer, M. J. 1971. Site-finding behaviour in helminths in intermediate and definitive hosts. In Fallis, A. M., editor. Ecology and physiology of parasites. University of Toronto Press, Toronto. pp. 123-160.

81. Vernberg, W. B., and F. J. Vernberg. 1971. Respiratory metabolism of a trematode metacercaria and its host. In Cheng, T. C., editor. Aspects of the biology of symbiosis. University Park Press, Baltimore. pp. 91-102.

82. Watts, S. D. M. 1970. Transamination in homogenates of rediae of *Cryptocotyle lingua* and of sporocysts of *Cercaria emasculans* Pelseneer, 1900. Parasitology 61:499-504.

83. Wesenberg-Lund, C. 1931. Biology of *Leuco-chloridium*. Kongl. Danske Vidensk. Selsk. Skr. Nat. Math. Afd., ser. 9, 4:89-142.
84. Wilson, R. A. 1968. The hatching mechanism of the egg of *Fasciola hepatica* L. Parasitology 58:79-89.
85. Wilson, R. A. 1969. Fine structure of the tegument of the miracidum of *Fasciola hepatica* L. J. Parasitol. 55:124-133.
86. Yamaguti, S. 1971. Synopsis of digenetic trematodes of vertebrates, vol. 1. Keigaku, Tokyo.

SUGGESTED READING

Acholonu, A. D. 1964. Freshwater cercariae of northern Colorado together with a checklist of the species described in the United States and their molluscan hosts. Dissertation Abstracts 25(6).

Baer, J. G. 1951. Ecology of animal parasites. Illinois University Press, Urbana, Ill. (Contains interesting examples of trematodes.)

Baer, J. G. and C. Joyeux. 1961. Classe des Trematodes (Trematoda Rudolphi). In Grassé, P. P., editor. Traité de zoologie. Anatomie, systematique, biologie. T. IV., F. I. Plathelminthes, Mesozoaires, Acanthocephales, Nemertiens. Masson et Cie, Paris. pp. 561-692. (A well-illustrated overview of trematodes.)

Buttner, A. 1951. La progénèse chez les trematodes digenetiques, sa signification, ses manifestations. Contribution à l'étude de son determinism. Ann. Parasitol. 25:376-434.

Calde, R.M. 1972. Behaviour of digenetic trematodes. Zool. J. Linn. Soc. 51(Suppl. 1):1-18.

Dawes, B. 1956. The Trematoda, with special reference to British and Other European Forms. Cambridge University. Press, Cambridge. (A classic reference work, of value to all interested in trematodes.)

Ginetsinskaya, T. A. 1968. Trematodes and their life cycle, biology and evolution. Acad. Nauk SSSR, Leningrad. (An important theoretical book.)

Holliman, R. B. 1961. Larval trematodes from the Apalache Bay area, Florida, with a checklist of known marine cercariae arranged in a key to their superfamilies. Tulane Stud. Zool. 9:2-74.

Hyman, L. H. 1951. The invertebrates: Platyhelminthes and Rhynchocoela. The acoelomate bilateria, vol. II. McGraw-Hill Book Co., New York. (A standard reference to all aspects of Trematoda.)

Kurashvili, B. E. 1967. The role of phylogeny and ecology of helminth and host in the formation of helminth fauna in animals. Helminthologia 8:283-288.

LaRue, G. R. 1957. The classification of digenetic Trematoda: a review and a new system. Exp. Parasitol. 6:306-349. (The classic system on which modern classification of flukes is based.)

Lim, H., and D. Heynemen. 1972. Intramolluscan intertrematode antagonism: a review of factors influencing the host-parasite system and its possible role in biological control. In Dawes, B., editor. Advances in parasitology, vol. 10. Academic Press, Inc., New York. pp. 192-268.

Lumsden, R.D. 1975. Surface ultrastructure and cytochemistry of parasitic helminths. Exp. Parasitol. 37:267-339. (Covers a wide variety of helminths in addition to trematodes.)

Mansour, T. E. 1967. Effect of hormones on carbohydrate metabolism of invertebrates. Fed. Proc. 26:1179-1185.

Manter, H. W. 1963. The zoogeographical affinities of trematodes of South American fishes. Syst. Zool. 12:45-70.

Pappas, P. W., and C. P. Read. 1975. Membrane transport in helminth parasites: a review. Exp. Parasitol. 37:469-530. (Covers a wide variety of helminths.)

Schell, S. C. How to know the trematodes. William C. Brown, Publishers, Dubuque. (An excellent key to the trematodes of North America.)

Stunkard, H. W. 1963. Systematics, taxonomy and nomenclature of the Trematoda. Q. Rev. Biol. 38:221-233.

Stunkard, H. W. 1970. Trematode parasites of insular and relict vertebrates. J. Parasitol. 56:1041-1054.

Chapter 17

SUPERORDER ANEPITHELIOCYSTIDIA: ORDER STRIGEATA

Of the several superfamilies in this order, only two, Strigeoidea and Schistosomatoidea, are of much economic or medical significance. The latter, however, contains some of the most important disease agents of humans.

SUPERFAMILY STRIGEOIDEA

Strigeoidea are bizarre in appearance, with their bodies divided into two portions (Fig. 17-1). The anterior portion usually is spoon- or cup-shaped, with accessory **pseudosuckers** on each side of the oral sucker. Behind the acetabulum is a spongy, pad-like organ referred to as the **adhesive** or **tribocytic organ.** This structure secretes proteolytic enzymes that digest host mucosa, probably functioning both as an accessory holdfast and as a digestive-absorptive organ. The hindbody contains most of the reproductive organs, although vitelline follicles often extend into the forebody. The genital pore is located at the posterior end.

Most strigeids are quite small and are found commonly in the digestive tracts of fish-eating vertebrates. Their cercariae are easily recognized by the fact that they have both a pharynx and a forked tail. No adult strigeids are known to parasitize humans, but they are so ubiquitous and their biology so interesting that we will briefly consider a few species.

Family Diplostomatidae
Alaria americana

The genus *Alaria* contains several very similar species, all of which mature in the small intestines of carnivorous mammals. *Alaria americana* is found in various species of Canidae in northern North America. They are about 2.5 to 4 mm long, with the forebody longer than the hindbody. The forebody has a pair of ventral flaps that are narrowest at the anterior end (Fig. 17-2). A pointed process flanks each side of the oral sucker. The tribocytic organ is relatively large and elongated and has a ventral depression in its center.

The life cycle of *Alaria* is remarkable in that the worms may require four hosts before they can develop to maturity (Fig. 17-3). The eggs are unembryonated when laid, and they hatch in about 2 weeks. The miracidium swims actively and will attack and penetrate any of several species of helisomid snails.[14] Mother sporocysts develop in the renal veins and produce daughter sporocysts in about 2 weeks. Daughter sporocysts migrate to the digestive gland and need about a year to mature and begin producing cercariae. The furcocercous cercaria leaves the snail during daylight hours and swims to the surface, where it hangs upside down. Occasionally it sinks a short distance and then returns to the surface. If a tadpole swims by, the resulting water currents stimulate the cercaria to swim after it. If it contacts the tadpole, the cercaria will quickly attack, drop its tail, penetrate the skin, and begin wandering within the amphibian. In about 2 weeks the cercaria has transformed into a mesocercaria, an unencysted form between a cercaria and a metacercaria. It is then infective to the next host, which may be the definitive host or a paratenic host. If a canid eats an infected tadpole (or adult frog), the mesocercariae are freed by digestion, penetrate into the coelom, and then move to the diaphragm and lungs. After about 5 weeks in the lungs, the mesocercaria has transformed into a **diplostomulum metacercaria.** Diplostomules migrate up the trachea and then to the intestine, where they mature in about a month.

Tadpoles, however, are not always available to terrestrial canids and, furthermore,

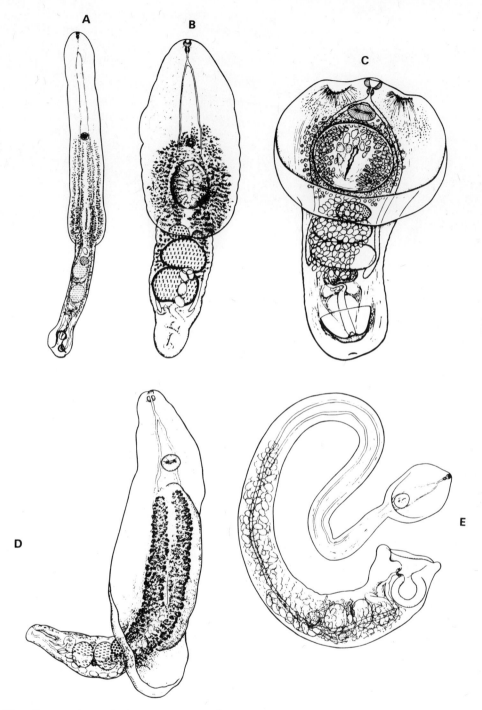

Fig. 17-1. Typical strigeid trematodes, illustrating body forms. **A,** *Mesodiplostomum gladiolum* Dubois, 1936. **B,** *Pseudoneodiplostomum thomasi* (Dollfus, 1935). **C,** *Proalarioides serpentis* Yamaguti, 1933. **D,** *Petalodiplostomum ancyloides* Dubois, 1936. **E,** *Cerocotyla cerylis* Yamaguti, 1939. (From Yamaguti, S. 1971. Synopsis of digenetic trematodes of vertebrates, vol. 2. Keigaku, Tokyo.)

are distasteful to all but the hungriest carnivores. This ecological barrier is overcome when a water snake eats the infected tadpole or frog and thereby becomes a paratenic host. A snake (or other animal) can accumulate large numbers of mesocercariae in its tissues, rendering a heavy infection to the definitive host when it is eaten. The mesocercariae then migrate, develop into diplostomulae, and mature in the intestine, as do those from tadpoles. Life cycles of other species of *Alaria* are similar.

Mature *Alaria* species are quite pathogenic, causing severe enteritis that often kills the definitive hosts in severe infections. Also, the mesocercaria is quite pathogenic, especially when accumulated in large numbers. Figure 17-4 is from a fatal infection of mesocercariae in a human.

Uvulifer ambloplitis

There are several species of strigeid trematodes that cause black spots in the skin of fishes, one of which is *Uvulifer ambloplitis*. This is a parasite of kingfishers, a fish-eating bird that is widely distributed across the United States. The spoon-

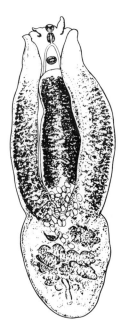

Fig. 17-2. *Alaria alata* (Goeze, 1782).

Fig. 17-3. Life cycle of *Alaria americana*. **A,** Ventral view of adult fluke; **B,** miracidium; **C,** mother sporocyst; **D,** daughter sporocyst; **E,** fork-tailed cercaria; **F,** mesocercaria showing some internal organs; **G,** external view of mesocercaria; **H,** diplostomulum; **I,** fox definitive host; **J,** planorbid snail *(Helisoma)*, first intermediate host; **K,** tadpole *(Rana, Bufo)*, second intermediate host; **L,** paratenic hosts (snakes, frogs, mice). *1,* Forebody; *2,* hindbody; *3,* lappet or pseudosucker; *4,* holdfast (tribocytic) organ; *5,* oral sucker; *6,* pharynx; *7,* cecum; *8,* ventral sucker; *9,* testes; *10,* ovary; *11,* eggs in uterus; *12,* vitelline glands; *13,* common genital pore; *14,* cilia; *15,* eyespot; *16,* flame cell; *17,* germ cells; *18,* excretory opening; *19,* daughter sporocyst with germ balls or developing cercariae; *20,* birth pore; *21,* cercaria; *22,* penetration glands; *23,* duct of penetration glands; *24,* genital primordium; *25,* excretory bladder; *26,* forked tail; *27,* body spines.

a, Adult fluke in small intestine; *b,* egg passing out of body in feces; *c,* unembryonated egg; *d,* embryonated egg; *e,* egg hatching; *f,* miracidium penetrating *Helisoma* snail; *g,* young mother sporocyst; *h,* mature mother sporocyst; *i,* daughter sporocyst; *j,* cercaria free in water in characteristic resting position; *k,* cercaria penetrating tadpole, casting tail as it enters; *l,* mesocercaria; *m,* mesocercaria in snake, frog, and mouse paratenic hosts; *n,* infection of definitive host by swallowing infected tadpole, second intermediate host; *o,* infection of definitive host by swallowing infected paratenic host; *p,* mesocercariae migrate through gut wall into coelom; *q,* mesocercariae enter hepatic portal vein, but it has not been shown that they reach the lungs via the blood; *r,* mesocercariae pass through the diaphragm and penetrate the lungs; *s,* in the lungs, the mesocercariae transform to a diplostomulum stage; *t,* diplostomulae migrate up trachea; *u,* diplostomulae are swallowed, go to small intestine, and develop to maturity *(a)* in 5 to 6 weeks. (From Olsen, O. W. 1974. Animal parasites. Their life cycles and ecology. University Park Press, Baltimore.)

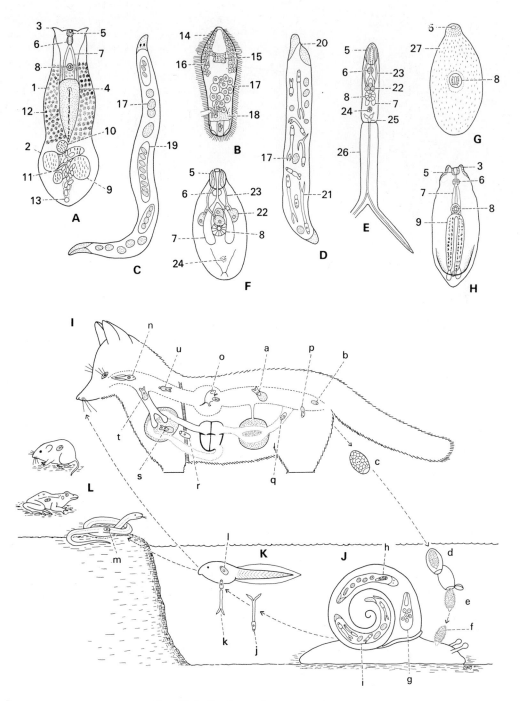

Fig. 17-3. For legend see opposite page.

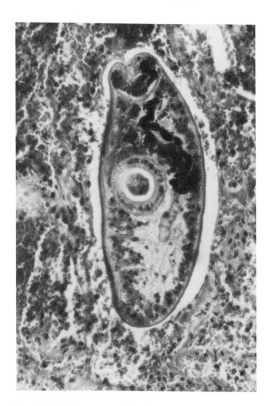

Fig. 17-4. Mesocercaria of *Alaria americana* in human lung biopsy. The case proved to be fatal, with nearly every organ of the body infected, presumably as a result of eating undercooked frog's legs. (From Freeman, R., and others. 1976. Am. J. Trop. Med. Hyg. 25:803-807.)

shaped forebody of the parasite is separated from the longer hindbody by a slender constriction. Adults are 1.8 to 2.3 mm long.

The eggs, which are unembryonated when laid, hatch in about 3 weeks. The miracidium will penetrate snails of the genus *Helisoma* and will transform into mother sporocysts that retain the eyespots of the first larva. Daughter sporocysts invade the digestive gland and produce cercariae in about 6 weeks. The cercariae escape from the tissues of the snail and rise to the surface of the water, where they are sensitive to the passing of fish. If they contact a fish, especially a centrarchid or percid, they drop their tails and penetrate the skin. Once inside the dermis, the flukes metamorphose into **neascus metacercariae** and secrete a delicate, hyaline cyst

wall around themselves. The fish host responds to the neascus by laying down layers of melanin granules. The result is a conspicuous "black spot" indicating the presence of a metacercaria. When such fish are heavily infected they are often discarded as diseased by fishermen. Kingfishers become infected when they eat such a fish. The flukes mature in 27 to 30 days.

Other, related flukes also cause "black spot" in a wide variety of fishes and have similar life cycles (Fig. 17-5).

Family Strigeidae
Cotylurus flabelliformis

This is a very common parasite of wild and domestic ducks in North America. Adult flukes are 0.55 to 1.0 mm long. The forebody is cup-shaped with the acetabulum and tribocytic organ located at its depths. The hindbody is short and stout and is curved dorsad (Fig. 17-6).

Adult worms live in the small intestines of ducks and lay unembryonated eggs. These hatch in about 3 weeks, and the miracidia attack snails of the family Lymnaeidae *(Lymnaea, Stagnicola)*. The mother sporocyst produces daughter sporocysts that migrate to the digestive gland and grow into slender, worm-like bodies. After about 6 weeks they begin to release furcocercous cercariae, which work their way free into the water. The cercariae are very active, and if they contact a snail of the same family Lymnaeidae, they penetrate, migrate to the ovotestis, and transform into **tetracotyle metacercariae.** When the snail is eaten by ducks, the flukes are released by digestion and mature in about 1 week.

On the other hand, if the cercaria enters a snail of the families Planorbidae or Physidae, it will attack sporocysts or rediae of other species of flukes already present and will develop into a tetracotyle metacercaria within them. This is an example of hyperparasitism among larval parasites. These, too, will mature in about 1 week, if eaten by a duck.

Strigeoid trematodes, then, exhibit complex life cycles that involve several unrelated hosts. Their adaptability is amazing, when one considers the differences in environments provided by snails, pond water, fish, amphibians, reptiles, and birds or

Fig. 17-5. A minnow, *Pimephales* sp., infected with neascus-type metacercariae, "black spot." (Photograph by John S. Mackiewicz.)

Fig. 17-6. *Cotylurus flabelliformis*, a common parasite of waterfowl. Lateral view. (Photograph by Warren Buss.)

mammals. It is also remarkable that a parasite that may require more than a year to complete its larval development can become sexually mature in its definitive host in a week or less and die a few days later. Such is the pattern in the life cycles of the strigeoids.[15]

SUPERFAMILY SCHISTOSOMATOIDEA

Flukes of the superfamily Schistosomatoidea are peculiar in that they have no second intermediate host in their life cycles and also in that they mature in the blood vascular system of their definitive hosts. Most species are dioecious. They are parasites of fishes, turtles, birds, and mammals throughout the world. Several species are parasites of humans, causing misery and death wherever they are distributed[18]

(Table 2). The families Aporocotylidae, Sanguinicolidae, and Spirorchidae parasitize fish and turtles and are of little economic importance. The family Schistosomatidae, however, includes species that are among the most dreaded parasites of humans. To date, ten species and two varieties of schistosomes have been reported in Africa. The taxonomy of this group and the validity of some species of parasites have been subjects of controversy for years, and the final decision on species recognition awaits in-depth studies on the basic biology of schistosomes.

Family Schistosomatidae
Schistosoma spp. and schistosomiasis

Three species of schistosomes are of vast medical significance: *Schistosoma haematobium, S. mansoni,* and *S. japonicum*—all parasites of humans since antiquity. Bloody urine was a well-recognized disease symptom in northern Africa in ancient times. At least 50 references to this condition have been found in surviving Egyptian papyri, and calcified eggs of *S. haematobium* have been found in Egyptian mummies dating from about 1200 B.C. Hulse has presented a well-reasoned theory that the curse that Joshua placed on Jericho can be explained by the introduction of *Schistosoma haematobium* into the communal well by the invaders. The removal of the curse occurred after the abandonment of Jericho and subsequent droughts had eliminated the snail host, *Bulinus truncatus.*[4] Today, Jericho (Ariha, Jordan) is

Table 2. World distribution of schistosomiasis in 1972: total regional population, population exposed, and population infected.[19]

Region	Total population	Population exposed	Population infected
Africa	308,021,000	226,102,740	91,200,310
Mascarene Islands (Mauritius)	741,000	370,500	66,690
Southwest Asia	87,003,000	10,745,050	2,271,020
Southeast Asia	868,531,000	337,051,500	25,223,650
The Americas	98,339,000	18,199,750	6,144,130
TOTAL	1,362,635,000	592,469,540	124,905,800

well-known for its fertile lands and healthy, well-nourished people.

The first Europeans to record contact with *S. haematobium* were surgeons with Napoleon's army in Egypt (1799-1801). They reported that **hematuria** (bloody urine) was prevalent among the troops, although the cause, of course, was quite unknown. Nothing further was learned about *schistosomiasis haematobia* for over 50 years, until a young German parasitologist, Theodor Bilharz, discovered the worm that caused it. He announced his discovery in letters to his old teacher, Von Siebold, naming the parasite *Distomum haematobium*.[3] Tragically Bilharz died of typhus at the age of 37. During the next few years it was discovered that 30% to 40% of the population in Egypt was infected with *S. haematobium,* and it even was found in an ape dying in London. The peculiar morphology of the worm made it apparent that it could not be included in the genus *Distomum,* so in 1858 Weinland proposed the name *Schistosoma*. Three months later Cobbold named it *Bilharzia,* after its discoverer. This latter name became widely accepted throughout the world, and was even given the slang name "Bill Harris" by British soldiers serving in Europe during the World War I. Today, however, the strict rules of zoological nomenclature decree that *Schistosoma* has priority and is thus the current name for the parasite. Even so, health officers in many parts of the world erect signs next to ponds and streams that warn prospective bathers of the dangers of "bilharzia." None the less, schistosoma is an apt name, referring to the "split body" (gynecophoral canal) of the male.

While information was accumulating on the biology of *Schistosoma haematobium,* doubts were being raised as to whether *S. haematobium* was a single species or whether two or more species were being confused. The problem was confounded by the occurrence of terminally spined eggs in both urine and feces. Whenever eggs with lateral spines were noticed, they were ignored as being "abnormal." Sir Patrick Manson, in 1905, decided that intestinal and vesicular (urinary bladder) schistosomiasis usually are distinct diseases, caused by distinct species of worms. He reached this conclusion when he examined a man from the West Indies who had never been to Africa and who passed laterally spined eggs in his feces but none at all in his urine.[10] Sambon argued in favor of the two-species concept in 1907, and he named those parasites producing laterally spined eggs *Schistosoma mansoni*. (Japanese zoologists had already detected still another species by this time, but their reports were generally unknown to Europeans.) But the eminent German parasitologist Looss disagreed and brought the full sway of his reputation and dialectic against the theory and even stated that he had seen a female worm with both kinds of eggs in its uterus. Sambon, undaunted, replied that until Professor Looss could "show me an actual specimen, I am bound to place the worm capable of producing the two kinds of eggs with the phoenix, the chimaera and other mythical monsters."[16]

The question was finally resolved by Leiper in 1915. He first visited Japan to acquaint himself with the work of Miyairi and Suzuki on *S. japonicum*. Then, working in Egypt, he discovered that cercariae

Table 3. Comparative morphology of human schistosomes

	S. haematobium	*S. mansoni*	*S.japonicum*
Tegumental papillae	Small tubercles	Large papillae with spines	Smooth
Size			
Male			
Length	10-15 mm	10-15 mm	12-20 mm
Width	0.8-1.0 mm	0.8-1.0 mm	0.5-0.55 mm
Female			
Length	ca. 20 mm	ca. 20 mm	ca. 26 mm
Width	ca. 0.25 mm	ca. 0.25 mm	ca. 0.3 mm
Number of testes	4-5	6-9	7
Position of ovary	Near midbody	In anterior half	Posterior to midbody
Uterus	With 20-100 eggs at one time, average 50	Short, few eggs at one time	Long, may contain up to 300 eggs, average 50
Vitellaria	Few follicles, posterior to ovary	Few follicles, posterior to ovary	In lateral fields, posterior quarter of body
Egg	Elliptical, with sharp terminal spine	Elliptical, with sharp lateral spine	Oval to almost spherical, rudimentary lateral spine
	112-170 μm × 40-70 μm	114-175 μm × 45-70 μm	70-100 μm × 50-70 μm

emerging from the snail *Bulinus* could infect the vesicular veins of various mammals, and they always produced eggs with terminal spines. Those emerging from a different snail, *Biomphalaria,* infected the intestinal veins and produced laterally spined eggs.[6] It was soon determined that *S. mansoni* had a broad distribution in the world, having been widely scattered by the slave trade. It is now widespread in Africa and the Middle East and is the only blood fluke of humans in the New World, with the possible exception of a small focus of *S. haematobium* in Surinam.[8] The original endemic area of *S. mansoni* is thought to have been the Great Lakes region of central Africa.

While Cobbold, Weinland, Bancroft, Sambon, and others were wrestling with the problem of *S. haematobium* and *S. mansoni,* Japanese researchers were investigating a similar disease in their country. For years physicians in the provinces of Hiroshima, Saga, and Yamanachi had recognized an endemic disease characterized by an enlarged liver and spleen, ascites, and diarrhea. At autopsy they noted eggs of an unknown helminth in various organs, especially in the liver. In 1904, Professor Katsurada of Okayama recognized that the larvae in these eggs resembled those of *S. haematobium.* Because he was unable

to make a postmortem examination of an infected person, he began examining local dogs and cats, in hopes that they were reservoirs for the parasite. He soon found adult worms containing eggs identical to those from humans and named them *Schistosoma japonicum.* The experimental elucidation of the life cycle by various Japanese researchers was a milestone in the history of parasitology and formed the basis for Leiper's work on blood flukes in Egypt. The distribution of this parasite is limited to Japan, China, Taiwan, the Philippines, Celebes, Laos, Cambodia, and Thailand.

Morphology. Though the three species are generally similar structurally, several differences in detail are listed in Table 3. There is considerable sexual dimorphism in the genus *Schistosoma,* the males being shorter and stouter than the females (Fig. 17-7). The males have a ventral, longitudinal groove, the **gynecophoric canal,** where the female normally resides. The mouth is surrounded by a strong oral sucker, and the acetabulum is near the anterior end. There is no pharynx. The paired intestinal ceca converge and fuse at about the midpoint of the worm, then continue as a single gut to the posterior end. The male possesses five to nine testes, according to species, each of which has a del-

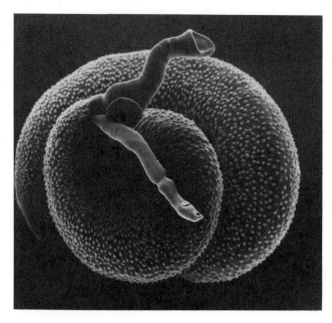

Fig. 17-7. Scanning electron micrograph of male and female *Schistosoma mansoni*. The female is lying in the gynocophoric groove in the ventral surface of the male. (Photograph by Sue Carlisle Ernst.)

icate vas efferens, and these combine to form the vas deferens. The latter dilates to become the seminal vesicle, which opens ventrally through the genital pore immediately behind the ventral sucker. Cirrus pouch, cirrus, and prostate cells are absent.

The suckers of the females are smaller and not so muscular as those of the males, and the tegumental tubercles (Fig. 17-8), if any, are confined to the ends of the female. The ovary is anterior or posterior to or at the middle of the body, and the uterus is correspondingly short or long, depending on the species. On the basis of differences between the species, La Roux proposed that the genera *Afrobilharzia* and *Sinobilharzia* be erected for *S. mansoni* and *S. japonicum*, respectively, but his suggestion has not been generally accepted.

Biology. Adult worms live in the veins that drain certain organs of their host's abdomen, and the three species have distinct preferences: *S. haematobium* lives principally in the veins of the urinary bladder plexus; *S. mansoni* prefers the portal veins draining the large intestine; and *S. japonicum* is more concentrated in the veins of the small intestine. The female worm is often found in the gynecophoric canal of the male worm, where copulation takes place; there may be other physiological reasons for this habitus as well. The worms work their way "upstream" into smaller veins, and the female may leave the gynecophoric canal to reach still smaller venules to deposit eggs. The eggs (Fig. 17-9) must then traverse the wall of the venule, some intervening tissue, and the gut or bladder mucosa before they are in a position to be expelled from the host. The mechanism by which this "escape" is achieved is not at all clear and has been the subject of much speculation. The spines on the eggs are often credited with contributing to the expulsion, but one must remember that the feat is accomplished by *S. japonicum*, which has only the most rudimentary spine. The most likely explanation has been provided by Kuba (1963).[5] He observed that when eggs are lodged adjacent to the wall of the venule, a small blood thrombus (clot) forms around each of them. As the thrombus is slowly infiltrated with fibroblasts (organized) and overgrown with endothelial cells, the eggs are isolated from the blood flow, and the venule wall thins. The eggs

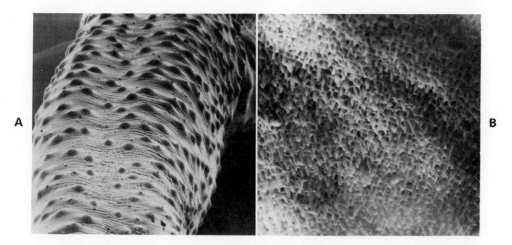

Fig. 17-8. Tegumental tubercles of *Schistosoma haematobium*. **A,** Dorsal surface; **B,** toothlike tubercles in oral sucker. (Photographs by Robert E. Kuntz.)

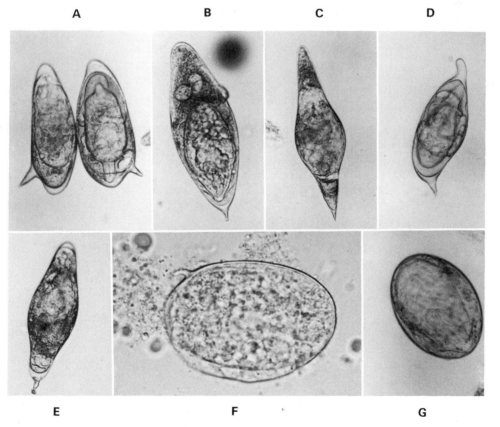

Fig. 17-9. Eggs of schistosome flukes. **A,** *Schistosoma mansoni;* **B,** *Schistosoma intercalatum;* **C,** *Schistosoma bovis;* **D,** *Schistosoma rhodhaini;* **E,** *Schistosoma mattheei;* **F,** *Schistosoma japonicum;* **G,** *Schistosomatium douthitti.* (Photographs by Robert E. Kuntz and Jerry A. Moore.)

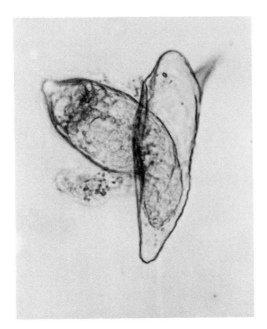

Fig. 17-10. Miracidium of *Schistosoma mansoni* escaping from its eggshell. (Photograph by Robert E. Kuntz.)

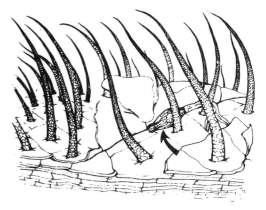

Fig. 17-12. Diagram of a cercaria of *Schistosoma* in exploring position on skin (arrow). (From Stirewalt, M. A. 1971. In Cheng, T. C., editor. Aspects of the biology of symbiosis. University Park Press, Baltimore.)

then pass through as the host repair processes continue. How the egg gets from there to the lumen of the gut or bladder is less clear, but it can only be accomplished if the extravasation is extremely close to the mucosa or a crypt. In any case, about two-thirds of the eggs do not make it, and large numbers build up in the gut or bladder wall, particularly in chronic cases where the wall is toughened by a great amount of connective (scar) tissue. Of course, many eggs are never expelled from the venules but are swept away by the blood, eventually to lodge in the liver or capillary beds of other organs. By the time the eggs reach the outside via the urine or feces, they are completely embryonated and hatch when exposed to the lower tonicity of fresh water.

The mechanism of hatching is poorly understood. The first indication of hatching is activation of the cilia on the miracidium. This increases until the miracidium is a veritable spinning ball. Then suddenly, an operculum-like **stigma** flies open on the side of the egg, and the miracidium emerges (Fig. 17-10). The miracidium usually contracts a few times to completely clear the shell, after which it rapidly swims away. However, some eggs will not hatch, no matter how active the miracidium becomes, and others will hatch before the larva becomes activated.

The miracidium is typical and swims ceaselessly during its short life. If hatching from an old egg, it will live only 1 or 2

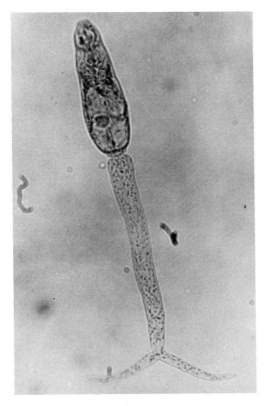

Fig. 17-11. Cercaria of *Schistosoma mansoni.* (Photograph by Warren Buss.)

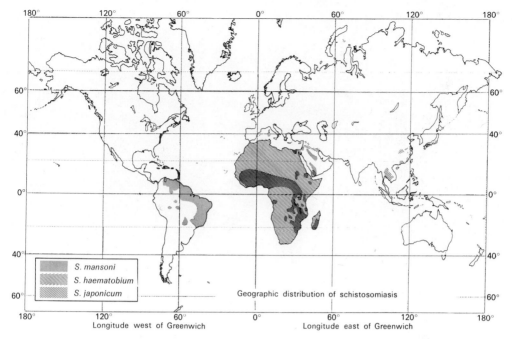

Fig. 17-13. Geographic distribution of schistosomiasis. (AFIP neg. no. 68-4866-3.)

hours, while in optimum conditions it will survive for 5 or 6 hours. Though the miracidia of schistosomes do not have eyespots, they apparently have photoreceptors, and they are positively phototropic.[2] When miracidia enter the vicinity of a snail host, they are stimulated to swim more rapidly and change direction much more frequently, thus increasing their chances of encountering the host. The most important snails are: for *S. haematobium,* several species of *Bulinus* and *Physopsis,* possibly also *Planorbarius;* for *S. mansoni, Biomphalaria alexandrina* in northern Africa, Saudi Arabia, and Yemen and *B. sudanica, B. rupellii, B. pfeifferi,* and others in the genus in other parts of Africa and *B. glabrata* being the most important in the Western Hemisphere, with *Tropicorbis centrimetralis* having been found naturally infected in Brazil; for *S. japonicum,* several species of *Oncomelania.*

After penetration of the snail, the miracidium sheds its epithelium and begins development into a mother sporocyst, usually near its point of entrance. After about 2 weeks the mother sporocyst, which has four protonephridia, gives birth to daughter sporocysts, which usually migrate to other organs of the snail, if there is room.

The mother sporocyst continues producing daughter sporocysts for up to 6 or 7 weeks.[11] There is no redial generation.

The furcocercous cercariae (Fig. 17-11) start to emerge from the daughter sporocysts and the snail host about 4 weeks after initial penetration by the miracidium. The cercaria has a body 175 to 240 μm long by 55 to 100 μm wide, and a tail 175 to 250 μm long by 35 to 50 μm wide, bearing a pair of furci 60 to 100 μm long. The oral sucker is comparatively large, and the ventral sucker is small and covered with minute spines. Four types of glands open through bundles of ducts at the anterior margin of the oral sucker (see p. 251).

There is no second intermediate host in the life cycle. The cercariae alternately swim to the surface of the water and slowly sink toward the bottom, continuing to live this way for 1 to 3 days. If they come into contact with the skin of a prospective host, such as a human, they attach and creep about for a time as if seeking a suitable place to penetrate (Fig. 17-12). They require only half an hour or less to completely penetrate the epidermis, and they can disappear through the surface in 10 to 30 seconds. Penetration is accompanied by a vigorous wiggling, together with se-

Fig. 17-14. Government-encouraged pit latrines in the Philippines serve as one means of prevention of schistosomiasis. (Photograph by Robert E. Kuntz.)

cretion of the products of the penetration glands. The tail is cast off in the process. The worms are somewhat smaller now that the penetration glands have emptied their contents. Within 24 hours the **schistosomules,** as they are now called, enter the peripheral circulation and are swept off to the heart. Some of the schistosomules may migrate via the lymphatics to the thoracic duct and thence to the subclavian veins and heart. Leaving the right side of the heart, the small worms wriggle their way through the pulmonary capillaries to gain access to the left heart and systemic circulation. It appears that only the schistosomules that enter the mesenteric arteries, traverse the intestinal capillary bed, and reach the liver by the hepatic portal system can continue to grow. Some parasitologists maintain that *S. mansoni* can complete the entire circuit of the circulatory system several times before reaching the portal ven-

ules.[7] After undergoing a period of about 3 weeks of development in the liver sinusoids, the young worms migrate to the walls of the gut or bladder (according to species), copulate, and begin producing eggs. The entire prepatent period is about 5 to 8 weeks.

Epidemiology. Human waste in water containing intermediate hosts of the *Schistosoma* is the single most important epidemiological factor in schistosomiasis, and the availability of suitable species of snail host will determine the endemicity of the particular species of *Schistosoma*. The latter is well illustrated by the fact that though both *S. mansoni* and *S. haematobium* are widespread in Africa, only *S. mansoni* became established in the New World by the slave trade, almost certainly because snails suitable for only that species were present there (Fig. 17-13). Survival of these parasites depends on human insistence on pol-

Fig. 17-15. Typical native dwelling in a schistosomiasis area of the Philippines. *Oncomelania,* the snail host of *S. japonicum,* is found in the stream near the house, which is also likely to be contaminated by human feces. (Photograph by Robert E. Kuntz.)

luting water with their organic wastes. Adequate sewage treatment is sufficient to eliminate schistosomiasis as a disease of humans (Fig. 17-14). However, really adequate sewage treatment has not yet been realized in the most advanced civilizations, and any treatment at all is lacking in many areas where the flukes are prevalent. Tradition, at once the salvation and the bane of culture, prompts people to use the local waterway for sewage disposal instead of foul-smelling outhouses (Fig. 17-15). A bridge across a small stream becomes a convenient toilet; a grove of mango trees over a rivulet is a haven for children who bombard the area with their feces (Fig. 17-16). Especially vulnerable to infection are farmers who wade in their irrigation water, fishermen who wade in their lakes and streams, children who play in any contaminated body of water, and women who wash clothes in streams. A focus of infec-

tion in Brazil was a series of ditches where watercress was grown for food. In some Moslem countries, the religious requirement of ablution, that is, washing the anal or urethral orifices after urination or defecation, is an important factor in transmission. Not only is the convenient water source to perform ablution likely to be a contaminated river or canal, the deposition of additional feces and urine in its vicinity is ensured.

Clearly, the economic and education level of the population will influence the transmission of the disease, and age and sex are important factors as well. Males usually show the highest rates of infection and the most intense infections, and the most hazardous age is the second decade of life. This appears to reflect occupational and recreational differences, rather than sex or age resistance to infection. In Surinam, where both sexes work in the

Fig. 17-16. Slow-running streams and protecting tropical vegetation provide ideal habitats for *Biomphalaria,* a snail host for *S. mansoni* in Puerto Rico. (Photograph by Robert E. Kuntz.)

fields, the highest (and equal) prevalence is found in adults of both sexes. Certain other factors, such as immunity and the cessation of egg release in chronic infections, must be considered when a survey of a population is carried out and transmission is studied. The fact that build-up of granulation tissue in the gut or bladder wall prevents release of eggs into the feces or urine will mask infection unless immunodiagnostic or biopsy methods are used. Further, even though such persons may be gravely disabled, they are removed from the cycle of transmission.

During the course of the infection, some protective immunity to superinfection is elicited, either by repeated exposures to the cercariae or by the presence of the adults, though the adult worms themselves are not affected by the immune response (see p. 26). This amount of protective immunity may prevent the disease from becoming an even worse scourge than it is.

It is of utmost importance to recognize that agricultural projects intended to increase food production in underdeveloped countries have, in many cases, created more misery than they have alleviated, by extending snail habitats. A $10 million irrigation project in southern Rhodesia had to be abandoned 10 years after it was started because of schistosomiasis.[13] The Aswan High Dam in Egypt, much acclaimed in its inception, may have its benefits canceled out by the increase in disease it has caused. Restraint of the wide fluctuations in the water level in the Nile, while making possible four crops per year by perennial irrigation, has also created conditions vastly more congenial to snails.[17] Before the dam construction, perennial irrigation was already practiced in the Nile

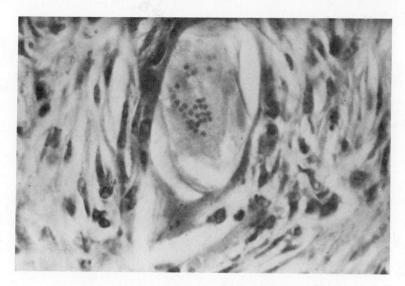

Fig. 17-17. Egg of *Schistosoma mansoni* embedded in intestinal wall. (Photograph by David F. Oetinger.)

delta region, and the prevalence of schistosomiasis was about 60%; in the 500 miles of river valley between Cairo and Aswan, where the river was subject to annual floods, the prevalence was only about 5%. Four years after the dam was completed, the prevalence of *S. haematobium* ranged from 19% to 75%, with an average of 35%, between Cairo and Aswan, or an average seven-fold increase! In the area above the dam, prevalence was very low before its construction; in 1972 76% of the fishermen checked in the impounded area were infected.

The cost, in terms of productivity, of *S. haematobium* to a native may well equal the worker's per capita income. Even oil production in some countries of the Middle East may be slightly, but directly, affected by the presence of schistosomes in oil refinery employees.

Only in Japan, Venezuela, and Puerto Rico has there been a reduction in transmission and pathological manifestations of the disease in recent years; in *all other centers of endemicity,* schistosomiasis has remained at the same level or has increased.[8]

The role of reservoir hosts and of strains of the parasite have some importance as epidemiological factors, depending on the species. Members of no less than seven mammalian orders have been successfully infected experimentally with *S. mansoni;* but certain monkeys and a variety of rodents are probably important natural reservoir hosts in Africa and tropical America. *S. haematobium* is more host specific than *S. mansoni,* and it is thought that there are no natural reservoir hosts for it. The opposite is true of *S. japonicum,* which seems to be the least host-specific. It can develop in dogs, cats, horses, swine, cattle, caribou, rodents, and deer, but there seems to be more than one race of this worm, and the susceptibility of a given host varies. For example, *S. japonicum* is widely prevalent in rats in Taiwan, but it is rare in humans there.

Pathogenesis. The pathogenesis of schistosomiasis differs somewhat according to the species of the fluke involved, mainly because of the preferred site of each. Progression of the disease caused by all three is commonly divided into three phases: (1) the initial phase, 4 to 10 weeks postinfection, characterized by fever and toxic or allergic phenomena; (2) the intermediate stage, 2.5 months to several years postinfection, with pathological changes in intestinal or urinary tracts and eggs in excreta; and (3) the final phase, with complications involving the gastrointestinal, renal, and other systems and often with no

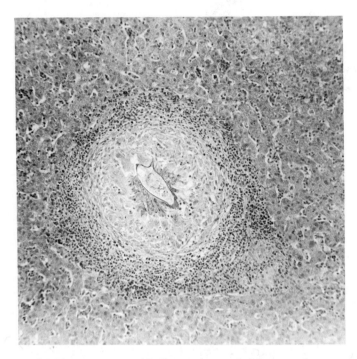

Fig. 17-18. Egg of *Schistosoma mansoni* in liver, surrounded by granulomatous cells. Note leukocytic infiltration around the granuloma. (AFIP neg. no. 64-6532. From Hopps, H. G., and D. L. Price. 1971. In Marcial-Rojas, R. A., editor. Pathology of protozool and helminthic diseases with clinical correlation. The Williams & Wilkins Co., Baltimore.)

eggs being passed. The initial phase is similar for all species: intermittent fever, frequently a skin rash, abdominal pain, bronchitis, enlargement of liver and spleen, and diarrhea. The most serious damage is done by the eggs in all three species. With *S. mansoni*, the large intestine is most notably affected, especially the sigmoid colon and rectum. The eggs lodged in the venules and submucosa (Fig. 17-17) act as foreign bodies and cause inflammatory reactions with leukocytic and then fibroblastic infiltration. These finally become small fibrous nodules called **pseudotubercles,** so called because of their resemblance to the localized nodules of tissue reaction (tubercles) in tuberculosis. Small abscesses occur, and the occlusion of small vessels leads to necrosis and ulceration. Often a high eosinophilia (high leukocyte count in blood, predominately of eosinophils) is followed by leukopenia (lowered white cell count). The clinical symptoms include abdominal pain and diarrhea with blood, mucus, and pus. Many eggs are swept back

into the liver where they lodge in the hepatic capillary bed (Fig. 17-18) and cause a similar foreign body reaction with pseudotubercle formation. As the eggs accumulate and the fibrotic reactions in the liver continue, a periportal cirrhosis and portal hypertension ensue. A marked enlargement of the spleen (splenomegaly) occurs, partly because of eggs lodged in it and partly because of the chronic passive congestion of the liver. Ascites (accumulation of fluid in the abdominal cavity) is common at this stage (Fig. 17-19). Some eggs pass the liver, lodging in the lungs, nervous system, or other organs, and produce pseudotubercles there.

S. japonicum causes similar pathological changes in the intestine and liver as *S. mansoni*, except that the small intestine is more extensively involved and the large intestine less so. Frequently fibrous nodules containing nests of eggs are found on the serosal and peritoneal surfaces. Eggs of *S. japonicum* seem to reach the brain more often than those of the other species, and

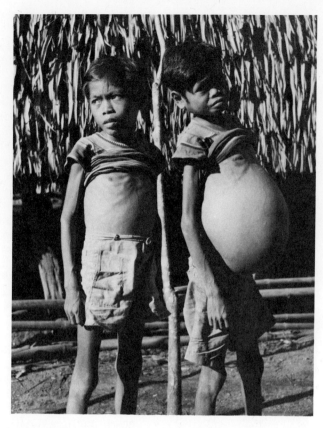

Fig. 17-19. Ascites in advanced schistosomiasis japonica, Leyte, Philippines (right). This is an example of dwarfing caused by schistosomiasis. The male on the left is 13 years old, the one on the right is 24 years old. (Photograph by Robert E. Kuntz.)

neurological pathology, including coma and paralysis, is shown in about 9% of cases. **Schistosomiasis japonica** is the most grave of the three, and the prognosis is unfavorable in heavy infections without early treatment.

Infection with *S. haematobium* is considered the least serious. Since the adults' of this species live in the venules of the urinary bladder, the chief symptoms are associated with the urinary system: cystitis (inflammation of the bladder), hematuria (bloody urine), and pain on urination. The onset of hematuria is usually gradual and becomes marked as the disease develops and the bladder wall becomes more ulcerated. Pain is most intense at the end of urination. The changes in the bladder wall (Fig. 17-20) are associated with the foreign body reactions around the eggs, that is,

pseudotubercles, fibrous infiltration, thickening of the muscularis layer, and ulceration. Malignant changes sometimes occur. Chronic cases lead to ureteral and kidney involvement and to lesions in other parts of the body, as with other species.

Diagnosis and treatment. As in diagnosing many other helminths, the demonstration of eggs in the excreta is the most straightforward mode of diagnosis. However, the number of eggs produced per female schistosome, even for *S. japonicum,* is far smaller than most other helminth parasites of humans; therefore, direct smears must be augmented by concentration techniques and other diagnostic methods, such as biopsy and immunodiagnosis. In the case of *S. mansoni,* which produces the smallest number of eggs, only 47% of patients could be diagnosed after

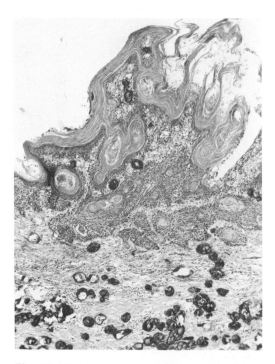

Fig. 17-20. Schistosomiasis of the urinary bladder. In this case, many eggs of *S. haematobium* can be seen in all the layers of the bladder. Many of the eggs are calcified. The epithelium has undergone squamous metaplasia. Note also leukocytic infiltration, granulosis, and ulceration. (AFIP neg. no. 65-6779.)

three direct smears.[9] With concentration techniques such as gravity or centrifugal sedimentation, over 90% of the coprologically demonstrable cases could be diagnosed. Nevertheless, it must be remembered, particularly in chronic cases, that few or no eggs may be passed. In such cases, rectal, liver, or bladder biopsies may be of great value, but these require the services of specialists and the availability of appropriate surgical facilities. Hence, a substantial research effort has been directed at finding sensitive, accurate, and reliable immunodiagnostic methods. Of these, the simple intradermal test is perhaps the most valuable. There is no immune protection against the adult worms; the phrase "immune serum" is used in the immunological sense of serum that contains antibodies.

The intradermal test becomes positive from 4 to 8 weeks after infection and remains so for years, even after the patient may be cured. The reaction is a histamine or immediate hypersensitivity type (see Chapter 2). A small amount of antigen is injected intradermally, and the area of the wheal produced is measured exactly 15 minutes later and compared with the area of the control injection (antigen vehicle alone). With good antigen and careful administration, this test has a 97% efficiency. Another valuable test is the **complement fixation reaction,** possibly the best test in the field. Troublesome cross-reactions with syphilis and *Paragonimus* have been eliminated by more recent modifications. The complement fixation test is 100% sensitive in infections of less than 3 years duration, and it decreases to 47% in infections that are 10 years old. The **circumoval precipitin** reaction depends on the formation of a precipitate around lyophilized eggs incubated in immune serum. The value of this test is that it becomes negative about 8 months after the eggs have died; therefore, it can be used to confirm cure. Both the circumoval test and the **cercarienhüllenreaktion** (CHR) are so specific that they can be used to determine the species of schistosome in an infection. The CHR is so-called because cercariae incubated in immune serum are inactivated, and an amorphous envelope forms around them. The CHR, or a related phenomenon, may be the mechanism providing some protection against superinfection. This may be the reaction that kills newly invading cercariae or schistosomules in vivo.

Difficulty in treating **schistosomiasis** is a major factor contributing to the disease as a world health problem. The most effective drugs in current use are the organic trivalent antimonials, but these are quite toxic to humans and must be given carefully—in small doses over a period of from 2 to 6 weeks, depending on the drug. Problems inherent in treating large numbers of people over wide areas in underdeveloped countries with such drugs are obvious. Some other drugs (lucanthone hydrochloride and niridazole) are less toxic but also less effective. They may inhibit egg production and cause the worms to move back into the liver for a time, but there is evidence that the worms can recover and begin producing eggs again. By the stage in the disease when liver damage is extensive, all chemotherapeutic treatment is contraindicated, and surgical intervention may be necessary. At the late stages, prognosis is poor, and the treatment can only be supportive.

Control of schistosomiasis is exceedingly difficult, depending ultimately on the almost intractable task of persuading masses of uneducated, poor people to change their customs and traditions. Although draining snail habitats is of value, it was pointed out before that many more good habitats are now being created by efforts to increase agricultural production. Use of chemical molluscicides has met with some success, but problems here include determination and application of the proper quantity in a given body of water, dilution, effects on other organisms in the environment, and errors in estimating the physical and chemical characteristics of the water. Molluscicidal control of *S. japonicum* is virtually ineffective because *Oncomelania* is amphibious and only visits water to lay its eggs. In Puerto Rico some success in control of *Biomphalaria* has been achieved with predatory snails *(Marisa cornuarietis, Tarebia granifera maniensis)*. Snail-eating fish have been cultured and released in infected waters with some success.

Other schistosomes of lesser medical importance

Schistosoma mattheei-intercalatum "complex." In Africa hundreds of cases are known in which terminal-spined eggs are recovered from the stool only. The worms have been named *Schistosoma intercalatum,* but they are morphologically indistinguishable from *S. mattheei,* a natural parasite of African ruminants and primates. If only one species is involved, *S. mattheei* is the correct name. To further confuse the issue, cases are known in which eggs of this species are passed in both stool and urine by human patients.[1]

Schistosome cercarial dermatitis ("swimmer's itch"). Several species in the genus *Schistosoma* are known to cause a severe rash when their cercariae penetrate the skin of an unsuitable host. Hence, *S. spindale,* a parasite of ruminants in India, Ma-

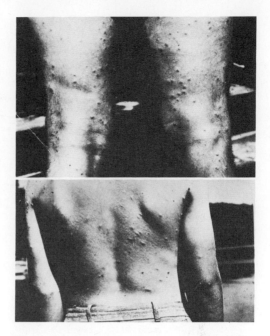

Fig. 17-21. Cercarial dermatitis, or "swimmer's itch," caused by cercariae of avian blood flukes. (AFIP neg. no. 77203.)

laya, Africa, and Sumatra, and *S. bovis* from ruminants, equines, and primates in Europe, Africa, and the Middle East are agents of dermatitis in humans throughout their range.

More importantly, several species of bird schistosomes are distributed throughout the world and cause "swimmer's itch" when their cercariae attack anyone who gets them on their skin. The genera *Trichobilharzia, Gigantobilharzia, Ornithobilharzia, Microbilharzia,* and *Heterobilharzia* are the guilty parties.

For the most part, the skin reaction is a product of sensitization, with repeated infections causing increasingly severe reactions.[12] When the cercaria penetrates the skin and is unable to complete its migration, the host's immune responses rapidly kill it. At the same time, the cercariae release allergic substances that cause inflammation, and typically, a pus-filled pimple (Fig. 17-21). The reaction may also be general, with an itching rash produced over much of the body. The condition is not a serious threat to health but is a terrific annoyance, much like poison ivy, which interrupts a summer vacation, for instance, or decreases the income of someone who rents lake front cottages to the summer crowd. In the United States the the problem is most serious in the Great Lakes area, but it has been reported from nearly all states.

Control depends mainly on molluscicides, but their usefulness is limited because they threaten sport fishing by poisoning the fish, and, of course, because of the other problems previously mentioned. Ocean beaches are occasionally infested with avian schistosome cercariae, for which no control has yet been devised.

REFERENCES

1. Blair, D. M. 1966. The occurrence of terminal spined eggs, other than those of *Schistosoma haematobium,* in human beings in Rhodesia. Cent. Afr. J. Med. 12:103-109.
2. Brooker, B. E. 1972. The sense organs of trematode miracidia. In Canning, E. U., and C. A. Wright, editors. Behavioural aspects of parasite transmission. Linnean Society. Academic Press, Inc., London. pp. 171-180.
3. Foster, W. D. 1965. A history of parasitology. E. S. Livingstone, Edinburgh.
4. Hulse, E. V. 1971. Joshua's curse and the abandonment of ancient Jericho: schistosomiasis as a possible medical explanation. Med. Hist. 15:376-386.
5. Kuba, N. 1963. Histopathological study on mechanism of extrusion of schistosome ova from blood vessels. Jpn. J. Vet. Sci. 25:289-297.
6. Leiper, R. T. 1915. Observations on the mode of spread and prevention of vesicle and intestinal bilharziosis in Egypt, with additions to August, 1916. Proc. R. Soc. Med. 9:145-172.
7. Mackie, T. T., G. W. Hunter III, and C. B. Worth. 1976. A manual of tropical medicine. W. B. Saunders Co., Philadelphia.
8. Maldonado, J. F. 1967. Schistosomiasis in America. Editorial Científico-Medica, Barcelona.
9. Maldonado, J. F., J. Acosta-Matienzo, and F. Velez-Herrera. 1954. Comparative value of fecal examination procedures in the diagnosis of helminth infections. Exp. Parasitol. 3:403-416.
10. Manson, P. 1905. Lectures on tropical diseases. Constable, London. p. 54.
11. Okabe, K. 1964. Biology and epidemiology of *Schistosoma japonicum* and schistosomiasis. In Morishita, K., Y. Komiya, and H. Matsubayashi, editors. Progress of medical parasitology in Japan, vol. I. Meguro Parasitological Museum, Tokyo. pp. 185-218.
12. Olivier, L. J. 1949. Schistosome dermatitis, a sensitization reaction. Am. J. Hyg. 49:209-301.
13. Osmundsen, J. A. 1965. Science: battle is on against a dread crippler. New York Times, Aug. 22, 1965, p. 8E.
14. Pearson, J. C. 1956. Studies on the life cycles and morphology of the larval stages of *Alaria orisaemoides* Augustine and Uribe, 1927 and *Alaria canis* La Rue and Fallis. Can. J. Zool. 34:295-387.
15. Pearson, J. C. 1959. Observations on the morphology and life cycle of *Strigea elegans* Chandler and Rausch, 1947 (Trematoda: Strigeidae). J. Parasitol. 45:155-170, 171-174.
16. Sambon, L. W. 1909. What is *Schistosoma mansoni* Sambon, 1907? J. Trop. Med. 12:1-11.
17. Van der Schalie, H. 1974. Aswan Dam revisited. Environment 16(9):18-20, 25-26.
18. Wright, W. H. 1968. Schistosomiasis as a world problem. Bull. N.Y. Acad. Med. 44:301-302.
19. Wright, W. H. 1972. A consideration of the economic import of schistosomiasis. Bull. WHO 197:559-566.

SUGGESTED READING

Ansari, N., editor. 1973. Epidemiology and control of Schistosomiosis (Bilharziasis). S. Karger, Basel. (This is an official publication of the World Health Organization, outlining latest advances in the area.)
Berrie, A. D. 1970. Snail problems in African schistosomiasis. In Dawes, B., editor. Advances in parasitology, vol. 8. Academic Press, Inc., New York. pp. 43-96.
Chu, G. W. T. C. 1958. Pacific area distribution of fresh-water and marine cercarial dermatitis. Pacific Sci. 12:299-312.
Erasmus, D. A. 1973. A comparative study of the reproductive system of mature, immature and "unisexual" female *Schistosoma mansoni*. Parasitology 67:165-183.
Kuntz, R. E. 1953. Demonstration of the "spreading factor" in the cercariae of *Schistosoma mansoni*. Exp. Parasitol. 2:397-402.
Smithers, S. R., and R. J. Terry. 1969. The immunology of schistosomiasis. In Dawes, B., editor. Advances in parasitology, vol. 7. Academic Press, Inc., New York. pp. 41-93.
Warren, K. S. 1973. The pathology of schistosome infections. Helm. Abstr. Series A 42:591-633. (A complete summary of the subject.)

Chapter 18

SUPERORDER ANEPITHELIOCYSTIDIA: ORDER ECHINOSTOMATA

Members of the order Echinostomata often show little resemblance to one another in the adult stages, but studies on their embryology have shown common ancestries through developmental similarities. Often, though by no means always, the tegument bears well-developed scales or spines, especially near the anterior end. The acetabulum is near the oral sucker. In many cases a second intermediate host is absent, and the metacercaria encyst on underwater vegetation or debris. Most species are parasitic in wild animals, but a few are important as agents of disease in humans and/or their domestic animals.

SUPERFAMILY ECHINOSTOMATOIDEA

Parasites of the Superfamily Echinostomatoidea infect all classes of vertebrates and are found in marine, freshwater, and terrestrial environments. Some are among the most common parasites encountered, and a few cause devastating losses to agriculture.

Family Echinostomatidae

Echinostomes are easily recognized by their circumoral collar of peg-like spines, hence their name (Fig. 18-1). The spines are arranged either in a single, simple circle or in two circles, one slightly lower than, and alternating with the other. The collar is interrupted ventrally and at each end has a group of "corner spines." The size, number, and arrangement of these spines are of considerable taxonomic importance in both cercariae and adults. Echinostomes typically are slender worms with large preequatorial acetabula, pretesticular ovaries, and tandem testes, although there are exceptions. The vitellaria are voluminous and mainly postacetabular. These worms are parasites of the intestine or bile duct of reptiles, birds, and mammals, especially those frequenting aquatic environments.

Echinostoma revolutum (Fig. 18-2)

Echinostoma revolutum is a cosmopolitan parasite that is one of the most common and abundant of all trematodes of semi-aquatic vertebrates. It shows little host specificity and appears to be at home in any kind of bird or mammal that eats the metacercariae. It is especially common in ducks, geese, muskrats, beavers, and shore birds, and it has been found in many terrestrial birds. Experimentally it develops well in rabbits, rats, dogs, and guinea pigs. It is a fairly common parasite of humans in the Orient, especially in Formosa and Indonesia.[2] Routine inspections of snails of the genera *Physa, Lymnaea, Helisoma, Paludina,* and *Segmentina*—all common genera —usually reveal infections with *E. revolutum.* Metacercariae encyst in molluscs, planaria, fish, and tadpoles. Infection of the definitive host is accomplished when the definitive host eats one of these. Humans are usually infected by eating raw mussels or snails.

Morphologically, the worm is easily identified. Its circumoral collar bears 37 spines in a double circle, of which five are corner spines at each end. The operculate eggs are large, 90 to 126 μm by 54 to 71 μm, and there are few of them in the uterus at any one time. The genital pore is median and preacetabular, and the cirrus pouch is large, passing dorsal to the voluminous acetabulum. The short uterus has an ascending limb only. Overall size varies greatly.

Echinostoma ilocanum

The eggs of *Echinostoma ilocanum* were first found in the stool of a prisoner in Manila in 1907. It has since been found

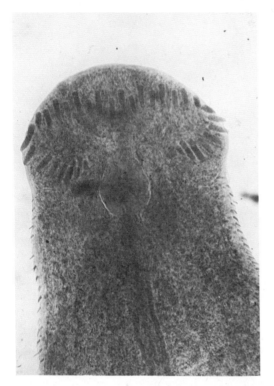

Fig. 18-1. Anterior end of *Echinostoma*, showing the double crown of peg-like spines on the circumoral collar. (Photograph by Warren Buss.)

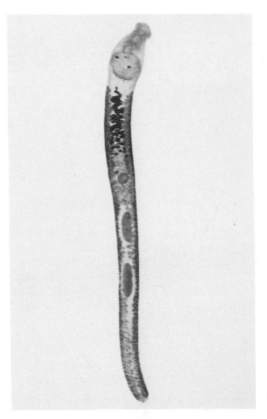

Fig. 18-2. *Echinostoma revolutum*, a common intestinal parasite of aquatic birds and mammals. (Photograph by Larry Jensen.)

commonly throughout the East Indies and China. Tubangui found that the Norway rat was an important reservoir of infection.[8]

The morphology of *E. ilocanum* is similar to that of *E. revolutum,* but it differs in collar spine number and arrangement. *Echinostoma ilocanum* has 49 to 51 spines, with five or six corner spines at each end. The double row of spines is continuous dorsally, and the testes are deeply lobate.

Its biology is similar to that of *E. revolutum,* with the metacercaria encysting in any freshwater mollusc. Infected snails are eaten raw by native peoples, who thereby become infected. The worms cause inflammation at their sites of attachment within the small intestine. Intestinal pain and diarrhea may develop in severe cases.

Other echinostomatid species reported from humans

Several species of echinostomes in different genera, which normally parasitize wild animals, have been reported from humans. These include *Echinostoma lindonese* in Celebes and possibly Brazil; *E. malayanum* from India (Fig. 18-3), southeast Asia, and the East Indies; *E. cinetorchis* from Japan, Taiwan, and Java; *E. melis,* which is circumboreal; and *Hypoderaeum conoidum* in Thailand. Others are *Himasthla muehlensi* in New York; *Paryphosotomum sufrartyfex* in Asia Minor, and *Echinochasmus perfoliatus* from eastern Europe and from Asia. These are only some of the species of echinostomes of wild animals that have found their ways into humans, but, considering their lack of host specificity, others probably do so fairly often and remain undetected. This points to the practical importance of systematic surveys of animal parasites; for, when zoonotic infections are found, it immediately becomes necessary to identify the pathogen and determine how the person became infected. Faunal surveys are the only way to fulfill this need.

Family Fasciolidae

Members of the family Fasciolidae are large, leaf-shaped parasites of mammals, mainly of plant-eaters. They have a tegument covered with scale-like spines, and the acetabulum is close to the oral sucker. The testes and ovary are dendritic, and the vitellaria are extensive, filling most of the postacetabular space. There is no second intermediate host in the life cycle; the metacercariae encyst on submerged objects or free in the water. One important species lives in the intestinal lumen, but most parasitize the liver of mammals.

Fasciola hepatica (Fig. 18-4)

Although *Fasciola hepatica* is rare in humans, it has been known as an important parasite of sheep and cattle for hundreds of years. Because of its size and economic importance, it has been the subject of many scientific investigations and is probably the best known of any trematode species. The first published record of it is that of Jean de Brie in 1379. He was well aquainted with a disease of sheep called "liver rot," in which the liver of a diseased animal is infected with large, flat worms. In 1668, the great pragmatist Francisco Redi was the first to illustrate this fluke, thereby stimulating others to investigate its biology. Leeuwenhoek was interested in the organism but apparently was distracted by all of the other beings he found with his microscope.

The cercaria and redia of *F. hepatica* were described in 1737 by Jan Swammerdam, a man who had a remarkable ability to see and understand microscopic objects without the use of a microscope. Linnaeus gave the worm its name in 1758 but considered it to be a leech. Pallas, in 1760, was the first to find it in a human. Professor C. L. Nitzsch, in 1816, was the first to recognize the similarity of cercariae and adult liver flukes. Thus the history of *Fasciola hepatica* parallels the history of trematodology itself. In 1844, Johannes Steenstrup published a landmark book on *Alternation of Generations,* in which he postulated that trematodes have two generations, one adult and one not.

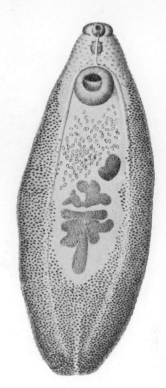

Fig. 18-3. *Echinostoma malayanum,* a parasite of humans in southern Asia. (From Odhner, T. 1913. Zool. Anz. 41:577-582.)

By the mid-1800s circumstantial evidence indicated that molluscs were involved in the transmission of *F. hepatica.* In 1880, George Rolleston, professor of anatomy and physiology at Oxford, was convinced that a common slug was the intermediate host of *F. hepatica.* Although he was wrong in this assumption, he recommended that A. P. Thomas undertake an investigation to determine the life cycle of this parasite. Thomas was a 23-year-old demonstrator at the time, but took on this formidable task with zeal. He soon found the snail *Limnaea truncatula* to be infected with rediae and cercariae that were similar in many regards to *Fasciola.* Then he successfully infected this snail with miracidia and followed its development through the sporocyst, redia, and cercarial stages.

At the same time as this "lowly" Oxford demonstrator was investigating liver rot, fascioliasis was also engrossing the mind of the greatest parasitologist then living, Rudolph Leuckart. After a series of false starts, Leuckart traced the development of

Fig. 18-4. *Fasciola hepatica,* the sheep liver fluke. (Courtesy Turtox/Cambosco.)

this parasite through the same species of snail and, as a final irony, published his results 10 days before Thomas published his. Credit is given to both men equally, but one can scarcely refrain from lending sympathy to the young Englishman who elucidated the first trematode life cycle in a truly scientific manner, without the advantages of a large budget and long experience.

Neither Thomas nor Leuckart determined the mode of infection of the definitive host. This was done by Adolph Lutz, a Brazilian working in Hawaii, who demonstrated in 1892-1893 that ruminants become infected by eating larvae encysted on vegetation. That Lutz was actually working with a different species, *F. gigantica,* is immaterial, for the biology of both species is the same.

Morphology. *Fasciola hepatica* is one of the largest flukes of the world, reaching a length of 30 mm and a width of 13 mm. It is rather "leaf-shaped," pointed posteriorly, and wide anteriorly, though the shape varies somewhat. The oral sucker is small but powerful and is located at the end of a cone-shaped projection at the anterior end. The marked widening of the body at the base of the so-called oral cone gives the worm the appearance of having shoulders. The combination of an oral cone and "shoulders" is an immediate means of identification. The acetabulum is somewhat larger than the oral sucker and is quite anterior, at about the level of the shoulders. The tegument is covered with large, scale-like spines, reminding one of echinostomes, to which they are closely related. The intestinal ceca are highly dendritic and extend to near the posterior end of the body.

The testes are large and greatly branched, arranged in tandem behind the ovary. The smaller, dendritic ovary lies on the right side, shortly behind the acetabulum, and the uterus is short, coiling between the ovary and the preacetabular cirrus pouch. The vitelline follicles are extensive, filling most of the lateral body and becoming confluent behind the testes. The operculate eggs are 130 to 150 μm by 63 to 90 μm.

Biology (Fig. 18-5). Adult *F. hepatica* live in the bile passages of the liver of many kinds of mammals, especially of ruminants. Humans are occasionally infected. The flukes feed on the lining of the biliary ducts. Their eggs are passed out of the liver with the bile and into the intestine to be voided with the feces. If they fall into water, they will complete their development into miracidia and hatch in 9 or 10 days, during warm weather. Colder water retards their development. Upon hatching, the miracidium has about 24 hours in which to find a suitable snail host, which in the United States is *Fossaria modicella* or *Stagnicola bulimoides.* In other parts of the world different but related snails are the important first intermediate hosts. Mother sporocysts produce first generation rediae, which in turn produce daughter rediae that develop in the snail's digestive gland. Cercariae begin emerging 5 to 7 weeks after infection. If the water in which the

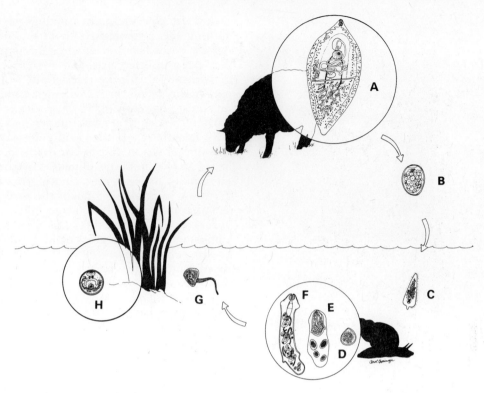

Fig. 18-5. Life cycle of *Fasciola hepatica*. **A,** Adult worm in bile duct of sheep or other mammal; **B,** egg; **C,** miracidium; **D,** mother sporocyst; **E,** mother sporocyst with developing rediae; **F,** redia with developing cercariae; **G,** free-swimming cercaria; **H,** metacercaria, encysted on aquatic vegetation. (Courtesy Carol Eppinger.)

snails live dries up, the snails burrow into the mud and survive, still infected, for months at a time. When water is again present, the snails emerge and rapidly shed many cercariae.

The cercaria has a simple, club-shaped tail about twice its body length. Once in the water the cercaria quickly attaches to any available object, drops its tail, and produces a transparent cyst around itself. If it does not encounter an object within a short time it will drop its tail and encyst free in the water. When a mammal eats metacercariae encysted on vegetation or in water, juvenile flukes excyst in the small intestine. They immediately penetrate the intestinal wall, enter the coelom, and creep over the viscera until contacting the capsule of the liver. Then they burrow into the liver parenchyma and wander about for nearly 2 months, feeding and growing and finally entering bile ducts.[4] The worms become sexually mature in another month

and begin producing eggs. Adult flukes are known to live as long as 11 years.

Epidemiology. Infection begins when metacercaria-infected aquatic vegetation is eaten or when water containing metacercariae is drunk. Humans can be infected by eating watercress, a common green plant. Human infection is common in parts of Europe, northern Africa, Cuba, South America, and other locales. Surprisingly, only one case is known in a human in the United States, although the worm is fairly common in parts of the South and West. Sheep, cattle, and rabbits are the most common reservoirs of infection.

Pathology. Little damage is done by larvae penetrating the intestinal wall and the capsule surrounding the liver (Glisson's capsule), but much necrosis results from the migration of flukes through the parenchyma of the liver. During this time, they feed on liver cells and blood.[3]

Worms in the bile ducts cause inflam-

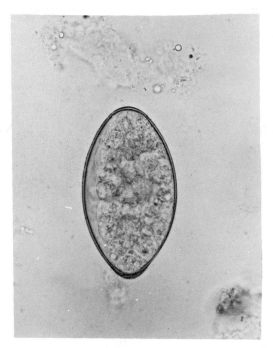

Fig. 18-6. Egg of *Fasciola hepatica*. (Photograph by Jay Georgi.)

mation and edema, which in turn stimulate the production of fibrous tissue in the walls of the ducts (pipestem fibrosis) (see Fig. 19-27). Thus thickened, the ducts can handle less bile and are less responsive to the needs of the liver. Back pressure causes atrophy of the liver parenchyma, with concomitant cirrhosis and possibly icterus. In heavy infections the gallbladder is damaged, and the walls of the bile ducts are eroded completely through, with the worms then reentering the parenchyma, causing large abesses. Migrating larvae frequently produce ulcers in ectopic locations, such as the eye, brain, skin, and lungs.

Ingestion of raw sheep or goat liver in the Middle East may result in adult worms establishing in the nasopharynx. The resulting respiratory blockage is called **halzoun** in Arabic. Recent research casts doubt on this disease entity, placing the full blame on pentastomid larvae and leeches. At this writing, *F. hepatica* is still suspect.

Diagnosis and treatment. Whenever liver blockage coincides with a history of watercress consumption, fascioliasis should be suspected. Specific diagnosis depends on finding the eggs (Fig. 18-6) in the stool. A false record can result when the patient has eaten infected liver and *Fasciola* eggs pass through with the feces. Daily examination during a liver-free diet will unmask this false diagnosis. Early diagnosis is important before irreparable damage to the liver occurs. Several drugs are effective in chemotherapy of fascioliasis, both in humans and in domestic animals. Prevention in humans depends on eschewing raw watercress. In domestic animals the problem is much more difficult to avoid. Snail control is always a possibility, although this is almost impossible in areas of high precipitation, where nearly every cow track is filled with water and snails. Reservoir hosts, especially rabbits, can maintain infestation of a pasture when pasture rotation is attempted as a control measure. *Fasciola hepatica* is one of the most important disease agents of domestic stock throughout the world and shows promise of remaining so for years to come. Losses are enormous because of mortality, condemned livers, reduction of milk and meat production, secondary bacterial infection, and expensive anthelmintic treatment.

Other fasciolid trematodes

Fasciola gigantica (Fig. 18-7) is a species very similar to *F. hepatica* found in Africa, Asia, and Hawaii, being relatively common in herbivorous mammals, especially cattle, in these areas. The morphology, biology, and pathology are nearly identical to those of *F. hepatica* although different snail hosts are necessary.

Fascioloides magna is a giant in a family of large flukes, reaching nearly 3 inches in length and 1 inch width. Formerly strictly an American species, it was first discovered in Italy, in an American elk in a zoo. Now the fluke has become established in Europe, mainly in game reserves. It is easily distinguished from *Fasciola* spp. by its large size and the absence of a cephalic cone and "shoulders." The life cycle is similar to that of *F. hepatica,* except that the adults live in the liver parenchyma rather than in the bile ducts. Because of their large size, they cause extensive damage. Often, but not always, they become encased in a calcareous cyst of host origin

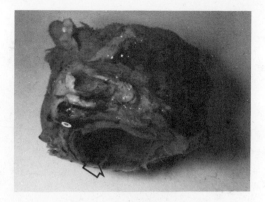

Fig. 18-8. A large calcareous cyst from the liver of a steer (the arrow points to its opening). This cyst contained two *Fascioloides magna.* (Photograph by Warren Buss.)

Fig. 18-7. *Fasciola gigantica,* a liver fluke. (Photograph by Warren Buss.)

(Fig. 18-8). Their excretory system produces great amounts of melanin, which fills their excretory canals and also the cyst containing them. The normal hosts are probably elk and other Cervidae, but cattle are commonly infected in endemic areas. Human infections have not been found.

Fasciolopsis buski (Fig. 18-9) is a common parasite of humans and pigs in the Orient. Stoll estimated 10 million human infections in 1947.[7] While basically a typical fasciolid, it is peculiar because it lives in the small intestine of its definitive host rather than in the liver. It is elongate-oval, reaching a length of 20 to 75 mm and a width of up to 200 mm. There is no cephalic cone or "shoulders." The acetabulum is larger than the oral sucker and is located close to it. Another difference from "typi-

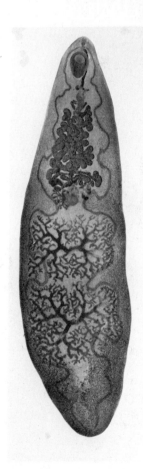

Fig. 18-9. *Fasciolopsis buski,* an intestinal fluke in the Fasciolidae. (Photograph by Robert E. Kuntz.)

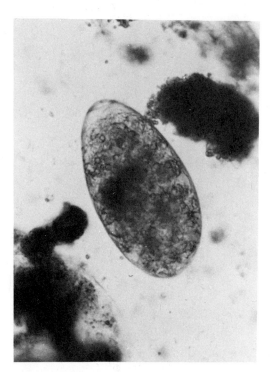

Fig. 18-10. An egg of *Fasciolopsis buski* in a human stool in Taiwan. (Photograph by Robert E. Kuntz.)

cal" fasciolids is the presence of unbranched ceca. The dendritic testes are tandem in the posterior half of the worm. The ovary is also branched and lies in the midline anterior to the testes. Vitelline follicles are extensive, filling most of the lateral parenchyma all the way to the caudal end. The uterus is short, with an ascending limb only. The eggs are nearly identical to those of *F. hepatica*

The life cycle of *F. buski* parallels that of *F. hepatica*. Each worm produces about 25,000 eggs (Fig. 18-10) per day, which take up to 7 weeks to mature and hatch at 80° to 90°F. Several species of snails of the genera *Segmentina* and *Hippeutis* (Planorbidae) serve as intermediate hosts. Cercariae encyst on underwater vegetation, including cultivated water chestnut, water caltrop, lotus, bamboo, and other edible plants (Fig. 18-11). Metacercariae are swallowed when these plants are eaten raw or when they are peeled or cracked with the teeth prior to eating. The worms excyst in the small intestine, grow, and mature in about 3 months without further migration. Infection, then, depends on human or pig feces being introduced directly or indi-

Fig. 18-11. A woman harvesting water caltrop, which serves as a medium for transport of metacercariae of *Fasciolopsis buski* to humans in Taiwan. (Photograph by Robert E. Kuntz.)

rectly into bodies of water where edible plants grow.[6]

Pathology resulting from *F. buski* is toxic, obstructive, and traumatic. Inflammation at the site of attachment provokes excess mucous secretion, which is a typical symptom of infection. Heavy infections block the passage of food and interfere with normal digestive juice secretions. Ulceration, hemorrhage, and abscess of the intestinal wall result from long-standing infections. Chronic diarrhea is symptomatic. Another aspect of pathology is a typical profound toxemia caused by absorption of the worm's metabolites. This **verminous intoxication** results from sensitization phenomena and may eventually cause the death of the patient. Treatment is usually effective in early or lightly infected cases. Late cases do not fare so well. Prevention is easy. Immersion of vegetables in boiling water for a few seconds will kill the metacercariae. Snail control should be attempted whenever it is impractical to prevent the use of nightsoil as a fertilizer.

SUPERFAMILY PARAMPHISTOMOIDEA

The superfamily Paramphistomoidea contains the "amphistomes," those flukes in which the acetabulum is located at or near the posterior end. Usually, they are thick, fleshy worms with the genital pore preequatorial and the ovary usually posttesticular. Species are found in fishes, amphibians, reptiles, birds, and mammals. Of several families in this group, we will consider three.

Fig. 18-12. *Stichorchis subtriquetrus,* a stomach parasite of the American beaver. (Photograph by Warren Buss.)

Family Paramphistomidae

Members of the family Paramphistomidae parasitize mammals, especially herbivores. Several species in different genera of this family parasitize sheep, goats, cattle, cervids, water buffalo, elephants, and other important animals. One species was found once in humans.

Paramphistomum cervi

The species *Paramphistomum cervi* lives in the rumen of domestic animals throughout most of the world. Adults are almost conical in shape and are pink when living. The testes are slightly lobated.

The life cycle is quite similar to that of *Fasciola hepatica,* and, in North America at least, they develop in the same snail hosts. The cercariae are large and pigmented and have eyespots. There is no second intermediate host; the metacercariae encyst on aquatic vegetation. When eaten, the worms excyst in the duodenum, penetrate the mucosa, and migrate anteriad through the tissues. On reaching the abomasum, or true stomach, they return to the lumen and creep farther forward to the rumen. There, they attach among the villi and mature in 2 to 4 months.

Paramphistomum cervi is a particularly pathogenic species. Migrating larvae cause severe enteritis and hemorrhage, often killing the host. Secondary bacterial infection often complicates the problem. No adequate prevention or treatment is known.

Stichorchis subtriquetrus (Fig. 18-12) is a parasite of beavers, occurring throughout their range. Like *P. cervi,* the metacercariae encyst on underwater objects, including sticks that beavers embed in the bottom of their ponds and streams. When they later eat the bark off those sticks, they swallow any attached metacercariae. A peculiarity of this worm's early embryogenesis is the development of a mother redia within the miracidium. This is released in the snail immediately after penetration, and the remainder of the miracidium then disintegrates. Adult worms live in the cecum and have been reportedly causing mortality in beavers in Russia.

Family Diplodiscidae

Flukes in the family Diplodiscidae have a pair of posterior diverticulae in the oral sucker.

Megalodiscus temperatus

Megalodiscus temperatus and other genera and species of amphistomes are common parasites of the rectum and urinary bladder of frogs. They measure up to 6 mm long and 2.25 mm wide at the posterior end. The posterior sucker is equal to about the greatest width of the body.

The life cycle of this species is similar to other amphistomes in that no second intermediate host is required. Miracidia hatch soon after the eggs reach water and penetrate snails of the genus *Helisoma*. Cercaria have eyespots and swim toward lighted areas. If they contact a frog, they will encyst almost immediately on its skin, especially on its dark spots. Frogs molt the outer layers of their skin regularly and not infrequently will eat the sloughed skin. Metacercariae excyst in the rectum and mature in 1 to 4 months. If a tadpole eats a cercaria, the worm will encyst in the stomach and excyst when it reaches the rectum. At metamorphosis, when the amphibian's intestine shortens considerably; the flukes migrate anteriad as far as the stomach and then return to the rectum. These parasites are of no economic importance, but they are an easily obtained amphistome for general studies.

Family Gastrodiscidae

The morphology of the family Gastrodiscidae is essentially similar to that of Paramphistomidae and perhaps should not be separate from it. One of its species is a common parasite of humans in restricted areas of the world.

Gastrodiscoides hominis

A typical amphistome, *G. hominis,* is cone-shaped, fleshy, and pink. It is an important parasite of humans in Assam, India, southeast Asia, and the Philippines, inhabiting the lower small intestine and the upper colon. Rodents and primates are reservoirs.

Adult worms are 5 to 8 mm long by 5 to 14 mm wide at the ventral disc, which occupies about two-thirds of the ventral surface. There is a conspicuous posterior notch in the rim of the ventral sucker.

The complete life cycle of this worm is unknown, but the planorbid snail *Helicorbus coenosus* serves as an experimental host in India.[5] Presumably, humans are infected by eating uncooked aquatic plants. An adult worm draws a mass of mucosal tissue into the ventral sucker and remains attached for some time, causing a nipple-like projection on the intestinal lining.[1] The most common symptom is a mucoid diarrhea. Treatment and prevention have not been well studied.

REFERENCES

1. Ahluwalia, S. S. 1960. *Gastrodiscoides hominis* (Lewis and McConnell) Leiper 1913. Indian J. Med. Res. 48:315-325.
2. Bonne, C., G. Bras, and Lie Kian Joe. 1948. Five human echinostomes in the Malayan Archipelago. Med. Monandbl. 23:456-465.
3. Dawes, B. 1961. On the early stages of *Fasciola hepatica* penetrating into the liver of an experimental host, the mouse: a histological picture. J. Helminthol., R.T. Leiper Supplement. pp. 41-52.
4. Dawes, B., and D. C. Hughes. 1970. Fascioliasis: the invasion stages in mammals. In Dawes, B., editor. Advances in parasitology, vol. 8. Academic Press, Inc., New York. pp.259-274.
5. Dutt, S. C., and H. D. Srivastava. 1966. The intermediate host and the cercaria of *Gastrodiscoides hominis* (Trematoda: Gastrodiscidae). J. Helminthol. 40:45-52.
6. Sadun, E. H., and C. Maiphoom. 1953. Studies on the epidemiology of the human intestinal fluke, *Fasciolopsis buski* (Lankester) in central Thailand. Am. J. Trop. Med. Hyg. 2:1070-1084.
7. Stoll, N. R. 1947. This wormy world. J. Parasitol. 33:1-18.
8. Tubangui, M. A. 1931. Worm parasites of the brown rat, *Rattus norvegicus,* in the Philippine Islands, with special reference to those that may be transmitted to human beings. Phil. J. Sci. 46:537-591.

SUGGESTED READING

Boray, J. C. 1969. Experimental fascioliasis in Australia. In Dawes, B., editor. Advances in parasitology, vol. 7. Academic Press, Inc., New York. pp. 96-210. (A very interesting general account of fascioliasis, together with many experimental approaches. Accent is on special problems of Australia.)

Dawes, B., and D. L. Hughes. 1964. Fascioliasis: the invasive stages of *Fasciola hepatica* in mammalian hosts. In Dawes, B., editor. Advances in parasitology, vol. 2. Academic Press, Inc. New York. pp. 97-168.

Horak, I. G. 1971. Paramphistomiasis of domestic ruminants. In Dawes, B., editor. Advances in parasitology, vol. 9. Academic Press, Inc., New York. pp. 33-72. (A thorough discussion of paramphistomiasis, mainly outside North America.)

Kendall, S. B. 1965. Relationships between the species of *Fasciola* and their molluscan hosts. In Dawes, B., editor. Advances in parasitology, vol. 3. Academic Press, Inc., New York. pp. 59-98.

Kendall, S. B. 1970. Relationships between the species of *Fasciola* and their molluscan hosts. In Dawes, B., editor. Advances in parasitology, vol. 8. Academic Press, Inc., New York. pp. 251-258. (This short paper brings the earlier one by this author up to date.)

Leuckart, R. 1882. The developmental history of the liver fluke, second part. English translation by J. H. Hoogewey. Ill. Vet. 2:8-10.

Pantelouris, E. M. 1965. The common liver fluke, *Fasciola hepatica* L. Pergamon Press, Oxford. (A general reference to *F. hepatica*.)

Reinhard, E. G. 1957. Landmarks of parasitology. I. The discovery of the life cycle of the liver fluke. Exp. Parasitol. 6:208-232.

Chapter 19

SUPERORDER EPITHELIOCYSTIDIA: ORDERS PLAGIORCHIATA AND OPISTHORCHIATA

Order Plagiorchiata

Even more than in the superorder Anepitheliocystidia, adults of the order Plagiorchiata often show little resemblance to each other. They do, however, have many larval and juvenile similarities. The wall of the excretory bladder is epithelial and of mesodermal origin (although some workers are beginning to doubt this). The cercaria has a simple tail and an oral stylet is common.

SUPERFAMILY PLAGIORCHIOIDEA

Parasites of the superfamily Plagiorchoidea parasitize fishes, amphibians, reptiles, birds, and mammals and thereby are among the most commonly encountered flukes. They inhabit hosts in marine, freshwater, and terrestrial environments. Most are unimportant parasites of wild animals, but a few are important disease agents of humans and domestic animals. Cercariae possess an oral stylet, and the metacercaria usually encysts in an invertebrate intermediate host. Of the many families in the superfamily, we will discuss three.

Family Dicrocoeliidae

This is one of the three major families of liver flukes that will be considered in this book (see Fasciolidae, Opisthorchiidae). Some species, though, parasitize the gallbladder, pancreas, or intestine. All are parasites of terrestrial or semiterrestrial vertebrates and use land snails as first intermediate hosts. All dicrocoeliids are medium-sized and flattened, with a subterminal oral sucker and a powerful acetabulum in the anterior half of the body. The body is usually pointed at both ends. The ceca are simple. Testes are pretequatorial, and the ovary is posttesticular. The voluminous uterus has a descending and an ascending limb, commonly filling most of the medullary parenchyma.

Most dicrocoeliids parasitize amphibians, reptiles, birds, and wild mammals and as such are of little economic importance. One cosmopolitan species is an important parasite of domestic mammals and, occasionally, of humans.

Dicrocoelium dendriticum (Fig. 19-1)

Dicrocoelium dendriticum is common in the bile ducts of sheep, cattle, goats, pigs, and cervids and rarely is found in humans. It is common throughout most of Europe and Asia and has foci in North America and Australia, being a recent introduction. It is commonly known as the **lancet fluke** because of its blade-like shape.

Morphology. *Dicrocoelium dendriticum* is 6 to 10 mm long by 1.5 to 2.5 mm at its greatest width, near the middle. Both ends of the body are pointed. The ventral sucker is larger than the oral sucker and is located near it. The large, lobate testes lie in tandem directly behind the acetabulum, and the small ovary lies immediately behind them. Loops of the uterus fill most of the body behind the ovary. The vitellaria are lateral and restricted to the middle third of the body. The operculate eggs are 36 to 45 μm by 22 to 30 μm.

Biology. *Dicrocoelium dendriticum* is an interesting example of a trematode that has dispensed with the aquatic environment at all stages of its life cycle. Adult *D. dendriticum* live in the bile ducts within the liver, much the same as *F. hepatica*. When laid the eggs contain miracidia and must be eaten by land snails before they will hatch (Fig. 19-2). *Cionella lubrica* appears to be the most important snail host in the United

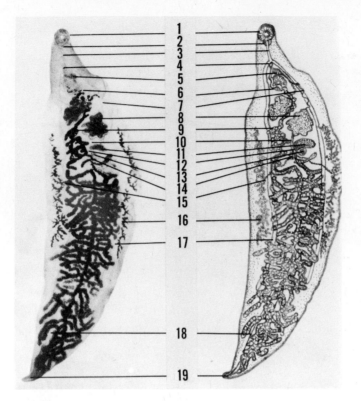

Fig. 19-1. Photograph and drawing of *Dicrocoelium dendriticum. 1,* Oral sucker; *2,* pharynx; *3,* esophagus; *4,* genital pore; *5,* cirrus pouch; *6,* acetabulum; *7,* vas deferens; *8,* testis; *9,* uterus; *10,* Laurer's canal; *11,* ovary; *12,* oviduct; *13,* seminal receptacle; *14,* ootype and Mehlis's gland; *15,* vitelline duct; *16,* vitellaria; *17,* cecum; *18,* excretory bladder; *19,* excretory pore. (AFIP neg. no. 67-7113.)

States, while other species serve in other lands. Upon hatching in the snail's intestine, the miracidium penetrates the gut wall and transforms into a mother sporocyst in the digestive gland. Mother sporocysts produce daughter sporocysts, which in turn produce xiphidiocercariae (stylet-bearing). The fact that these cercariae possess well-developed tails probably indicates a recent aquatic origin. About 3 months after infection, cercariae may accumulate in the "lung" (mantle cavity) of the snail or on its body surface. The snail surrounds this irritant with thick mucus and eventually deposits these cercaria-containing **slime balls** as it crawls along. The slime ball may be expelled from the snail's pneumostome with some force. Slime balls are most abundantly produced during a wet period immediately following a drought. Individual slime balls may con-

tain up to 500 cercariae each. Drying of the slime ball surface retards desiccation of the interior and thereby prolongs the lives of the cercariae within.

Continued development of the fluke depends on its ingestion by an ant, which becomes the second intermediate host. The common brown ant *Formica fusca* is the arthropod host in North America. Upon eating the delectable slime balls or feeding them to their larvae, the ants become host to metacercariae, most of which encyst in the hemocoel and are then infective to the definitive host. Over 100 metacercariae may occur in a single ant. One or two, however, emigrate to the subesophageal ganglion and encyst there.[9] These will not become infective, but they alter the ant's behavior in a most remarkable way. When the temperature drops in the evening, ants thus infected climb to the

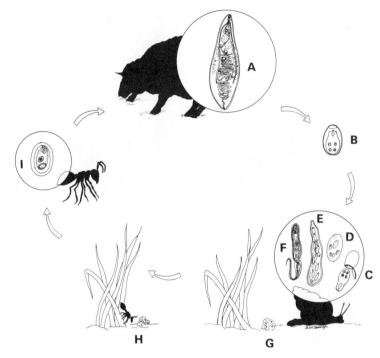

Fig. 19-2. The life cycle of *Dicrocoelium dendriticum*. **A,** Adult, in bile duct of sheep or other plant-eating mammal; **B,** egg; **C,** miracidium hatching from egg after being eaten by snail; **D,** mother sporocyst; **E,** daughter sporocyst; **F,** cercaria; **G,** slime balls containing cercariae; **H,** ant, eating slime balls; **I,** metacercaria encysted in ant; the definitive host becomes infected when it accidentally eats the ant. (Diagram by Carol Eppinger.)

tops of grasses and other plants and grasp them firmly in their mandibles, while the uninfected nest-mates return to warmer digs. They remain attached until later the next day when they warm up and seemingly resume normal behavior. [2] This behavioral pattern keeps the infected ants near the tops of vegetation during the active periods of grazing by ruminants during the evening and morning hours, but allows them to retreat to cooler places during the hot hours of the day. The parasite thus influences its intermediate host to behave in a manner encouraging passage to the definitive host. A similar phenomenon is found when the metacercariae of the dicrocoeliid of *Brachylecithum mosquensis*, a parasite of American robins, encyst near the supraesophageal ganglion of carpenter ants, *Campanotus* spp. Instead of retreating from brightly lighted areas, as is normal for these ants, infected individuals actually seek such places and wander aimlessly or

in circles on exposed surfaces. This makes them much more attractive to the bird definitive host. [4]

Upon being eaten by a definitive host, *Dicrocoelium dendriticum* excysts in the duodenum. It is apparently attracted by bile and quickly migrates upstream to the common bile duct and thence into the liver. The flukes mature in sheep in 6 or 7 weeks and begin producing eggs about a month later. Up to 50,000 *D. dendriticum* have been found in a single sheep.

Pathology of dicrocoeliiasis is basically the same as that for fascioliasis, except that there is no trauma to the gut wall or liver parenchyma resulting from migrating larvae. General biliary dysfunction, with its several symptoms, is typical.

Numerous cases of *D. dendriticum* in humans have been reported. Most of these were false infections. That is, the eggs that were detected in the stool were actually part of a liver repast that the person had

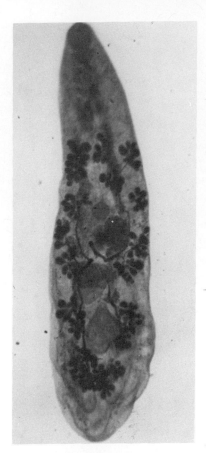

Fig. 19-3. *Haematoloechus medioplexus,* a common parasite in the lungs of frogs. The eggs have been extruded from the uterus of this specimen. (Photograph by Warren Buss.)

enjoyed a few hours earlier. A few genuine infections in humans have been diagnosed, however, mainly in Russia, Europe, Asia, and Africa. More than 15 cases of human infection with a related species, *Dicrocoelium hospes,* have been reported from Africa.[12]

Adequate drug treatment is available for both humans and domestic animals, but control promises to be difficult in the foreseeable future because of the ubiquity of land snails and ants.

Family Haematoloechidae

The flukes of the family Haematoloechidae are parasitic in the lungs of frogs and toads. They are of no economic or medical importance to humans, but because of their large size and easy availability, they are often the first live parasites seen by beginning students of biology. Their transparent beauty, enigmatic location, and fascinating biology have led more than one novitiate into a career in parasitology.

Haematoloechus medioplexus is a typical species among the more than 40 known species that are found in all parts of the world where amphibians occur (Fig. 19-3). This is a flat, soft worm up to 8 mm long and 1.2 mm wide. The acetabulum is small and inconspicuous in this and related species because of their undisturbed site in the lung. The uterus is voluminous, with a descending limb reaching near the posterior end and then ascending with wide loops to the genital pore near the oral sucker. So many eggs fill the uterus that most internal organs are obscured. However, when living worms are placed in tap water, they will expel most of the eggs and thereby become transparent enough to study.

Adult flukes lay prodigious numbers of eggs, which are carried out of the respiratory tract by ciliary action and thence through the gut to the outside. When swallowed by a scavanging *Planorbula armigera* snail, the miracidium hatches and migrates to the hepatic gland, where it develops into a sporocyst. Cercariae escape the snail by night and live the free life for up to 30 hours. When sucked into the anal "lung" of a dragonfly nymph the cercaria penetrates the thin cuticle and encysts in nearby tissues. When the insect metamorphoses into a teneral and then into an adult, the metacercariae remain in the posterior end of the abdomen, to be eaten, along with the rest of the luckless dragonfly, by a frog or toad. Excystment occurs in the stomach. The little flukes creep through the stomach, up the esophagus, through the glottis, and into the respiratory tree. As many as 75 worms have been found in a single lung, although two or three are average.

The known life cycles of other haematoloechids are quite similar.

Family Plagiorchiidae

Another family of flukes often encountered in wild animals is Plagiorchiidae. They parasitize vertebrate classes from amphibians through birds and mammals

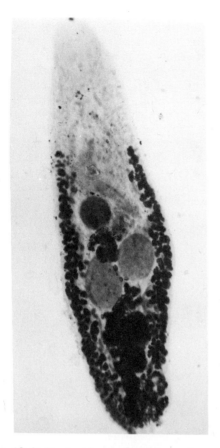

Fig. 19-4. *Plagiorchis maculosus,* from the small intestine of a cliff swallow. Note the oblique testes and intratesticular uterus typical of the genus. Birds are infected when they eat mayflies containing metacercariae. (Photograph by Warren Buss.)

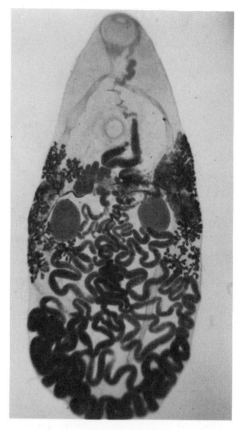

Fig. 19-5. *Prosthogonimus macrorchis,* an oviduct fluke of birds. (Photograph by Warren Buss.)

and are the "typical" flukes in the animal world. One species is extremely unusual for a trematode in that it develops to maturity in freshwater leeches.[20] All species use aquatic snails as first intermediate hosts, and insects, such as mayflies and dragonflies, are the second intermediate hosts. The literature on plagiorchiids is replete with variations of life cycles and descriptions of the many species of worms encountered. Apparently this is a plastic group, adaptable to different situations. Representative species are *Plagiorchis maculosus,* cosmopolitan in swallows (Fig. 19-4); *P. nobeli* in American blackbirds; *Ochetosoma* in the mouth and esophagus of snakes; *Astiotrema* in fishes; and *Neoglyphe* in shrews. Several species of *Plagiorchis* have been reported from humans in Japan, Java, and the Philippines, but these no doubt were zoonotic infections.[6]

Family Prosthogonimidae

Most prosthogonimids are parasites living in the oviduct, bursa of Fabricius, or gut of birds. They are remarkably transparent, stain well, and make good examples for classroom studies of trematode morphology.

Prosthogonimus macrorchis (Fig. 19-5) is the oviduct fluke of domestic fowl and various wild birds in North America. It causes considerable damage to the oviduct and can decrease or even prevent egg laying. Many have been found within eggs after being trapped in the membranes formed by the oviduct, presumably giving the cook an unexpected surprise.

Its life cycle is similar to that of *Haematoloechus* spp. When the embryonated eggs are passed into water, they sink to the bottom. They do not hatch until eaten by a snail *(Amnicola),* where they burrow into the digestive gland, become sporocysts, and produce short-tailed xiphidiocercariae. When these are sucked into the rectal branchial chamber of a dragonfly nymph, they attach and penetrate into the hemocoel and encyst in the muscles of the body wall, remaining infective after the insect metamorphoses. When eaten by a bird, they excyst in the intestine, migrate downstream to the cloaca and into the bursa of Fabricius or the oviduct, and mature in about 1 week. In male birds the infection is lost when the bursa atrophies. Over 30 species of *Prosthogonimus* are known from various areas around the world.

SUPERFAMILY ALLOCREADIOIDEA

Allocreadiids are mainly parasites of fishes, although species are found in amphibians, reptiles, and rarely mammals. Most are of no known importance to humans but one family has species of great medical significance. The cercariae usually have eyespots; metacercariae encyst in arthropods or fishes.

Family Troglotrematidae

The Troglotrematidae are oval, thick flukes with a spiny cuticle and dense vitellaria. They are parasites of the lungs, intestine, nasal passages, cranial cavities, and various ectopic locations of birds and mammals in many parts of the world. We will illustrate the biology of this interesting group with discussions of two species.

Paragonimus westermani

Paragonimus westermani was first described from two Bengal tigers that had died in zoos in Europe in 1877. During the next 2 years, infections by this worm in humans were found in Formosa. It was very quickly found in the lungs, brain, and viscera of humans in Japan, Korea, and the Philippines. The life cycle was worked out by Kobayashi[13] and Yokagawa.[25] The major focus of infection today remains in

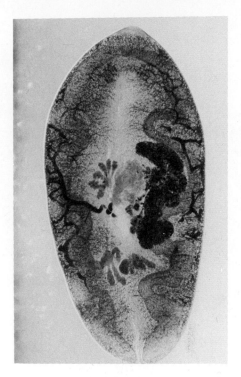

Fig. 19-6. *Paragonimus westermani,* adult. (Photograph by Robert E. Kuntz and Jerry A. Moore.)

the Orient, including India and the Philippines. It also appears to be endemic throughout the East Indies, New Guinea, the Solomon Islands, Samoa, western Africa, Peru, Colombia, and Venezuela. The taxonomy of the genus is difficult, and some of these reports may be of other, closely related species. Paragonimiasis is an excellent example of a zoonosis.

Morphology. Adult worms (Fig. 19-6) are 7.5 to 12 mm long and 4 to 6 mm at their greatest width. They are very thick, however, measuring 3.5 to 5 mm in the dorsoventral axis. In life, they are reddish brown in color, lending the worms the overall size, shape, and color of coffee beans. The tegument is densely covered with scale-like spines. The oral and ventral suckers are about equal in size, with the latter placed slightly preequatorially. The excretory bladder extends from the posterior end to near the pharynx. The lobated testes are at the same level, located at the junction of the posterior fourth of the

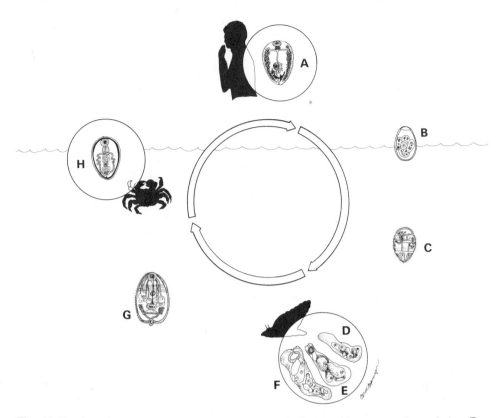

Fig. 19-7. The life cycle of *Paragonimus westermani*. **A,** Sexually mature lung fluke; **B,** the egg is coughed up, swallowed, and passed out with the feces; **C,** miracidium; **D,** sporocyst; **E,** mother redia; **F,** daughter redia; **G,** cercaria; **H,** metacercaria in the second intermediate host. (Diagram by Carol Eppinger.)

Fig. 19-8. The lung of a cat with two cysts containing adult *Paragonimus westermani* (arrows). (Photograph by Robert E. Kuntz.)

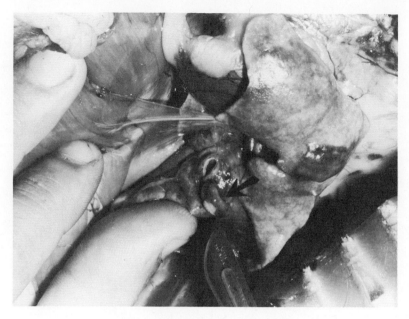

Fig. 19-9. The same specimen as in Fig. 19-8, with a worm dissected out of its cyst (arrow). (Photograph by Robert E. Kuntz.)

body. A cirrus and cirrus pouch are absent. The genital pore is postacetabular.

The ovary is also lobated and is found to the left of midline, slightly postacetabular. The uterus is tightly coiled into a "rosette" at the right of the acetabulum and opens into the common genital atrium with the vas deferens. Vitelline follicles are extensive in the lateral fields, from the level of the pharynx to the posterior end. The eggs are ovoid and have a rather flattened operculum set into a rim. They measure 80 to 118 μm by 48 to 60 μm.

Identification of the 30 or so species of *Paragonimus* is difficult, with much emphasis being placed on the characters of the metacercaria and the shape of the tegumental spines.[15] Several nominal species probably will be synonymized after they have been properly studied.

Biology (Fig. 19-7). Adult *P. westermani* usually live in the lungs, encapsulated in pairs by layers of connective tissues (Fig. 19-8 and 19-9). They have been found in many other organs of the body, however (Fig. 19-10). Cross fertilization usually occurs. The eggs (Fig. 19-11) are often trapped in surrounding tissues and cannot leave the lungs, but those that escape into the air passages are moved up and out by the ciliary epithelium. Arriving at the

pharynx, they are swallowed and passed through the alimentary canal to be voided with the feces. The larvae require from 16 days to several weeks in water before development of the miracidium is complete. Hatching is spontaneous, and the miracidium must encounter a snail in the family Thieridae if it is to survive. As these snails usually live in swift-flowing streams, the chances of survival of any miracidium are slight. This is offset by the numbers of eggs produced by the adult. Upon entering a snail, it forms a sporocyst that produces rediae, which in turn develop many cercariae. These cercariae (Fig. 19-12) are microcercous, with spined, knob-like tails and minute oral stylets.

After escaping from the snail, cercariae become quite active, creeping over rocks in inch-worm fashion, and attack crabs and crayfish of at least 11 species, encysting in the viscera and muscles. A common second intermediate host in Taiwan is *Eriocheir japonicus* (Fig. 19-13). Recent evidence suggests that crustaceans may become infected by eating infected snails.[16] The metacercariae (Fig. 19-14) are pearly white in life, and, through microscopic examination, can be recognized to species by an expert. When the crustacean is eaten by a proper definitive host, the worms ex-

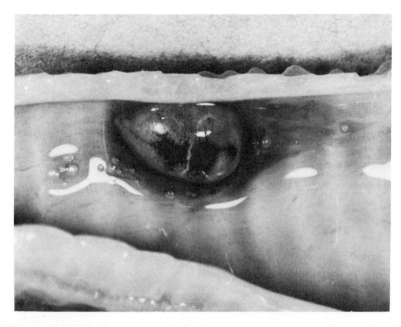

Fig. 19-10. An adult *Paragonimus westermani* in the trachea of an experimentally infected cat. (Photograph by Robert E. Kuntz.)

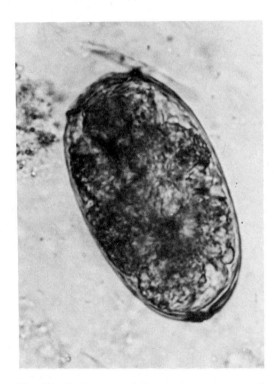

Fig. 19-11. An egg of *Paragonimus westermani* from the feces of a cat. (Photograph by Robert E. Kuntz.)

cyst in the duodenum, pierce its wall, and embed themselves in the abdominal wall. Several days later they reenter the coelom, penetrate the diaphragm (Fig. 19-15) and pleura, and enter the bronchioles of the lungs. They mature in 8 to 12 weeks. Wandering larvae may locate in ectopic locations, such as the brain, mesentery, pleura, or skin.

Epidemiology. The natural definitive hosts, and therefore reservoir hosts, of *P. westermani* are several species of carnivores, including felids, canids, viverrids, and mustelids, as well as some rodents and pigs. Humans are probably a lesser source of infective eggs than other mammals, but like the others, humans become infected when they eat raw or insufficiently cooked crustaceans. Crab collectors in some countries distribute their catch miles from their source, effectively propagating paragonimiasis (Fig. 19-16). Completely raw crab or crayfish is not as commonly eaten in the Orient as that prepared by marination in brine, vinegar, or wine, which coagulates the protein in the muscles, giving it a cooked appearance and taste but not affecting the metacercariae. Exposure commonly is effected by contamination of

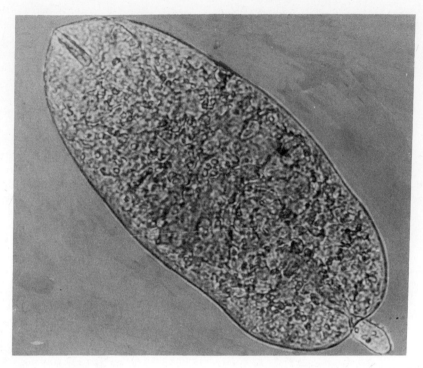

Fig. 19-12. A microcercous cercaria of *Paragonimus westermani*. (Photograph by Robert E. Kuntz.)

Fig. 19-13. *Eriocheir japonicus,* the second intermediate host for *Paragonimus westermani* in Taiwan. (Photograph by Robert E. Kuntz.)

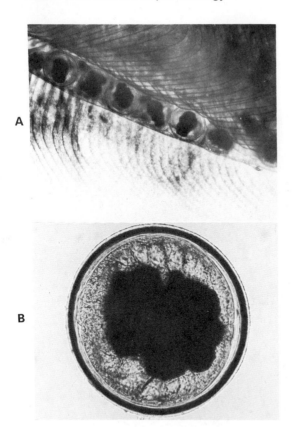

Fig. 19-14. Metacercaria of *Paragonimus westermani.* **A,** Several metacercariae in a gill filament of a crab. **B,** A single metacercaria. The opaque mass is characteristic for this genus. (Photograph by Robert E. Kuntz.)

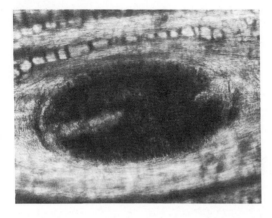

Fig. 19-15. A juvenile *Paragonimus westermani* penetrating the diaphragm of a civet (a small, carnivorous mammal). (Photograph by Robert E. Kuntz.)

fingers or cooking utensils during food preparation (Fig. 19-17).[22] It is even possible that persons accidentally become infected when they smash rice-eating crabs in the paddies, splashing themselves with juices that contain metacercariae. Another factor of possible epidemiological significance in some ethnic groups is the medicinal use of juices strained from crushed crabs or crayfish.

Pathology. The early, invasive stages of paragonimiasis cause little or no symptomatic pathology. Once in a lung or an ectopic site, the worm stimulates connective tissue proliferation that eventually will enshroud it in a brownish or bluish capsule. Such capsules often ulcerate and heal slowly. Eggs in surrounding tissues will themselves become centers of pseudotubercles. Worms in the spinal cord are known to cause paralysis, which sometimes is total. Fatal cases of *Paragonimus* in the heart have been recorded. Cerebral cases have the same results as cerebral cysticercosis. Pulmonary cases usually cause chest symptoms, with breathing difficulties, chronic cough, and sputum containing blood or brownish streaks (fluke eggs). Fatal cases are common.

Diagnosis and treatment. The only sure diagnosis, aside from surgical discovery of the adult worm, is by finding the highly characteristic eggs in sputum, aspirated pleural fluid, feces, or matter from a *Paragonimus*-caused ulcer. The pulmonary type is easily mistaken for tuberculosis, pneumonia, spirochaetosis, and other such illnesses, and x-ray examination may be incorrectly interpreted. Cerebral involvement requires differentiation from tumors, cysticercosis, hydatids, encephalitis, and others. Seroimmunological diagnosis is useful and particularly valuable in detecting ectopic infection. The intradermal test is practiced for surveys but must be followed by a complement fixation test on those testing positive. This is because the dermal reaction persists for long periods after recovery from the disease.[24]

Several drugs show promise in treatment of paragonimiasis, although a widely accepted regimen of therapeutics has not yet been developed. Clinical symptoms decrease after 5 or 6 years of infection, but worms are known to live for 20 years

Fig. 19-16. Crab collectors stringing crabs for trip to market in Taiwan. This practice distributes *Paragonimus* far from its source. (Photograph by Robert E. Kuntz.)

Fig. 19-17. Children "cooking" fresh-caught crabs on an open fire. Such practices contribute to the widespread incidence of infection in the Orient. (Photograph by Robert E. Kuntz.)

or more. Infection can be avoided by cooking crustaceans before eating them and by avoiding contamination with their juices.

Paragonimus kellicotti closely resembles *P. westermani.* It is common in mink and similar mammals in North America east of the Rocky Mountains. The first intermediate host in the United States is *Pomatiopsis lapidaria.* Crayfish of the common genus *Cambarus* serve as second intermediate hosts, with the metacercariae usually encysting on the heart. One case of infection in a human has been reported. Several species of *Paragonimus* are known from a variety of wild animals.

Nanophyetus salmincola (Fig. 19-18)

After 1814 it was known that dogs that ate raw salmon were prone to a disease so severe that scarcely one in 10 survived.[8] It was not until 1926 that this disease was shown to be associated with a minute fluke, whose metacercariae were common in the flesh and viscera of salmon.[5] It is now known that the disease itself is caused by the rickettsia, *Neorickettsia helminthoeca,* which is transmitted to the dogs by the

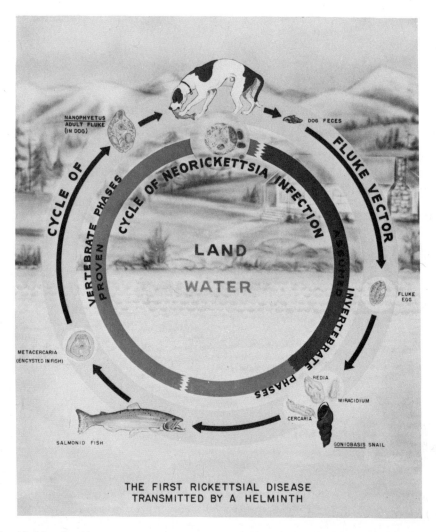

NANOPHYETUS
ADULT FLUKE
(IN DOG)

DOG FECES

CYCLE OF

FLUKE VECTOR

VERTEBRATE PHASES

CYCLE OF NEORICKETTSIA INFECTION

PROVEN

ASSUMED

INVERTEBRATE PHASES

LAND

WATER

FLUKE
EGG

METACERCARIA
(ENCYSTED IN FISH)

REDIA

MIRACIDIUM

CERCARIA

GONIOBASIS SNAIL

SALMONID FISH

THE FIRST RICKETTSIAL DISEASE
TRANSMITTED BY A HELMINTH

Fig. 19-18. The life cycle of *Nanophyetus salmincola* and of the neorickettsia it harbors. (From Philip, C. B. 1959. Archives de l'Institut Pasteur de Tunis 36:595-603.)

flukes. Infected dogs can be treated effectively with sulfanilamides and antibiotics.

The identity of the fluke was almost as elusive as that of the rickettsia it carried. When first described, it was placed in the family Heterophyidae, whose members it superficially resembles. Witenberg, in 1932, restudied the original specimens and found a cirrus pouch but no genital sucker or seminal receptacle. He concluded that it was actually a species of *Troglotrema* and transferred it to the Troglotrematidae. Wallace in 1935 returned it to *Nanophyetus* and established the subfamily Nanophyetinae within the Troglotrematidae, where the species resides today.

Morphology. Adult worms are 0.8 to 2.5 mm long and 0.3 to 0.5 mm wide. The oral sucker is slightly larger than the midventral acetabulum. The testes are side by side in the posterior third of the body. A cirrus pouch is present but there is no cirrus. The small ovary is lateral to the acetabulum, and the uterus is short, containing only a few eggs at a time.

Biology. Adult *N. salmincola* live deeply embedded in crypts in the wall of the small intestine of at least 32 species of mammals, including humans, as well as in fish-eating birds. They produce unembryonated eggs that hatch in water after 87 to 200 days. The snail host in northwestern United

States is *Oxytrema silicula,* an inhabitant of fast-moving streams. Experimental infection of snails in the laboratory has not been accomplished. Sporocysts have not been found, but rediae are well known, occurring in nearly all tissues of the snail. The xiphidiocercaria is microcercous. It penetrates and encysts in at least 34 species of fish, but salmonid fishes are more susceptible than fish of other families. Metacercariae can be found in nearly any tissue of the fish, but they are most numerous in the kidneys, muscles, and fins. Young fish have a high rate of mortality in heavy infection.[7] A variety of mammals and even two bird species (heron and merganser) can be infected with the trematode, but the racoon and spotted skunk are clearly the main definitive hosts in nature.[19] The worm has not been reported from humans in North America, but it occurs in natives of eastern Siberia.

Pathology. Adult flukes themselves cause surprisingly little pathology. Philip noted that the inflammatory changes in the intestines of dogs carrying hundreds of *Nanophyetus* were no more extensive than in animals infected with salmon poisoning disease by injection with lymph node suspensions.[18] Salmon poisoning disease is restricted to dogs, coyotes, and other canids and does not affect humans. Its course in dogs is rapid and severe. After an incubation period of 6 to 10 days, the dog's temperature rises to 40 to 42°C, often with edematous swelling of the face and discharge of pus from the eyes. The dog exhibits depression, loss of appetite, and increased thirst, then vomiting and diarrhea by 4 to 7 days after onset of symptoms. The fever usually lasts from 4 to 7 days, and the dog can be expected to die about 10 days to 2 weeks after onset, but those that recover are immune for the rest of their lives.

Much remains to be learned about the biology of this fluke and the rickettsia it harbors. Dogs are extremely susceptible, and, when untreated, the mortality is about 90%. The disease can be transmitted experimentally by injection of lymph node preparations from other infected dogs or by injecting eggs (evidence of transovarial transmission in the fluke), metacercariae, or adult flukes and digestive glands from infected snails.[17] The geographic range of the disease coincides with the distribution of the snail host of the fluke, and the proportion of salmonids within that range that are infected is extremely high. In light of these facts and the mortality in dogs, it is assumed that there must be some reservoir of the rickettsia, but the identity of that reservoir is not at all clear. Racoons do not seem to be susceptible; after fluke infection or injection with infected lymph nodes, they have a transitory, low-grade fever, but attempts to transmit the disease from them to dogs via lymph node preparations were unsuccessful.[18]

Order Opisthorchiata

The order Opisthorchiata contains three superfamilies. The Hemiuroidea and Isoparorchioidea parasitize only fishes (and a few amphibians) in the adult stage, and, while they are interesting, there is not room in this book to discuss them. The remaining superfamily is of more importance to humans and domestic animals. It will serve, then, to illustrate the biology of this order of parasites.

SUPERFAMILY OPISTHORCHIOIDEA

These are medium to small flukes, often spinose and with poorly developed musculature. The testes are at or near the posterior end, and a cirrus pouch is absent. A seminal receptacle is present, and the metraterm and ejaculatory duct unite to form a common genital duct. Eggs are embryonated when passed, but hatching occurs only after ingestion by a suitable snail. Adults live in the intestine or biliary system of fishes, reptiles, birds, and mammals. Metacercariae are in fishes.

Family Opisthorchiidae

Opisthorchiids are delicate, leaf-shaped flukes with weakly developed suckers. Most are exceptionally transparent when prepared for study and so are popular subjects for parasitology classes. Adults are in the biliary system of reptiles, birds, and mammals. Two species are of substantial consequence to humans.

Clonorchis sinensis

Clonorchis sinensis was first discovered in the bile passages of a Chinese carpenter in Calcutta, in 1875. Other infections were

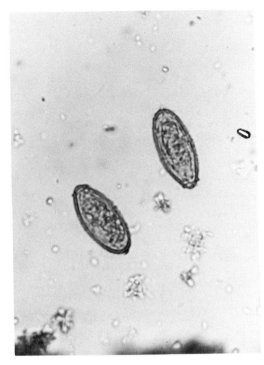

Fig. 19-20. Eggs of *Clonorchis sinensis* from a human stool. (Photograph by Robert E. Kuntz and Jerry A. Moore.)

Fig. 19-19. The Chinese liver fluke, *Clonorchis sinensis.* (Photograph by Robert E. Kuntz.)

quickly discovered in Hong Kong and Japan. Today it is known that the "Chinese liver fluke" is widely distributed in Japan, Korea, China, Taiwan, and Vietnam, where it causes untold suffering and economic loss. Stoll estimated that 19 million persons were infected in eastern Asia in 1947. The number is probably higher today. Reports of this parasite outside the Orient involve infections acquired while visiting there or by eating frozen, dried, or pickled fish imported from endemic areas.

Morphology. Adults (Fig. 19-19) measure 8 to 25 mm long by 1.5 to 5 mm wide. The tegument lacks spines, and the musculature is weak. The oral sucker is slightly larger than the acetabulum, which is about one-fourth of the way from the anterior end.

The male reproductive system consists of two large, branched testes in tandem near the posterior end and a large, serpentine seminal vesicle leading to the genital pore. A cirrus and cirrus pouch are absent. The pretesticular ovary is relatively small and has three lobes. The seminal receptacle is large and transverse and is located just behind the ovary. The uterus ascends in broad, tightly packed loops and joins the ejaculatory duct to form a short, common genital duct. The genital pore is median, just anterior to the acetabulum. Vitelline follicles are small and dense and are confined to the level of the uterus. Laurer's canal is conspicuous.

Biology. Chinese liver flukes mature in the bile ducts and produce up to 4,000 eggs per day for at least 6 months. The mature egg (Fig. 19-20) is yellow-brown in color, 26 to 30 μm long and 15 to 17 μm wide. The operculum is large and fits into a broad rim of the egg shell. There is usually a small knob or curved spine on the aboperucular end that helps to distinguish the eggs of this species. When passed, the egg contains a well-developed miracidium that is rather asymmetrical in its internal organization.

Hatching of the miracidium will occur only after the egg is eaten by a suitable snail, of which *Parafossarulus manchouricus* is the most common and, therefore, the most important, first intermediate host throughout the Orient. The miracidium transforms into a sporocyst in the wall of the intestine or in other organs within 4 hours after infection. These produce rediae within 17 days. Each redia produces from five to 50 cercariae. The cercaria (Fig. 19-21) has a pair of eyespots and is beset with delicate bristles and tiny spines. The entire cercaria is brownish in color. The tail has dorsal and ventral fins **(pleurolophocercous cercaria).**

The cercaria hangs upside down in the water and slowly sinks to the bottom. When contacting any object, it rapidly swims upward toward the surface and again begins to sink. Even a slight current of water will also cause this reaction. Thus, when a fish swims by, the cercaria is stimulated to react in a way favoring its contact with its next host. Upon touching the epithelium of the fish, the cercaria attaches with its suckers, casts off its tail, and bores through the skin, coming to rest and encysting under a scale or in a muscle (Fig. 19-22). Nearly a hundred species of fishes, mostly in Cyprinidae (Fig. 19-23), have been found naturally infected with metacercaria of *C. sinensis,* although some species are more susceptible than others. Thousands of metacercariae may accumulate in a single fish, but the number usually is much smaller. Metacercariae will also develop in the crustaceans *Caridina, Macrobrachium,* and *Palaemonetes,* and such metacercariae have been shown to be infective to guinea pigs.[23] The definitive host is infected when it eats raw or undercooked fish or crustaceans.

Mammals other than humans that have been found infected with adult *Clonorchis sinensis* are pigs, dogs, cats, rats, and camels.[14] Experimentally, rabbits and guinea pigs are highly susceptible. Possibly any fish-eating mammal can become infected. Dogs and cats undoubtedly are important reservoir hosts. Birds may possibly be infected.

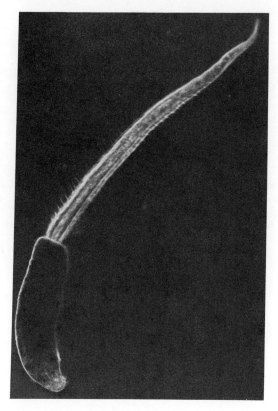

Fig. 19-21. A cercaria of *Clonorchis sinensis.* (From Gibson, J. B., and Sun, T. 1971. In Marcial-Rojas, R. A., editor. Pathology of protozool and helminthic diseases with clinical correlation. The Williams & Wilkins Co., Baltimore.)

The young flukes excyst in the duodenum. The route of migration to the liver is not clear, for conflicting reports have been published. It seems probable to us that the juveniles migrate up the common bile duct to the liver. Young flukes have been found in the liver 10 to 40 hours after infection of experimental animals. The worms mature and begin producing eggs in about 1 month. The entire life cycle can be completed in 3 months under ideal conditions. Adult worms can live at least 8 years in humans.

Epidemiology. It is easy to see why clonorchiasis is common in countries where raw fish is considered a delicacy (Fig. 19-24). In some areas the most heavily infected people are wealthy epicures who can afford beautifully cut and arranged slices of raw fish. On the other hand, the

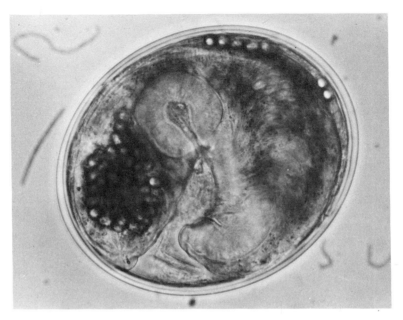

Fig. 19-22. Encysted metacercaria from fish muscle. The oral and ventral suckers are clearly seen; the round bodies are excretory corpuscles. (From Gibson, J. B., and Sun, T. 1971. In Marcial-Rojas, R. A., editor. Pathology of protozool and helminthic diseases with clinical correlation. The Williams & Wilkins Co., Baltimore.)

Fig. 19-23. Grass carp, *Ctenopharyngodon idellus.* A common intermediate host of *Clonorchis sinensis;* this fish is widely cultivated in the Orient. (From Gibson, J. B., and Sun, T. 1971. In Marcial-Rojas, R. A., editor. Pathology of protozool and helminthic diseases with clinical correlation. The Williams & Wilkins Co., Baltimore.)

poor are also afflicted, for fish is often their only source of animal protein. The prevalence may range from an average of 14% in cities such as Hong Kong to 80% in some endemic rural areas.[21] Though complete protection is achieved simply by cooking fish, it would be a futile exercise to try to get millions of people to change centuries-old eating habits. Even so, to educate these people to cook their fish would not change matters, for fuel is commonly a luxury that many cannot afford. Kim and Kuntz (1964) discuss the epidemiology of clonorchiasis in Taiwan.[11]

Fish farming is a mainstay of protein production throughout the Orient, as in

Fig. 19-24. "Yue-shan chuk," thin slices of raw carp with rice soup, vegetable garnishing, and soya sauce—a Cantonese delicacy. (From Gibson, J. B., and Sun., T. 1971. In Marcial-Rojas, R. A., editor. Pathology of protozool and helminthic diseases with clinical correlation. The Williams & Wilkins Co., Baltimore.)

Fig. 19-25. A privy over a fish-culture pond in Hong Kong. The Chinese characters on the structure advertise a worm medicine. (From Gibson, J. B., and Sun, T. 1971. In Marcial-Rojas, R. A., editor. Pathology of protozool and helminthic diseases with clinical correlation. The Williams & Wilkins Co., Baltimore.)

Europe and, increasingly, in America. More protein in the form of fish can be harvested from an acre of pond than in the form of beef, beans, or corn from an acre of the finest farmland. The fastest growing fish are primary consumers of algae and other plants. Such ponds typically are fertilized with human feces (Fig. 19-25), which increases the growth rate of water plants and thereby that of the fish. And of course this abets the life cycle of *Clonorchis sinensis.* Where fish farming is not so important, dogs and cats serve as reservoirs of infection, contaminating streams and ponds with their feces.

Metacercariae will withstand certain

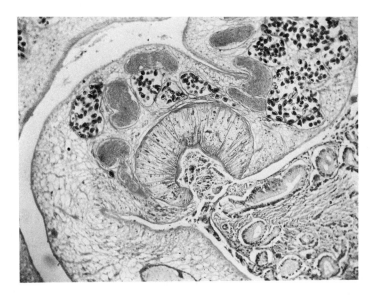

Fig. 19-26. Adult *Clonorchis sinensis* attached by its ventral sucker to biliary epithelium in a human. (From Gibson, J. B., and Sun, T. 1971. In Marcial-Rojas, R. A., editor. Pathology of protozool and helminthic diseases with clinical correlation. The Williams & Wilkins Co., Baltimore.)

types of preparation of fish, such as salting, pickling, drying, and smoking. Because of this, people can become infected thousands of miles from an endemic area when they eat such imported fish.[3]

Pathology. The basic pathogenesis of *Clonorchis* infection is erosion of the epithelium lining the bile ducts (Fig. 19-26). The ultimate effect depends mainly on the intensity and duration of infection, and, fortunately, worm burdens are usually small. The mean intensity of infection in most endemic areas is 20 to 200 flukes, but as many as 21,000 have been removed at a single autopsy. Chronic defoliation of the bilary epithelium leads to gradual thickening and occlusion of the ducts (Fig. 19-27). Pockets form in the walls of bile ducts, and complete perforation into surrounding parenchyma may result. Infiltrating eggs become surrounded by connective tissues, thereby interfering with liver function.

Ascites nearly always occurs in fatal cases, but its relationship to *Clonorchis* infection is uncertain. Jaundice is found in a small percentage of cases and is probably caused by bile retention when ducts are obstructed. Eggs, and sometimes entire worms, often become nuclei of gallstones. Cancer of the liver is more prevalent in Japan than elsewhere, and its relationship to clonorchiasis should be investigated.

Diagnosis and treatment. Diagnosis is based on the recovery of the characteristic eggs in the feces. Liver abnormalities described above should suggest clonorchiasis in endemic areas, but care must be taken to exclude cancer, hydatid disease, beriberi, amebic abscess, and other types of hepatic disease. Intradermal tests seem promising in making a diagnosis. A variety of drugs have been tried against *Clonorchis* infections, but no satisfactory agent has yet been found.

Opisthorchis felineus

The cat liver fluke is very similar to *Clonorchis sinensis* but has a more European distribution. Originally described from a domestic cat in Russia, it is known to be common throughout southern, central, and eastern Europe, Turkey, southern Russia, Vietnam, India, and Japan. Besides cats and other carnivores, it parasitizes humans, probably infecting more than a million humans within its range.

The most obvious difference in morphology from *C. sinensis* is the shape of its testes, which are slightly lobed in *O. felineus* but greatly branched in the Chinese

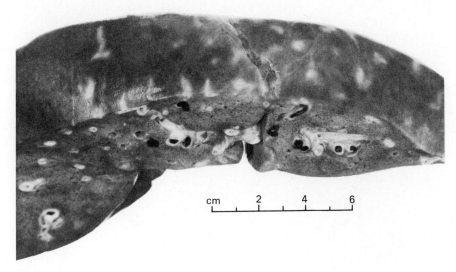

Fig. 19-27. Severe clonorchiasis with "pipestem fibrosis" in a human. The dilated, thick-walled bile ducts are full of flukes. (From Gibson, J. B., and Sun, T. 1971. In Marcial-Rojas, R. A., editor. Pathology of protozool and helminthic diseases with clinical correlation. The Williams & Wilkins Co., Baltimore.)

liver fluke. Their life cycles also are nearly identical. The only known snail first intermediate host is *Bithynia leachi*. The pleuro-lophocercous cercaria encysts within the muscles of several species of cyprinid fishes. Its epidemiology and pathogenesis parallel those of *C. sinensis*.

Other, related species of opisthorchiids are *Opisthorchis viverrini* in wild and domestic carnivores (and possibly 3 million cases in humans) in southeast Asia and *Amphimerus pseudofelineus* in cats and wild carnivores in North America.

Family Heterophyidae

Heterophyids are tiny, teardrop-shaped flukes, usually maturing in the small intestine of fish-eating birds and mammals. The suckers are usually feeble, with the acetabulum enclosed inside a sucker-like genital sinus, or **gonotyl**, which is greatly modified in different species. A cirrus pouch is absent. The tegument is scaly, especially anteriorly. This is a large family with several subfamilies. Several species are important parasites of humans.

Heterophyes heterophyes (Fig. 19-28)

Heterophyes heterophyes is a minute fluke that was first discovered in an Egyptian in Cairo in 1851. It is common in north Africa, Asia Minor, and the Far East, including Korea, China, Japan, Taiwan, and the Philippines. Because the eggs cannot be differentiated from other, related species, an accurate estimate of human infections cannot be made.

Morphology. The adults are 1.0 to 1.7 mm long and 0.3 to 0.4 mm at their greatest width. The entire body is covered with slender scales, most numerous near the anterior end. The oral sucker is only about 90 μm in diameter, while the acetabulum is around 230 μm wide and is located at the end of the first third of the body. The two oval testes lie side by side near the posterior end of the body. The vas deferens expands to form a sinuous seminal vesicle, which constricts again, becoming a short ejaculatory duct. The ovary is small, medioanterior to the testes, at the beginning of the last fourth of the body. A seminal receptacle and Laurer's canal are present. The uterus coils between the ceca and constricts before joining the ejaculatory duct to form a short common genital duct, which then opens into the genital sinus. The gonotyl is about 150 μm wide and has 60 to 90 toothed spines on its margin. Lateral vitelline follicles are few in number

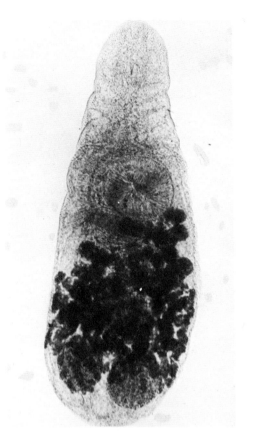

Fig. 19-28. *Heterophyes heterophyes.* (Photograph by Warren Buss.)

and are confined to the posterior third of the worm. The eggs are 28 to 30 μm by 15 to 17 μm.

Biology. Adult worms live in the small intestine, burrowed between the villi. The egg contains a fully developed miracidium when laid, but hatches only when eaten by an appropriate fresh- or brackish-water snail (*Pironella conica* in Egypt, *Cerithidia cingula* in Japan). After penetrating the gut of the snail, the miracidium transforms into a sporocyst that produces rediae. A second generation redia gives birth to cercariae with eyespots and finned tails **(ophthalmolophocercous cercaria),** which emerge from the snail. Like the cercaria of *Clonorchis sinensis,* this species swims toward the surface of the water and slowly drifts downward. Upon contacting a fish it penetrates the epithelium, creeps beneath a scale, and encysts in muscle tissue. Metacercariae are most abundant in

various species of mullet, which are exposed when they enter estuaries or brackish-water shorelines. Several thousand metacercariae have been found in a single, small fish. The definitive host becomes infected when it eats raw or undercooked fish.

Epidemiology. For eggs to be available to estuarine and brackish-water snails, pollution must occur in those waters. Therefore, boatmen, fishermen, and others who live by or on the water are often the main reservoirs of infection. Infected fish are distributed widely in fish markets. Other fish-eating mammals, such as cats, foxes, and dogs, serve as natural reservoirs of infection.

Pathology. Each worm elicites a mild inflammatory reaction at the site of contact with the intestine. Heavy infections (which are common) cause damage to the mucosa and produce intestinal pain and mucous diarrhea. Perforation of the mucosa and submucosa sometimes occurs and allows eggs to enter the blood and lymph vascular systems and to be carried to ectopic sites in the body.[1] The heart is particularly affected, with tissue reactions in the valves and myocardium leading to heart failure. Kean and Breslau reported that 14.6% of cardiac failure in the Philippines resulted from heterophyid myocarditis.[10] Eggs in the brain or spinal cord lead to neurological disorders that are sometimes fatal. Two bizarre cases are known where adult *H. heterophyes* were found in the brains of humans, and in another case an adult worm was found in the myocardium.[1]

Diagnosis is difficult when adult worms are not available. The eggs closely resemble those of several other heterophyids and are not very different from those of *Clonorchis sinensis.* Tetrachloroethylene has been found effective in treatment.

Other heterophyid parasites of humans

Heterophyes katsuradai is very similar to *H. heterophyes.* It has been found in humans near Kobe, Japan. Infection is acquired by eating raw mullet.

Metagonimus yokagawai is a very common heterophyid in the Far East, Russia, and the Balkan region, where it infects humans. It superficially resembles *Heterophyes heterophyes,* but its acetabulum is displaced to the left, where it is fused with the gonotyl. The biology of *M. yokagawai* is identical with that of *H. heterophyes* except that a different snail host is required (*Semisulcospira* spp.), and the second intermediate hosts are freshwater fishes of several species. The definitive host becomes infected when it eats uncooked fish. Various fish-eating mammals are natural reservoirs, and even pelicans have been incriminated in this regard. Pathogenesis, diagnosis, and treatenk are as for *Heterophyes heterophyes*.

Until proved otherwise, all species of Heterophyidae should be considered potential parasites of humans. More than a dozen species have been proved infective to date.

REFERENCES

1. Africa, C. M., W. de Leon, and E. Y. Garcia. 1937. Heterophyidiasis: VI. Two more cases of heart failure associated with the presence of eggs in sclerosed veins. J. Philippine Isl. Med. Assoc. 17:605-609.
2. Anokhin, I. A. 1966. Daily rhythm in ants infected with metacercariae of *Dicrocoelium lanceatum*. Dokl. Akad. Nauk SSSR 166:757-759.
3. Binford, C. H. 1934. Clonorchiasis in Hawaii. Report of cases in natives of Hawaii. Public Health Rep. 49:602-604.
4. Carney. W. P. 1969. Behavioral and morphological changes in carpenter ants harboring dicrocoeliid metacercariae. Am. Midland Nat. 82:605-611.
5. Donham, C. R., B. T. Simms, and F. W. Miller. 1926. So-called salmon poisoning in dogs (progress report). J. Am. Vet. Med. Assoc. 68:701-715.
6. Faust, E. C., P. F. Russell, and R. C. Jung. 1970. Craig and Faust's clinical parasitology, ed. 8. Lea & Febiger, Philadelphia.
7. Gebhardt, G. A., R. E. Millemann, S. E. Knapp, and P. A. Nyberg. 1966. "Salmon poisoning" disease. II. Second intermediate host susceptibility studies. J. Parasitol. 52:54-59.
8. Henry's Astoria Journal. 1814. In The Oregon country under the Union Jack. A reference book of historical documents for Scholars and Historians. Rayette Radio Ltd., Montreal. 1962.
9. Hohorst, W. 1964. Die Rolle der Ameisen in Entwicklungsangan des Lanzettegels (*Dicrocoelium dendriticum*). Z. Parasitenk. 22:105-106.
10. Kean, B. H., and R. C. Breslau. 1964. Parasites of the human heart. Grune & Stratton, Inc., New York. pp. 95-103.
11. Kim, D. C., and R. E. Kuntz. 1964. Epidemiology of helminth diseases: *Clonorchis sinensis* (Cobbold, 1875) Looss, 1907 on Taiwan (Formosa). Chinese Med. J. 11:29-47.
12. King, E. V. J. 1971. Human infection with *Dicrocoelium hospes* in Sierra Leone. J. Parasitol. 57:989.
13. Kobayashi, H. 1918. Studies on the lung-fluke in Korea. Mitteil. Med. Hochschule Keijo. 2:95-113.
14. Komiya, Y., and N. Suzuki. 1964. Biology of *Clonorchis sinensis*. In Morishita, K., and others, editors. Progress of medical parasitology in Japan, vol. 1. Meguro Parasitol. Mus., Tokyo. pp. 551-645.
15. Miyazaki, I. 1965. Recent studies on *Paragonimus* in Japan with special reference to *P. ohirai* Miyazaki, 1939, *P. iloktsuenensis* Chen, 1940 and *P. miyazakii* Kamo, Nishida, Hatsushika et Tomimura, 1961. In Morishita K., and others, editors. Progress of medical parasitology in Japan, vol. 2. Meguro Parasit. Mus., Tokyo. pp. 349-345.
16. Noble, G. A. 1963. Experimental infection of crabs with *Paragonimus*. J. Parasitol. 44:352.
17. Nyberg, P. A., S. E. Knapp, and R. E. Millemann. 1967. "Salmon poisoning" disease. IV. Transmission of the disease to dogs by *Nanophyetus salmincola* eggs. J. Parasitol. 53:694-699.
18. Philip, C. B. 1955. There's always something new under the "parasitological" sun (the unique story of the helminth-borne salmon poisoning disease). J. Parasitol. 41:125-148.
19. Schlegel, M. W., S. E. Knapp, and R. E. Millemann. 1968. "Salmon poisoning" disease. V. Definitive hosts of the trematode vector, *Nanophyetus salmincola*. J. Parasitol. 54:770-774.
20. Schmidt, G. D., and K. Chaloupka. 1969. *Alloglossidium hirudicola* sp. n., a neotenic trematode (Plagiorchiidae) from leeches, *Haemopis* sp. J. Parasitol. 55:1185-1186.
21. Spencer, H., and others. 1973. Tropical pathology. Springer-Verlag, Berlin.
22. Suzuki, Z. 1958. Epidemiological studies on paragonimiasis in South Izu District, Shizuoka Prefecture, Japan.
23. Tang, C. C., and others. 1963. Clonorchiasis in South Fukien with special reference to the discovery of crayfishes as second intermediate hosts. Chinese Med. J. 82:545-618.
24. Yokagawa, M. 1969. *Paragonimus* and paragonimiasis. In Dawes, B., editor. Advances in parasitology, vol. 7. Academic Press, Inc., New York. pp. 375-387.
25. Yokagawa, S. 1919. A study of the lung distoma. Third report, Formosan Endoparasitic Disease research.

SUGGESTED READING

Holmes, J. C., and W. M. Bethel. 1972. Modification of intermediate host behaviour by parasites. In Canning, E. U., and C. A. Wright, editors. Behavioral aspects of parasite transmission. Zool. J. Linn. Soc. 51(Suppl. 1):123-149. (An outstanding summary of the subject. Should be required reading for all students of parasitology.)

Komiya, Y. 1966. *Clonorchis* and clonorchiasis. In Dawes, B., editor. Advances in parasitology, vol. 4. Academic Press, Inc., New York. pp. 53-106.

Milleman, R. E., and S. E. Knapp. 1970. Biology of *Nanophyetus salmincola* and "salmon poisoning" disease. In Dawes, B., editor. Advances in parasitology, vol. 8. Academic Press, Inc. New York. pp. 1-41.

Miyata, I. 1965. The development of *Eurytrema pancreaticum* and *Eurytrema coelomaticum* in the intermediate host snails. In Morishita, K., and others, editors. Progress of medical parasitology in Japan, vol. 2. Meguro Parasit Mus., Tokyo. pp. 348-357.

Travassos, L. 1944. Revisào da familia Dicrocoeliidae Odhner, 1910. Inst. Oswaldo Cruz Monogr. 2:1-357. (The definitive monograph on Dicrocoeliidae.)

Yokagawa, M. 1965. *Paragonimus* and paragonimiasis. In Dawes, B., editor. Advances in parasitology, vol. 3. Academic Press, Inc., New York. pp. 99-158. (A complete summary of paragonimiasis. Required reading by all who are interested in the subject.)

Chapter 20

CLASS CESTOIDEA: FORM, FUNCTION, AND CLASSIFICATION OF THE TAPEWORMS

Fear and superstition still abound among laypeople, who generally hold tapeworms to be the lowliest and most degenerate of creatures (Fig. 20-1). Most of the repugnance with which most people hold these animals derives from the fact that the tapeworms live in the intestine and are only seen when they are passed with the feces of the host. Further, they seem to be generated spontaneously, and mystery is nearly always accompanied by fear. Finally, in a few instances, their presence initiates disease conditions that traditionally have been difficult to cure. Be that as it may, a scientific approach to cestodology has increased understanding of tapeworms and shown that they are one of the most fascinating groups of organisms in the animal kingdom. Their complex life cycles and intricate host-parasite relationships are rivaled by few known organisms. Observations of cestodes no doubt began in earliest times and extends today into sophisticated laboratories throughout the world.

Historically, Hippocrates, Aristotle, and Galen appreciated the animal nature of tapeworms.[14] The Arabs suggested that the segments passed with the feces were a separate species of parasite from tapeworms; they called these segments the cucurbitini, after their similarity to cucumber seeds.[15] Andry, in 1718, was the first to illustrate the scolex of a tapeworm from a human (Fig. 20-2). Three common species in humans, *Taeniarhynchus saginata*, *Taenia solium*, and *Diphyllobothrium latum*, were confused by all scientists until the brilliant efforts of Küchenmeister, Leuckart, Mehlis, Siebold, and others in the nineteenth century determined both the external and internal anatomy of these and

other common species. These researchers also proved conclusively that bladderworms, hydatids, and coenuri were larval tapeworms and not separate species or degenerate forms in improper hosts. Although these organisms have been removed from the realms of ignorance and superstition within the past 150 years, much misconception persists.

Sexually mature tapeworms live in the intestine or its diverticulae (rarely in the coelom) of all classes of vertebrates. Only one form, *Archigetes* (Order Caryophyllidea), is known to mature in an invertebrate, that is, in the coelom of a freshwater oligochaete.

FORM AND FUNCTION
Strobila

Typically, a mature tapeworm possesses a chain of **segments** or **proglottids,** each of which contains one or more sets of reproductive organs in various stages of development. A few genera, such as *Spathebothrium* and *Cyathocephalus,* have numerous sets of reproductive organs, but there is no trace of external segmentation. Cestodes with more than one proglottid are said to be **polyzoic,** and the entire chain is called the **strobila** (Fig. 20-3). In many polyzoic tapeworms the proglottids are continuously produced near the anterior end by a process of asexual budding called **strobilization.** Each bud moves toward the posterior end as a new one takes its place and, during the process, becomes sexually mature. By the time they approach the posterior end of the strobila, the proglottids will have copulated and produced eggs. A given proglottid can copulate with itself, with others in the same strobila, or with those in other worms.

Fig. 20-1. Advertisements of this kind illustrate the low regard most people have for cestodes. (Courtesy SANE, Washington, D.C.)

After the proglottid contains fully developed eggs or shelled embryos, it is then said to be **gravid.** When it reaches the end of the strobila, it often detaches and passes intact out of the host with the feces, as *Taenia* does, or disintegrates en route, releasing the eggs, as *Hymenolepis* does. This process of detaching is called **apolysis.** In some species the eggs are released from the gravid proglottid via a uterine pore, such as in *Diphyllobothrium,* or through tears or slits in the proglottid (as in Trypanorhyncha); and the proglottid only detaches when it is senile or exhausted (**pseudopolysis** or **anapolysis**). In some forms the proglottids may be shed while immature and lead an independent existence in the gut while developing to matu-

Fig. 20-2. A *Taenia*, probably *T. saginata*, illustrated for the first time by Andry in 1718.

rity **(hyperapolysis),** as in Tetraphyllidea. If the posterior margin of a proglottid overlaps the anterior of the following one, the strobila is said to be **craspedote;** if not, it is called **acraspedote** (Fig. 20-4).

The number of proglottids in a single worm varies from two or three to several thousand, depending on the species. Some groups (such as Amphilinidea and Caryophyllidea), have only one set of reproductive organs, never budding off new proglottids. Such worms are said to be **monozoic** (Fig. 20-5). Caryophyllideans are common parasites of some freshwater fish, especially of cyprinids and catastomids (for a review, see Mackiewicz[20]).

Scolex

Most tapeworms bear a "head" or **scolex** at the anterior end that may be equipped with a variety of holdfast organs to maintain the position of the animal in the gut (Fig. 20-6). The scolex may be provided with suckers, grooves, hooks, spines, glands, or combinations of these (Fig. 20-7). However, the scolex can be simple or

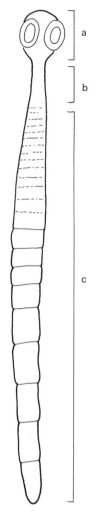

Fig. 20-3. Generalized diagram of a tapeworm, showing scolex *(a)*, neck *(b)*, and strobila *(c)*. (From Schmidt, G. D. 1970. How to know the tapeworms. William C. Brown Co. Publishers, Dubuque, Iowa.)

absent altogether. In some forms, the holdfast function of the scolex is lost early in life, and a holdfast organ is formed by a distortion of the anterior end of the strobilia called a **pseudoscolex** (Fig. 20-8). Some species penetrate the gut wall of the host to a considerable distance, with the scolex and a portion of the strobila then encapsulated by reacting host tissues.

Sucker-like organs on scolices of tapeworms can be divided roughly into three types: acetabula, bothria, and bothridia. The acetabulum is more or less cup-shaped, circular or oval in outline, and

with a heavy muscular wall. There are normally four acetabula on a scolex, spaced equally around it. Bothridia usually are in groups of four and are quite muscular, projecting sharply from the scolex and have highly mobile, leaf-like margins. Bothria are usually two in number, although up to six may occur, and take the form of shallow pits or longer grooves. Accessory suckers sometimes occur, and most cestodes have a variety of keratinaceous hooks for anchoring the scolex to the host gut. In acetabulate worms, the hooks often are arranged in one or more circles anterior to the suckers, borne on a protrusible, dome-shaped area on the apex of the scolex called the **rostellum.** Both the presence or absence and the shape and arrangement of the hooks is of great taxonomic value. If the rostellum is armed with hooks, it is supplied internally with a heavy muscular pad, which becomes flat and disc-shaped when the hooks attach to the host's gut wall. Retraction of the central area of the pad allows withdrawal of the hooks.

Various kinds of gland cells have been reported in the scolices of tapeworms. In *Echinococcus granulosus,* a cyclophyllidean with an armed rostellum, these cells are located near the anterior surface of the rostellar pad and secrete a substance to the outside that is thought to be a lipoprotein. The value of the secretion to the tapeworm is unknown. *Hymenolepis diminuta,* a cyclophyllidean with an unarmed rostellum, has secretory cells within a muscular rostellar capsule. The secretion of these cells stains as if it were neurosecretory material,[11] and it is possible that its function is regulatory. Within the anteriormost portion of the rostellum of *H. diminuta* is found a pocket-like apical organ, formed by an invagination of the apical tegument and of unknown function (Fig. 20-9). Such apical organs are found in many other cestodes, although their structural similarity and homology to that of *H. diminuta* is unknown. They are particularly prominent in the Proteocephalata where, at least in certain cases, their secretion has proteolytic activity and probably functions in penetration.[41]

The scolex contains the chief neural ganglia of the worm (as will be discussed further), and it bears numerous sensory

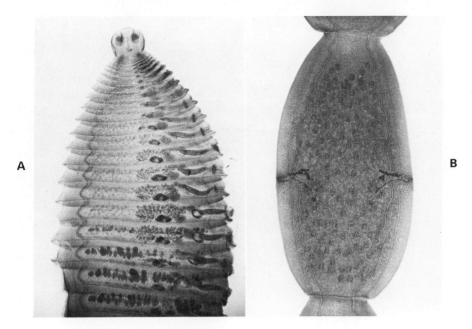

Fig. 20-4. A, *Paranoplocephala mamillana,* a craspedote cestode; **B,** *Dipylidium caninum,* an acraspedote species. (Photographs by Jay Georgi.)

endings on its anterior surface, probably detecting both physical and chemical stimuli. Such sensory input may allow optimal placement of the scolex and entire strobila with respect to the gut surface and physicochemical gradients within the intestinal milieu. The apical organ previously mentioned has been thought to function in this capacity,[36] but there is little evidence for this assumption.

Commonly, between the scolex and the strobila lies a relatively undifferentiated zone called the **neck,** which may be long or short. It contains germinal cells that apparently are responsible for budding off new proglottids. In the absence of a neck, similar cells may be present in the posterior portion of the scolex.

Tegument

Cestodes lack any trace of a digestive tract and must, therefore, absorb all required substances via their external covering. Because of this fact, the structure and function of the body covering have been of great interest to parasitologists, who have used electron microscopy and radioactive tracers to contribute much to this area of cestodology. Before 1960 the body covering of cestodes and trematodes was

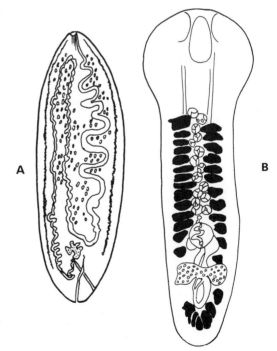

Fig. 20-5. Examples of monozoic tapeworms. **A,** *Amphilina foliacea* (Amphilinidea); **B,** *Penarchigetes oklensis* (Caryophyllidea). (From Schmidt, G. D. 1970. How to know the tapeworms. William C. Brown Co. Publishers, Dubuque, Iowa.)

Fig. 20-6. Two types of holdfast organs. **A,** Bothridea of a tetraphyllidean; **B,** spiny tentacles of a trypanorhynchan. (Photographs by Frederick H. Whittaker.)

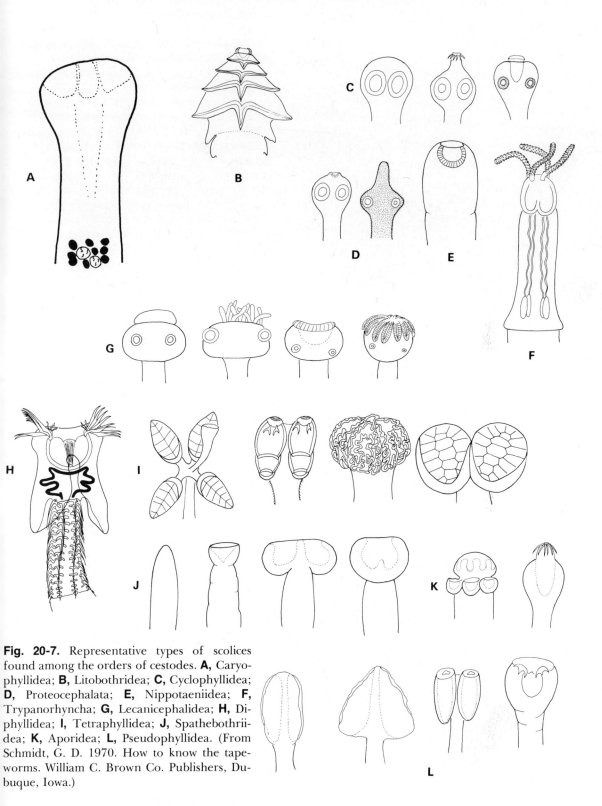

Fig. 20-7. Representative types of scolices found among the orders of cestodes. **A,** Caryophyllidea; **B,** Litobothridea; **C,** Cyclophyllidea; **D,** Proteocephalata; **E,** Nippotaeniidea; **F,** Trypanorhyncha; **G,** Lecanicephalidea; **H,** Diphyllidea; **I,** Tetraphyllidea; **J,** Spathebothriidea; **K,** Aporidea; **L,** Pseudophyllidea. (From Schmidt, G. D. 1970. How to know the tapeworms. William C. Brown Co. Publishers, Dubuque, Iowa.)

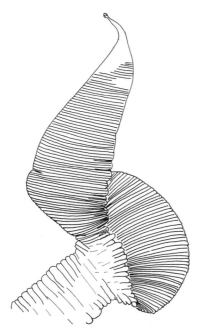

Fig. 20-8. *Fimbriaria fasciolaris,* a tapeworm with a pseudoscolex in addition to a tiny true scolex. (Adapted from Mönnig, H. O. 1934. Veterinary helminthology and entomology. William Wood, London.)

commonly referred to as a "cuticle," but it is now known that it is a living tissue with high metabolic activity, and most parasitologists prefer the term **tegument.**

Tegumental structure is generally similar in all cestodes studied, differing in details according to species. The tegument is covered by minute projections called **microtriches** (singular: microthrix) that are underlaid by the tegumental distal cytoplasm (Fig. 20-10). The distal cytoplasm is connected to perikarya via trabeculae that run through superficial muscle layer (Fig. 20-11). The microtriches are similar in some respects to the microvilli found on gut mucosal cells and other vertebrate and invertebrate transport epithelia, and they completely cover the worm's surface, including the suckers. They have a dense distal portion set off from the base by a multilaminar plate. The cytoplasm of the base is continuous with that of the rest of the tegument, and the entire structure is covered by a plasma membrane. A layer of carbohydrate-containing macromolecules, the **glycocalyx,** is found on the membrane. The microtriches serve to increase the absorptive area of the tegument,

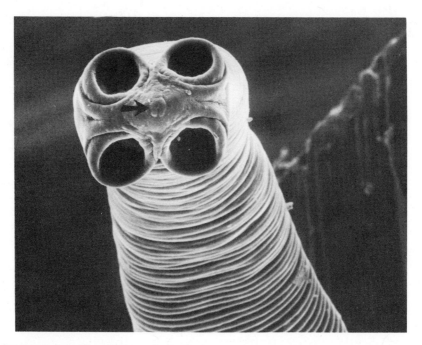

Fig. 20-9. The scolex of *Hymenolepis diminuta.* Note the apical organ (arrow). (From Ubelaker, J. E., V. F. Allison, and R. D. Specian. 1973. J. Parasitol. 59:667-671.)

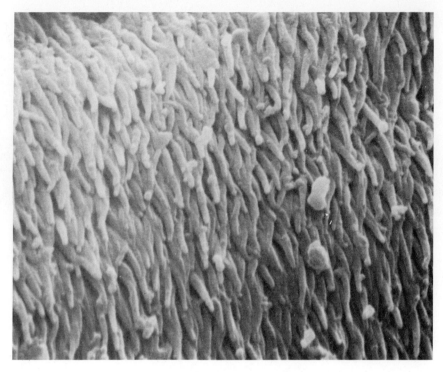

Fig. 20-10. Posteriorly-directed microtriches on the surface of a proglottid of *Hymenolepis diminuta*. (From Ubelaker, J. E., V. F. Allison, and R. D. Specian. 1973. J. Parasitol. 59:667-671.)

but the microtriches also may help the worm maintain its position in the host gut.[34] A number of phenomena, apparently depending on the absorption of certain molecules to the glycocalyx, have been reported recently; enhancement of host amylase activity; inhibition of host trypsin, chymotrypsin, and pancreatic lipase; adsorption of cations; and adsorption of bile salts. The functional value of these phenomena to the worm is uncertain, but interaction with nutrient absorption, protection against digestion by host enzymes, and maintanence of the integrity of the worm's surface membrane may be involved.

The distal cytoplasm beneath the microtriches contains abundant vesicles and electron-dense bodies, as well as numerous mitochondria. The tegumental nuclei are not found in this layer but lie in the perikarya. The vesicles are secreted in the perikarya, are passed to the distal cytoplasm via the trabeculae, and at least some of them contribute to microthrix and hook

formation.[18,22] Although each perikaryon contains but one nucleus, the distal cytoplasm is continuous, with no intervening cell membranes; therefore, the tegument of cestodes is a syncytium.

Muscular system

Just internal to the distal cytoplasm of tapeworms are circular and longitudinal muscle layers called superficial muscles. More powerful musculature lies below the superficial muscles. These longitudinal muscles are usually arranged around a central mesenchymal area, largely free of longitudinal fibers. They may extend to the tegumental perikarya or they may leave a zone of cortical mesenchyme, which is also free of longitudinal fibers. There are numbers of dorsoventral and transverse fibers, and sometimes radial fibers as well. The pattern and relative development of muscle bundles are highly variable in the Cestoidea, but constant within a species; therefore, they are often valuable taxonomic characters. Internal

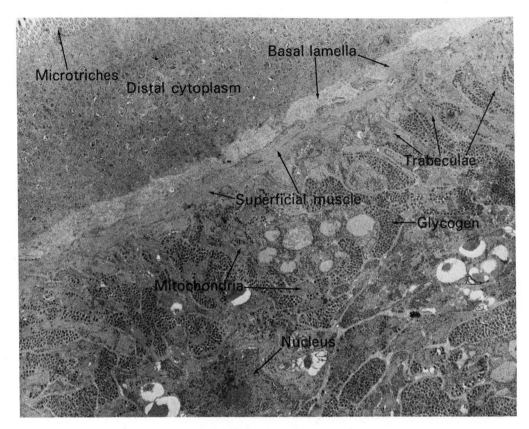

Fig. 20-11. Low magnification electron micrograph of *Hymenolepis diminuta* tegument. Small area of surface with microtriches is shown at upper left. Note the large number of vesicular inclusions in the distal cytoplasm, the basal lamella interrupted by trabeculae from perikarya, the large number of glycogen particles in the parenchymal cells, and the ribosomes and endoplasmic reticulum in the tegumental perikarya. The nucleus indicated is in a tegumental perikaryon. (×6,000.) (Photograph by R. D. Lumsden.)

musculature of the scolex is very complex, making it extraordinarily mobile.

Like those of other platyhelminths, cestode muscles are nonstriated and lack transverse sarcolemmal tubules (T-tubules),[17] as might be expected of muscles with slow contraction.

Nervous system

The main nerve center of the cestode is in its scolex, and the complexity of ganglia, commissures, and motor and sensory innervation there depends on the number and complexity of other structures on the scolex. Among the simplest are the bothriate cestodes such as *Bothriocephalus*, which have only a pair of lateral cerebral ganglia united by a single ring and a transverse commissure. Arising from the cerebral ganglia are a pair of anterior nerves, supplying the apical region of the scolex; four short posterior nerves; and a pair of lateral nerves that continue on posteriorly through the strobila. The bothria are innervated by small branches from the lateral nerves. In contrast, worms with bothridia or acetabula and hooks, rostellum, and so on may have a substantially more complex system of commissures and connectives in the scolex, with three to five longitudinal nerves running posteriorly from the cerebral ganglia through the strobila (Fig. 20-12). In addition to the motor innervation of the scolex, there may

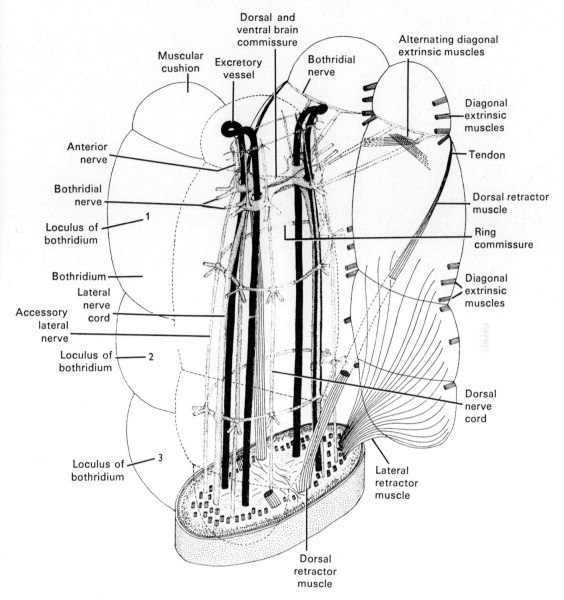

Fig. 20-12. *Acanthobothrium coronatum.* Reconstruction of nervous system of scolex. Excretory vessels and some muscles included. (From Rees, G., and H. H. Williams. 1965. Parasitology 55:617-651.)

be many sensory endings, particularly at the apex of the tegument. Stretch receptors have been described.[26a]

As the longitudinal nerves proceed posteriad, they are connected by interproglottidal commissures in a ladder-like fashion. Smaller nerves emanate from them to supply the general body musculature and sensory endings. The cirrus and vagina are richly innervated, and sensory endings around the genital pore are more abundant than in other areas of the strobilar tegument.[40] Such an arrangement has obvious value.

Study of the neuroanatomy of cestodes is difficult because the nerves are unmyelinated and do not stain well with conventional histological stains. However, certain

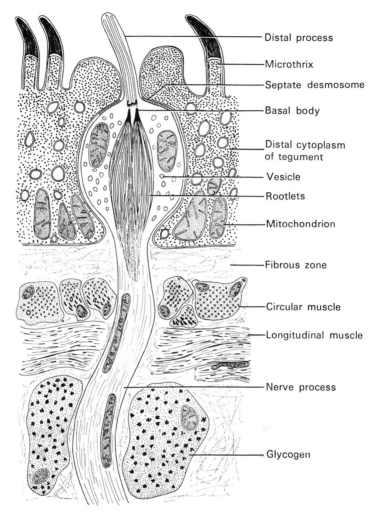

Distal process

Microthrix

Septate desmosome

Basal body

Distal cytoplasm of tegument

Vesicle

Rootlets

Mitochondrion

Fibrous zone

Circular muscle

Longitudinal muscle

Nerve process

Glycogen

Fig. 20-13. Schematic drawing of a longitudinal section through a sensory ending in the tegument of *Echinococcus granulosus*. (From Morseth, D. J. 1967. J. Parasitol. 53:492-500.)

histochemical techniques that show sites of acetylcholinesterase activity delineate the nervous systems of cestodes beautifully, even the sensory endings in the tegument. This fact, as well as studies on the actions of cholinesterase inhibitors, lends support to the conclusion that neural transmission in cestodes is essentially cholinergic.[34]

Sensory function probably includes both tactoreception and chemoreception, and cestodes possess at least two, and perhaps more, morphologically distinct types of sensory endings in their tegument.[6] One of these has a modified cilium projecting as a terminal process (Fig. 20-13), as is common in such cells in invertebrates.

Osmoregulatory system

In many families of cestodes, the main "osmoregulatory" canals run the length of the strobila from the scolex to the posterior end. These are usually in two pairs, one ventrolateral and the other dorsolateral on each side (Fig. 20-14). Most often the dorsal pair is smaller in diameter than the ventral pair, a useful criterion for determining the dorsal and ventral sides of a tapeworm. The canals may branch and rejoin throughout the strobila or may be independent. Usually a transverse canal joins the ventral canals at the posterior margin of each proglottid. The dorsal and ventral canals unite in the scolex, often

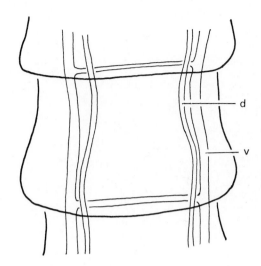

Fig. 20-14. Diagram showing the typical arrangement of dorsal *(d)* and ventral *(v)* osmoregulatory canals. (From Schmidt, G. D. 1970. How to know the tapeworms. William C. Brown Co. Publishers, Dubuque, Iowa.)

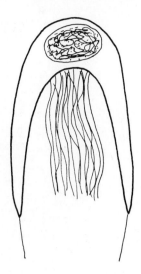

Fig. 20-15. Diagrammatic representation of a flame cell protonephridium. (From Schmidt, G. D. 1970. How to know the tapeworms. William C. Brown Co. Publishers, Dubuque, Iowa.)

with some degree of branching. Posteriorly, the two pairs of canals merge into an excretory bladder with a single pore to the outside. When the terminal proglottid of a polyzoic species is detached, the canals empty independently at the end of the strobila. Rarely, the major canals also empty through short, lateral ducts. In some orders, such as Caryophyllidea and Pseudophyllidea, the canals form a network that lacks major dorsal and ventral ducts.

Embedded throughout the parenchyma are flame cell protonephridia (Fig. 20-15), whose ductules feed into the main canals. The cilia of the flame cell presumably provide some motive force to the fluid in the system.

Although the system is often called "osmoregulatory," it is also commonly referred to as "excretory," and there is some evidence that the latter is more descriptive. In fact, the cestodes tested so far have little ability to osmoregulate; that is, they behave like osmometers, changing their body volume when placed in a medium with a different osmotic concentration. DeRycke maintained that *Hymenolepis microstoma* could regulate more effectively in hyperosmotic than hyposmotic conditions; when the worms were placed in hyposmotic solutions, there was an initial, rapid gain in

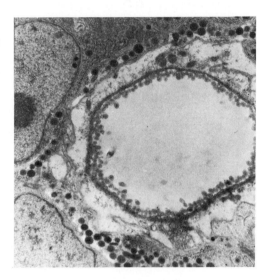

Fig. 20-16. Low magnification electron micrograph of excretory duct of *Hymenolepis diminuta* showing beadlike microvilli.

weight, followed by a small, much slower weight loss.[12] An analysis of fluid from the excretory canals of *Hymenolepis diminuta* demonstrated glucose, soluble proteins, lactic acid, urea, and ammonia; lipid was absent.[38]

In some cases, at least, the excretory ducts are lined with microvilli (Fig. 20-16),

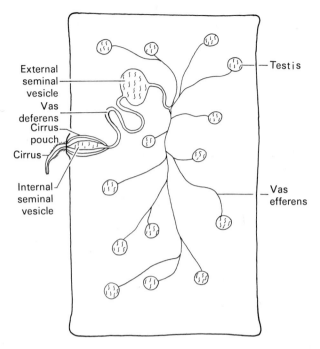

Fig. 20-17. Diagrammatic representation of a tapeworm male reproductive system. (From Schmidt, G. D. 1970. How to know the tapeworms. William C. Brown Co. Publishers, Dubuque, Iowa.)

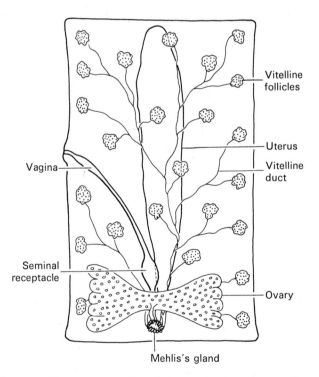

Fig. 20-18. Diagram of a tapeworm female reproductive system. (From Schmidt, G. D. 1970. How to know the tapeworms. William C. Brown Co. Publishers, Dubuque, Iowa.)

thus suggesting that the duct linings serve a transport function. Therefore, possible functions of the system might include active transport of excretory wastes and resorption of substances such as ions from the excretory fluid.

Reproductive systems

Tapeworms are monoecious with the exception of a few rare species from birds and one from a stingray that are dioecious. Usually each proglottid has one complete set of both male and female systems, but some genera have two sets of each system per segment, and a few species in birds have one male and two female systems in each proglottid.

As a segment moves toward the posterior of the strobila, the reproductive systems mature, sperm are transferred, and ova are fertilized. Usually the male organs mature first and produce sperm that are stored until maturation of the ovary; this is called **protandry,** or **androgyny.** In a few species the ovary matures first; this is called **protogyny** or **gynandry.** This may be an adaptation that avoids self-fertilization of the same proglottid. Many variations occur in structure, arrangement, and distribution of reproductive organs in tapeworms. These variations are useful at all levels of taxonomy.

The male reproductive system (Fig. 20-17) consists of one to many testes, each of which has a fine vas efferens. The vasa efferentia unite into a common vas deferens that channels the sperm toward the genital pore. The vas deferens may be a simple duct, or it may have sperm storage capacity in convolutions or in a spheroid external seminal vesicle. As the vas deferens leads into the cirrus pouch, which is a muscular sheath containing the terminal organs of the male system, it may form a convoluted ejaculatory duct or dilate into an internal seminal vesicle. The male copulatory organ is the muscular cirrus, which may or may not bear spines. It can invaginate into the cirrus pouch and evaginate through the cirrus pore. Commonly the reproductive pores of both sexes open into a common sunken chamber, the genital atrium, which may be simple or equipped with spines, stylets, glands, or accessory pockets. The cirrus pore may open on the mar-

gin or somewhere on the flat surface of the proglottid. If two male systems are present, they open on margins opposite from one another.

The female reproductive system (Fig. 20-18) consists of an ovary and associated structures, which are variable in size, shape, and location, depending on the genus. The entire complex is called the oogenotop. Vitelline cells, which contribute yolk and eggshell material to the embryo, may be arranged into a single, compact vitellarium or they may be scattered as follicles in various patterns. Ova are called **ectolecithal,** for they do not produce their own yolk. As oocytes mature, they leave the ovary through a single oviduct, which usually has a controlling sphincter, the **oocapt.** Fertilization occurs in the proximal oviduct. One or more cells from the vitelline glands pass through a common vitelline duct, sometimes equipped with a small vitelline reservoir, and join with the zygote. Together they pass into an area of the oviduct known as the ootype. This zone is surrounded by unicellular glands, called Mehlis's glands, which appear to secrete a thin membrane around the zygote and its associated vitelline cells. Eggshell formation is then completed from within by the vitelline cells. Some eggs, such as those in the Pseudophyllidea, Tetraphyllidea, and Trypanorhyncha, are covered by a thick capsule of sclerotin. These embryonate in water after passing from the host and usually hatch to release a free-swimming larval stage that is eaten by the aquatic intermediate host. Other egg types, as found in the Cyclophyllidea, have a thin capsule or no capsule at all but may have a thick embryophore or other covering. They pass from the host fully embryonated and must be eaten by their intermediate host before they hatch. Pseudophyllidean type eggs (Fig. 20-19), designated as group I, are homologous with trematode eggs and formed in quite a similar manner. Group II, cyclophyllidean eggs, are of three different types: (1) *Dipylidium* type with a thin capsule and an embryophore (as in *Dipylidium, Moniezia,* and *Hymenolepis*); (2) *Taenia* type with no capsule but with a thick embryophore (as in *Taenia* and *Echinococcus*); and (3) *Stilesia* type, formed by species with no vitellaria,

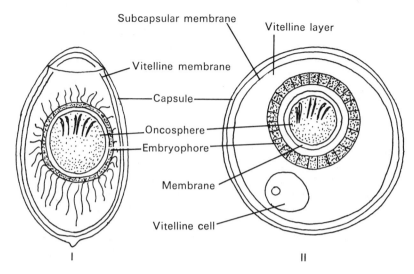

Fig. 20-19. Morphology of group I (Pseudophyllidean) and group II (Cyclophyllidean) eggs.

with cellular covering laid down by the uterine wall.[34]

In both *Dipylidium* and *Taenia* type eggs of group II, a single vitelline cell envelopes the zygote; in the *Dipylidium* type, a globule of material, which becomes the capsule, is released from the vitelline cell. Some parasitologists believe the capsule is sclerotin, though there is disagreement on this point. No capsule is formed by *Taenia* type eggs, but there is a thin, fragile outer membrane. In both types there is an embryophore of keratin formed by peripheral embryonic cells.[31]

As the vitelline cell (or cells) becomes associated with the zygote, the incipient egg passes through the ootype where the secretions of Mehlis's gland are added. In the past Mehlis's gland was often called the "shell gland" because it was thought to contribute shell material. But since later work has shown that this is not so, the function of Mehlis's gland is still uncertain. The mucous cells in Mehlis's gland may produce material to form the ground lamella of the capsule or embryophore, and serous cells produce a surface-active substance that helps spread the presclerotin globules released from the vitelline cells. Leaving the ootype, the developing larva passes into the uterus where embryonation is completed. The form of the uterus varies considerably between groups. It

may be reticulated, lobulated, or circular; or it may be a simple sac or a simple or convoluted tube; or it may be replaced by other structures. In some tapeworms the uterus disappears and the eggs, either singly or in groups, are enclosed within hyaline **egg capsules** embedded within the parenchyma. In some species one or more fibromuscular structures called **paruterine organs** form, attached to the uterus. In this case, the eggs pass from the uterus into the paruterine organ, which then assumes the functions of a uterus. The uterus then usually disintegrates.

GENERAL BIOLOGY OF TAPEWORMS

Nearly every tapeworm whose life cycle is known has an **indirect life cycle.** One notable exception with a **direct life cycle** is *Hymenolepis nana,* a cyclophyllidean parasite of mice and humans, which can complete its larval stages within the definitive host (see p. 371). Complete life cycles are known for only a comparatively few species of tapeworms. In fact, there are several orders in which not a single life cycle has been determined. Among the life cycles that are known, much variety exists in larval forms and patterns of development.

Sexually mature tapeworms live in the intestine or its diverticulae or, rarely, in the coelom of all classes of vertebrates. As

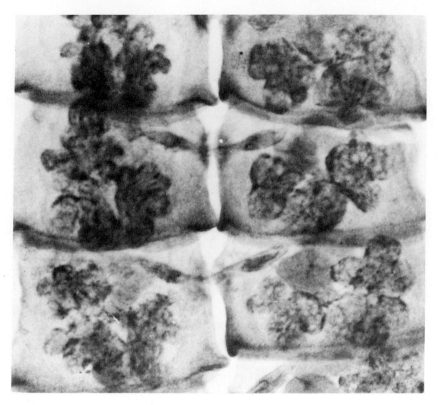

Fig. 20-20. Cross-insemination of two strobilas of *Sobelevacanthus coloradensis*. (Photograph by William B. Ahern.)

stated earlier, the genus *Archigetes* (Caryophyllidea, Caryophyllaeidae) can mature in the coelom of freshwater oligochaetes. A mature tapeworm may live for a few days or up to many years, depending on the species. During its reproductive life, a single worm produces from a few to millions of eggs, each with the potential of developing into the adult form. Because of the great hazards obstructing the course of transmission and development of each worm, mortality is very high.

Most tapeworms are hermaphroditic and are capable of fertilizing their own eggs. Sperm transfer is usually from the cirrus to the vagina of the same segment or between adjacent strobilae, if the opportunity affords (Fig. 20-20). A few species are known in which a vagina is absent; **hypodermic impregnation** has been observed in some of these. In such cases, the cirrus is forced through the body wall, and the sperm are deposited within the paren-

chyma (Fig. 20-21). It is not known how they find their way into the seminal receptacle.

In many apolytic species, especially those in elasmobranchs, the proglottids detach when mature and lead independent existences within the gut of the host, copulating with each other on contact. The eggs then develop and are shed, as in anapolytic species.

A few species of tapeworms are known to be dioecious. In these cases, it is not clear what triggers the sexuality of a given strobila, for it appears that each has the potential of maturing as either male or female. Interaction between two or more strobilae is important in sex determination of dioecious forms. For example, in *Shipleya* (Cyclophyllidea, Dioecocestidae), if a single strobila is present in a host, it is usually female; if two are present one is nearly always a male. Similar phenomena are known in certain flukes and nema-

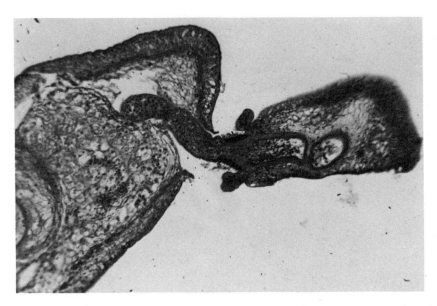

Fig. 20-21. Hypodermic impregnation of *Dioecotaenia cancellatum*. A cross-section view: the smaller male proglottid (right) has pierced the female proglottid (left) with its cirrus.

Fig. 20-22. Diagram of a ciliated coracidium. (From Schmidt, G. D. 1970. How to know the tapeworms. William C. Brown Co. Publishers, Dubuque, Iowa.)

todes. Much opportunity for research exists in this aspect of cestode biology.

Both invertebrates and vertebrates serve as intermediate hosts of tapeworms. Nearly every group of invertebrates has been discovered harboring larval cestodes, but the most common are crustaceans, insects, molluscs, mites, and annelids. As a general rule, when a tapeworm occurs in an aquatic definitive host, the larval forms are found in aquatic intermediate hosts. A similar assumption can be made for terrestrial hosts.

Vertebrate intermediate and paratenic hosts are found among fishes, amphibians, reptiles, and mammals. Tapeworms found in these hosts normally mature within predators whose diets include the intermediary.

Life cycles

Among the life histories that are known, much variety exists in larval forms and details of development, but there seem to be two basic patterns; the others apparently being variations on these themes. Since cestodes with one or the other life cycle types have eggs corresponding with groups I and II, we will adopt these designations for the developmental patterns; that is, group I, typified by Pseudophyllidea and several other orders, and group II, containing the Cyclophyllidea.

Eggs of cestodes in group I must reach water after passing from the host in the feces. They embryonate and hatch. The larva emerging from the egg has a ciliated epithelium and is called a **coracidium** (Fig. 20-22). The coracidium must be eaten by the first intermediate host, usually an arthropod, within a short time. There it sheds its ciliated epithelium and actively

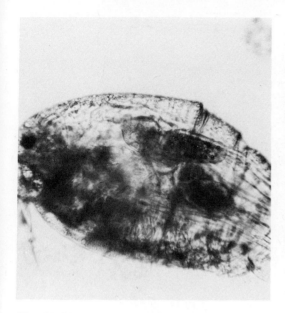

Fig. 20-23. A copepod with two procercoids in its hemocoel. Note the pale-staining cercomer on the right end of the upper worm. (Photograph by Robert W. Mead.)

uses its six hooks to penetrate the gut of its host. In the hemocoel it metamorphoses into a **procercoid** (Fig. 20-23). This involves some degree of elongation and reorganization, while the embryonic hooks are relegated to the posterior in a structure known as the **cercomer.** The procercoid is defined as the stage in which the embryonic hooks are still present but the definitive holdfast has not developed. When the first intermediate host is consumed by the second intermediate host—often a fish—the procercoid penetrates the host gut into the peritoneal cavity and mesenteries and then commonly into the skeletal muscles. It undergoes a gradual metamorphosis to a **plerocercoid** (Fig. 20-24), that is, it loses the embryonic hooks (and cercomer, if present) and develops rudiments of the holdfast it will have as an adult. It should be pointed out that what has been presented is a "typical," perhaps ancestral, group I life cycle, and there is a considerable telescoping and tendency toward neoteny in various worms in this group. For example, several proteocephalans have been described in which the holdfast develops in the first intermediate host before the embryonic hooks are lost. In pseudophyllideans *Ligula* and *Schistocephalus,* development as plerocercoids proceeds so far that little growth occurs when these worms reach the definitive host, and the gonads mature within 72 hours and start producing eggs within 36 hours.[4] Most caryophyllideans could probably be regarded as neotenic plerocercoids (no proglottidization), and *Archigetes,* mentioned before, could be considered a neotenic procercoid (reproduction in the invertebrate host with the cercomer still present).

Group II life cycles are found in that large order, the Cyclophyllidea. In these, the eggs are fully embryonated and infective when they pass from the definitive host, but they do not hatch until eaten by an intermediate host. The larva that hatches from the egg is called an **oncosphere** (Fig. 20-25), or **hexacanth,** but is not a coracidium. It penetrates the gut of its host and metamorphoses and grows into a juvenile called the **cysticercoid** (Fig. 20-26 and 20-27). The cysticercoid is a solid-bodied organism with a fully developed scolex invaginated into its body. It is surrounded by cystic layers, and the cercomer, which contains the larval hooks, is outside the cyst. If not displaced mechanically, the cercomer will be digested away, along with parts of the cyst, in the gut of the definitive host. A few cysticercoids have been described that undergo asexual reproduction by budding. When cestode larvae are infective to a definitive host, they are often called **metacestodes.** Several schemes for naming different types of cestode larvae have been proposed. We prefer to retain a simple, classical approach.

Contrasting juveniles are produced in vertebrate intermediate hosts by the cyclophyllidean family Taeniidae. These are encysted forms called **cysticerci** (Fig. 21-10). They differ from the cysticercoid in that the scolex is *introverted* as well as invaginated, and the scolex forms on a germinative membrane enclosing a fluid-filled bladder. Several variations from the simple cysticercus in the Taeniidae undergo asexual reproduction by budding (as will be discussed further). They are of considerable medical and veterinary importance.

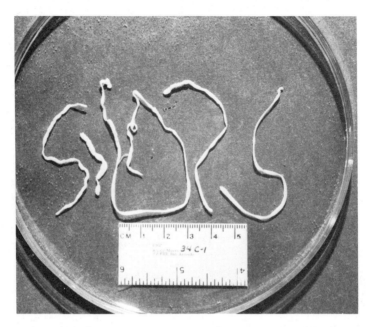

Fig. 20-24. Pleroceroids from the musculature of a vervet, an African primate. (Photograph by Robert E. Kuntz.)

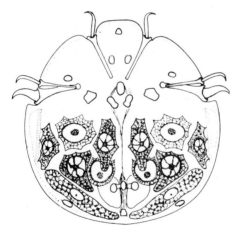

Fig. 20-25. Diagram of oncosphere of *Hymenolepis diminuta,* dorsal view. (From Ogren, R. E. 1967. Trans. Am. Microsc. Soc. 86:250-260.)

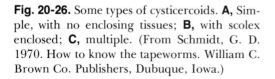

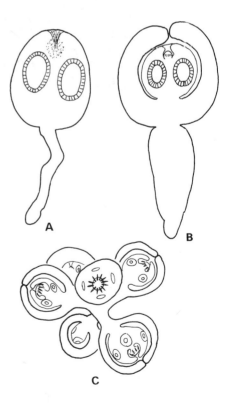

Fig. 20-26. Some types of cysticercoids. **A,** Simple, with no enclosing tissues; **B,** with scolex enclosed; **C,** multiple. (From Schmidt, G. D. 1970. How to know the tapeworms. William C. Brown Co. Publishers, Dubuque, Iowa.)

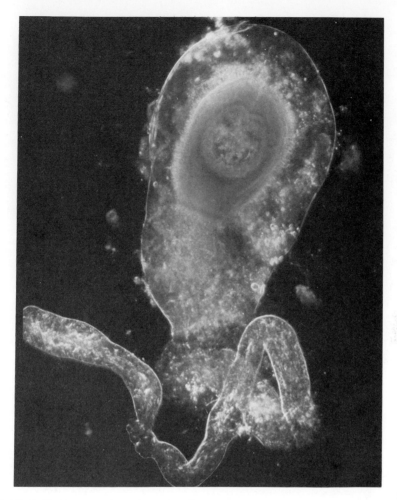

Fig. 20-27. Fully developed cysticercoid of *Hymenolepis diminuta.* (From Voge, M. 1969. In Schmidt, G. D., editor. Problems in systematics of parasites. University Park Press, Baltimore.)

There are numerous other sorts of larvae and juveniles that can be distinguished from the typical forms described previously, but they are, for the most part, simply modifications of the types:

1. **Decacanth** or **lycophora** (Fig. 20-28) —the ten-hooked larva that hatches from the egg in Cestodaria, often ciliated (for example, Amphilina).

2. **Sparganum**—a term originally proposed to be applied to any pseudophyllidean plerocercoid of unknown species, but now usually used for plerocercoids of the genus *Spirometra.*

3. **Plerocercus** (Fig. 20-29)—a modified plerocercoid found in some Trypanorhyncha, in which the posterior forms a bladder, the blastocyst, into which the rest of the body can withdraw (as in *Gilquinia*).

4. **Strobilocercoid**—a cysticercoid that undergoes some strobilization, found only in *Schistotaenia.*

5. **Tetrathyridium** (Fig. 20-30)—a fairly large solid-bodied juvenile that can be regarded as a modified cysticercoid, developing in vertebrates that have ingested the cysticercoid encysted in the invertebrate host. It is known only in the atypical cyclophyllidean *Mesocestoides.*

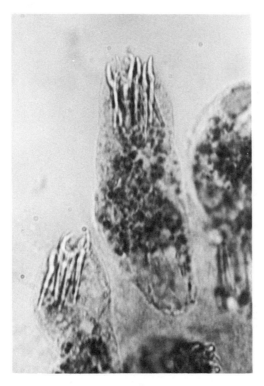

Fig. 20-28. Lycophora larvae, found only in the Cestodaria. (Photograph by Warren Buss.)

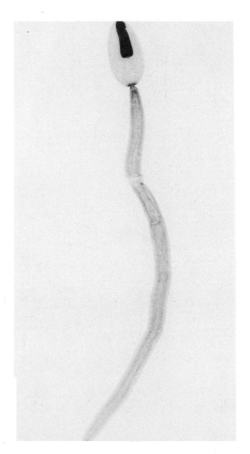

Fig. 20-29. A plerocercus larva of a trypanorhynchan cestode. In this species the blastocyst, which contains the scolex, bears a long caudal appendage. (Photograph by Warren Buss.)

6. Variations on cysticercus.
   a. **Strobilocercus** (Fig. 20-31)—a simple cysticercus in which some strobilization occurs within the cyst (for example *Taenia taeniaeformis*).
   b. **Coenurus** (Figs. 20-32 and 21-18) —budding of a few to many scolices (called **protoscolices**) from the germinative membrane of the cyst, each on a simple stalk invaginated into the common bladder (as in *Taenia multiceps*).
   c. **Unilocular hydatid** (Fig. 21-20)— up to several million protoscolices present; occasional · sterile specimens. Usually there is an inner or **endogenous budding** of brood cysts, each with many protoscolices inside. **Exogenous budding** rarely occurs, resulting in two more hydatids called **daughter cysts.** This form may grow very large, sometimes containing several quarts of fluid. Occasionally, many proto-

scolices break free and sink to the bottom of the cyst, forming **hydatid sand** (Fig. 21-22), but this is probably rare in the living, normal cyst. This larval form is known only for the cyclophyllidean genus *Echinococcus*.
   d. **Multilocular** or **alveolar hydatid** (Fig. 21-24)—known only for *Echinococcus multilocularis*, exhibiting extensive exogenous budding, when in an abnormal host such as humans; resulting in an infiltration of host tissues by numerous cysts. It forms a single mass with many little pockets that contain protoscolices when in a normal host.

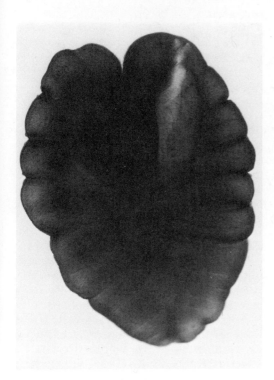

Fig. 20-30. A tetrathyridium larva, typical of the family Mesocestoididae. (Photograph by Warren Buss.)

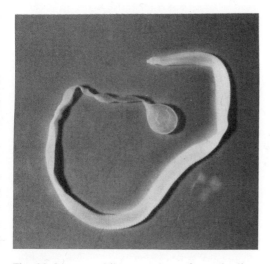

Fig. 20-31. A strobilocercus larva from the liver of a rat. Note the small bladder at the posterior end. (Photograph by James Jensen.)

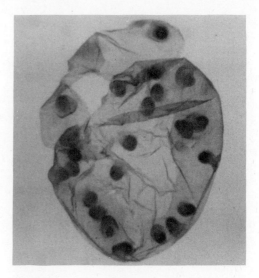

Fig. 20-32. A coenurus larva. Each round body in the bladder is an independent scolex. (Photograph by Warren Buss.)

Development in the definitive host

As with many other areas of parasitology, generalizations regarding this phase of development may be ill advised because detailed studies of relatively few species are available. However, substantial data have accumulated for those species that have been examined.

When the juvenile tapeworm reaches the small intestine of its definitive host, certain stimuli cause it to excyst, evaginate, or both, and begin growth and sexual maturation. In encysted forms, action of digestive enzymes in the host's gut may be necessary to at least partially free the organism from its cyst. In *Hymenolepis diminuta,* most of the cyst wall may be removed by treatment with pepsin then with trypsin, but few worms will evaginate and emerge from the cyst unless bile salts are present.[30] In fact, bile salts, for example, sodium taurocholate or glycocholate, appear to furnish an almost essential stimulant for excystment in this species; a high proportion of worms will excyst in 5 mM sodium taurocholate alone, without any enzymatic treatment. The effect does not seem to be the result of the surface-active properties of the bile salt, since a variety of detergents do not stimulate excystment. This is not the case in all tapeworms, however, since Campbell reported that

Haemo-Sol, a commercial detergent, stimulated cysticerci of *Taenia pisiformis* to evaginate.[10]

Smyth has emphasized the possible role of bile salts in determining host specificity.[33] He found that the protoscolices of *E. granulosus* rapidly lysed in solutions of deoxycholic acid, and bile from rabbits and sheep, rich in deoxycholic acid, had the same effect. Bile from normal definitive hosts of *E. granulosus,* such as foxes and dogs, has relatively little deoxycholic acid and does not lyse the protoscolices. Although deoxycholate is a fairly harsh compound and is often used as a lytic agent in the laboratory, rabbits and sheep are well supplied with their own tapeworms, and these parasites clearly are unharmed by the bile of their hosts. Without doubt a combination of factors interact to determine whether a tapeworm can develop in a given species of host. Among these are conditions which stimulate strobilization and maturation. The identity of such conditions is largely unknown, though in some instances they may be quite simple. In some pseudophyllideans with a well-developed strobila in the plerocercoid (for example, *Ligula, Schistocephalus*), an increase in temperature to that of their definitive host is all that is required for them to mature.[5] In *Echinococcus* it has been shown that contact of the rostellum with a suitable protein substrate is necessary to induce strobilar growth.[35]

As strobilar development begins, subsequent events are influenced by a variety of conditions, including size of the infecting juvenile, species of the worm and host, size of the host, diet of the host, presence of other worms, and so on. Under optimal conditions, certain species have a burst of growth that must surely rival growth rates found anywhere in the animal kingdom. *Hymenolepis diminuta* (a tapeworm of rats and one of the best studied of all cestodes) can increase its weight by up to 1.8×10^6 times within 15 to 16 days.[27] Such rapid growth, accompanied by strictly organized differentiation, makes this worm a fascinating system for the study of development, particularly so since the course of the growth may be altered experimentally. The growth of the worm is especially sensitive to the composition of the host diet with respect to carbohydrate. The situation is best known for *H. diminuta,* but the findings can be extended to other tapeworms, to some extent at least. *H. diminuta* apparently has a high carbohydrate requirement, but it can only absorb glucose and, to a lesser degree, galactose across its tegument. This is true for other cestodes tested, though some can absorb a limited number of other monosaccharides and disaccharides.[9] For optimal growth, the carbohydrate must be supplied in the host diet in the form of *starch,* so that the glucose will be released as the digestion proceeds in the host gut. If glucose per se—or a disaccharide containing glucose, such as sucrose—is furnished in the host diet, the worm is placed at a competitive disadvantage for glucose with respect to the gut mucosa, physiological conditions in the gut are altered, or both, such that the worm's growth is substantially restrained.

Another important condition affecting worm growth is the presence of other tapeworms in the gut, the so-called crowding effect. This may be viewed as an interesting adaptation by which the parasite biomass is adjusted to the carrying capacity of the host. Again, though best known in *H. diminuta,* there is evidence that the crowding effect occurs in at least several other species.[26] Within certain limits, the weight of the individual worms in a given infection is, on the average, inversely proportional to the number of worms present. In consequence, the total worm biomass and the number of eggs produced is the same and is maximal for that host, regardless of the number of worms present. In the case of *H. diminuta,* whether the rat is infected with a large number of worms all at once or acquires them a few at a time is immaterial; earlier worms will reduce their size as later worms establish and population density increases.[28] The operational mechanism of the crowding effect is of considerable biological interest as a mode of developmental control. The most widely held view has been that the individual worms compete for available host dietary carbohydrate. However, the means by which the competition might be translated into lower rates of cell division and cell growth have not been elucidated, and simple starvation may not be an adequate explanation.[7]

As the worm approaches maximal size,

growth rate decreases, and production of new proglottids is only sufficient to replace those lost by apolysis. Although some species, such as *Hymenolepis nana,* characteristically become senescent and pass out of the host after a period, others may be limited only by the length of their host's life. *Taeniarhynchus saginata* has been known to live in a human for over 30 years, and *H. diminuta* lives as long as the rat it inhabits. In fact, Read reported an "immortal" worm that he kept alive for 14 years by periodically removing it from its host, severing the strobila in the region of the germinative area, then surgically reimplanting the scolex in another rat.[25]

Finally, it should be noted that some tapeworms manage a surprising degree of mobility within their host's intestine. It has been known for some time that cestodes may establish initially in one part of the gut, then move to another as they grow. In fact, *Diphyllobothrium dendriticum* in rats passes all the way to the large intestine within a few hours of infection, but less than 24 hours later it moves back to the duodenum to start growth.[4] More recently, it has been discovered that *H. diminuta* actually undergoes a diurnal migration in the rat's gut (Fig. 20-33). This migration is related to the nocturnal feeding habits of rats and can be reversed by giving the rat food only in the daytime.

METABOLISM
Acquisition of nutrients

All nutrient molecules of a tapeworm are absorbed across its tegument by active transport, mediated diffusion, or simple diffusion, but probably not by pinocytosis.[19] Absorption of nutrients across the body surface has been reviewed recently by Pappas and Read (1975).

Glucose seems to be the most important nutrient molecule to fuel energy processes in tapeworms. As noted before, the only carbohydrates that most cestodes can absorb are glucose and galactose, and, though some tapeworms can absorb other monosaccharides and disaccharides, only glucose and galactose are known actually to be metabolized. Although the question has not been thoroughly investigated, the primary fate of galactose seems to be incorporation into membranes or other structural components, such as glycocalyx.[23] Both glucose and galactose are actively transported and accumulated in the worm against a concentration gradient. Of the two sugars, glucose has been studied more extensively. Glucose influx in a number of species is coupled to a sodium pump mechanism, that is, the maintenance of a sodium concentration difference across the membrane. The accumulation of glucose, in *H. diminuta* at least, is also sodium dependent. At least two transport sites for glucose are kinetically distinct in the tegument of *H. diminuta,* and the relative proportion of these sites changes during development (Starling, 1975).

Amino acids are also actively transported and accumulated, though less is known about them than about glucose. However, efflux of amino acids from the worm is stimulated by the presence of other amino acids in the ambient medium; therefore, the worm pool of amino acids rapidly comes to equilibrium with the amino acids in the intestinal milieu.

Purines and pyrimidines are absorbed by facilitated diffusion, and the transport locus is distinct from the amino acid and glucose loci.[19a]

The actual mechanism of lipid absorption has not been investigated, but it is likely to be a form of diffusion. Fatty acids, monoglycerides, and sterols are absorbed at a considerably greater rate when they are in a micellar solution with bile salts.[5a]

Only two cases of the requirements for external supplies of the various vitamins are known. Investigations of vitamin requirements are difficult, as they are in some other parasites, because of limitations in in vitro cultivation techniques, because the worm may be less sensitive than its host to a vitamin-deficient diet, or both. In any case, the pathogenesis of vitamin deficiency in the host may have indirect effects on the worm. The necessity for an external supply of a vitamin has been demonstrated unequivocally in only one case, that of pyridoxine and *H. diminuta.*[24,29] By inference we can assume that *Diphyllobothrium latum* has a requirement for vitamin B_{12}, since the worm accumulates unusually large amounts of it.[8] In some cases *D. latum* can compete so successfully with its host for the vitamin that the worm causes pernicious anemia (see Chapter 21).

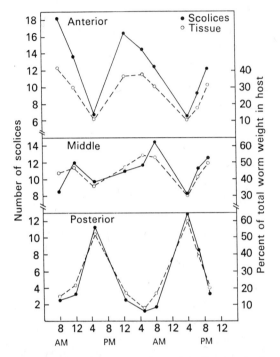

Fig. 20-33. The distribution of scolices and of wet tissue of *H. diminuta* in the host intestine at various times of the day. "Anterior" refers to the first 10 inches, "middle" to the second 10 inches, and "posterior" to the remainder of the small intestine. Each point is the mean of determinations from four host animals, representing 110 to 120 worms. (From Read, C. P. 1970. J. Parasitol. 56:643-652.)

Fig. 20-34. A branched chain electron transport system with cytochrome o, facultatively transporting electrons to fumarate or oxygen. There is evidence that a similar system operates in *Moniezia, Taenia,* and probably other cestodes. (Solid lines represent the major pathway; dotted lines represent the minor pathway.) (Adapted from Bryant, C. 1970. In Dawes, B., editor. Advances in parasitology, vol. 8. Academic Press, Inc., New York.)

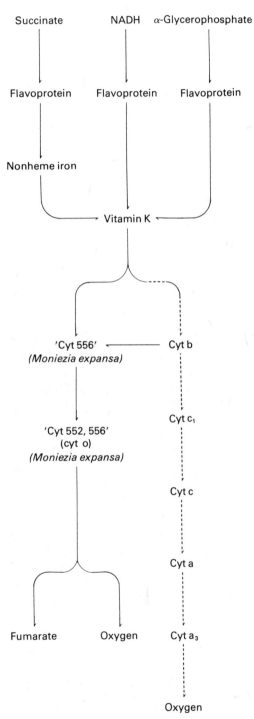

Energy metabolism

As in a number of other endoparasites that have been investigated, cestodes apparently derive their energy predominantly by anaerobic processes, even in the presence of oxygen. They do take up oxygen when it is available, but it is not clear that the function of oxygen is as a terminal electron acceptor in an energy-producing series of reactions (for example, oxidative phosphorylation via the "classical" cytochrome system). Although earlier research indicated that some cytochromes (b, c, and cytochrome oxidase) might be present in some cestodes, later research failed to confirm that a cytochrome system was operating, and the function of such cytochromes as were present was a mystery. Use of more sensitive techniques now has provided evidence that a classical mammalian type electron transport system is present in at least some cestodes but that the classical chain is probably of minor importance (Fig. 20-34) and that the major cytochrome system was a so-called o-type, similar to that reported in many bacteria. The significance of this system is that it seems to be an adaptation to facultative anaerobiosis. The terminal oxidase can transfer electrons to either fumarate or oxygen, depending on whether conditions are aerobic or anaerobic, and the products are either succinate or hydrogen peroxide, respectively. A peroxidase destroys the hydrogen peroxide before it reaches toxic levels. Under anaerobic conditions, the succinate formed in this pathway would be excreted; another mechanism for succinate formation and excretion, even under aerobic conditions, will be described later.

A complete Krebs tricarboxylic acid cycle apparently does not function in adult cestodes. Certain of the enzymes and intermediates in the Krebs cycle occur (for example, succinate, oxaloacetate, succinic dehydrogenase), but their metabolic significance lies elsewhere (as will be discussed further). Only in protoscolices of *Echinococcus granulosus* does there seem to be evidence for a complete Krebs cycle.[3]

There is some evidence for the existence of a pentose phosphate pathway in some cestodes but not in *H. diminuta*.[32] The complete pathway probably operates in *E. granulosus* protoscolices.[1,2]

There seems little doubt that the most important series of reactions to produce energy in adult cestodes is the Embden-Meyerhof sequence, or glycolysis. In general, the total energy derived from degradation of 1 mole of glucose is that required to form 4 moles, or possibly 6 (Fig. 20-35), of ATP from ADP. The products, which would then be further catabolized to carbon dioxide and water in vertebrate tissues, are excreted by the tapeworm. This seems a very inefficient usage of energy since much more of it could be derived from a mole of glucose by its complete oxidation. However, some facts of parasitic life must be kept in mind: (1) oxygen is likely to be absent from any given locus of worm tissue in the gut, or at least often severely limited, (2) a living host must furnish the worm an abundant and continuing food supply, and (3) the products excreted by the worm can be absorbed by the host and further catabolized. For every mole of glucose absorbed by the worm, almost 90% of the energy in it is returned to the host, assuming lactate or succinate is excreted.

The chemical composition of a variety of juvenile and adult pseudophyllideans and cyclophyllideans has been investigated, and all have relatively high concentrations of stored glycogen from about 20% to over 50% of dry weight. While the tissue-dwelling juveniles will be exposed to a reasonably constant glucose concentration maintained by the homeostatic mechanisms of the host, the adults must survive between host feeding periods. The large amount of stored glycogen serves at these times as an effective cushion. *H. diminuta* consumes 60% of its glycogen during 24 hours of host starvation and another 20% during the next 24 hours. When glucose is again available, the glycogen stores are rapidly replenished.[13]

The terminal phases of the glycolytic sequence in cestodes are strikingly different from those in vertebrate tissues, many bacteria, and other organisms but bear similarities to those of several other parasites and some other invertebrates that are facultative anaerobes. The difference involves the fixation of carbon dioxide and

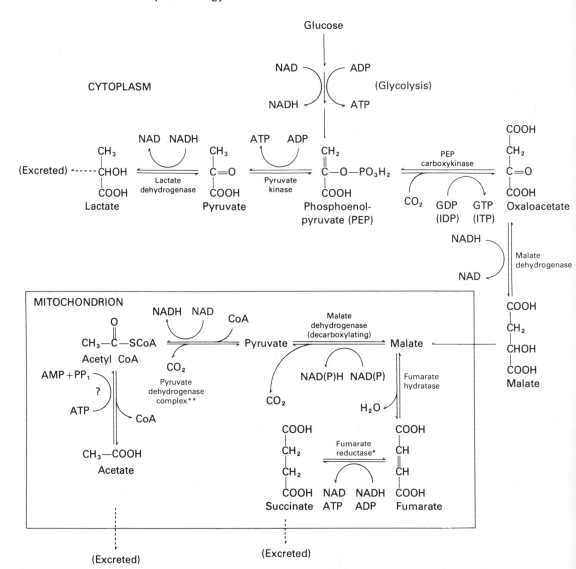

Fig. 20-35. Reactions forming the major end products of energy metabolism from phosphoenolpyruvate in *Hymenolepis diminuta* (adapted and proposed from various sources). These reactions yield additional ATP above that from classical glycolysis, with a balanced cytoplasmic oxidation-reduction and a balanced mitochondrial oxidation-reduction (ratio of succinate : acetate excreted, approximately 2:1). *This enzyme (fumarate reductase) usually is referred to as succinate dehydrogenase, but it acts in an opposite direction from mammalian systems, that is, as a fumarate reductase. **(Watts and Fairbairn[37] and Hochachka and Mustafa [1972].)

the ultimate excretion of large amounts of succinate. As in vertebrate tissue, glucose is degraded through the Embden-Meyerhof sequence to phosphoenolpyruvate (PEP). At that point, however, rather than converting PEP to pyruvate and then, through the Krebs cycle, to carbon dioxide, the cestode *fixes* carbon dioxide and produces oxaloacetate (OAA) from the PEP (Fig. 20-35). The reaction is catalyzed by PEP carboxykinase. The OAA is subsequently reduced through malate and fumarate to succinate and is excreted. Thus, the presence of carbon dioxide in relatively high concentrations is very important to normal worm metabolism. Glucose uptake and glycogen synthesis are severely inhibited in the absence of carbon dioxide.[13] Other major end-products of carbohydrate dissimulation are acetate and lactate, at least in *H. diminuta*. Lactate is derived from PEP by the actions of pyruvate kinase and lactic dehydrogenase, while acetate appears to be formed from pyruvate via a coenzyme A intermediate.[37] Formation of either succinate or lactate serves the dual role of reoxidizing the NADH formed in glycolysis and conserving the high energy phosphate bond in PEP. Lactate is the strongest acid of the three major products and is, therefore, the most toxic. The excretion of lactic acid by small helminths is probably related to the relative ease with which the substance can be excreted when the surface-volume ratio is high, and interestingly, the metabolism of *H. diminuta* shifts from production of a high proportion of lactate to a high proportion of succinate as the worm grows.

Energy derived by cestodes from lipids and proteins appears to be extremely limited. *H. diminuta* has only a modest capacity for carrying out transaminations and can degrade only four amino acids.[39] The availability of energy from degradation of lipid is even less. In fact, the function served by much of the lipid in cestodes remains a mystery since no one has been able to show that lipids are depleted at all during starvation, even though they may comprise up to about 20% of worm dry weight or over 30% in the parenchyma of gravid proglottids. Lipids in cestodes may represent metabolic end products, since they are relatively nontoxic to store, and the parenchyma of gravid proglottids is discarded during apolysis.

Nitrogenous end products excreted by *H. diminuta* include considerable quantities of ammonia, α-amino nitrogen, and urea.[13]

Synthetic metabolism

Little need be said of the protein and nucleic acid synthetic abilities of cestodes. It is clear that they can absorb amino acids, purines, pyrimidines, and nucleosides from the intestinal milieu and synthesize their own proteins and nucleic acids (see, for example, Bolla and Roberts, 1971[7]). *Moniezia expansa* cannot synthesize carbamyl phosphate; therefore, it is dependent on its host for both pyrimidine and arginine (Kurelec, 1972).

In contrast, capacity for synthesis of lipids appears minimal. The worm cannot synthesize fatty acids de novo from acetyl-CoA, nor can it introduce double bonds into the fatty acids it absorbs.[16] *H. diminuta* rapidly hydrolyzes monoglycerides after absorption, and it can then resynthesize triglycerides. It can lengthen the chain of fatty acids provided that the acid already contains 16 or more carbons. Similar observations have been reported for *Spirometra mansonoides*.[21]

Finally, *H. diminuta* cannot synthesize cholesterol, a biosynthesis that is known to require molecular oxygen in other systems. The normal precursors of cholesterol (acetate, hydroxymethylglutarate, and mevalonate) are converted by *H. diminuta* into 2-*cis*,6-*trans* farnesol and lesser amounts of 2-*trans*, 6-*trans* farnesol, the latter being a normal precursor of cholesterol in mammals. These compounds are normally present in this cestode, and that fact is of interest because both isomers are mimics of juvenile hormone in insects.[14a]

Classification of Class Cestoidea

Subclass CESTODARIA

Monozoic, with single set of reproductive organs. No scolex present, but there may be a small, proboscis-like organ at the anterior end. Posterior end single, rounded, genital pores near posterior end. Testes in two lateral, preovarian fields. Ovary posterior. Vitelline glands follicular, lateral. Uterus N-shaped or looped. Uterine pore near anterior end.

Families: Amphilinidae, Austramphilinidae

Subclass EUCESTODA

Polyzoic (except orders Caryophyllidea and Spathebothriidea), with one or more sets of reproductive systems per proglottid. Scolex usually present. Shelled embryo with six hooks. Parasites of fishes, amphibians, reptiles, birds, and mammals.

Order CARYOPHYLLIDEA

Scolex unspecialized or with shallow grooves or loculi or shallow bothria. Monozoic. Genital pores midventral. Testes numerous. Ovary posterior. Vitellaria follicular, scattered or lateral. Uterus a coiled median tube, opening, often together with vagina, near male pore. Parasites of teleost fishes and aquatic annelids.

Family: Caryophyllaeidae

Order SPATHEBOTHRIIDEA

Scolex feebly developed, undifferentiated or with funnel-shaped apical organ or with one or two hollow, cup-like organs. External segmentation absent, proglottids distinguished internally. Genital pores and uterine pore ventral or alternating dorsal and ventral. Testes in two lateral bands. Ovary dendritic. Vitellaria follicular, lateral or scattered. Uterus coiled. Parasites of teleost fishes.

Families: Spathebothriidae, Cyathocephalidae

Order TRYPANORHYNCHA

Scolex elongate, with two or four bothridia and four eversible (rarely atrophied) tentacles armed with hooks. Each tentacle invaginates into internal sheath provided with muscular bulb. Neck present or absent. Strobila apolytic, anapolytic, or hyperapolytic. Genital pores lateral, rarely ventral. Testes numerous. Ovary posterior. Vitellaria follicular, cortical, and encircling other reproductive organs. Uterine pore present or absent. Parasites of elasmobranchs.

Families: Paranybeliniidae, Tentaculariidae, Hepatoxylidae, Sphyriocephalidae, Dasyrhynchidae, Lacistorhynchidae, Pterobothriidae, Gymnorhynchidae, Otobothriidae, Gilquiniidae, Eutetrarhynchidae, Hornelliellidae, Mustelicolidae

Order PSEUDOPHYLLIDEA

Scolex with two bothria, with or without hooks. Neck present or absent. Strobila variable. Proglottids anapolytic. Genital pores lateral, dorsal, or ventral. Testes numerous. Ovary posterior. Vitellaria follicular, as in Trypanorhyncha; occasionally in lateral fields but not interrupted by interproglottidal boundaries. Uterine pore present, dorsal or ventral. Egg usually operculate, containing a coracidium. Parasites of fishes, amphibians, reptiles, birds, and mammals.

Families: Cephalochlamydidae, Haplobothriidae, Diphyllobothriidae, Ptychobothriidae, Bothriocephalidae, Echinophallidae, Amphicotylidae, Triaenophoridae, Parabothriocephalidae

Order LECANICEPHALIDEA

Scolex divided into anterior and posterior regions by horizontal groove. Anterior portion cushion-like or with unarmed tentacles, capable of being withdrawn into posterior portion, forming a large sucker-like organ. Posterior portion usually with four suckers. Neck present or absent. Testes numerous. Ovary posterior. Vitellaria follicular, lateral or encircling proglottid. Uterine pore usually present. Parasites of elasmobranchs.

Families: Balanobothriidae, Disculiceptidae, Lecanicephalidae, Adelobothriidae

Order APORIDEA

Scolex with simple suckers or grooves and armed rostellum. External segmentation absent, proglottids distinguished internally or separate proglottids not evident. Genital ducts and pores, cirrus, ootype, and Mehlis's gland absent. Hermaphroditic, rarely dioecious. Vitelline cells mixed with ovarian cells. Parasites of Anseriformes.

Family: Nematoparataeniidae

Order TETRAPHYLLIDEA

Scolex with highly variable bothridia, sometimes also with hooks, spines, or suckers. Myzorhynchus present or absent. Proglottids commonly hyperapolytic. Hermaphroditic, rarely dioecious. Genital pores lateral, rarely posterior. Testes numerous. Ovary posterior. Vitellaria follicular (condensed in *Dioecotaenia*), usually medullary in lateral fields. Uterine pore present or not. Vagina crosses vas deferens. Parasites of elasmobranchs.

Families: Phyllobothriidae, Onchobothriidae, Dioecotaeniidae, Triloculariidae

Order DIPHYLLIDEA

Scolex with armed or unarmed peduncle. Two spoon-shaped bothridia present, lined with minute spines, sometimes divided by median, longitudinal ridge. Apex of scolex with insignificant apical organ or with large rostellum bearing dorsal and ventral groups of T-shaped hooks. Strobila cylindrical, acraspedote. Genital pores posterior, midventral. Testes numerous, anterior. Ovary posterior. Vitellaria follicular, lateral, or surrounding segment. Uterine pore absent. Uterus tubular or saccular. Parasites of elasmobranchs.

Families: Echinobothriidae, Ditrachybothridiidae

Order LITOBOTHRIDEA

Scolex a single, well-developed apical sucker. Anterior proglottids modified, cruciform in cross section. Neck absent. Strobila dorsoventrally flattened, with numerous proglottids, each with single set of medullary reproductive organs. Segments laciniated and craspedote; apolytic or anapolytic. Testes numerous, preovarian. Genital pores lateral. Ovary two- or four-lobed, posterior. Vitellaria follicular, encircling medullary parynchyma. Parasites of elasmobranchs.

Family: Litobothridae

Order PROTEOCEPHALATA

Scolex with four suckers, often with prominant apical organ, occasionally with armed rostellum. Neck usually present. Genital pores lateral. Testes numerous. Ovary posterior. Vitelline glands follicular, usually lateral, cortical or medullary. Uterine pore present or absent. Parasites of fishes, amphibians, and reptiles.

Family: Proteocephalidae

Order CYCLOPHYLLIDEA

Scolex usually with four suckers. Rostellum present or not, armed or not. Neck present or absent. Strobila usually with distinct segmentation, monoecious or rarely dioecious. Genital pores lateral (ventral in Mesocestoididae). Vitelline gland compact, single (double in Mesocestoididae), posterior to ovary (anterior or beneath ovary in Tetrabothriidae). Uterine pore absent. Parasites of amphibians, reptiles, birds, and mammals.

Families: Triplotaeniidae, Mesocestoididae, Tetrabothriidae, Nematotaeniidae, Dioecocestidae, Progynotaeniidae, Taeniidae, Diploposthidae, Amabiliidae, Davaineidae, Hymenolepididae, Catenotaeniidae, Anoplocephalidae, Dilepididae

Order NIPPOTAENIIDEA

Scolex with single sucker at apex, otherwise simple. Neck short or absent. Strobila small. Proglottids each with single set of reproductive organs. Genital pores lateral. Testes anterior, ovary posterior. Vitelline gland compact, single, between testes and ovary. Osmoregulatory canals reticular. Parasites of teleost fishes.

Family: Nippotaeniidae

REFERENCES

1. Agosin, M., and L. Aravena. 1960. Studies on the metabolism of *Echinococcus granulosus*. IV. Enzymes of the pentose phosphate pathway. Exp. Parasitol. 10:28-38.
2. Agosin, M., and Y. Repetto. 1961. Studies on the metabolism of *Echinococcus granulosus*. VI. Pathways of glucose ^{14}C metabolism of *Echinococcus granulosus* scolices. Biologica 32:33-38.
3. Agosin, M., and Y. Repetto. 1963. Studies on the metabolism of *Echinococcus granulosus*. VII. Reactions of the tricarboxylic acid cycle in *E. granulosus* scoleces. Comp. Biochem. Physiol. 8:245-261.
4. Archer, D. M., and C. A. Hopkins. 1958. Studies on cestode metabolism. III. Growth pattern of *Diphyllobothrium* sp. in a definitive host. Exp. Parasitol. 7:125-144.
5. Arme, C. 1966. Histochemical and biochemical studies on some enzymes of *Ligula intestinalis* (Cestoda: Pseudophyllidea). J. Parasitol. 52:63-68.
5a. Bailey, H. H., and D. Fairbairn. 1968. Lipid metabolism in helminth parasites. V. Absorption of fatty acids and monoglycerides from micellar solution by *Hymenolepis diminuta* (Cestoda). Comp. Biochem. Physiol. 26:819-836.
6. Blitz, N. M., and J. D. Smyth. 1973. Tegumental ultrastructure of *Raillietina cesticillus* during the larval-adult transformation, with emphasis on the rostellum. Int. J. Parasitol. 3:561-570.
7. Bolla, R. I., and L. S. Roberts. 1971. Developmental physiology of cestodes. X. The effect of crowding on carbohydrate levels and on RNA, DNA, and protein synthesis in *Hymenolepis diminuta*. Comp. Biochem. Physiol. 40A:777-787.
8. von Bonsdorff, B. 1956. *Diphyllobothrium latum* as a cause of pernicious anemia. Exp. Parasitol. 5:207-230.
9. von Brand, T., P. McMahon, E. Gibbs, and H. Higgins. 1964. Aerobic and anaerobic metabolism of larval and adult *Taenia taeniaeformis*. II. Hexose leakage and absorption; tissue glucose and polysaccharides. Exp. Parasitol. 15:410-429.
10. Campbell, W. C. 1963. The efficacy of surface-active agents in stimulating the evagination of cysticerci in vitro. J. Parasitol. 49:81-84.
11. Davey, K. G., and W. R. Breckenridge. 1967. Neurosecretory cells in a cestode, *Hymenolepis diminuta*. Science 158:931-932.
12. DeRycke, P. H. 1972. Osmoregulation of *Hymenolepis microstoma*. I. Influence of the osmotic pressure on the gravid adult. Zeitschr. Parasitenk. 38:141-146.
13. Fairbairn, D., G. Wertheim, R. P. Harpur, and E. L. Schiller. 1961. Biochemistry of normal and irradiated strains of *Hymenolepis diminuta*. Exp. Parasitol. 11:248-263.
14. Foster, W. D. 1965. A history of parasitology. E. & S. Livingston, Edinburgh.
14a. Frayha, G. J., and D. Fairbairn. 1969. Lipid metabolism in helminth parasites. VI. Synthesis of 2-*cis*,6-*trans* farnesol by *Hymenolepis diminuta* (Cestoda). Comp. Biochem. Physiol. 28:1115-1124.
15. Hoeppli, R. J. C. 1959. Parasites and parasitic infections in early medicine and science. University of Malaya Press, Singapore.
16. Jacobsen, N. S., and D. Fairbairn. 1967. Lipid metabolism in helminth parasites. III. Biosynthesis and interconversion of fatty acids by *Hymenolepis diminuta* (Cestoda). J. Parasitol. 53:355-361.
17. Lumsden, R. D., and J. Byram III. 1967. The

ultrastructure of cestode muscle. J. Parasitol. 53:326-342.

18. Lumsden, R. D., J. A. Oaks, and J. F. Mueller. 1974. Brush border development in the tegument of the tapeworm, *Spirometra mansonoides*. J. Parasitol. 60:209-226.

19. Lumsden, R. D., L. T. Threadgold, J. A. Oaks, and C. Arme. 1970. On the permeability of cestodes to colloids: an evaluation of the transmembranosis hypothesis. Parasitology 60:185-193.

19a.MacInnis, A. J., F. M. Fisher, Jr., and C. P. Read. 1965. Membrane transport of purines and pyrimidines in a cestode. J. Parasitol. 51:260-267.

20. Mackiewicz, J. S. 1972. Caryophyllidea (Cestoidea): a review. Exp. Parasitol. 31:417-512.

21. Meyer, F., S. Kimura, and J. F. Mueller. 1966. Lipid metabolism in the larval and adult forms of the tapeworm *Spirometra mansonoides*. J. Biol. Chem. 241:4224-4232.

22. Mount, P. M. 1970. Histogenesis of the rostellar hooks of *Taenia crassiceps* (Zeder, 1800) (Cestoda). J. Parasitol. 56:947-961.

23. Oaks, J. A., and R. D. Lumsden. 1971. Cytological studies on the absorptive surfaces of cestodes. V. Incorporation of carbohydrate-containing macromolecules into tegument membranes. J. Parasitol. 57:1256-1268.

24. Platzer, E. G., and L. S. Roberts. 1969. Developmental physiology of cestodes. V. Effects of vitamin deficient diets and host coprophagy prevention on development of *Hymenolepis diminuta*. J. Parasitol. 55:1143-1152.

25. Read, C. P. 1967. Longevity of the tapeworm, *Hymenolepis diminuta*. J. Parasitol. 53:1055-1056.

26. Read, C. P., and J. E. Simmons. 1963. Biochemistry and physiology of tapeworms. Physiol. Rev. 43:263-305.

26a.Rees, G. 1966. Nerve cells in *Acanthobothrium coronatum* (Rud.) (Cestoda: Tetraphyllidea). Parasitology. 56:45-54.

27. Roberts, L. S. 1961. The influence of population density on patterns and physiology of growth in *Hymenolepis diminuta* (Cestoda: Cyclophyllidea) in the definitive host. Exp. Parasitol. 11:332-371.

28. Roberts, L. S., and F. Mong. 1968. Developmental physiology of cestodes. III. Development of *Hymenolepis diminuta* in superinfections. J. Parasitol. 54:55-62.

29. Roberts, L. S., and F. N. Mong. 1973. Developmental physiology of cestodes. XIII. Vitamin B_6 requirement of *Hymenolepis diminuta* during in vitro cultivation. J. Parasitol. 59:101-104.

30. Rothman, A. H. 1959. Studies on the excystment of tapeworms. Exp. Parasitol. 8:336-364.

31. Rybicka, K. 1964. The embryonic envelopes in cyclophyllidean cestodes. Acta Parasitol. Polonica 13:25-34.

32. Schiebel, L. W., and H. J. Saz. 1966. The pathway for anaerobic carbohydrate dissimulation in *Hymenolepis diminuta*. Comp. Biochem. Physiol. 18:151-162.

33. Smyth, J. D. 1962. Lysis of *Echinococcus granulosus* by surface-active agents in bile and the role of this phenomenon in determining host specificity in helminths. Proc. R. Soc. B 156:553-572.

34. Smyth, J. D. 1969. The physiology of cestodes. W. H. Freeman and Co. Publishers, San Francisco.

35. Smyth, J. D., J. Morgan, and A. B. Howkins. 1967. Further analysis of the factors controlling strobilization, differentiation and maturation of *Echinococcus granulosus* in vitro. Exp. Parasitol. 21:31-41.

36. Wardle, R. A., and J. A. McLeod. 1952. The zoology of tapeworms. University of Minnesota Press, Minneapolis.

37. Watts, S. D. M., and D. Fairbairn. 1974. Anaerobic excretion of fermentation acids by *Hymenolepis diminuta* during development in the definitive host. J. Parasitol. 60:621-625.

38. Webster, L. A., and R. A. Wilson. 1970. The chemical composition of protonephridial canal fluid from the cestode *Hymenolepis diminuta*. Comp. Biochem. Physiol. 35:201-209.

39. Wertheim, G., R. Zeledon, and C. P. Read. 1960. Transaminases of tapeworms. J. Parasitol. 46:497-499.

40. Wilson, V. C. L. C., and E. L. Schiller. 1969. The neuroanatomy of *Hymenolepis diminuta* and *H. nana*. J. Parasitol. 55:261-270.

41. Wood, D. E. 1965. Nature of the end organ in *Ophiotaenia filaroides* (La Rue). J. Parasitol. 51:541-544.

SUGGESTED READING

All-Union Institute of Helminthology. 1951. Essentials of Cestodology, vols. 1-7. Academy of Science, Moscow. (In Russian, some volumes translated into English. This important series of volumes on cestode taxonomy is authored by several Soviet authorities. It is indispensable to the student in classifying tapeworms.)

Bryant, C. 1970. Electron transport in parasite helminths and Protozoa. In Dawes, B., editor. Advances in parasitology, vol. 8. Academic Press, Inc., New York. pp. 139-172.

Dollfus, R. P. 1942. Étude critiques sur le tétrarhynques du Museum de Paris. Arch. Mus. Hist. Nat. 19:1-466. (This is the outstanding monograph on the taxonomy of the order Trypanorhyncha.)

Fairbairn, D., 1970. Biochemical adaptation and loss of genetic capacity in helminth parasites. Biol. Rev. 45:29-72.

Florkin, M., and B. T. Scheer, editors. 1968. Chemical zoology, vol. 2. Academic Press, Inc., New York. (Several contributions are presented by different authors on the subject of cestode physiology.)

Freeman, R. 1973. Ontogeny of cestodes and its bearing on their phylogeny and systematics. In Dawes, B., editor. Advances in Parasitology, vol. 11. Academic Press, Inc., New York. pp. 481-557.

Fuhrmann, O. 1930. Dritte Klasse des Cladus Platyhelminthes. Cestoidea. In Kükenthal's Handbuch der Zoologie, vol. 2. pp. 141-416. (A classic reference on all aspects of cestodology.)

Hyman, L. H. 1951. The Invertebrates, vol. 2. McGraw-Hill Book Co., New York. (A complete summary of knowledge of cestodes up to 1951.)

Joyeux, C., and J. G. Baer. 1936. Cestodes. In Faune de France, vol. 30. Kraus Reprint, Nadeln. (A summary of classification of European tapeworms.)

Joyeux, C., and J. G. Baer. 1961. Classe des Cestodes. Cestoidea Rudolphi. In Grasse's Traite de Zoologie, vol. IV. Masson et Cie., Paris. (A summary of cestode biology with an abbreviated classification.)

Kurelec, B. 1972. Lack of carbamyl phosphate synthesis in some parasitic platyhelminthes. Comp. Biochem. Physiol. 43B:769-780.

Pappas, P. W., and C. P. Read. 1975. Membrane transport in helminth parasites: a review. Exp. Parasitol. 37:469-530.

Read, C. P. 1959. The role of carbohydrates in the biology of cestodes. VIII. Exp. Parasitol. 8:365-382.

Read, C. P. 1973. Contact digestion in tapeworms. J. Parasitol. 59:672-677.

Schmidt. G. D. 1970. How to know the tapeworms. William C. Brown, Co. Publishers Dubuque, Iowa. (An introduction to cestodology with keys to genera of tapeworms of the world.)

Southwell, T. 1925. A monograph on the Tetraphyllidea. Liverpool University Press, Liverpool. Mem. II. (The standard reference on the order Tetraphyllidea.)

Southwell, T. 1930. Cestoda. In Fauna of British India, including Ceylon and Burma. Taylor and Francis, London. (A useful reference to the tapeworms of the British Empire in India.)

Starling, J. A. 1975. Tegumental carbohydrate transport in intestinal helminths: correlation between mechanisms of membrane transport and the biochemical environment of absorptive surfaces. Trans. Am. Microsc. Soc. 94:508-523.

Voge, M. 1969. Systematics of cestodes—present and future. In Schmidt, G. D., editor. Problems in systematics of parasites. University Park Press, Baltimore. (An interesting account of current events in cestode systematics, with a view to the future of the science.)

Wardle, R. A., and J. A. McLeod. 1952. The zoology of tapeworms. Hafner Publishing Co., New York. (This monograph is the classic in its field. No student of tapeworms should be without it.)

Williams, H. H. 1968. The taxonomy, ecology, and host-specificity of some Phyllobothriidae (Cestoda: Tetraphyllidea), a critical revision of *Phyllobothrium* Beneden, 1849, and some comments on some allied genera. Phil. Trans. R. Soc. London 786:231-307. (A useful revision of a troublesome group and a model of revisionary taxonomy.)

Chapter 21

TAPEWORMS OF PARTICULAR IMPORTANCE TO HUMANS

Although most species of cestodes are parasites of wild animals, a few are of particular interest because they infect humans or their domestic animals. For this reason the following species have been selected to illustrate tapeworm biology. All tapeworms of humans are in the two largest orders, Pseudophyllidea and Cyclophyllidea.

ORDER PSEUDOPHYLLIDEA

Pseudophyllidean cestodes typically have a scolex with dorsal and ventral longitudinal grooves called **bothria.** These may be deep or shallow, smooth or fimbriated, and in some cases they are fused along all or part of their length, forming longitudinal tubes. Sclerotized hooks accompany the bothria in some species. The genital pores may be lateral or medial, depending on the species. The vitellaria are always follicular and scattered throughout the segment. The testes are numerous. Some species are fairly small, but the largest tapeworms known are in the Pseudophyllidea. For example, *Polygonoporus* from the sperm whale measures over 30 meters long. In addition, each segment has five to 14 complete sets of reproductive organs. *Tetragonoporus,* also a parasite of sperm whales, has up to 45,000 segments, each with four complete sets of reproductive organs. The reproductive capacity of such animals is staggering. Generally, the life cycles of pseudophyllideans involve crustacean first intermediate hosts and fish second intermediate hosts.

Family Diphyllobothriidae
Diphyllobothrium latum

Usually called the broad fish tapeworm, this cestode is common in fish-eating carnivores, particularly in northern Europe. It appears to exhibit a striking lack of host

specificity, occurring in many canines and felines, mustelids, pinnipeds, bears, and humans. Most of these records, however, are misidentifications. In northeastern North America humans may also become infected with *D. ursi,* a closely related species. Humans seem to be quite suitable as hosts; *D. latum* is so common in some small areas of the world that nearly 100% of the human population are infected.[5] It is most abundant in Scandinavia, the Baltic states, and Russia and is present in the Arctic and Great Lakes areas of North America.[5] It has also been found in Africa, Japan, South America, Ireland, and Israel, although some of these records were probably erroneous.

Morphology. The adult worm (Fig. 21-1) may attain a length of 30 feet and shed up to a million eggs a day. The species is anapolytic and characteristically releases long chains of spent proglottids, usually the first indication that the infected person has a secret guest.

The scolex (Fig. 21-2) is finger-shaped and has dorsal and ventral bothria. Proglottids (Fig. 21-3) are usually wider than long. There are numerous testes and vitelline follicles scattered throughout the proglottid, except for a narrow zone in the center. The male and female genital pores open midventrally. The bilobed ovary is near the rear of the segment. The uterus consists of short loops and extends from the ovary to a midventral uterine pore.

Biology. The ovoid eggs measure about 60 to 40 μm and have a lid-like operculum at one end and a small knob on the other (Fig. 21-4). When released through the uterine pore, the shelled embryo is at an early stage of development and, as in Group I life cycles, it must be deposited in water for development to continue. Completion of development to coracidium

Fig. 21-1. *Diphyllobothrium latum.* The scolex is at the tip of the thread-like end at upper left. (Photograph by Warren Buss.)

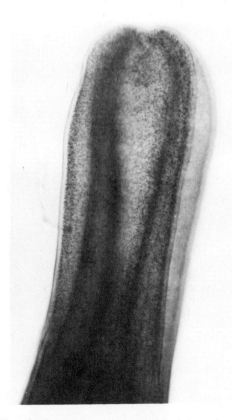

Fig. 21-2. Scolex of *Diphyllobothrium latum.* Note the dorsal bothrium (right side of photo). (Photograph by Warren Buss.)

takes from 8 days to several weeks, depending on the temperature. Emerging through the operculum (Fig. 21-5), the ciliated coracidium swims randomly about, where it may attract the attention of predaceous copepods of the genus *Diaptomus.* Soon after being eaten, the coracidium loses its ciliated epithelium and immediately begins to attack the wall of the midgut with its six tiny hooks. Once through the intestine and into the crustacean's hemocoel, it becomes parasitic, absorbing nourishment from the surrounding blood. In about 3 weeks it increases its length to around 500 μm, becoming an elongate, undifferentiated mass of parenchyma with a cercomer at the posterior end. It is now a procercoid (Fig. 21-6), incapable of further development until eaten by a suitable second intermediate host—any of several species of freshwater fishes, especially pike and related fish, or any of the salmon family. The cercomer may be lost while still in the copepod or soon after the procercoid enters a fish. Large, predaceous fish eat comparatively few microcrustaceans, but can still become infected by eating smaller fish containing pleroceroids, which then migrate into the new host.

When the infected copepod is eaten by a fish, the procercoid is released in the

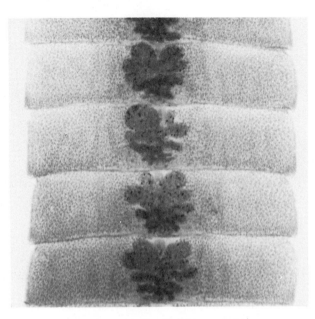

Fig. 21-3. Gravid proglottids of *Diphyllobothrium latum.* (Photograph by Larry Jensen.)

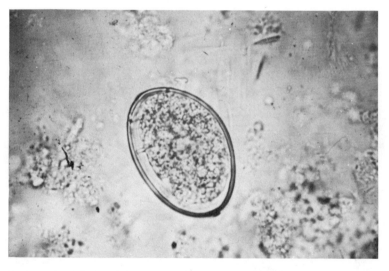

Fig. 21-4. Egg of *Diphyllobothrium latum* in a human stool; note operculum at upper end and small knob at opposite end. (Photograph by David Oetinger.)

fish's intestine. From here it bores its way through the intestinal wall and into the body muscles. Here it again becomes parasitic, absorbing nutrients and growing rapidly into a plerocercoid. Mature plerocercoids vary from a few millimeters to several centimeters in length. They are still mainly undifferentiated, but there may be evidence of shallow bothria at the anterior end. Usually they are found unencysted and coiled up in the musculature, although they may be encysted in the viscera. They are easily seen as white masses in uncooked fish (Fig. 21-7), but, when the flesh is cooked, the worms are seldom noticed. Plerocercoids of other pseudophyllideans, as well as those of proteocephalideans and trypanorhychans, are also

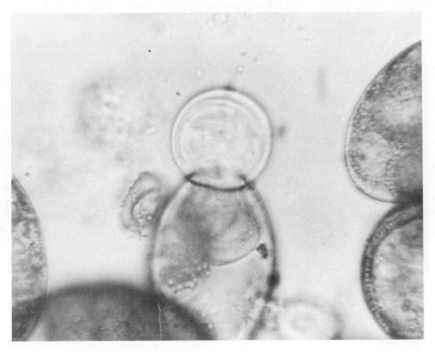

Fig. 21-5. A diphyllobothriid coracidium emerging from its eggshell. The cilia are moving so fast they appear as a blur. (Photograph by Justus F. Mueller.)

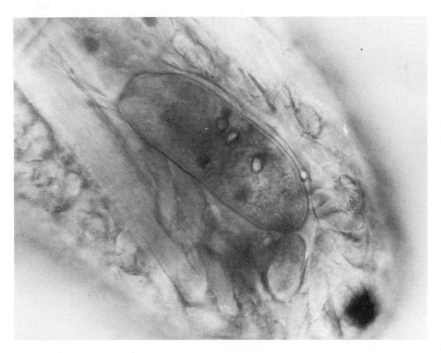

Fig. 21-6. Procercoid in the hemocoel of a copepod. Note the anterior pit, the posterior cercomer, and the internal calcareous granules. (Photograph by Justus F. Mueller.)

Fig. 21-7. Two plerocercoids in the flesh of a perch. (From Vik, R. 1971. In Marcial-Rojas, R. A., editor. Pathology of protozool and helminthic diseases with clinical correlation. The Williams & Wilkins Co., Baltimore.)

found in fish and are often mistaken for those of *D. latum.* When the plerocercoid is ingested by a suitable host, it survives the digestive fate of its late host and begins a close relationship with a new one. The worms grow rapidly and may begin egg production by 7 to 14 days. Little of the growth may be attributed to the production of new proglottids but is caused by growth in primordia already in the plerocercoid. As much as 70% of the strobila may mature on the same day.[1]

Epidemiology. Obviously, persons become infected when they eat raw or undercooked fish. Hence, infection rates are highest in countries where raw fish are eaten as a matter of course. Communities that dispose of sewage by draining it into lakes or rivers without proper treatment create an opportunity for a massive build-

up in local fishes. These fishes may be harvested for local consumption or shipped thousands of miles by refrigerated freight to distant markets. There an unsuspecting customer may gain infection in a restaurant or at home by tasting such dishes as gefilte fish during preparation.

Pathogenesis. Many cases of diphyllobothriasis are apparently asymptomatic or have poorly defined symptoms associated with other tapeworms, such as vague abdominal discomfort, diarrhea, nausea, and weakness. However, in a small number of cases, the worm causes a serious megaloblastic anemia, and virtually all of these are in Finnish people. It has been estimated that almost one-fourth of the population of Finland may be infected with *D. latum,* and about 1,000 of these will have pernicious anemia.[4] It was thought originally that toxic products of the worm produced the anemia, but it is now known that the large amount of vitamin B_{12} absorbed by the cestode, in conjunction with some degree of impairment of the patient's normal absorptive mechanism for vitamin B_{12}, is responsible for the disease. Nyberg reported that an average of 44% of a single oral dose of ^{60}Co-labeled vitamin B_{12} was absorbed by *D. latum* in otherwise healthy patients, but, in patients with tapeworm pernicious anemia, 80% to 100% of the dose was absorbed by the cestode.[19] The clinical symptoms of tapeworm pernicious anemia are similar in many respects to "classical" pernicious anemia (caused by a failure in intestinal absorption of vitamin B_{12}) except that expulsion of the worm generally brings a rapid remission of the anemia.

Diagnosis and treatment. Demonstration of the characteristic eggs or proglottids passed with the stool gives positive diagnosis. In the past, a variety of drugs have been used against *D. latum* and other tapeworms, aspidium oleoresin (extract of male fern), mepacrine, dichlorophen, and even extracts of fresh pumpkin seeds (*Cucurbita* spp.) have anticestodal properties.[7] However, the drug of choice at this time seems to be niclosamide (Yomesan). Its mode of action seems to be an inhibition of an inorganic phosphate–ATP exchange reaction associated with the worm's anaerobic electron transport system.

Other pseudophyllideans found in humans

Several other species of *Diphyllobothrium* have been reported from humans in different parts of the world. These include *D. chordatum* and *D. pacificum*, parasites of pinnipeds in the northern and southern hemispheres, respectively, and *D. ursi* of bears. Other species of *Diphyllobothrium* reported from humans are probably synonyms of *D. latum*. An expert can clearly differentiate the species in animals from *D. latum*.

Spirometra haughtoni (S. erinacei), Digramma brauni, and *Ligula intestinalis* have also been reported from humans, but such occurrences must be considered rare. *Diplogonoporous grandis (D. balaenopterae)* has been reported numerous times from humans in Japan.[12] A parasite of whales, its plerocercoid occurs in marine fishes, the mainstay of the Japanese protein diet. The proglottid is easily recognized, for it has two sets of male and female reproductive organs in each segment.

Sparganosis. With the exception of the forms with scolex armature, it is impossible to distinguish the species of plerocercoids by examining their morphology. When procercoids of some species are ingested accidentally, usually by swallowing an infected copepod in drinking water, they can migrate from the gut and develop into plerocercoids, sometimes reaching a length of 14 inches. The infection is called **sparganosis** and may cause severe pathological consequences. Cases have been reported from most countries of the world but are most common in the Orient. Yamane and others reported a living sparganum that had infected a breast of a woman for at least 30 years.[26]

Other means of infection are by ingestion of insufficiently cooked amphibians, reptiles, birds, or even mammals such as pigs.[6] Plerocercoids present in these animals may then infect the person indulging in such delicacies. Many Chinese are infected in this way because of their tradition of eating raw snake to cure a panoply of ills.[14]

A third method of infection results from the Oriental treatment of skin ulcers, inflamed vagina, or inflamed eye (Fig. 21-8), by poulticing the area with a split frog or flesh of a vertebrate that may be inciden-

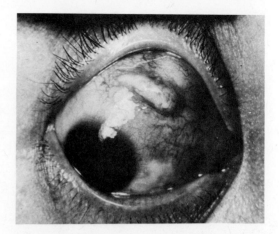

Fig. 21-8. Right eye of a patient with sparganosis. Note the protruding mass in the upper conjunctiva. (From Wang, L. T., and J. H. Cross. 1974. J. Formosan Med. Assoc. 73:173-177.)

tally infected with spargana. The active worm then crawls into the orbit, vagina, or ulcer and establishes itself. Most cases of sparganosis in the Orient are probably caused by *Spirometra erinacei*, a parasite of carnivores.

In North America, most spargana are probably *Spirometra mansonoides*, a parasite of cats.[17] It usually does not proliferate, except by occasionally breaking transversly, and may live up to 10 years in a human.[25] The current public awareness of the symptoms of cancer has led to an increase in reported cases of sparganosis in this country. Subdermal lumps are no longer ignored by the average person, and more than one physician has been shocked to find a gleaming, white worm in a lanced nodule. Wild vertebrates are commonly infected with spargana (Fig. 21-9).

Rarely, a sparganum will be proliferative, splitting longitudinally and budding profusely. Such cases are very serious, for many thousands of worms can result, with the infected organs becoming honeycombed.

Treatment of sparganosis is usually by surgery, but some success has been obtained by chemotherapy.[13]

ORDER CYCLOPHYLLIDEA

The most characteristic morphological features of cyclophyllideans are a single

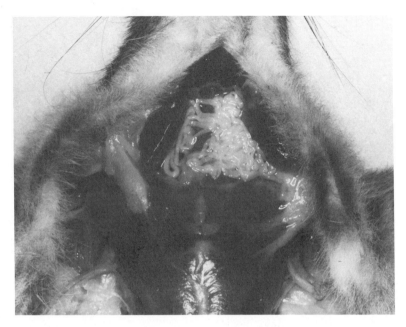

Fig. 21-9. Spargana in subcutaneous connective tissues of a wild rat in Taiwan. (Photograph by Robert E. Kuntz.)

compact vitelline gland and a scolex with four suckers. A rostellum, which usually is armed with hooks, is commonly present. The genital pores are lateral in all except the family Mesocestoididae, in which they are midventral. The single vitellarium is usually postovarian but may be preovarian. The number of testes varies from one to several hundred, depending on the species. Most species are rather small, although some are giants of over 30 feet in length. Most tapeworms of birds and mammals belong to this order.

Family Taeniidae

The largest cyclophyllideans are in the family Taeniidae, as are the most medically important tapeworms of humans. A remarkable morphological similarity occurs between most species in the family; a striking exception is *Echinococcus*, which is very much smaller than cestodes of the other genera. An armed rostellum is present on most species and, when present, is not retractable. The testes are numerous, and the ovary is a bilobed mass near the posterior margin of the proglottid. The larval forms are various types of bladderworms (Fig. 21-10), and mammals serve as their intermediate hosts.

Taeniarhynchus saginatus

Taeniarhynchus saginatus is by far the most common taeniid of humans, occurring in nearly all countries where beef is eaten. The beef tapeworm, as it is usually known lacks a rostellum or any scolex armature (Fig. 21-11). An exceptionally large species, individuals may attain a length of over 75 feet, but 10 to 15 feet is much more common. Even the smaller specimens may consist of as many as 2,000 proglottids.

Morphology. The scolex, with its four powerful suckers, is followed by a long, slender neck. Mature proglottids are slightly wider than long, while gravid ones are much longer than wide. Usually, it is the gravid proglottid passed in the feces that is first noticed and taken to a physician for diagnosis. Because the eggs of this species cannot be differentiated from those of *Taenia solium*, the next most common taeniid of humans, accurate diagnosis depends on a critical examination of a gravid uterus. Of course, if the entire worm is passed, the combination of unarmed scolex and taeniid-type proglottid (Fig. 21-12) leads to an unmistakable diagnosis.

The spherical eggs are characteristic of the Taeniidae (Fig. 21-13). There is a thin,

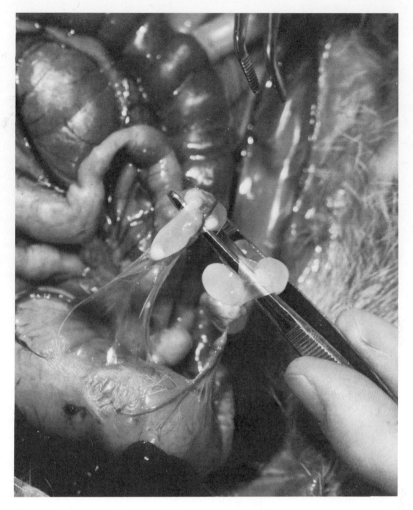

Fig. 21-10. Bladderworms of *Taenia pisiformis* in the mesenteries of a rabbit. (Photograph by John Mackiewicz.)

hyaline, outer membrane, which is usually lost by the time the egg is voided with the feces. The embryophore is very thick and riddled with numerous tiny pores, giving it a striated appearance in optical section. Unfortunately, the egg sizes of several taeniids in humans overlap, making diagnosis to species impossible on this character alone.

Biology. When gravid, the proglottids detach and either pass out with the feces or migrate out of the anus. Each segment behaves like an individual worm, crawling actively about, as if searching for something. The segments are easily mistaken for trematodes or even nematodes at this stage. As a segment begins to dry up, a rupture occurs along the midventral body

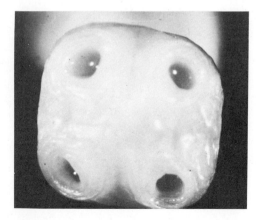

Fig. 21-11. En face view of the scolex of *Taeniarhynchus saginatus*. Note the absence of a rostellum or armature. (AFIP neg. no. 65-12073-2.)

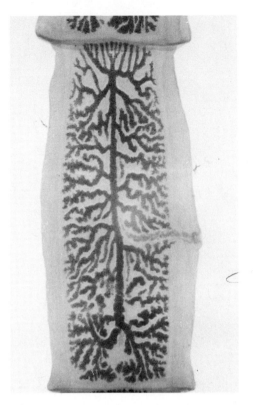

Fig. 21-12. A gravid proglottid of *Taenia* sp. The uterus is characteristic for this genus, consisting of a median stem with lateral branches. (Photograph by Warren Buss.)

wall, allowing eggs to escape. The eggs are fully embryonated and infective to the intermediate host at this time; they remain viable for many weeks. Cattle are the usual intermediate host, although cysticerci have also been reported from llamas, goats, sheep, giraffes, and possibly reindeer (perhaps incorrectly).

When eaten by a suitable intermediate host, the egg hatches in the duodenum under the influence of gastric and intestinal secretions. The released hexacanth quickly penetrates the mucosa and enters an intestinal venule, to be carried throughout the body. Typically, it leaves a capillary between muscle cells, enters a muscle fiber, and becomes parasitic, developing into an infective cysticercus in about 2 months. This larva is white, pearly, about 10 mm at its greatest diameter and contains a single, invaginated scolex. Humans are probably an unsuitable intermediate host, and the few records of *T. saginatus* cysticerci in humans are most likely misidentifications. Before the beef cysticercus was known to be a larval form of *T. saginatus,* it was placed in a separate genus under the name of *Cysticercus bovis.* The disease produced in cattle is thus known as **cysticercosis bovis,** and flesh riddled with the larvae is called **measly beef.**

A person who eats infected beef, cooked

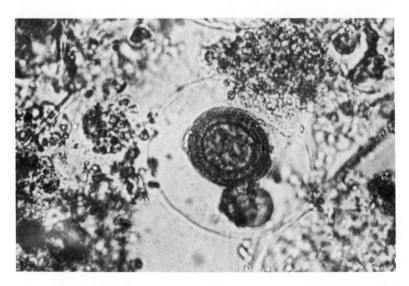

Fig. 21-13. A *Taenia* egg in human feces. The thin outer membrane is often lost at this stage. (Photograph by David Oetinger.)

insufficiently to kill the larvae, becomes infected. The invaginated scolex and neck of the cysticercus evaginate in response to bile salts. The bladder is digested by the host or absorbed by the scolex, and budding begins. Within 2 to 12 weeks the worm will begin shedding gravid proglottids.

Epidemiology. Human infection is highest in areas of the world where beef is a major food and sanitation is of little concern. Thus, in several developing nations of Africa and South America, for instance, ample opportunity exists for cattle to eat tapeworm eggs and for people to eat infected flesh. Many people are content to eat a chunk of meat that is cooked in a campfire, charred on the outside and raw on the inside. Local custom may have profound effect on infection rates. Hence, in India there may be a high rate of infection in a Moslem population, while Hindus, who do not eat beef, are unaffected. In the United States, federal meat inspection laws and a high degree of sanitation combine to keep the incidence of infection low. However, only 80% of the cattle slaughtered in the United States is federally inspected, and studies have shown that standard inspection procedures fail to detect one-fourth of infected cattle.[8] So, one wonders if the current fad of backyard cookery and the increasing popularity of steak tartare might not contribute to an increase in taeniiasis in the near future.

Despite the high level of sanitation in any country, it still is possible for cattle to be exposed to the eggs of this parasite. One infected person who defecates in a pasture or cattle-feeding area can quickly infect an entire herd. The use of human feces as fertilizer can have the same effect. The eggs are known to remain viable in liquid manure for 71 days, in untreated sewage for 16 days, and on grass for 159 days.[11] Cattle are coprophagous and often will eat human dung, wherever they find it. In India, where cattle roam at will, it is common for a cow to follow a person into the woods, in hopes of obtaining a fecal meal.

Prevention of human infection is easy; when meat is cooked until it is no longer pink in the center, it is safe to eat, for cysticerci are killed at 56° C. Further, meat is also rendered safe by freezing at −5° C for at least a week.

Pathogenesis. Disease characteristics of *T. saginatus* infection are similar to those of infection by any large tapeworm, except that the avitaminosis B_{12} found in association with *D. latum* is unknown. Verminous intoxication, caused by absorption of the worm's excretory products, is common, with the characteristic symptoms of dizziness, abdominal pain, headache, localized sensitivity to touch, and nausea. Delirium is rare but does occur. Neither diarrhea nor intestinal obstruction is uncommon. Hunger pains, universally accepted by laypeople as a symptom of tapeworm infection, are not common, but *loss* of appetite is frequent. Additionally, it is difficult to estimate the psychological effects of observing continued migration of proglottids out of the intestine of an infected person.

Diagnosis and treatment. Identification of taeniid eggs to species is impossible. Therefore, accurate diagnosis depends on examination of a scolex or a gravid proglottid. The latter is characterized by having 15 to 20 lateral branches on each side. Because these branches tend to fuse in old segments freshly passed specimens must be obtained for reliable results.

Numerous taeniicides have been used in the past. Today quinacrine and niclosamide appear to be the drugs of choice.

Taenia solium

The most potentially dangerous adult tapeworm of humans is the pork tapeworm, *T. solium,* because of the possibilities of self-infection with cysticerci. Further, it is possible to infect others in the same household with the larval forms of this parasite, often with grave results.

Morphology. The scolex of the adult (Fig. 21-14) bears a typical, nonretractable taeniid rostellum, armed with two circles of 22 to 32 hooks measuring 130 to 180 μm long. Whereas the scolex of *T. saginata* is cuboidal and up to 2 mm in diameter, that of *T. solium* is spheroid and only half as large. The strobila has been reported as being up to 30 feet long, but 6 to 10 feet is much more common. Mature proglottids are wider than long and are nearly identical with those of *T. saginata,* differing in number of testes (150 to 200 in *T.*

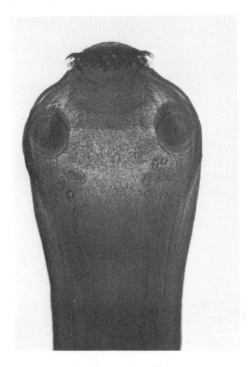

Fig. 21-14. Scolex of *Taenia solium*. Note the large rostellum with two circles of hooks. (Photograph by David Oetinger.)

solium, 300 to 400 in *T. saginatus*). Gravid proglottids are longer than wide and have the typical taeniid uterus, a medial stem with only seven to 13 lateral branches.

Biology. The life cycle of *T. solium* (Fig. 21-15) is in most regards quite like that of *T. saginatus,* except that the intermediate hosts are pigs instead of cattle. Gravid proglottids passed in the feces are laden with eggs infective to swine. When eaten, the oncospheres develop into cysticerci *(Cysticercus cellulosae)* in the muscles and other organs. A person easily becomes infected when a bladderworm is eaten along with insufficiently cooked pork. Evaginating by the same process as *T. saginatus,* the worm attaches to the mucosa of the small intestine and matures in 5 to 12 weeks. Specimens of *T. solium* have been known to live up to 25 years. Pathogenesis caused by the adult worm is similar to that in taeniiasis saginata.

Cysticercosis

Unlike most other species of *Taenia,* the cysticerci of *T. solium* will develop readily in humans. Infection occurs when embryonated eggs pass through the stomach and hatch in the intestine. People who are infected by adult worms may contaminate their households or food with eggs that are accidentally eaten by themselves or others. Possibly, a gravid proglottid may migrate from the lower intestine to the stomach or duodenum, or it may be carried there by reverse peristalsis. Subsequent release and hatching of many eggs at the same time results in a massive infection by cysticerci. Fortunately, this parasite is not common anywhere in the world.

Virtually every organ and tissue of the body may harbor cysticerci. Most commonly they are found in the subcutaneous connective tissues. The second most common site is the eye, followed by the brain, muscles, heart (Fig. 21-16), liver, lungs and coelom. A fibrous capsule of host origin surrounds the larva, except when it develops in the chambers of the eye. The effect of any larva on its host depends on where it is located. In skeletal muscle, skin, or liver, little noticeable pathogenesis usually results, except in massive infection. Ocular cysticercosis may cause irreparable damage to the retina, iris, uvea, or choroid. A developing larva in the retina may be mistaken for a malignant tumor, resulting in the unnecessary surgical removal of the eye. Removal of the larva by fairly simple surgery is usually successful.

Cysticerci are rare in the spinal cord but common in the brain. Symptoms of infection are vague and rarely diagnosed except at autopsy. Pressure necrosis may cause severe CNS malfunction, blindness, paralysis, disequilibrium, obstructive hydrocephalus, or disorientation. Perhaps the most common symptom is epilepsy of sudden onset. When this occurs in an adult with no family or childhood history of epilepsy, cysticercosis should be suspected.[9]

When a cysticercus dies, it elicits a rather severe inflammatory response. Many of them may rapidly prove to be fatal to the host, particularly if the worms are located in the brain. This was observed frequently in former British soldiers of whom a high proportion who had served in India became infected. Other types of cellular reaction also occur, usually resulting in eventual calcification of the parasite (Fig.

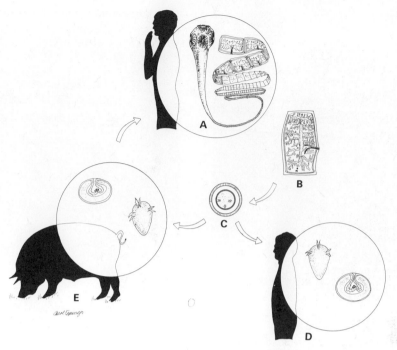

Fig. 21-15. The life cycle of *Taenia solium*. **A,** Adult tapeworm in the small intestine of a human. **B,** Gravid proglottids detach from the strobila and migrate out of the anus or passed with feces. **C,** Egg. **D,** If eaten by a human the oncosphere hatches, migrates to some site in the body, and develops into a cysticercus. **E,** Cysticerci will also develop if the eggs are eaten by a pig. The life cycle is completed when a person eats pork containing live cysticerci. (Diagram by Carol Eppinger.)

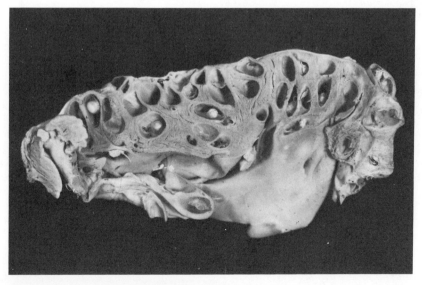

Fig. 21-16. A human heart containing numerous cysticerci of *Taenia solium (Cysticercus cellulosae).* (Photograph by Warren Buss.)

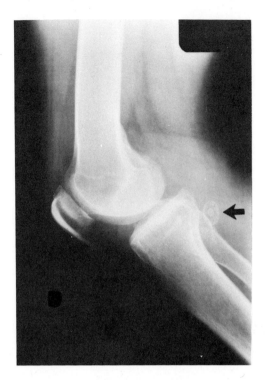

Fig. 21-17. *Cysticercosis cellulosae:* partially calcified cyst found in a routine x-ray of a human leg. (From Raudabush, R. L., and G. A. Ide. 1975. J. Parasitol. 61:512.)

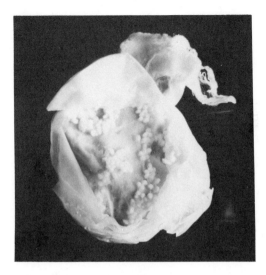

Fig. 21-18. A coenurus larva of *Taenia serialis.* It has been opened to show the numerous protoscolices arising from the germinal epithelium. (Photograph by James Jensen.)

21-17). If this occurs in the eye, there is little chance of corrective surgery.

A cysticercus will rarely become proliferative, developing branching extensions that destroy even more of the host's tissues. The increased danger of such an organism is obvious; futhermore, there is less chance of successful surgery in such cases.

Prevention of cysticercosis depends on early detection and elimination of the adult tapeworm and a high level of personal hygiene. Fecal contamination of food and water must be avoided, and the use of untreated sewage on vegetable gardens eschewed.

Curiously, swine cysticercosis is fairly common, even in countries where the adult is rare. It is evident that the multitudes of eggs produced by even one adult worm overcome great odds in infecting pigs and thereby continue the species.

Other taeniid species of medical importance

Taeniarhynchus confusa is very similar to *T. saginatus.* The gravid segments are about twice as long as those of *T. saginatus,* while the uterus has branched arms that are fewer in number. It has been reported from humans in Africa and the United States; cattle are satisfactory intermediate hosts.

Taenia multiceps, T. glomeratus, T. brauni, and *T. serialis* are all characterized by developing a coenurus type of bladderworm (Fig. 21-18). This is similar to a cysticercus but has many rather than a single scolex. Such coenuri occasionally occur in humans, particularly in the brain, eye, muscles, or subcutaneous connective tissue, where they often grow to be longer than 40 mm. The resulting pathogenesis is similar to that of cysticercosis. The adults are parasites of carnivores, particularly dogs, with herbivorous mammals serving as intermediate hosts. Accidental infection of humans occurs when the eggs are ingested. Coenuriasis of sheep causes a characteristic vertigo called "gid," or "staggers."

Echinococcus granulosus

The genus *Echinococcus* contains the smallest tapeworms in the Taeniidae. Yet their larval forms are often huge and are capable of infecting humans, resulting in hydatidosis, a very serious disease in many parts of the world. An interesting review

Fig. 21-19. Adult *Echinococcus granulosus* from the intestine of a dog. (Courtesy Ann Arbor Biological Center.)

of the ecology and distribution of this and other species of *Echinococcus* was given by Rausch.[21]

Echinococcus granulosus uses carnivores as definitive hosts, especially dogs and other canines. Many mammals may serve as intermediate hosts, but herbivorous species are most likely to become infected by eating the eggs on contaminated herbage.

The adult (Fig. 21-19) lives in the small intestine of the definitive host. It measures 3 to 6 mm long when mature and consists of a typically taeniid scolex, a short neck, and usually only three proglottids. The nonretractable rostellum bears a double crown of 28 to 50 (usually 30 to 36) hooks. The anteriormost proglottid is immature, the middle is usually mature, and the terminal one is gravid. The gravid uterus is an irregular longitudinal sac. The eggs cannot be differentiated from those of *T. solium* and *T. saginatus*. The ripe segment detaches and develops a rupture in its wall, releasing the eggs, which are fully capable of infecting an intermediate host.

Hatching and migration of the onco-

sphere are the same as previously described for *T. saginatus*, except that the liver and lungs are the usual sites of development. By a very slow process of growth, the oncosphere metamorphoses into a type of bladderworm called a **unilocular hydatid** (Fig. 21-20). In about 5 months the hydatid has developed a thick outer, laminated, noncellular layer and an inner, thin, nucleated germinal layer. The inner layer eventually produces the protoscolices that are infective to the definitive host. Protoscolices (Fig. 21-21) are usually produced singly into the lumen of the bladder as in a coenurus (though probably not in naturally infected hosts) and also within **brood capsules.** The latter are smaller cysts, containing ten to 30 protoscolices, which usually are attached to the germinal layer by a slender stalk; they may break free and float within the hydatid fluid. Similarly, individual scolices and brood capsules may break free and sink to the bottom of the bladder, where they are known as **hydatid sand** (Fig. 21-22) (although this may happen only in dead cysts). Rarely, germinal cells penetrate the laminated layer and form **daughter capsules.** When eaten by a carnivore, the cyst wall is digested away, freeing the protoscolices, which evaginate and attach among the villi of the small intestine. A small percentage of hydatids lack protoscolices and are sterile, being unable to infect a definitive host. The worm matures in about 56 days and may live for 5 to 20 months.

Epidemiology. The life cycle of *E. granulosus* in wild animals may involve a wolf-moose, wolf-reindeer, dingo-wallaby, or other carnivore-herbivore relationship. This is known as **sylvatic echinococcosis.** Humans are seldom involved as accidental intermediate hosts in these cases. However, ample opportunities do exist for human infection in situations where domestic herbivores are raised in association with dogs. For example, hydatid disease is a very serious problem in sheep-raising areas of Australia, New Zealand, North and South America, Europe, Asia, and Africa. Similarly, goats, camels, reindeer, and pigs, together with dogs, maintain the cycle in various parts of the world. Dogs are infected when they feed on the offal of butchered animals, and herbivores are infected when they eat herbage contami-

Fig. 21-20. Several unilocular hydatid larvae in the lung of a sheep. Each hydatid contains many protoscolices. (Photograph by James Jensen.)

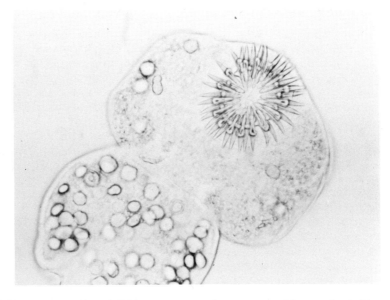

Fig. 21-21. A protoscolex of *Echinococcus granulosus*, removed from a hydatid cyst. (Photograph by Sharon File.)

nated with dog dung. Humans are infected with hydatids when they accidentally ingest *Echinococcus* eggs, usually as a result of fondling dogs.

Local traditions may contribute to massive infections. Some primitive tribes of Kenya, for instance, are said to relish dog intestine roasted on a stick over a camp-fire. Because cleaning of the intestine may involve nothing more than squeezing out its contents, and cooking may entail nothing more than external scorching, these people probably have the highest rate of infection with hydatids in the world. Nelson and Rausch discussed how some infants in Kenya become infected when their

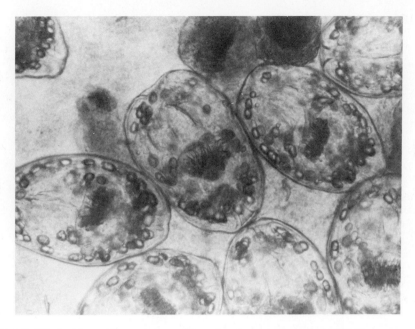

Fig. 21-22. "Hydatid sand," consisting of free-floating protoscolices of *Echinococcus granulosus*. (Photograph by Robert E. Kuntz.)

parents encourage the family dog to clean up the mess when the child vomits or defecates, by licking its face and anal area.[18] A different set of circumstances leads to infection in tanners in Lebanon, where dog feces are used as an ingredient of the tanning solution. Scats picked off the street are added to the vats, and any eggs present may infect their handler by contamination.[22] Sheep herders in the United States risk infection by living closely with their dogs. Surveys of cattle, hogs, and sheep in abattoirs reveal that *E. granulosus* is distributed throughout most of the United States, with greatest concentrations in the deep South and far West. Recent outbreaks have been diagnosed in California and Utah. This disease can be eliminated from an endemic area only by interrupting the life cycle by denying access by dogs to offal, by destroying stray dogs, and by a general education program.

Pathogenesis. The effects of a hydatid may not become apparent for many years after infection because of its slow growth. Up to 20 years may elapse between infection and overt pathogenesis. If infection occurs early in life, the parasite may be almost as old as its host.[3]

The type and extent of pathology depend on the location of the cyst in the

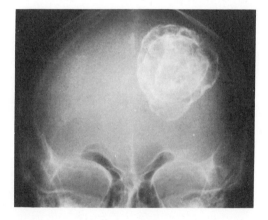

Fig. 21-23. A partially calcified hydatid cyst in the brain. (AFIP neg. no. 68-2740.)

host. As the size of the hydatid increases, it crowds adjacent host tissues and interferes with their normal functions (Fig. 21-23). The results may be very serious. Clinical effects may be manifested relatively early in the infection, before much growth occurs, if the parasite is lodged in the nervous system. When bone marrow is affected, the growth of the hydatid is restricted by lack of space. Chronic internal pressure caused by the parasite usually causes necrosis of the bone, which becomes thin and fragile; characteristically, the first sign of such an infection is a spontaneous fracture of an arm or leg. When the hydatid grows in an unrestricted location, it may become enormous, containing over 15 quarts of fluid and millions of protoscolices. Even if it does not occlude a vital organ, it can still cause sudden death if it ruptures. Hydatid fluid is quite proteinaceous, inducing an adverse host reaction called anaphylactic shock. Unconsciousness and death are nearly instantaneous in such instances.

Diagnosis and treatment. When hydatids are found, it is often during routine medical x-rays. Or, they may be encountered in exploratory surgery. An intradermal immunological test (Casoni's test) is available for use in suspected cases. The antigen, which is manufactured from the proteins in hydatid fluid, is inoculated into the skin, and if a hydatid is or has been in the patient, a characteristic wheal develops at the site of injection. If all signs of the inoculation disappear almost immediately, the results are negative for hydatidosis. Other tests have been developed, utilizing complement fixation, hemagglutination, fluorescent antibody, latex slide agglutination, bentonite flocculation, and precipitin. None is 100% accurate, but they are of sufficient accuracy to make tests most useful on a probability basis.

Surgery remains the only routine method of treatment and then only when the hydatid is located in an unrestricted location. A high rate of surgical success is obtained on ocular hydatidosis. The typical procedure involves incising the surrounding adventitia until the capsule is encountered and aspirating the hydatid fluid with a large syringe. Considerable delicacy is required at this point, for fluid spilled into a body cavity can quickly cause fatal anaphylactic shock. Following aspiration of the cyst contents, 10% formalin is injected into the hydatid to kill the germinal layer. This fluid is withdrawn after 5 minutes, and the entire cyst is then excised.

Echinococcus multilocularis

Echinococcus multilocularis is primarily boreal in its distribution. It is known from Europe, Asia, and North America, having been recently discovered as far south as

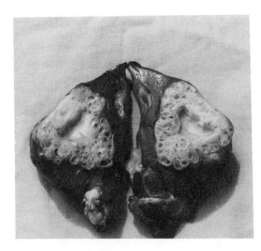

Fig. 21-24. An alveolar hydatid cyst in the liver of a vole, *Clethrionomys*. (Photograph by Robert Rausch.)

South Dakota and Iowa.[15] Cases have also been reported from South America and New Zealand. The adult is mainly a parasite of foxes, but dogs, cats, and coyotes may also serve as definitive hosts. The hydatid develops in several species of small rodents such as voles, lemmings, and mice.

The adult is very similar to *E. granulosus*, differing from it in the following characteristics: (1) *E. granulosus* is 3.0 to 6.0 mm long, while *E. multilocularis* is only 1.2 to 3.7 mm long; (2) the genital pore of *E. granulosus* is about equatorial, but it is preequatorial in *E. multilocularis;* (3) *E. granulosus* has 45 to 65 testes with a few located anterior to the cirrus pouch; *E. multilocularis* has 15 to 30 testes, all located posterior to the cirrus pouch.

The larval form (Fig. 21-24) differs in several respects from that of *E. granulosus*. Instead of developing a thick, laminated layer and growing into large, single cysts, this parasite has a thin, outer wall that grows and infiltrates processes into the surrounding host tissues like a cancer. Each process may have several small, fluid-filled pockets containing several protoscolices. This is particularly true in humans; in natural intermediate hosts the cyst is more regular. In humans, pieces of the cyst sometimes break off and metastasize to other parts of the body. Because of its type of construction, this larval form is called an **alveolar** or **multilocular hydatid.** Some authorities, especially in the Soviet Union, place this species in a separate genus, *Alveococcus,* because of its unique larval form.

Human infection with alveolar hydatid is rare because the normal life cycle is sylvatic rather than urban. Anyone handling wild foxes, however, may be exposed to infection. Thus, this disease is most common among professional trappers and among handlers of sled dogs, where the dogs catch and eat wild mice as a regular part of their diet. Dogs appear to be the only important source of infection in humans.

Diagnosis of alveolar hydatid is difficult; even at necropsy it may be mistaken for a malignant tumor. As a result of the difficulties of liver surgery, excision is usually practical only when the hydatid is localized near the tip of a lobe of the liver; infections of the hilar area are inoperable. The infiltrative nature of the cyst and its slow rate of growth may advance the disease to an inoperable state without its presence being detected.

Alveolar hydatidosis can be prevented only by avoiding dogs and their feces in endemic regions, by carefully washing all strawberries, cranberries, and the like that may be contaminated by dung, and by regularly worming dogs that may be liable to infection.

Echinococcus oligarthrus

Echinococcus oligarthrus of wild Felidae of South and Central America is rarely known to cause hydatidosis in humans.[23] Adult parasites of this species have been found in the jaguarundi, jaguar, and puma. The hydatid is alveolar but less so than that of *E. multilocularis.*

OTHER TAPEWORMS OF HUMANS
Family Hymenolepididae
Hymenolepis nana (Fig. 21-25)

Commonly called the dwarf tapeworm, *Hymenolepis nana* is a cosmopolitan species that is the most common cestode of humans in the world, especially among children. Rates of infection run from 1% in the southern United States to 9% in Argentina.

As its name implies, this is a small species, seldom exceeding 40 mm long and 1 mm wide. The scolex bears a retractable

rostellum armed with a single circle of 20 to 30 hooks. The neck is long and slender, and the proglottids are wider than long. The genital pores are unilateral, and each mature segment contains three testes. After apolysis, the gravid segments disintegrate, releasing the eggs, which measure 30 to 47 μm in diameter. The oncosphere (Fig. 21-26) is covered with a thin, hyaline, outer membrane and an inner, thick membrane with polar thickenings that bear several filaments. The pores that give taeniid eggs their characteristic striated appearance are lacking in this and the other families of tapeworms infecting humans.

The life cycle of *H. nana* is unique in that an intermediate host is optional (Fig. 21-25). When eaten by a person or a rodent, the eggs hatch in the duodenum, releasing the oncospheres, which penetrate the mucosa and come to lie in the lymph channels of the villi. Here each develops into a cysticercoid (Fig. 21-27). In 5 or 6 days the cysticercoid emerges into the lumen of the small intestine, where it attaches and matures.

This direct life cycle is doubtless a recent modification of the ancestral two-host cycle, found in all other species of *Hymenolepis,* for the cysticercoid of *H. nana* can still develop normally within larval fleas and beetles (Fig. 21-28). One reason for the facultative nature of the life cycle is that *H. nana* cysticercoids can develop at higher temperatures than can those of other hymenolepidids. Direct contaminative infection by eggs is probably the most common route in human cases, but accidental ingestion of an infected grain beetle or flea cannot be ruled out.

Besides humans, domestic mice and rats also serve as suitable hosts for *H. nana.* Some authors contend that two subspecies exist: *H. nana nana* in humans and *H. nana fraterna* in murine rodents. Differences do seem to exist in the physiological host-parasite relationships of these two subspecies, for higher rates of infection result from eggs obtained from the same host species than from the other.[20] This is a good example of **allopatric speciation** in action.

Pathological results of infection by *H. nana* are rare and usually occur only in massive infections. Heavy infections can occur through autoinfection,[10] and the

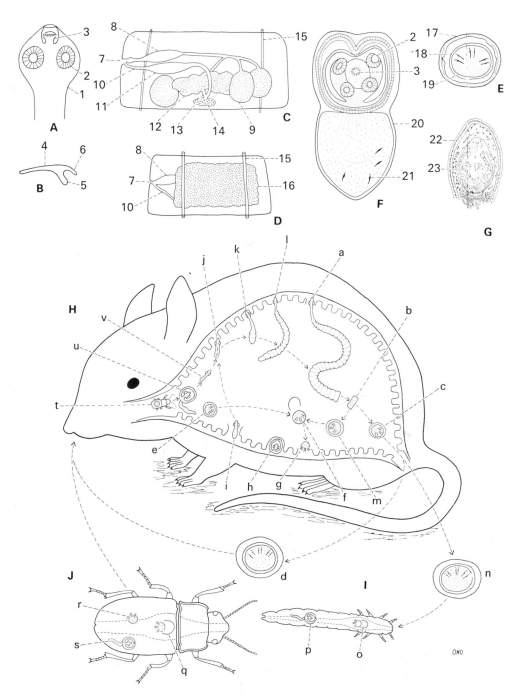

Fig. 21-25. For legend see opposite page.

Fig. 21-25. The life cycle of *Hymenolepis nana,* the dwarf tapeworm. **A,** Scolex; **B,** rostellar hook; **C,** mature proglottid; **D,** gravid proglottid; **E,** embryonated egg; **F,** cysticercoid; **G,** cysticercoid in villus of intestine; **H,** mouse definitive host; **I,** larval intermediate beetle host *(Tenebrio molitor, Tribolium confusum)*; **J,** adult beetle.

1, Scolex; *2,* sucker; *3,* armed rostellum; *4,* handle of hook; *5,* guard; *6,* blade; *7,* common genital pore; *8,* cirrus pouch; *9,* testis; *10,* vagina; *11,* seminal receptacle; *12,* ovary; *13,* oviduct; *14,* vitelline gland; *15,* longitudinal excretory canal; *16,* gravid uterus; *17,* shell or outer membrane of egg; *18,* thick inner membrane with terminal filaments; *19,* oncosphere; *20,* tail of cystercoid; *21,* oncospheral hooks; *22,* intestinal villus; *23,* cysticercoid.

a, Adult worm in small intestine; *b,* gravid proglottid detached from strobilus; *c,* egg in feces passing out of intestine; *d,* egg free; *e,* egg swallowed and returned to small intestine; *f,* egg hatches in small intestine; *g,* oncosphere burrows into intestinal wall; *h,* cysticercoid develops in villus; *i,* cysticercoid breaks out of intestinal wall and evaginates; *j-l* and *a,* cysticercoid attaches to intestinal wall and grows to sexually mature cestode; *m,* some infective eggs do not leave the body, but hatch in the intestine, initiating internal autoinfection in which development proceeds as in *e-l; n,* eggs passed in feces are infective to arthropod intermediate hosts; *o,* eggs swallowed by larva of beetle or flea hatch in intestine; *p,* oncospheres enter hemocoel and develop into cysticercoids; *q,* eggs swallowed by adult beetles hatch in intestine; *r,* oncospheres migrate from intestine into hemocoel; *s,* tailed cysticercoids develop; *t,* infected beetles swallowed by definitive hosts; *u,* cysticercoids released from beetle; *v,* cysticercoids evaginate and shed tail; *j-l* and *a,* cysticercoids develop to adult worms. (From Olsen, O. W. 1974. Animal parasites, their life cycles and ecology, ed. 3. University Park Press, Baltimore.)

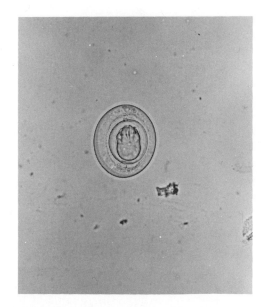

Fig. 21-26. An egg of *Hymenolepis nana.* Note the polar filaments on the inner membrane and the well-developed oncosphere. (Photograph by Jay Georgi.)

symptoms are similar to those already described for *Taeniarhynchus saginatus* intoxication. Treatment with a vermifuge is efficacious but may have to be repeated in a month to remove the worms that were developing in villi at the time of treatment.

Hymenolepis diminuta

Hymenolepis diminuta is a cosmopolitan worm that is primarily a parasite of domestic rats, but many cases of human infection have been reported. It is a much larger species than *H. nana* (up to 90 cm) and differs from it in lacking an armed rostellum. Typical of the genus, it has unilateral genital pores and three testes per proglottid. The eggs (Fig. 21-29) are easily differentiated from those of *H. nana,* for they are at least twice as large and have no polar filaments. It has been demonstrated experimentally that more than 90 species of arthropods can serve as suitable intermediate hosts. Stored-grain beetles (*Tribolium* spp.) are probably most commonly involved in infections of both rats and humans. A household shared with rats is also likely to have its cereal foods infested with beetles.

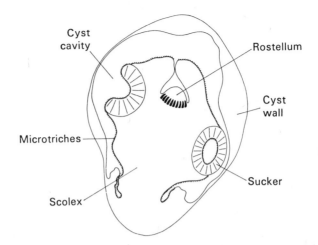

Fig. 21-27. *Hymenolepis nana:* a diagrammatic representation of a longitudinal section through a cysticercoid from a mouse villus. (From Caley, J. 1975. Zeitschrift für Parasitenkunde 47:218-228.)

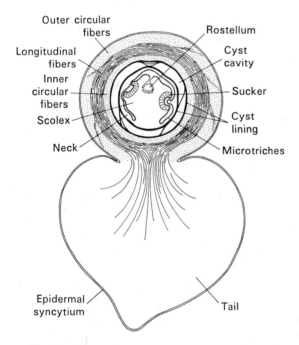

Fig. 21-28. A diagrammatic representation of a longitudinal section through a cysticercoid of *Hymenolepis nana* from the insect host. (From Caley, J. 1975. Zeitschrift für Parasitenkunde 47:218-228.)

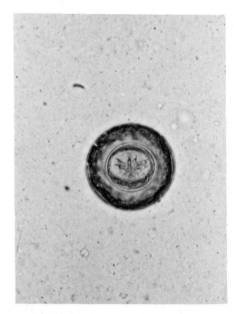

Fig. 21-29. The egg of *Hymenolepis diminuta.* (Photograph by Jay Georgi.)

The ease with which this parasite is maintained in laboratory rats and beetles makes it an ideal model for many types of experimental studies; its physiology has been more thoroughly examined than that of any other tapeworm.

Treatment is as recommended for *H. nana.*

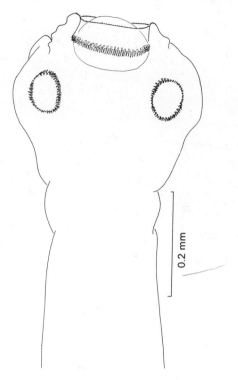

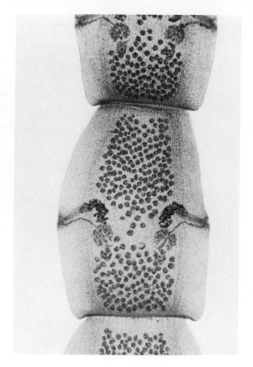

Fig. 21-30. Scolex of *Raillietina*. The suckers are weak and have a double circle of spines, and the massive rostellum has many hammer-shaped hooks. (Drawing by Thomas Deardorff.)

Fig. 21-31. A mature segment of *Dipylidium caninum*, the "double-pored tapeworm" of dogs and cats. The two vitelline glands are directly behind the larger ovaries. The smaller spheres are testes. (Courtesy Ann Arbor Biological Center.)

Family Davaineidae
Raillietina spp.

The following species of *Raillietina* have been reported from humans: *R. siriraji, R. asiatica, R. garrisoni, R. celebensis,* and *R. demarariensis.* All normally parasitize domestic rats and possibly represent no more than two actual species. The genus is easily recognized by its large rostellum with hundreds of tiny, hammer-shaped hooks and by its spiny suckers (Fig. 21-30). The life cycle is known for none of them, but because several other species of *Raillietina* are known to use ants or other insects as intermediate hosts, it seems probable that the epidemiology of this infection is similar to that of *Hymenolepis diminuta.* The clinical pathology of raillietiniasis is unknown.

Raillietina cesticillus is one of the most common poultry cestodes in North America, and a wide variety of grain, dung, and ground beetles serve as intermediate hosts.

Family Dilepididae
Dipylidium caninum

A cosmoplitan, common parasite of domestic dogs and cats, *Dipylidium caninum* has been found many times in children.[16] It is easily recognized because each segment has two sets of male and female reproductive systems and a genital pore on each side (Fig. 21-31). The scolex has a retractable, rather pointed rostellum with several circles of rosethorn-shaped hooks. The uterus disappears early in its development and is replaced by hyaline, noncellular egg capsules, each containing eight to 15 eggs. Gravid proglottids detach and either wander out of the anus or are passed with feces. They are very active at this stage and are the approximate size and

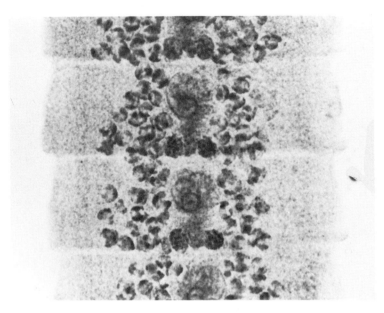

Fig. 21-32. *Mesocestoides*, a cyclophyllidean cestode with a midventral genital pore and a bilobed vitellarium. (Photograph by Larry Shults.)

Fig. 21-33. Tetrathyridea larvae of *Mesocestoides* sp. in the mesenteries of a baboon, *Papio cyanocephalus*. (Photograph by Robert E. Kuntz.)

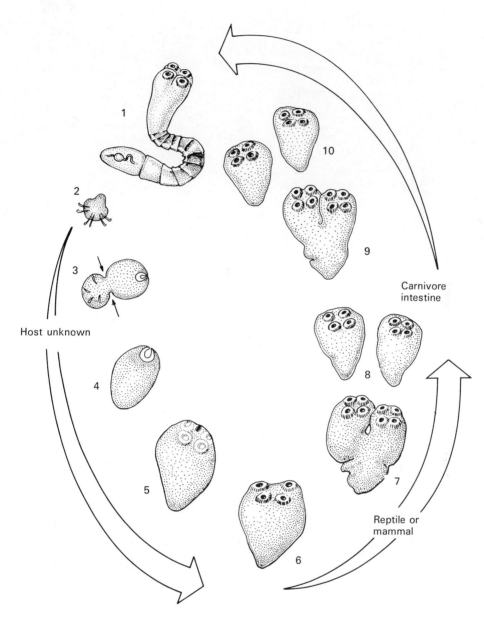

Fig. 21-34. Developmental sequence of *Mesocestoides corti,* illustrating early developmental stages *(2-5),* tetrathyridium and asexual multiplication in second intermediate host *(6-8),* and asexual multiplication with subsequent formation of adult worms in intestine of definitive host *(9-10).* Not illustrated here is the potential reinvasion of tissues from intestinal lumen of carnivore with continuing asexual multiplication of tetrathyridial stage. (From Voge, M. 1969. In Schmidt, G. D. Problems in systematics of parasites. University Park Press, Baltimore.)

shape of cucumber seeds. As the detached segments begin to desiccate, the egg capsules are released. Fleas are the usual intermediate hosts, although chewing lice have also been implicated. Unlike the adult, a larval flea has simple, chewing mouth-parts and feeds on organic matter, which may include *Dipylidium* egg capsules. The resulting cysticercoids survive their host's metamorphosis into the parasitic adult stage, when the flea may be nipped or licked out of the fur of the dog or cat, thereby completing the life cycle. This, by the way, is an example of **hyperparasitism,** since the flea is itself a parasite.

Nearly every reported case of infection of humans has involved a child. This may reflect adult resistance or may simply be a result of the familiarity of a child with a dog, with increased chances of a flea being accidentally swallowed. The symptoms and treatment are the same as for *Hymenolepis nana.*

Family Anoplocephalidae
Bertiella studeri

Normally a parasite of wild, Old World primates, *Bertiella studeri* has been reported many times from humans, especially in southern Asia, the East Indies, and the Philippines. The scolex is unarmed, and the proglottids are much wider than they are long, with the ovary located between the middle of the segment and the cirrus pouch. The egg is characteristic: 45 to 50 μm in diameter, with a bicornuate **piriform apparatus** on the inner shell.

Ripe segments are shed in chains of about a dozen at a time. The intermediate hosts are various species of oribatid mites.[24] These free-living animals feed on organic detritus and readily ingest eggs of *B. studeri*. Accidental ingestion of mites infected with cysticercoids completes the life cycle within primates. No pathology has been ascribed to infection by *B. studeri*. Treatment is as for *Hymenolepis nana.*

Bertiella mucronata is very similar to *B. studeri* and has also been reported from humans. It appears to be a parasite of New World monkeys, and children may become infected when living with a pet monkey and the ubiquitous oribatid mite. Distinguishing this species from *B. studeri* normally requires a specialist.

Inermicapsifer madagascariensis

Inermicapsifer madagascariensis is normally parasitic in African rodents, but it has been reported repeatedly in humans in several parts of the world, including South America and Cuba. Baer concluded that humans are the only definitive host outside of Africa.[2]

The scolex is unarmed. The strobila is up to 42 cm long. Mature proglottids are somewhat wider than long and bear a centrally located ovary. The uterus is replaced by egg capsules in ripe segments, each capsule containing six to 10 eggs, which do not possess a piriform apparatus.

The life cycle of this parasite is unknown but undoubtedly involves an arthropod intermediate host. Clinical pathology has not been studied, and treatment is similar to that described for other species.

Family Mesocestoididae
Mesocestoides sp.

Unidentified specimens of the genus *Mesocestoides* have occasionally been reported from humans in Denmark, Africa, United States, Japan, and Korea. The ventromedial location of the genital pores is clearly diagnostic of the genus (Fig. 21-32). The complete life cycle is not known for any species in this difficult family, but many have a rodent or reptile intermediate host, in which a cysticercoid-type larva known as a **tetrathyridium** (Figs. 20-30 and 21-33) develops. Neither mammals nor reptiles can be infected directly by eggs, so a first host must be involved. As yet, such a host has not been identified (Fig. 21-34). Pathology and treatment of humans have not been studied.

REFERENCES

1. Archer, D. M., and C. A. Hopkins. 1958. Studies on cestode metabolism, III. Growth pattern of *Diphyllobothrium* sp. in a definitive host. Exp. Parasitol. 7:125-144.
2. Baer, J. G. 1956. The taxomomic position of *Taenia madagascariensis* Davaine, 1870, a tapeworm parasite of man and rodents. Ann. Trop. Med. Parasitol. 50:152-156.
3. Barnett, L. 1939. Hydatid disease: errors in teaching and practice. Br. Med. J. 2:593-599.
4. von Bonsdorff, B. 1956. *Diphyllobothrium latum* as a cause of pernicious anemia. Exp. Parasitol. 5:207-230.
5. Chandler, A. C., and C. P. Read. 1961. Introduc-

tion to parasitology, ed. 10, John Wiley & Sons, Inc., New York.

6. Corkum, K. C. 1966. Sparganosis in some vertebrates of Louisiana and observations on human infection. J. Parasitol. 52:444-448.
7. Davis, A. 1973. Drug treatment in intestinal helminthiases. World Health Organization, Geneva.
8. Dewhirst, L. W., J. D. Cramer, and J. J. Sheldon. 1967. An analysis of current inspection procedures for detecting bovine cysticercosis. J. Am. Vet. Med. Assoc. 150:412-417.
9. Dixon, H. B. F., and D. W. Smithers. 1934. Epilepsy in cysticercosis *(Taenia solium)*. A study of seventy-one cases. Q. J. Med. 3:603-616.
10. Heyneman, D. 1962. Studies of helminth immunity: I. Comparison between lumenal and tissue phases of infection in the white mouse by *Hymenolepis nana* (Cestoda: Hymenolepididae). Am. J. Trop. Med. Hyg. 11:46-63.
11. Jepsen, A., and H. Roth. 1952. Epizootiology of *Cysticercus bovis*—resistance of the eggs of *Taenia saginata*. Report 14. Int. Vet. Cong. 22:43-50.
12. Kamegai, S., A. Ichihara, H. Nonobe, M. Machida, and T. Hara. 1968. A case of human infection with the immature worm of *Diplogonoporus grandis*. Res. Bull. Meguro Parasitol Mus. 2:1-8.
13. Keller, M. 1937. Sur une nouvelle méthode de traitmente du la sparganose oculaire. Ann. École Super. Med. Pharm. Indochine. 1:77-89.
14. Kuntz, R. E. 1963. Snakes of Taiwan. Q. J. Taiwan Mus. 16:1-79.
15. Leiby, P. D., W. P. Carney, and C. E. Woods. 1970. Studies on sylvatic echinococcosis. III. Host occurrence and geographic distribution of *Echinococcus multilocularis* in the north central United States. J. Parasitol. 56:1141-1150.
16. Moore, D. V. 1962. A review of human infections with the common dog tapeworm, *Dipylidium caninum,* in the United States. Southwestern Vet. 15:283-288.
17. Mueller, J. F. 1974. The biology of *Spirometra*. J. Parasitol. 60:3-14.
18. Nelson, G. S., and R. L. Rausch. 1963. *Echinococcus* infections in man and animals in Kenya. Ann. Trop. Med. Parasitol. 57:136-149.
19. Nyberg, W. 1958. Absorption and excretion of vitamin B_{12} in subjects infected with *Diphyllobothrium latum* and in noninfected subjects following oral administration of radioactive B_{12}. Acta Haematol. 19:90-98.
20. Pampiglione, S. 1962. Indagine sulla diffusion dell'imenolepiasi nella Sicilia occidentale. Parassitologia 4:49-58.
21. Rausch, R. L. 1967. On the ecology and distribution of *Echinococcus* spp. (Cestoda: Taeniidae), and characteristics of their development in the intermediate host. Ann. Parasitol. 42:19-63.
22. Schwabe, C. W., and K. A. Daoud. 1961. Epidemiology of echinococcosis in the Middle East. I. Human infection in Lebanon, 1949-1959. Am. J. Trop. Med. Hyg. 10:374-381.
23. Sousa, O. E., and J. D. Lombardo Ayala. 1965.

Informe de un caso de hidatidosis en sujeto nativo panameño. Primer caso autoctono. Arch. Med. Panameños 14:79-86.
24. Stunkard, H. 1940. The morphology and life history of the cestode *Bertiella studeri*. Am. J. Trop. Med. 20:305-332.
25. Swartzwelder, J. C., P. C. Beaver, and M. W. Hood. 1964. Sparganosis in southern United States. Am. J. Trop. Med. Hyg. 13:43-48.
26. Yamane, Y., N. Okada, and M. Takihara. 1975. On a Case of long term migration of *Spirometra erinacei* larva in the breast of a woman. Yonago Acta Medica. 19:207-213.

SUGGESTED READING

von Brand, T. 1966. Biochemistry of parasites. Academic Press, Inc., New York. (The standard reference on parasite physiology.)
Faust, E. L., P. F. Russell, and R. C. Jung. 1970. Craig and Faust's Clinical parasitology, ed. 8. Lea and Febiger, Philadelphia. (A standard reference, useful to clinician and zoologist alike.)
Fuhrmann, O. 1932. Les ténias des oiseoux. Mem Univ. Neuchâtel, vol. 8. (A synopsis of the genera and species of bird tapeworms up to 1932.)
Hoffman, G. L. 1967. Parasites of North American freshwater fishes. University of California Press, Berkeley. (Keys to all genera of tapeworms of North American fishes with lists of species and hosts.)
Jackson, G. J., R. Herman, and I. Singer. 1970. Immunity to parasitic animals, vol. 2. Appleton-Century-Crofts, New York. pp. 295-1217.
Olsen, O. W. 1974. Animal parasites: their life cycles and ecology, ed. 3. University Park Press, Baltimore. (An excellent atlas of cestode life cycles.)
Read, C. P. 1970. Parasitism and symbiology. Ronald Press, New York. (An interesting treatise on symbiotic relationships. Useful to beginner and professional alike.)
Soulsby, E. J. L. 1965. Textbook of veterinary clinical parasitology, vol. 1. F. A. Davis Co., Philadelphia. (This is a useful reference to tapeworms of veterinary significance.)
Spasskaya, L. P. 1966. Cestodes of birds of SSSR. Akad, Nauk SSSR, Moscow. (A useful, illustrated survey of tapeworms of northern birds.)
Taylor, A. E. R., and J. R. Baker. 1968. The cultivation of parasites in vitro. Blackwell, Oxford. (This is a compendium of knowledge of in vitro cultivation up to 1968.)
Stunkard, H. W. 1962. The organization, ontogeny and orientation of the cestodes. Q. Rev. Biol. 37:23-34.
Voge, M. 1967. The post-embryonic developmental stages of cestodes. In Dawes, B., editor, Advances in parasitology, vol. 5. Academic Press, Inc., New York. pp. 247-297.
Ward, H. B. 1930. The introduction and spread of the fish tapeworm *(Diphyllobothrium latum)* in the United States. The Williams & Wilkins Co., Baltimore.

Chapter 22

PHYLUM NEMATODA: FORM, FUNCTION, AND CLASSIFICATION

Nematodes are among the most abundant animals on earth. Of course, more species of insects have been described, but when one realizes that nearly every kind of insect examined harbors at least one species of parasitic nematode, and when one further calculates the number of kinds of nematodes parasitic in the rest of the animal kingdom, there is no contest. There are also many species of nematodes that parasitize plants. Finally, the species of free-living marine, freshwater, and soil-dwelling nematodes probably far outnumber those that are parasitic. Numbers of individuals are often extremely high; 90,000 were found in a single rotting apple; 1,074 individuals, representing 36 species, were counted in 6 to 7 ml. of mud; and 3 to 9 billion per acre may be found in good farmland in the United States.[44]

Obviously, these animals have a lot going for them. We hope that this chapter will give some insight into their enormous success.

Most nematodes are small, inconspicuous, and apparently unimportant to humans, and therefore attract the attention only of specialists. A few, however, cause diseases of extreme importance to humans and domestic and wild plants and animals. These are the roundworms that attract the attention of parasitologists.

HISTORICAL ASPECTS

Ancient people were probably familiar with the larger nematodes, which they encountered when they slew game or disemboweled an opponent at war. The earliest records mention the worms or contain recognizable allusions to them. Aristotle discussed the worm we now call *Ascaris lumbricoides*, and the Ebers Papyrus of 1550 B.C. Egypt described clinical hookworm disease, as did Hippocrates, Lucretius, and the ancient Chinese. Moses wrote of a scourge that probably was caused by the guinea worm. The Arabians Avicenna and Avenzoar, who kept parasitology alive during the Dark Ages in Europe, studied elephantiasis, differentiating it from leprosy.[24]

Linnaeus, in 1758, placed the roundworms in his class Vermes, along with all other worms and worm-like animals. Goeze, Zeder, and Rudolphi made great advances in recognition of various nematodes, although they still believed the worms arose by spontaneous generation. Further work by Gegenbaur, Huxley, Hatschek, Leuckart, Beneden, Diesing, Linstow, Looss, Railliet, Stossich, and many others established the nematodes as a distinct and important group of animals.

The name Nematoda is a modification of Rudolphi's Nematoidea and was placed in Nemathelminthes, itself first considered a class in phylum Vermes, by Gegenbaur in 1859 and later was elevated to phylum status. Although still employed by several authors today, the name Nemathelminthes is little used. The name Aschelminthes was proposed by Grobben in 1910 as a superphylum to contain several divergent groups of worm-like animals that had in common a pseudocoelomic body cavity. Hyman resurrected the name Aschelminthes for use as a phylum containing the classes Nematoda, Rotifera, Priapulida, Gastrotricha, Nematomorpha, and Kinorhyncha.[29] Today, most authorities consider each of these groups as separate phyla, often placing them in the superphylum Aschelminthes, recognizing their rather distant phyletic relationship. Attempts by a few students of free-living nematodes to change Nematoda to Nema or Nemata have not been widely accepted.

Curiously, studies of nematodes have

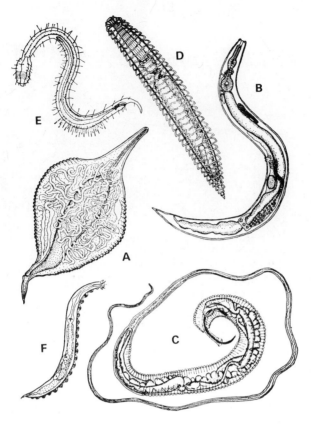

Fig. 22-1. Variety of form in the nematodes: **A,** *Tetrameres;* **B,** *Rhabditis;* **C,** *Trichuris;* **D,** *Criconema;* **E,** *Chaetosoma;* **F,** *Bunonema.* (From Crofton, H. D. 1966. Nematodes. Hutchinson University Library, London.)

developed along two separate lines, with parasitologists claiming the parasites of vertebrates and nematologists accounting for free-living and plant and invertebrate-parasitic roundworms. This doubtless is because of the parasitologists' historical concern for parasites of medical and veterinary importance. Exceptions occur, of course, for a few individuals, such as Chitwood, Bird, Crites, Inglis, Mawson, Schuurmanns-Steckhoven, and others have made significant contributions to both areas. Each discipline publishes in its own journals, uses its own terminology and taxonomic formulas, and to a large extent employs different techniques for handling and preparing specimens for study. Recent trends suggest that the two disciplines are beginning to merge. If so, the science of nematodology will assuredly benefit, for each school of thought will profit from the special knowledge of the other, and the waste engendered by lack of communication will be minimized.

FORM AND FUNCTION

Typical nematodes are elongate, tapered at both ends, bilaterally symmetrical, and possess a pseudocoel, that is, a body cavity derived from the embryonic blastocoel. There are many varieties of this basic shape, however (Fig. 22-1). The digestive system is complete, with a mouth at the extreme anterior end and an anus near the posterior tip. The lumen of the pharynx is characteristically triradiate. The body is covered with a noncellular cuticle that is secreted by an underlying hypodermis and is shed four times during ontogeny. The muscles of the body wall are only one layer thick, and a distinguishing feature is that they are all *longitu-*

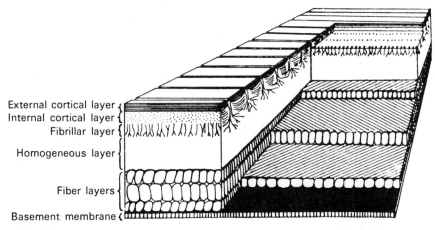

External cortical layer
Internal cortical layer
Fibrillar layer
Homogeneous layer
Fiber layers
Basement membrane

Fig. 22-2. Diagram showing transverse, longitudinal, and tangential sections of the cuticle of *Ascaris*. The strands of each of the three fiber layers run at an angle of about 75° to the longitudinal axis of the worm, and the strands of the middle layer run about 135° from those of the inner and outer layers. (From Bird, A. F., and K. Deutsch. 1957. Parasitology 47:319-329.)

dinally arranged with no separate circular layer. The excretory system consists of either lateral canals or ventral glands, or both, which open near the anterior end through a ventral excretory pore. Except for some sensory endings of modified cilia, neither cilia nor flagella are present, even in the male gamete. Most nematodes are dioecious and show considerable sexual dimorphism, with the females usually larger and the tail of the male being the more curled. Some free-living and plant-parasitic species are hermaphroditic and others are parthenogenetic. The female reproductive system opens through a ventral genital pore, and the male system opens into a cloaca, together with the digestive system. Adult nematodes vary in size from less than 1 mm, as in *Caenorhabditis*, to over a meter, as in *Dracunculus*.

A considerable body of knowledge has accumulated on the function and structure (both at the light and electron microscope levels) of nematodes, far beyond our ability to review within the confines of this chapter. Many reviews and literature references are available.*

Body wall

The body wall of nematodes is comprised of the cuticle, hypodermis, and

_____
*See references 6, 10-12, 18, 21, 35, 37, 40, 45, 51, 52, and 68.

body wall musculature. The outermost covering is the **cuticle,** a complex structure of great functional significance to the animals. The cuticle also lines the stomodeum, proctodeum, excretory pore, and vagina. It is basically of three regions; the cortical, the middle (also called homogeneous or matrix), and the inner fibrous layers, and these are commonly subdivided. The cuticle in *Ascaris* seems to be fairly typical, and a total of nine regions can be distinguished (Fig. 22-2). The cortex is covered by what appears to be a thin layer of lipid less than 0.1 μm thick, although some evidence suggests that this may be a trilaminar membrane.[7] The main body of the cortex is divided into an **external** and an **internal cortical layer.** The external layer is a stabilized protein related to keratin, recognized because disulfide cross-links have been detected (see trematode eggshell formation, Chapter 16). In some species quinone-tanning may play a role in stabilizing the protein in this layer. The inner cortical layer, as well as the other layers in the cuticle, are primarily of collagen, a protein type also abundant in vertebrate connective tissue. The external layer is amorphous and electron dense and is interrupted by a series of fine, transverse grooves. These are punctuated by a series of apparent pores leading into canals extending through the inner cortical layer to the **fibrillar layer.** The pore canals may function as means to transport substances to the surface of the cuticle, as skeletal supports, or both. Even under the electron microscope one can distinguish little structure in the **middle layer,** though some researchers have reported fine striations. Three **fiber layers,** each of parallel strands of collagen-like protein, run at an angle of about 75 degrees to the longitudinal axis of the worm. Strands of the middle layer run at an angle of about 135 degrees to those of the inner and outer layers, which are parallel to each other, thus forming a lattice-like arrangement. The fibrous layers form an important component of the hydrostatic skeleton; the strands themselves are not extensible but allow for longitudinal expansion and contraction of the overlying cuticle by change in the angles between the layers. The innermost layer of the cuticle is the **basal lamella,** a layer of fine fibrils that merges with the underlying hypodermis. Transverse annulations occur in the cuticle, aiding flexibility of the animal. These striations are more prominent in some species than in others.

Cuticular markings and ornamentations of many types occur in various kinds of nematodes. These include shallow **punctations,** deeper **pores,** and **spines** of varying complexity.[30,78] Lateral or sublateral cuticular thickenings called **alae** are present in many species. Cervical alae (Fig. 22-3) are found on the anterior part of the body, caudal alae are on the tail ends of some males, and longitudinal alae, when present, extend the entire length of both sexes. The lateral alae may be of value to the animal when it is swimming or lend greater stability on solid substrate, when the nematode is crawling on its side by dorsoventral undulations, as in the larva of *Nippostrongylus brasiliensis* (Fig. 22-4).[38] Longitudinal ridges occur in many adult trichostrongylids, and a species in which cuticular ultrastructure has been studied is *N. brasiliensis.* Lee reported that the ridges are supported by a series of struts or skeletal rods in the middle layer of the cuticle (Fig. 22-5).[36] The struts are held erect by collagenous fibers inserted in the cortical and fibrous layers, but the middle layer itself is fluid-filled and contains hemoglobin. The function of the longitu-

Fig. 22-3. Scanning electron micrograph of *Toxocara cati,* illustrating cervical alae. (Photograph by John Ubelaker.)

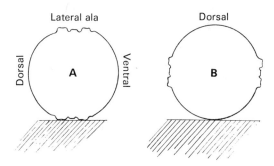

Fig. 22-4. Silhouette of a transverse section through the third-stage larva to show the more stable position of the larva when it lies on its side **(A),** as it does when moving by two-dimensional undulatory propulsion, and the less stable position, when it lies on its ventral surface **(B).** (From Lee, D. L. 1964. In Taylor, A. E. R., editor. *Nippostrongylus* and *Toxoplasma.* Blackwell Scientific Publications, Oxford.)

dinal ridges in trichostrongylids is both to aid in locomotion, as the worm moves between villi with a corkscrew-type motion (Fig. 22-6), and to abrade the microvillar surface, thus helping to obtain food in the absence of biting or piercing mouthparts.[38]

The **hypodermis** lies just beneath the basal lamella of the cuticle. It is usually syncytial in adult worms, and the nuclei lie in four thickened portions that project

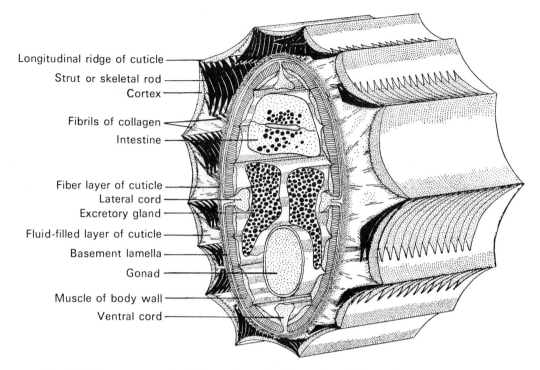

Longitudinal ridge of cuticle
Strut or skeletal rod
Cortex
Fibrils of collagen
Intestine
Fiber layer of cuticle
Lateral cord
Excretory gland
Fluid-filled layer of cuticle
Basement lamella
Gonad
Muscle of body wall
Ventral cord

Fig. 22-5. Stereogram of a thick section taken from the middle region of an adult *Nippostrongylus brasiliensis* to show the arrangement of the various layers in the cuticle. (From Lee, D. L. 1965. Parasitology 55:174.)

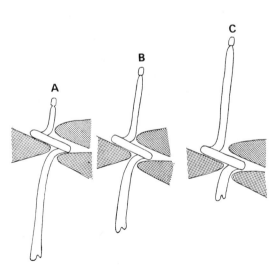

A   B   C

Fig. 22-6. The locomotion of an adult *Nippostrongylus* removed from the intestine of the host and placed among moist sand grains. It is probable that similar movements are performed by the nematode among the villi of the host intestine. (From Lee, D. L. 1965. The physiology of nematodes. Oliver and Boyd Ltd., Edinburgh.)

into the pseudocoel, the **hypodermal cords.** The hypodermal cords run longitudinally and divide the somatic musculature into four quadrants. On large nematodes, these may be discernible with the naked eye as pale lines. The dorsal and ventral cords contain longitudinal nerve trunks, while the lateral cords contain the lateral canals of the excretory system in most species. Especially in the regions of the cords, the hypodermis contains mitochondria and endoplasmic reticulum. An important function of the hypodermis is secretion of the cuticle, described in the section on development. In at least two genera of trichuroids, *Trichuris* and *Capillaria,* specialized areas of the hypodermis, the **bacillary bands,** occur. They open through pores lateral to the esophagus in *Trichuris* and extend the length of the body in *Capillaria.* The function of the bacillary bands is unknown, but they contain both apparently glandular and nonglandular cells.[77]

Musculature and pseudocoel

The somatic musculature is technically a part of the body wall, but it is convenient

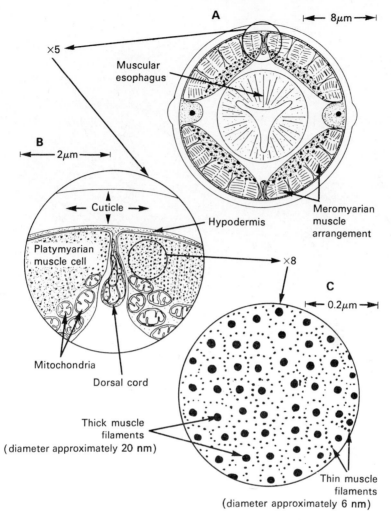

Fig. 22-7. Diagrams depicting a typical meromyarian-platymyarian muscle type at different levels of magnification. **A,** Whole transverse section; **B,** portion of muscle cells on either side of dorsal nerve cord; and **C,** two types of muscle filaments as seen at high resolution with the aid of the electron microscope. (From Bird, A. F. 1971. The structure of Nematodes. Academic Press, Inc., New York.)

to consider its function along with that of the pseudocoel, and indeed they, along with the cuticle, function together as a **hydrostatic skeleton.** The hydrostatic skeleton will be discussed further.

Schneider (1866) ascribed taxonomic significance to certain aspects of muscle morphology and coined several terms.[64] The terms are still in use because they are convenient to describe forms and arrangements of muscle cells, though it is recognized now that these arrangements are of little taxonomic value. Schneider's terms describe muscle cells with respect to rows

per quadrant and shape and distribution of contractile sarcoplasm. If there are no rows or only two rows of cells between adjacent cords, the arrangement is **holomyarian.** If there are a few rows (two to five) per quadrant, the arrangement is described as **meromyarian,** and, if there are many, then it is called **polymyarian.** The **platymyarian** muscle cell is rather ovoid in cross section, contains its contractile fibrils at one side, adjacent to the hypodermis, and has a noncontractile portion of about the same width bulging into the pseudocoel (Fig. 22-7). The noncontractile

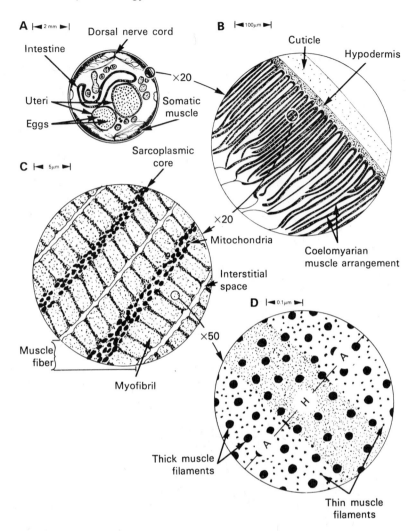

Fig. 22-8. Diagrams depicting the typical polymyarian-coelomyarian muscle type of *Ascaris lumbricoides* over a wide range of magnifications. **A,** Whole transverse section; **B,** part of the muscle quadrant between the dorsal nerve and lateral hypodermal cord; **C,** fibers of two muscle cells; **D,** an *H* and two *A* bands and the two types of muscle filaments at high resolution. (From Bird, A. F. 1971. The structure of Nematodes. Academic Press, Inc., New York.)

portion contains the nuclei, large mito- chondria with numerous cristae, ribo- somes, endoplasmic reticulum, glycogen, and lipid. The **coelomyarian** cell is more spindle-shape, with the contractile portion at the distal end in the shape of a narrow U (Fig. 22-8). The distal end of the U is placed against the hypodermis, the con- tractile fibrils extend up along its sides, and the space in the middle is tightly packed with mitochondria. In some cases the elongate contractile portion does not

sandwich the mitochondria, but these or- ganelles are concentrated in the distal por- tion of the cell body close to the contractile fibrils.[75] The "cell body," or perikaryon, bulges medially into the pseudocoel. It contains the nucleus, some mitochondria, endoplasmic reticulum, a Golgi body, and a large amount of glycogen. One impor- tant function of the coelomyarian peri- karyon seems to be as a glycogen storage depot. In the **circomyarian** cell, the con- tractile fibrils at the periphery entirely en-

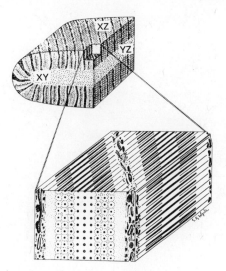

Fig. 22-9. Diagram of a muscle fiber showing the pattern of striation in three planes. In the XZ plane, the myofilaments are staggered with the result that the striations are oblique rather than transverse. A second consequence of the stagger is that the adjacent rows of myofilaments do not reach the XY plane in phase resulting in the appearance of striation in this plane also. The YZ plane shows cross-striation. (From Rosenbluth, J. 1965. J. Cell Biol. 25:510.)

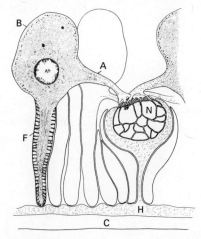

Fig. 22-10. Diagram of muscle cells and myoneural junctions in transverse section. The belly *(B)*, containing the nucleus of the muscle cell, is continuous with the core of the striated fiber *(F)* and with the elongated arm *(A)*. The arm subdivides as it approaches the nerve cord *(N)*. The individual axons comprising the nerve cord are embedded in a trough-like extension of the hypodermis *(H)*, which underlies the animal's cuticle *(C)*. (From Rosenbluth, J. 1965. J. Cell Biol. 26:580.)

circle the noncontractile portion of the cell.

The foregoing classification of muscle types seems to have little taxonomic or evolutionary significance other than that meromyarian-platymyarian is usual in small nematodes and larvae, while polymyarian-coelomyarian is associated with larger, specialized parasites. The circomyarian form is often shown by specialized muscles, such as those functioning in ingestion, digestion, defecation, and reproduction. They may be circomyarian in only one area, becoming coelomyarian then platymyarian closer to their origin.[6] Functioning of these muscles will be discussed further in reference to the digestive and reproductive tracts.

The myofilaments seem to be essentially similar in all muscle types and are of two sizes; thick filaments of about 23 nm in diameter and thin filaments of about 8 nm. The thick filaments are made up of subunits of about 5 nm. It is believed that con-

traction occurs in a manner similar to the Hanson-Huxley model for vertebrate striated muscle, the thick filaments containing myosin and the thin filaments of actin. The actin filaments slide past the myosin filaments in contraction. The A, H, and I band typical of striated muscle can be distinguished, but Z lines are absent. Thus the structure of nematode muscle is similar to vertebrate striated muscle and insect flight muscle and is referred to as "obliquely striated" (Fig. 22-9).[53-55]

Nematode muscles are very unusual, for processes run from the perikarya to the nerves, rather than processes from the nerve perikarya running to the muscles (Fig. 22-10). (This condition is also found in the myotome of *Amphioxus.*) Another unusual feature of nematode muscle cells is their frequent muscle-muscle connectives, at least in coelomyarian types.[76] These occur most often in the anterior regions of the worms and between the innervation processes of the muscle cells,

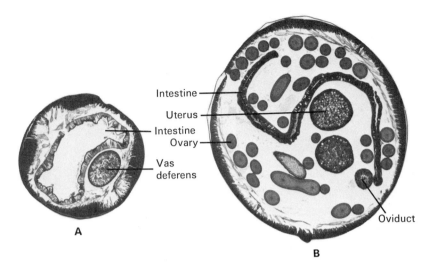

Intestine
Uterus
Intestine
Ovary
Vas
deferens
Oviduct

A

B

Fig. 22-11. Cross sections of male and female *Ascaris suum*. Space between organs is pseudocoel **A,** Male; **B,** female. (Photograph by Warren Buss.)

though they may be between perikarya. There is sometimes cytoplasmic continuity between the cells (cytoplasmic bridges) and sometimes continuity of the external layers (sarcolemmal bridges). A higher degree of muscular coordination is thought to result from transmission of nerve impulses between muscle cells that are so connected.

The somatic musculature and the rest of the body wall enclose a fluid-filled cavity, the **pseudocoelom** or **pseudocoel** (Fig. 22-11). The pseudocoel differs from the true coelom in that it is derived from a persistent embryonic blastocoel, rather than being a cavity within the endomesoderm, and, therefore, has no peritoneal (mesodermal) lining. The nematode coelom functions as a hydrostatic skeleton. Hydrostatic skeletons are widespread in invertebrates. Their function depends on the enclosure of a volume of fluid, the ability of muscle contraction to apply pressure to that fluid, and the transmission of the pressure in all directions in the fluid as the result of its incompressibility. Thus, in a simple case, simultaneous contraction of circular muscles and relaxation of longitudinal muscles will cause an animal to become thinner and longer, while relaxation of circular muscles and a contraction of longitudinal muscles make an animal shorter and thicker. However, in nematodes the somatic musculature is entirely composed of longitudinal fibers, and the muscles act

not against other antagonistic muscles but against forces exerted by the internal pressure on the cuticle.[28] An increase in efficiency and strength of this system in locomotion can only be achieved by an increase in the pressure of the pseudocoelomic fluid. Measurements on the hydrostatic pressure (turgor) of the fluid of *Ascaris* have shown that the pressure can average from 70 to 120 mm Hg and vary up to 210 mm Hg.[26,28] This is an order of magnitude higher than the pressure in the body fluids of animals with hydrostatic skeletons in other phyla. Thus it seems highly probable that the limitations imposed by this high internal pressure determine many features of the nematode body plan. The mechanism of body movement can be summarized as follows: the longitudinal muscles in an area on one side of the body contract, those on the other side relax, forming a curve; the high internal pressures, plus the elastic cuticle being compressed on the concave side and stretched on the convex side of the curve, antagonize the contracted muscles to bring the body back to its resting state upon relaxation. Thus, the worm moves in its characteristic undulating fashion.

The pseudocoelomic fluid is known as **hemolymph.** In *Ascaris,* and probably in other nematode parasites of animals, it is a clear, pink, almost cell-free, complex solution. Aside from its structural signifi-

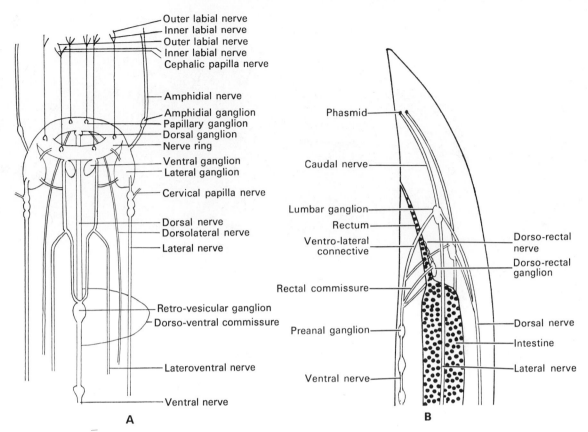

Fig. 22-12. A diagrammatic representation of the nervous system of a nematode. **A,** Anterior; **B,** posterior. (From Crofton, H. D. 1966. Nematodes. Hutchinson University Library, London.)

cance, it almost certainly is important in transport of solutes from one tissue to another. These solutes include a variety of electrolytes, proteins, fats, and carbohydrates. The proteins include albumins, globulins, hemoglobin, and several enzymes. Curiously, the fluid has far less chloride than would be required to balance the cations present, and the anion deficiency is made up mostly of volatile and nonvolatile organic acids.[18]

A peculiar and unique cell type found in the pseudocoel is the **coelomocyte.** Usually two, four, or six such cells, ovoid or many-branched, lie in the pseudocoel, attached to surrounding tissues. Although often small, in some species they are enormous; in *Ascaris* the coelomocytes are 5 mm by 3 mm by up to 1 mm thick. Their function is still obscure, though they may have a role in accumulation and storage of vitamin B_{12} and in protein synthetic and secretory function.[8]

Nervous system

The nervous system of nematodes is relatively simple. Because the mechanical coordination of movement in different parts of the body is accomplished by local changes in hemolymph volume, there is no need for local reflex networks.

There are two main concentrations of nerve elements in nematodes, one in the esophageal region and one in the anal area, connected by longitudinal nerve trunks. The most prominent feature of the anterior concentration is the **nerve ring, or circumesophageal commissure.** In *Ascaris* the nerve ring is composed of eight cells, four of which are nerve cells and four

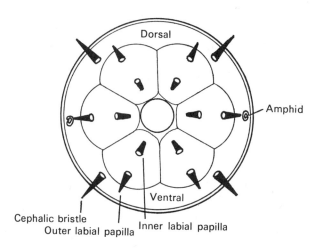

Cephalic bristle
Outer labial papilla
Inner labial papilla

Fig. 22-13. Diagram of the anterior end of a hypothetical primitive nematode to show the arrangement of the sense organs. (Adapted from de Conick. From Crofton, H. D. 1966. Nematodes. Hutchinson University Library, London.)

of which are supporting, or **glial,** cells. The ring lies close to the outer wall of the esophagus and is fairly easily seen in most species. Because its location is constant within a species, it is a good taxonomic character. The ring serves as a commissure for the **ventral, lateral,** and **dorsal** cephalic **ganglia** (Fig. 22-12, *A*). These are usually paired; the ventral ganglia are largest, the dorsal are the smallest. Emanating from each ganglion posteriorly are the **longitudinal nerve trunks,** which become embedded in the hypodermal cords, and, again, the ventral nerve is largest. Proceeding anteriorly from the lateral ganglia are two **amphidial nerves.** Six **papillary nerves,** which are derived directly from the nerve ring, innervate the cephalic sensory papillae surrounding the mouth.

The ventral nerve trunk runs posteriorly as a chain of ganglia, the last of which is the **preanal ganglion.** The preanal ganglion gives rise to two branches that proceed dorsally into the pseudocoel to encircle the rectum, thus forming the **rectal commissure,** or **posterior nerve ring.** Other posterior nerves and ganglia are depicted in Fig. 22-12, *B*. Posterior innervation in the male is usually not much more complex than in the female, in spite of the presence of the copulatory apparatus, except in the bursate nematodes

(Chapter 25). The peripheral nervous system consists of a latticework of nerves that interconnect with fine commissures and supply nerves to sensory endings within the cuticle.

The main sense organs are the cephalic and caudal papillae, the amphids, the phasmids, and in certain free-living species, the ocelli. The pattern of sensory papillae on the head of a nematode is a very important taxonomic character. The primitive pattern of lips surrounding the mouth of the ancestral nematodes is thought to have been two lateral, two dorsolateral, and two ventrolateral, each of which were supplied with sensory papillae (Fig. 22-13). In addition to the papillae forming the **inner** and **outer labial circles,** there were four **cephalic papillae,** one located behind the lips in each of the dorsolateral and ventolateral quadrants. Most parasitic nematodes are modified from this basic form. Labial papillae are often lost or fused together, and cephalic papillae usually are quite reduced in size. However, some papillae are found on all species, and careful study will reveal all 16 nerve endings on most species, even those that have lost all semblances of lips. The pattern of lips and papillae on nematodes is studied by slicing the anterior tip from the worm with a sharp blade and orienting the end

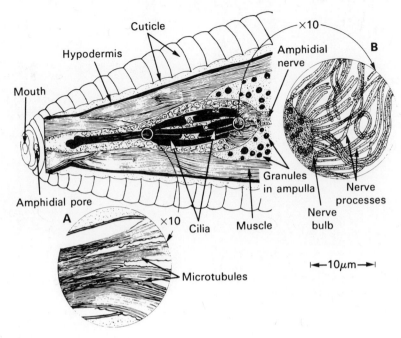

Fig. 22-14. Scale diagram of part of the tip of the head of *Meloidogyne* cut open to reveal one of the two amphids. (From Bird, A. F. 1971. The structure of Nematodes. Academic Press, Inc., New York.)

for an en face view on a microscope slide. Studies with the electron microscope have shown that the sensory endings of the papillae are modified cilia.[6] The papillae are probably tactile receptors.

The **amphids** are a pair of somewhat more complex sensory organs that open on each side of the head at about the same level as the cephalic circle of papillae. They are most conspicuous in marine, free-living forms and usually are reduced in animal parasites. The amphidial opening, which usually is at the tip of a papilla, leads into a deep, cuticular pit, at the base of which is a nerve bulb with several nerve processes (Fig. 22-14). The sensory endings are modified cilia, up to 23 in one amphid, in contrast to the one to three per papilla. Until modified cilia were discovered in the sense organs of nematodes, it was thought that these worms had no cilia. Of course, their structure is rather different from ordinary kinetic cilia. They have no kinetosomes, and the microtubules usually diverge from the normal 9 + 2 pattern, for example to 9 + 4, 8 + 4, or 1 + 11 + 4. The amphids are considered chemoreceptors but may have a se-

cretory function in some species; extracts of hookworm amphids inhibit clotting of vertebrate blood.[70]

Most parasitic nematodes have a pair of cuticular papillae, the **deirids** or **cervical papillae,** at about the level of the nerve ring, and other sensory papillae are found at different levels along the body of many species. **Caudal papillae** (Fig. 22-15) are more elaborately developed in the male, aiding in copulation. The pattern of distribution is an important taxonomic character. These papillae reach maximum development in the order Strongylata, where they form a complex copulatory bursa (Chapter 25). Also near the posterior end of most parasitic nematodes is a bilateral pair of cuticle-lined organs, the phasmids. The phasmids are similar in structure to the amphids and are of two basic types: one glandular, with an excretory function, and one sensory, involved in chemoreception.[46] These enigmatic structures are not found in most free-living nematodes or in the parasitic Dioctophymata and Trichurata. They are used to separate the classes Phasmidia (with phasmids) and Aphasmidia (without phasmids). While

Fig. 22-15. Ventral view of *Toxascaris* sp., showing caudal papillae. (Photograph by Jay Georgi.)

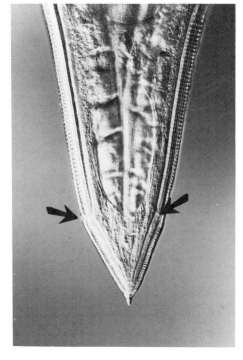

Fig. 22-16. Ventral view of female *Toxascaris* sp., showing ducts (arrows) leading to phasmids. (Photograph by Jay Georgi.)

difficult to see in some species, in most they are easily recognized by their cuticle-lined ducts (Fig. 22-16) that open at the apices of papillae near the tip of the tail.

Neural and neuromuscular transmission in nematodes is predominately cholinergic,[35] because the neurotransmitter at the synapses is acetylcholine. In *Ascaris*, the noncontractile part of the muscle cell has a membrane resting potential of 20 to 30 mV, and the contractile part has a resting potential of 40 to 60 mV. The muscle cell undergoes spontaneous depolarization in the innervation arm, then generation of action potential in a repeated or oscillatory manner.[14] This spontaneous rhythmic spike production is contrasted with vertebrate skeletal muscle, in which an action potential (spike) is initiated by a transmission of a nerve impulse across the neuromuscular junction. It is similar to cardiac muscle, in which there is a myogenic, rhythmic spike production. The rate of

spontaneous firing is increased with lowered resting potential and decreased with higher resting potential. The role of the nerve fibers is primarily one of modulation, and there are both excitatory and inhibitory fibers. Stimulation of excitatory fibers releases acetylcholine at the neuromuscular junction, depolarizes the muscle membrane, and increases the rate of spikes. The inhibitory fibers release γ-aminobutyric acid (or a pharmacologically similar compound), hyperpolarize the muscle, and decrease the rate of action potentials. Rhythmic spikes disappear altogether at resting potentials above 40 mV. Interestingly, it has been shown that a very efficient ascaricidal drug, piperazine, acts by hyperpolarizing the muscle membrane to about 45 mV, effectively paralyzing the worm, which then passes out of the host. The worms are not killed; the muscles still respond to electrical stimulation, and motility and tone of expelled worms return

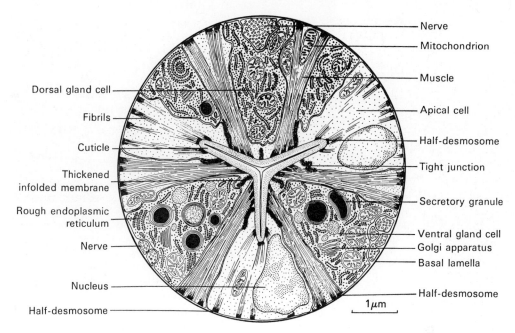

Nerve

Mitochondrion

Muscle

Apical cell

Half-desmosome

Tight junction

Secretory granule

Ventral gland cell

Golgi apparatus

Basal lamella

Half-desmosome

Dorsal gland cell

Fibrils

Cuticle

Thickened infolded membrane

Rough endoplasmic reticulum

Nerve

Nucleus

Half-desmosome

1 μm

Fig. 22-17. Diagram of a transverse section through the posterior part of the esophagus of *Nippostrongylus brasiliensis* to show the arrangement of various cells, cell membranes, and cellular organelles. Reconstructed from several electron micrographs. (From Lee, D. L. 1968. J. Zool. 154:9-18.)

if they are immersed in balanced saline at 37° C.[14]

Digestive system and acquisition of nutrients

The digestive system is complete in most nematodes, with mouth, gut, and anus, although in mermithids and a few filariids the anus is atrophied. The stomodeum (buccal cavity and esophagus) and proctodeum (rectum) are lined with cuticle, and the cuticular lining is shed with the molting of the exterior cuticle.

The mouth is usually a circular opening surrounded by a maximum of six lips. Few parasitic nematodes possess as many as six lips; in some they have fused in pairs to form three. In many species the lips are absent altogether, while in others two lateral lips develop as new structures derived from the inner margin of the mouth. Regardless of the morphology of a given species, it is a variation of the primitive, six-lipped form.

A buccal cavity lies between the mouth and esophagus of most nematodes. The size and shape of this area vary between species and are important taxonomic characters. In some species, the cuticular lining is quite thick, forming a rigid structure known as a **buccal capsule;** in others the lining is thin. The cavity may be elongate, reduced, or absent altogether, with a mouth that opens almost immediately into the lumen of the esophagus. Buccal armament is often present in parasitic and predaceous nematodes. The elements arise from the cavity wall or as anterior projections of the esophagus. Some nematodes have teeth of both types.

Food ingested by a nematode moves into a muscular region of the digestive tract known as the esophagus or pharynx. This is a pumping organ that sucks food into the alimentary canal and forces it into the intestine. It appears to be necessary because of the high turgor of the coelom. The esophagus assumes a variety of shapes, depending on the order and species of nematode, and for this reason is an important taxonomic character. It is highly muscular, cylindrical, often with

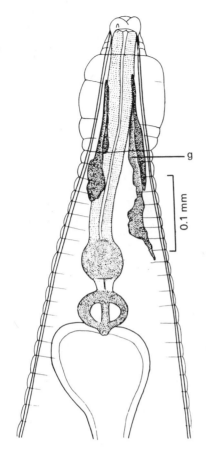

g

0.1 mm

Fig. 22-18. *Syphacia,* a rodent pinworm with enlarged esophageal glands *(g).* (From Schmidt, G. D., and R. E. Kuntz. 1968. Parasitology 58:845-854.)

one or more enlargements **(bulbs).** The lumen of the esophagus is lined with cuticle and is triadiate in cross section, with one radius directed ventrad and the other two pointed laterodorsad (Fig. 22-17). Radial muscles insert on the cuticular lining in the interradii and run the length of the esophagus. Interspersed among the muscles are three esophageal glands, one in each of the interradial zones. The dorsal gland is usually more extensive than the ventrolaterals. Each gland is usually uninucleate and opens independently into the lumen of the esophagus, although the dorsal one commonly opens farther anteriad. In some species the dorsal gland opens into the buccal cavity or even on the margin of the mouth. The secretions produced by these glands are digestive, for amylase, proteases, pectinases, chitinases, and cellu-

lases have been detected in them. In hookworms the secretions have anticoagulant properties.[7] In some species the glands fuse together near the posterior end of the esophagus, and in some nematodes, such as the Spiruroidea and many filariids, the posterior portion of the esophagus is mostly glandular. In some species the glands, especially the dorsal gland, are so extensive that much of their mass lies outside the esophagus proper (Fig. 22-18). In the specialized esophagus of the Trichuroidea, the anterior portion is a thin-walled, muscular tube, while the posterior portion is a very thin tube surrounded by a column of single cells, the **stichocytes,** the entire structure being referred to as the **stichosome.** The ultrastructure of the stichocytes suggests that they are secretory, and they communicate with the esophageal lumen by small ducts.[65]

Rapid contraction of the buccal muscles and the anterior esophageal muscles opens the mouth and dilates the anterior end of the esophagus, sucking food in (Fig. 22-19). Internal hydrostatic pressure closes the mouth and esophageal lumen when the muscles relax. The food is passed down the esophagus by the posteriorly progressing wave of muscle contraction opening the lumen for it until it reaches the intestine. The posterior bulb of many species appears to function as a one-way, nonregurgitation valve for food in the intestine. Thus, the mechanism is a kind of peristalsis in which the force moving the food is not the contraction of circular muscles but the closure of the esophageal lumen by hydrostatic pressure behind the food. The frequency of pumping has been recorded as two to 24 per second.[16]

In a few ascaroids (*Contracaecum, Multicaecum, Polycaecum,* and others), there are one to five posteriorly directed esophageal ceca that originate from a short, glandular **ventriculus** between the body of the esophagus and the intestine (Fig. 22-20).

The intestine is a simple, tube-like structure, extending from the esophagus to the proctodeum, and is constructed of a single layer of intestinal cells.

In females there is a short terminal, cuticle-lined rectum between the anus and intestine. In the male, the rectum is further specialized in its terminal portion to receive the products of the reproductive

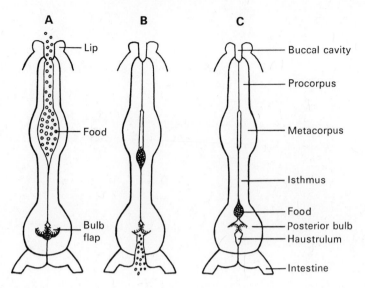

Fig. 22-19. Diagrams to show the structure and function of the *Rhabditis*-type pharynx during feeding. Food particles, small enough to pass through the buccal cavity, are drawn into the lumen of the metacorpus by sudden dilation of the pro- and metacorpus **(A)**. Closure of the lumen of the pharynx in these regions expels excess water **(B)**, and the mass of food particles is passed backwards along the isthmus **(B, C)**. Food is drawn between the bulb flaps of the posterior bulb by dilation of the haustrulum, which inverts the bulb flaps **(A)**, and is passed to the intestine by closure of haustrulum and by dilation, followed by closure of the pharyngeal-intestinal valve **(B)**. The bulb flaps contribute to the closure of the valve in the posterior bulb and, when they invert **(A)**, also crush food particles. (From Lee, D. L. 1965. The physiology of nematodes. Oliver & Boyd Ltd., Edinburgh.)

system and, therefore, is a cloaca. The dorsal wall of the cloaca is usually invaginated into two pouches, the spicule sheaths, which contain the copulatory spicules to be described along with the reproductive system. The vas deferens opens into the ventral wall of the cloaca.

The intestine is nonmuscular. Its contents are forced posteriad by the action of the esophagus as it adds more food to the front end of the system and perhaps by locomotor activity of the worm. Internal pressure of the coelom causes the intestine to be bilaterally flattened when empty. Between the dorsal wall of the cloaca and the body wall is a powerful muscle bundle called the **depressor ani.** This is a misnomer, for when it contracts, the anus is opened; it is therefore a dilator rather than a depressor. Defecation is also caused by hydrostatic pressure, and fecal matter is expelled by pressure surrounding the intestine when the anus is opened. The force of the hydrostatic pressure is demonstrated by the fact that *Ascaris* can project its feces nearly 2 feet, when lifted from the saline solution maintaining it.[16]

The wall of the intestine consists of tall, simple columnar cells with prominent brush borders. It was often suggested that cilia covered the free surface of these cells, such as in an earthworm, but electron microscope studies have shown these cilia-like structures to be microvilli.[66] Each microvillus is about 0.1 μm wide and consists of several core filaments surrounded by a filamentous coat of mucoprotein (Fig. 22-21). It seems probable that the resultant increased surface area is more important in absorption than secretion.[6] Although several digestive enzymes have been identified in the intestinal lumen, intestinal digestion is probably of minor importance in most forms because of the rapid rate of food movement through the intestine.

The number of intestinal cells varies from about 30 in some free-living species to over a million in the larger parasitic

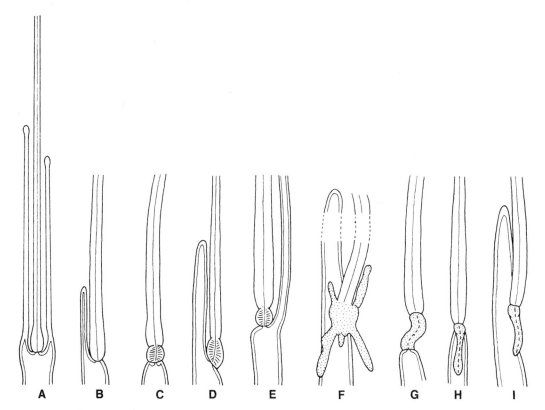

Fig. 22-20. Variations in esophagi in some ascaroid nematodes: **A,** *Crossophorus;* **B,** *Angusticaecum;* **C,** *Toxocara;* **D,** *Porrocaecum;* **E,** *Paradujardinia;* **F,** *Multicaecum;* **G,** *Anisakis;* **H,** *Raphidascaris;* **I,** *Contracaecum.* (Redrawn from Hartwich, G. 1974. CIH keys to the Nematode parasites of vertebrates, no. 2. Commonwealth Agricultural Bureaux, Herts.)

forms. These cells rest on a basement membrane, which is attached to random extensions of the body wall musculature. It is probable that the intestine serves as the primary means of excretion of nitrogenous waste products, in addition to its function in nutrient absorption. Crofton states that the intestine of *Ascaris lumbricoides* is emptied by defecation every 3 minutes under experimental conditions.[16] Such a rapid turnover of materials must surely limit the amount of enzymatic action possible in the intestinal lumen, but would favor the excretion of water-soluble waste products.

The food of parasitic nematodes, especially those in the intestine of their hosts, is high in amino acids and sugars. Nematodes feed extravagantly and wastefully, and the thin-walled intestine with its brush border is an efficient absorptive mechanism.

Excretion and osmoregulation

An excretory system has been observed in all parasitic nematodes except the aphasmidians Trichurata and Dioctophymata. Although the several types of excretory systems obviously are derived from one or two basic organs, and all have an external opening called the excretory pore, it has not been completely proved that the systems are excretory in function. There is strong evidence that most excretion occurs within the intestine.[58] As in the evolution of excretion in other animal groups, the excretory system of nematodes probably originated as an osmoregulatory system, any excretion of metabolic wastes by means of this system being secondarily acquired. Following tradition, we shall refer to it as an excretory system.

The presence of an excretory system is apparently primitive and probably evolved first in fresh-water forms. There are no

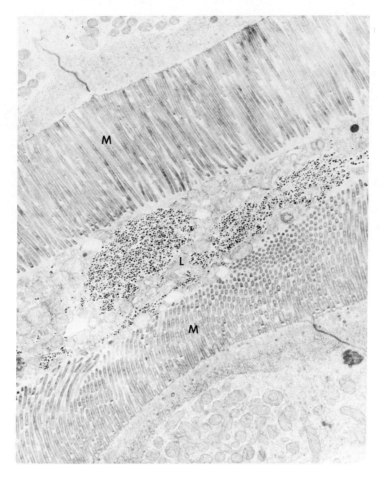

Fig. 22-21. Cross section of intestine showing microvilli *(M)* of dorsal and ventral sides. Glycogen, mitochondria, and other cellular debris fill the lumen *(L)*. (×10,800.) (From Sheffield, H. G. 1964. J. Parasitol. 50:365-379.)

flame cells or nephridia; in fact, the nematode excretory system seems to be unique in the animal kingdom. The two basic types are **glandular** and **tubular.** The glandular type is typical of the free-living Aphasmidea and may be involved in secretion of enzymes, proteins, or mucoproteins; it will not be considered here.

Several varieties of tubular excretory systems occur in parasitic forms (Fig. 22-22). Basically there are two long canals in the lateral hypodermis that connect to each other by a transverse canal near the anterior end. This transverse canal opens to the exterior by means of a median, ventral duct and pore, the excretory pore. This pore is conspicuous in most species; its location is fairly constant within a species and therefore is a useful taxonomic character.

Several variations of this basic, H-shaped system are common. The arms anterior to the transverse canal may be absent, forming a U-shaped system. In some species one entire lateral half is missing, resulting in an asymmetrical system. The posterior crura may be short, resulting in an inverted-U system. Many parasitic nematodes have a pair of large, granular, subventral gland cells associated with the transverse ducts. These are probably secretory in function.

The ability to osmoregulate varies greatly among nematodes and is correlated generally with the requirements of their habitats. The body fluids of species

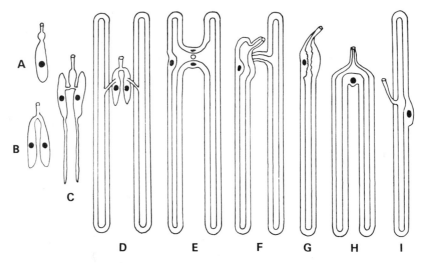

Fig. 22-22. Excretory systems. **A,** Single renette in a dorylaimid; **B,** two-celled renette in *Rhabdias;* **C,** larval *Ancylostoma;* **D,** rhabditoid type; **E,** oxyuroid type; **F,** *Ascaris;* **G,** *Anisakis;* **H,** *Cephalobus;* **I,** *Tylenchus.* (From Crofton, H. D. 1966. Nematodes. Hutchinson University Library, London.)

parasitic in animals may be somewhat different in tonicity from the tissues they inhabit but not dramatically so. For example, *Ascaris* hemolymph is about 320 to 350 mOsm, while pig intestinal contents are around 400 mOsm.[27] *Ascaris* clearly can control its electrolyte concentrations to some degree: chloride ion concentration of host intestinal contents varies between 34 and 102 mM, but *Ascaris* hemolymph is fairly constant at around 52 mM. Adults of most parasitic species cannot tolerate media much different in osmotic pressure from their hemolymph; when placed in tap water, they will burst, sometimes within minutes, from addition of the imbibed water to the already high internal pressure. Of course, freshwater and terrestrial nematodes, including larvae of many parasitic species, must withstand (and regulate in) very hypotonic conditions.

Details of water and ion excretion are poorly known. Contractions of excretory canals and the ampulla near the excretory pore have been observed in several species. Contractions of the ampulla in free-living, third-stage larvae of *Ancylostoma* and *Nippostrongylus* are inversely proportional to the salt concentration of the solution in which they are maintained.[74] Information

on ultrastructure of the excretory system strongly suggests that the system functions in osmoregulation and perhaps in excretion of waste products and in secretion as well.[43] The surface area of the peripheral cell membrane may be greatly increased by numerous bulbular invaginations, and, on the interior, the lumen is perforated by drainage ductules or canaliculi (Fig. 22-23). Filaments that are presumably contractile may surround the lumen of the duct. The hydrostatic pressure in the pseudocoel is thought to provide filtration pressure to excrete substances through the canals embedded in the hypodermal cords.

The ultrastructure of the gland cells clearly suggests secretory function. Enzymes responsible for exsheathment (shedding the old cuticle at ecdysis) are produced there by various strongyle larvae. A variety of nematodes excrete substances antigenic for their hosts via the excretory pores. Lee suggested that digestive enzymes were secreted by adult *Nippostrongylus* to act in conjunction with the abrading action of the cuticle.[38,39]

The major nitrogenous waste product of nematodes is ammonia. In normal saline, *Ascaris* excretes 69% of the total nitrogen excreted as ammonia and 7% as urea. Under conditions of osmotic stress, these

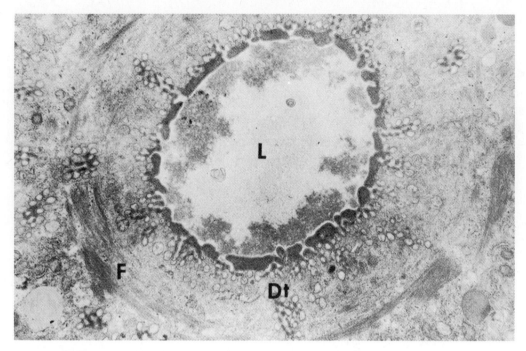

Fig. 22-23. *Anisakis* excretory gland. Transverse section through main excretory canal, showing round canal with interrupted dense material lining the lumen *(L)*, ramifying drainage tubules *(Dt)*, and congregated vesicles surrounding main canal and drainage tubules. Filaments *(F)* appear in circular arrangement around main canal. (×10,000.) (From Lee, H. 1973. J. Parasitol. 59:289-298.)

proportions can be changed to 27% ammonia and 52% urea. Amino acids, peptides, and amines may be excreted by nematodes. Other excretory products include carbon dioxide and a variety of fatty acids. The fatty acids are end products of energy metabolism and will be considered further. The role of the so-called excretory system in the elimination of the foregoing substances is not well established. Larval *Nippostrongylus* excrete several primary aliphatic amines via their excretory pore. It has been shown that a large proportion of nitrogenous waste products can be excreted via the intestine and anus by *Ascaris*,[58] and it would seem that the cuticle must play a major role in ammonia excretion in most nematodes.

Reproduction

Most nematodes are dioecious, although a few monoecious species are known, none of them important parasites. Parthenogenesis also exists in some. Morphological dimorphism usually attends dioecious forms,

with females growing larger than the males. Further, males have a more coiled tail and often have associated external features, such as bursae, alae, and papillae. Such dimorphism achieves the ultimate in the Tetrameridae and the plant-parasitic Heteroderidae, where the males have typical nematode anatomy, but the females are little more than swollen bags of uteri.

The gonads of nematodes are solid cords of cells that are continuous with the ducts that lead to the external environment. This allows the reproductive systems to function in spite of the high turgor of the pseudocoelom.

Male reproductive system

Usually, there is a single testis in nematodes, although two have been found in a few species. This organ may be relatively short and uncoiled, but in the larger animal parasites it appears as a long, threadlike structure that is coiled around the intestine and itself at various levels of the body. Two zones usually can be distin-

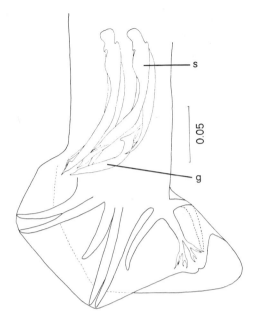

Fig. 22-24. A bursate nematode *Molineus,* showing complex spicules *(s)* and a gubernaculum *(g)*. (From Schmidt, G. D. 1965. J. Parasitol. 51:164-168.)

guished: the **germinal zone,** incorporating the blind end and in which spermatogonial divisions take place, and the **growth zone.** The end of the growth zone merges with a more tubular structure, the **seminal vesicle,** which is a sperm storage organ. The seminal vesicle merges into the vas deferens, which is usually divided into an anterior, glandular region and a posterior, muscular region, the ejaculatory duct. The ejaculatory duct opens into the cloaca. Some species have a pair of cement glands near the ejaculatory duct that secrete a hard, brown material to plug the vulva after copulation.

Nearly all nematodes have a pair of sclerotized, acellular, copulatory spicules (Fig. 22-24). They originate within dorsal outpocketings of the cloacal wall and are controlled by proximal muscles. Each spicule is surrounded by a fibrous spicule sheath. The spicule structure varies between species but is fairly constant among individuals within a species, making the size and morphology of the spicules two of the most important taxonomic characters. A dorsal sclerotization of the cloacal wall, the **gubernaculum,** occurs in many species. It guides the exsertion of the spicules from the cloaca at copulation. In several strongyloid genera, an additional ventral sclerotization of the cloaca, the **telemon,** has the same general function as the gubernaculum. Both structures are important taxonomic characters. The spicules are inserted into the vulva at copulation. They are not true intromittent organs, for they do not conduct the sperm, but are another adaptation to cope with the high internal hydrostatic pressure. The spicules must hold the vulva open while the ejaculatory muscles overcome the hydrostatic pressure in the female and rapidly inject sperm into her reproductive tract.

Nematode spermatozoa are unique among those studied in the animal kingdom in that they lack a flagellum and acrosome. Further, internal organization of organelles differs markedly from all other sperm previously described. Once inside the female reproductive tract (where they usually are studied) they become rounded, triangular, or ameboid. It is possible that they are incapable of fertilizing ova before this transformation.[67] Furthermore, they are rather diverse between species in cytological characteristics; Foor recognized at least four types among those so far described.[23] As mature sperm in the seminal vesicle of the male, the types range from small, rounded structures to ameboid types with distinct anterior and posterior cytoplasm to elongate, tadpole-shaped structures, with "head" and "tail." In all types, the nucleus is not bounded by a nuclear membrane. Some types undergo further morphological development after insemination of the female; for example, the "tadpole" becomes ameboid, and it is thought that the sperm may not be able to fertilize an ovum until these changes have occurred.[67] Motility has not been easy to discern, though clearly the gametes must travel up the female tract. The "tail" of the tadpole-shaped sperm apparently is nonmotile, but the "head" can put out pseudopodia.[41] Good ultrastructural evidence for pseudopodial movement is available (Fig. 22-25).[23]

An interesting tropism of the female in some species causes her to seek the coiled posterior end of the male, which she

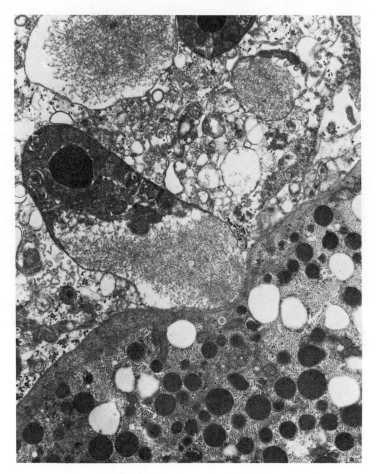

Fig. 22-25. *Angiostrongylus cantonensis* sperm in contact with an oocyte *(O)* in the uterus of the female worm. Note large pseudopod *(P)* and continuity between the membrane specialization and the sperm plasma membrane. (×14,500.) (From Foor, W. E. 1970. Biol. Reprod. 2[Suppl.]:177-202.)

enters. The caudal papillae of the male detect the vulva; this excites a probing response of the spicules, leading to sperm transfer. Females of some species have **vulvar papillae,** which no doubt aid in the mating reflexes. Curiously, if no males are present within a host, females of some species tend to wander, seeking a constriction to squeeze through. This may result in dire consequences to the host if a bile duct, for example, is selected for exploration. Other unexpected results of this behavior have been recorded (Fig. 22-26). Serial copulations of a female with several males have been observed,[1] and copulation also has been described.[4,67]

Female reproductive system

Most female nematodes have two ovaries, although some have from one to more than six. The general pattern of structure of the reproductive system in the female is quite similar to that in the male, except that the gonopore is independent of the digestive system: a linear series of structures, with the gonad at the internal or proximal end, followed by developmental, storage, and ejective areas. The number of tracts per female and their disposition relative to each other are given descriptive terms. If a species has only one ovary and uterus, it is called **monodelphic.** Monodelphic species nearly always have

Fig. 22-26. A female *Ascaris lumbricoides* strangled by a shoe-eyelet. This illustrates the tropism of female nematodes to seek the coiled tail of males. (From Beaver, P. C. 1964. Am. J. Trop. Med. Hyg. 13:295-296.)

the vulva near the anus. The more common situation, in which there are two ovaries and uteri, is termed **didelphic.** More than two uteri and associated structures is a condition called **polydelphic.** If the two uteri converge from opposite directions at their junction with the vagina, they are called **amphidelphic.** When they are parallel and converge from an anterior direction, they are **prodelphic;** if they converge from a posterior direction, they are **opisthodelphic.**

The ovaries are solid cords of cells that produce gametes and move them distally into the terminal portion of the system. The proximal end of the ovary is the **germinal zone,** which produces oogonia; the oogonia become oocytes and move into the **growth zone** of the ovary, toward the oviduct. In the large ascarids the oocytes are attached to a central supporting structure, the **rachis.** In *Ascaris* the germinal zone is very short, and most of the 200 to 250 cm length of the ovary is comprised of oocytes attached in a radial man-

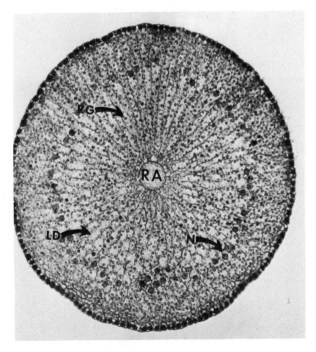

Fig. 22-27. *Ascaris lumbricoides:* transverse section through growth zone of ovary. *LD,* lipid droplet; *N,* nucleus; *RA,* rachis; *RG,* refringent granule. (×440.) (From Foor, W. E. 1967. J. Parasitol. 53:1245-1261.)

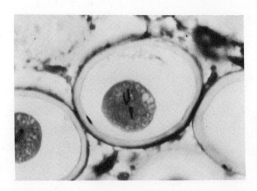

Fig. 22-28. A zygote of *Parascaris equorum* in prophase, showing two chromosomes. The diploid number of this species varies from 2 to 12.[71] (Courtesy Ann Arbor Biological Center.)

ner by cytoplasmic bridges to the rachis (Fig. 22-27).[22]

The haploid chromosome number of oocytes varies somewhat between families: Ascaridae (n = 1-12) (Fig. 22-28), Oxyuridae (n = 2-4), Strongyloididae (n = 3), Strongylata (n = 6).[71] The oocytes increase in size as they move down the rachis, and, about 3 to 5 cm from the oviduct, they become detached from it. The proximal end of the oviduct in most nematodes is a distinct **spermatheca,** or sperm storage area. As the oocytes enter the oviduct (spermathecal area), they are penetrated by sperm, and only then do they undergo their meiotic divisions. A polar body is extruded at the first division, and the second division is followed by expulsion of another polar body. Concurrent with these events, shell formation is occurring, which will be described in the section on development.

The wall of the uterus has well-developed circular and oblique muscle fibers, and these move the developing embryos ("eggs") distally by peristaltic action. The shape of the eggs may be molded by the uterus, and uterine secretory cells may contribute additional material to the eggshells. The distal end of the uterus is usually quite muscular, and is known as the **ovijector.** The ovijectors of the uteri fuse to form a short vagina that opens through a ventral, transverse slit in the body wall, the vulva. The vulva may be located anywhere from near the mouth to immediately in front of the anus, depending on the species. The vulva never opens posterior to the anus and only very rarely into the rectum to form a cloaca. When the lips of the vulva protrude, they are said to be salient. The muscles of the vulva act as dilators, and constriction of the circular muscles of the ovijector both expels the eggs and restrains more proximal, undeveloped eggs from being expelled because of hydrostatic pressure.

DEVELOPMENT

Historically, studies on the development of nematodes have led to fundamental discoveries in zoology. For example, van Beneden, in 1883, was the first to elucidate the meiotic process and realize that equal amounts of nuclear material were contributed by sperm and egg after fertilization.[5] Boveri (1899) first demonstrated the genetic continuity of chromosomes and determinate cleavage, that is, embryogenesis in which the fate of the blastomeres is determined very early.[9] Both men based their insights on studies of nematode material.

Not surprisingly, in such a successful group as nematodes, details of development and life history differ greatly between the various groups. However, the general pattern is remarkably similar for all species known. There are four juvenile stages and the adult, each separated from the succeeding one by an ecdysis, or molting of cuticle. The juvenile stages are referred to conventionally as "larvae," and we shall often follow convention in this respect. It should be noted, however, that the first-stage larva is quite similar in body form to the adult. No real metamorphosis occurs during ontogeny, and, with certain possible exceptions, all of the somatic cells of the adult may be present in the embryo![16]

Eggshell formation

Penetration of the ovum by the sperm initiates the process by which protective layers are produced around the zygote and developing embryo. The fully formed shell in most nematodes consists of three layers: (1) an outer vitelline layer, often not detectable by light microscopy, (2) a chitinous layer, and (3) a lipid layer, inner-

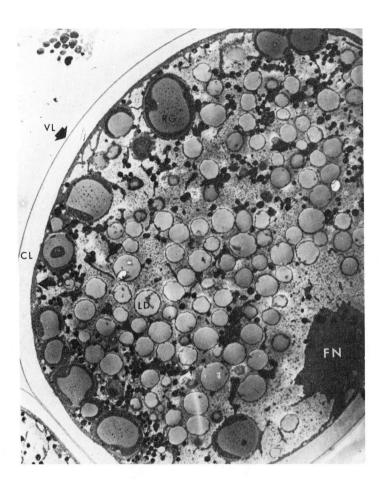

Fig. 22-29. Low magnification of newly fertilized egg, showing vitelline layer *(VL)*; incipient chitinous layer *(CL)*; dense, particulate, cortical cytoplasm *(DC)*; and numerous lipid droplets *(LD)*. Female nucleus *(FN)* lies near surface, and refringent granules *(RG)* have migrated to position just beneath cortical cytoplasm. After extrusion, the contents of the refringent granules will become the lipid layer. (×3,800.) (From Foor, W. E. 1967. J. Parasitol. 53:1245-1261.)

most and called the "vitelline membrane" in older literature.[6] A fourth, **proteinaceous** layer is contributed by uterine cell secretions in some nematodes *(Ascaris, Thelastoma, Meloidogyne)* and consists of an acid mucopolysaccharide–tanned protein complex. Formation of the shell layers has been best studied in *Ascaris*, but it seems likely that the process is similar in other nematodes. Just after sperm penetration a new oolemma forms beneath the original; the old oolemma becomes the vitelline layer and separates from the peripheral cytoplasm, and the cytoplasm shrinks back, leaving an electron-lucid space within

which the chitinous layer forms (Fig. 22-29).[22,42] Refringent bodies, previously dispersed throughout the cytoplasm, migrate to the periphery and extrude their contents, the fusion of which forms the lipid layer. The so-called chitinous layer is probably supportive or structural in function and also contains protein; the proportion of chitin present varies between groups from great (ascaroids, oxyuroids) to very small (strongyloids). Resistance to desiccation and to penetration of polar substances is conferred by the lipid layer. At least in *Ascaris*, this layer is composed of 25% protein and 75% **ascarosides.**

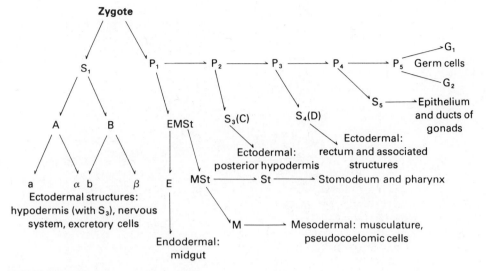

Fig. 22-30. Cell lineage of nematodes. The two cells produced at the first cleavage of the zygote are P_1 and S_1. The diagram indicates the progeny of those cells and the tissues to which they give rise.

Ascarosides are very interesting and unique glycosides (compounds with a sugar and an alcohol joined by a glycosidic bond). In ascarosides, the sugar is **ascarylose** (3,6-dideoxy-L-arabinohexose), and the alcohols are a series of secondary monols and diols containing 22 to 37 carbon atoms.[31] The ascarosides render the eggshell virtually impermeable to substances other than gases and lipid solvents; further discussion of the resistance of *Ascaris* eggs will be found in Chapter 26. Whether ascarosides are present in the lipid layers of nematode eggs other than ascaroid is not known.

Embryogenesis

Molecular biology of early development has been studied in *Ascaris,* and some data indicate similar phenomena in *Parascaris.*[20] During oogenesis in other well-studied systems (such as amphibians and echinoderms), there is considerable synthesis of ribosomal and informational RNA (rRNA, mRNA). These are conserved in the unfertilized egg, and little or no RNA synthesis follows fertilization and early cleavage. In contrast, there is almost no RNA in mature *Ascaris* oocytes, and fertilization is followed by a burst of rRNA synthesis in the *male* pronucleus. This may reflect an adaptation to the fact that the female nuclei are otherwise occupied at this time, undergoing their maturation divisions, and the burden of RNA production in preparation for protein synthesis falls on the male pronucleus.[20]

The determinate cleavage of the nematode embryo is the clearest and best documented example of germinal lineage in the animal kingdom.[16] Because of the early determination of the fate of each cell (blastomere) in the cleaving embryo, names or letter designations can be given to each blastomere, and the tissues that will develop from each are known (Fig. 22-30). At the first cleavage, the zygote produces one cell that will give rise to somatic tissues and one cell whose progeny will comprise more somatic cells and the germinal cells. The early cleavages of nematodes are marked by a very curious phenomenon called **chromatin diminution.** The chromosomes fragment, and only the middle portions are retained, the ends being extruded to the cytoplasm to degenerate. Since this diminution only occurs in the somatic cells, the germ line can be recognized by its full chromatin complement. Finally, the only cells left with complete chromosomes are G_1 and G_2, which will give rise to the gonads. Interestingly, further differentiation of the nuclei in the various tissues seems to go in both directions with respect to chromatin content. Some, such as muscle and ganglia nuclei, further diminish until DNA can no longer be detected, while others, particularly those with protein synthetic activity as excretory and pharyngeal glands and uterine cells, exhibit polyploidy.[69]

Rather typical morula and blastula stages are formed. Gastrulation is by invagination and also by epiboly (movement of the micromeres down over the macromeres).

In the fully formed embryo, the nuclei other than the germinal cells cease to divide; thus, all cells of the adult are present at this time. The phenomenon is known as **cell** or **nuclear constancy,** or **eutely,** and it is characteristic of several aschelminth phyla. There are some exceptions: cells of the intestine and hypodermis of the larger nematodes divide further, but in most species growth after embryogenesis is a matter of cell enlargement rather than cell division. The number of cells per individual is fairly constant within a species and varies between species.

The timing, site, and physical requirements for embryogenesis vary greatly between the species. In some, the cleavage will not begin until the egg reaches the external environment and oxygen is available. Others begin (or even complete) embryogenesis before the egg passes from the host, while in some the larvae complete development and hatch within the female nematode (ovoviviparity).

Studies on embryonation of *Ascaris* eggs have revealed a most fascinating sequence of biochemical epigenetic adaptation: adaptive appearance and disappearance of biochemical pathways through ontogeny, based on repression and derepression of genes.[20] The energy metabolism of adult *Ascaris* is anaerobic but that of the embryonating eggs is obligately aerobic. Dependence on pathways such as glycolysis would not only be wasteful of the limited stored nutrient in the embryos, but also a toxic concentration of acidic end products would soon build up as the result of impermeability of the eggshell. Eggs survive temporary anaerobiosis, but they do not develop unless oxygen is present. They are completely embryonated and infective after 20 days at 30°C,

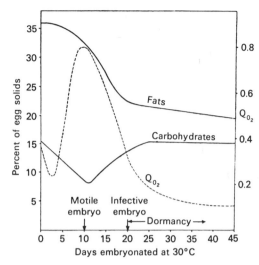

Fig. 22-31. Changes in oxygen consumption, fat, and carbohydrates during embryonation of *Ascaris* eggs. Fats are connected to carbohydrates via the glyoxylate cycle, which requires the enzymes isocitrate lyase and malate synthase. These enzymes are present in the embryo but are apparently absent in the adult. (From Fairbairn, D. 1960. In Sasser, J. N., and W. R. Jenkins, editors. Nematology. University of North Carolina Press, Chapel Hill, N.C.)

and throughout this time, a tricarboxylic acid cycle and cytochrome *c*-cytochrome oxidase electron transport system are present. The infective stage is the second-stage larva (peculiar to ascaroids; most nematodes are infective in the third stage), having undergone one molt in the egg, which hatches in the host intestine and goes through a tissue migration. Third-stage larvae break into the lung alveoli, travel up the trachea, then are swallowed to gain access to the intestine, where they go through the fourth stage and become adults. Cytochrome oxidase is still present in the third stage recovered from the lungs, and third-stage larvae require oxygen for motility. Oxidase activity disappears from the fourth stage and is essentially repressed through adult life. A similar phenomenon has been observed with regard to the enzymes of the glyoxylate cycle.[2] It was shown some years ago that embryonating *Ascaris* eggs consumed both lipid and carbohydrate reserves during the first 10 days of embryonation, then *resynthesized carbohydrate* (glycogen and trehalose) from fat (Fig. 22-31).[47] Derivation of energy from lipids normally requires degradation to two-carbon fragments in the form of acetyl CoA. Then the acetyl-CoA is oxidized in the tricarboxylic acid cycle. Most higher animals cannot synthesize carbohydrates from the acetyl-CoA, but many plants and microorganisms can accomplish this feat because they have two essential enzymes, **isocitrate lyase** and **malate synthase,** to perform the glyoxylate cycle. Other enzymes necessary for the complete cycle are normally associated with the tricarboxylic acid cycle (Figs. 3-14 and 22-36). Isocitrate lyase and malate synthase have been demonstrated in several other nematodes and in the trematode *Fasciola hepatica.* However, *Ascaris* is the only metazoan in which the glyoxylate cycle and its role in the conversion of fat to carbohydrate has been established.[2] Finally, all activity of the two critical enzymes seems to be repressed in the adult muscle. It may be supposed that the synthesized trehalose may play a role in egg-hatching, and the glycogen accumulation may be in anticipation of the larvae during their tissue migration.

Egg hatching

Hatching of nematodes whose larvae are free-living before becoming parasitic occurs spontaneously. This is probably a result of synthesis of lipid-hydrolyzing enzymes in the subventral esophageal glands, these structures only becoming active in the terminal phases of embryogenesis.[6] A number of species, however, will only hatch after being swallowed by a prospective host. Upon reaching the infective stage, such eggs remain dormant until the proper stimulus is applied, and this requirement has the obvious adaptive value of preventing premature hatching. Ascarid eggs require a combination of conditions: temperature about 37°C, a moderately low oxidation-reduction potential, a high carbon dioxide concentration, and a pH of about 7.0. These conditions are present in the gut of many warm-blooded vertebrates, and, indeed, *Ascaris* will hatch in a wide variety of mammals and even in some birds, but all four conditions are unlikely to be present simultaneously in the

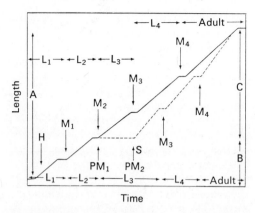

Fig. 22-32. An idealized form of the basic life cycle of nematodes. The life cycle of a free-living nematode is represented by a solid line. Hatching *(H)* is "spontaneous" and there are four molts $(M_1$-$M_4)$. The broken line represents a life cycle in which a change in environment is necessary to stimulate *(S)* the completion of the second molt (PM_2). A, B, and C are different environments. (From Rogers, W. P. and R. I. Sommerville. 1963. In Dawes, B., editor. Advances in parasitology, vol. I. Academic Press, Inc., New York.)

external environment. The first change detectable upon application of the stimulus is a rapid change in permeability; trehalose from the perivitelline fluid leaks from the eggs. The lipid layer is now permeable to chitinase secreted by the larva. Esterases and proteinases also are secreted, and these enzymes attack the hard shell, digesting it sufficiently for the larva to force a hole in it and escape.[18]

Growth and ecdysis

Unlike most arthropods, there is growth in body dimensions of nematodes between molts of their cuticle (Fig. 22-32). After the fourth molt in large nematodes such as *Ascaris,* there is considerable increase in size, and the cuticle itself continues to grow after the last ecdysis. The molting process has been studied in several species. First the hypodermis detaches from the basal lamella of the old cuticle and starts to secrete a new one, beginning with the cortical layers. By the time the new cuticle is secreted, it may be substantially folded under the old cuticle, to be stretched out later after ecdysis. In some cases the old cuticle up to the cortical layer is dissolved and resorbed through the new cuticle. This is particularly important when conservation of materials and space are considerations, such as in the first molt of *Ascaris* and less so when there is plenty of food and the old cuticle is very complex in structure, as in the fourth molt of *Nippostrongylus*.[39] Escape from the old cuticle seems to be facilitated by the enzyme leucine aminopeptidase, which is secreted in the excretory gland and released through the excretory pore.[49] The leucine aminopeptidase attacks the proteins in the old cuticle and, interestingly, has a marked substrate specificity, so that the enzyme from one species will only hydrolyze the cuticular sheath of that species.

Growth and ecdysis are controlled to an as yet undetermined degree by neurosecretory mechanisms. **Neurosecretion** is a process in which substances with endocrine function are secreted by nerve cells, or modified nerve cells, and its action is well-documented in vertebrates and some arthropods. On the basis of staining reactions, neurosecretory activity has been suggested in several nematodes, but Davey and Kan have shown that *Terranova decipiens* fails to undergo ecdysis in the absence of neurosecretory cycle.[17] Furthermore, extracts of isolated neurosecretory cells stimulate isolated excretory cells to synthesize leucine aminopeptidase.

A common adaptation in many parasites is a resting stage at some point in their development, enabling them to survive adverse conditions while awaiting access to a new host. Such **developmental arrest** is of particular interest in nematodes, not only because of the variety of stages and situations in which it takes place, but also because fundamentally similar processes are demonstrated in some free-living species.[51] An example is *Rhabditis dubia*, which lives in cow dung. It may go through an indefinite number of generations, developing normally, but when unsuitable conditions occur, special third-stage juveniles called **dauer larvae** are produced. The dauer larvae develop no further but await access to psychodid flies, to which they attach. When the fly obligingly transports them to a new pile of cow dung, they detach and proceed with development. An-

Fig. 22-33. Third-stage infective larvae of *Nippostrongylus brasiliensis*, illustrating the typical behavior of crawling up on pebbles, blades of grass, or the like and waving their anterior ends to and fro. In this photograph of living worms, they have mounted granules of charcoal and even each other.

other species is *Rhabditis coarctata*, which uses dung beetles for transport and in which dauer larvae are produced every generation. Several other examples could be cited. Typically, the dauer larvae do not feed but have stored reserves in their intestinal cells and are more resistant to desiccation than normal worms. One reason for their resistance seems to be an incomplete second ecdysis. When the third-stage cuticle is secreted, the old second-stage cuticle is retained in place as a sheath. A wide variety of parasitic nematodes produces infective third-stage juveniles that are quite comparable to dauer larvae. They develop no further until a new host is available, remaining ensheathed in the second-stage cuticle. They live on stored food reserves and usually exhibit behavior patterns that enhance the likelihood of reaching a new host. For example, third-stage larvae of *Haemonchus* and *Trichostrongylus* migrate out of the fecal mass and onto vegetation that is eaten by the host. Third-stage larvae of species that penetrate the host skin, such as hookworms and *Nippostrongylus,* migrate onto small objects (sand grains, leaves, and others) and move their anterior ends freely back and forth, in exactly the same manner as do some dauer larvae (Fig. 22-33). In both dauer larvae and infective larvae, a more or less specific stimulus is required for development in completion of the ecdysis of the second-stage cuticle. Skin penetrators usually exsheath in the processes of penetration, but the stimulus for exsheathment of swallowed larvae (*Haemonchus, Trichostrongylus,* and others) is very similar to that required for hatching of *Ascaris* eggs, including carbon dioxide, temperature, redox potential, and pH. In fact, Rogers and Sommerville considered infective eggs fundamentally the same as infective larvae and dauer larvae.[51] Nematodes with intermediate hosts normally undergo developmental arrest at the third stage and remain dormant until they reach the definitive host. Some species are astonishingly plastic in their capacities to sustain more than one developmental arrest in their ontogenies. For example, if some species of hookworms and ascarids infect an unsuitable host, they enter another developmental arrest and lie dormant in the host tissues until they receive another stimulus to resume.[45] In several of these that are known, the older animal is an unsuitable host, and the worms lie dormant until they are stimulated by the hormones of host pregnancy. They then migrate to the uterus or mammary glands and infect the infant via the placenta in utero or the milk after birth. More examples of adaptational arrests in development will be found in the nematode life cycles described in the chapters to follow.

METABOLISM
Energy metabolism

More probably is known about the nematodes than about any other group of parasitic helminths.* *Ascaris* was one of the first organisms in which cytochrome was demonstrated.[33] Nevertheless, numerous questions await resolution.

Parasitic nematodes are a very diverse

---

*See references 11, 12, 19, 50, 59, and 60.

group, occupying a number of different habitats, and, not surprisingly, their energy metabolisms, although basically similar, exhibit a wide variety of minor variations. As indicated in Table 4, a number of different compounds appear to be end products of anerobic carbohydrate metabolism. However, these compounds apparently are all derived from the products of a similar pathway (Fig. 22-34). The scheme is best documented for the pig roundworm, *Ascaris suum.* Glucose is converted to phosphoenolpyruvate (PEP) through classical Embden-Meyerhof glycolysis. Pyruvate kinase activity (see Fig. 20-35) is low, and the PEP is converted to oxaloacetate by the carbon dioxide–fixing enzyme, PEP carboxykinase, rather than to pyruvate. Since cytoplasmic malate dehydrogenase activity is high, the oxaloacetate is rapidly converted to malate. In addition, this reaction oxidizes the NADH, formed previously in glycolysis, to NAD and maintains the oxidation-reduction balance of the system, a role normally played by lactate dehydrogenase.

Cytoplasmic malate enters the mitochondria and is utilized by a dismutation reaction, that is, the reduction of one metabolite by another, which requires that NAD shuttle between the surfaces of the two substrate specific dehydrogenases. Half of the malate is oxidized to pyruvate and carbon dioxide by the action of malic enzyme and generates intramitochondrial reducing power in the form of NADH. The NADH shuttles to fumarate reductase, which reduces the remaining malate, via fumarate, to succinate. This last reaction results in a site I, electron transport –associated, phosphorylation of ADP to ATP. The importance of the ATP generated and the maintenance of redox balance in these reactions is indicated by the observation that some antinematodal agents (tetramisole and thiabendazole) are effective because they block the fumarate reductase.

A wide array of other end products are excreted besides succinate, of which the most abundant quantitatively in *Ascaris* are α-methylbutyrate and α-methylvalerate. These branched chain acids are formed by condensations of propionate units or a propionate and acetate (Fig. 22-35), the

Table 4. Examples of some end products of energy metabolism excreted by some nematodes*

Substance excreted†	Ascaris	Trichinella larvae	Heterakis	Litosomoides	Dracunculus	Dirofilaria	Caenorhabditis‡	Ancylostoma	Trichuris
Lactate	T	T	T	+	+	+	T	−	+
Propionate	+	+	+	−	−	−	T	+	+
Acetate	+	+	+	+	−	−	T	+	−
Pyruvate	−	−	+	−	−	−	T	−	−
Succinate	−	−	+	−	−	−	T	−	−
α-Methylbutyrate	+	−	−	−	−	−	T	+	−
n-Valeric acid	+	+	−	−	−	−	T	−	+
Isocaproic acid	+	−	−	−	−	−	T	−	−
n-Caproic acid	+	+	−	−	−	−	T	−	−
Acetylmethyl carbinol	+	−	−	+	−	−	T	−	−
Isobutyric acid	−	−	−	−	−	−	T	+	−
n-Butyric acid	+	+	−	−	−	−	T	−	T
Tiglic acid	+	−	−	−	−	−	T	−	−
C₆ acids (unidentified)	−	+	−	−	−	−	T	T	−

* Adapted from Lee, D. L. 1965. The physiology of nematodes. Oliver & Boyd Ltd., Edinburgh.
†T, trace; +, present;−, absent or not investigated.
‡Free-living nematode.

propionate and acetate arising from decarboxylations of succinate and pyruvate, respectively.[62,63] The exact significance of such reactions and their products remains unclear, but some general roles in anaerobic electron transport may be ascribed as probable. Many synthetic reactions are oxidative, and formation of such reduced end products would offer a means to reoxidize coenzymes reduced in synthetic reactions. Certain steps may be sites of additional electron transport–associated phosphorylations, that is, ATP production. Also, the worms may generate less toxic end products by these routes.

Other electron transport reactions are present in *Ascaris,* but their importance and sequence are still not known with certainty. Mitochondria contain a-, b-, and c-type cytochromes and a very low concentration of functional cytochrome a_3.[15] Other researchers have held that terminal electron transport is via flavoproteins.[11] Succinate and malate are oxidized in mitochondrial preparations with hydrogen peroxide as an end product. The toxicity of hydrogen peroxide, and the fact that it is produced in terminal electron transport by this worm, may account for the observation that oxygen at atmospheric concentrations drastically reduces survival time of *Ascaris* in vitro.[25] Though oxygen is unnecessary in the energy metabolism of these "anaerobic" nematodes and may be toxic in moderate concentrations, small amounts may be necessary in certain biosynthetic reactions.

In addition to the foregoing, other nematodes have been studied that survive and metabolize carbohydrates in the absence of oxygen for extended periods. Examples of these include *Heterakis gallinae* and *Trichuris vulpis.* These organisms serve as particularly good examples of how some parasites have solved the metabolic problem of reoxidizing NADH (see Chapter 3) in the absence of oxygen as a terminal electron acceptor. However, some other adult nematodes seem to be obligate aerobes with respect to their energy metabolism, requiring the presence of at least low concentrations of oxygen for survival and motility. Nonetheless, even among these, glucose is not oxidized completely to carbon dioxide and water, and substantial quantities of various reduced end products are ex-

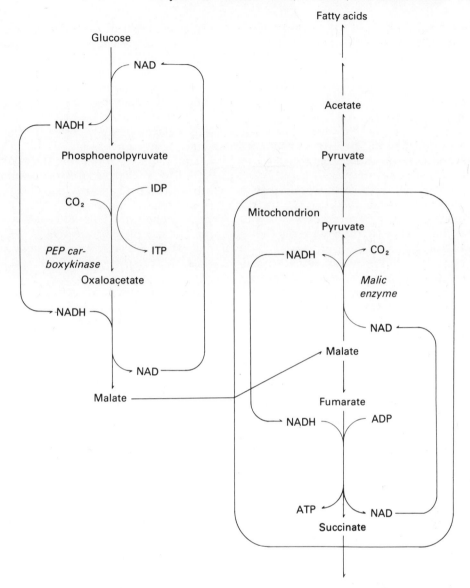

Fig. 22-34. Anaerobic oxidation of glucose by *Ascaris* (modified from Saz[61] and Bryant[12]). Reactions that occur in mitochondria are within box, others occur in cytosol.

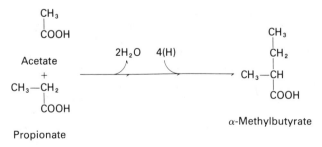

Fig. 22-35. Mode of α-methylbutyrate formation in *Ascaris* muscle; α-methylvalerate is formed in a similar manner except that the α-carbon of one propionate unit condenses with the carboxyl carbon of another propionate unit (Saz and Weil[62,63]).

creted. Some species apparently have a classical cytochrome system, or, alternatively, the electrons may be transported via a flavoprotein and terminal flavin oxidase to oxygen, producing hydrogen peroxide.

Examples of obligate aerobes are *Nippostrongylus brasiliensis* and *Litomosoides carinii*. *Nippostrongylus brasiliensis* is a trichostrongyle parasite in the intestine of rats, and *L. carinii* is a filarial worm found in the pleural cavity of cotton rats. Both species exhibit a **Pasteur effect,** that is, they consume more glucose in the absence of oxygen than in its presence. This suggests that they derive at least some energy from reaction sequences with oxygen as the terminal electron acceptor. Both can survive short periods of anaerobiosis but are killed by longer periods (a few hours).[48,59] *Nippostrongylus* excretes moderate quantities of lactate and succinate, plus small amounts of pyruvate, aerobically, and the excretion of lactate and succinate is markedly increased in the absence of oxygen. *Litomosoides* produces lactate, acetate, and carbon dioxide in the presence of oxygen, and anaerobiosis causes a dramatic shift toward much higher lactate production and lower acetate production. The carbon dioxide produced by *Litomosoides* comes almost entirely from the decarboxylation of pyruvate and not from the tricarboxylic acid cycle.[60] The character of the terminal electron transport reactions is still unclear; *L. carinii* has both cytochrome *c* and cytochrome oxidase, while *N. brasiliensis* has cytochrome oxidase but apparently no cytochrome *c*. Nevertheless, oxygen consumption of *L. carinii* is inhibited by cyanide and by drugs called **cyanine dyes.** The cyanine dyes have no inhibitory effect on mammalian cytochrome systems, and they have chemotherapeutic activity against *L. carinii* in vivo. Unfortunately, these drugs are ineffective against the filarial parasites of humans, presumably because these species are less dependent on oxygen than is *L. carinii*.

All of this discussion has been in reference to adult nematodes. Different stages in the life cycles may show dramatic biochemical adaptations in energy metabolism, as, for example, the aerobic embryonating eggs and the anaerobic adult of *Ascaris*. *Strongyloides* is another interesting example of biochemical epigenetic adaptation.[34] *Strongyloides* has a complex life cycle with free-living adults (males and females) and parasitic adults (parthenogenetic females only). The first three larval stages of both types are free-living, but those destined to become parasitic undergo developmental arrest at the third stage, until penetration of the host (see Chapter 24). All free-living stages are subjected to a selective pressure common to other free-living animals, namely, to utilize as completely as possible the energetic value in their nutrient molecules, and they have a complete tricarboxylic acid cycle and probably a cytochrome system. In contrast, the parasitic females do not have a complete tricarboxylic acid cycle or cytochrome system, a situation similar to many other intestinal helminths. Larvae of several other parasitic species have apparently functional tricarboxylic acid cycles,[73] although in some species the significance of the cycle may lie in regulation of four-carbon intermediates rather than energy production.

The normal pathway of fatty acid oxidation is referred to as β-oxidation, so called because the β-carbon of the fatty acetyl-CoA is oxidized, and the two-carbon fragment, acetyl-CoA, is cleaved off to enter the tricarboxylic acid cycle. It would be expected, therefore, that the presence of β-oxidation enzymes would be correlated with a functional tricarboxylic acid cycle (though not necessarily so), and, in the few cases investigated, this is the case. β-oxidation of fatty acids has been found in embryonating *Ascaris* eggs and in free-living *Strongyloides* larvae and adults.[34,72] It has been observed that tissue lipids gradually disappear (are consumed?) by infective eggs or larvae of several species. Interestingly, β-oxidation could not be demonstrated in *Ascaris* muscle or in parasitic females of *Strongyloides*.

Synthetic metabolism

Synthetic metabolism of nematodes has not been as intensively studied as has energy metabolism, probably because, in contrast to prokaryotes, energy pathways of helminths usually offer the better sites for chemotherapy. However, there are several points of interest. In light of the enormous number of progeny produced by an organism such as *Ascaris,* protein and nucleic acid synthetic ability must be correspondingly great. In this connection, the RNA metabolism of fertilized *Ascaris* eggs deserves further comment (see embryogenesis, p. 405). The young oocytes have nucleoli and large amounts of cytoplasmic RNA, and these presumably are responsible for the very large amount of yolk protein synthesized in the developing oocyte. By the time the oocyte matures, the nucleoli and most of the cytoplasmic RNA have disappeared.[32] At the same time, the sperm contains little or no RNA. Immediately after fertilization, there is a massive ribosomal RNA synthesis in the male pronucleus, along with a smaller amount of informational RNA, while the female pronucleus is going through its maturation divisions. Kaulenas and Fairbairn suggested that the female genome, therefore, is responsible for the high rate of oocyte production and yolk synthesis, while ribosomes provided by the male genome largely support shell formation and cleavage. Though the sperm brings with it no ribosomes, it does carry a protein of uncertain function called ascaridine.[18] Ascaridine contains no phosphorus, sulfur, or purines, and two amino acids, aspartic acid and tryptophane, account for 35% and 15% of the total nitrogen, respectively. The protein is contained in refringent granules that coalesce during sperm formation to become the **refringent body.** It is clear that the protein in the refringent body is directly related to the ribosome formation just after fertilization, and the ascaridine may well be precursor material for the ribosomes.[23]

As mentioned, the developing oocytes in the ovary are sites of much protein synthesis. Presumably, much of the amino acid supply is furnished by the nearby intestinal absorption, but there is evidence that some amino acids are synthesized in the ovaries as well. The ovaries contain active transaminases, which form amino acids from the corresponding α-keto acids derived from carbohydrate metabolism. In addition, the ovaries can condense pyruvate with ammonia to form the amino acid alanine. In this connection it is interesting to note that some free-living nematodes can synthesize a wide variety of amino acids from a simple substrate, such as acetate. When incubated in a medium containing glycine, glucose, and acetate, *Caenorhabditis briggsae* synthesized an array of "nonessential" and "essential" amino acids. This was the first metazoan known that could synthesize "essential" amino acids, and it was later found that *C. briggsae* could synthesize glycine by a transamination of glyoxylate, the glyoxylate having been produced by the action of isocitrate lyase (Fig. 22-36).[57]

At least some nematodes can synthesize polyunsaturated fatty acids de novo but apparently are unable to synthesize sterols de novo.[56] *Ascaris* incorporates acetate into long chain fatty acids, probably by the malonyl-CoA pathway as found in vertebrates.[3] The nonsugar parts of the ascarosides (the alcohols) in *Ascaris* are synthesized from long chain fatty acids. This involves a condensation in which the carboxyl carbon of one fatty acid condenses with carbon number 2 of another, with the elimination of 1 mole of carbon dioxide.[19] The ascarylose is freely synthesized by *Ascaris* ovaries from glucose or glucose-1-phosphate, and the end product of the synthesis is probably ascarylose-dinucleotidephosphate, which then condenses with the nonsugar moiety to give the ascaroside.

As noted in the discussion on the body wall, much collagen is found in the cuticle of *Ascaris.* These stabilized proteins are important factors in the resistance and strength of the nematode's cuticle. Collagens are stabilized by bonds between lysine residues in the subunits, and they are unusual in that they contain around 12% proline and 9% hydroxyproline; hydroxyproline is an amino acid rarely found in other proteins. In addition to cuticle, collagens exist in muscle, intestine, and reproductive organs of *Ascaris,* and these collagens differ in their hydroxyproline content. Collagen synthesis in *Ascaris* has been studied by several authors.[13] In all systems, the normal collagen precursor is a polypeptide called protocollagen, and the proline in protocollagen is hydroxylated to hydroxyproline by the enzyme **protocollagen proline hydroxylase (PPH or proline monooxygenase).** Among other cosub-

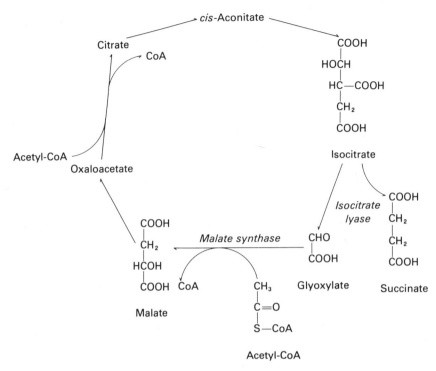

Fig. 22-36. Glyoxylate cycle. A pathway found in plants, microorganisms, and nematodes that can convert fatty acids or acetate to carbohydrates. Each turn of the cycle requires 2 moles of acetyl CoA and yields 1 mole of succinate. Enzymes required other than isocitrate lyase and malate synthase are ones normally found in the tricarboxylic acid cycle.

strates, this enzyme requires molecular oxygen to carry out the hydroxylation of proline. Here we have an example of a biosynthetic reaction that requires oxygen in an organism that is anaerobic with respect to its energy metabolism. PPH activity appears in the embryonating *Ascaris* egg, rises to a peak at the time of the larval molt, decreases, then rises to another peak coinciding with infectivity. In comparing PPH from adult muscle with that from embryonating eggs, Cain and Fairbairn found that oxygen concentration above 5% inhibited the enzyme from muscle but not the PPH from the embryos.[13]

CLASSIFICATION OF PHYLUM NEMATODA
Parasites of vertebrates

Recently the trend has been to divide the phylum into the classes Secernentia and Adenophorea rather than Phasmidia and Aphasmidia, since Phasmidia is occupied for an order of orthopteran insects (walking sticks). As originally conceived, the Se-

cernentia have lateral canals opening into the excretory system, while the Adenophorea do not. Because this is an even more cryptic character than the presence or absence of phasmids, we choose Phasmidia and Aphasmidia as the more useful taxonomic concept. There are other, correlating characters that serve to support this division.[46] It is commonly accepted that the Aphasmidia are more primitive and that the Phasmidia evolved from them.

Class APHASMIDEA

Amphids generally well developed (except in parasitic forms), well behind the lips, often with complex pores. Caudal and hypodermal glands common. Phasmids absent. Excretory system lacking lateral canals, basically formed of a single, ventral, glandular cell or entirely absent. Deirids always absent. Mostly free-living, some parasitic on plants or in invertebrates or vertebrates.

✳ Order TRICHURATA

Anterior end more slender than posterior end. Lips and buccal capsule absent or much reduced.

Esophagus a very slender, capillary-like tube, embedded within one or more rows of large, glandular cells (stichocytes) along its posterior portion. Bacillary band present. Both sexes with a single gonad. Males with one spicule or none. Eggs with polar plugs. Parasites of nearly all organs of all classes of vertebrates.

Families: Anatrichosomatidae, Capillariidae, Cystoopsidae, Trichinellidae, Trichosomoididae, Trichuridae.

Order DIOCTOPHYMATA

Stout worms, often very large. Anterior cuticle spinose in some species. Esophageal glands highly developed, multinucleate. Lips and buccal capsule reduced, replaced by muscular oral sucker in Soboliphymatidae. Esophagus cylindrical. Nerve ring far anterior. Anus at posterior end in both sexes. Male with bell-shaped muscular copulatory bursa without rays. Both sexes with a single gonad. Male with single spicule. Eggs deeply sculptured or pitted. Parasites of birds and mammals.

Families: Dioctophymatidae, Eustrongylidae, Soboliphymatidae.

Class PHASMIDEA

Amphids generally poorly developed, with small, simple pores near or on the lips. Caudal and hypodermal glands absent. Phasmids present. Excretory system with one or two lateral canals, with or without associated glandular cells. Deirids commonly present. Free-living or parasitic in plants or invertebrates or vertebrates.

Order RHABDITATA

Tiny to small worms, commonly with six small lips. Esophagus muscular, divided into anterior corpus, median isthmus, and posterior bulb. Pseudobulb often present between corpus and isthmus. Bulb usually absent in parasitic stages. Buccal capsule small or absent. Tail conical in both sexes, spicules equal, gubernaculum usually present. Parasitic generations parthenogenetic hermaphroditic (males usually unknown), alternating with gonochoristic, free-living generations. Parasites of lungs of amphibians and reptiles or of intestines of amphibians, reptiles, birds, and mammals.

Families: Rhabdiasidae, Strongyloididae.

Order STRONGYLATA

Commonly long, slender worms. Esophagus usually swollen posteriorly but lacking definite bulb. Male with well-developed copulatory bursa supported by sensory rays. Usually oviparous. Eggs thin-shelled, rarely developed beyond morula when laid. Parasites of all classes of vertebrates (rare in fishes).

Families: Ancylostomatidae, Amidostomatidae, Cloacinidae, Cyathostomidae, Deletrocephalidae, Diaphanocephalidae, Dictyocaulidae, Heligmosomatidae, Metastrongylidae, Oesophagostomatidae, Pharyngostrongylidae, Protostrongylidae, Pseudaliidae, Stephanuridae, Strongylidae, Syngamidae, Trichostrongylidae, Angiostrongylidae, Filaroididae, Strongylacanthidae, Ichthyostrongylidae, Ollulanidae.

Order ASCARIDATA

Commonly large, stout worms. Usually three lips present, less often two or none. Esophagus simple, muscular, lacking specializations; occasionally with posterior ventriculus, with or without appendix. Intestine occasionally with one or more appendixes at esophagointestinal junction. Preanal sucker present in a few species. Eggs usually with sculptured uterine layer, unembryonated when laid; larvae infective to host in second stage. Parasites of all classes of vertebrates.

Families: Acanthocheilidae, Angusticaecidae, Anisakidae, Ascaridiidae, Ascaridae, Crossophoridae, Goeziidae, Heterocheilidae, Inglisonematidae, Oxyascarididae, Toxocaridae.

Order OXYURATA

Medium-sized to small worms, commonly with sharply pointed tails. Lips, when present, usually three in number; often reduced or absent. Esophagus with posterior bulb. Caudal alae often well developed. Preanal sucker present in some species. Eggs thin-shelled, usually fully embryonated when laid, and life cycle usually direct. Parasites of arthropods and all classes of vertebrates.

Superfamily OXYUROIDEA

Tails of both sexes usually sharply pointed. Spicules small; one or both absent in several species. Eggs often with operculum. No intermediate host in life cycle.

Families: Heteroxynematidae, Oxyuridae, Ozolaimidae, Pharyngodonidae, Syphaciidae.

Superfamily ATRACTOIDEA

Head usually with complex ornamentation. Intestinal cecum present in *Cruzia*. One ovary present (except *Probstmayria*) . Viviparous.

Families: Atractidae, Cruziidae, Crossocephalidae, Hoplodontophoridae, Labiduridae, Schrankianidae, Travnematidae.

Superfamily COSMOCERCOIDEA

Three or six lips present. Males with two equal spicules; caudal papillae numerous.

Families: Cosmocercidae, Gyrinicolidae, Lauroiidae.

Superfamily HETERAKOIDEA

Three well-defined lips. Preanal sucker present on males. Two spicules. Eggs not embryonated when laid.

Families:Aspidoderidae, Heterakidae, Spinicaudidae, Strongyluridae.

Superfamily KATHLANIOIDEA

Mouth complex, with three or six lips. Buccal capsule well developed, sometimes with teeth. Preanal sucker or powerful preanal musculature present.

Family: Kathlaniidae.

Superfamily SUBULUROIDEA

Lips absent or quite reduced. Anterior end of esophagus commonly with teeth. Preanal sucker present.

Families: Maupasinidae, Parasubuluridae, Subuluridae.

Order SPIRURATA

Mouth surrounded by six small lips, or surrounded by a cuticular ring, or with two lateral pseudolabia. Buccal capsule present. Cephalic ornamentation common. Esophagus usually divided into an anterior, muscular portion and a posterior, glandular portion; never with posterior bulb. Spicules usually unequal in size and shape.

Families: Acuariidae, Ascaropsidae, Cobboldinidae, Crassicaudidae, Desmidocercidae, Gnathostomatidae, Gongylonematidae, Habronematidae, Haplonematidae, Hedruridae, Physalopteridae, Pneumospiruridae, Rhabdochonidae, Rictulariidae, Salobrellidae, Schistorophidae, Seuratidae, Spinitectidae, Spirocercidae, Spiruridae, Streptocaridae, Tetrameridae, Thelaziidae.

Order CAMALLANATA

Lips absent. Buccal capsule present or absent or replaced with large bilateral, sclerotized valves. Esophagus long, distinctly divided into anterior muscular and posterior glandular portions. Spicules unequal and dissimilar or equal and similar. Ovoviviparous. Anus and vulva may be atrophied in females. Parasites of tissue, coelom, air bladder, circulatory system or digestive system of aquatic and terrestrial vertebrates, including humans.

Families: Anguillicolidae, Camallanidae, Dracunculidae, Philometridae, Phlyctainophoridae, Skrjabillanidae, Tetanonematidae.

Order FILARIATA

Mouth simple, lacking lips. Buccal capsule absent in most species. Esophagus usually divided into anterior muscular and posterior glandular portions. Spicules usually unequal and dissimilar. Oviparous or ovoviviparous. Parasites of tissues or respiratory system of terrestrial vertebrates.

Families: Aproctidae, Desmidocercidae, Diplotriaenidae, Filariidae, Onchocercidae, Setariidae.

REFERENCES

1. Anderson, R. V., and H. M. Darling. 1964. Embryology and reproduction of *Ditylenchus destructor* Thorne, with emphasis on gonad development. Proc. Helm. Soc. Wash. 31:240-256.
2. Barrett, J., C. W. Ward, and D. Fairbairn. 1970. The glyoxylate cycle and the conversion of triglycerides to carbohydrates in developing eggs of *Ascaris lumbricoides*. Comp. Biochem. Physiol. 35:577-586.
3. Beames, C. G., Jr., B. G. Harris, and F. A. Hopper, Jr. 1967. The synthesis of fatty acids from acetate by intact tissue and muscle extract of *Ascaris lumbricoides suum*. Comp. Biochem. Physiol. 20:509-521.
4. Beaver, P. C., Y. Yoshida, and L. R. Ash. 1964. Mating of *Ancylostoma caninum* in relation to blood loss in the host. J. Parasitol. 50:286-293.
5. van Beneden, E. 1883, Recherches sur la maturation de l'oeuf et la fécondation *(Ascaris megalocephala)*. Arch. Biol. 4:265-641.
6. Bird, A. F. 1971. The structure of nematodes. Academic Press, Inc., New York.
7. Bird, A. F., and J. Bird. 1969. Skeletal structures and integument of Acanthocephala and Nematoda. In Florkin, M., and B. T. Scheer, editors. Chemical zoology, vol. III. Academic Press, Inc., New York. pp. 253-288.
8. Bolla, R. I., P. P. Weinstein, and G. D. Cain. 1972. Fine structure of the coelomocyte of adult *Ascaris suum*. J. Parasitol. 58:1025-1036.
9. Boveri, T. 1899. Die Entwicklung von *Ascaris megalocephala* mit besonderer Rücksicht auf die Kernverhältnisse. Festschr. f. C. Von Kupffer, Jena.
10. von Brand, T. 1966. Biochemistry of parasites. Academic Press, Inc., New York.
11. Bryant, C. 1970. Electron transport in parasitic helminths and protozoa. In Dawes, B., editor. Advances in parasitology, vol. 8. Academic Press, Inc., New York. pp. 139-172.
12. Bryant, C. 1975. Carbon dioxide utilization, and the regulation of respiratory metabolic pathways in parasitic helminths. In Dawes, B., editor. Advances in parasitology, vol. 13. Academic Press, Inc., New York. pp. 36-69.
13. Cain, G. D., and D. Fairbairn. 1971. Protocollagen proline hydroxylase and collagen synthesis

in developing eggs of *Ascaris lumbricoides*. Comp. Biochem. Physiol. 40B:165-179.

14. del Castillo, J. 1969. Pharmacology of Nematoda. In Florkin, M., and B. T. Scheer, editors. Chemical zoology, vol. III. Academic Press, Inc., New York. pp. 521-554.

15. Cheah, K.S., and B. Chance. 1970. The oxidase systems of *Ascaris*-muscle mitochondria. Biochim. Biophys. Acta 223:55-60.

16. Crofton, H. D. 1966. Nematodes. Hutchinson University Library, London.

17. Davey, K. G., and S. P. Kan. 1968. Molting in a parasitic nematode, *Phocanema decipiens*. IV. Ecdysis and its control. Can. J. Zool. 46:893-898.

18. Fairbairn, D. 1960. The physiology and biochemistry of nematodes. In Sasser, J. N., and W. R. Jenkins, editors. Nematology. University of North Carolina Press, Chapel Hill, N.C. pp. 267-296.

19. Fairbairn, D. 1969. Lipid components and metabolism of Acanthocephala and Nematoda. In Florkin, M., and B. T. Scheer, editors. Chemical zoology, vol. III. Academic Press, Inc., New York. pp. 361-378.

20. Fairbairn, D. 1970. Biochemical adaptation and loss of genetic capacity in helminth parasites. Biol. Rev. 45:29-72.

21. Florkin, M., and B. T. Scheer, editors. 1969. Chemical zoology, vol. III. Academic Press, Inc., New York.

22. Foor, W. E. 1967. Ultrastructural aspects of oocyte development and shell formation in *Ascaris lumbricoides*. J. Parasitol. 53:1245-1261.

23. Foor, W. E. 1970. Spermatozoan morphology and zygote formation in nematodes. Biol. Reprod. 2 (Suppl.):177-202.

24. Foster, W. D. 1965. A history of parasitology. E. & S. Livingstone, Edinburgh.

25. Harpur, R. P. 1962. Maintenance of *Ascaris lumbricoides* in vitro: a biochemical and statistical approach. Can. J. Zool. 40:991-1011.

26. Harpur, R. P. 1964. Maintenance of *Ascaris lumbricoides* in vitro—III. Changes in the hydrostatic skeleton. Comp. Biochem. Physiol. 13:71-85.

27. Harpur, R. P., and J. S. Popkin, 1965. Osmolality of blood and intestinal contents in the pig, guinea pig, and *Ascaris lumbricoides*. Can. J. Biochem. 43:1157-1169.

28. Harris, J. E., and H. D. Crofton. 1957. Structure and function in the nematodes: internal pressure and cuticular structure in *Ascaris*. J. Exp. Biol. 34:116-130.

29. Hyman, L. H. 1951. The invertebrates: Acanthocephala, Aschelminthes, and Entoprocta, the pseudocoelomate bilateria, vol. III. McGraw-Hill Book Co., New York.

30. Inglis, W. G. 1964. The structure of the nematode cuticle. Proc. Zool. Soc. Lond. 143:465-502.

31. Jezyk, P. F., and D. Fairbairn. 1967. Metabolism of ascarosides in the ovaries of *Ascaris lumbri-coides* (Nematoda). Comp. Biochem. Physiol. 23:707-719.

32. Kaulenas, M. S., and D. Fairbairn. 1968. RNA metabolism of fertilized *Ascaris lumbricoides* eggs during uterine development. Exp. Cell Res. 52:233-251.

33. Keilin, D. 1925. On cytochrome, a respiratory pigment common to animals, yeasts and higher plants. Proc. R. Soc. Ser. B, 98:312-339.

34. Körting, W., and D. Fairbairn. 1971. Changes in beta-oxidation and related enzymes during the life cycle of *Strongyloides ratti* (Nematoda). J. Parasitol. 57:1153-1158.

35. Lee, D. L. 1965a. The physiology of nematodes. Oliver & Boyd Ltd. Edinburgh.

36. Lee, D. L. 1965b. The cuticle of adult *Nippostrongylus brasiliensis*. Parasitology 55:173-181.

37. Lee, D. L. 1966. The structure and composition of the helminth cuticle. In Dawes, B., editor. Advances in parasitology, vol. 4. Academic Press, Inc., New York. pp. 187-254.

38. Lee, D. L. 1969. *Nippostrongylus brasiliensis:* some aspects of the fine structure and biology of the infective larva and the adult. In Taylor, A. E. R., editor. *Nippostrongylus* and *Toxoplasma*. Symposia of the British Society of Parasitology, vol. 7. Blackwell Scientific Publications, Oxford. pp. 3-16.

39. Lee, D. L. 1970. Moulting in nematodes: the formation of the adult cuticle during the final moult of *Nippostrongylus brasiliensis*. Tissue Cell 2:139-153.

40. Lee, D. L. 1972. The structure of the helminth cuticle. In Dawes, B., editor. Advances in parasitology, vol. 10. Academic Press, Inc., New York. pp. 347-379.

41. Lee, D. L., and A. O. Anya. 1967. The structure and development of the spermatozoon of *Aspiculuris tetraptera* (Nematoda). J. Cell Sci. 2:537-544.

42. Lee, D. L., and P. Leštan. 1971. Oogenesis and eggshell formation in *Heterakis gallinarum* (Nematoda). J. Zool. (Lond.) 164:189-196.

43. Lee, H., I. Chen, and R. Lin. 1973. Ultrastructure of the excretory system of *Anisakis* larva (Nematoda: Anisakidae). J. Parasitol. 59:289-298.

44. Meglitsch, P. A. 1972. Invertebrate zoology, ed.2. Oxford University Press, New York.

45. Michel, J. F. 1974. Arrested development of nematodes and some related phenomena. In Dawes, B., editor. Advances in parasitology, vol. 12. Academic Press, Inc., New York. pp. 280-366.

46. Paramonov, A. A. 1954. On the structure and function of the phasmids. Trudy Gelmint. Lab. Akad. Nauk SSSR 7:19-49. (Translation available from G. D. Schmidt.)

47. Passey, R. F., and D. Fairbairn. 1957. The conversion of fat to carbohydrate during embryonation of *Ascaris* eggs. Can. J. Biochem. Physiol. 35:511-525.

48. Roberts, L. S., and D. Fairbairn. 1965. Metabolic studies on adult *Nippostrongylus brasiliensis* (Nematoda: Trichostrongyloidea). J. Parasitol. 51:129-138.

49. Rogers, W. P. 1965. The role of leucine aminopeptidase in the moulting of nematode parasites. Comp. Biochem. Physiol. 14:311-321.

50. Rogers, W. P. 1969. Nitrogenous components and their metabolism: Acanthocephala and Nematoda. In Florkin, M., and B. T. Scheer, editors. Chemical zoology, vol. III. Academic Press, Inc. New York. pp. 379-428.

51. Rogers, W. P., and R. I. Sommerville. 1963. The infective stage of nematode parasites and its significance in parasitism. In Dawes, B., editor. Advances in parasitology, vol. 1. Academic Press, Inc., New York. pp. 109-177.

52. Rogers, W. P., and R. I. Sommerville. 1968. The infectious process, and its relation to the development of early parasitic stages of nematodes. In Dawes, B., editor. Advances in parasitology, vol. 6. Academic Press, Inc., New York. pp. 327-348.

53. Rosenbluth, J. 1965a. Ultrastructural organization of obliquely striated muscle fibers in *Ascaris lumbricoides*, J. Cell Biol. 25:495-515.

54. Rosenbluth, J. 1965b. Ultrastructure of somatic cells in *Ascaris lumbricoides*. II. Intermuscular junctions, neuromuscular junctions, and glycogen stores. J. Cell Biol. 26:579-591.

55. Rosenbluth, J. 1967. Obliquely striated muscle. III. Contraction mechanism of *Ascaris* body muscle. J. Cell Biol. 34:15-33.

56. Rothstein, M. 1970. Nematode biochemistry. XI. Biosynthesis of fatty acids by *Caenorhabditis briggsae* and *Panagrellus redivivus*. Int. J. Biochem. 1:422-428.

57. Rothstein, M., and H. Mayoh. 1964. Glycine synthesis and isocitrate lyase in the nematode, *Caenorhabditis briggsae*. Biochem. Biophys. Res. Comm. 14:43-47.

58. Savel, J. 1955. Études sur la constitution et le métabolism protéiques d'*Ascaris lumbricoides* Linné, 1758. Rev. Path. Comp. Hyg. Gen. Comp. 55:52-121.

59. Saz, D. K., T. P. Bonner, M. Karlin, and H. J. Saz. 1971. Biochemical observations on adult *Nippostrongylus brasiliensis*. J. Parasitol. 57:1159-1162.

60. Saz, H. J. 1969. Carbohydrate and energy metabolism of nematodes and Acanthocephala. In Florkin, M., and B. T. Scheer, editors. Chemical zoology, vol. III. Acad. Press, Inc., New York. pp. 329-360.

61. Saz, H. J. 1972. Comparative biochemistry of carbohydrates in nematodes and cestodes. In Van den Bossche, H., editor. Comparative biochemistry of parasities. Academic Press, Inc., New York. pp. 33-47.

62. Saz, H. J., and A. Weil. 1960. The mechanism of the formation of α-methylbutyrate from carbohydrate by *Ascaris lumbricoides* muscle. J. Biol. Chem. 235:914-918.

63. Saz, H. J., and A. Weil. 1962. Pathway of formation of α-methylvalerate by *Ascaris lumbricoides*. J. Biol. Chem. 237:2053-2056.

64. Schneider, A. 1866. Monographie der Nematoden. Gregg, Berlin.

65. Sheffield, H. G. 1963. Electron microscopy of the bacillary band and stichosome of *Trichuris muris* and *T. vulpis*. J. Parasitol. 49:998-1009.

66. Sheffield, H. G. 1964. Electron microscope studies on the intestinal epithelium of *Ascaris suum*. J. Parasitol. 50:365-379.

67. Sommerville, R. I., and P. P. Weinstein. 1964. Reproductive behavior of *Nematospiroides dubius* in vivo and in vitro. J. Parasitol. 50:401-409.

68. Soulsby, E. J. L. 1966. Biology of parasites. Academic Press, Inc., New York.

69. Swartz, F. J., M. Henry, and A. Floyd. 1967. Observations on nuclear differentiation in *Ascaris*. J. Exp. Zool. 164:297-307.

70. Thorson, R. E. 1956. The effect of extracts of the amphidial glands, excretory glands, and esophagus of adults of *Ancylostoma caninum* on the coagulation of dog's blood. J. Parasitol. 42:26-30.

71. Walton, A. C. 1959. Some parasites and their chromosomes. J. Parasitol. 45:1-20.

72. Ward, C. W., and D. Fairbairn. 1970. Enzymes of β-oxidation and their function during development of *Ascaris lumbricoides* eggs. Dev. Biol. 22:366-387.

73. Ward, C. W., and P. J. Schofield. 1967. Comparative activity and intracellular distribution of tricarboxylic acid cycle enzymes in *Haemonchus contortus* larvae and rat liver. Comp. Biochem. Physiol. 23:335-359.

74. Weinstein, P. P. 1952. Regulation of water balance as a function of the excretory system of the filariform larvae of *Nippostrongylus muris* and *Ancylostoma caninum*. Exp. Parasitol. 1:363-376.

75. Wright, K. A. 1964. The fine structure of the somatic muscle cells of the nematode *Capillaria hepatica* (Bancroft, 1893). Can. J. Zool. 42:483-490.

76. Wright, K. A. 1966. Cytoplasmic bridges and muscle systems in some polymyarian nematodes. Can. J. Zool. 44:329-340.

77. Wright, K. A. 1968. Structure of the bacillary band of *Trichuris myocastoris*. J. Parasitol. 54:1106-1110.

78. Wright, K. A., and W. D. Hope. 1968. Elaborations of the cuticle of *Acanthonchus duplicatus* Wieser, 1959 (Nematoda: Cyatholaimidae) as revealed by light and electron microscopy. Can. J. Zool. 46:1005-1011.

SUGGESTED READING

Anderson, R. C., A. G. Chabaud, and S. Willmott. 1976. C. I. H. keys to the nematode parasites of vertebrates. Commonwealth Agricultural Bureaux, Farnham Royal, Bucks, England. (A continuing, up-to-date series of keys for the identification of nematode parasites.)

Baylis, H. A., and R. Daubney. 1926. A synopsis of the families and genera of Nematoda. British Museum of Natural History, London. (The first complete summary of nematodes. Still a useful reference, although no illustrations are included.)

Bird, A. F. The structure of nematodes. Academic Press, Inc., New York. (This is the most up-to-date summary of nematode morphology. Indispensable for students of nematodology.)

Crites, J. L. 1969. Problems in systematics of parasitic nematodes. In Schmidt, G. D., editor. Problems in systematics of parasites. University Park Press, Baltimore. pp. 77-87. (A philosophical discussion of nematode systematics.)

Crofton, H. D. 1966. Nematodes. Hutchinson University Library, London. (A very useful summary of nematode characteristics.)

Florkin, M., and B. T. Scheer editors. 1968. Chemical zoology. vol. III. Echinodermata, Nematoda and Acanthocephala. Academic Press, Inc., New York. (Several contributions are presented by different authors on the subjects of nematode anatomy, physiology, and classification.)

Grassé, P. P. 1965. Traité de zoologie. Anatomie, systématique, biologie, némathelminthes (Nématodes-Gordiacés) rotifères-gastrotriches, kinorinques, vol. 4, parts 2 and 3. Masson et Cie., Paris. (An indispensable reference for serious students of nematodes.)

Hyman, L. H. 1951. The invertebrates: Acanthocephala, Aschelminthes, and Entoprocta. The pseudocoelomate bilateria, vol. 3. McGraw-Hill Book Co., New York. (An excellent reference to the morphology and biology of nematodes.)

Lee, D. L. 1965. The physiology of nematodes. University reviews in biology, vol. 3. Oliver & Boyd Ltd., Edinburgh. (This is a good reference to nematode physiology. Readers who desire an in-depth account should consult this work.)

Levine, N. D. 1968. Nematode parasites of domestic animals and of man. Burgess Publishing Co., Minneapolis. (An excellent general reference. The introductory chapter is useful for anatomy; Levine's proposed standard endings for taxa are used throughout.)

Skrjabin, K. I., and others. 1949-1952. Key to parasitic nematodes, vols. 1-3. Academiya Nauk SSSR, Moscow. (Excellent taxonomic presentation of several groups of parasitic nematodes. Volumes 1 and 3 have been translated into English.)

Skrjabin, K. I., and others. 1953-1972. Essentials of nematodology, vols. 1-22. Academiya Nauk SSSR, Moscow. (The most complete taxonomic compilation of information on parasitic nematodes in the world. Several volumes have been translated into English.)

Stone, W., and F. W. Smith. 1973. Infection of mammalian hosts by milk-borne nematode larvae: a review. Exp. Parasitol. 34:306-312.

York, W., and P. A. Maplestone. 1926. The nematode parasites of vertebrates. (Reprint, 1962.) Hafner Publishing Co., New York. (Still a valuable reference work.)

Chapter 23

ORDERS TRICHURATA AND DIOCTOPHYMATA: THE APHASMIDIAN PARASITES

Nematodes of particular importance to humans

Some of the most dreaded, disfiguring, and debilitating diseases of humans are caused by nematodes. Further, agriculture suffers mightily from attacks by these animals. Nematodes normally parasitic in wild animals can occasionally infect humans and domestic animals, causing mystifying diseases. Further, nonparasitic nematodes may accidentally find their way into a vertebrate and become a short-lived, but pathogenic, parasite. Many thousands of nematodes are known to parasitize vertebrates; many still are unknown. A few examples are presented here, selected for their interest as parasites of humans and as illustrations of parasitism as exemplified by nematodes.

ORDER TRICHURATA

This order of nematodes contains, among others, three genera of medical importance: *Trichuris, Capillaria,* and *Trichinella.* They have morphological and biological peculiarities that place them in the class Aphasmidia, a group of mostly nonparasitic worms. See p. 414 for details of classification.

Family Trichuridae

Whipworms, members of the family Trichuridae, are so-called because they are thread-like along most of their body, then they abruptly become thick at the posterior end, reminiscent of a whip with a handle (Fig. 23-1). The name *Trichocephalus* ("thread-head"), in widespread use in some countries, was coined when it was realized that the "thread" was the anterior end rather than the tail, but the term

Trichuris has priority. There are many species in a wide variety of mammalian hosts, and one is a very important parasite of humans.

Trichuris trichiura

Morphology. *Trichuris trichiura* measure from 30 to 50 mm long, with males being somewhat smaller than females. The mouth is a simple opening, lacking lips. The buccal cavity is tiny and is provided with a minute spear. The esophagus is very long, occupying about two-thirds of the body length and consists of a thin-walled tube surrounded by large, unicellular glands, the stichocytes. The entire structure often is referred to as the stichosome (see Chapter 22). The anterior end of the esophagus is somewhat muscular and lacks stichocytes. Both sexes have a single gonad, and the anus is near the tip of the tail. Males have a single spicule that is surrounded by a spiny spicule sheath. The ejaculatory duct joins the intestine anterior to the cloaca. In the female, the vulva is near the junction of the esophagus and the intestine. The uterus contains many unembryonated, lemon-shaped eggs, each with a prominent plug at each end (Fig. 23-2).

The excretory system is absent. The ventral surface of the esophageal regions bears a wide band of minute pores, leading to underlying glandular and nonglandular cells.[12,18] This bacillary band is typical of the order. Though the function of the cells in the bacillary band is unknown, their ultrastructure suggests that the gland cells may have a role in osmotic or ion regulation, and the nongland cells may function in cuticle formation and food storage.

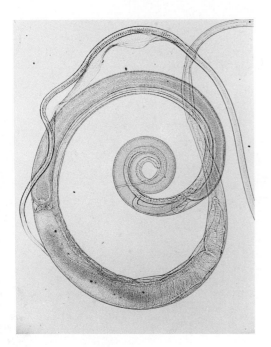

Fig. 23-1. A male *Trichuris*. Note the slender anterior end and the stout posterior end with a single, terminal spicule. (Photograph by Jay Georgi.)

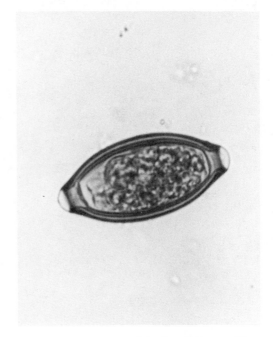

Fig. 23-2. Egg of *Trichuris trichiura*. (Photograph by Robert E. Kuntz.)

Biology. Estimates of egg production range from 1,000 to 7,000 per day. Embryonation is completed in about 21 days in soil, which must be moist and shady. When swallowed, the infective larva hatches in the small intestine and enters the crypts of Lieberkühn. After a short period of development, it reenters the intestinal lumen and migrates to the ileocecal area, where it matures in about 3 months. Adults live for several years, so large numbers may accumulate in a person, even in areas where the rate of new infection is low.

Epidemiology. The two requirements for *Trichuris* to become a serious health problem are poor standards of sanitation in which human feces are deposited on the soil and the combination of physical conditions that allows the worm's survival and development. These conditions are a warm climate, high rainfall and humidity, moisture-retaining soil, and dense shade. Though generally coextensive in distribution with *Ascaris, Trichuris* is more sensitive to the effects of desiccation and direct sunlight. Appropriate physical conditions

exist in much of the world, including parts of the southeastern United States, where the incidence of infection may reach 20% to 25%, mainly in small children. Stoll calculated the world prevalence at 355 million.[16] It is doubtful that the figure is lower today, considering the population increase in tropical areas of the world; *T. trichiura* is perhaps the most common nematode of humans after *Ascaris* and *Enterobius*. Small children are most commonly infected, either by drinking contaminated water or by placing egg-contaminated fingers in their mouths.

Pathology. Fewer than 100 worms rarely cause clinical symptoms, and the majority of infections are symptomless. A heavier burden may result in a variety of conditions, occasionally terminating in death. The anterior ends of the worms burrow in the mucosa (Fig. 23-3), where the worms consume blood cells, although blood loss by that mechanism is negligible. Trauma to the intestinal epithelium and underlying submucosa, however, can cause a chronic hemorrhage that may result in anemia. Secondary bacterial infection, possibly

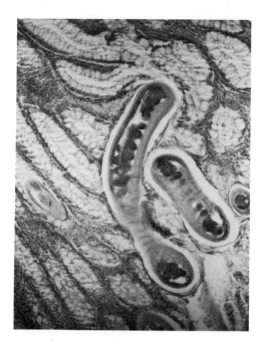

Fig. 23-3. A section of large intestine. Note the sections of *Trichuris* embedded in the mucosa. (Photograph by Robert E. Kuntz.)

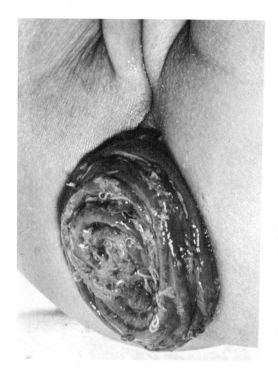

Fig. 23-4. Prolapse of the rectum caused by whipworm infection. (Courtesy University of Miami, School of Medicine. From Beck, J. W., and J. E. Davies. 1976. Medical parasitology, ed. 2. The C. V. Mosby Co., St. Louis.)

coupled with allergic responses, results in colitis, proctitis, and in extreme cases, prolapse of the rectum (Fig. 23-4). Inflammation of the appendix incriminates whipworm as a possible cause of appendicitis, although their mere presence in an infected appendix is not proof of a causal relationship. Amebic dysentery may be complicated by *Trichuris*. Among other symptoms of infection are insomnia, nervousness, loss of appetite, vomiting, urticaria, prolonged diarrhea, constipation, flatulence, and verminous intoxication. It is believed, however, that the more severe "toxic" effects are not solely the result of the worm infection but involve other factors, such as malnutrition.[15]

Diagnosis and treatment. Specific diagnosis depends on demonstrating a worm or egg in the stool. The eggs, with distinctive bipolar plugs, are 50 to 54 μm by 22 to 23 μm and have smooth outer shells. Clinical symptoms may be confused with hookworm, amebiasis, or acute appendicitis. Species of *Trichuris*, which are morphologically identical to *T. trichiura,* are found in pigs and wild primates.

Because of their location in the cecum, appendix, or lower ileum, whipworms are difficult to reach with oral drugs or medicated enemas. Thiabendazole and dithiazanine (not always available in many countries) are effective orally, while an enema of hexylresorcinol crystals in distilled water will remove worms in the rectum and lower colon.[5] Training of children and adults in sanitary disposal of feces and washing of hands is necessary to prevent reinfection.

Family Capillariidae

Members of the genus *Capillaria* look very much like *Trichuris*, except that the transition between the anterior, filiform portion and the posterior, stout portion is gradual, rather than sudden. Other morphological features are similar. A large genus, *Capillaria* includes species that are parasitic in nearly all organs and tissues of all classes of vertebrates.

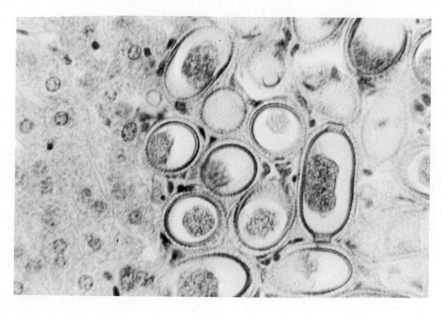

Fig. 23-5. Eggs of *Capillaria hepatica* in liver. Note the extensive damage to hepatic parenchyma. (Photograph by Warren Buss.)

Capillaria hepatica

Biology. *Capillaria hepatica* is a parasite of the liver, mainly of rodents, but it has been found in a wide variety of mammals, including humans. The female deposits eggs in the liver parenchyma, where they have no means of egress until eaten by a predator or until the liver decomposes after death. The eggs cannot embryonate while in the liver, so a new host cannot be infected when it eats an egg-laden liver. The eggs merely pass through the digestive tract of the predator with the feces. Embryonation occurs on the soil, and new infection is by contamination. After hatching in the small intestine, the larvae migrate to the liver, where they mature.

Epidemiology. As in *Trichuris trichiura,* infection occurs when contaminated objects, food, or water are ingested. Unlike whipworm, however, human feces are not the source of contamination; more likely, the feces of carnivores or flesh-eating rodents are involved.

Pathology. Wandering of adult *Capillaria* through the host liver causes loss of liver cells and thereby loss of normal function. Large areas of parenchyma may be replaced by masses of eggs (Fig. 23-5). Rarely, eggs will be carried to the lungs or other organs by the bloodstream.

Diagnosis and treatment. Verified cases of this parasite in humans are rare, partly because of difficulties of diagnosis. Most cases have been determined post mortem, but liver biopsy has uncovered others. Clinical symptoms resemble numerous liver disorders, especially hepatitis with eosinophilia. Specific diagnosis depends on demonstrating the eggs, which closely resemble those of *Trichuris* except that they measure 51 to 67 μm by 30 to 35 μm and have deep pits in the outer shells. There is no known treatment for this disease.

Discovery of *C. hepatica* eggs in human feces probably indicates the presence of a spurious infection caused by eating an infected liver.

Capillaria philippinensis

Capillaria philippinensis was discovered in 1963 as a parasite of humans in the Philippines. In contrast to *C. hepatica, C. philippinensis* is an intestinal parasite. Its appearance as a human pathogen was sudden and unexpected. One or two isolated cases were followed by an epidemic in Luzon in 1967 that killed several dozen persons.[3] Probably a zoonotic disease, it is possibly transmitted between humans by fecal contamination. The original animal host remains unknown, but circumstantial evidence points to fish as the source of human infection.

Morphology. This parasite is very small, males measuring 2.3 to 3.17 mm and females 2.5 to 4.3 mm long. The male has small caudal alae and a nonspined spicule sheath. The esophagus of the female is about half as long as the body. The female produces typical *Capillaria*-type eggs that lack pits.

Epidemiology. Humans remain the only known host of this parasite, although it probably is shared by some species of wild animals. Intensive surveys of the Philippine fauna have so far failed to identify any reservoir host. Once established in a village, human transmission is readily maintained, particularly during the rainy season when soil contaminated with feces abounds in the form of mud.

Pathology. The worms repeatedly penetrate the mucosa of the small intestine and reenter the lumen, especially in the jejunum, leading to progressive degeneration of the epithelium and submucosa. The two primary results are pathological malabsorption of nutrients and violent diarrhea. Abdominal distention and pain usually occur. Death appears to be caused by emaciation and ion imbalance, leading to fatal shock.[2]

Diagnosis and treatment. Both adults and eggs, as well as larvae, are abundant in feces of heavily infected persons, and at least one of them is necessary for specific diagnosis. An intradermal test is now also available.

Mebendazole is very effective in curing this disease.[13] Control is unsatisfactory because the source of infection in nature is unknown. Epidemics can be prevented by a high standard of hygiene.

Other Capillaria species

Several species of *Capillaria* are important parasites of domestic animals. *Capillaria aerophila* is a lung parasite of cats, dogs, and other carnivores and has been reported several times from humans in Russia. It is probably the most destructive parasite of commercial fox farms.

Capillaria annulata and *C. caudinflata* infect the esophagus and crop of chickens, turkeys, and several other species of birds. Unlike most species in the genus, an inter-mediate host (earthworm) is required in the life cycle.

Few species of terrestrial vertebrates, wild or domestic, are free from at least one species of *Capillaria*.

Anatrichosoma spp.

Species of *Anatrichosoma* are very similar to *Capillaria,* except they lack a spicule and spicule sheath. They have been reported from the tissues of a wide variety of Asian and African monkeys and gerbils and from the North American opossum. *Anatrichosoma ocularis* (Fig. 23-6) lives in the corneal epithelium of tree shrews, *Tupaia glis.* The eggs of this genus (Fig. 23-7) have the polar plugs characteristic of the order. No species is known to parasitize humans, but the species in monkeys, at least, should be considered as possible potential zoonoses.

Family Trichinellidae
Trichinella spiralis

Curiously, the smallest nematode parasite of humans, which exhibits the most unusual life cycle, is one of the most widespread and clinically important organisms in the world. *Trichinella spiralis* is the only species in the family (although physiological subspecies may occur) and is responsible for the disease variously known as trichinosis, trichiniasis, or trichinelliasis. It is common in carnivorous mammals, including rodents and humans, primarily on the circumboreal continents. *T. spiralis* is less common in tropical regions but is well known in Mexico, parts of South America, Africa, southern Asia, and the Middle East. Incidence of infection is always higher than suspected, because of the vagueness of symptoms, which usually suggests other conditions; over 50 different diseases have been diagnosed incorrectly as trichinosis.

Morphology. The males (Fig. 23-8) measure 1.4 to 1.6 mm long and are more slender at the anterior end than at the posterior end. The anus is nearly terminal and has a large papilla on each side of it. A copulatory spicule is absent. Like other members of the order Trichurata, stichocytes are arranged in a row behind a short muscular esophagus. Females are about twice the size of males, also tapering

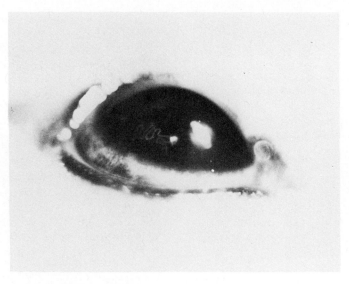

Fig. 23-6. *Anatrichosoma ocularis* in the eye of a tree shrew, *Tupaia glis.* (From File, S. K. 1974. J. Parasitol. 60:985-988.)

toward the anterior end. The anus is nearly terminal. The vulva is located near the middle of the esophagus, which is about one-third the length of the body. The single uterus is filled with developing eggs in its posterior portion, while the anterior portion contains fully developed, hatched larvae.

Biology. The biology of this organism is unusual in that one animal serves as both definitive and intermediate hosts, with the larvae and the adults located in different organs.

Although her mate dies shortly after copulation, the female burrows into the mucosa and submucosa, sometimes entering and following the lymphatic ducts to the mesenteric lymph nodes. During and subsequent to this migration, she gives birth to about 1,500 larvae over a period of 4 to 16 weeks. Eventually, the spent female dies and is absorbed by the host. Some larvae may escape into the intestinal lumen where they can remain infective to another intermediate host if passed in the feces.[10] Any females remaining in the intestine 2 weeks after infection are spontaneously expelled by immune reactions.

Most larvae are carried away by the hepatic portal system through the liver, then to the heart, lungs, and the arterial system,

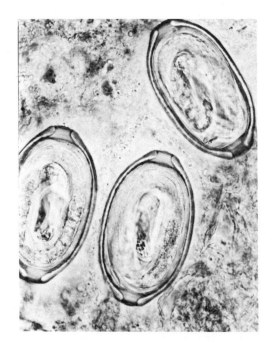

Fig. 23-7. Eggs of *Anatrichosoma ocularis* from eye secretions. (Courtesy Sharon K. File.)

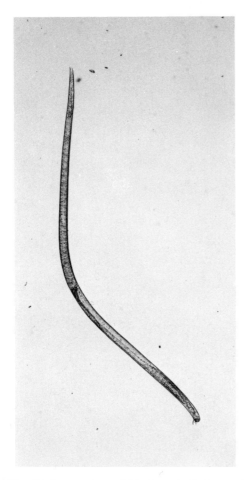

Fig. 23-8. A male *Trichinella spiralis* from the intestine of a rat. (Photograph by Jay Georgi.)

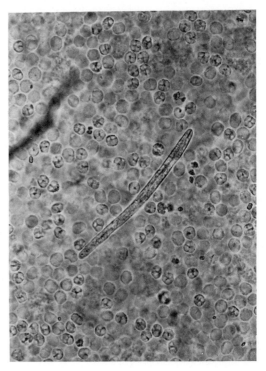

Fig. 23-9. Live *Trichinella spiralis* larva, migrating in the blood of a rat. (Photograph by Jay Georgi.)

which distributes them throughout the body (Fig. 23-9). During this migration, they have been found in literally every kind of tissue and space in the body. When they reach skeletal muscle, they penetrate individual fibers and begin to grow, eventually forming a spiral and becoming encysted by infiltrating leukocytes (Fig. 23-10). Although one worm per cyst is most common, up to seven have been observed in a single cyst within a single muscle cell.

Some muscles are much more heavily invaded than others, but the reasons for this are not understood. Most susceptible are muscles of the eye and tongue and masticatory muscles, then diaphragm and intercostals, and finally the heavy muscles of the arms and legs. The larvae absorb their nu-

trients from the enclosing muscle cell and increase in length to about 1.0 mm in about 8 weeks, at which time they are infective to their next host.

The cyst wall becomes gradually thicker during this time, and finally achieves a length of 0.25 to 0.5 mm. The enclosing muscle cell degenerates, as the result of movements of the larva, loss of nutrients, and toxicity of the parasite's metabolic wastes. The larvae enter a developmental arrest and can live for months while encysted. Gradually, after about 10 months, the host reactions begin to calcify the cyst walls and, eventually, the worms themselves. However, some are believed still to be viable up to 30 years.[8]

It was the presence of such calcified cysts in human cadavers that led to the discovery of this species in 1835 by James Paget, a medical student in London. Noticing that his subject had gritty particles in its muscles that tended to dull his scalpels, he studied the particles and demonstrated

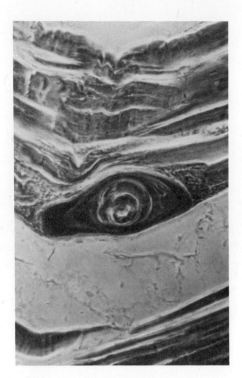

Fig. 23-10. A larva of *Trichinella spiralis* encysted in a muscle fiber.

their worm-like nature to his fellow students. He then showed them to the eminent anatomist Richard Owen, who reported on them further and gave them their scientific name. It was another 25 years before it was determined that these minute animals cause disease.

As in other nematodes, there are four larval stages, but various authors disagree on whether the infective larva is second, third, or fourth stage.[7] After ingestion, the enclosing muscle cells and cyst wall are digested off, freeing the worm, which enters intestinal crypts, molts at least once, and matures within 40 hours. Most mammals are susceptible to infection. **Sylvatic trichinosis** occurs between wild carnivores and their prey or carrion. Hence, bears, badgers, foxes, and even walruses are commonly heavily infected. **Urban trichinosis** occurs primarily as an unfortunate triangle between humans, rats, and pigs.

Trichinosis may best be considered a zoonotic disease, for humans can scarcely be important in the life cycle of the parasite. Unless an infected person is eaten by a carnivorous predator or finds himself in a cannibal's pot, both unlikely eventualities these days, the parasites are at a dead-end course.

Epidemiology. Because sylvatic trichinosis involves wild mammals, humans are infected only when they interject themselves into the sylvatic food chain. Eskimos, Indians, and others who rely on wild carnivores for food, or urban dwellers who return home with the spoils of the hunt are all subject to infection with *Trichinella*. Fatal cases of trichinosis are common among those who eat undercooked or underfrozen bear, wild pig, cat, dog, or walrus meat. Theoretically, any wild mammal may be a source of infection, but of course most rarely find their way to the dinner table. In Alaska, polar bears and black bears are common sources of infection.[9,11] Arctic explorers have been killed by *Trichinella* acquired from uncooked polar bear meat. The cause of death of the three members of the ill-fated André polar expedition of 1897 was determined *50 years later* by finding *Trichinella* larvae in museum specimens of the polar bear meat that the men had been eating before they died.[17]

Urban trichinosis is epidemiologically more important to humans because of the close relationship between rats, pigs, and people. Infected pork is our most common source of infection. Pigs become infected by eating offal or trichinous meat in garbage or by eating rats, which are ubiquitous in pig farms. It is usually concluded that garbage containing raw pork scraps is the usual source of infection for pigs, but it should not be overlooked that pigs will greedily devour dead or dying rats or even live rats when they can catch them. The rats probably maintain their infections by cannibalism, although it has been demonstrated that rat and mouse feces can contain larvae capable of infecting rats, pigs, or humans.[7,10] Infection can spread from pig to pig when they nip off and eat each other's tails, a common practice in crowded piggeries.[14] It has been suggested that prenatal infection is possible.[1]

The importance of cooking pork thoroughly before it is eaten cannot be overstated. A roast or other piece of solid meat is safe when all traces of pink have disappeared. Many persons are careful about this but become careless when cooking sausage, which is equally dangerous. Raw sausage is a delicacy among many peoples of the world, particularly in the areas where trichinosis is a chronic health problem. Even a casual taste to determine proper seasoning can be fatal: heavily infected pork may contain more than 100,000 larvae per ounce. If half of these are females, and each produces 1,500 larvae, 1.5 million larvae can theoretically result from a single bite of meat. Five larvae per gram of body weight is usually fatal for human beings. Particularly important in transmission is meat processed by "backyard butchers" (individuals who slaughter their own stock) or by very small packing houses. Sausage from such sources is unlikely to be diluted with uninfected meat. In December, 1975, 30 symptomatic cases (five confirmed) around Springfield, Massachusetts, were attributed to eating sausage prepared by a "backyard butcher."

Nevertheless, a taste for rare pork can still be satisfied, for freezing at −15° C for 20 days destroys all parasites.

Survivors of trichinosis have varying degrees of immunity to further infection. Duckett, Denham, and Nelson demonstrated that this immunity in mice can be passed by a mother to young by suckling.[4]

Pathogenesis. The pathogenesis of *Trichinella* infection can be considered in three successive stages: penetration of adult females into the mucosa, migration of larvae, and penetration and encystment in muscle cells.

First symptoms may appear between 12 hours and 2 days after ingestion of infected meat. Commonly, this phase is clinically inapparent because of low-grade infection or is misdiagnosed because of the vagueness of symptoms. When the gravid females penetrate the intestinal epithelium, they cause traumatic damage to the host tissues; they begin poisoning the host with their waste products, and they introduce enteric bacteria into the wounds they cause. These result in intestinal inflammation and pain, with symptoms of food poisoning, such as nausea, vomiting, sweating, and diarrhea. Respiratory diffi-

culties may occur, and red blotches erupt on the skin in some cases. This period usually terminates with facial edema and fever 5 to 7 days after the first symptoms.

During migration, the newborn larvae damage blood vessels, resulting in localized edema, particularly in the face and hands. Wandering larvae may also cause pneumonia, pleurisy, encephalitis, meningitis, nephritis, deafness, peritonitis, brain or eye damage, and subconjunctival or sublingual hemorrhage. Death resulting from myocarditis (inflammation of the heart muscle) may occur at this stage. Though the larvae do not stay in the heart, they migrate through its muscle, causing local areas of necrosis and infiltration of leukocytes.

By the tenth day after the first symptoms appear, the larvae begin penetration of muscle fibers. Attendant symptoms are again varied and vague: intense muscular pain, difficulty in breathing or swallowing, swelling of masseter muscles (occasionally leading to a misdiagnosis of mumps), weakening of pulse and blood pressure, heart damage, and various nervous disorders, including hallucination. Extreme eosinophilia is common but may not be present in even severe cases. Death is usually caused by heart failure, respiratory complications, toxemia, or kidney malfunction.

Diagnosis and treatment. Most cases of trichinosis, particularly subclinical cases, go undetected. Even extreme infections of 1,000 larvae or more per gram of body weight may go undetected if the attending physician does not suspect trichinosis. Routine examinations rarely detect larvae in feces, blood, milk, or other secretions. Although muscle biopsy is seldom employed, it remains an accurate diagnostic if trichinosis is suspected. Pressing the tissue between glass plates and examining it under low-power microscopy is useful, although digestion of the muscle in artificial gastric enzymes for several hours provides a much more reliable diagnostic technique. **Xenodiagnosis,** feeding suspected biopsies to laboratory rats, may be employed. Several immunodiagnostic techniques have been developed, none of which is 100% effective but which are useful nonetheless.

No really satisfactory treatment for

trichinosis is known. Treatment is basically that of relieving the symptoms by use of analgesics and corticosteroids. Purges during the initial symptoms may dislodge females that have not yet begun penetrating the intestinal epithelium. Thiabendazole has been shown effective in experimental animals, but results in clinical cases have been variable.

Despite an immense amount of research, trichinosis remains an important disease of humans, one that has the potential of striking anyone, anywhere. One hopeful note: for unknown reasons the incidence of infection has slowly but steadily declined throughout the world.

ORDER DIOCTOPHYMATA

The few members of the order Dioctophymata are parasites of aquatic birds and terrestrial mammals. Most are of no economic or medical importance to humans. Of the three families in the order, Soboliphymidae has the single genus *Soboliphyme,* parasites of shrews and mustelid carnivores (Fig. 23-11), and Eustrongylidae has the genera *Eustrongylides* and *Hystrichis,* both parasites of wild birds. The other family is of more concern and will be considered more fully.

Family Dioctophymatidae

The family Dioctophymatidae has three genera, *Dioctowittius, Mirandonema,* and *Dioctophyme,* each with a single species. The first is known only from a snake, the second from a Brazilian carnivore, while the third has been reported from a very wide variety of mammals, including humans, in many parts of the world.

Dioctophyme renale

Morphology. *Dioctophyme renale* is truly a giant among nematodes, with the males up to 20 cm long and 6 mm wide and the females up to 100 cm long and 12 mm wide. They are blood-red in color and rather blunt at the ends. The male has a conspicuous, bell-shaped copulatory bursa that lacks any supporting rays or papillae (Fig. 23-12). There is a single, simple spicule 5 to 6 mm long. The vulva is near the anterior end. Eggs (Fig. 23-13) are lemon-shaped, with deep pits in the shells, except at the poles.

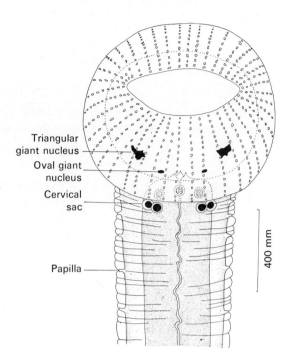

Triangular giant nucleus

Oval giant nucleus

Cervical sac

Papilla

400 mm

Fig. 23-11. *Soboliphyme baturini* from a short-tailed weasel. Note the swollen mouth capsule typical of this genus. (From Schmidt, G. D., and J. M. Kinsella. 1965. Trans. Am. Microsco. Soc. 84:413-415.)

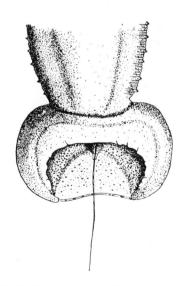

Fig. 23-12. Posterior end of a male *Dioctophyme renale.* Note the single spicule and the powerful copulatory bursa. (From Stefanski, W. 1928. Ann. Parasitol. Hum. Comp. 6:93-100.)

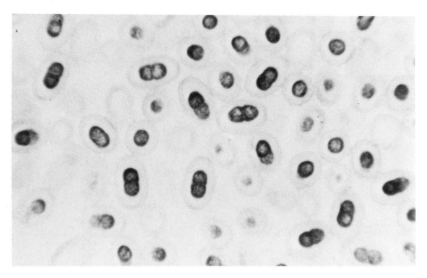

Fig. 23-13. Eggs of *Dioctophyme renale,* showing the corrugated shell and the two-cell embryo. This is the stage released by the female worm. (Photograph by Arthur E. Woodhead. Courtesy Ann Arbor Biological Center.)

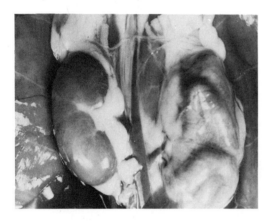

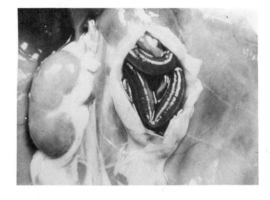

Fig. 23-14. Ferret dissection: the kidney on the left is normal, while that on the right is distended by an adult *Dioctophyme renale.* (Photograph by Arthur E. Woodhead. Courtesy Ann Arbor Biological Center.)

Fig. 23-15. Same specimen as in Fig. 23-14, but with infected kidney opened to reveal the worm. The organ is reduced to a hollow shell. (Photograph by Arthur E. Woodhead. Courtesy Ann Arbor Biological Center.)

Biology. The life cycle of this parasite and a description of the history of its elucidation are given by Karmanova.[6] The thick-shelled eggs require about 6 months in water to embryonate. The larva, which bears an oral spear, hatches when eaten by the aquatic oligochaete annelid *Lumbriculus variegatus.* There, it penetrates into the abdominal blood vessels, where it develops into the fourth-stage larva. When the small annelid is swallowed, the larva migrates to a kidney of the new host, where it matures. If the annelid is eaten by any of several species of fish, the larva will encyst in the muscles or viscera, using the fish as a paratenic host. When swallowed by a definitive host, the larva penetrates the duodenum and enters a kidney. Usually the right kidney is infected, as it is near the loop of the duodenum. The worms mature in the kidney, and the eggs are voided from the host in its urine.

Epidemiology. Probably any species of large mammal can serve as definitive host. Because of their fish diets, mustelids, canids, and bears are particularly susceptible, as are humans. But even such non–fish-eating mammals as cows, horses, and pigs can become infected, when they accidentally ingest an infected annelid. Thorough cooking of fish and drinking of only pure water will prevent infection in people.

Pathology. Pressure necrosis caused by the growing worm, together with its feeding activities, reduces the infected kidney to a thin-walled, ineffective organ (Figs. 23-14 and 23-15). Loss of kidney function is compounded by uremic poisoning. The worms will sometimes penetrate the renal capsule and wander in the coelomic cavity.

Diagnosis and treatment. The rarity of this parasite makes physicians unlikely to suspect its presence. Demonstration of the characteristic eggs in the urine is the only positive means of diagnosis, aside from surgical discovery of the worm itself. Surgical removal of the worm is the only treatment known.

REFERENCES

1. Bourns, T. K. R. 1952. The discovery of trichina cysts in the diaphragm of a six-week old child. J. Parasitol. 38:367.
2. Canlas, B. D., Jr., B. O. Cabrera, and U. Diaz. 1967. Human intestinal capillariasis. II. Pathological features. Acta Med. Philippina 4:84-91.
3. Diaz, U., B. O. Cabrera, and B. D. Canlas, Jr. 1967. Human intestinal capillariasis. I. Clinical features. Acta Med. Philippina 4:72-83.
4. Duckett, M. G., D. A. Denham, and G. S. Nelson. 1972. Immunity to Trichinella spiralis: V. Transfer of immunity against the intestinal phase from mother to baby mice. J. Parasitol. 58:550-554.
5. Faust, E. C., P. C. Beaver, and R. C. Jung. 1968. Animal agents and vectors of human disease, ed. 3. Lea & Febiger, Philadelphia. pp. 222-225.
6. Karmanova, E. M. 1968. Essentials of nematodology, vol. 20. Dioctophymidea of animals and man and the diseases caused by them. Academiya Nauk USSR, Moscow.
7. Levine, N. D. 1968. Nematode parasites of domestic animals and of man. Burgess Publishing Co., Minneapolis.
8. Mackie, T. T., G. W. Hunter III, and C. B. Worth. 1954. A manual of tropical medicine. W. B. Saunders Co., Philadelphia.
9. Maynard, J. E., and F. P. Pauls. 1962. Trichinosis in Alaska. A review and report of two outbreaks due to bear meat, with observations of serodiagnosis and skin testing. Am. J. Hyg. 76:252-261.
10. Olsen, O. W. and H. A. Robinson. 1958. Role of rats and mice in transmitting Trichinella spiralis through their feces. J. Parasitol. 44(Sect. 2):35.
11. Rausch, R. 1953. Animal-borne diseases. Publ. Health Rep. 68:533.
12. Sheffield, H. G. 1963. Electron microscopy of the bacillary band and stichosome of Trichuris muris and T. vulpes. J. Parasitol. 49:998-1009.
13. Singston, C. N., T. C. Banzon, and J. H. Cross. 1975. Mebendazole in the treatment of intestinal capillariasis. Am. J. Trop. Med. Hyg. 24:932-934.
14. Smith, H. J. 1975. Trichinae in tail musculature of swine. Can. J. Comp. Med. 39:362-363.
15. Spencer, H. 1973. Nematode diseases I. In Spencer, H., editor. Tropical pathology. Springer-Verlag, Berlin. pp. 457-509.
16. Stoll, N. R. 1947. This wormy world. J. Parasitol. 33:1-18.
17. Sundman, P. O. 1970. The flight of the eagle. Random House, Inc., New York.
18. Wright, K. A. 1968. Structure of the bacillary band of Trichuris myocastoris. J. Parasitol. 54:1106-1110.

SUGGESTED READING

Gould, S. E. 1945. Trichinosis. Charles C Thomas, Publisher, Springfield, Ill. (A classic, standard reference.)

Kagan, I. G. 1959. Trichinosis in the United States. Publ. Health. Rep. 74:159-162. (An accurate assessment of the problem in the United States to 1959.)

Madsen, H. 1945. The species of Capillaria (Nematoda: Trichinelloidea) parasites in the digestive tract of Danish gallinaceous and anatine game birds with a revised list of species of Capillaria in birds. Danish Rev. Game Biol. 1:1-112.

Zimmerman, W. J. 1971. The trichiniasis problem: facts, fallacies, and future. Iowa State Univ. Vet. 33:93-97.

Chapter 24

ORDER RHABDITATA: PIONEERING PARASITES

The tiny worms in the order Rhabditata appear to bridge the gap between free-living and parasitic modes of life, for several species alternate between free-living and parasitic generations. Most species inhabit decaying organic matter and are common in soil, foul water, decaying fruit, and so on. For this reason they often find their way into the bodies of larger animals; the digestive, reproductive, respiratory, and excretory tracts are particularly susceptible, as are open wounds. Once in such locations, they may become facultatively parasitic for a time or may simply pass through the body. Since most species are similar to each other, it usually requires the services of a specialist to differentiate the pathogenic species from nonpathogenic ones. Species are more difficult to identify during free-living phases than during parasitic stages.

Some of the behavioral and developmental adaptations of free-living species (*Rhabditis* spp.) in this order were mentioned in Chapter 22 (dauer larvae). It is not difficult to visualize how analogous adaptations for transfer to new food supplies in the ancestors of the frankly parasitic species could have been preadaptations to parasitism. The diversity in life cycles among the parasites in the Rhabditata further illustrates this evolutionary opportunism. Thus, within the order, one can observe completely free-living species, species with preadaptations for parasitism, facultative parasites, obligate parasites, and even species that seem to produce obligate free-living or obligate parasitic forms, depending on conditions. Families that have obligate parasites or parasite interspersed with free-living adults are the Rhabdiasidae and the Strongyloididae. Although the latter family is of far more medical and veterinary importance, the

Rhabdiasidae is of interest in illustrating the diversity of life cycles in the order. It is often encountered in frog dissections.

FAMILY RHABDIASIDAE

Rhabdias bufonis, a common parasite in the lungs of toads and frogs, has a very curious life cycle. The parasitic adult is a **protandrous hermaphrodite,** that is, an individual that is a functional male before it becomes a female. Sperm are produced in an early male phase and stored in a seminal receptacle. Then, the gonad produces functional ova that are fertilized by the stored sperm. The resulting shelled zygotes are passed up the trachea of the host, then swallowed, embryonating along the way. The larvae hatch in the intestine of the frog, and the first-stage larvae accumulate in the cloaca to be voided with the feces. The first-stage larvae often are referred to as **rhabditiform** because the posterior end of their esophagus has a prominent **bulb** that is separated from the anterior portion **(corpus)** by a narrower region **(isthmus)** (Fig. 24-1). These larvae undergo four molts to produce a generation of free-living males and females, a dioecious generation. This nonparasitic generation feeds on bacteria and other inhabitants of humus soil. The progeny of this generation hatch in utero and proceed to consume the internal organs of their mother, destroying her. "How sharper than a serpent's tooth!" They escape from the female's body and, by now, have become third-stage infective juveniles. They are referred to as filariform at this stage because the esophagus has no terminal bulb or isthmus. The filariform larvae undergo developmental arrest unless they penetrate the skin of a toad or frog. After penetration, they lodge in various tissues; those that reach the lungs mature into her-

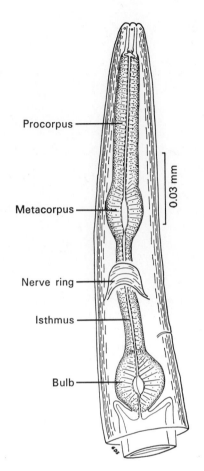

Procorpus

Metacorpus

0.03 mm

Nerve ring

Isthmus

Bulb

Fig. 24-1. A typical rhabditiform esophagus. (From Schmidt, G. D., and Robert E. Kuntz. 1972. Proc. Helminth. Soc. Wash. 39:189-191.)

maphroditic adults, while the rest apparently expire. This type of life cycle is **indirect** or **heterogonic,** which is to say, a free-living generation is interspersed between parasitic generations.

Rhabdias fuscovenosa is a common parasite in the lungs of some kinds of aquatic snakes (Natricinae). The life cycle differs somewhat from that of *R. bufonis* in that most of the eggs from the parasitic forms yield filariform larvae. Very few free-living adults are found[2]; therefore, the life cycle in this species is predominately **direct** or **homogonic,** and the worm is an obligate parasite.

FAMILY STRONGYLOIDIDAE

Some species of this family are among the smallest nematode parasites of hu-

mans, the males being even smaller than *Trichinella. Strongyloides stercoralis* is the most common, widespread species, while *S. fuelleborni* occurs commonly in parts of Africa.[6] *Strongyloides stercoralis* also infects other primates, as well as dogs, cats, and some other mammals, and races from various geographic areas vary in infectivity for different hosts.[4] *Strongyloides fuelleborni* has been reported from a variety of primates. Other species are parasites of other mammals, for example, *S. ratti* in rats, *S. ransomi* in swine, and *S. papillosus* in sheep. They also are common in birds, amphibians, and reptiles. The species of *Strongyloides* are remarkable in their ability, at least in some cases, to maintain homogonic, parasitic life cycles or repeat free-living generations indefinitely, depending on conditions. The parasitic generation apparently consists only of parthenogenetic females, for sperm have not been found in the seminal receptacles of the females.[1] Parasitic males, though rare, have been reported. The free-living generation consists of both males and females. The following description pertains to *Strongyloides stercoralis* (Fig. 24-2).

Morphology. Parthenogenetic females reach a length of 2.0 to 2.5 mm, while parasitic males (if published reports are accurate) are about 0.7 mm long and appear identical to free-living males. The buccal capsule of both sexes is small, and they possess a long, cylindrical esophagus that lacks a posterior bulb. The vulva is in the posterior third of the body; the uteri are divergent and contain only a few eggs at a time. The free-living adults both have a rhabditiform esophagus. The male is up to 0.9 mm long and 40 to 50 μm wide. The male has two simple spicules and a gubernaculum; its pointed tail is curved ventrad. The female is stout and has a vulva that is about equatorial; the uteri generally contain more eggs than those of the parasitic female.

Biology (Fig. 24-3). The parasitic females anchor themselves with their mouths to the mucosa of the small intestine or burrow their anterior ends into the submucosa. They are found occasionally in the respiratory, biliary, or pancreatic system. They produce several dozen, thin-shelled, partially embryonated eggs a day and re-

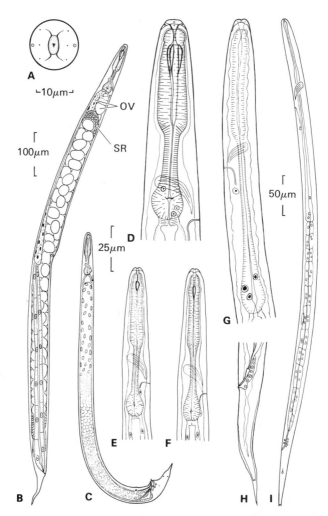

Fig. 24-2. *Strongyloides stercoralis.* **A,** Free-living female, en face view. **B,** Free-living female, lateral view (*OV*, ovary; *SR*, seminal receptacle containing sperm). **C,** Free-living male. **D,** Anterior end of free-living female, showing details of esophagus. **E,** Newly hatched, first-stage larva obtained by duodenal aspiration from human. **F,** First-stage larva from freshly passed feces of same patient as was larva in **E. G,** Second-stage larva developing to the filariform stage; cuticle is separating at anterior end. **H,** Tail of same larva as shown in **G;** notched tail is developing and cuticle is separating at tip and in rectum. **I,** Filariform larva. (From Little, M. D. 1966. J. Parasitol. 52:69-84.)

lease them into the gut lumen or submucosa. The eggs measure 50 to 58 μm by 30 to 34 μm. They hatch during passage through the gut or within the submucosa, and the larvae escape to the lumen. These first-stage larvae are 300 to 380 μm long, and they usually are passed with the feces. The larvae go on either to develop into free-living adults or to become infective, filariform larvae with a developmental ar-

rest at the third stage; they are now 490 to 630 μm long. The filariform larvae develop no further unless they gain access to a new host by skin penetration or ingestion. If by skin penetration, they are carried by the blood to the lungs, where they exit into the alveoli, travel up the trachea, are swallowed, and mature in the small intestine. If they are ingested, a lung migration appears unnecessary. The free-living

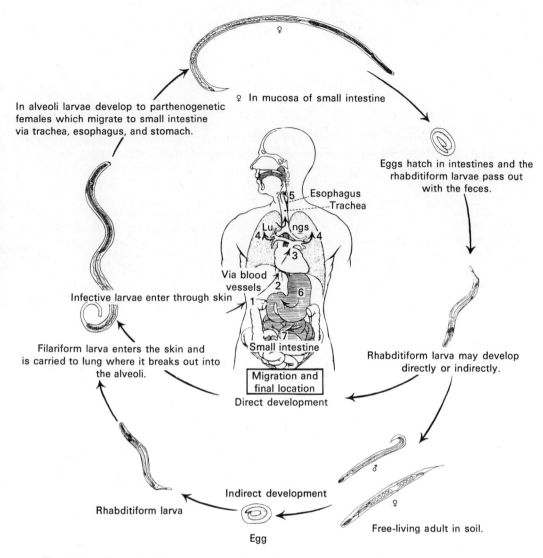

In alveoli larvae develop to parthenogenetic females which migrate to small intestine via trachea, esophagus, and stomach.

♀ In mucosa of small intestine

Eggs hatch in intestines and the rhabditiform larvae pass out with the feces.

Esophagus
--Trachea

Lungs

Via blood vessels

Infective larvae enter through skin

Small intestine

Filariform larva enters the skin and is carried to lung where it breaks out into the alveoli.

Rhabditiform larva may develop directly or indirectly.

Migration and final location
Direct development

Indirect development

Rhabditiform larva

Egg

Free-living adult in soil.

Fig. 24-3. Life cycle of *Strongyloides stercoralis* in humans. (Adapted from Medical protozoology and helminthology. 1959. U.S. Naval Medical School.)

adults, in turn, can produce successive generations of free-living adults. Both parasitic and free-living females can produce larvae that will become filariform, infective juveniles and larvae that will mature into free-living adults. In other words, the homogonic and heterogonic life cycles seem to be mixed in a random fashion.

The mechanism that determines whether a given embryo will become a free-living male or female or a parasitic female is still unclear. Some evidence suggests that the parasitic females are haploid.[1] If this is so,

then all haploid embryos will produce parasitic females, and diploid embryos will produce free-living adults. Environmental conditions that appear to encourage production of one or the other type of larva may *in fact* strongly favor the *survival* of either the free-living adults or filariform larvae.[3] Some important differences in the energy metabolism of free-living and parasitic *Strongyloides* have already been mentioned (Chapter 22), and such factors may have bearing on the question.

Still other permutations of the life cycle

are possible. If the larvae have time to molt twice during their transit down the digestive tract, they may penetrate the lower gut mucosa or perianal skin, go through their migration, and mature. This process is called **autoinfection.** Some researchers have held, further, that filariform larvae could repenetrate the small gut mucosa in a process of **hyperinfection,** but others have questioned whether this can occur. Nevertheless, whether by maintenance of the original adult worms, by autoinfection, or otherwise, cases are known in which patients have had *Strongyloides* infections for as long as 36 years.[7]

Epidemiology. People typically become infected by contacting the larvae in contaminated soil or water. This is chiefly a parasite of tropical regions but extends well into temperate zones in several continents. Like most filth-borne diseases, it is most prevalent under conditions of low sanitation standards. While traditionally a disease of poor, uneducated persons in depressed areas of the world, it could become important among the affluent with "vacation hideways," who rely on inadequate sewage disposal facilities, such as in mountain environments.

Pathology. The effects of strongyloidiasis may be described in three stages: **invasion, pulmonary,** and **intestinal.**

Penetration of the skin by invasive larvae results in slight hemorrhage and swelling, with intense itching at the site of entry. If pathogenic bacteria are introduced with the larvae, inflammation may result also.

During migration through the lungs, damage to lung tissues results in massive host-cell reactions, which often delay or prevent further migration. When this happens, the worms may establish themselves in the pulmonary tissues and begin reproducing as if they were in the intestine, again demonstrating their extraordinary adaptability. A burning sensation in the chest, a nonproductive cough, and other symptoms of bronchial pneumonia may accompany this phase. The pulmonary phase has been suspected of reactivating quiescent pulmonary tuberculosis.[5]

After being swallowed, the juvenile females enter the crypts of the intestinal mucosa, where they rapidly mature and invade the tissues. They rarely penetrate deeper than the muscularis mucosae, but some cases of deeper penetration have been reported. The worms migrate randomly through the mucosa, each depositing several eggs per day. An intense, localized burning sensation or aching pain in the abdomen usually is felt at this time. Destruction of tissues by adult worms and larvae results in sloughing of patches of mucosa, with fibrotic changes in chronic cases, sometimes with death resulting from septicemia (bacterial infection of the blood) following ulceration of the intestine. Host resistance lowered by other diseases or by immunodepressant drugs increases chances of a fatal outcome.[7]

Diagnosis and treatment. Demonstration of rhabditiform (or occasionally filariform) larvae in freshly passed stools is a sure means of diagnosis. A direct fecal smear is often effective in cases of massive infections, and various concentration techniques, such as Baermann isolation or zinc floatation with centrifugation, will increase chances in cases where fewer worms are present. Rarely, after purgation or in severe diarrhea, embryonating eggs may be seen in the stool. These resemble hookworm eggs but are more rounded.

Difficulties arise, however, because of day-to-day variability in numbers of larvae in the feces. Further, once autoinfection or hyperinfection becomes established, the number of larvae passed in the feces may decrease. Duodenal aspiration is a very accurate technique, but only applies to duodenal infections: larvae farther down the intestine cannot be obtained.

Once larvae are obtained, there are difficulties of identification. First-stage larvae are similar to rhabditiform hookworm larvae, which may be present if the stool was constipated or had remained at room temperature long enough for hookworm eggs to hatch. Two morphological features can be useful to separate the two: *Strongyloides* has a short buccal cavity and a large genital primordium, while hookworm larvae have a long buccal cavity and a tiny genital primordium (see p. 439).

If the stool has been exposed to soil or water, species of *Rhabditis* or related genera may invade it, compounding the confusion. Further, filariform larvae appear in cases of constipation or autoinfection.

These, however, are easily recognized by their notched tails. Although there is no clinically available immunological test, there is promise that such tests will become available in the near future.

Several drugs are effective in treatment of strongyloidiasis, but most have undesirable side effects. Thiabendazole currently is the drug of choice, with a high percentage of cure and minimal side effects.

REFERENCES

1. Bolla, R. I., and L. S. Roberts. 1968. Gametogenesis and chromosomal complement in *Strongyloides ratti* (Nematoda: Rhabdiasoidea). J. Parasitol. 54:849-855.
2. Chu, T. 1936. Studies on the life cycle of *Rhabdias fuscovenosa* var. *catanensis* (Rizzo, 1902). J. Parasitol. 22:140-160.
3. Hansen, E. L., B. J. Buecher, and W. S. Cryan. 1969. *Strongyloides fülleborni:* environmental factors and free-living generations. Exp. Parasitol. 26:336-343.
4. Levine, N. D. 1968. Nematode parasites of domestic animals and of man. Burgess Publishing Co., Minneapolis.
5. Palmer, E. D. 1944. A consideration of certain problems presented by a case of strongyloidiasis. Am. J. Trop. Med. 24:249-254.
6. Pampiglioni, S., and M. L. Ricciardi. 1971. The presence of *Strongyloides fülleborni* von Linstow, 1905, in man in central and east Africa. Parasitologia 13:257-269.
7. Spencer, H. 1973. Nematode diseases. I. In Spencer, H. editor. Tropical pathology. Springer Verlag, Berlin. pp. 457-509.

SUGGESTED READING

Cable, R. M. 1971. Parthenogenesis in parasitic helminths. Am. Zool. 11:267-272. (Contains a short discussion of parthenogenesis in nematodes, with emphasis on *Strongyloides*.)
Little, M. D. 1966. Comparative morphology of six species of *Strongyloides* (Nematoda) and redefinition of the genus. J. Parasitol. 52:69-84.

Chapter 25

ORDER STRONGYLATA: THE BURSATE PHASMIDIANS

The large order Strongylata is of great economic and medical importance. It seems to have evolved directly from rhabditoid-type ancestors. One feature in common with nearly all species is a broad copulatory bursa on the posterior end of the males. Most, but not all, species are parasites of the intestine of vertebrates and have a direct life cycle, requiring no intermediate hosts. Of the numerous families in the order, the Ancylostomidae, Trichostrongylidae, and superfamily Metastrongyloidea are the most important to humans. Examples from these illustrate the parasitological significance of the order.

FAMILY ANCYLOSTOMIDAE

Members of the family Ancylostomidae are commonly known as hookworms. They live in the intestine of their hosts, attaching to the mucosa and feeding on blood and tissue fluids sucked from it.

Morphology. There is much similarity of morphology and biology between the numerous species in this family, so we will first give them a general consideration. Most species are rather stout and the anterior end is curved dorsad, giving the worm a hook-like appearance. The buccal capsule is large and heavily sclerotized and is usually armed with cutting plates, teeth, lancets, or a dorsal cone (Fig. 25-1). A **dorsal gutter** extends along the middorsal wall of the buccal capsule, emptying the dorsal esophageal gland into it. Lips are reduced or absent. In one subfamily (Arthrostominae), the buccal capsule is subdivided into several articulated plates.

The esophagus is stout, with a swollen posterior end, giving it a club shape. It is mainly muscular, corresponding to its action as a powerful pump. The esophageal glands are extremely large and are mainly outside of the esophagus, extending posteriad into the body cavity. Cervical papillae are present near the rear level of the nerve ring.

Males have a conspicuous copulatory bursa, consisting of two broad **lateral lobes** and a smaller **dorsal lobe,** all supported by fleshy rays (Fig. 25-2). These rays follow a common pattern in all species, varying only in relative size and point of origin; consequently, they are important taxonomic characters. The spicules are simple, needle-like, and similar. A gubernaculum and cement gland are present.

Females have a simple, conical tail. The vulva is postequatorial, and two ovaries are present. About 5% of the daily output of eggs is found in the uteri at any one time; the total production is several thousand per day for as long as 9 years.

Biology (Fig. 25-3). Hookworms mature and mate in the small intestine of their host. The embryos develop into the two- or four- or several-cell stage by the time they are passed with the feces (Fig. 25-4). The species infecting humans cannot be diagnosed reliably by the egg alone. The eggs require warmth, shade, and moisture for continued development. Coprophagous insects may mix the feces with soil and air, thus hastening embryonation, which may occur in 24 to 48 hours in ideal conditions (Fig. 25-5). Urine is fatal to developing embryos. Newly hatched first-stage larvae have a rhabditiform esophagus with its characteristic constriction at the level of the nerve ring, such as that occurring in the rhabditiform larvae of *Strongyloides*. In fact, differentiation of hookworm larvae from *Strongyloides* larvae is difficult for the beginner.

The larvae live in the feces, feeding on fecal matter, and molt their cuticle in 2 or 3 days. The second-stage larva, which also has a rhabditiform esophagus, continues

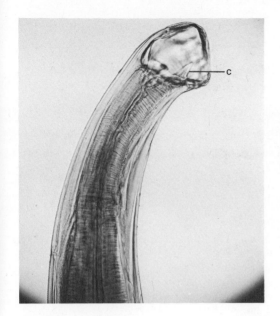

Fig. 25-1. Lateral view of the anterior of *Bunostomum* sp., a hookworm of ruminants. Note the large buccal capsule typical of hookworms, and the dorsal cone *(c)*. The dorsal flexure of the head is also typical of hookworms. (Photograph by Jay Georgi.)

to feed and grow and, after about 5 days, molts to the third stage, which is infective to the host. The second-stage cuticle may be retained as a loose-fitting sheath until penetration of a new host, or it may be lost earlier. The third-stage filariform larva has an **strongyliform esophagus,** that is, it has a posterior bulb, but the bulb is not separated from the corpus by an isthmus. The intestine is filled with food granules that sustain the worm through the nonfeeding, third stage. It is similar to the filariform larva of *Strongyloides* but can be distinguished from it by the tail, which is pointed in hookworms and notched in *Strongyloides* (see p. 434).

Third-stage larvae live in the upper few millimeters of soil, remaining in the capillary layer of water surrounding soil particles. They are killed quickly by freezing or desiccation. There is a short, vertical migration in the soil, depending on the weather or time of day; when the surface of the ground is dry, they migrate a short distance into the soil, following the retreat-

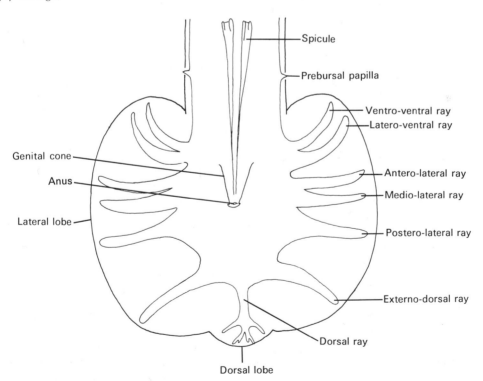

Fig. 25-2. Ventral view of a typical strongyloid copulatory bursa. The basic pattern is found in all of the strongylata.

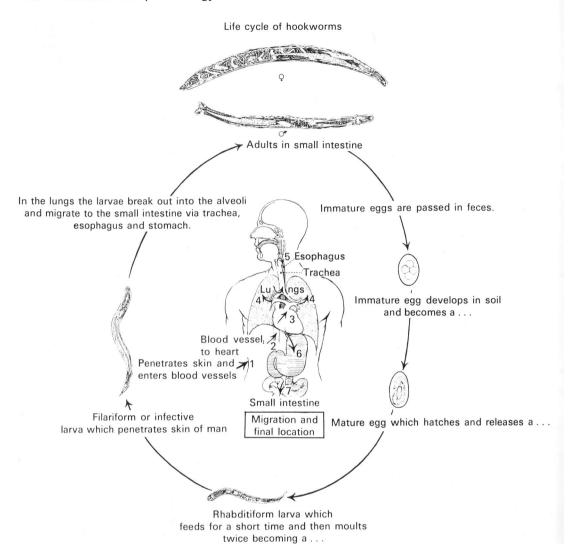

Life cycle of hookworms

♀

♂

Adults in small intestine

In the lungs the larvae break out into the alveoli and migrate to the small intestine via trachea, esophagus and stomach.

Immature eggs are passed in feces.

5 Esophagus

Trachea

Lungs

4 4

3

Immature egg develops in soil and becomes a . . .

Blood vessel to heart

2 6

Penetrates skin and enters blood vessels 1

7

Small intestine

| Migration and final location |

Mature egg which hatches and releases a . . .

Filariform or infective larva which penetrates skin of man

Rhabditiform larva which feeds for a short time and then moults twice becoming a . . .

Fig. 25-3. The life cycle of hookworms. (From Medical protozoology and helminthology. 1959. U.S. Naval Medical School.)

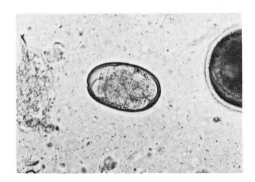

Fig. 25-4. A hookworm egg from the feces of a cat. (Photograph by Jay Georgi.)

ing water. When the surface of the ground is wet, after rain or morning dew, they move to the surface and extend themselves snake-like, waving back and forth in an attitude that allows maximum opportunity for contact with a host. Thousands of larvae often group together, crawling over each other and waving rhythmically in unison (see Fig. 22-33). They can live for several weeks under ideal conditions.

Infection usually occurs when third-stage larvae contact the skin and burrow into it, although transplacental transmission and transmission in mother's milk occur in some species. Any epidermis can

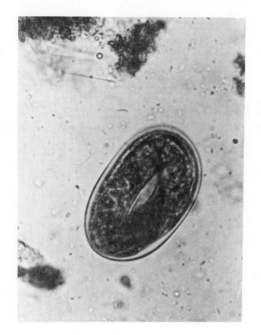

Fig. 25-5. A fully embryonated hookworm egg. (Photograph by R. E. Kuntz and J. Moore.)

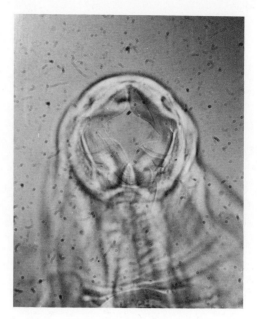

Fig. 25-6. En face view of the mouth of *Necator americanus*. Note the two broad cutting plates in the ventrolateral margins (top of illustration).

be penetrated by larvae, although those parts most often in contact with the soil, such as hands, feet, and buttocks, are most often attacked. If larvae are introduced into the mouth, they usually pierce the oral mucosa, although some species can survive if they are swallowed.

Larvae that wander through subcuticular tissues soon die and are absorbed, but those that enter vessels are carried to the heart and thereby to the lungs. After breaking into the air spaces, they are carried by ciliary action up the respiratory tree to the glottis and are swallowed. Arriving in the small intestine, they attach to the mucosa, grow, and molt to the fourth stage, which has an enlarged buccal capsule. After further growth and a final molt, the worm becomes sexually mature. At least 5 weeks are required from the time of infection to egg production.

Several genera and many species of hookworm plague humans and domestic and wild mammals. The following few species are among those most important to humans.

Necator americanus

Necator americanus, the "American killer," was first discovered in Brazil, then Texas, but was later found indigenous in Africa, India, southeast Asia, China, and the southwest Pacific islands. It probably was introduced into the New World with the slave trade. The worm has had an important impact on the development of the southern United States, as well as other regions of the world where it occurs. The image of the lazy, "poor white trash" of the southern United States is now known to be, in large measure, the result of hookworm disease. In 1947, Stoll estimated 384.3 million infected persons in the world, with 1.8 million cases in North America.[18] Those numbers are probably not much different today, considering the increase in world population since 1947. Recent surveys in the southeast United States show average prevalence of about 4%, with up to 15% in certain areas.[21]

Necator americanus has a pair of dorsal and a pair of ventral cutting plates surrounding the anterior margin of the buccal capsule (Fig. 25-6). In addition, there are a pair of subdorsal and a pair of subventral teeth near the rear of the buccal capsule.

Males are 7.0 to 9.0 mm long and have a bursa diagnostic for the genus (Fig. 25-7).[13] The needle-like spicules have minute

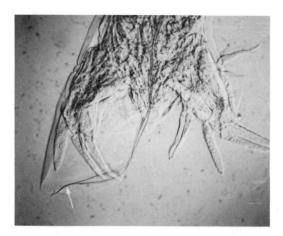

Fig. 25-7. The copulatory bursa and spicules of *Necator americanus*. The spicules are fused at their distal ends (arrow) and form a characteristic hook.

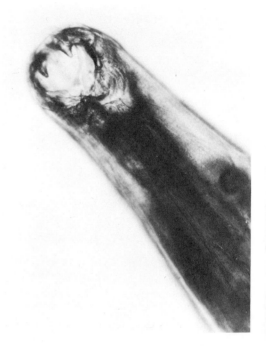

Fig. 25-8. *Ancylostoma duodenale,* dorsal view. Notice the powerful ventral teeth. (AFIP neg. no. N-41730-2.)

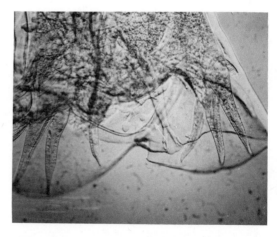

Fig. 25-9. The copulatory bursa and spicules of *Ancylostoma duodenale.* The tips of the spicules are not fused into a hook, as in *Necator americanus.*

barbs at their tips and are occasionally fused distally. Females are 9.0 to 11.0 mm long and have the vulva located in about the middle of the body. The life cycle is generally as described previously, although infection apparently cannot occur by swallowing larvae.[22] Dönges and Madecki could find no evidence of transmission through mother's milk in Nigeria.[6] Female worms produce about 9,000 eggs per day.

Primarily a tropical parasite, *N. americanus* is the dominant species in humans in most of the world. About 95% of the hookworms in the southern United States are this species.

Ancylostoma duodenale

The parasite *Ancylostoma duodenale* is abundant in southern Europe, northern Africa, India, China and southeastern Asia, as well as in other scattered locales, including small areas of the United States, Caribbean Islands, and South America. It is known in mines as far north as England and Belgium; since Lucretius, in the first century, it was known to cause a serious anemia in miners. Generally speaking, it is never as abundant as *N. americanus.*

The anterior margin of the buccal capsule has two ventral plates, each with two large teeth that are fused at their bases (Fig. 25-8). A pair of small teeth are found in the depths of the capsule.

Adult males are 8.0 to 11.0 mm long and

have a bursa characteristic for the species (Fig. 25-9). The needle-like spicules have simple tips and are never fused distally. Females are 10.0 to 13.0 mm long, with the vulva located about one-third of body length from the posterior end. A single female can lay from 25,000 to 30,000 eggs per day.

This is the first hookworm for which the life cycle was elucidated. In 1896 Artur Looss, working in Egypt, was dropping cultures of *Ancylostoma* larvae into the mouths of guinea pigs when he spilled some of the culture onto his hand. He noticed that it produced an itching and redness and wondered if infection would occur this way. He began examining his feces at intervals and, after a few weeks, found that he was passing hookworm eggs. He next placed some larvae on the leg of an Egyptian boy who was to have his leg amputated within an hour. Subsequent microscopic sections showed larvae penetrating the skin. Looss' monograph on the morphology and life cycle of *A. duodenale* remains one of the most elegant of all works on helminthology.[14]

It is possible for swallowed larvae to develop normally without a migration through the lungs, but this is probably a fairly rare means of infection.

Other hookworms reported from humans

Ancylostoma ceylanicum is normally a parasite of carnivores in Sri Lanka, southeast Asia, and the East Indies, but it has been reported from humans in the Philippines.[19] A very similar species, *A. braziliense,* is found in domestic and wild carnivores in most of the tropics. Though it has been reported from humans in Brazil, Africa, India, Sri Lanka, Indonesia, and the Philippines, these infections probably were *A. ceylanicum. Ancylostoma malayanum,* a parasite of bears in Malaysia and India, has been found once in a human.

Ancylostoma caninum (Fig. 25-10) is the most common hookworm of domestic dogs and cats, especially in the Northern Hemisphere. It has been found in humans on five occasions, and the worm is a common cause of creeping eruption (see p. 446). The species has proved to be a useful tool in studying hookworm biology because of the ease of maintaining it in the laboratory.

Hookworm disease

An important distinction is to be made between hookworm infection and hookworm disease. Far more people are infected with the worm than exhibit symptoms of the disease. Whether disease is manifested depends strongly on two factors: the number of worms present and the nutritional condition of the infected person. In general fewer than 25 *Necator americanus* in a person will cause no symptoms, 25 to 100 worms lead to light symptoms, 100 to 500 produce considerable damage and moderate symptoms, 500 to 1,000 worms result in severe symptoms and grave damage, and with more than 1,000 worms, very grave damage may be accompanied by drastic and often fatal consequences. Because *Ancylostoma* sucks more blood, fewer worms cause greater disease; for example, 100 worms may cause severe symptoms. However, the clinical disease is intensified by the degree of malnutrition, corresponding impairment of the host's immune response, and other considerations.

Epidemiology. From the discussion on the biology of hookworms, it is obvious that the combination of poor sanitation with appropriate environmental conditions is necessary for high endemicity.

Environmental conditions conducive to transmission of the disease have been well studied, and they are the conditions that favor the development and survival of the larvae. The disease is restricted to warmer parts of the world (and to specialized habitats such as mines in more severe climates) because the larvae will not develop to maturity at less than 17° C, with 23° to 30° C being optimum. Frost kills eggs and larvae. Oxygen is necessary for hatching of eggs and larval development because their metabolism is aerobic. This means, among other things, that the larvae will not develop in undiluted feces or water-logged soil. Therefore, a loose, humusy soil that has reasonable drainage and aeration is favorable to larvae. Both heavy clay and coarse sandy soils are unfavorable for the parasite, the latter because the larvae are also sensitive to desiccation. Alternate drying and moistening are particularly damaging to the larvae; hence, very sandy soils become noninfective after brief periods of frequent rainfall. However, the larvae live in the film of water surrounding soil particles, and even apparently dry soil may have enough moisture to enable survival, particularly below the surface. The larvae are quite sensitive to direct sunlight and survive best in shady locations, as, for ex-

Fig. 25-10. *Ancylostoma caninum*, male. (Courtesy Ann Arbor Biological Center.)

ample, on coffee, banana, or sugarcane plantations. Workers and other adults in such situations often have preferred defecation sites, not out in the open where the larvae would be killed by the sun, of course, but in shady, cool, secluded spots beneficial for larval development. Repeated return of people to the defecation site exposes them to continual reinfection. Further, the use of preferred defecation sites makes it possible for hookworm to be endemic in otherwise quite arid areas. Another physical condition conducive to survival of larvae is pH; they develop best near neutrality, and acid or alkaline soils inhibit development, as does the acid pH of undiluted feces (pH 4.8 to 5.0). Chemical factors have an influence. Urine mixed with the feces is fatal to the eggs, and several strong chemicals that may be added to feces as disinfectants of fertilizers are lethal to the free-living stages. Salt in the water or soil inhibits hatching and is fatal to juveniles.

The longevity of the worms is important in transmission to new hosts, continuity of infection in a locality, and introduction to new areas. The larvae can survive in reasonably good environmental conditions for about 3 weeks, except in protected sites like mines, where they can last for a year. There is some dispute as to the life span of the adult, but a good estimate is about 2 years. If a person is removed from an endemic area, the infection is lost in about that time.

The degree of soil contamination is an important factor in transmission. Obviously, a higher average number of worms per individual will seed the soil with more eggs. Promiscuous defecation, associated with poverty and ignorance, keeps soil contamination high. The use of nightsoil as fertilizer for crops is an especially important factor in the Orient.

Because the worm penetrates the skin, habits of going barefoot in tropical countries make an elemental contribution to transmission.

Race is another important epidemiological factor, white people being about ten times more susceptible than black people to hookworm. Prevalence and numbers of worms per individual tend to be higher among whites than blacks. The image of the lazy, apathetic poor whites in the Old South, as contrasted with the more industrious blacks on comparable socioeconomic levels, is generally attributed to the differential effects of hookworm.

Pathogenesis. Hookworm disease manifests three main phases of pathogenesis: the cutaneous or invasion period, the migration or pulmonary phase, and the intestinal phase. Another pathogenic condition, caused when a larva enters an unsuitable host, will be discussed separately.

The cutaneous phase begins when lar-

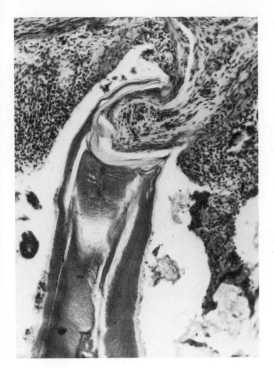

Fig. 25-11. A hookworm attached to intestinal mucosa. Notice how the ventral tooth in the depth of the buccal capsule lacerates the host tissue. (AFIP neg. no. N-33818.)

vae penetrate the skin. They do little damage to the surficial layers, for they seem to slip through tiny cracks between skin scales or to penetrate hair follicles. Once in the dermis, however, their attack on blood vessels initiates a tissue reaction that may isolate and kill the worms. If, as usually happens, pyogenic bacteria are introduced into the skin with the invading larva, an urticarial reaction will result, causing a condition known as **ground itch.**

The pulmonary phase occurs when the larvae break out of the lung capillary bed into the alveoli and progress up the bronchii to the throat. Each site hemorrhages slightly, with serious consequences in massive infections, though this is rare. The phase is usually asymptomatic, though there may be some dry coughing and sore throat. A pneumonitis may result in severe infections.

The intestinal phase is the most important period of pathogenesis. On reaching the small intestine, the juvenile worm attaches to the mucosa with its strong buccal capsule and teeth, and it begins to feed

on blood (Fig. 25-11). In heavy infections, worms are found from the pyloric stomach to the ascending colon, but usually they are restricted to the anterior one-third of the small intestine. The worms move from place to place, but bleeding quickly ceases at a lesion that a worm has left, contrary to prevalent belief.[17] The worms pass substantially more blood through their digestive tracts than would appear necessary for their nutrition alone, but the reason for this is unknown. Recent estimates of blood loss per worm are about 0.03 ml per day for *Necator* and around 0.15 ml per day for *Ancylostoma*. Up to 200 ml of blood may be lost by patients with heavy infections,[13] but around 40% or so of the iron may be reabsorbed by the patient before it leaves the intestine.[11] Nevertheless, a moderate hookworm infection will gradually produce an iron-deficiency anemia as the body reserves of iron are used up. The severity of the anemia depends on the worm load and the dietary iron intake of the patient. Slight, intermittent abdominal pain, loss of normal appetite, and desire to eat soil (geophagy) are common manifestations of moderate hookworm disease. (Certain areas in the southern United States became locally famous for the quality of their clay soil, and people traveled for miles to eat it. In fact, in the early 1920s an enterprising person began a mail order business, shipping clay to hookworm sufferers throughout the country!)

In very heavy infections, patients suffer severe protein deficiency, with dry skin and hair, edema, and "pot-belly" in children and with delayed puberty, mental dullness, heart failure, and death. Intestinal malabsorption is not a marked feature of infection with hookworms, but hookworm disease is usually manifested in the presence of malnutrition and is often complicated by *Ascaris, Trichuris,* and/or malaria infections. The drain of protein and iron is catastrophic to one who is subsisting on a minimal diet. Such chronic malnutrition, particularly in the young, often causes irreversible damage, resulting in stunted growth and below average intelligence. Impairment in ability to produce gamma globulin results in lowered antibody response to the hookworm as well as to other infectious agents. No living organism could be expected to live up to its potential under such conditions, and it is small wonder that development has been so difficult for many tropical countries.

Diagnosis, treatment, and control. Demonstration of hookworm eggs or the worms themselves in the feces is, as usual, the only definitive diagnosis of the disease. Demonstration of eggs in direct smears may be difficult, however, even in clinical cases, and one of the several concentration techniques should be used. If estimation of worm burden is necessary, there are techniques that give reliable data on egg counts (Beaver and Stoll techniques).

It is not necessary or possible to distinguish *Necator* eggs from those of *Ancylostoma,* but care should be taken to differentiate *Strongyloides* infections. This is not a problem unless some hours pass between time of defecation and time of examination of feces, or unless the feces were obtained from a constipated patient. Then the hookworm eggs may have hatched, and the larvae must be distinguished from those of *Strongyloides.* It is desireable, nevertheless, to distinguish *Necator* and *Ancylostoma* in studies on the efficacy of various drugs or chemotherapeutic regimes because the two species are not equally sensitive to particular drugs. Differentiation can be accomplished by recovery of adults after anthelmintic treatment or by culturing the larvae from the feces.

There is no 100% effective treatment known for hookworm.[5] Hexylresorcinol was once widely employed but has fallen into disuse because of its irritation to the gastrointestinal tract. Tetrachloroethylene is well tolerated and easily administered and is the cheapest drug available, being especially valuable in mass treatment campaigns. It is more effective against *N. americanus* than against *A. duodenale,* but it is ineffective against *Ascaris.* Tetrachloroethylene may irritate *Ascaris,* causing it to cluster into knots with resultant intestinal blockage. Therefore, when *Ancylostoma* or *Ascaris* are present, preferable drugs are bephenium hydroxynaphthoate (Alcopar), pyrantel embonate, and mebendazole. All three are active against both species of hookworm, as well as against *Ascaris.*

Treatment for hookworm disease should always include dietary supple-

mentation. In many cases provision of an adequate diet alleviates the symptoms of the disease without worm removal, but treatment for the infection should be instituted, if only for public health reasons.

Control of hookworm disease depends on lowering worm burdens in a population to the extent that remaining worms, if any, can be sustained within the nutritional limitations of the people, without causing symptoms. Mass treatment campaigns do not eradicate the worms but certainly lower the "seeding" capacity of their hosts. Education and persuasion of the population in the sanitary disposal of feces are also vital. The economic dependence on nightsoil in family gardens remains one of the most persistent of all problems in parasitology.

Recognizing these factors, the Sanitary Commission of the Rockefeller Foundation initiated a hookworm campaign in 1913. Beginning in Puerto Rico, where it was estimated that one-third of all deaths resulted from hookworm disease, then extending throughout the southeastern United States, the Commission would first survey an area. Residents of the area were examined for infection, then treated with anthelminthics. Thousands of latrines were provided, together with instructions on how to use and maintain them. It is a study in human nature to note that many persons refused to use latrines and could be persuaded only with great difficulty. As the result of the efforts of this and other similar hygiene commissions, hookworm prevalence has been reduced greatly from the earlier levels in the southern United States, most of the Caribbean, and a few other areas of the world. Nevertheless, hookworm remains one of the most important parasites of humans in the world today.

Creeping eruption

Also known as **cutaneous larva migrans,** creeping eruption is caused by invasive larval hookworms of species or strains normally maturing in animals other than humans. The larvae manage to penetrate the integument of humans but are incapable of successfully completing migration to the intestine. However, before they are overcome by immune responses, they re-

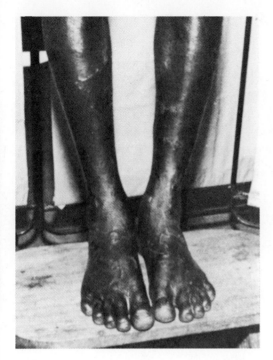

Fig. 25-12. Creeping eruption caused by accidental infection with *Ancylostoma braziliense.* (From Hopps, H. C., and D. C. Price. 1971. In Marcial-Rojas, R. A. Pathology of protozoal and helminthic diseases. The Williams & Wilkins Co., Baltimore. WRAIR neg. no. 13190-8.)

sult in distressing and occasionally serious complications of the skin (Fig. 25-12). Possibly any species of hookworm can cause this condition, but those of cats, dogs, and other domestic animals are most likely to come into contact with people. *Ancylostoma braziliense* appears to be the most common agent, followed by *A. caninum.* Both are very common parasites of domestic dogs and cats.

After entering the top layers of epithelium the larvae are usually incapable of penetrating the basal layer (stratum germinativum), so they begin an aimless wandering. As they tunnel through the skin, they leave a red, itchy wound that usually becomes infected by pyogenic bacteria. The worms may live for weeks or months. The larvae can attack any part of the skin, but because people's feet and hands are more in contact with the ground, they are most often affected. Thiabendazole has revolutionized treatment of creeping eruption, and topical application of a thiabendazole ointment has supplanted all other forms of treatment.[5]

FAMILY TRICHOSTRONGYLIDAE

Many genera and an immense number of species comprise the family Trichostrongylidae. They are parasites of the small intestine of all classes of vertebrates, causing great economic losses in domestic animals, especially ruminants, and in a few cases causing disease in humans.

Trichostrongylids (Fig. 25-13) are small, very slender worms with a rudimentary buccal cavity in most cases. Lips are reduced or absent and teeth rarely are present. The cuticle of the head may be inflated. Males have a well-developed bursa, and the spicules vary from simple to complex, depending on the species. Females are considerably larger than males. The vulva is located anywhere from preequatorial to near the anus, according to the species. Thin-shelled eggs are laid while in the morula stage.

Life cycles are similar in all species. No intermediate host is required; the eggs hatch in soil or water and develop directly to infective third-stage larvae. Some infections may occur through the skin, but as a rule the larvae must be swallowed with contaminated food or water. Enormous numbers of larvae may accumulate on heavily grazed pastures, causing serious or even fatal infections in ruminants and other grazers. Since their life cycles are similar, a given host usually is infected with several species, and the severe pathogenesis results from the cumulative effects of all the worms.

Haemonchus contortus

Haemonchus contortus lives in the "fourth stomach," or abomasum, of sheep, cattle, goats, and wild ruminants of many species. The species has been reported in humans in Brazil and Australia. It is one of the most important nematodes of domestic animals, causing a severe anemia in heavy infections.

The small buccal cavity contains a single well-developed tooth, which pierces the mucosa of the host. Blood is sucked from the wound, giving the transparent worms a reddish color. The larger females have

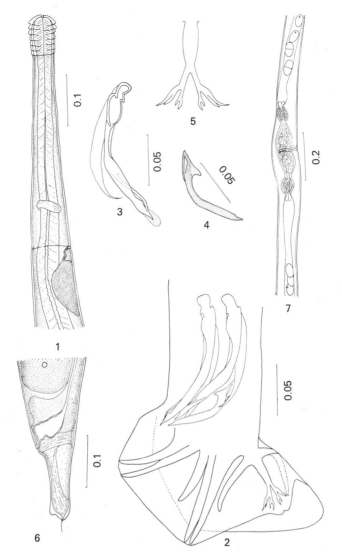

Fig. 25-13. *Molineus mustelae,* showing the characters typical of the Trichostrongylidae. **1,** Anterior end, lateral view; **2,** posterior end of male; **3,** complex spicules, lateral view; **4,** gubernaculum, lateral view; **5,** dorsal ray; **6,** posterior end of female, lateral view; **7,** midregion of female, showing ovijectors. All scales are in millimeters. (From Schmidt, G. D. 1965. J. Parasitol. 51:164-168.)

the white ovaries wrapped around the red intestine, lending it a characteristic red and white appearance, leading to its common names—"twisted stomach worm" and "barber-pole worm." Prominent cervical papillae are found near the anterior end. The male bursa is powerfully developed, with an asymmetrical dorsal ray (Fig. 25-14). The spicules are 450 to 500 μm long, each with a terminal barb. The vulva has

a conspicuous anterior flap in many individuals, but not in all. The frequency of occurrence of the vulvar flap seems to vary according to strain.

Infection occurs when the third-stage larva, still wearing the loosely fitting second-stage cuticle, is eaten with forage. Exsheathment takes place in the forestomachs. Arriving in the abomasum or upper duodenum, the worm molts within 48

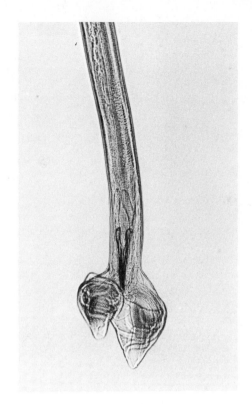

Fig. 25-14. *Haemonchus contortus:* ventral view of male, showing asymmetrical copulatory bursa. (Photograph by Jay Georgi.)

hours, becoming a fourth-stage larva with a small buccal capsule. It feeds on blood, which forms a clot around the anterior end of the worm. The worm molts for a final time in 3 days and begins egg production about 15 days later.

The anemia, emaciation, edema, and intestinal disturbances caused by these parasites result principally from loss of blood and injection of hemolytic proteins into the host's system. The host often dies with heavy infections, but those that survive usually develop an immunity and effect a self cure.

Ostertagia spp.

Ostertagia spp. are similar to *Haemonchus* in host and location but differ in color, being a dirty brown—hence their common name, the "brown stomach worm." The buccal capsule is rudimentary and lacks a tooth. Cervical papillae are present. The male bursa is symmetrical. The vulva has a large anterior flap, and the tip of the female's tail bears several cuticular rings.

The life cycle is similar to that of *Haemonchus* except that the third-stage larva burrows into the abomasal mucosa, where it molts. Returning to the lumen, it feeds, molts, and begins producing eggs about 17 days after infection. *Ostertagia* spp. suck blood, but not as much as does *Haemonchus.*

Some common species of *Ostertagia* are *O. circumcincta* in sheep, *O. ostertagi* in cattle and sheep, and *O. trifurcata* in sheep and goats. *Ostertagia ostertagi* and *O. circumcincta* have been reported from humans in Russia. It is possible that the infection followed eating insufficiently cooked abomasum of sheep, cattle, or goats.

Trichostrongylus spp.

Trichostrongylus spp. are the smallest members of the family, seldom exceeding 7 mm in length. Many species exist, parasitizing the small intestine of ruminents, rodents, pigs, horses, birds, and humans.

They are colorless, lack cervical papillae, and have a rudimentary, unarmed buccal cavity. The male bursa is symmetrical, with a poorly developed dorsal lobe. Spicules are brown and distinctive in size and shape in each species. The vulva lacks an anterior flap.

The life cycle is similar to that of *Ostertagia* except that the worms pass through the abomasum and burrow into the mucosa of the duodenum, where they molt. After returning to the lumen, they bury their heads in the mucosa and feed, grow, and molt for the last time. Egg production begins about 17 days after infection.

Common species of *Trichostrongylus* are *T. colubriformis* in sheep; *T. tenuis* in chickens and turkeys; *T. capricola, T. falcatus,* and *T. rugatus* in ruminants; *T. retortaeformis* and *T. calcaratus* in rabbits; and *T. axei* in a wide variety of mammals.

Eight species of *Trichostrongylus* have been reported in humans, with records from nearly every country of the world: six species in Armenia alone. Rate of infection varies from very low to as high as 69% in southwest Iran[16] and 70% in a village in Egypt.[10] Stoll estimated 5.5 million cases of human infection in the world.[18]

Pathology is identical in humans and

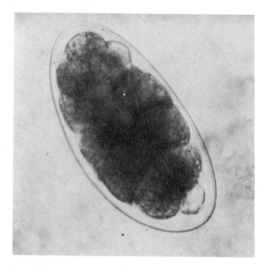

Fig. 25-15. An egg of *Trichostrongylus* sp., found in a human stool. (Photograph by David Oetinger.)

other infected animals. Traumatic damage to the intestinal epithelium may be produced by burrowing larvae and feeding adults. Systemic poisoning by metabolic wastes of the parasites and possible thyroid deficiency, hemorrhage, emaciation, and mild anemia may develop in severe infections.

Diagnosis can be made by finding the characteristic eggs (Fig. 25-15) in the feces or by culturing the larvae in powdered charcoal. The larvae are very similar to the larvae of hookworms and *Strongyloides*, and careful differential diagnosis is required.

Treatment with thiabendazole or with bephenium hydroxynaphthoate have proved effective. Cooking vegetables adequately will prevent infection in humans.

Other Trichostrongylids

In addition to the species from ruminants already mentioned, *Cooperia curticei*, *Nematodirus spathiger*, and *N. filicollis* should be included in that group of trichostrongyles that often occur in the same host and cause so much damage. *Hyostrongylus rubidus* is a serious pathogen of swine and can cause death, when present in large numbers. *Amidostomum anseris* burrows under the horny lining of the gizzard in ducks and other waterfowl, and it seems to be one of the main causes of mortality in overwintering Canadian geese. *Nema-*

tospiroides dubius in mice and *Nippostrongylus brasiliensis* in rats are of no direct medical or veterinary significance, but because they are easily kept in the laboratory, they serve as important tools for investigation of trichostrongyle nematodes.

Superfamily Metastrongyloidea

Metastrongyloidea are the bursate nematode lungworms of mammals. Most species for which the life cycles are known require an invertebrate intermediate host, and some also utilize a vertebrate or invertebrate transport host. Most species mature in terrestrial mammals, although several species in numerous genera are important parasites of marine mammals. Taxonomy of the group is unsettled, with several schemes in use. The metastrongyles are fairly homogeneous, morphologically, with buccal cavities reduced or absent and bursal lobes and rays reduced. Most are parasitic in the bronchioles, but some inhabit the pulmonary arteries, heart, muscles, and frontal sinuses.

Dictyocaulus filaria

This important parasite of sheep and goats shows close relationship to the Trichostrongylidae; for example, the life cycle does not involve an intermediate host. Adults live in the bronchii and bronchioles, where the females produce embryonated eggs. The eggs hatch while being carried out of the respiratory tree by ciliary action. First-stage larvae appear in the feces and develop to the third stage in contaminated soil, without feeding. The cuticles of both the first and second stages are retained by the third stage until the worm is eaten by a definitive host, then cuticles of all these stages are shed together. Fourth-stage larvae penetrate the mucosa of the small intestine, enter the lymphatic system, mix with the blood, and are carried to the lungs, where they enter alveoli and migrate to the bronchioles. They commonly cause death to their host.

These worms are slender and quite long, males reaching 80 mm and females 100 mm. The bursa is small and symmetrical; the spicules are short and boot-shaped in lateral view. The uterus is near the middle of the body.

Dictyocaulus arnfieldi in horses and *D.*

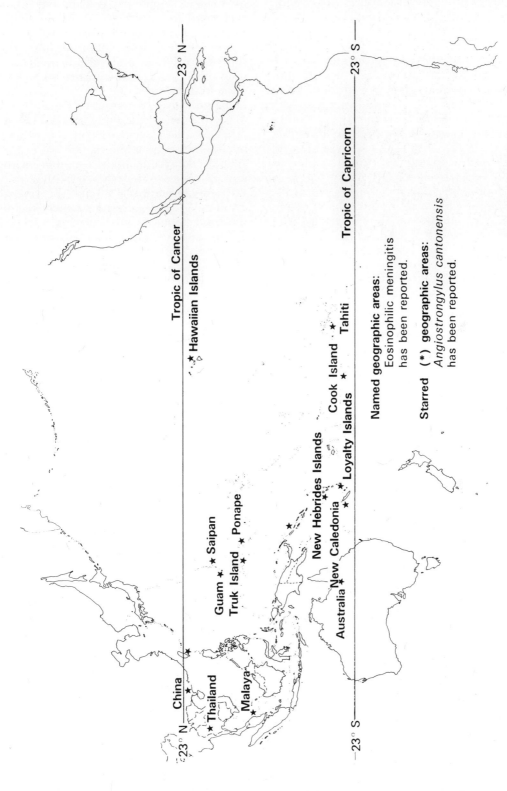

Fig. 25-16. The known geographic distribution of *Angiostrongylus cantonensis*. (AFIP neg. no. 68-5861.)

viviparus in cattle are similar to *D. filaria* in morphology and biology.

Angiostrongylus cantonensis

Angiostrongylus cantonensis was first discovered in the pulmonary arteries and heart of domestic rats in China, in 1935. Later the worm was found in many species of rats and bandicoots, and it may mature in other mammals throughout southeast Asia, the East Indies, Madagascar, and Oceanica, with infection rates as high as 88%. As a parasite of rats, it attracted little attention, but 10 years after its initial discovery, it was found in the spinal fluid of a 15-year-old boy in Taiwan. It has been discovered since in humans in Hawaii, Tahiti, the Marshall Islands, New Caledonia, Thailand, New Hebrides, and the Loyalty Islands (Fig. 25-16). This illustrates the value of basic research in parasitology to medicine, for when the medical importance of the parasite was realized, the reservoir of infection in rats already was known. Surveys of parasites endemic to the wild fauna of the world remain the first step in understanding the epidemiology of zoonotic diseases.

Angiostrongylus cantonensis is a delicate, slender worm with a simple mouth and no lips or buccal cavity. Males are 15.9 to 19.0 mm long, while females attain 21 to 25 mm. The bursa is small and lacks a dorsal lobe. Spicules are long, slender, and about equal in length and form. An inconspicuous gubernaculum is present. In the female, the intertwining of the intestine and uterine tubules give the worm a conspicuous barber-pole appearance. The vulva is about 0.2 mm in front of the anus. The eggs are thin-shelled and unembryonated when laid.

Biology. The eggs are laid in the pulmonary arteries, carried to the capillaries, and break into the air spaces, where they hatch. The larvae migrate up the trachea, are swallowed, and are expelled with the feces.

A number of types of molluscs serve as satisfactory intermediate hosts, including slugs and aquatic and terrestrial snails. Terrestrial planarians, freshwater shrimp, land crabs, and coconut crabs serve as paratenic hosts. Frogs have been found naturally infected with infective larvae.[2] Experimentally, Cheng infected American oysters and clams,[4] and Wallace and Rosen succeeded in infecting crabs.[20] All larvae thus produced were infective to rats.

When eaten by a definitive host, the third-stage larvae undergo an obligatory migration to the brain, which they leave 4 weeks after ingestion. Maturation occurs in the pulmonary arteries in about 6 weeks. Many wander in the body and mature in other locations, primarily in the central nervous system, meninges, and the eye.

Epidemiology. Humans or other mammals become infected when third-stage larvae are ingested. There may be several avenues of human infection, depending on the food habits of particular peoples. In Tahiti, where freshwater shrimp are known to be infected, it is a common practice to catch and eat them raw or to make sauce out of their raw juices. It is possible to eat slugs or snails accidentally with raw vegetables or fruit, so this route of infection should be considered. In Thailand and Taiwan, raw snails are often considered a delicacy. Heyneman and Lim showed that infective larvae escape from slugs and are left behind in their mucous trail.[7] These investigators also found larvae on lettuce sold in a public market in Malaya. Thus, while the epidemiology of angiostrongyliasis is not completely known, ample opportunities for infection exist.

Pathology. For many years, a disease of unknown etiology was recognized in tropical Pacific islands and was named **eosinophilic meningoencephalitis.** Patients with this condition have high eosinophil counts in peripheral blood and spinal fluid in about 75% of the cases and increased lymphocytes in cerebrospinal fluid. Nervous disorders commonly accompany these symptoms, occasionally followed by death. It is now known that *A. cantonensis* is at least one cause of this condition.

The presence of worms in blood vessels of the brain and meninges, as well as free-wandering worms in the brain tissue itself, results in serious damage. Some effects of such infection are severe headache, fever in some cases, paralysis of the fifth cranial nerve, stiff neck, coma, and death. Destruction of brain and spinal cord cells by trauma and immune responses evoked by dead worms results in vague symptoms for which the cause is most difficult to diagnose.

Diagnosis and treatment. When the symptoms described above appear in a patient in areas of the world where *A. cantonensis* exists, angiostrongyliasis should be suspected. It should be kept in mind that many of these symptoms can be produced by hydatids, cysticerci, flukes, *Strongyloides, Trichinella,* various larval ascarids, and possibly other lungworms. Alicata and Ash differentiate the larvae of several species of metastrongyles that could be confused with *A. cantonensis.*[1,3]

Thiabendazole shows promise in treating early, invasive stages, but little is known of treating the adults. Dead worms in blood vessels and the central nervous system may be more dangerous than live ones. A spinal tap to relieve headache may be recommended.

Angiostrongylus costaricensis parasitizes the mesenteric arteries of rats in Costa Rica. Morera and Céspedes found more than 70 human cases of infection with this parasite.[15] The worms matured in the mesenteric arteries and their intramural branches. Most damage was to the wall of the intestine, especially the cecum and appendix, which had become thickened and necrotic, with massive eosinophilic infiltrations. Abdominal pain and high fever were the most evident symptoms. Evidently, intestinal pathology was caused by blockage of the arterioles supplying the area by eggs and larvae of the parasites. No symptoms of meningoencephalitis, typical of *A. cantonensis,* were noted.

Other species of *Angiostrongylus* are potential parasites of humans, perhaps actually causing a disease that has yet to be diagnosed. North American species are *A. michiganensis* and *A. blarina* in shrews, *A. schmidti* from rice rats, and *A. gubernaculatus* in mustelid carnivores. A key to species of the world is given by Kinsella.[9]

Other Metastrongylids

Protostrongylus rufescens parasitizes the bronchioles of ruminants in many parts of the world. Its intermediate hosts are terrestrial snails, in which it develops to the third stage, and the definitive host is infected when the snail is eaten along with forage. Mountain sheep in America are seriously threatened by this and related species, which take a high toll of lambs every spring. Hibler, Lange, and Metzgar demonstrated transplacental transmission of *Protostrongylus* spp. in bighorn sheep.[8]

Muellerius capillaris lives in nodules in the parenchyma of the lungs of sheep and goats in most areas of the world. The life cycle is similar to that of *Protostrongylus rufescens,* involving a snail intermediate host.

Metastrongylus apri mainly infects swine, but sheep, cattle, and three cases in humans have been reported. Adults live in the bronchioles, and the eggs may hatch as in *Dictyocaulus,* or they may appear in the feces before hatching. An earthworm intermediate host is required for development to the infective third stage. This lungworm is also known to serve as a vector for the virus that causes **swine influenza.**[12] The nematodes serve as reservoirs for the disease, for they may live up to 3 years while encysted in an earthworm, all the while carrying the virus within their bodies.

REFERENCES

1. Alicata, J. E. 1963. Morphological and biological differences between the infective larvae of *Angiostrongylus cantonensis* and those of *Anafilaroides rostratus.* Can. J. Zool. 41:1179-1183.
2. Ash, L. R. 1968. The occurrence of *Angiostrongylus cantonensis* in frogs of New Caledonia with observations on paratenic hosts of metastrongyles. J. Parasitol. 54:432-436.
3. Ash, L. R. 1970. Diagnostic morphology of the third-stage larvae of *Angiostrongylus cantonensis, Angiostrongylus vasorum, Aelurostrongylus abstrusus,* and *Anafilaroides rostratus* (Nematoda: Metastrongyloidea). J. Parasitol. 56:249-253.
4. Cheng, T. C. 1965. The American oyster and clam as experimental intermediate hosts of *Angiostrongylus cantonensis.* J. Parasitol. 51:296.
5. Davis, A. 1973. Drug treatment in intestinal helminthiases. World Health Organization. Geneva.
6. Dönges, J., and O. Madecki. 1968. The possibility of hookworm infection through breast milk. German Med. Monthly 13:391-392.
7. Heyneman, D., and B. L. Lim. 1967. *Angiostrongylus cantonensis:* proof of direct transmission with its epidemiological implications. Science 158:1057-1058.
8. Hibler, C. P., R. E. Lange, and C. J. Metzger. 1972. Transplacental transmission of *Protostrongylus* spp. in bighorn sheep. J. Wildl. Dis. 8:389.
9. Kinsella, J. M. 1971. *Angiostrongylus schmidti* sp. n. (Nematoda: Metastrongyloidea), from the rice rat, *Oryzomys palustris,* in Florida, with a key to

the species of *Angiostrongylus* Kamensky, 1905. J. Parasitol. 57:494-497.

10. Lawless, D. K., R. E. Kuntz, and C. P. A. Strome. 1956. Intestinal parasites in an Egyptian village of the Nile Valley with emphasis on the protozoa. Am. J. Trop. Med. Hyg. 5:1010-1014.
11. Layrisse, M., A. Paz, N. Blumenfeld, and M. Roche. 1961. Hookworm anemia: iron metabolism and erythrokinetics. Blood 18:61-72.
12. Lee, D. L. 1971. Helminths as vectors of microorganisms. In Fallis, A. M., editor. Ecology and physiology of parasites. University of Toronto Press, Toronto. pp. 104-122.
13. Levine, N. D. 1968. Nematode parasites of domestic animals and of man. Burgess Publishing Co., Minneapolis.
14. Looss, A. 1898. Zur Lebensgeschichte des *Ankylostoma duodenale.* Cbt. Bakt. 24:441-449, 483-488.
15. Morera, P., and R. Céspedes. 1971. *Angiostrongylus costaricensis* n. sp. (Nematoda: Metastrongyloidea), a new lungworm occurring in man in Costa Rica. Rev. Biol. Trop. 18:173-185.
16. Sabha, G. H., F. Arfaa, and H. Bijan. 1967. Intestinal helminthiasis in the rural area of Khuzestan, south-west Iran. Ann. Trop. Med. Parasitol. 61:352-357.
17. Spencer, H. 1973. Nematode diseases. I. In Spencer, H., editor. Tropical pathology. Springer Verlag, Berlin. pp. 457-509.
18. Stoll, N. R. 1947. This wormy world. J. Parasitol. 33:1-18.
19. Velasquez, C., and B. C. Cabrera. 1968. *Ancylostoma ceylanicum* (Looss), in a Filipino woman. J. Parasitol. 54:430-431.
20. Wallace, G. D., and L. Rosen. 1966. Studies on eosinophilic meningitis. 2. Experimental infection of shrimp and crabs with *Angiostrongylus cantonensis.* Am. J. Epidemiol. 84:120-141.
21. Warren, K. S. 1974. Helminthic diseases endemic in the United States. Am. J. Trop. Med. Hyg. 23:723-730.
22. Yoshida, Y., K. Okamoto, A. Higo, and K. Imai. 1960. Studies on the development of *Necator americanus* in young dogs. Jap. J. Parasitol. 9:735-743.

SUGGESTED READING

Andrews, J. 1942. Modern views on the treatment and prevention of hookworm disease. Ann. Intern. Med. 17:841-901. (An interesting assessment of the problem to 1942.)
Beaver, P. C. 1953. Persistence of hookworm larvae in soil. Am. J. Trop. Med. Hyg. 2:102-108.
Looss, A. 1898. Zur Lebensgeschichte des *Ankylostoma duodenale.* Cbt. Bakt. 24:441-449, 483-488. (For the parasite historian: the first account of a hookworm life cycle.)
Skrjabin, K. I., N. P. Shikhobolova, R. S. Schulz, T. I. Popova, S. N. Boev, and S. L. Delyamure. 1952. Key to parasitic nematodes, vol. 3. *Strongylata.* Acad. Nauk SSSR, Moscow (English translation, 1961.) (Identification keys to all genera known to 1952.)
Stoll, N. R. 1962. On endemic hookworm. Where do we stand today? Exp. Parasitol. 12:241-252.
Stoll, N. R. 1972. The osmosis of research: example of the Cort hookworm investigations. Bull. N.Y. Acad. Med. 48:1321-1329.
Travassos, L. 1937. Revisão da familia Trichostrongylidae Leiper, 1912. Monogr. Inst. Oswaldo Cruz. I. (Out of date but still a valuable reference, especially the 295 plates.)

Chapter 26

ORDER ASCARIDATA: THE LARGE INTESTINAL ROUNDWORMS

The ascaroid worms are typically large, stout, intestinal parasites with three large lips. However, there are minute species with small or no lips, large species with two lips, and small species with well-defined lips. A preanal sucker is found on males of some. In one large group the esophageal-intestinal junction is highly specialized, with muscular or glandular appendages. Usually, however, the esophagus is simple and muscular. The life cycle is usually simple, lacking an intermediate host, although such a host is required in a few species. When the life cycle is direct, it is peculiar in that the second-stage larva is infective rather than the third-stage, which is typical in parasitic nematodes. It is as if a two-host cycle has been compressed into one host.

Of the several families in this order, we will emphasize the Ascaridae and Toxocaridae, which have the most medical importance. Other families will be discussed briefly.

FAMILY ASCARIDAE

The ascarids are among the largest of nematodes, some species achieving a length of 18 inches or more. Cervical, lateral, and caudal alae are absent, as are any esophageal ceca or ventriculi. Three large rounded or trapezoidal lips are present; interlabia are absent. Spicules are simple and equal. This family contains one of the oldest associates of people—*Ascaris,* the large intestinal roundworm.

Ascaris lumbricoides

Because of its great size, abundance, and cosmopolitan distribution, this may well have been the first parasite known to humans. Certainly the ancient Greeks and the Romans were familiar with them, and they were mentioned in the Ebers Papyrus.

It is probable that *A. lumbricoides* was originally a parasite of pigs that adapted to humans when swine were domesticated and began to live in close association with humans—or perhaps it was a human parasite that we gave to pigs. This is not surprising, for the physiologies of people and swine are remarkably similar, as, on occasion, are their eating and social habits. Today, two populations of this parasite exist, one in humans and one in pigs. They show a strong host specificity, but the two forms are so close morphologically that they were long considered the same species. Sprent pointed out slight differences in the tiny denticles on the dentigerous ridges along the inner edge of the lips.[11] This difference seems consistent and is much clearer when the structures are viewed with the scanning electron microscope[15]; therefore, we can now consider the two as separate species, *Ascaris suum* from pigs and *A. lumbricoides* from humans. This seems to be a good example of evolution in action. Each species may diverge even further with time, now that they have been reproductively isolated in many parts of the world where pigs no longer enjoy the homes of their masters.

Aside from the host specificity and the characteristics of the denticles, there are few, if any, other differences in the two species, and the following remarks on morphology and biology apply to both equally.

Morphology. In addition to its great size (Fig. 26-1), this species is characterized by having three prominent lips, each with a dentigerous ridge (Fig. 26-2), and no interlabia or alae. Lateral lines are visible grossly.

Males are 15 to 31 cm long and 2 to 4 mm at greatest width. The posterior end is curved ventrad, and the tail is bluntly

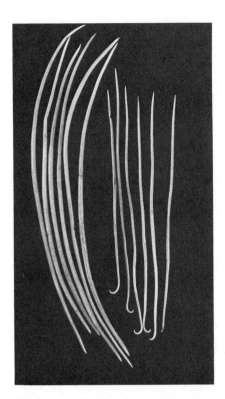

Fig. 26-1. *Ascaris lumbricoides*, males (right) and females (left). (Courtesy Ann Arbor Biological Center.)

pointed. Spicules are simple, nearly equal, and measure 2 to 3.5 mm long. No gubernaculum is present.

Females are 20 to 49 cm long and 3 to 6 mm wide. The vulva is about one-third the body length from the anterior end. The ovaries are extensive and the uteri may contain up to 27 million eggs at a time, with 200,000 being laid per day. Fertilized eggs (Fig. 26-3) are oval to round, 45 to 75 μm long by 35 to 50 μm wide, with a thick, lumpy outer shell (**mammillated,** uterine, or proteinaceous layer) that is contributed by the uterine wall. When the eggs are passed in the feces, the mammillated layer is bile-stained to a golden brown. The eggs are usually uncleaved when laid and passed in the feces. An unfertilized female, or one in early stages of oviposition, commonly deposits unfertilized eggs (Fig. 26-4) that are longer and narrower, measuring 88 to 94 μm long by 44 μm wide. The shell layers of unfertilized eggs are thinner and less distinct.

Biology. A period of 9 to 13 days is the minimal time required for the embryo to develop into an active first-stage larva. Although extremely resistant to low temperatures, desiccation, and strong chemicals, embryonation is retarded by such factors. Sunlight and high temperatures are lethal in a short time. The larva molts to the second stage before hatching through an indistinct operculum (Fig. 26-5). Contrary to the pattern of most parasitic nematodes, the second-stage larva is infective to the host (see discussion of embryonation in Chapter 22).

Infection occurs when embryonated eggs are swallowed with contaminated food and water. They hatch in the duodenum, where they penetrate the mucosa and submucosa and enter lymphatics of venules. After passing through the right heart, they enter the pulmonary circulation and break out of capillaries into the air spaces. Many larvae get lost during this migration and accumulate in nearly every organ of the body, causing acute tissue reactions.

While in the lungs, the larvae molt twice during a period of about 10 days. They then move up the respiratory tree to the pharynx, where they are swallowed. Many larvae make this last step of their migration before moulting to the fourth stage, but these cannot survive the gastric juices in the stomach. Fourth-stage larvae are resistant to such a hostile environment and pass through the stomach to the small intestine, where they mature. In 60 to 65 days after being swallowed, they begin producing eggs. It seems curious that the worms embark on such a hazardous migration only to end up where they began. One theory to account for it suggests that the migration simulates an intermediate host, which normally would be required for the larva of the ancestral form to develop to the third stage. Another possibility is that the ancestor was a skin penetrator for which the migration was a developmental necessity.

Epidemiology. The dynamics of *Ascaris* infection are essentially the same as for *Trichuris*. Indiscriminate defecation, particularly near habitations, "seeds" the soil with eggs that remain viable for many months or even years. The resistance of

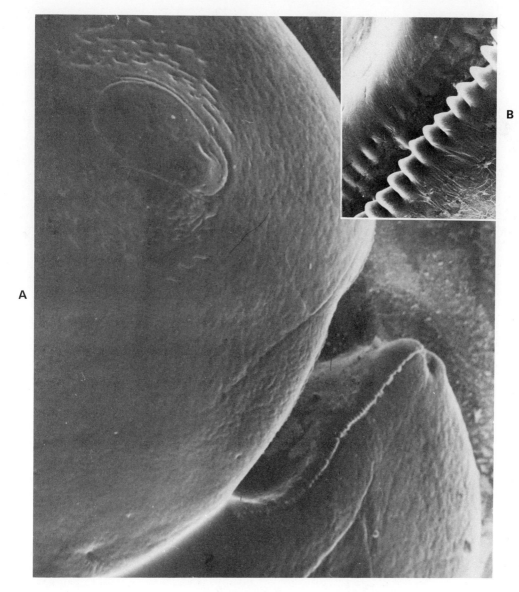

Fig. 26-2. A, Lips of *Ascaris lumbricoides;* note the large double papilla on the upper lip and the dentigerous ridge on the lower one. **B,** Enlarged view of the denticles of *Ascaris suum.* (Photographs by John Ubelaker.)

Ascaris eggs to chemicals is almost legendary. They can embryonate successfully in 2% formalin, in potassium dichromate, and in 50% solutions of hydrochloric, nitric, acetic, and sulfuric acid, among other similarly inhospitable substances.[10] This extraordinary chemical resistance is the result of the lipid layer of the eggshell, which contains the ascarosides.

The longevity of *Ascaris* eggs also con-

tributes to the success of the parasite. Brudastov and others infected themselves with eggs kept for 10 years in soil at Samarkand, Russia.[4] Of these eggs 30.7% to 52.7% were still found to be infective. Because of this longevity, it is impossible to prevent reinfection when houseyards have been liberally seeded with eggs, even when proper sanitation habits are initiated later.

Contamination, then, is the typical

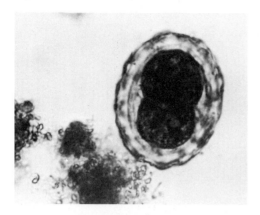

Fig. 26-3. A fertilized egg of *Ascaris lumbricoides* from a human stool. (Photograph by Robert E. Kuntz.)

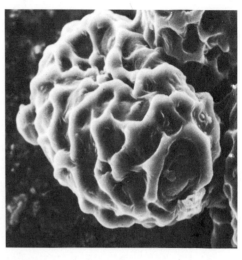

Fig. 26-5. Scanning electron micrograph showing egg of *Ascaris lumbricoides*. An operculum is visible at one end. (Photograph by John Ubelaker.)

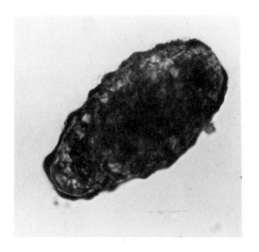

Fig. 26-4. An unfertilized egg of *Ascaris lumbricoides* from a human stool. (Photograph by Robert E. Kuntz.)

means of infection. Children are the most likely to become infected (or reinfected) by eating dirt or placing soiled fingers and toys in their mouths. In regions where nightsoil is used as fertilizer, principally the Orient, Germany, and certain Mediterranean countries, uncooked vegetables become important vectors of *Ascaris* eggs. Experimental support for this came from Mueller, who seeded a strawberry plot with eggs; he and volunteers ate unwashed strawberries from this plot every year for 6 years and became infected each year.[9]

Even windborne dust can carry *Ascaris*

eggs, when conditions permit. Bogojawlenski and Demidowa found *Ascaris* eggs in nasal mucus of 3.2% of school children examined in Russia.[3] From the nasal mucosa to the small intestine is a short trip in children. Dold and Themme found *Ascaris* eggs on 20 German banknotes in actual circulation.[5]

Pathogenesis. Little damage is caused by the penetration of intestinal mucosa by newly hatched worms. Larvae that become lost and wander and die in anomalous locations, such as the spleen, liver, lymph nodes or brain, often elicit an inflammatory response that may cause vague symptoms that are difficult to diagnose and may be confused with other diseases. This is apparently the fate of most *A. suum* larvae in humans. Transplacental migration into a developing fetus is also known.

When the larvae break out of lung capillaries into the respiratory system, they cause a small hemorrhage at each site. Heavy infections will cause small pools of blood to accumulate, which then initiate edema with resultant clogging of air spaces. Accumulations of white blood cells and dead epithelium add to the congestion, which is known as *Ascaris* **pneumonitis.** Large areas of lung can become diseased, and when bacterial infections

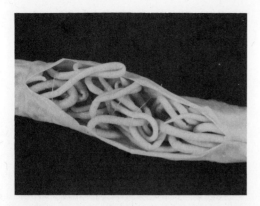

Fig. 26-6. The intestine of a pig, nearly completely blocked by *Ascaris suum* (threads were inserted to hold worms in place). Such heavy infections are also fairly common with *A. lumbricoides* in humans.

become superimposed, death can result. One instance is known in which a (perhaps unbalanced) parasitology graduate student vented his ire on his roommates by "seeding" their dinner with embryonated *Ascaris* eggs. They almost died before their malady was diagnosed.[1]

Pathogenesis of adult worms can be discussed conveniently in two categories: normal worm activities in the small intestine and wandering worms.

Although it is probable that *Ascaris* occasionally sucks blood from the intestinal wall, its main food is liquid contents of the intestinal lumen. In moderate and heavy infections the resulting theft of nourishment can cause malnutrition and underdevelopment in small children. Abdominal pains and sensitization phenomena—including rashes, eye pain, asthma, insomnia, and restlessness—often result as allergic responses to metabolites produced by the worms.

A massive infection can cause fatal intestinal blockage (Fig. 26-6). Why, in one case, do large numbers of worms cause no apparent problem, while in another, the worms knot together to form a mass that completely blocks the intestine? It is known that certain drugs, such as tetrachlorethylene used to treat hookworm, can aggravate *Ascaris* to knot up, but other factors are still unknown. Penetration of the intestine or appendix is not uncommon.

The resulting peritonitis is usually quickly fatal. According to Louw, 35.5% of all deaths in acute abdominal emergencies of children in Capetown were caused by *Ascaris.*[8]

Wandering adult worms cause various conditions, some serious, some bizarre, all unpleasant. The tropism of a female to squirm through the coiled tail of a male causes her to wander if no males are present. A similar restlessness can be noted in even higher forms of animals. Overcrowding may also lead to wandering. A downstream wandering leads to the appendix, which can be clogged or penetrated, or to the anus, with an attendant surprise for the unsuspecting host. Upstream wandering leads to the pancreatic and bile ducts, possibly occluding them with grave results. Worms reaching the stomach are aggravated by the acidity and writhe about, often causing nausea. The psychological trauma induced in one who vomits an 18-inch ascarid is difficult to quantify. Worms that reach the esophagus, usually while the host is asleep, may crawl into the trachea, causing suffocation or lung damage, or into the eustachian tubes and middle ears, causing extensive damage, or may simply exit through the nose or mouth, causing a predictable consternation.

Diagnosis and treatment. Accurate diagnosis of migrating larvae is impossible at this time. Demonstration of larvae in sputum is definitive, provided the technician can identify them.

Most diagnoses are made by identifying the characteristic, mammillated eggs in the stool or by an appearance of the worm itself. So many eggs are laid each day by one worm that one or two direct fecal smears are usually sufficient to demonstrate at least one. Otherwise, diagnosis is difficult if not impossible. *Ascaris* should be suspected when any of the previously listed pathogenic conditions are noted. Most light infections are asymptomatic and presence of worms may be determined only by spontaneous elimination of spent individuals from the anus.

There are several safe and effective drugs against *Ascaris,* of which piperazine has become the most commonly used. Others are levamisole, pyrantel, and mebendazole. Bephenium is less effective, but it is useful in concomitant infection with hookworm. No efficient treatment of migrating larvae has been discovered.

Parascaris equorum

This large nematode is the only ascaroid found in horses. *Parascaris equorum* is a cosmopolitan species that also infects the mule, ass, and zebra. It is very similar in gross appearance to *Ascaris lumbricoides* but is easily differentiated by its huge lips, which give it the appearance of having a large, round head.

The life cycle is similar to that of *A. lumbricoides,* involving a lung migration. Resulting pathogenesis is especially important in young animals, with pneumonia, bronchial hemorrhage, colic, and intestinal disturbances resulting in unthriftiness and morbidity. Intestinal perforation or obstruction is common. Older horses are usually immune to infection. Prenatal infection is not known to occur. Piperazine is the drug of choice, usually administered with a purgative to prevent intestinal obstruction.

FAMILY TOXOCARIDAE
Toxocara canis and Toxocara cati

These two species are cosmopolitan parasites of domestic dogs and cats and their relatives, and they have been found as adults in humans on several occasions. The biology and morphology of the two species are similar and the following remarks apply generally to both.

It is not uncommon for 100% of puppies and kittens to be infected in enzootic areas. As the result of prenatal infections (to be discussed further), one may expect even puppies in well-cared-for kennels to be infected at birth, and they are treated accordingly. The casual owner of a new puppy or kitten is likely to be startled by the vomiting by the pet of a number of large, active worms. Older dogs and cats seem to develop strong immunity to further infection, and they only rarely harbor adult worms.

Adults look basically like *Ascaris,* only they are much smaller. Three lips are present (Fig. 26-7). Unlike *Ascaris,* however, *Toxocara* has prominent cervical alae in both sexes. Males are 4 to 6 cm, and

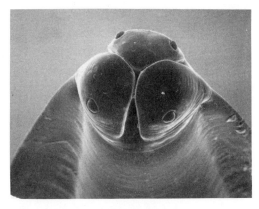

Fig. 26-7. *Toxocara cati:* Scanning electron micrograph, en face view illustrating the three lips with sensory papillae. (Photograph by John Ubelaker.)

females are 6.5 to over 15 cm long. The brownish eggs are almost spherical, with surficial pits, and are unembryonated when laid. *Toxocara canis* and *T. cati* have slightly different eggs.

Biology. Adult worms live in the small intestine of their host (Fig. 26-8), producing prodigious numbers of eggs, which are passed with the host's feces. Development of the second-stage larva takes 5 or 6 days under optimal conditions. The fate of ingested larvae depends on the age and immunity of the host. If the puppy is young and has had no prior infection, the worms hatch and migrate through the portal system and lungs and back to the intestine, as in *Ascaris lumbricoides.*

If the host is an older dog, particularly with some immunity acquired from past infection, the larvae do not complete the lung migration. They wander through the body, eventually becoming inactive but remaining alive for a long period of time. If a bitch becomes pregnant, the dormant larvae apparently are activated by host hormones and reenter the circulatory system, where they are carried to the placentas. There they can penetrate through to the fetal bloodstream, where they complete a lung migration en route to the intestine. Thus, a puppy can be born with an infection of *Toxocara,* even though the dam has shown no sign of infection.

A third option in the life cycle of *Toxocara canis* is offered when a rodent eats em-

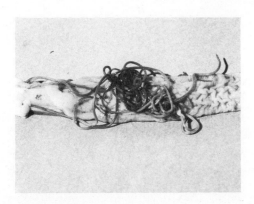

Fig. 26-8. Intestine of a domestic cat, opened to show numerous *Toxocara cati.* (Photograph by Robert E. Kuntz.)

bryonated eggs. In this unfavorable host, the larva begins to migrate but becomes lost and dormant. If the rodent is eaten by a dog that is not immune, the worms promptly migrate through the lungs to the intestine. Although this adaptability favors survival of the parasite, it bodes ill for certain accidental hosts, such as humans.

According to Sprent, the life cycle of *T. cati* varies from that of *T. canis* in an apparent absence of prenatal infection.[12] Some development occurs in the intestinal or stomach wall, and the lung migration may be omitted. Also, transport hosts may play a greater role.

Visceral larva migrans

When eggs of *Toxocara canis,* and probably also of *T. cati,* are eaten by an improper host, the larvae hatch and begin the typical liver-lung-intestine migration. However, they do not complete the normal migration but undergo developmental arrest and begin an extended, random wandering through the body. The resulting disease entity is known as **visceral larva migrans,** in contrast to dermal larva migrans (see p. 446). In its widest sense, visceral larva migrans can be caused by a variety of spirurid, filariid, strongylid, and other nematodes in addition to *Toxocara.* Thus, Sprent listed 36 species known to occur in humans in Australia and southeast Asia.[13] Even hookworm larvae that cause dermal larva migrans occasion-

ally enter the deep tissues of the body, thereby initiating a visceral disease. In some of these, humans are true intermediate hosts, for some development of the worms occurs.

In a narrower sense, visceral larva migrans only occurs when the larva maintains an extended migration but *does not undergo further development itself.* In these cases, the infected animal is a paratenic host, and the infection reflects a normal element in the life cycle of the parasite. The behavior of the larvae in humans is essentially the same as in a natural paratenic host.[2]

Therefore, while several genera of nematodes can cause visceral larva migrans, we will emphasize *Toxocara,* which is by far the most common cause of the condition in human beings, with *T. canis* seeming to be the most important species.

Epidemiology. Only a few years ago it was generally assumed that dog and cat worms could not infect humans or were not dangerous to them. We have come to realize since the early 1950s that the assumption was not true, particularly in the case of *T. canis.* Actually, very few cases of visceral larva migrans have been reported worldwide since then, but "most observers believe that this is the very small tip of a very large iceberg."[7] About 20% of adult dogs and 98% of puppies in the United States are infected with *T. canis;* therefore, the risk of exposure is very high. Most cases are either unrecognized or unreported. One of us knows of two local cases, one of which terminated fatally, the other with loss of an eye, but neither case found its way into the literature.

Dogs and cats defecating on the ground seed the area with eggs, which embryonate and become infective to any mammal eating them, including children. Considering the crawling-walking age of small children as the time when virtually every available object goes into the mouth for a taste, it is not surprising that the disease is most common in children between 1 and 3 years old. That favorite outdoor playground, the children's sandbox, unfortunately also constitutes an ideal cat toilet and *Toxocara* embryonation site. In the urban setting, the dog owner looks on the city park as the perfect place to "walk" the dog, while the parent at the same time brings young

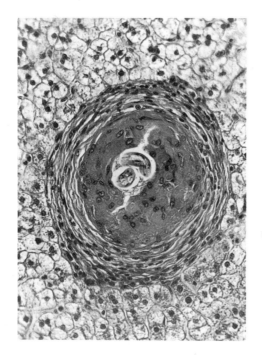

Fig. 26-9. *Toxocara canis* larva in liver of a monkey at 9 months infection. The larva rests in a matrix of epithelioid cells surrounded by a fibrous capsule lacking intense inflammatory reaction. (From Beaver, P. C. 1955. J. Parasitol. 55:3-12.)

children to play on the "seeded" grass. An especially unhappy fact in the epidemiology of larva migrans is the high risk to children by exposure to the environment of puppies and kittens. Finally, a factor to contemplate in light of the foregoing is the durability and longevity of *Toxocara* eggs, which are comparable to those of *Ascaris* (discussed before).

Pathogenesis. Characteristic symptoms of visceral larva migrans include fever, pulmonary symptoms, hepatomegaly, and eosinophilia. Extent of damage usually is related to the number of larvae present and their ultimate homestead in the body. Deaths have occurred when larvae were especially abundant in the brain; however, relatively few larvae can be life-threatening in the presence of a severe allergic reaction.[2] There seems little doubt that most cases result in rather minor, transient symptoms, which are undiagnosed or misdiagnosed. The most common site of larval invasion is the liver (Fig. 26-9), but no

organ is exempt. Eventually, each larva is surrounded by a granulomatous host reaction that blocks further migration. Beaver suggested that this reaction might actually be advantageous to the parasite, for nourishment and protection against further defenses of the host are gained.[2]

Larvae in the eye cause chronic inflammation of the inner chambers or retina or provoke dangerous granulomas of the retina. These reactions can lead to blindness in the affected eye. Ocular involvement has been reported in 245 patients with an average age of 7.5 years.[17] Other lesions destroy lung, liver, kidney, muscle, and nervous tissues.

Diagnosis and treatment. Diagnosis is difficult. A liver biopsy may demonstrate the characteristic granuloma surrounding the larva, but to obtain a biopsy containing a larva may be a matter of luck. A high eosinophilia is suggestive, especially if the possibility of other parasitic infections can be eliminated. Various serological and cutaneous tests are being perfected for diagnostic purposes, but unreliability and cross-reactions with *Ascaris lumbricoides* limit the usefulness of the tests developed so far.

No dependably effective treatment is known at this time, although several drugs have been used.[7] Control consists of periodic worming of household pets, especially young animals, and proper disposal of the animal's feces. Dogs and cats should be restrained, if possible, from eating available transport hosts. Children's sandboxes should be covered when not in use.

Toxocara vitulorum (Neoascaris vitulorum)

The only ascarid that occurs in cattle is *T. vitulorum*, which is nearly cosmopolitan in calves. It has also been reported in zebu, water buffalo, and sheep.

Adults appear much like *Ascaris*, except that there is a small, glandular ventriculus at the posterior end of the esophagus, typical of *Toxocara*. Lateral alae are absent. Males are 15 to 26 cm long, while females reach a length of 22 to 30 cm. Like those of *Toxocara canis*, the eggs have a pitted outer layer.

The life cycle is direct. Transplacental infection has been suggested, but infection by ingestion of larvae in mother's milk has

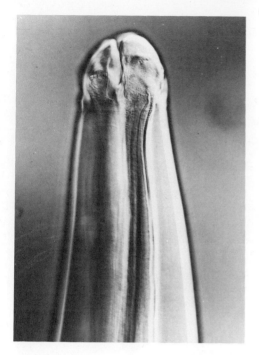

Fig. 26-10. The anterior end of *Toxascaris leonina*, an intestinal parasite of dogs and cats. Note the narrow cervical alae as compared with the broad alae of *Toxocara cati*. (Photograph by Jay Georgi.)

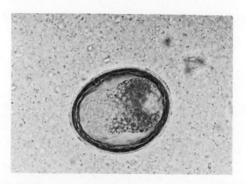

Fig. 26-11. The egg of *Toxascaris leonina*. (Photograph by Jay Georgi.)

now been proved.[16] Adult hosts are refractory to intestinal infection, but the second-stage larvae invade their tissues. At parturition, the larvae migrate to the mammary glands and are swallowed by the nursing calf. Calves can pass eggs as early as 23 days after birth.[6]

Young calves may succumb to verminous pneumonia during the migratory stages of the parasites. Diarrhea or colic in later stages result in unthriftiness and subsequent economic loss to the owner. Piperazine is effective in diagnosed cases, but prevention of infection is difficult.

Toxascaris leonina

Toxascaris leonina is a cosmopolitan parasite of dogs and cats and related felids and canids. It is similar in appearance to *Toxocara*, being recognized in the following ways: (1) the body tends to flex dorsally in *Toxascaris*, ventrally in *Toxocara;* (2) alae of *T. cati* are short and wide, while they are long and narrow in *T. canis* and *T.*

leonina (Fig. 26-10); (3) the surface of the egg of *T. leonina* is smooth (Fig. 26-11) but pitted in *Toxocara;* and (4) the tail of male *Toxocara* is constricted abruptly behind the anus, while it gradually tapers in *Toxascaris.*

The life cycle of *Toxascaris* is simple. Ingested eggs hatch in the small intestine, where the larvae penetrate the mucosa. After a period of growth, they molt and return directly to the intestinal lumen, where they mature.

Although they are mildly pathogenic, their main importance is a diagnostic one: it may be useful to separate *T. leonina* from *Toxocara* in identification procedures because *T. leonina* is considered of little importance as a source of visceral larva migrans.

Lagochilascaris minor

Lagochilascaris minor is evidently closely related to *Toxocara* and *Toxascaris,* differing only in minor morphological details (Fig. 26-12). Little is known about its biology in nature. It has been found in the stomach, pharynx, and trachea of various wild cats in South America and the Caribbean, with three related species in other felids and opossums in the same areas and in Africa.

Lagochilascaris minor has been reported in humans a dozen or so times, usually in the tonsils, nose, or neck.[14] When present, they cause abscesses that may contain from one to over 900 individuals (Fig. 26-13). They can mature in those locations, and they produce pitted eggs, much like those

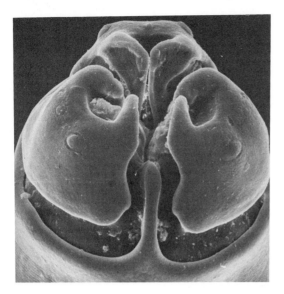

Fig. 26-12. *Lagocheilascaris turgida.* Note the prominent cleft in the tip of each lip, typical of the genus (lagocheil means harelip). (Photograph by John Sprent.)

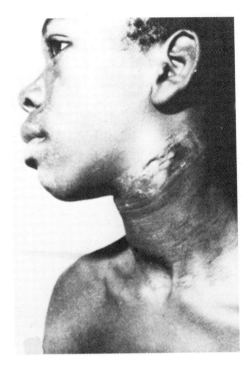

Fig. 26-13. An abscess in the neck of a 15-year-old native of Surinam. It contained numerous adults, larvae, and eggs of *Lagocheilascaris.* After treatment with thiabendazole, the fistula closed and the abscess healed, leaving only a small scar. (From Oostburg, B. F., 1971. Am. J. Trop. Med. Hyg. 20:580-583.)

of *Toxocara.* It is common for human infections to last many years or to rapidly kill the infected person. These worms epitomize the zoonotic infection. How humans become infected is unknown.

FAMILY ANISAKIDAE
Anisakis spp.

The many species in the genus *Anisakis* are parasites of the stomachs of marine fishes, birds, and mammals. While the life cycles generally are unknown, fishes of many species serve as paratenic hosts. Humans enter the picture when such larvae are eaten. Two aspects of the situation are important to humans: esthetics and public health. The first relates to the disgust experienced by persons who find large, stout worms in the flesh of the meal they are preparing or eating. Many a finnan haddie has ended up in the garbage pail when *Anisakis* larvae were discovered, a rather common occurrence at that!

More importantly, *Anisakis* larvae can produce a variety of pathological conditions in humans, when eaten in raw, salted, or pickled fish. Intestinal obstruction, colic, abscesses, and peritonitis commonly result from infections with these worms. The stomach commonly is afflicted with a peculiar, tumor-like growth at the site of attachment. Most cases have been reported from Japan, Europe, and Scandinavia, where raw fish is relished. Fatalities have been recorded.

The taxonomy of this group of worms is confused and contradictory. For the sake of convenience, most parasitologists refer to the larval stages as "anisakis-type" and let it go at that. The extreme abundance of such larvae in marine fishes the world over suggests that fish dishes must always be well cooked or deep frozen before eating.

FAMILY ASCARIDIIDAE

The family Ascaridiidae is of particular interest because it appears to represent an evolutionary link between the Ascaridata and the Oxyurata. The esophagus is typically ascaridoid, while the male posterior end is typically heterakoid (see p. 470).

The type genus *Ascaridia* has numerous species in birds, including the large nematode of chickens, *A. galli.*

Ascaridia galli is a cosmopolitan parasite of the small intestine of domestic fowl. Males reach a length of 77 mm; females are up to 115 mm.

Typical of ascarids, the second-stage larva hatches from the egg after it is ingested with contaminated food or water. No extensive migration is involved in the life cycle. Instead, at 8 or 9 days postinfection, the larvae molt to the third stage and begin to burrow into the mucosa, where they generally remain with their tails still in the intestinal lumen. After molting to the fourth stage at about 18 days, they return to the lumen where they undergo their final molt and mature. Probably the majority of larvae complete their three molts and attain maturity without ever leaving the lumen. Those that do attack the mucosa, however, cause extensive damage, which may result in unthriftiness or even death. Adult chickens seem to be refractory to infection.

REFERENCES

1. Anonymous. 1970. LIer sought in roommates' poisoning. Newsday, Feb. 27.
2. Beaver, P. C. 1969. The nature of visceral larva migrans. J. Parasitol. 55:3-12.
3. Bogojawlenski, N. A., and A. Demidowa. 1928. Ueber den Nachweis von Parasiteneiern auf der menschlichen Nasenschleimhaut. Russian J. Trop. Med. 6:153-156.
4. Brudastov, A. N., V. R. Lemelev, S. K. Kholnuk-hanedov, and L. N. Krasnos. 1971. The clinical picture of the migration phase of ascariasis in self-infection. Medskaya Parazitol. 40:165-168.
5. Dold, H., and H. Themme. 1949. Ueber die Möglichkeit der uebertragung der Askaridiasis durch Papiergeld. Deutsch. Med. Wochenschr. 74:409.
6. Herlich, H., and D. A. Porter. 1954. Experimental attempts to infect calves with *Neoascaris vitulorum.* Proc. Helm. Soc. Wash. 21:75-77.
7. Levine, N. D. 1968. Nematode parasites of domestic animals and of man. Burgess Publishing Co., Minneapolis.
8. Louw, J. H. 1966. Abdominal complications of *Ascaris lumbricoides* infestation in children. Brit. J. Surg. 53:510-521.
9. Mueller, G. 1953. Untersuchungen ueber die Lebensdauer von Askarideiern in Gartenerde. Zentralbl. Bakt. I. Orig. 159:377-379.
10. Schwartz, B. 1960. Evolution of knowledge concerning the roundworm *Ascaris lumbricoides.* Smithsonian Report for 1959. pp. 465-481.
11. Sprent, J. F. A. 1952. Anatomical distinction between human and pig strains of *Ascaris.* Nature 170:627-628.
12. Sprent, J. F. A. 1956. The life history and development of *Toxocara cati* (Schrank, 1788) in the domestic cat. Parasitology 46:54-78.
13. Sprent, J. F. A. 1969. Nematode *larva migrans.* New Zealand Vet. J. 17:39-48.
14. Sprent, J. F. A. 1971. Speciation and development in the genus *Lagocheilascaris.* Parasitology 62:71-112.
15. Ubelaker, J. E., and V. F. Allison. 1972. Scanning electron microscopy of the denticles and eggs of *Ascaris lumbricoides* and *Ascaris suum.* In Arceneaux, C. J., editor. Thirtieth Annual Proceedings of the Electron Microscopy Society of America. Claitor's Publishing Division, Baton Rouge.
16. Warren, E. G. 1971. Observations on the migration and development of *Toxocara vitulorum* in natural and experimental hosts. Int. J. Parasitol. 1:85-99.
17. Warren, K. S. 1974. Helminthic diseases endemic in the United States. Am. J. Trop. Med. Hyg. 23:723-730.

SUGGESTED READING

Chabaud, A. G. 1974. Keys to subclasses, orders and superfamilies. In Anderson, R. C., A. G. Chabaud, and S. Willmott, editors. CIH keys to the nematode parasites of vertebrates. Commonwealth Agricultural Bureaux, Bucks.

Hartwich, G. 1974. Keys to genera of the Ascaridoidea. In Anderson, R. C., A. G. Chabaud, and S. Willmott, editors. CIH keys to the nematode parasites of vertebrates. Commonwealth Agricultural Bureaux, Bucks.

Willmott, S. 1974. General introduction, glossary of terms. In Anderson, R. C., A. G. Chabaud, and S. Willmott, editors. CIH keys to the nematode parasites of vertebrates. Commonwealth Agricultural Bureaux, Bucks.

Chapter 27

ORDER OXYURATA: THE PINWORMS

Members of the Oxyurata are called "pinworms" because they typically have slender, sharp-pointed tails, especially the females. All pinworms have one feature in common: a conspicuous muscular bulb on the posterior end of the esophagus (Fig. 27-1). Three lips are present around the mouth of the more primitive species (Fig. 27-2), but the lips are reduced or absent in more advanced forms. Caudal and cervical alae are common, and males of many species have a preanal sucker. Life cycles are typically direct, with no intermediate host required. Pinworms are common in mammals, birds, reptiles, and amphibians but are rare in fishes. Most domestic birds and mammals harbor pinworms, but, curiously, they are absent in dogs and cats. Terrestrial arthropods, especially insects and millipedes, are commonly infected. One species, *Enterobius vermicularis*, may be the most common nematode parasite of humans. Because pinworms usually inhabit the large intestine and apparently feed only on bacteria and other intestinal contents, it has been suggested that they are not parasitic. The following discussion shows that the pinworm of humans, at least, qualifies for the status of parasite, as defined in the introductory chapter of this book.

The oxyurids will be illustrated by examples from humans, rodents, and domestic fowl.

FAMILY OXYURIDAE
Enterobius vermicularis

In some ways, pinworms are rather paradoxical among the nematode parasites of humans. For one thing, they are not tropical in their distribution, thriving best in the temperate zones of the world. Further, pinworms often are found in families at high socioeconomic levels, where, after introduction into the premises by one member, they rapidly become a "family affair." It is fair to say, however, that the greatest pinworm problems are among institutionalized persons, such as in orphanages and mental hospitals, where conditions facilitate transmission and reinfection.

The fact that this worm inhabits at least 500 million persons is perhaps less surprising than the fact that practically nothing is being done to eliminate it. At least part of the reason is simple and practical: pinworms cause no obvious, debilitating or disfiguring effects. Their presence is an embarrassment and an irritation, like acne or dandruff. Resources of democracies, kingdoms, and dictatorships could scarcely be expected to mobilize to combat such an innocuous foe, particularly when they seem to have so much trouble mounting efforts against more disabling infectious agents.

And yet, is enterobiasis so unimportant after all? Certainly it is important to the millions of persons who suffer the discomforts of infection. Further, a great deal of money is spent in efforts to be rid of pinworms. The frantic efforts by persons to rid their households of the tiny worms often lead to what has been called a "pinworm neurosis." The mental stress and embarrassment suffered by families who know they harbor parasites are unmeasurable, but very real consequences of infection, especially when multiplied by the vast number of persons involved. Finally, the pathogenesis of these worms may be greatly underrated.[1]

Morphology. Both sexes have three lips surrounding the mouth, followed by a cuticular inflation of the head (Fig. 27-3). The male *Enterobius vermicularis* is 2.0 to 5.0 mm long and has the posterior end strongly curved ventrad. It has a single, simple spicule about 70 μm long, and conspicuous caudal alae supported by papillae (Fig. 27-4).

Females measure 8 to 13 mm long and have the posterior end extended into a

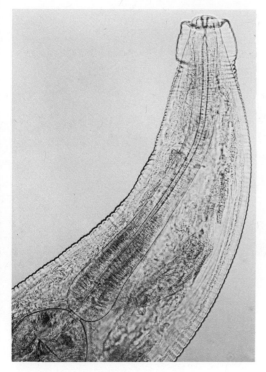

Fig. 27-1. *Enterobius* sp., a pinworm from a monkey. Note the conspicuous bulb at the posterior end of the esophagus. Such a bulb is typical of the order Oxyurata. (Photograph by Jay Georgi.)

Fig. 27-2. *Pharyngodon*, a pinworm of reptiles, showing the primitive, three-lipped condition. (Photograph by John Ubelaker.)

long, slender point (Fig. 27-5), giving pinworms their name. The vulva opens between the first and second thirds of the body. When gravid, the two uteri contain thousands of eggs, which are elongate-oval and flattened on one side (Fig. 27-6), measuring 50 to 60 μm by 20 to 30 μm.

Biology. Adult worms congregate mainly in the ileocecal region of the intestine, but they commonly wander throughout the gastrointestinal tract from the stomach to the anus. They attach themselves to the mucosa where they presumably feed on epithelial cells and bacteria. Gravid females begin migrating within the lumen of the intestine, commonly passing out of the anus onto the perianal skin. As they crawl about, both within the bowel and on the outer skin, they leave a trail of eggs. One worm may deposit from 4,600 to 16,000 eggs.[5] Females die soon after oviposition, while males die soon after copulation.

Consequently, it is usual to find many more females than males within a host.

When laid, each egg contains a partially developed larva, which can develop to infectivity within 6 hours at body temperature.[3] Ovic embryos are resistant to putrefaction and disinfectants but succumb to dehydration in dry air within a day.

Reinfection occurs by two routes. Most often the eggs, containing third-stage larvae, are swallowed and hatch in the duodenum. They slowly move down the small intestine, molting twice to become adults by the time they arrive at the ileocecal junction. Total time from ingestion of eggs to sexual maturity of the worms is 15 to 43 days.

If the perianal folds are unclean for long periods of time, the attached eggs may hatch and the larvae wander into the anus and hence to the intestine, a process known as **retrofection.** Hatching of the eggs while still inside the intestine apparently does not occur, except, perhaps, during constipation.

Epidemiology. Clothing and bedding

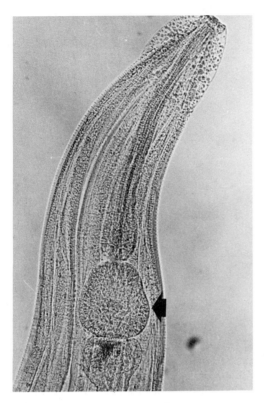

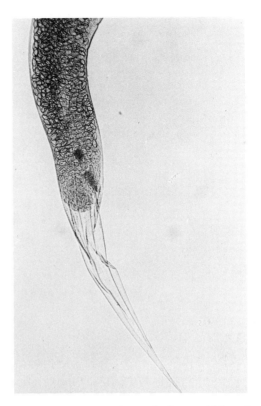

Fig. 27-3. Anterior end of the pinworm, *Enterobius vermicularis.* Note the large esophageal bulb (arrow) and the swollen cuticle at the head end, typical of this genus. (Photograph by Warren Buss.)

Fig. 27-5. The posterior end of a gravid female *Enterobius vermicularis.* The long pointed tail lends this species the name "pinworm." (Photograph by Warren Buss.)

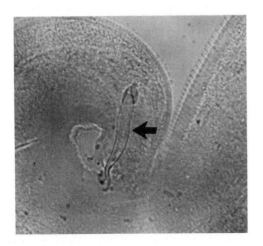

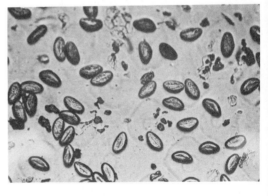

Fig. 27-6. Eggs of *Enterobius vermicularis.* (Photograph by David Oetinger.)

Fig. 27-4. The posterior end of a male *Enterobius vermicularis,* illustrating the single spicule (arrow). (Photograph by Warren Buss.)

rapidly become seeded with eggs when an infection occurs. Even curtains, walls, and carpets become sources of subsequent infection (or reinfections). The microscopic eggs are very light in weight and are wafted about by the slightest air currents, where they are deposited throughout the building. The eggs remain viable in cool, moist conditions for up to a week.

The most common means of infection is by placing soiled fingers or other objects into the mouth, as well as by use of contaminated bedding, towels, and so on. Obviously, it becomes next to impossible to avoid contamination when eggs are abundant. And it remains impossible to avoid reinfection when retrofection occurs.

Airborne eggs may be inhaled and subsequently swallowed, or they may remain in the nose until they hatch. This, together with nose-picking, accounts for the occasional case of pinworm in the nose. Contrary to popular belief, pinworms cannot be transmitted by dogs and cats, for these animals are free of pinworms.

White people seem more susceptible to pinworms than are black people.[2]

Pathogenesis. About one-third of infections are completely asymptomatic, and in many more, clinical symptoms are negligible. Nevertheless, very large numbers of worms may be present and lead to more serious consequences. Pathogenesis has two aspects: damage caused by worms within the intestine and damage resulting from egg deposition around the anus. Minute ulcerations of the intestinal mucosa from attachment of adults may lead to mild inflammation and bacterial infection.[7] Very rarely, pinworms will penetrate into the subserosa with fatal results. The movements of the females out the anus to deposit eggs, especially when the patient is asleep, lead to a tickling sensation of the perianus, causing the patient to scratch. The subsequent vicious circle of bleeding, bacterial infection, and intensified itching can lead to a nightmare of discomfort.

It is common for worms to wander into the vulva where they may remain for several days, causing a mild irritation. Cases have been reported where pinworms have wandered up the vagina, uterus, and oviducts into the coelum, to become encysted in the peritoneum.

Children with heavy pinworm infection are often nervous, restless and irritable and may suffer from loss of appetite, nightmares, insomnia, weight loss, and perianal pain.

Diagnosis and treatment. Positive diagnosis can be made only by finding eggs or worms on or in the patient. Ordinary fecal examinations are usually unproductive because few eggs are deposited within the intestine and passed in the feces. Heavy infections can be discovered by examining the perianus closely under bright light, during the night or early morning. Wandering worms glisten and can be seen easily. When adults cannot be found, eggs often can be, for they are left behind in the perianal folds. A short piece of cellophane tape, held against a flat, wooden applicator or similar instrument, sticky-side out, is pressed against the junction of the anal canal and the perianus. The tape is then reversed and stuck to a microscope slide for observation. If a drop of xylene or toluene is placed on the slide before the tape, it will dissolve the glue on the tape and clear away bubbles, simplifying the search for the characteristic, flat-sided eggs. It is desirable for the physician to teach the parent how to prepare the slide, since it should be done just after awakening in the morning, certainly before bathing the child for a trip to the doctor's office.

Numerous home remedies and over-the-counter medications have been in use for many years, with results ranging from poor to completely ineffective. Today, the drugs of choice are piperazine citrate (Antepar), pyrvinium pamoate (Povan). and mebendazole (Vermox). All are highly effective, inexpensive, and safe. Treatment should be repeated after about 10 days to kill worms acquired after the first dose, and sanitation procedures should be instituted concurrently. All members of the household should be treated simultaneously, regardless of whether the infection has been diagnosed in all.

While diagnosis and cure of enterobiasis are easy, preventing reinfection is more difficult. Personal hygiene is most important. Completely sterilizing the household is a gratifyingly difficult activity, but of limited usefulness. Nevertheless, at time of treatment all bed linens, towels, and the like should be washed in hot water, and the household should be cleaned as well as possible to lower the prevalence of infective eggs in the environment. If all persons are undergoing chemotherapy while reasonable care is taken to avoid reinfection, the family infection can be eradicated —until the next time a child brings it home from school.

FAMILY SYPHACIIDAE
Syphacia spp.

The tiny parasites of the family Syphaciidae (*Syphacia* spp.) are rarely found in humans but are commonly encountered in their natural hosts, wild and domestic rodents. Laboratory rats and mice are frequently infected by *Syphacia obvelata,* which lives in the cecum and has been reported from humans.[6]

Male *Syphacia* are easily recognized by their **mamelons,** two or three ventral, serrated projections (Fig. 27-7). Females are typical pinworms, with long, pointed tails. The eggs are operculated. The life cycle is direct, with the worms maturing in the cecum or large intestine.[4] No migration within the host is known.

FAMILY HETERAKIDAE
Heterakis gallinarum

The large pinworms *Heterakis gallinarum* are cosmopolitan in domestic chickens and related birds. They live in the cecum, where they feed on its contents.

Three large lips and a bulbar esophageal swelling are found in this genus, as are lateral alae. Males are up to 13 mm long and possess wide caudal alae supported by 12 pairs (usually) of papillae (Fig. 27-8). The tail is sharply pointed, and there is a prominent preanal sucker. The spicules are strong and dissimilar, and there is no gubernaculum.

Females are typical pinworms, with the vulva near the middle of the body, and a long, pointed tail.

Many species of *Heterakis* are known in birds, particularly in ground-feeders, and one species, *Heterakis spumosa,* is cosmopolitan in rodents.

Biology. The eggs of *H. gallinarum* are in the zygote stage when laid. They develop to the infective stage in 12 to 14 days at 72°F and can remain infective for 4

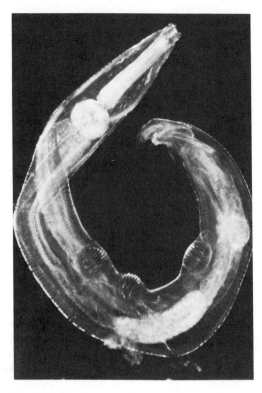

Fig. 27-7. *Syphacia,* a pinworm of rodents. Note the three corrugated mamelons on the ventral surface of the male. (Photograph by Warren Buss.)

years in soil. Infection is contaminative: when embryonated eggs are eaten, the second-stage larvae hatch in the gizzard or duodenum and pass down to the ceca. Most complete their development in the lumen, but some penetrate the mucosa, where they remain for 2 to 5 days without further development. Returning to the lumen, they then mature, about 14 days after infection.

If eaten by an earthworm, the larva may hatch and become dormant in the worm's tissues, remaining infective to chickens for at least a year. Since the nematodes do not develop further until eaten by a bird, the earthworm is a paratenic host.

Epidemiology. As the result of the longevity of the eggs, it is difficult to eliminate *Heterakis* from a domestic flock. Thus, although adult chickens may effect a self-cure, infective eggs are still available the following spring, when new chicks are hatched. Further, as earthworms feed in contaminated soil, they accumulate large

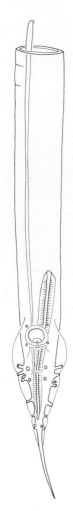

Fig. 27-8. The posterior end of *Heterakis varia-bilis,* a parasite of pheasants that is similar to *H. gallinarum.* Note the conspicuous preanal sucker. (From Inglis, W. G., G. D. Schmidt, and R. E. Kuntz. 1971. Rec. S. Aust. Mus. 16:1-14.)

numbers of larvae, which in turn causes massive infections in the unlucky birds that eat them.

Pathogenesis. In heavy infections, there may be a thickening of the cecal mucosa, with slight bleeding. Generally speaking, *Heterakis* is not highly pathogenic in itself.

However, a flagellate protozoan, *Histomonas meleagridis,* is transmitted between birds within eggs of *Heterakis gallinarum.* This protozoan is the etiological agent of **histomoniasis,** a particularly serious disease in turkeys. The protozoan is eaten by the nematode and multiplies in the worm's intestinal cells, in the ovaries, and, finally, in the embryo within the egg (see p. 94, Chapter 5). Hatching of the worm within a new host releases *Histomonas.* Hence we encounter the curious phenomenon of one parasite acting as a true intermediate host and vector of another.

Diagnosis and treatment. *Heterakis* can be diagnosed by finding the eggs in the feces of its host. The worms are effectively eliminated with phenothiazine. Usually, a flock of birds is routinely fed this or other drugs in its feed or water. This, together with rearing the birds on hardware cloth, will eliminate the parasite from the flock. Birds that are allowed to roam the barnyard are usually infected.

REFERENCES

1. Bijlmer, E. 1946. Exceptional case of oxyuriasis of the intestinal wall. J. Parasitol. 32:359-366.
2. Cram, E. B. 1941. Studies on oxyuriasis. The familial nature of pinworm infestation. Med. Ann. D.C. 10:39-48.
3. Hulinská, D. 1968. The development of the female *Enterobius vermicularis* and the morphogenesis of its sexual organ. Folia Parasitol. 15:15-27.
4. Prince, M. J. R. 1950. Studies on the life cycle of *Syphacia obvelata,* a common nematode parasite of rats. Science 111:66-67.
5. Reardon, L. 1938. Studies on oxyuriasis. XVI. The number of eggs produced by the pinworm, *Enterobius vermicularis,* and its bearing on infection. Publ. Health Rep. 53:978-984.
6. Riley, W. A. 1920. A mouse oxyurid, *Syphacia obvelata,* as a parasite of man. J. Parasitol. 6:89-92.
7. Shubenko-Gabuzova, I. N. 1965. Appendicitis in enterobiasis. Med. Parasitol. Parasitol. Dis. 34:563-566.

SUGGESTED READING

Beaver, P. C., J. J. Kriz, and T. J. Lau. 1973. Pulmonary nodule caused by *Enterobius vermicularis.* Am. J. Trop. Med. Hyg. 22:711-713.
Little, M. D., C. J. Cuello, and A. D'Alessandro. 1973. Granuloma of the liver due to *Enterobius vermicularis.* Report of a case. Am. J. Top. Med. Hyg. 22:567-569.
Skrjabin, K. I., N. P. Schikhobalova, and A. A. Mosgovoi. 1951. Key to parasitic nematodes, vol. 2. *Oxyurata* and *Ascaridata.* Akad. Nauk SSSR, Moscow. (A useful key to genera, with lists of species.)
Skrjabin, K. I., N. P. Schikhobalova, and E. A. Lagodovskaya. 1960-1967. Essentials of nematodology, vols. 8, 10, 13, 15, and 18. Oxyurata. Akad. Nauk SSSR, Moscow. (Indispensable reference works for the oxyurid taxonomist.)

Chapter 28

ORDER SPIRURATA: A POTPOURRI OF NEMATODES

Spirurids are parasitic in all classes of vertebrates and utilize an intermediate host in their development, usually an arthropod. It is a very large, heterogeneous group, with many species. The many variations of morphology in this order make it difficult to generalize, but most have two lateral lips, called pseudolabia, and an esophagus that is divided into anterior muscular and posterior glandular portions. The lips do not represent the fusion of primitive lips but are evolutionarily new structures that originate in an anterior shifting of tissues from within the buccal walls. Though the esophagus usually has both muscular and glandular portions, there are species whose esophagus is primarily muscular and others in which it is mainly glandular. Some of those, however, are of uncertain taxonomic position. Spirurid spicules are usually dissimilar in size and shape.

Spirurids seldom parasitize humans, and when they do it is only as zoonoses. Several, however, are important in domestic livestock; the rest are parasites of wild animals. When a nematode is found in a wild animal, the chances are 50% that it will be a spirurid, especially if insects are a preferred part of the host's diet.

Of the many families in this order, a few that demonstrate the diversity in the group will be examined briefly.

FAMILIES ACUARIIDAE AND SCHISTOROPHIDAE

Nematodes of these two families, all parasites of birds, exhibit very peculiar morphological structures at their head ends. Acuariids have four grooves or ridges, called cordons, which begin two dorsally and two ventrally at the junctions of the lateral lips, and proceed posteriad for varying distances (Fig. 28-1). The cordons may be straight, sinuous, recurving, or even anastomosing in pairs.

The schistorophids, which are very closely related to the acuariids, do not have cordons but instead possess four extravagant cuticular projections, sometimes simple, sometimes serrated, or even feathered (Fig. 28-2).

Both specializations, cordons and cuticular projections of the head, seem to correlate with the parasite's location within the host, the stomach. Most mature under the koilon, or gizzard lining, where they cause considerable damage to the underlying epithelium. How these anterior modifications aid the parasite is not known.

Common genera of acuariids are *Acuaria* and *Cheilospirura* in terrestrial birds and *Echinuria*, *Skrjabinoclava*, *Chevreuxia*, and *Cosmocephalus* in aquatic birds.

Some genera of schistorophids are *Torquatella*, *Viquiera*, and *Serticeps* in terrestrial birds and *Schistorophus*, *Ancyracanthopsis*, and *Sciadiocara* in aquatic birds.

With the exception of *Echinuria* in ducks, geese, and swans, these parasites are of little or no economic importance. They do, however, represent an interesting example of adaptive radiation.

FAMILY GNATHOSTOMATIDAE

Family Gnathostomatidae contains the genera *Tanqua,* found in reptiles, *Echinocephalus* in elasmobranchs, and *Gnathostoma* in the stomachs of carnivorous mammals (Fig. 28-3). These distinctive nematodes have two powerful, lateral lips, followed by a swollen "head," which is separated from the rest of the body by a constriction. Internally, four peculiar, glandular, cervical sacs, reminiscent of acanthocephalan lemnisci, hang into the coelom from their attachments near the anterior end of the esophagus. The head

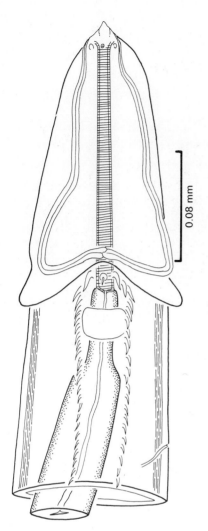

Fig. 28-1. *Cordonema venusta,* from the stomach of an aquatic bird (dipper). Note the helmet-like inflation of the cuticle, which bears two cordons on each side. The cordons of each side join at their posterior ends in this genus. (From Schmidt, G. D., and R. E. Kuntz. 1972. Parasitology 64:235-244.)

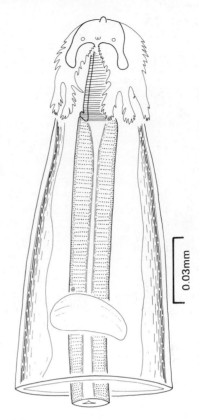

Fig. 28-2. *Sobolevicephalus chalcyonis* from the stomach of a kingfisher. The head cuticle has four feathered projections. (From Schmidt, G. D., and R. E. Kuntz. 1972. Parasitology 64:264-278.)

bulb is divided internally into four hollow areas called **ballonets.** Each cervical sac has a central canal, which is continuous with a ballonet. The functions of these organs are unknown.

Gnathostoma spp. are particularly interesting because of their widespread distribution and their peculiar biology and because they often cause disease in humans in some areas of the world. In the United States, *G. procyonis* is common in the stomachs of raccoons and opossums, while *G. spinigerum* has been reported from a wide variety of carnivores. *Gnathostoma doloresi* is common in pigs in the Orient (Fig. 28-4). Of the 20 or so species that have been described, *G. spinigerum* has most consistently been demonstrated as a cause of disease in humans, so we will examine this parasite in more detail.

Gnathostoma spinigerum

In 1836, Richard Owen discovered *Gnathostoma spinigerum* in the stomach wall of a tiger that had died in the London Zoo. Since then, it has been found in many kinds of mammals in several countries, although it is most common in southeast Asia.

Morphology. The body is stout and pink

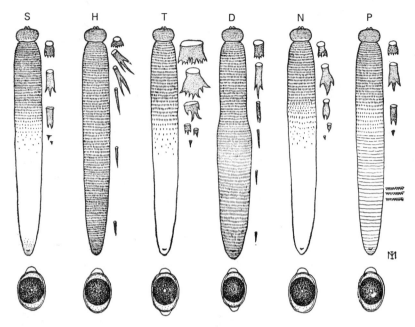

Fig. 28-3. Morphological comparison among six species of female *Gnathostoma:* **S,** *G. spinigerum;* **H,** *G. hispidum;* **T,** *G. turgidum;* **D,** *G. doloresi;* **N,** *G. nipponicum;* **P,** *G. procyonis.* This figure indicates the arrangement and shape of the cuticular spines and fresh fertilized uterine eggs, which at times may show various developmental stages when preserved. (From Miyazaki, I. 1966. In Morishita, K., Y. Komiga, and H. Mitsubayashi, editors. Progress of medical parasitology in Japan, vol. 3. Meguro Parasitological Museum, Tokyo.)

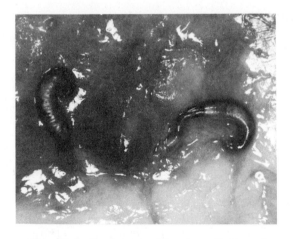

Fig. 28-4. *Gnathostoma doloresi* attached to the stomach mucosa of a pig. (Photograph by Robert E. Kuntz.)

in life. The swollen head bulb is covered with four circles of stout spines. The anterior half of the body is covered with transverse rows of flat, toothed spines, followed by a bare portion. The posterior tip of the body has numerous tiny cuticular spines.

Males are 11 to 31 mm long and have a bluntly rounded posterior end. The anus is surrounded by four pairs of stumpy papillae. The spicules are 1.1 mm and 0.4 mm long and are simple with blunt tips.

Females are 11 to 54 mm long and also have a blunt posterior end. The vulva is slightly postequatorial in postion. The eggs are unembryonated when laid, 65 to 70 µm by 38 to 40 µm in size, and have a polar cap at only one end. The outer shell is pitted.

Biology (Fig. 28-5). The eggs complete embryonation and hatch in about 1 week at 27° to 31° C.[7] The actively swimming larva is eaten by a cyclopoid copepod, where it penetrates into the hemocoel and

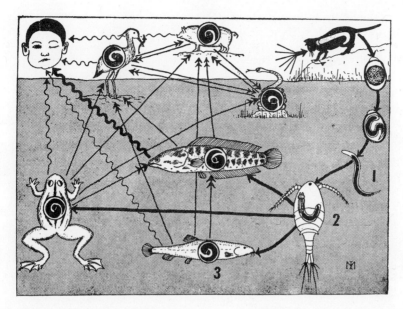

Fig. 28-5. The life history of *G. spinigerum: 1,* The first-stage larva (averaging 0.27 mm long) swimming in the water; *2,* the second-stage larva (averaging 0.5 mm long) parasitic in cyclops; *3,* encysted third-stage larva (averaging 3 to 4 mm long) in the muscle of the second intermediate host. Double arrow in the figure indicates "secondary infection." Humans are susceptible to the infection from a variety of second intermediate hosts. (From Miyazaki, I. 1966. In Morishita, K., Y. Komiya, and H. Matsubayashi, editors. Progress of medical parasitology in Japan, vol. 3. Meguro Parasitological Museum, Tokyo.)

develops further into a second-stage larva in 7 to 10 days. The second-stage larva already has a swollen head bulb covered with four transverse rows of spines.

When the infected crustacean is eaten by a vertebrate second intermediate host, the second-stage larva penetrates the intestine of its new host and migrates to muscle or connective tissue, where it molts to the third stage. The third stage is infective to a definitive host. But if it is eaten by the wrong host, it may wander in that animal's tissues without further development. More than 35 species of paratenic hosts are known, including crustaceans, freshwater fishes, amphibians, reptiles, birds, and mammals, including humans. The biology of this parasite in Japan was reviewed by Miyazaki.[8]

Adult worms are found embedded in tumor-like growths in the stomach wall of the definitive host. They begin producing eggs about 100 days after infection.

Epidemiology. Human infection results from eating a raw or undercooked intermediate or paratenic host containing third-stage larvae. In Japan this is most often a fish, while in Thailand, domestic duck and chicken are probably the most important vectors.[5] However, any amphibian, reptile, or bird may harbor larvae and thereby contribute an infection if eaten raw.

Pathology. In humans, the third-stage larvae usually migrate to the superficial layers of the skin, causing **gnathostomiasis externa.** They may become dormant in abscessed pockets in the skin, or they may wander, leaving swollen red trails in the skin behind them. This creeping eruption resembles larva migrans caused by hookworms or fly larvae.

If the worms remain in the skin with little wandering, they cause relatively little disease. Often, they will erupt out of the skin spontaneously. However, erratic migration may take them into an eye, the brain, or the spinal cord, with serious results that may cause death.

Diagnosis and treatment. Diagnosis depends on recovery and accurate identification of the worm. An intradermal test, using an antigen prepared from *Gnathostoma,* has been employed with success in Japan. Gnathostomiasis should be suspected in an endemic area when a localized edema is accompanied by leukocytosis with a high percentage of eosinophils.

Chemotherapy has not been effective against this zoonosis. The only effective treatment is surgical removal of the worm.

Prevention is the most realistic means of controlling this disease. Cooling by any means will kill the worms. In regions where ritualistic consumption of raw fish is an important tradition, the fish should be well frozen before preparation, or marine fish should be used. Consumption of raw, previously unfrozen fish in any area of the world is dangerous for a variety of parasitological reasons.

FAMILY PHYSALOPTERIDAE

Members of the family Physalopteridae are mostly rather large, stout worms that live in the stomachs or intestines of all classes of vertebtates. All have two large, lateral pseudolabia, usually armed with teeth. The head papillae are on the pseudolabia. The cuticle at the base of the lips is swollen into a "collar" in some genera. Caudal alae are well developed on males. Spicules are equal or unequal, and a gubernaculum is absent. This family has a tendency toward **polydelphy,** or many ovaries and uteri. Of the several genera in this family, we will briefly consider *Physaloptera.*

Physaloptera spp.

In the genus *Physaloptera,* the triangular pseudolabia are armed with varying numbers of teeth, and a conspicuous cephalic collar is present. The male has numerous pedunculated caudal papillae and caudal alae that join anterior to the anus. In a few species, the cuticle of the posterior end is inflated into a prepuce-like sheath, which encloses the tail. There are three species found in Amphibia, around 45 spe-

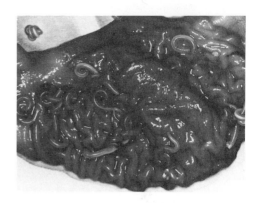

Fig. 28-6. *Physaloptera praeputialis* in the stomach of a domestic cat. (Photograph by Robert E. Kuntz.)

cies in reptiles, 24 in birds, and nearly 90 in mammals.

Physaloptera praeputialis lives in the stomachs of domestic and wild dogs and cats throughout the world (Fig. 28-6), except Europe. It is common in dogs, cats, coyotes, and foxes in the United States. A flap of cuticle covers the posterior ends of both sexes. Its life cycle is incompletely known, but development has been experimentally obtained in cockroaches.

Physaloptera rara is the most common physalopterid of carnivores in North America, to which it is apparently restricted. It is similar to *P. praeputialis* but lacks the posterior cuticular flap. The life cycle involves an insect intermediate host, usually a field cricket, in which it develops to the third stage. A paratenic host, such as a snake, is commonly necessary in the life cycle because of the feeding habits of the definitive hosts.

Physaloptera caucasica is the only species recorded from humans.[10] It is normally parasitic in African monkeys. Most recorded cases in humans were from Africa, although several records, some based only on eggs found in patients' feces, have been reported from South and Central America, India, and the Middle East. It is possible that some of these were misidentified.

The life cycle is unknown but most likely involves insect intermediate hosts and a vertebrate paratenic host. Humans may become infected by eating either of these hosts.

Symptoms include vomiting, stomach

pains, and eosinophilia. Tentative diagnosis can be made by demonstrating the eggs in a fecal sample or by obtaining an adult specimen for accurate identification.

FAMILY TETRAMERIDAE

Members of the family Tetrameridae are bizarre in their degree of sexual dimorphism. While males exhibit typical nematoid shape and appearance, the females are greatly swollen and often colored bright red. There are three genera in this family, all parasitic in the stomachs of birds. *Geopetitia* is represented by five rare species, which live in cysts on the outside of the proventriculus or gizzard of birds, where they communicate through the enteric lumen through a tiny pore. The posterior end of the female is distorted and swollen, while the anterior portion is normal. Both sexes are colorless.

The other two genera in the family are very common parasites, which live in the branched secretory glands of the proventriculus, although males can be found wandering throughout the organ.

Tetrameres (Fig. 28-7) is a large genus of about 50 species, which are mainly parasites of aquatic birds. A well-developed, sclerotized buccal capsule is present in both sexes. Males are typically nematoid in form, lacking caudal alae and possessing spicules that are vastly dissimilar in size. Lateral, longitudinal rows of spines are present on many species. Females, however, are greatly swollen, with only the front and back ends retaining the appearance of a nematode. In addition, females are blood-red in color, while the sac-like intestine is black. They are easily seen as reddish spots in the wall of the proventriculus, where they mature with the tail end near the lumen of that organ. The vulva is near the anus and thus is available to males who chance upon it. The eggs are embryonated when laid. The intermediate hosts are crustaceans or insects. Definitive hosts are water birds, chickens, and hawks.

The terrestrial counterpart of *Tetrameres* is *Microtetrameres* (Fig. 28-8). About 40 species have been described in this genus, all of which live in the proventricular glands of insectivorous birds. Females of this genus are also swollen and, in addition, are twisted into a spiral. Morphological and bi-

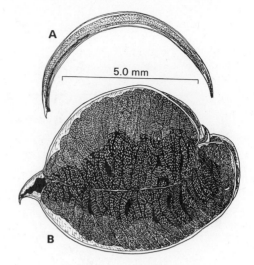

Fig. 28-7. *Tetrameres strigiphila* from owls. **A,** Male; **B,** female. (From Pence, D. B. 1975. J. Parasitol. 61:494-498.)

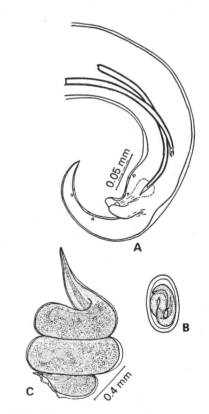

Fig. 28-8. *Microtetrameres aguila,* a parasite of the proventriculus of eagles. **A,** Posterior end of male, lateral view; **B,** embryonated egg; **C,** female. (From Schell, S. C. 1953. Trans. Am. Microsc. Soc. 72:227-236.)

ological characteristics are otherwise similar to those of *Tetrameres*, with terrestrial crustaceans and insects serving as intermediate hosts.

The quaint little worms in this family are familiar to all who survey the parasites of birds, for they are common in many species of hosts. Because of the difficulties of taxonomy of the group, only a small percentage of actual species have been described. The economic importance of these nematodes is slight. Even though 100 or more females are commonly embedded in the proventriculus of a single duck, for example, they seem to have little effect on the overall health of their host.

FAMILY GONGYLONEMATIDAE

Gongylonematidae contains the single genus *Gongylonema*, which has several species that are found in the upper digestive tracts of birds and mammals. Morphologically, they resemble several spiruriids, except that the cuticle of the anterior end is covered with large bosses, or irregular scutes, arranged in eight longitudinal rows (Fig. 28-9). Cervical alae are present, as are cervical papillae. The posterior end of the male bears wide caudal alae, which are supported by numerous pedunculated papillae.

Of the 25 or so species in this genus, *Gongylonema pulchrum* is probably the best known. Primarily a parasite of ruminants and swine, the worm has also been reported from monkeys, hedgehogs, bears, and humans.[6,9] It has been demonstrated experimentally that the life cycle involves an insect intermediate host, either a dung beetle or a cockroach. While these would appear rather unpalatable fare for people, they have been ingested often enough, for numerous cases of human gongylonemiasis have been reported. In normal hosts, the worms invade the esophageal epithelium, where they burrow stitch-like in shallow tunnels. In an abnormal host, such as humans, they behave similarly but do not mature and seem to wander further, being found often in the epithelium of the tongue, gums, or buccal cavity. Their active movements, together with resulting irritation and bleeding, soon make their presence known. Treatment is surgical removal of worms that can be seen. Chemotherapy is seldom employed.

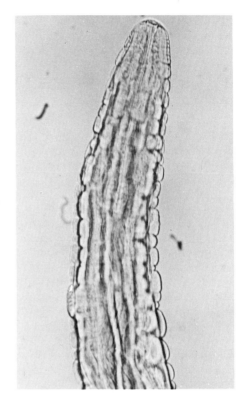

Fig. 28-9. Anterior end of *Gongylonema*, demonstrating the cuticular bosses typical of the genus. (Photograph by Warren Buss.)

Gongylonema neoplasticum and *G. orientale* in domestic rats are thought to induce neoplastic tumors. They are not known to infect humans.

FAMILY SPIROCERCIDAE

The family Spirocercidae is closely related to Spiruridae and *Cyathospirura*; all are parasites of mammals. Of these, *Spirocerca lupi* is the most interesting, because of its complex life cycle and its relationship to esophageal cancer in dogs.[1,2]

Spirocerca lupi

These stout worms are bright pink to red in color when alive. The mouth is surrounded by six rudimentary lips, and the buccal capsule is well developed, with thick walls. A short muscular portion of the esophagus is followed by a longer glandular portion. Males are 30 to 54 mm long, with a left spicule 2.45 to 2.8 mm long and a right spicule 475 to 750 μm long. Females are 50 to 80 mm long, with the vulva 2 to 4 mm from the anterior end. The eggs are cylindrical and embryonated when laid.

Biology. Adults are normally found in clusters, entwined within the wall of the upper digestive tract of dogs, although they have been reported in other organs, mostly the dorsal aorta, and in a wide variety of carnivorous hosts. Hounds seem to be the breeds most frequently infected in the United States. This is probably related to their opportunities for exposure rather than to a breed susceptibility.

Embryonated eggs pass out of the host with its feces. Any of several species of scarabeid dung beetles can serve as intermediate host. A wide variety of paratenic hosts are known, including birds, reptiles, and other mammals. Dogs can become infected by eating dung beetles or infected paratenic hosts. Domestic dogs are probably most often infected by eating the offal of chickens that have third-stage larvae encysted in their crops.

Once in the stomach of a definitive host, the larvae penetrate its wall and enter the wall of the gastric artery, migrating up to the dorsal aorta and forward to the area between the diaphragm and the aortic arch. They remain in the wall of the aorta for $2^1/_2$ to 3 months, after which they emerge and migrate to the nearby esophagus, which they penetrate. After establishing a passage into the lumen of the esophagus, they move back into the submucosa or muscularis where they complete their development about 5 to 6 months after infection. Eggs pass into the esophageal lumen through the tiny passage formed by the worm.

Many worms get lost during migration and may be found in abnormal locations, including lung, mediastinum, subcutaneous tissue, trachea, urinary bladder, and kidney.

Epidemiology. This parasite is most common in warm climates, but has been found in Manchuria and northern Russia. Many questions are still unanswered, such as: Why is the parasite distributed so sporadically throughout the United States and the world? What factors influence the change in prevalence of the infection in a given area? The attractiveness of dog feces to susceptible beetle species is a factor, as is the application of pesticides in an endemic area. Certainly, the successful transfer of third-stage larvae from one paratenic host to another increases the parasite's chances for survival.

Pathology. When the third-stage larva penetrates the mucosa of the stomach, it causes a small hemorrhage in the area. This irritation quite commonly causes the dog to vomit. The lesions in the aorta caused by the migrating worms are often quite severe, with hemorrhage accounting for death in some dogs with heavy infections. Destruction of tissues in the wall of the aorta with subsequent scarring is typical of this disease and may lead to numerous aneurysms.

Worms that leave the aorta and migrate upward come in contact with the tissues surrounding the vertebral column, where they frequently cause, by a mechanism not yet elucidated, a condition known as **spondylosis.** This deformation may be so severe as to cause adjacent vertebrae to fuse. Hypertrophic pulmonary osteoathropathy, with inflamed and swollen joints, is a common sequel to this disease (Fig. 28-10).

The most striking lesion associated with spirocercosis is in the wall of the esophagus, where the worm matures. Here, their presence stimulates the formation of a **reactive granuloma,** made up of fibroblasts. The granulomas are rather more loosely organized than is usually the case in granulomatous reactions, and the cell structure is characteristic of incipient **neoplasia** (cancer). Some of the granulomas change to **sarcomas,** true cancerous growths (Fig. 28-11). The worms may continue to live inside these tumors for some time, or they may be extruded or compressed and killed by the rapidly growing tissue. The precise oncogenic factor responsible for stimulation of neoplasia is still unknown. While several other helminths have been thought to be associated with malignancy, in no other instance is there as strong evidence for a cause-effect relationship as in canine spirocercosis.

The only known case of spirocercosis in humans was that of a fatal, prenatal infection reported in Italy.[4] The baby was born prematurely and died 12 days later. Mature worms were found in the wall of the terminal ileum. The mother may have become infected by eating a coprophagous beetle or undercooked chicken.

Diagnosis and treatment. Diagnosis in

Fig. 28-10. Hound with severe hypertrophic pulmonary osteoarthropathy associated with esophageal sarcoma. (From Bailey, W. S. 1963. Ann. N.Y. Acad. Sci. 108:890-923.)

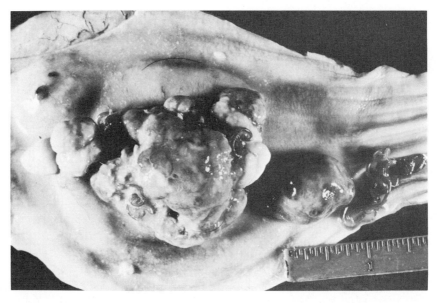

Fig. 28-11. An esophageal sarcoma associated with *Spirocerca lupi* infection. Pedunculated masses protruding into the lumen; adult *S. lupi* are partially embedded in the neoplasm. (From Bailey, W. S. 1972. J. Parasitol. 58:3-22. Photograph courtesy Department of Pathology and Parasitology, School of Veterinary Medicine, Auburn University, Auburn, Ala.)

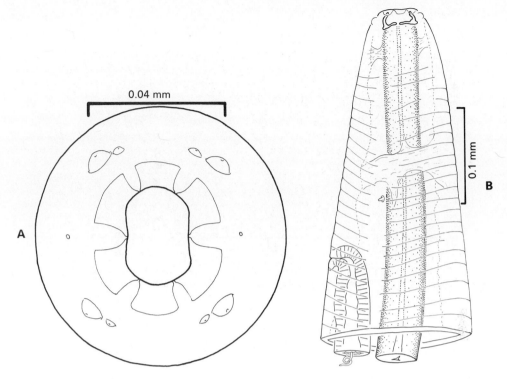

Fig. 28-12. *Thelazia digiticauda* from under the nictitating membrane of the eye of a kingfisher. **A,** En face view; **B,** female, lateral view of anterior end. (From Schmidt, G. D. and R. E. Kuntz. 1971. Parasitology 63:91-99.)

dogs is usually by demonstrating the characteristic eggs, which measure 40 μm by 12 μm and have nearly parallel sides, in a fecal examination. At necropsy, aortic scarring and aneurysms and esophageal granuloma or sarcoma are considered diagnostic, even if worms are no longer present.

Disophenol is effective against *Spirocerca lupi,* but if extensive aneurysms or a sarcoma have already developed, the treatment will not affect these conditions.

FAMILY THELAZIIDAE

Members of the family Thelaziidae live on the surface of the eye in birds and mammals, usually remaining in the lacrimal ducts or conjunctival sacs or under the nictitating membrane. Most are parasites of wild animals, but two species of *Thelazia* have been reported from humans.

These worms lack lips but show evidence of the primitive condition in having a hex-agonal mouth (Fig. 28-12, *A*). The buccal capsule is well developed, with thick walls. Alae and cuticular ornamentations are absent, except for conspicuous transverse striations near the anterior end (Fig. 28-12, *B*). These are deep, and their overlapping edges ostensibly aid movements across the smooth surface of the cornea.

Thelazia callipaeda is a parasite of dogs and other mammals in southeast Asia, China, and Korea, and *Thelazia californiensis* parasitizes deer and other mammals in western North America. Both species have been reported from humans several times.[3]

Little is known of the biology of these worms except that filth flies (*Musca* and *Fannia*) are capable of serving as intermediate hosts. Probably, when an infected fly is swallowed, the third-stage larvae migrate up the esophagus to the pharynx and then up the lacrimal ducts to the orbits.

REFERENCES

1. Bailey, W. S. 1963. Parasites and cancer: sarcoma in dogs associated with *Spirocerca lupi*. Ann. N.Y. Acad. Sci. 108:890-923.
2. Bailey, W. S. 1971. *Spirocerca lupi:* a continuing enquiry. J. Parasitol. 58:3-22.
3. Bhaibulaya, M., S. Prasertsilpa, and S. Vajrasthira. 1970. *Thelazia callipaeda* Railliet and Henry, 1910, in man and dog in Thailand. Am. J. Trop. Med. Hyg. 19:476-479.
4. Biocca, E. 1959. Infestazione umana prenatale da *Spirocerca lupi* (Rud. 1809). Parassitologia 1:137-142.
5. Daengsvang, S., P. Thienprasitthi, and P. Chomcherngpat. 1966. Further investigations on natural and experimental hosts of larvae of *Gnathostoma spinigerum* in Thailand. Am. J. Top. Med. Hyg. 15:727-729.
6. Feng, L. C., M. S. Tung, and S. C. Su. 1955. Two Chinese cases of *Gongylonema* infection. A morphological study of the parasite and clinical study of the case. Chin. Med. J. 73:149-162.
7. Miyazaki, I. 1954. Studies on *Gnathostoma* occurring in Japan (Nematoda: Gnathostomidae). II. Life history of *Gnathostoma* and morphological comparison of its larval forms. Kyushu Mem. Med. Sci. 5:123-140.
8. Miyazaki, I. 1966. Gnathostoma and gnathostomiasis in Japan. In Morishita, K., Y. Komiya, and H. Matsubayashi, editors. Progress of medical parasitology in Japan, vol. 3. Meguro Parasitological Museum, Tokyo. pp. 529-586.
9. Thomas, L. J. 1952. *Gongylonema pulchrum,* a spirurid nematode infecting man in Illinois, U.S.A. Proc. Helm. Soc. Wash. 19:124-126.
10. Vandepitte, G., J. Michaux, J. L. Fain, and F. Gatti. 1964. Premieres observations congolaises de physaloptérose humaine. Ann. Soc. Belg. Med. Trop. 44:1067-1076.

SUGGESTED READING

Chabaud, A. G. 1954. Valeur des charactèrs biologiques pour la systématique des nématodes spirurides. Vie Milieu 5:299-309.

Chabaud, A. G. 1975. Keys to the order Spirurida, part 2. In Anderson, R. C., A. G. Chabaud, and S. Willmott, editors. CIH keys to the nematode parasites of vertebrates. Commonwealth Agricultural Bureaux, Bucks.

Chitwood, B. G., and E. E. Wehr. 1934. The value of cephalic structures as characters in nematode classification, with special reference to the superfamily Spiruroidea. Zeit. Parasitol. 7:293-335.

Skrjabin, K. I. 1949. Key to parasitic nematodes, vol. 1. Spirurata and Filariata. Akad. Nauk SSSR, Moscow. (English translation, 1968.) (Useful keys to genera with lists of species.)

Skrjabin, K. I., A. A. Sobolev, and V. M. Ivaskin. 1963-1967. Essentials of Nematodology, vols. 11, 12, 14, 16, and 19. Spirurata of animals and man and the diseases they cause. Akad. Nauk SSSR, Moscow. (The most complete monographs on the subject.)

Chapter 29

ORDER CAMALLANATA: GUINEA WORMS AND OTHERS

The order Camallanta appears to be transitional between the Spirurata, which are primarily parasites of the digestive tract, and the Filariata, mainly parasites of tissues. The morphology of several families, such as the Dracunculidae, is strikingly filariid, while that of others, such as Camallanidae and Cucullanidae, is quite spirurid. All but Cucullanidae are ovoviviparous, like most filariids, but the arthropod intermediate host must be eaten to complete transmission, as in the spirurids. Most species are uncommon parasites of economically unimportant vertebrate hosts and, therefore, will not be discussed here. Two families, Camallanidae and Philometridae, are commonly encountered in fishes, and a third, Dracunculidae, has a species of great medical importance to humans. These three families will serve to illustrate the order.

FAMILY CAMALLANIDAE

Included in the family Camallanidae are several similar genera that inhabit the intestines of fishes, amphibians, and reptiles. Their most conspicuous character is the head, in which the buccal capsule has been replaced with a pair of large bilateral, sclerotized valves (Fig. 29-1). The complex ornamentation of these valves (Fig. 29-2) is a useful taxonomic character.

The genus *Camallanus* is common in freshwater fishes and turtles in the United States. *Camallanus oxycephalus* is often seen as a bright red worm extending from the anus of a crappie or other warmwater panfish. The life cycles of all species that have been investigated involve a cyclopoid copepod crustacean as intermediate host. Development proceeds to maturity in the intestine of the vertebrate with no tissue migration.

FAMILY PHILOMETRIDAE

Two common genera in the family Philometridae are *Philometra,* with a smooth cuticle, and *Philometroides,* with a cuticle covered with bosses. Each is a tissue parasite of fishes. The mouth is small, there is no sclerotized buccal capsule, and the esophagus is short. Males of many species are unknown. Gravid females live under the skin, in the swimbladder, or in the coelom of fishes, where they release first-stage larvae. After reaching the external environment, the larvae develop further if they are eaten by a cyclopoid crustacean. The microcrustacean, containing third-stage larvae, must be eaten by the definitive host if the worm is to survive. Development to the adult is not well known. Males and females mate in the deep tissues of the body, and the males die soon after. The females then migrate to their definitive site, where the young are released. *Philometra oncorhynchi,* a parasite of salmon in the western United States and Canada, apparently passes out with the fish's eggs when it spawns, bursts in the fresh water, and, thus, releases its larvae.[10]

Philometroides, under the skin of the head and fins of suckers (Catostomidae), are familiar sights to those who work with these fishes in the United States (Fig. 29-3).

FAMILY DRACUNCULIDAE

Members of the family Dracunculidae are tissue-dwellers of reptiles, birds, and mammals. All have life cycles involving aquatic intermediate hosts. Morphological characteristics of the several genera and species are remarkably similar, with small differences between those in reptile hosts, for example, and those in mammals.

Several species of *Dracunculus* are known from snakes, and one is common

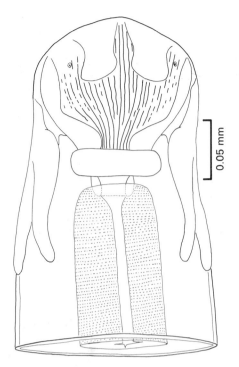

Fig. 29-1. The head of *Camallanus marinus,* lateral view. In this genus the buccal cavity is replaced by large, sclerotized valves with various markings. (From Schmidt, G. D. and R. E. Kuntz. 1969. Parasitology 59:389-396.)

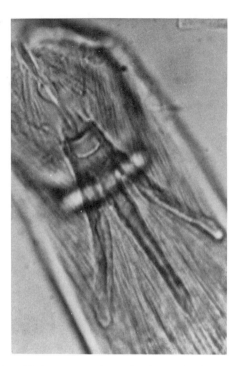

Fig. 29-2. Dorsal view of the head of *Camallanus marinus,* showing the large, sclerotized trident characteristic of this genus.

Fig. 29-3. *Philometroides* sp. in the skin of a fin of a white sucker, *Catostomus commersoni.* (Photograph by John S. Mackiewicz.)

in snapping turtles in the United States. *Micropleura* is found in crocodilians and turtles in South America and India, while *Avioserpens* has species in aquatic birds.

Dracunculus is also known from mammals. In the Americas, a species known as *Dracunculus insignis* is common in muskrats, opossums, and raccoons and other carnivores, especially those occupying semiaquatic environments. *Dracunculus medinensis* is prevalent in circumscribed areas of Africa, India, and the Middle East. It has been reported from humans in the United States several times, but these cases may have been caused by *Dracunculus insignis*. In fact, *D. insignis* may well be *D. medinensis,* perhaps in a form attenuated in its pathogenicity to humans. This might explain the scarcity of reports of it in humans in this country. It has been shown to be infective to rhesus monkeys.[2]

However, *D. medinensis* is not scarce in humans in all countries of the world, so we will examine it in greater detail.

Dracunculus medinensis

Dracunculus medinensis has been known since antiquity, particularly in the Middle East and Africa, where it causes great suffering even today. Because of its large size and the conspicuous effects of infection, it is not surprising that the parasite was mentioned by classical authors. The Greek, Agatharchidas of Cnidus, who was tutor to one of the sons of Ptolemy VII in the second century B.C., gave a lucid description of the disease: " . . . the people taken ill on the Red Sea suffered many strange and unheard of attacks, amongst other worms, little snakes, which came out upon them, gnawed away their legs and arms, and when touched retracted, coiled themselves up in the muscles, and there gave rise to the most unsupportable pains."[4] The Greek and Roman writers Paulus Aegineta, Soranus, Aetius, Actuarus, Pliny, and Galen all described the disease, although most of them probably never saw an actual case. The Arabian scholars Avicenna, Avenzoar, Rhazes, and Albucasis also discussed this parasite, probably from firsthand observations. In 1674 Velschius described winding the worm out on a stick as a cure. European parasitologists remained ignorant of this

worm until about the beginning of the nineteenth century, when British army medical officers began serving in India. Information about *Dracunculus* slowly accumulated, but it remained for a young Russian traveler and scientist, Aleksej Fedchenko, to give the first detailed account of the morphology and life cycle of the worm in 1869 to 1870.[5] His discovery that humans become infected by swallowing infected *Cyclops* pointed the way to a means of prevention of dracunculiasis. However, the disease is still common in some areas today, and certain details of the worm's biology are still unknown. An excellent review of this parasite is given by Muller.[9]

Morphology. *Dracunculus medinensis* is one of the largest nematodes known. Adult females have been recorded up to 800 mm long, although the few males known do not exceed 40 mm. The mouth is small and triangular and is surrounded by a quadrangular, sclerotized plate. Lips are absent. Cephalic papillae are arranged in an outer circle of four double papillae at about the same level as the amphids and an inner circle of two double papillae, which are peculiar in that they are dorsal and ventral (Fig. 29-4). The esophagus has a large glandular portion that protrudes and lies alongside the thin muscular portion.

In the female, the vulva is about equatorial in young worms; it is atrophied and nonfunctional in adults. The gravid uterus has an anterior and a posterior branch, each of which is filled with hundreds of thousands of embryos. The intestine becomes squashed and nonfunctional as the result of the pressure of the uterus.

A major difficulty in the taxonomy of dracunculids is the sparsity of discovered males. The few specimens known range from 12 to 40 mm long; the spicules are unequal and 490 to 730 μm long. The gubernaculum ranges from 115 to 130 μm long. Genital papillae vary considerably in published descriptions. In fact, in monkeys, at least, males taken from a single animal have varying numbers of papillae. It is quite possible that more than one species is responsible for dracunculiasis, or there may be a complex of subspecies. Because of the technical difficulties of obtaining many specimens, it remains for an

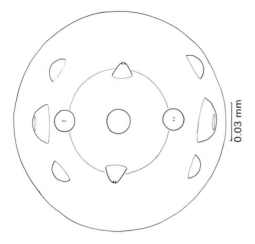

Fig. 29-4. En face view of *Dracunculus,* showing the arrangement of papillae and amphidial pores.

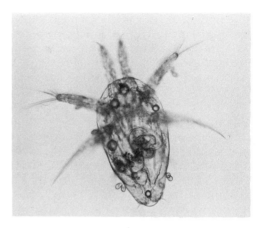

Fig. 29-6. A living nauplius of *Cyclops vernalis* with a larva of *Dracunculus medinensis* in its hemocoel. (Photograph by Ralph Muller.)

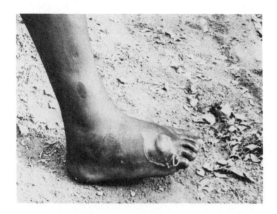

Fig. 29-5. A blister, caused by a female *Dracunculus medinensis,* in the process of bursting. There has been an unusually severe tissue reaction resulting in a very large blister. A loop of the worm can be seen protruding through the skin. (From Muller, R. 1971. In Dawes, B., editor. Advances in parasitology, vol. 9. Academic Press, Inc., New York.)

experimental approach to solve the taxonomy of the species.

Biology. *Dracunculus medinensis* is ovoviviparous. When gravid, the thousands of embryos in the uteri cause a high internal pressure. At this stage, the female has migrated to the skin of the host. Usually the legs and feet are infected, but nearly any portion of the body is susceptible. Internal pressure and progressive senility cause the body wall and uterus of the parasite to burst, forcing a loop of the uterus through, freeing many larvae. These larvae cause a violent immune reaction that causes a blister in the skin of the host (Fig. 29-5). This eventually ruptures, forming an exit for the embryos, which trickle out onto the surface of the skin. Sometimes, instead of the body wall rupturing, the uterus forces itself out of the mouth of the worm. Muscular contractions of the body wall force embryos out in periodic spurts, with over half a million ejected at a single time. These contractions are instigated by cool water, which causes the worm and its uterus to protrude through the wound. As portions of the uterus empty, they disintegrate and adjacent portions move into the ulcer. Eventually all of the worm will be "used up," and the wounds will heal.

The first-stage larva must enter directly into water, after leaving its mother and host, in order to survive. It can live for 4 to 7 days but is able to infect an intermediate host for only 3 days. To develop further, they must be eaten by cyclopoid crustaceans. Once in the intestine of their new host, the juveniles penetrate into the hemocoel, especially dorsad to the gut, where they develop to the infective third stage in 12 to 14 days at 25° C (Fig. 29-6).

Infection of the definitive host is effected when infected copepods are swallowed with drinking water. The released larvae penetrate the duodenum, cross the abdominal mesenteries, pierce the abdominal muscles, and enter the subcutaneous connective tissues, where they migrate to the axillary and inguinal regions. The third molt occurs about 20 days postinfection, and the final one at about 43 days. Females are fertilized by the third month postinfection. Males die between the third and seventh months, become encysted, and degenerate. Gravid females migrate to the skin of the extremities between the eighth and tenth months, by which time the embryos are fully formed. Between 10 to 14 months after initial infection, the female causes a blister in the skin.

Little is known about the physiology of this parasite, but the gut is often filled with a dark brown material, suggesting that the worms feed on blood. Glycogen is stored in several tissues of the mature female. Glucose utilization and the rate of formation of lactic acid are not affected by the presence or absence of oxygen.[3] The blister formation in the definitive host is an immunological response to parasite antigens.

Epidemiology. To become infected, a person must swallow a copepod that had been exposed to larvae previously released from the skin of a definitive host. Thus, three conditions must be met before the parasite's life cycle can be completed: the skin of an infected individual must come in contact with water, the water must contain the appropriate species of microcrustaceans, and the water must be used for drinking. It is curious that a parasite life cycle that is so dependent on water is most successfully completed under conditions of drought.

In some areas of Africa, for instance, people depend on rivers for their water. During periods of normal river flow, few or no new cases of dracunculiasis occur. During the dry season, however, rivers are reduced to mere trickles with occasional deep pools, which are sometimes enlarged and deepened by those who depend on them as a water source (Fig. 29-7). Planktonic organisms flourish in this warm, semistagnant water, and a cyclopean population explosion occurs. At the same time, any bathing, washing, and water-drawing bring infected persons in contact with the

Fig. 29-7. A pond in the Mabauu area of Sudan, in the Sahel savannah zone. Conditions such as these favor the transmission of the guinea worm. (Photograph by J. Bloss; from Muller, R. 1971. In Dawes, B., editor. Advances in parasitology, vol. 9. Academic Press, Inc., New York.)

water, where larvae are shed. When such water is drunk, many infected copepods may be downed at a quaff.

In areas of India, the step well (Fig. 29-8) is a time-honored method of exposing ground water. These wells, often centuries old, have steps leading into the water, upon which water-bearers enter the well to fill their jars and, incidentally, release larvae into the water at the same time.

In many desert areas, the populace depends on deep wells, which are crustacean-free, during the dry season. Most villages also have one or more ponds that fill during the rainy season and become a source of infection with *Dracunculus*. Most villagers prefer the pond water because they have to pay for well water, and, moreover, the well water is usually saline.

Fig. 29-8. A step well at Kantarvos, near Kherwara, India, infected with *Dracunculus medinensis.* (Photograph by A. Banks; from Muller, R. 1971. In Dawes, B., editor. Advances in parasitology, vol. 9. Academic Press, Inc., New York.)

With these examples in mind, it is no wonder that a parasite with an aquatic life cycle should thrive in a desert environment, for all animals, humans and beasts alike, depend on isolated waterholes for their existence. So does *Dracunculus medinensis.*

Pathogenesis. Dracunculiasis may result in three major disease conditions: the emergence of adult worms, secondary bacterial infection, and nonemergent worms.

At the onset of migration to the skin, the female worm elicits an allergic reaction caused by the release of metabolic wastes into the host's system. The reaction may produce a rash, nausea, diarrhea, dizziness, and localized edema. The worms remain just under the skin for about a month before a reddish papule develops.

This rapidly becomes a blister, caused mainly by larvae escaping from a rupture in the worm. The feet and legs are most often affected, although the blister may appear nearly anywhere on the surface of the body. On rupture of the blister, the allergic reactions usually subside. The site of the blister becomes abscessed, but this heals rapidly if serious secondary complications do not occur. A tiny hole remains, through which the worm protrudes. When the worm is removed or is expelled, healing is completed.

Serious complications can result from the introduction of bacteria under the skin by the retreating worm. In parts of Africa, this is the third most common mode of entry of tetanus spores.[6] Other complications are abscesses, synovitis, arthritis, bubo, and other infections.

Worms that fail to reach the skin often cause complications in deeper tissues of the body, although many die and are absorbed or calcified, with no apparent affect on the host. Chronic arthritis, with a calcified worm in or alongside the joint, is common. More serious symptoms, such as paraplegia, result from a worm in the central nervous system. Adult worms have also been found in the heart and urogenital system.

Commonly, when worms do not emerge, they eventually begin to degenerate and release powerful antigenic substances. These cause aseptic abscesses, which can also lead to arthritis. These abscesses can be large, with up to half a liter of fluid containing leukocytes and, frequently, numerous embryos. Usually, however, they become calcified.

Diagnosis and treatment. The appearance of an itchy, red papule that rapidly transforms into a blister is the first strong symptom of dracunculiasis. On a few occasions, the patient can feel or see the worm in the skin before papule formation. After the blister ruptures, embryos can be obtained by placing cold water on the wound; when mounted on a slide they can be seen actively moving about under a low power microscope.

When a part of the worm emerges, diagnosis is fairly evident, although the drying, disintegrating worm does not show the typical morphology of a nematode. An oc-

Fig. 29-9. An ancient woodcut showing removal of a guinea worm by winding it on a stick. (Velschius, 1674.)

casional sparganum may be diagnosed as dracunculiasis. Immunological tests show promise but are not yet perfected.

Pulling out guinea worms by winding them on a stick is a treatment used successfully since antiquity (Fig. 29-9).

And they journeyed from Mount Hor by way of the Red Sea, to compass the land of Edom . . .

And the Lord sent fiery serpents among the people, and they bit the people; and much people of Israel died . . . And the Lord said unto Moses, "Make thee a fiery serpent and set it upon a pole: and it shall come to pass that everyone that is bitten, when he looketh upon it, shall live."

NUMBERS 21:6

This exerpt from the Old Testament is a pretty fair account of dracunculiasis and its treatment. A serpent on a pole and a worm on a stick are not all that different, after all. Moses and his people were, at the time, near the Gulf of Akaba, where *Dracunculus* is still endemic. Also, the Israelites had for some time been in a drought area, existing on water where they could find

it. This is consistent with the epidemiology of dracunculiasis.

The staff with serpents carried by Aesculapius, the Roman god of medicine, adopted today as the official symbol of medicine (and the double-serpent caduceus of the military), may well depict the removal of *Dracunculus* (Fig. 29-10). This form of cure is still widely used (Fig. 29-11). If cold water is applied to the worm, she will expel enough embryos to allow about 5 cm of her body to be pulled out. The procedure is repeated once a day, complete removal requiring about 3 weeks. In some areas of India, the worms are said to be sucked out by native doctors using a crude aspirator!

An alternate method is removal of the complete worm by surgery. This is often successful when the entire worm is near the skin and also in the case of deep abscesses containing worms that failed to reach the skin. However, if the worm is threaded through a tendon or deep fascia

Fig. 29-10. The seal of the American Medical Association and the double-serpent caduceus of the military medical profession. Might the serpent on a staff originally have depicted the removal of guinea worm?

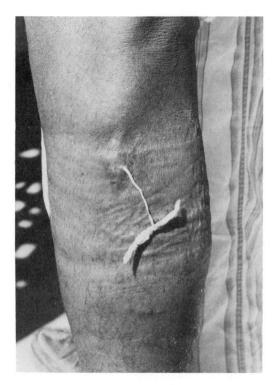

Fig. 29-11. An uncomplicated case of dracunculiasis. The worm is being pulled out through a small hole left after the ulcer is mostly healed. (From Muller, R. 1971. In Dawes, B., editor. Advances in parasitology, vol. 9. Academic Press, Inc., New York.)

or is broken into several pieces, it may be impossible to remove completely.

Several chemotherapeutic agents show promise against this disease organism, particularly niridazole, thiabendazole, and metronidazole.[1] After a short time, the worms are spontaneously ejected, or they may be pulled out with a minimum of effort. Muller found that worms treated with these drugs in vivo showed no histological changes that would indicate that the drugs killed them.[8] As these drugs have an anti-inflammatory effect, he suggested that this may be the sole factor in expelling the worms. He supported this idea by applying 2.5% hydrocortisone cream, an anti-inflammatory, to new blisters, and the worms were pulled out easily after 5 days, leaving little tissue reaction behind. Larvae in the uterus were still infective to copepods.

Prevention and control lie in interrupting the life cycle. The most logical means of doing this is by chemical treatment of ponds and wells to eliminate copepods. On a cost/effectiveness basis, DDT would probably be the most effective compound. Those who fear DDT will have to pay a higher price, either by using more expensive compounds or by living with *Dracunculus medinensis*. One compound that shows promise, among others, is Abate (O,O,O',O'-tetramethyl O,O'-thio-di-p-phenylene phosphorothioate).[7]

REFERENCES

1. Antani, J., H. V. Srinivas, K. R. Krishnamurthy, and B. R. Jahagirdar. 1970. Metronidazole in dracunculiasis. Am. J. Trop. Med. Hyg. 19:821-822.
2. Beverly-Burton, M., and V. F. J. Chrichton. 1973. Identification of guinea-worm species. Trans. R. Soc. Trop. Med. Hyg. 67:152.
3. Bueding, E., and J. Oliver-Gonzalez. 1950. Aerobic and anaerobic production of lactic acid by the filarial worm *Dracunculus insignis*. Br. J. Pharmacol. Chemother. 5:62-64.
4. Cobbold, T. S. 1864. Entozoa. Groombridge, London.
5. Fedchenko, A. P. 1870. Concerning the structure and reproduction of the guinea worm *(Filaria medinensis)*. Proc. Imp. Soc. Friends Nat. Sci., Anthropol., Ethnograph. 8:columns 71-81. (Translation Am. J. Trop. Med. Hyg. 20:511-523.)
6. Lauckner, T. R., A. M. Rankin, and F. C. Adi. 1961. Analysis of medical admissions to University College Hospital, Ibadan. W. Afr. Med. J. 10:3.
7. Muller, R. 1970. Laboratory experiments on the control of *Cyclops* transmitting guinea worm. Bull. WHO 42:563-567.
8. Muller, R. 1971a. The possible mode of action of some chemotherapeutic agents in guinea worm disease. Trans. R. Soc. Trop. Med. Hyg. 65:843-844.
9. Muller, R. 1971b. *Dracunculus* and dracunculiasis. In Dawes, B., editor. Advances in parasitology, vol. 9. Academic Press, Inc., New York. pp. 73-151.
10. Platzer, E. G., and J. R. Adams. 1967. The life history of a dracunculoid *Philonema onchorhynchi*, in *Onchorhynchus nerka*. Can. J. Zool. 45:31-43.

SUGGESTED READING

Chabaud, A. G. 1975. Keys to genera of the order Spirurida, part 1. In Anderson, R. C., A. G. Chabaud, and S. Willmott, editors. CIH keys to the nematode parasites of vertebrates. Commonwealth Agricultural Bureaux, Bucks.

Ivashkin, V. N., A. A. Sobolev, and L. A. Hromova. 1971. Essentials of nematodology, vol. 22. Camallanata of animals and man and the diseases they cause. Akad. Nauk SSSR, Moscow. (A most valuable reference to all species in this order.)

Chapter 30

ORDER FILARIATA: THE FILARIAL WORMS

ORDER FILARIATA

The filariids are tissue-dwelling parasites, with the exception of the Diplotriaenidae, which live in the air sacs of birds. They are among the most highly evolved of the parasitic nematodes, appearing to have arisen from the Camallanata by way of the Spirurata. All species utilize arthropods as intermediate hosts, most of which deposit third stage larvae on the skin with their bite (Fig. 30-1). They are parasitic in all classes of vertebrates except fishes. Generally speaking, filariids are slender worms with reduced lips and buccal capsule. Most are parasites of wild animals, especially of birds, but several are very important disease organisms of humans and domestic animals. The majority of these belong to the large family Onchocercidae.

FAMILY ONCHOCERCIDAE

Members of the family Onchocercidae live in the tissues of amphibians, reptiles, birds, and mammals. Most are of no known medical or economic importance, but a few cause some of the most tragic, horrifying, and debilitating diseases in the world today. Of these, species of *Wuchereria, Brugia, Onchocerca,* and *Loa* will be considered in detail. Short mention will be made of others.

Wuchereria bancrofti

Perhaps the most striking disease of humans is the clinical entity known as **elephantiasis** (Fig. 30-2). The horribly swollen parts of the body afflicted with this condition have been known since antiquity. The ancient Greek and Roman writers likened the thickened and fissured skin of infected persons to that of the elephant, although they also confused leprosy with this condition. Actually, elephantiasis is a nonsense word, for literally translated it means "a condition caused by elephants."

The word is so deeply entrenched, however, that it is not likely ever to be abandoned. Classic elephantiasis is a rather rare consequence of infection by *Wuchereria bancrofti* and by at least two other species of filariids.

Infection by *W. bancrofti* and other filariids is best referred to as filariasis. **Bancroftian filariasis** *(W. bancrofti)* is the most widespread of the filariases of humans, extending throughout central Africa, the Nile Delta, Turkey, India and southeast Asia, the East Indies, the Philippine and Oceanic Islands, Australia, and parts of South America; in short, across a broad equatorial belt. It was probably brought to the New World by the slave trade. A nidus of infection remained in the vicinity of Charleston, South Carolina, until it died out spontaneously in the 1920s.

Filariasis was a cause of great psychological concern to American armed forces in the Pacific theater in World War II. Although thousands of cases of filariasis were in fact contracted by American servicemen, no single case of classical elephantiasis resulted. Some individuals experienced symptoms for up to 16 years.[9]

The history of the evolution of knowledge of this disease and its cause remains one of the classics of medical history.[4] Two species are known in the genus.

Morphology. Adult worms are long and slender with a smooth cuticle and bluntly rounded ends. The head is slightly swollen and bears two circles of well-defined papillae. The mouth is small; a buccal capsule is lacking.

The male is about 40 mm long and 100 μm wide. Its tail is finger-like. The female is 6 to 10 cm long and 300 μm wide. The vulva is near the level of the middle of the esophagus.

Biology. Adult *Wuchereria* live in the major lymphatic ducts of humans, tightly coiled into nodular masses. They are nor-

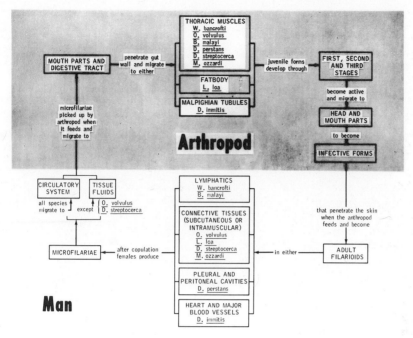

Fig. 30-1. Filarioid diseases. Relationships of life cycles of filarial parasites to mode of transmission. (AFIP neg. no. 67-19037-1.)

mally found in the afferent lymph channels near the major lymph glands in the lower half of the body. Rarely, they invade a vein. The females are ovoviviparous, producing thousands of larvae known as **microfilariae** (Fig. 30-3). Microfilariae are not as differentiated as normal first-stage larvae and are sometimes considered advanced embryos. The microfilariae of *W. bancrofti* retain the egg membrane as a "sheath" (not to be confused with the sheath of some third-stage strongyle larvae, which is the second-stage cuticle). The sheath is rather delicate and close-fitting but can be detected where it projects at the anterior and posterior ends of the microfilaria. When stained, several internal nuclei and primordia of organs can be seen in the microfilariae. The location of these and the presence or absence of a sheath are used to identify the several species of microfilariae found in humans.

The microfilariae are released into the surrounding lymph by the female. Some may wander into the adjacent tissues, but most are swept into the blood via the tho-

racic duct. Throughout much of the geographical distribution of this parasite, there is a marked **periodicity** of microfilariae in the peripheral blood, that is, they can be demonstrated at certain times of the day, while at other times, they virtually disappear from the peripheral circulation. The maximum number usually can be found between 10 P.M. and 2 A.M. For this reason, night-feeding mosquitoes are the primary vectors of *Wuchereria* in areas where microfilarial periodicity occurs. During the day, the microfilariae are concentrated in blood vessels of the deep tissues of the body, predominately in the pulmonary vessels proximal to the pulmonary arterioles.[8] The causes of the periodicity remain obscure, but they apparently do not involve daily release of a new generation of progeny by the adult female. Stimuli, such as arterial oxygen tension and body temperature, probably are involved. Administration of pure oxygen to a patient during peak microfilaremia can cause the microfilariae to localize in the deep tissues. Reversal of the patient's sleep schedule causes reversal of periodicity, so that mi-

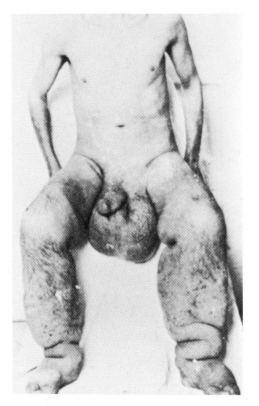

Fig. 30-2. Elephantiasis involving both legs, scrotum, and penis, as seen in chronic filariasis. (From Price, D. L., and H. C. Hopps. 1971. In Marcial-Rojas, R. A., editor. Pathology of protozoal and helminthic diseases with clinical correlation. The Williams & Wilkins Co., Baltimore.)

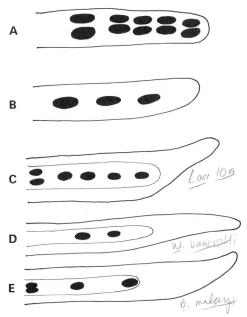

Fig. 30-3. The presence or absence of a sheath and the arrangement of nuclei in the tail are useful criteria in identifying microfilariae. **A,** *Dipetalonema perstans;* **B,** *Mansonella ozzardi;* **C,** *Loa loa;* **D,** *Wuchereria bancrofti;* **E,** *Brugia malayi.*

crofilaremia becomes diurnal. The adaptive value of the periodicity is difficult to explain. While it is clearly advantageous for the microfilariae to be present in the peripheral blood when the vector is likely to be feeding, what value is there in being absent when the vector is not feeding? Some strains of *Wuchereria* have a normal diurnal periodicity. The periodicity is unimportant clinically, but it has significant diagnostic and epidemiological implications.

In certain areas of the South Pacific, including Fiji, Samoa, the Philippines, and Tahiti, a strain of *Wuchereria* is common that shows a diurnal periodicity, referred to as **subperiodic.** The morphology of the adults is identical to those producing periodic microfilariae; most investigators believe that only one species is involved, although some designate the subperiodic type as a separate species named *W. pacifica.* Daytime-feeding mosquitoes are the major vectors of the subperiodic strain.

Microfilariae are ingested by the mosquito along with its blood meal. They lose their sheath in the first 2 to 6 hours in the insect's stomach, after which they penetrate the gut of the host and reach the thoracic muscles. The first cuticular molt occurs about 2 days later. The second-stage larva is a short, sausage-shaped worm **(sausage stage)** (Fig. 30-7) in which most of the organ systems are present. Within 2 weeks the second molt takes place. The larva is now an elongate, slender **filariform** third stage (Fig. 30-8), and development ceases. The filariform larvae are 1.4 to 2 mm long and are infective to the definitive host. They migrate throughout the hemocoel, eventually reaching the labium, or proboscis sheath, from which they escape when the mosquito is feeding.

They enter the skin through the wound made by the mosquito. After migrating through the peripheral lymphatics, the worms settle in the larger lymph vessels, where they mature.

Epidemiology. Many mosquito vectors of *Wuchereria* have a preference for human blood and often breed near human habitation. At least 77 species and subspecies of mosquitoes in the genera *Anopheles, Aedes, Culex,* and *Mansonia* are known intermediate hosts for *Wuchereria*. In areas where the periodic strain of *Wuchereria* is found, the mosquito vectors are primarily night feeders. The species of mosquito serving as vector in a particular area seems to depend more on coincidence (which species feeds when the larvae within the definitive host are available) than on physiological determinants of host specificity. Nevertheless, the periodicity has practical epidemiological significance because this fact determines which mosquito species must be controlled and, consequently, what control measures must be applied.

Suitable breeding sites for mosquitoes abound in tropical areas. Some sites are difficult or impossible to control, such as tree holes, hollows at the bases of palm fronds, and so on, while others can be controlled with a degree of effort. Hollow coconuts, killed while still green by rats gnawing holes in them, fall, fill with rain water, and become havens for developing mosquito larvae. These can be collected and burned. Even dugout canoes that are unused for a few days can partially fill with rainwater and become mosquito nurseries. Conditions for transmission of *Wuchereria* vary from locality to locality and country to country. The epidemiologist must consider each case independently within the framework of the biology of the vector and host, putting the economic and technical resources that are available to best advantage.

Pathogenesis. Pathogenesis in filariasis depends heavily on inflammatory and immune responses, and these are predominately responses to the adult worms, especially the female; little or no disease is caused by the microfilariae. There are three clinical phases: the **incubation** stage, the **acute** or **inflammatory** stage, and the **obstructive** phase or stage of complications caused by the chronic lymphoedema. The incubation phase is the time between infection and the appearance of microfilariae in the blood. It is largely symptomless, but there may be transient lymphatic inflammation with mild fever and malaise.

The acute inflammatory stage follows when the females reach maturity and start releasing microfilariae. There is intense lymphatic inflammation, usually in the lower half of the body, with chills, fever, and toxemia. The area of the affected lymphatic is swollen and painful, and the overlying skin may be reddened. The attack usually subsides after a few days, but this and the other manifestations described further often recur at frequent intervals.

Additional common symptoms in the acute stage of filariasis include **inguinal lymphadenitis** (inflammation of the lymph nodes in the inguinal region), **orchitis** (inflammation of the testes, usually with sudden enlargement and considerable pain), **hydrocele** (forcing of lymph into the tunica vaginalis of the testis or spermatic cord), and **epididymitis** (inflammation of the spermatic cord). Acute febrile episodes called **elephantoid fever** recur frequently. These are marked by sudden onset, rigors and sweating, and fever to 104° F and are from a few hours to several days duration. On the histological level, extensive proliferation of the lining cells exists in the lymphatics, with much inflammatory cell infiltration, especially of polymorphonuclear leukocytes and eosinophils, around the lymphatics and adjacent veins. The most prominent cells in the infiltration become lymphocytes, plasma cells, and eosinophils, as the most acute phase subsides. Abscesses around dead worms may exist, with accompanying bacterial infection. Microfilariae may disappear from the peripheral blood during and after the acute phase, presumably because the lymphatic vessel containing the female becomes blocked.

The obstructive phase is marked by **lymph varices, lymph scrotum, hydrocele, chyluria,** and **elephantiasis.** Lymph varices are "varicose" lymph ducts, caused when lymph return is obstructed and the lymph "piles up," greatly dilating the affected duct. This causes chyluria, or lymph in the urine, a common symptom of filariasis. The chyle gives the urine a milky appearance, and some blood is often present. A feature of the chronic obstructive phase is progressive infiltration of the affected areas with fibrous connective tissue, or "scar" formation, after inflammatory episodes. However, dead worms are sometimes calcified instead of absorbed, usually causing little further difficulty.

In a certain proportion of cases, thought to be associated with repeated attacks of acute lymphatic inflammation, the condition known as elephantiasis gradually develops. This is a chronic lymphoedema with much fibrous infiltration and thickening of the skin. In men, the organs most commonly afflicted with elephantiasis are scrotum, legs, and arms; in women, the legs and arms are usually afflicted, with vulva and breasts being affected more rarely. Elephantoid organs are composed mainly of fibrous connective tissues, granulative tissue, and fat. The skin becomes thickened and cracked, and invasive bacteria and fungi further complicate the matter. Microfilariae usually are not present.

Elephantiasis is thus seen to be a result of complex immune responses of long duration. After worms die and are absorbed, the symptoms gradually disappear. Repeated superinfections over many years are usually necessary to cause elephantiasis. Casual visitors into endemic areas may well become infected with the parasite, however, and suffer from localized edema and painful inflammation of the lymphatic system, but they may have no microfilariae in their peripheral blood. This condition may persist for many years, subsiding and recurring from time to time.[9]

One of the most perplexing problems of the disease is that of why microfilariae are so rarely found in the peripheral blood if a person is first infected as an adult. In World War II, 10,431 U.S. naval personnel were infected with *W. bancrofti,* yet only 20 showed a microfilaremia.[1] Yet, among adults indigenous to an endemic area, there is a high incidence of microfilaremia. It is possible that this phenomenon may be explained by a condition of immune tolerance. Transplacental infection has been demonstrated in some filarial worms and, if such occurred with *W. bancrofti,* invasion of the human embryo might result in later failure of the individual's immune system to recognize the microfilariae as foreign. It must be, however, that the tolerance is not complete because the person would soon be overwhelmed by the larvae, if a certain proportion were not destroyed. It seems likely that the clinical disease is determined by individual reactions to continual and sometimes massive antigenic stimulation. Also, it appears that associated bacterial infections play a role, but the nature of this role is not understood.

Diagnosis and treatment. Demonstration of microfilariae in the blood is a simple and fairly accurate diagnostic technique, provided that thick blood smears are made during the period when the larvae are in the peripheral blood. The technician must be able to distinguish this species from others that could be present. X-rays can detect dead, calcified worms. Because microfilariae often cannot be demonstrated, especially in newcomers to an endemic area, the intradermal skin test is valuable. A preparation of powdered *Dirofilaria immitis* (the heart filarial worm of dogs) in saline gives nearly 100% accuracy in diagnosis, although false positives result when the patient is infected with other species of filariids.[2] Filariasis should always be suspected if clinical symptoms occur about 3 months or more after arrival in an endemic area.

The drug of choice at this time is diethylcarbamazine (Hetrazan), which eliminates microfilariae from the blood but apparently does not kill the adults. Other, newer drugs show promise as chemotherapeutic agents against *Wuchereria,* but few are currently available.

Swollen limbs are sometimes successfully treated by applying pressure bandages, which force the lymph out of the swollen area. This may gradually reduce the size of the member, which returns to nearly normal. Any connective tissue proliferation that might have developed will not be affected, however. Surgical removal of elephantoid tissue is often possible.

Prevention remains primarily protection against the bite of mosquitoes when in endemic areas. Insect repellant, mosquito netting, and other preventatives should be rigorously used by persons temporarily visiting such places. Long-term protection requires mosquito control and mass chemotherapy of indigenous people to eliminate microfilariae from the circulating blood, where they are available to mosquitoes.

Brugia malayi

It was first noticed in 1927 that a microfilaria, different from that of *W. bancrofti,* occurred in the blood of natives of Celebes. It was not until 1940 that the adult form was found in India, and a year later it was discovered in Indonesia. It is now known to parasitize humans in China, Korea, Japan, Southeast Asia, India, Sri Lanka, the East Indies, and the Philippines. Much of its distribution overlaps that of *Wuchereria.*

The morphology of this parasite, *Brugia malayi,* is very similar to *W. bancrofti,* although it is only about half as large. The number of anal papillae of males differs slightly between the two species, and the left spicule of *B. malayi* is a little more complex than that of *W. bancrofti.* These are feeble differences on which to separate two genera, but because a large literature is accumulating on Malayan filariasis under the name of *Brugia,* we reluctantly follow common usage.

Morphology. Males are 13.5 to 20.5 mm long and 70 to 80 μm wide. The tail is curved ventrad and bears three or four pairs of adanal and three or four pairs of postanal papillae. The spicules are unequal and dissimilar, and a small gubernaculum is present.

Females are 80 to 100 mm long by 240 to 300 μm wide. The tail is finger-like and is covered with minute cuticular bosses. The vulva is near the level of the middle of the esophagus.

Biology and pathology. The life cycle of *B. malayi* is nearly identical to that of *W. bancrofti,* Mosquitoes of the genera *Mansonia, Aedes,* and *Culex* are intermediate hosts. Adults live in the lymphatics and cause the same disease symptoms as *W. bancrofti,* although elephantiasis, when it occurs, is more restricted to the legs; the genitalia are rarely affected.

The microfilaria is somewhat similar to that of *Wuchereria,* but can be differentiated from it by the presence of nuclei in the tail tip. There are both periodic and subperiodic strains.

Diagnosis and treatment are as for *Wuchereria.* Control is also primarily by mosquito eradication. Because *Mansonia* is the major vector in many areas, herbicides can be put to good advantage in eliminating the aquatic plants that the mosquito larvae depend on as their source of oxygen. Larvae of this genus pierce the stems of aquatic vegetation and tap oxygen, obviating the need for the wriggler to reach the surface regularly.

Two related species of filariids in humans are known from the microfilariae only. One was found in Timor, in the East Indies, and the other, *Wuchereria lewisi,* is known from Brazil. It has been common in the history of filaria research that the microfilaria was discovered years before the adult was found. It would not be surprising if more unknown species exist in humans in various parts of the world.

Onchocerca volvulus

River blindness is a disease caused by this large filarid worm in areas of Africa (where 30 million are infected), Arabia, Guatemala, Mexico, Venezuela, and Colombia. It probably has an even greater distribution than is presently known, for it is a cryptic disease that does not always manifest overt symptoms. When it does, the symptoms often are overlooked by health authorities, most of whom are busy with more pressing and immediate problems. An estimated 2,000 infected persons now live as immigrants in London.[10]

Onchocerciasis, as the condition is also known, is not a fatal disease. However, it does cause disfigurement and blindness in many cases; there are small communities in Africa and Central America where most of the people of middle age and over are blind. Eradication of this disease from the earth would not result in the "parasitologist's dilemma," for it would not increase the birthrate or increase the chances for infant survival. It would, instead, free hundreds of thousands of persons from a debilitating disease and thereby remove this economic burden from developing nations. Onchocerciasis was extensively reviewed by Nelson.[7]

Morphology. The morphology of *Oncho-*

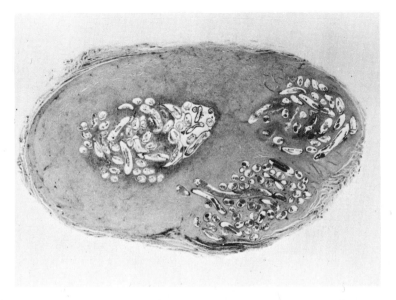

Fig. 30-4. A cross-section of a fibrous nodule **(onchocercoma)** removed from the chest of an African. It contained several worms bound together in a mass. (From Connor, D. H., and others. 1970. Hum. Pathol. 1:553-579. AFIP neg. no. 69-3625.)

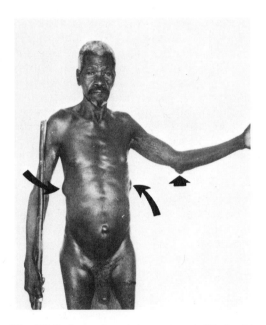

Fig. 30-5. Several nodules (arrows) filled with *Onchocerca volvulus* are found in the skin of this man. Note also the elephantoid scrotum and the depigmentation and wrinkling of the skin of the upper arms, also symptoms of onchocerciasis. (From Connor, D. H., and others. 1970. Hum. Pathol. 1:553-579. AFIP neg. no. 68-10071-3.)

cerca volvulus is not very different from that of *Wuchereria bancrofti*. The worms characteristically are knotted together in pairs or groups in the subcutaneous tissues (Fig. 30-4). They are slender and blunt at both ends. Lips and a buccal capsule are absent, and there are two circles of four papillae each surrounding the mouth. The esophagus is not conspicuously divided.

Males are 19 to 42 cm long by 130 to 210 μm wide; females are 33.5 to 50 cm long by 270 to 400 μm wide, with the vulva just behind the posterior end of the esophagus. The tail of the male is curled ventrad and lacks alae; it bears four pairs of adanal and six or eight pairs of postanal papillae. The microfilariae are unsheathed.

Biology. Adult worms locate under the skin, where they become encapsulated by host reactions. If this is over a bone, such as at a joint or over the skull, a prominent nodule appears (Fig. 30-5). The location of these nodules is correlated with the geographical area. In Africa, most infections are below the waist, while in Central America, they are usually above the waist. This is probably an adaptation to the biting preferences of the insect vectors, for the microfilariae are concentrated in the areas where the insects prefer to bite. Perhaps

Fig. 30-6. A blackfly, *Simulium damnosum*, biting the arm of a human. This insect is a major vector of *Onchocerca volvulus* in Africa. (From Connor, D. H., and others. 1970. Hum. Pathol. 1:553-579. AFIP neg. no. 68-2763-1.)

the cause is simply that "bush country" Americans are more inclined to wear trousers than their African counterparts.

The unsheathed microfilariae remain in the skin, where they can be ingested by the blackfly intermediate hosts, *Simulium* spp. (Fig. 30-6). These ubiquitous pests become infected when they take a blood meal. Their mouthparts are not adapted for deep piercing; so, much of their food consists of tissue juices, which contain numerous microfilariae in infected persons. The first-stage larva migrates from the intestinal tract of the fly to its thoracic muscles. There it molts to the sausage stage (Fig. 30-7), then molts again to the infective, filariform stage (Fig. 30-8). The filariform larva moves to the labium of the fly and can infect a new host when the insect next feeds. Mature worms appear in the skin in less than a year.

Onchocerca volvulus was probably introduced to the Americas with African slaves. It became established in Central America and has since mutated sufficiently to cause different clinical symptoms in its definitive host and to differ in its infectability to various vectors and laboratory animals. That the species has done this within about 400 years is an indication of the mutability

of dioecious parasites with high reproductive capacity. Humans appear to be the only natural definitive host for *O. volvulus*. The physiology of this parasite has not been studied.

Epidemiology. Generally speaking, onchocerciasis is a model system for the landscape epidemiologist. *Simulium* spp. live their larval stages only in clear, fast-running streams. The adult flies survive only where there is high humidity and plenty of streamside vegetation. It was long known by certain African natives that the disease was associated with rivers (and even with blackflies, although it was not officially "discovered" until 1926), and they gave it the name river blindness. Anyone who intrudes into such an area is viciously attacked by these insects. Wild-caught blackflies are often infected with a variety of species of filariids, most of which are still unidentified, but in areas endemic with *O. volvulus*, the larvae can often be recognized as that species.

Surprisingly, foci of onchocerciasis occur in the arid savannah of west Africa and the desert areas along the Nile near the Egypt-Sudan border. The epidemiology in these areas has not been thoroughly studied, but it is certain to depend on adaptations for survival of the blackfly vectors.

Pathogenesis. Two different elements contribute to the pathogenesis of onchocerciasis: the adult worms and the microfilariae. Of these, the adult is the least pathogenic, often causing no symptoms whatever and, at the worst, stimulating the growth of palpable subcutaneous nodules called *onchocercomas*, especially over bony prominences. In the African strain, these nodules are most frequent in the pelvic area, with a few along the spine, chest, and knees. The Venezuelan form is much like that in Africa, but in Central America the nodules are mostly above the waist, especially on the neck and head. These nodules are relatively benign, causing some disfigurement but no pain or ill health. The number of nodules may vary from one to well over 100. They consist mainly of collagen fibers surrounding one to several adult worms. Rarely, the nodule will degenerate to form an abscess, or the worm will become calcified.

True elephantiasis is sometimes caused

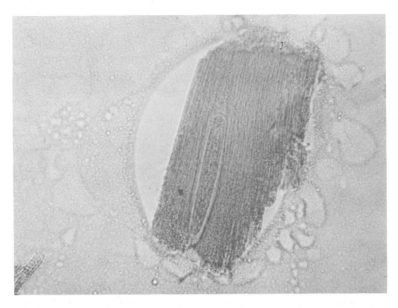

Fig. 30-7. A piece of thoracic muscle from *Simulium damnosum*. Note the second, or "sausage-stage," larva of *Onchocerca volvulus*. (Photograph by John Davies.)

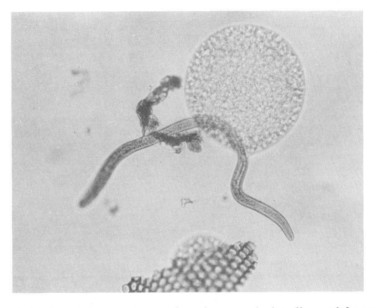

Fig. 30-8. A third, or filariform, larva of *Onchocerca volvulus,* dissected from the head of a *Simulium damnosum*. (Photograph by John Davies.)

by this worm (Fig. 30-9), and a similar condition, known as "hanging groin," is common in some areas of Africa. A loss of elasticity of the skin causes a sagging of the groin into pendulous sacs, often containing lymph nodes. The testes and scrotum are not affected, and hydrocele does not accompany the condition. Females are sim-ilarly affected (Fig. 30-10). Leonine face is a rare complication in Central America and Africa. Onchocerciasis frequently causes hernias, especially femoral hernia, in Africa.

The presence of microfilariae in the skin often causes a severe **dermatitis** caused either by allergic responses or toxic effects

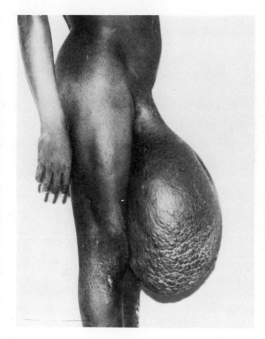

Fig. 30-9. A severe elephantoid scrotum on a native of Ubangi territory. It was removed surgically and a good cosmetic result was obtained. The scrotum weighed 20 kg and, when seen microscopically, was an edematous mass of interlacing collagen and smooth muscle fibers. (From Connor, D. H., and others. 1970. Hum. Pathol. 1:553-579. AFIP neg. no. 68-8582-9.)

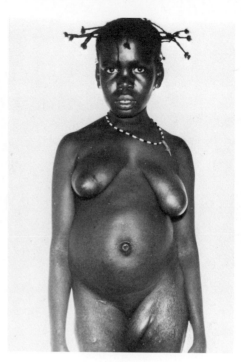

Fig. 30-10. A "hanging groin," or adenolymphocele. The tissue was excised and contained a group of lymph nodes embedded in subcutaneous tissue. The nodes contained many microfilariae of *Onchocerca volvulus*. (From Connor, D. H., and others. 1970. Hum. Pathol. 1:553-579. AFIP neg. no. 68-10066-1.)

after the death of the larvae. The first symptom is an intense itching, which may lead to secondary bacterial infection, often accompanied by dyspigmentation of the skin in small or extensive areas. This is followed by a thickening, discoloration, and cracking of the skin. These symptoms parallel those of avitaminosis A, and it has been suggested that they result from the parasite's competition for or interference with vitamin A metabolism. The last stage of the skin lesion is characterized by loss of elasticity, which gives the patient a look of premature age. **Depigmentation** is accentuated and may extend over large areas, especially of the legs (Fig. 30-11). Patients at this stage are often misdiagnosed as leprous. The distressing effects on the life-style of these persons can only be imagined.

Microfilariae in advanced cases often are located in the deeper part of the dermis and are not detected by skin-snip biopsy.

By far the most dreadful complications of onchocerciasis are those of the eyes. It has been calculated that the number of blind persons per 100,000 is over 1,500 in areas endemic for onchocerciasis, compared with 250 per 100,000 in a random sample in Europe.[7] The rate of impaired vision may reach 30% in some communities of Africa, where blindness exceeds 10% of the adult population. In these areas, and in similar areas of Guatemala, it is not unusual to see a child with good vision leading a string of blind adults to the local market.

Ocular complications are less common in the rain forest areas of Africa but are frequent in the savannah. The reason for this remains one of the enigmas of parasitology. It is possible that different strains of worms have different tropisms for the cornea.

Lesions of the eye take many years to develop, with most affected individuals

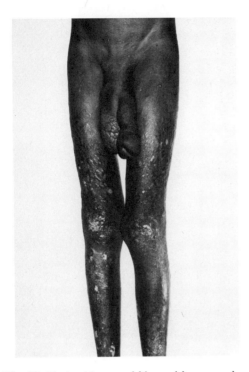

Fig. 30-11. An 11-year-old boy with severe dermatitis characterized by depigmentation, wrinkling, and thickening of the skin. He also has elephantoid changes of the penis and scrotum and onchocercomas over the knees. (From Connor, D. H., and others. 1970. Hum. Pathol. 1:553-579. AFIP neg. no. 68-7912-1.)

Fig. 30-12. A skin snip from a patient with onchocerciasis. Note the emerging microfilariae. (Photograph by Warren Buss.)

being over 40 years old. However, in Central America, with more worms concentrated on the head, young adults also show symptoms.

The earliest and most common complications begin when microfilariae invade the cornea. This causes inflammation of sclera, or white of the eye, followed by an invasion of fibrous tissue, leading to extensive vascularization of the cornea, which, in turn, severely impairs vision. Subsequent fibrosis may lead to complete blindness.

In some cases, there is damage to the retina, with complications of the optic nerve. The exact etiology of this condition is incompletely known, but it probably results from immune responses to dead microfilariae, perhaps in conjunction with toxic metabolites of the parasite. Microfilariae in the chambers of the eye are easily

demonstrated in onchocerciasis. In fact, it is often the ophthalmologist who first diagnoses the disease, during routine ocular inspection. Many aspects of ocular complications of river blindness parallel those of avitaminosis A.

A bizarre manifestation of onchocerciasis in Uganda is **dwarfism.** At first thought to be pygmies, these cases are now known to derive from normal parents and are always infected with *O. volvulus.* The symptoms are of pituitary deficiency, and there is little doubt but that the pituitary has been damaged, either directly or indirectly, by microfilariae.

Adult worms may live as long as 16 years.

Diagnosis and treatment. The best method of diagnosis is the demonstration of microfilariae in bloodless skin-snips. These are made by raising a small bit of skin with a needle and slicing it off with a razor or scissors. The bit of skin is then placed in saline on a slide and observed with a microscope for emerging microfilariae (Fig. 30-12). These must be differentiated from other species that might be present. Nodules may be aspirated, but no microfilariae will be found if only males or dead worms are present. Skin-snip biopsies can be taken anywhere, but if only a single snip is available it should be taken from the buttock. If the snip is so deep as to draw blood, it might be contaminated with other species of filariid. Also, in old cases the microfilariae may be so deep as to elude the snip. Each case may demonstrate other, overt symptoms that obviate the need for demonstrating the microfilariae.

In parts of Africa, there may be some periodicity of microfilariae in the skin, but this has not been adequately studied. No reliable immunological test has yet been devised, although there is promise that current research will refine tests that are still in the experimental stages.

Treatment of onchocerciasis is by two methods, surgical and chemotherapeutic. Excision of nodules, especially those around the head, may be effective in lowering both the rate of eye damage and the number of new infections within a population. It is a simple operation that can be performed under rather primitive conditions.

Various drugs will kill the microfilariae or the adults, but so far all are so toxic in their reactions as to make widespread use impractical. Diethylcarbamazine (Hetrazan) kills microfilariae very rapidly but does not affect adult worms. In fact, it kills the microfilariae too rapidly, for tissue reactions to dead larvae are usually violent, resulting in massive eruptions of the skin, extreme prostration, and sometimes death caused by anaphylactic shock. Melarsoprol and suramin sodium are arsenic compounds that kill adult worms, but they are so toxic that they are extremely dangerous to use for mass treatment. Under hospital conditions, these drugs can be administered, along with antihistamines and/or corticosteroids, to diminish the side effects, with good results.

Prevention may best be accomplished by eliminating blackflies from inhabited areas. It has been demonstrated clearly that DDT applications to swift-running streams will destroy all simulids.[5] Similar results have been obtained by spraying DDT along river banks from airplanes or helicopters. The widespread use of DDT has caused international uneasiness about its possible, and as yet largely unknown, effects on humans and other organisms. This should be cautiously weighed against the ultimate goal of eradicating one of humankind's most devastating diseases. Other, more biodegradable insecticides are being developed for use in place of DDT.

An ongoing program of blackfly control

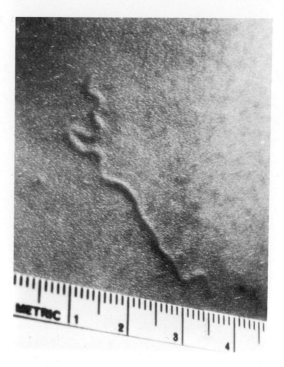

Fig. 30-13. An adult female *Loa loa* visible under the skin of a patient. (From Price, D. L., and H. C. Hopps. 1971. In Marcial-Rojas, R. A., editor. Pathology of protozoal and helminthic diseases with clinical correlation. The Williams & Wilkins Co., Baltimore. AFIP neg. no. 67-5366.)

in west Africa promises to free 10 million people otherwise condemned to an arduous battle for economic and social survival.[3]

Loa Loa

Loa Loa is the "eye worm" of Africa, which produces **loaiasis** or **fugitive** or **Calabar swellings.** It is distributed in the rain forest areas of west Africa and equatorial Sudan. Although it was established for a short time in the West Indies, where it was first discovered during slavery, it no longer exists there.

The morphology of *Loa* is typical of the family: a simple head with no lips and eight cephalic papillae, and a long, slender body and a blunt tail. The cuticle is covered with irregular, small bosses, except at the head and tail. Males are 20 to 34 mm long by 350 to 430 μm wide. The three

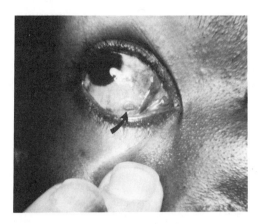

Fig. 30-14. An adult female *Loa loa* coiled under the conjunctival epithelium (arrow) of the eye of an African from the Congo. (From Price, D. L., and H. C. Hopps. 1971. In Marcial-Rojas, R. A., editor. Pathology of protozoal and helminthic diseases with clinical correlation. The Williams & Wilkins Co., Baltimore. AFIP neg. no. 67-5368-1.)

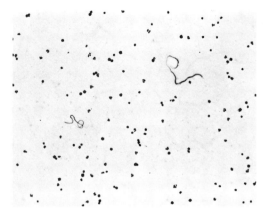

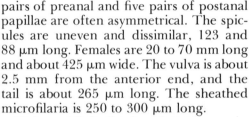

Fig. 30-15. The microfilariae of *Dipetalonema perstans* (left) and *Loa loa* (right). (From Price, D. L., and H. C. Hopps. 1971. In Marcial-Rojas, R. A., editor. Pathology of protozoal and helminthic diseases with clinical correlation. The Williams & Wilkins Co., Baltimore. AFIP neg. no. 67-10842.)

pairs of preanal and five pairs of postanal papillae are often asymmetrical. The spicules are uneven and dissimilar, 123 and 88 μm long. Females are 20 to 70 mm long and about 425 μm wide. The vulva is about 2.5 mm from the anterior end, and the tail is about 265 μm long. The sheathed microfilaria is 250 to 300 μm long.

Biology. Adults live in subcutaneous tissues (Fig. 30-13), including back, chest, axilla, groin, penis, scalp, and eyes in humans and several other primates. The microfilariae (Fig. 30-15) are periodic, appearing in the peripheral blood in maximum numbers during daylight hours and concentrating in the lungs at night. The intermediate host is any of several species of deerfly, genus *Chrysops*, which feeds by slicing the skin and imbibing the blood as it wells into the wound. The worms develop to the third-stage, filariform larvae, in the fat body of the fly, after which they migrate to the mouthparts. The prepatent period in humans is about 1 year, and adults may live at least 15 years.

Pathogenesis. These worms have a tendency to wander through the subcutaneous connective tissues, provoking inflammatory responses as they go. When they remain in one spot for a short time, the host reaction results in localized "Cala-

bar swellings," which disappear when the worm moves on. Localized inflammation of the area is most evident in Caucasian hosts. Adult worms also have an annoying habit of migrating through the conjunctiva and cornea (Fig. 30-14), with swelling of the orbit and psychosomatic results to the host. The overall pathology is rather benign compared with that of other filariids of humans.

Diagnosis and treatment. Demonstration of typical microfilariae in the blood (Fig. 30-15) is ample proof of loaiasis. The visual observation of a worm in the cornea or over the bridge of the nose is also indicative of this species. Finally, transient swellings of the skin are suspect, although spargana or *Onchocerca* may be confused with loaiasis before the parasite is excised and examined. Surgical removal is simple and effective, providing the worm is properly located, but most of the worms are inapparent. Chemotherapy is as in filariasis bancrofti. Control of deerflies, which breed in swampy areas of the forest, is extremely difficult.

Other filariids occasionally found in humans

Dipetalonema perstans exists in people in tropical Africa and South America. Several

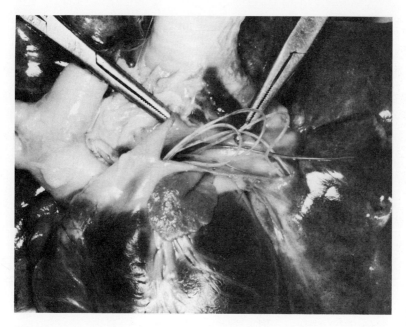

Fig. 30-16. *Dirofilaria immitis* protruding from the opened pulmonary artery of a German shepherd dog. (Photograph by Robert E. Kuntz.)

primates have been incriminated as reservoir hosts. Adults live in the coelom and produce unsheathed microfilariae (Fig. 30-15). Intermediate hosts are species of biting midges of the genus *Culicoides*. They appear to cause little pathological effect.

Dipetalonema streptocerca is a common parasite in the skin of humans in many of the rain forests of Africa.[6] It probably is a normal parasite of chimpanzees.

Dirofilaria immitis (Fig. 30-16) is parasitic in the right heart and pulmonary artery of dogs and other mammals throughout most of the world. It has been found in humans several times, including cases in the United States. The worm is transmitted by several species of mosquito. It often results in right heart failure and pulmonary complications in dogs, and it constitutes a dangerous pathogen for dogs in many areas of the United States. In humans the symptoms are vague and unpredictable.

Dirofilaria conjunctivae has been found in subcutaneous lesions of humans in several European and American countries. It probably is not an actual species but rather is a zoonotic expression of *D. repens*, a parasite of dogs in Europe, Asia, and South America, and *D. tenuis*, a raccoon parasite in North America.

Mansonella ozzardi is a filariid parasite of the New World, with distribution known to encompass northern Argentina, the Amazon drainage, the northern coast of South America, Central America, and several islands of the West Indies. It has never been found in the Old World. Adults live in the body cavity, threaded among the mesenteries and peritoneum. The intermediate host is *Culicoides* sp. in the West Indies, but *Simulium* sp. may transmit infection in Brazil. Males are known only from the posterior end of a single specimen. Females are better known and can be recognized by a pair of large caudal papillae, probably representing phasmidial pores. This parasite is most likely a zoonosis in humans.

REFERENCES

1. Beaver, P. C. 1970. Filariaris without microfilaremia. Am. J. Trop. Med. Hyg. 19:181-189.
2. Bozicevich, J., and A. M. Hutter. 1944. Intradermal and serological tests with *Dirofilaria immitis* antigen. Publ. Health Rep. 53:2130-2138.
3. Fatoyinbo, A. 1975. Initial success in battle against West African "river blindness." Trop. Med. Hyg. News 24:6-13.

4. Foster, W. D. 1965. A history of parasitology. E. S. Livingstone, Edinburgh.

5. Garnham, P. C. G., and J. P. McMahon. 1947. The eradication of *Simulium neavei* Raubaud, from an onchocerciasis area in Kenya Colony. Bull. Ent. Res. 37:619-628.

6. Meyers, W. M., and others. 1972. Human streptocerciasis. A clinico-pathologic study of 40 Africans (Zairians) including identification of the adult filaria. Am. J. Trop. Med. Hyg. 21:528-545.

7. Nelson, G. S. 1970. Onchocerciasis. In Dawes, B. editor. Advances in parasitology, vol. 8. Academic Press, Inc., New York. pp. 173-224.

8. Spencer, H. 1973. Nematode diseases. II. Filarial diseases. In Spencer, H. editor. Tropical pathology. Springer Verlag, Berlin. pp. 511-559.

9. Trent, S. 1963. Reevaluation of World War II veterans with filariasis acquired in the South Pacific. Am. J. Trop. Med. Hyg. 12:877-887.

10. Woodhouse, D. F. 1975. Tropical eye diseases in Britain. Practitioner 214:646-653.

SUGGESTED READING

Chabaud, A., and R. C. Anderson. 1959. Nouvel essai de classification des filaries (Superfamille des Filarioidea) II, 1959. Ann. Parasitol. 34:64-87. (An accurate, easy to use key to the genera of Filariata.)

Connor, D. H. and others. 1970. Onchocerciasis, onchocercal dermatitis, lymphadenitis, and elephantiasis in the Ubangi Territory. Human Pathol. 1:553-579.

Duke, B. O. L. 1971. The ecology of onchocerciasis in man and animals. In Fallis, A. M., editor. Ecology and physiology of parasites. University of Toronto Press, Toronto. pp. 213-222.

Duke, B. O. L. 1971. Onchocerciasis. Br. Med. Bull. 28:66-71.

Khanna, N. N., and G. K. Joshi. 1971. Elephantiasis of female genitalia. A case report. Plast. Reconstr. Surg. 48:374-381.

Sonin, M. D. 1966, 1968. Essentials of nematodology, vol. 17 and 21. Filarata of animals and man. Akad. Nauk SSSR, Moscow. (Comprehensive treatments of the Aproctoidea and Diplotriaenoidea.)

Chapter 31

PHYLUM ACANTHOCEPHALA: THE THORNY-HEADED WORMS

Few zoologists, and still fewer veterinarians and physicians, ever encounter a thorny-headed worm. Compared to parasitic platyhelminthes or nematodes, they are fairly rare. Still, representatives are to be found inhabiting the intestines of fishes, amphibians, reptiles (rarely), birds, and mammals, where they have established a parasitic relationship with their host and, occasionally, cause serious disease.

The first recognizable description of an acanthocephalan in the literature is that of Redi, who, in 1684, reported white worms with hooked, retractable proboscides in the intestines of eels. From the time of Linnaeus to the end of the nineteenth century, all species were placed in the collective genus *Echinorhynchus* Zoega in Mueller, 1776, although Koelreuther is credited with naming the genus *Acanthocephalus* in 1771. Hamann divided *Echinorhynchus,* which by then had become large and unwieldly, into *Gigantorhynchus, Neorhynchus,* and *Echinorhynchus,* thereby beginning the modern classification of the Acanthocephala.[10]

Lankester proposed elevating the order Acanthocephala, proposed by Rudolphi in 1808, to the level of phylum.[15] This suggestion was not widely accepted until Van Cleave convincingly argued in its favor.[30,31] Today, most students of the group consider the Acanthocephala to represent a separate phylum, though a few still hold it to be a class or order in the phyla Nemathelminthes or Aschelminthes.

MORPHOLOGY

The conservatism of the parasitologists who have developed the classification of the group may be a reflection of the conservatism in the structure of the worms themselves. Adaptation to parasitism has

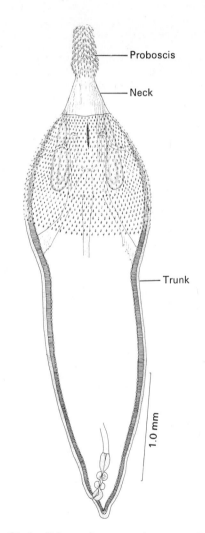

Fig. 31-1. *Polymorphus swartzi,* a parasite of ducks, showing the main body divisions. (From Schmidt, G. D. 1965. J. Parasitol. 51:809-813.)

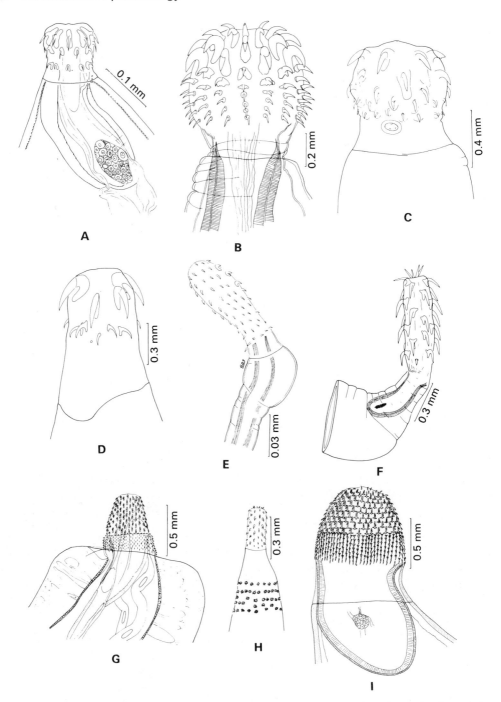

Fig. 31-2. Examples of different types of acanthocephalan proboscides. **A,** *Octospini-feroides australis;* **B,** *Sphaerechinorhynchus serpenticola;* **C,** *Oncicola spirula;* **D,** *Acanthosentis acanthuri;* **E,** *Pomphorhynchus yamagutii;* **F,** *Paracanthocephalus rauschi;* **G,** *Mediorhynchus wardae;* **H,** *Palliolisentis polyonca;* **I,** *Owilfordia olseni;* **J,** *Centrorhynchus falconis;* **K,** *Prosthen-orchis elegans;* **L,** *Paracavisoma impudica;* **M,** *Centrorhynchus spilornae;* **N,** *Andracantha phala-crocoracis;* **O,** *Southwellina dimorpha.* (**A, E, H,** and **L,** from Schmidt, G. D., and E. J. Hugghins. 1973. J. Parasitol. 59:829-838. **B,** from Schmidt, G. D., and R. E. Kuntz. 1966. J. Parasitol. 52:913-916. **C** and **K,** from Schmidt, G. D. 1972. In Fiennes, R. N.

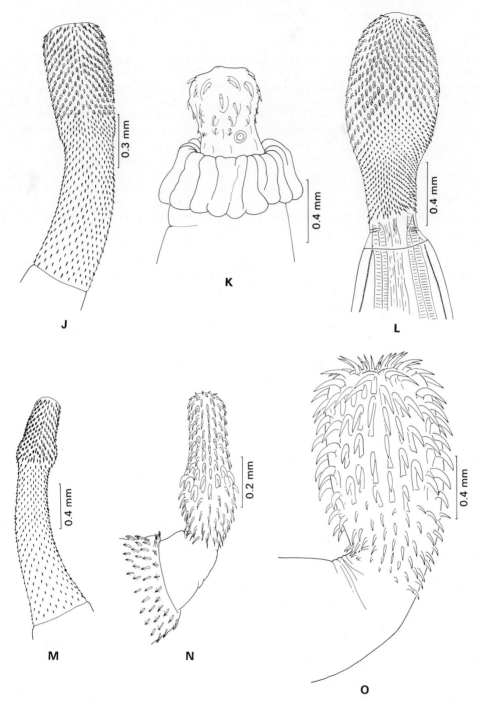

Fig. 31-2, cont'd. T. W., editor. Pathology of simian primates. S. Karger, Basel. pp. 144-156. **D,** from Schmidt, G. D. 1975. J. Parasitol. 61:865-867. **F,** from Schmidt, G. D. 1969. Can. J. Zool. 47:383-385. **G,** from Schmidt, G. D., and A. G. Canaris. 1967. J. Parasitol. 53:634-637. **I,** from Schmidt, G. D., and R. E. Kuntz. 1967. J. Parasitol. 53:130-141. **J** and **M,** from Schmidt, G. D., and R. E. Kuntz. 1969. J. Parasitol. 55:329-334. **N,** from Schmidt, G. D. 1975. J. Parasitol. 61:615-620. **O,** from Schmidt, G. D. 1973. J. Parasitol. 59:299-305.)

resulted in genetic reduction of muscular, nervous, circulatory, and excretory systems and to complete loss of a digestive system. The remaining animal is little more than a pseudocoelomate bag of reproductive organs with a spiny holdfast at one end. The worms range in size from the tiny *Octospiniferoides chandleri,* only 0.92 to 2.4 mm long, to *Oligocanthorhynchus longissimus,* exceeding a meter in length. Extensive reviews of acanthocephalan physiology have been published.[6,17,18]

General body structure

Superficially, the acanthocephalan body is seen to consist of an anterior proboscis, a neck, and a trunk (Fig. 31-1).

The proboscis is variable in shape, from spherical to cylindrical, depending on the species (Fig. 31-2). It is covered by a tegument and has a thin, muscular wall within which are embedded the roots of recurved, sclerotized hooks. The sizes, shapes, and numbers of these hooks are among the most useful characters in the taxonomy of the worms. The proboscis is hollow and fluid-filled. Attached to its inner apex is a pair of muscles, called **proboscis inverter muscles,** which extend the length of the proboscis and neck and insert in the wall of a muscular sac called the **proboscis receptacle.** The proboscis receptacle itself is attached to the inner wall of the proboscis. Its morphology varies somewhat depending on the family, but, generally speaking, it consists of one or two layers of muscle fibers. When the proboscis inverter muscles contract, the proboscis invaginates into the proboscis receptacle, with the hooks completely inside. When the proboscis receptacle contracts, it forces the proboscis to evaginate by a hydraulic system.[11] A nerve ganglion called the **brain** is located within the receptacle. The proboscis and its receptacle are sometimes referred to as the **praesoma.**

The neck is a smooth, unspined zone between the most posterior hooks of the proboscis and an infolding of the body wall. **Neck retractor muscles** attach this infolding of the body wall to the inner surface of the trunk. In some species, other muscles, called **protrusers,** attach to the proboscis receptacle. When the proboscis retractor and the neck retractor muscles contract, the entire anterior end is withdrawn into the trunk. Some species have a pit on each side of the neck, and these, presumably, are sensory in nature. Two similar pits are found on the tip of the proboscis of many species.

The rest of the body, posterior to the neck, is called the **trunk,** or **metasoma.** Like the proboscis and neck, it is covered by a tegument and has muscular internal layers. Many species have simple, sclerotized spines embedded in the trunk wall that maintain close contact with the mucosa of the host's intestine. The trunk contains the reproductive system (Fig. 31-3) and also functions in absorbing and distributing nutrients from the host's intestinal contents. In the living worm, the trunk is bilaterally flattened, usually with numerous transverse wrinkles, but when the worm is placed in a hypotonic solution, such as tap water, it swells and becomes turgid. This is desirable for ease of study of the specimen, for it places the internal organs in constant relationship with each other, and it usually forces the introverted proboscis to evaginate.

Body wall

The body wall is a complex syncytium containing nuclear elements and a series of internal, interconnecting canals called the **lacunar system.** In some species, the nuclei are gigantic (Fig. 31-3) but few in number. In others the nuclei fragment during larval development and are widely distributed throughout the trunk wall. When entire nuclei are present, their number is constant for each species, demonstrating the principle of **eutely,** or nuclear constancy. Development of the wall was described by Butterworth.[3]

The lacunar system is filled with a fluid and apparently serves as a circulatory system. It is present in the proboscis and neck as well as the trunk and connects with two structures called **lemnisci** that grow from the base of the neck into the pseudocoelom. Each lemniscus has a central canal that is continuous with the lacunar system. The function of the lemnisci is unknown. In most species there are two main longitudinal lacunar canals, either dorsal and ventral or lateral. These are connected by numerous irregular, transverse canals. The

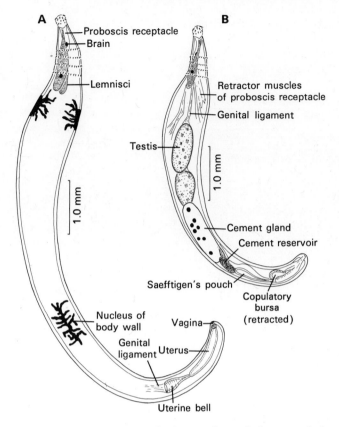

Fig. 31-3. *Quadrigyrus nickoli,* illustrating basic acanthocephalan morphology. **A,** Female; **B,** male. (From Schmidt, G. D., and E. J. Hugghins. 1973. J. Parasitol. 59:829-835.)

location and arrangement of the lacuni are used as taxonomic characters.

The tegument has no true layers, but several regions differ in their construction. These are, beginning with the outermost, the (1) surface coat, (2) striped zone, (3) vesicular zone, (4) felt zone, (5) radial fiber zone, and (6) basement lamina. Inside the tegument is a layer of irregular connective tissue, followed by circular and longitudinal muscle layers. Like the trematodes and cestodes, the tegument is syncytial, but unlike those groups, the nuclei are in the basal region of the tegument, not in perikarya separated from the distal cytoplasm. The **surface coat** or **glycocalyx,** a filamentous material, was formerly known as the epicuticle. It is, for instance, about 0.5 μm thick on *Moniliformis dubius,* an acanthocephalan of rats and the most commonly investigated species in the laboratory. The surface coat is composed of acid muco-

polysaccharides and neutral polysaccharides and/or glycoproteins.[37] The surface coat fits the definition of a glycocalyx, a carbohydrate-rich coat found on a variety of eukaryotic and prokaryotic cells. The stabilized system of polyelectrolytic filaments in the surface coat constitute an extensive surface for molecular interactions, including those involved in transport functions and enzyme-substrate interactions.

Immediately beneath the surface coat and limited by the trilaminate outer membrane is the **striped zone.** This zone is 4 to 6 μm thick and is punctuated by a large number of crypts about 2 to 4 μm deep that open to the surface by pores (Fig. 31-4).[4] These crypts give this zone a striped appearance *(Streifenzone)* under the light microscope. The crypts increase the surface area of the worm by 44 times that of a smooth surface. A filamentous molecular

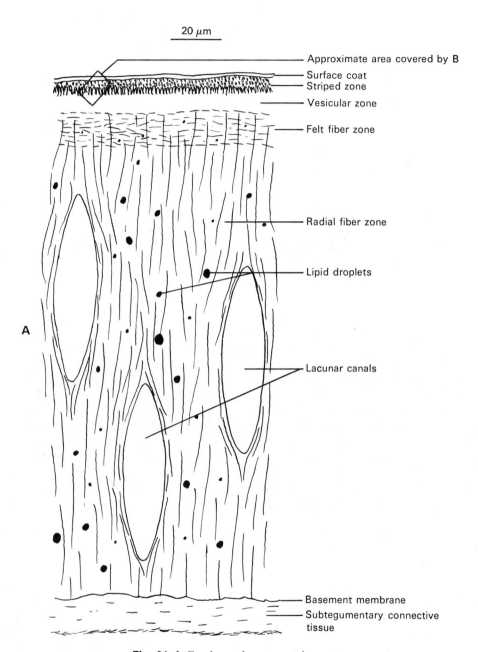

Fig. 31-4. For legend see opposite page.

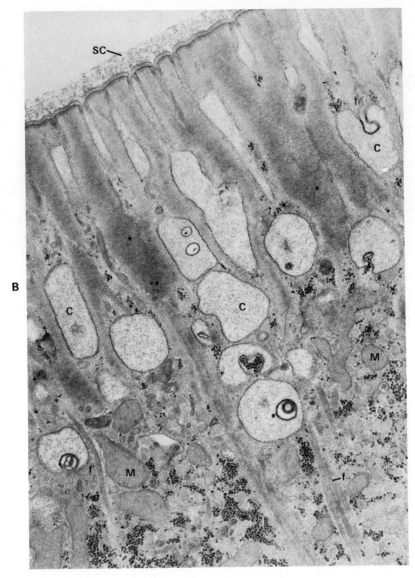

Fig. 31-4. Tegument of *Moniliformis dubius.* **A,** Diagram of transverse section to show layers. The vesicular zone is transitional between the striped zone and the felt fiber zone; it contains many vesicles and mitochondria with poorly developed cristae. The lacunar canals are in the radial fiber zone. **B,** Electron micrograph showing the major features of the striped zone. The worm is coated with a finely filamentous surface coat *(SC).* Numerous surface crypts *(C)* appear as large scattered vesicular structures with elements occasionally appearing to course to the surface of the helminth. The crypts are separated by patches of moderately electron-opaque material *(*),* giving the zone its striped appearance under the light microscope. Mitochondria *(M),* glycogen particles, microtubules, and other cytoplasmic details are evident in the inner portion of the striped zone. Bundles of fine cytoplasmic filaments *(f)* extend between this region and the deeper cytoplasm of the body wall. (×42,000.) (**A,** adapted from Nicholas, W. L. 1967. In Dawes, B., editor. Advances in parasitology, vol. 5. Academic Press, Inc., New York. **B,** from Byram, J. E., and F. M. Fisher. 1974. Tissue Cell 5:559.)

sieve is seen in the necks of the crypts, but particles of less than about 8.5 nm can gain access to the crypts and undergo pinocytosis by the crypt membrane.[4] The importance of pinocytosis in the acquisition of nutrients by the worms is unknown. In the deeper aspects of the striped zone, numerous lipid droplets, mitochondria, Golgi complexes, and lysosomes are found.

The striped zone grades into a region of numerous, closely packed, randomly arranged fibrils known as the **felt-fiber zone.** Mitochondria, numerous glycogen particles, vesicles, and occasional lipid droplets and lysosomes also are found in the felt-fiber zone. The **radial fiber zone** is just within the felt-fiber zone and makes up about 80% of the thickness of the body wall. It contains large bundles of filaments that course radially through the cytoplasm, large lipid droplets, and nuclei of the body wall. Here, too, are many glycogen particles, mitochondria, Golgi complexes, and lysosomes. Rough endoplasmic reticulum is found in the perinuclear cytoplasm. The nuclei have numerous nucleoli. The lacunar canals course through the radial fiber zone.

The structure of the proboscis wall is similar to that of the trunk, except with fewer crypts, a thinner radial zone, and the absence of a felt zone.

Reproductive system

Acanthocephalans are dioecious and usually demonstrate some degree of sexual dimorphism in size, with the female being larger (Fig. 31-3). In both sexes one or two thin **ligament sacs** are attached to the posterior end of the proboscis receptacle and extend to near the distal genital pore. Within these sacs are the gonads and some accessory organs of the reproductive systems. In some species, the ligament sacs are permanent; in others they break down as the worm matures.

Male reproductive system. Two testes normally occur in all known species, and their location and size are somewhat constant for each species. Spermiogenesis has been described.[35] Each testis has a vas efferens through which mature spermatozoa, which appear as slender, headless threads, travel to a common vas deferens and/or to a small penis. Several accessory organs also are present, the most obvious

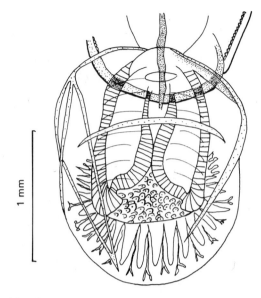

Fig. 31-5. Extended copulatory bursa of *Owilfordia olseni.* Note the numerous sensory papillae. (From Schmidt, G. D., and R. E. Kuntz. 1967. J. Parasitol. 53:130-141.)

of which are the cement glands. These syncytial organs, numbering from one to eight, contain one or more giant nuclei or several nuclear fragments. In many species they are joined in places by slender bridges. They secrete a **copulatory cement** of tanned protein, which is stored in a **cement reservoir,** in some species, until copulation occurs. At that time, the cement plugs the vagina after sperm transfer and rapidly hardens to form a **copulatory cap.** This remains attached to the posterior end of the female during subsequent development of the embryos within her body, but eventually disintegrates.

Another male accessory sex organ is the **copulatory bursa** (Fig. 31-5), a bell-shaped specialization of the distal body wall that is invaginated into the posterior end of the coelom except during copulation. A muscular sac, the **Saefftigen's pouch,** is attached to the base of the bursa. When it contracts, fluid is forced into the lacunar system of the bursa and, by hydrostatic pressure, it is everted. Many sensory papillae line the bursa; when it contacts the posterior end of a female, it clasps the female by muscular contraction, and sperm transfer is effected.

Female reproductive system. The ovary

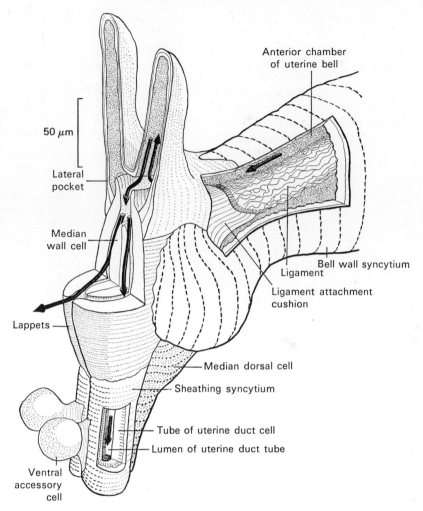

Fig. 31-6. Stereogram of mature uterine bell, cut away to reveal complex internal luminal system. Possible routes for egg translocation indicated by heavy arrows. (From Whitfield, P. J. 1968. Parasitology 58:671-682.)

of the female acanthocephalan is peculiar in that it fragments into **ovarian balls** early in life, often while the worm is still a juvenile in the intermediate host. These balls of oogonia float freely within the ligament sac, increasing slightly in size before insemination occurs. The posterior end of the ligament sac is attached to a muscular **uterine bell** (Fig. 31-6). This organ allows mature eggs to pass through into the uterus and vagina and out the genital pore, while returning immature eggs to the ligament sac. Its mechanism has been described.[34]

After copulation, the spermatozoa migrate from the vagina, through the uterus and uterine bell, and into the ligament sac. There they begin fertilizing the oocytes of the ovarian balls. After the first few cleavages, the embryos detach from the ovarian ball and float freely in the pseudocoelomic fluid. This exposes underlying oocytes for fertilization. Thus, several stages of early embryology may be found in a single female. Eventually, from this one copulation, many thousands or even millions of embryonated eggs are produced and released by each female. These, when sufficiently mature, pass from the host in its feces, where they may become available to the proper intermediate host.

As the shelled embryos are pushed into

the uterine bell by peristaltic action, two possible routes are available. They may pass back into the pseudocoelom through slits in the bell or on into the uterus. Fully developed embryos are slightly longer than immature ones and, therefore, cannot pass through the bell slits[34]: hence, they are passed on into the uterus, while immature eggs are retained for further maturation. The efficiency of the sorting is quite high, and apparently no immature forms are passed into the uterus.

Excretory system

Excretion in most species appears to be effected by diffusion through the body wall. However, members of one family in the class Archiacanthocephala, the Oligacanthorhynchidae, are unique in possessing two **protonephridial excretory organs.** Each is comprised of many anucleate flame bulbs with tufts of cilia and may be encapsulated or not, depending on the species. In the male, these organs are attached to the vas deferens and empty through it; in the female, they are attached to the uterine bell and empty into the uterus.

Acanthocephalans show little ability to osmoregulate, swelling in hypotonic, balanced saline or sucrose solutions and becoming flaccid in hypertonic solutions. The tonicity of their pseudocoelomic fluid is close to or somewhat above that of the intestinal contents. They take up sodium and potassium, swelling in hypertonic solutions of sodium chloride or potassium chloride at 37° C. In balanced saline, they lose sodium and accumulate potassium against a concentration gradient. Their hexose transport mechanism is not sodium-coupled.

Nervous system

The brain of an acanthocephalan consists of a fairly constant number of cell bodies, which lie in a mass within the proboscis receptacle. Several bundles of fibers lead from the brain; the two largest (the retinacula) pierce the sides of the receptacle and insert in the body wall, branching there into the trunk. Other bundles of fibers extend anteriad into the neck and proboscis, where they innervate the walls, retractor muscles, and apical and lateral sensory pores, if present. Actually, as is the case with most of the organ systems of the Acanthocephala, little is known of the nervous system.

DEVELOPMENT AND LIFE CYCLES

Each species of Acanthocephala uses at least two hosts in its life cycle. The first is an insect, crustacean, or myriapod, and the arthropod must eat an egg that was voided with the feces of a definitive host. Development proceeds through a series of stages to that which is infective to a definitive host. Many species, when eaten by a vertebrate that is an unsuitable definitive host, can penetrate the gut and encyst in some location where they survive without further development. This unsuitable vertebrate becomes a paratenic host; for if it is eaten by the proper definitive host, the parasite excysts, attaches to the intestinal mucosa, and matures. Such adaptability has survival value. For example, ecological gaps exist in the food chain between a microcrustacean and a large predaceous fish or between a grasshopper and an eagle. The paratenic host is one member in a food chain that bridges such a gap and, incidentally, ensures the survival of the parasite.

The manner of early embryogenesis is an unusual characteristic of the group (Fig. 31-7). Early cleavage is spiral, although this pattern is somewhat distorted by the spindle shape of the eggshell. At about the 4 to 34-cell stage, the cell boundaries begin to disappear, and the entire organism becomes syncytial. Gastrulation occurs by migration of nuclei to the interior of the embryo.[23] They continue to divide, but become smaller, until they form a dense core of tiny nuclei, the **inner nuclear mass.** These nuclei give rise to all the internal organ systems of the worm. In some species, the uncondensed nuclei remaining in the peripheral area give rise to the tegument; in some, the tegument is derived from a nucleus that separates from the inner mass, while in others, there are contributions from both.

The fully embryonated larva that is infective to the arthropod intermediate host is called the **acanthor.** The acanthor is an elongate organism that is usually armed at its anterior end with six or eight blade-like

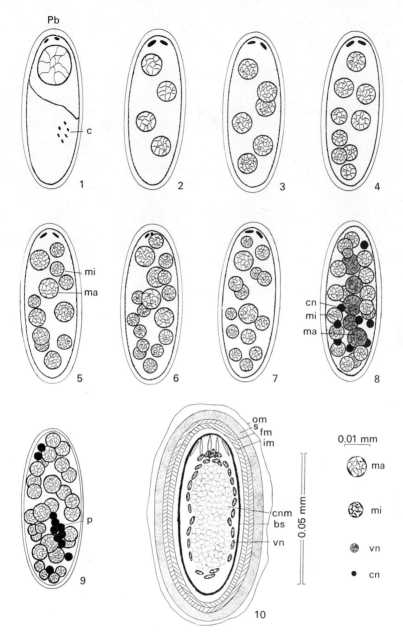

Fig. 31-7. Stages in the early embryological development of *Mediorhynchus grandis*. **1,** Prophase of division of posterior blastomere. **2,** Four-nuclei stage, showing spiral cleavage and absence of apparent cell envelopes. **3,** Six-nuclei stage. **4,** Nine-nuclei stage, with posterior quadrant of micromeres. **5,** 17-nuclei stage. **6** and **7,** Macromeres and micromeres, showing anterior quadrant of micromeres (some nuclei omitted). **8,** Formation of condensed nuclei. **9,** Inward migration of condensed nuclei, forming primordium of central nuclear mass. **10,** Mature acanthor, with enclosing envelopes (diagrammatic). *bs,* Body spines; *c,* chromosomes; *cn,* condensed nuclei; *cnm,* condensed nuclear mass; *fm,* fertilization membrane; *im,* inner envelope; *ma,* macromere; *mi,* micromere; *om,* outer membrane; *p,* primordium of inner nuclear mass; *pb,* polar bodies; *s,* shell; and *vn,* vesicular nucleus. (From Schmidt, G. D. 1973. Trans. Am. Microsc. Soc. 92:512-516.)

hooks. The hooks may be replaced by smaller spines in some species. The hooks or spines with their muscles are called the **aclid organ** or **rostellum.** The hooks aid in penetration of the gut of the intermediate host.[36] The acanthor is a resting, resistant stage and will undergo no further development until it reaches the intermediate host. Under normal environmental conditions, the acanthors may remain viable for months or longer. Acanthors of *Macracanthorhynchus hirudinaceus* can withstand subzero temperatures and desiccation, and they can remain viable for up to 3½ years in the soil. The acanthors of some species completely penetrate the gut, coming to lie in the host's hemocoel, while in others, they stop just under the serosa. In both cases, the worm then becomes parasitic on the arthropod, absorbing nutrients and enlarging, thus initiating the developmental stage known as the **acanthella.** The end of the acanthor that bears the aclid organ apparently becomes the anterior end of the adult in some species, while others exhibit a curious 90 degree change in polarity, in which the anterior end of the adult develops from the side of the acanthor. During the acanthella stage, the organ systems develop from the central nuclear mass and the hypodermal nuclei of the acanthor.

At termination of this development, the juvenile is an infective stage called a **cystacanth.** In most species, the anterior and posterior ends invaginate, and the entire cystacanth becomes encased in a hyaline envelope. The parasite then must be eaten by the definitive host before it can fulfill its potential. Obviously, mortality is very high, for only a tiny fraction of the immense number of eggs produced may survive the numerous hazards involved in completion of the life cycle.

Complete life cycles are known for only about 20 species in the phylum, although we have partial information on several more. The following examples illustrate the pattern followed in the life histories of the three major groups.

Class Eoacanthocephala

Neoechinorhynchus saginatus is an eoacanthocephalan parasite of various species of suckers and of creek chubs, *Semotilus atromaculatus,* a fish distributed from Maine to Montana. Its life cycle and embryology were described by Uglem and Larson[27] (Fig. 31-8). When the eggs are eaten by the common ostracod crustacean *Cypridopsis vidua,* they hatch within 1 hour and begin penetrating the gut within 36 hours. Following penetration, the unattached larva begins to enlarge and rearrange its nuclei, initiating the formation of internal organs. By 16 days postinfection, the acanthella has developed into an infective cystacanth. Time required for maturation within the fish has not been determined.

Other eoacanthocephalan life cycles are similar, although paratenic hosts are

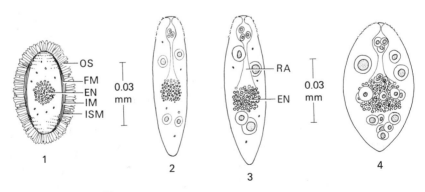

Fig. 31-8. For legend see opposite page.

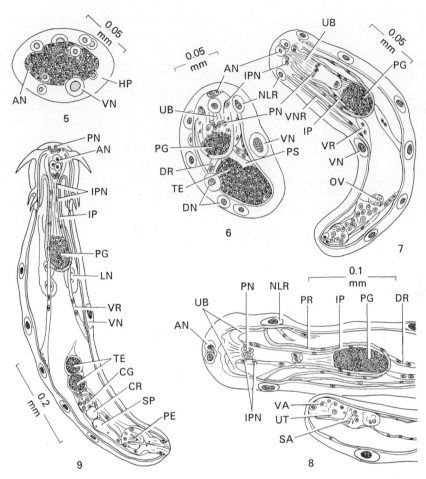

Fig. 31-8. Stages in development of *Neoechinorhynchus saginatus.* **1,** Shelled acanthor from body cavity of an adult female. **2,** Acanthor from gut of ostracod, 1 hour postfeeding. **3,** Acanthor, age 36 hours. **4,** Young acanthella from hemocoel 4 days postfeeding. **5,** Acanthella, age 7 days. **6,** Male acanthella, age 10 days. **7,** Female acanthella, age 11 days (one nucleus of future lemnisci omitted for clarity). **8,** Female acanthella, age 12 days (one lemniscal nucleus omitted). **9,** Late male acanthella, age 14 days (neck retractors omitted).

AN, Apical nuclei; *CG,* cement gland; *CR,* cement reservoir; *DN,* giant nuclei of dorsal trunk wall; *DR,* dorsal retractor of proboscis receptacle; *EN,* condensed nuclear mass; *FM,* fertilization membrane; *HP,* hypodermal primordium; *IM,* inner membrane; *IP,* proboscis inverter; *IPN,* proboscis inverter nuclei; *ISM,* inner shell membrane; *LN,* lemniscal nucleus; *NLR,* lemniscal ring nuclei; *OS,* outer shell; *OV,* ovary; *PE,* penis; *PG,* brain anlage; *PN,* proboscis nuclear ring; *PR,* proboscis receptacle muscle sheath; *PS,* pseudocoel; *RA,* retractor apparatus; *SA,* selector apparatus; *SP,* Saefftigen's pouch; *TE,* testes; *UB,* uncinogenous bands; *UT,* uterus; *VA,* vagina; *VN,* giant nucleus of ventral trunk wall; *VNR,* ventral neck retractor; *VR,* ventral retractor of proboscis receptacle. (From Uglem, G. L., and O. R. Larson. 1969. J. Parasitol. 55:1212-1217.)

known for *Neoechinorhynchus cylindratus*[33] and for *Neoechinorhynchus emydis*.[12]

Class Palaeacanthocephala

Plagiorhynchus formosus is a palaeacanthocephalan that is common in robins and other passerine birds in North America. Its life cycle and embryology were described by Schmidt and Olsen[24] (Fig. 31-9). When the eggs are eaten by the terrestrial isopod crustacean *Armadillidium vulgare*, they hatch in the midgut within 15 minutes to 2 hours. Active entrance of the acanthor into the gut wall occurs within 1 to 12 hours, and the acanthor lies within the tissues of the gut wall. After 15 to 25 days of apparent dormancy, it migrates to the outside of the gut, where it clings loosely to the serosa. Progressive changes follow in which the overall size increases, and the organs of the mature worm are delineated. The cystacanth appears fully developed in 30 to 40 days but is not infective to the definitive host until 60 to 65 days. On ingestion of an infected isopod by the definitive host, the proboscis of the cystacanth evaginates, pierces the cyst, and attaches to the gut wall, where development to maturity occurs.

Nickol and Oetinger found encapsulated *P. formosus* in the mesenteries of a shrew.[19] This illustrates how the interjection of a paratenic host into a life cycle may doom a parasite rather than serve it,

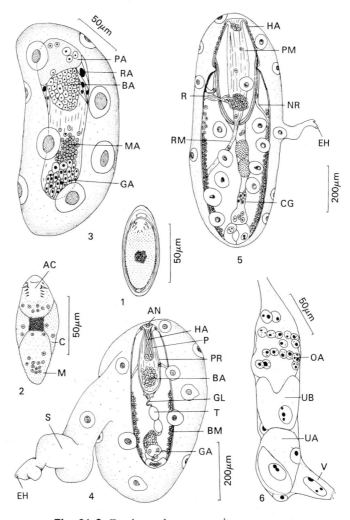

Fig. 31-9. For legend see opposite page.

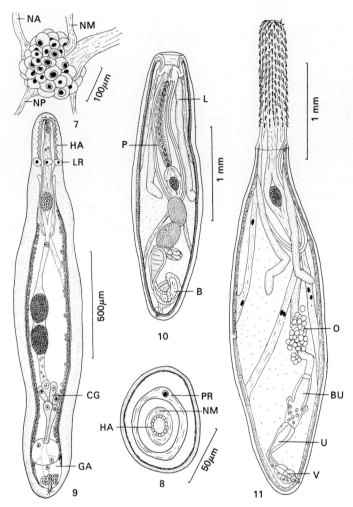

Fig. 31-9. *Plagiorhynchus formosus,* anatomy and development. **1,** Egg, containing mature acanthor. **2,** Acanthor after escape from egg (*AC,* aclid organ; *C,* cortex; *M,* medulla). **3,** Acanthella, 22 days (*BA,* brain anlage; *GA,* genital apparatus anlagen; *MA,* anlagen of body wall musculature, gonads, and genital ligament; *PA,* proboscis anlage; *RA,* anlagen of proboscis receptacle and retractor muscles). **4,** Acanthella, 25 days (*AN,* apical nuclei; *BA,* brain anlage; *BM,* anlagen of body wall musculature; *EH,* embryonal hooks; *GA,* anlagen of copulatory apparatus and cement glands; *GL,* genital ligament; *HA,* hook anlagen; *P,* proboscis; *PR,* proboscis receptacle; *S,* stalk; *T,* testis). **5,** Acanthella, 27 days (*CG,* cement glands; *EH,* embryonal hooks; *HA,* hook anlage; *NR,* neck retractor; *PM,* proboscis retractor; *R,* retinaculum; *RM,* retractor muscle). **6,** Anlagen of female reproductive system, 32 days (*OA,* ovary anlage; *UA,* uterus anlage; *UB,* uterine bell anlage; *V,* vagina anlage). **7,** Developing brain with principal nerve trunks, 28 days (*NA,* lateroanterior; *NM,* anteromedial; *NP,* lateroposterior). **8,** Cross-section at level of developing proboscis, 25 days (*HA,* hook anlage; *NM,* anteromedial nerve trunk; *PR,* proboscis receptacle). **9,** Acanthella, 30 days (*CG,* cement gland; *GA,* anlagen of copulatory apparatus; *HA,* hook anlage; *LR,* nucleus of lemniscal ring). **10,** Cystacanth, 37 days (*B,* bursa; *L,* lemniscus; *P,* proboscis). **11,** Cystacanth, 60 days (*BU,* uterine bell; *O,* ovarian balls; *U,* uterus; *V,* vagina). (From Schmidt, G. D., and O. W. Olsen. 1964. J. Parasitol. 50:721-730.)

for it is unlikely that a robin would eat a shrew. Chances of the worm invading a hawk or owl would be improved, however, and, if the parasite were preadapted to survive in such a host, its host range would thus be extended.

Class Archiacanthocephala

Macracanthorhynchus hirudinaceus is a cosmopolitan archiacanthocephalan parasite of pigs. Its life cycle has been known since 1868 and was more recently reported in detail by Kates.[13] When the eggs are eaten by white grubs (larvae of the beetle family Scarabaeidae), they hatch in the midgut within an hour and penetrate its lining. Within 5 to 20 days postinfection, the developing acanthellas are found free in the hemocoel or attached to the outer surface of the serosa. By 60 to 90 days postinfection, the cystacanth is infective to the definitive host. Pigs are infected by eating the grubs or the adult beetles, which have metamorphosed with their parasites intact.

Most archiacanthocephalans are parasites of predaceous birds and mammals, so paratenic hosts often are involved in life cycles within this class.

EFFECTS OF THE PARASITE ON ITS HOST

Because most acanthocephalans are parasitic in wild animals, their host-parasite relations have been little studied. Surveys of effects of these parasites on wild mammals and captive primates have been discussed.[20,22]

The nature of damage to intestinal mucosa is primarily traumatic, by penetration of the proboscis, and is compounded by the tendency of the worm to release its hold occasionally and reattach at another place. Complete perforation of the gut sometimes occurs, and in mammals, at least, the results are often rapidly fatal (Fig. 31-10). Great pain accompanies this phase: infected monkeys show evident distress, and Grassi and Calandruccio recorded the symptoms of pain and delirium experienced by Calandruccio after he voluntarily infected himself with cystacanths of *Moniliformis moniliformis*, common parasite of domestic rats.[9]

It is suspected that secondary bacterial infection is responsible for localized and

Fig. 31-10. Complete perforation of the large intestine of a squirrel monkey by *Prosthenorchis elegans*. (From Schmidt, G. D. 1972. In Fiennes, R. N. T. W., editor. Pathology of simian primates, vol. 2. S. Karger AG, Basel.)

generalized peritonitis, hemorrhage, pericarditis, myocarditis, arteritis, cholangiolitis, and other complications.

In view of the invasive nature of the parasites, it is surprising that they elicit so little inflammatory response in many cases. The reaction seems mainly a result of the traumatic damage, with granulomatous infiltration and sometimes collagenous encapsulation around the proboscis. Some species show evidence that antigens are released from pores on the proboscis hooks (as in *M. hirudinaceus*), and the inflammatory response is intense. It is clear that the pathogenesis caused by acanthocephalans can be severe but little consideration usually is given to the effects of this group of parasites as a controlling factor of wildlife populations.

Little chemotherapy has been developed for acanthocephalans. Various authors have proposed chenopodium and castor oil, calomel and santonin, carbon tetrachloride, and tetrachloroethylene for primates and pigs, with varying results. Oleoresin of aspidium has been used successfully in human cases but is not recommended for children. Mebendazole was used successfully in a 12-month-old child.[7] Control of intermediate hosts is helpful in preventing infection of domestic or captive animals.

ACANTHOCEPHALA IN HUMANS

Records of Acanthocephala in humans are few, no doubt because of the nature of the intermediate and paratenic hosts involved in the life cycles of the parasites. Few peoples of the world eat such animals as insects, microcrustaceans, toads, or lizards, at least without cooking them first. Yet, human infections with five different species have been reported.[21] *Macracanthorhynchus hirudinaceus* has been recognized as an occasional parasite of humans from 1859 to the present. *Moniliformis moniliformis* has been found repeatedly in people. *Acanthocephalus rauschi* is known only from specimens taken from the peritoneum of an Alaskan Eskimo, an obvious case of accidental parasitism, for the proper host is undoubtedly a fish. The zest of Eskimos for raw fish probably contributes to such zoonotic infections rather commonly. *Corynosoma strumosum*, a common seal parasite, also has been found in humans. More puzzling is a case of *Acanthocephalus bufonis*, a toad parasite, in an Indonesian. In this instance, it is probable that the man ate a raw paratenic host.

Thus, it seems that the Acanthocephala do not pose much of a threat to human health. They are much more important as parasites of wild and captive animals, where sudden epizootics have been known to kill a great number of individuals in a short time.

PHYLOGENETIC RELATIONSHIPS

While the phylogenies of most parasitic groups of animals seem reasonably clear, the thorny-headed worms are closely related to no known form. Their affinities with the nematodes and Nematomorpha seem indicated by the presence of a pseudocoel and exterior "cuticle." However, the tegument is quite different from the cuticle of nematodes; and the body wall structure, the eversible, spined proboscis with its accompanying mechanisms, and the complete lack of a digestive system set

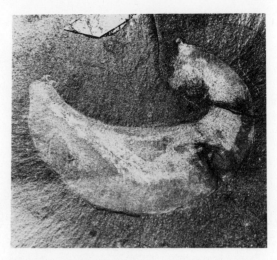

Fig. 31-11. *Ottoia prolifica,* a mid-Cambrian fossil from British Columbia. The similarities of this worm to modern-day Acanthocephala are striking. (Photograph by S. Conway-Morris. USNM cat. no. 57619; from Crompton, D. W. T. 1975. In Symbiosis. Symposia of the Society for Experimental Biology. Cambridge University Press, Cambridge.)

these worms well apart from other phyla in the superphylum Aschelminthes. They seem to represent a small relict of a larger population, most members of which have become extinct. This has left phylogenetic gaps between the Acanthocephala and other phyla and, indeed, between several groups within the phylum as well. A further paradox is the fact that the Archiacanthocephala, parasites of the most highly evolved vertebrates, are in many ways more primitive than the Eoacanthocephala, which are parasites of fishes. Golvan discussed the subject in detail and proposed a hypothetical ancestor, which he named *Protacanthocephala.*[8]

Very interesting fossils were found in the mid-Cambrian Burgess Shale of British Columbia in 1911.[32] The worms, named *Ottoia prolifica,* show many similarities to existing acanthocephalans (Figs. 31-11 and 31-12). Vertebrates had not yet evolved in the mid-Cambrian; so *Ottoia,* if a parasite, could only have parasitized invertebrates, such as trilobites, which were very abundant then. More likely it was a predator or scavenger.

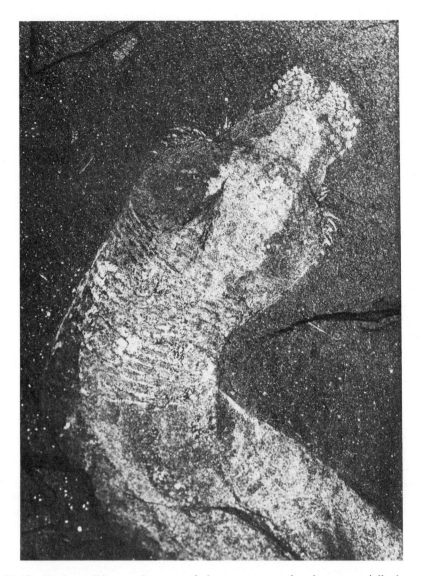

Fig. 31-12. *Ottoia prolifica:* a close-up of the praesoma, showing a partially inverted proboscis. Note the conspicuous trunk spines. (Photograph by H. B. Whittington. Geological Survey of Canada no. 40972; from Crompton, D. W. T. 1975. In Symbiosis. Symposia of the Society for Experimental Biology. Cambridge University Press, Cambridge.)

METABOLISM

Because of the availability of a good laboratory subject (*Moniliformis dubius* in rats), investigators have been able to accumulate some knowledge of acanthocephalan metabolism. However, the problem of assessing the general applicability of observations on *M. dubius* and the few others reported is acute. (*Moniliformis dubius* is probably a junior synonym of *M. moniliformis,* but we retain the name here because of the large literature that has accumulated on the physiology of the organism under this name.)

Acquisition of nutrients

Since the Cestoda and the Acanthocephala are both groups that must obtain all nutrient molecules through their body surfaces, comparisons between the two are quite interesting, particularly in light of their structural differences.[16] Some points of divergence will be pointed out in the brief account to follow. One difference, of course, is the ability of the acanthocephalan tegument to carry out pinocytosis,[4] mentioned before, though the importance of the process in acquisition of nutrients is uncertain.

Acanthocephalans can absorb at least some triglycerides, amino acids, nucleotides, and sugars. Amino acids are absorbed, at least partially, by stereospecific membrane transport systems in *M. dubius* and *M. hirudinaceus*.[28] The surface of *M. dubius* contains peptidases, which can cleave several dipeptides, and the amino acid products are then absorbed by the worm.[29] Absorbed thymidine is incorporated into DNA in the perilacunar regions and into the nuclei of the ovarian balls and testes. Nuclei in the body wall are not labelled by radioactive thymidine; therefore, it is assumed that the DNA synthesized there is mitochondrial.

Like the tapeworm *Hymenolepis diminuta*, *M. dubius* has an absolute dependence on host dietary carbohydrate for growth and energy metabolism as an adult.[25,26] The worm can absorb glucose, mannose, fructose, and galactose, as well as several glucose analogs. Absorption of glucose is through a single transport locus, while transport of mannose, fructose, and galactose is mediated both by the glucose locus and another site, referred to by Starling and Fisher as the "fructose site."[26] Maltose and glucose-6-phosphate (G6P) are absorbed also, but first they are hydrolyzed to glucose by enzymes in or on the tegumental surface. Acid phosphatase is localized in the crypts of the tegument[5]; kinetic evidence suggests that the hydrolysis of the G6P and part of the glucose absorption occurs there, while some of the glucose diffuses back out of the crypts and is absorbed by the outer surface.[26] A most interesting observation, and one in sharp contrast to tapeworm and other glucose transport systems, is that glucose absorption by *M. dubius* is not coupled to cotransport of sodium. Therefore, an alternative explanation for the mediated

transport of glucose is required. Part of the explanation may lie in the fact that glucose is rapidly phosphorylated, and this removal of free glucose from the vicinity of the tegumental transport loci would form a metabolic sink for the flow of additional hexose down its concentration gradient. However, substantial amounts of free glucose are found in the body wall. Evidence suggests the free glucose pool is not derived directly from absorbed glucose but is first "shuttled" through the nonreducing disaccharide, trehalose. The glucose would then be deposited, perhaps by intervention of a membrane-bound trehalase, in an internal membranous compartment that by some means can resist the efflux of the glucose it contains. The scheme offers an interesting possible metabolic role for trehalose, perhaps similar to that in insects.

Energy metabolism

Moniliformis dubius can ferment the hexoses it absorbs: glucose, fructose, galactose, and mannose. As in trematodes and cestodes, glycolysis is an important degradative pathway. The tricarboxylic acid cycle apparently does not operate in adult *M. dubius* or *M. hirudinaceus*, although there is evidence for it in *Echinorhynchus gadi*, a parasite of cod.

As in the other helminth parasites, the energy metabolism is adapted for facultative anaerobiosis. The terminal reactions of glycolysis derive more energy from glucose than would classical glycolysis alone and also reoxidize reduced coenzymes and produce less toxic end products. *Moniliformis* fixes carbon dioxide, and the principal enzyme of carbon dioxide fixation is phosphoenolpyruvate carboxykinase (see Fig. 20-35). It is likely that some or all of the reactions shown in Fig. 20-35 occur in acanthocephalans, though species vary. Lactate and succinate are the main end products of glucose degradation in *Polymorphus minutus*. Interestingly, the main end products of glycolysis in *M. dubius* are ethanol and carbon dioxide with a small amount of lactate and only traces of succinate, acetate, and butyrate. The presence of the enzymes necessary to produce pyruvate from phosphoenolpyruvate (pyruvate kinase), lactate from pyruvate (lactate dehydrogenase), and ethanol (and carbon dioxide) from pyruvate (pyruvate decarboxylase and alcohol oxidoreductase) in *M. dubius* has been confirmed.[14] Even though phosphoenolpyruvate carboxykinase activity is high, it must be regulated in such a way that the major end products are ethanol and lactate, rather than succinate. One possible explanation may lie in the fact that fumarate hydratase (see Fig. 20-35) is very low or absent. Thus, the terminal reactions of glycolysis in *M. dubius* resemble those of yeast rather than those of other intestinal helminths. However, the alcohol oxidoreductase of the acanthocephalan requires NADP rather than being NAD-dependent, as is the yeast enzyme.

Lipids apparently are not used as energy sources. Körting and Fairbairn found that endogenous lipids were not metabolized during in vitro incubation of *M. dubius*.[14] This was correlated with the fact that enzymes necessary for the β-oxidation of lipids were low in activity, and one of them seemed to be completely absent.

Electron transport in acanthocephalans has been studied very little. Oxidation of both succinate and NADH lead to reduction of cytochrome b.[2] Two pathways for reoxidation of this compound have been postulated, the major one independent of cytochrome c and cytochrome oxidase. This is somewhat similar to the branched chain electron transport postulated for the cestode, *Moniezia expansa* (see Fig. 20-34).[1] It could provide an additional means for reoxidation of NADH and generation of ATP by anaerobic mechanisms.

Classification of Phylum Acanthocephala

Class ARCHIACANTHOCEPHALA

Main longitudinal lacunar canals dorsal and ventral or just dorsal. Hypodermal nuclei few. Giant nuclei present in lemnisci and cement glands. Two ligament sacs persist in females. Protonephridia present in one family. Cement glands separate, piriform. Eggs oval, usually thick-shelled. Parasites of birds and mammals. Intermediate hosts are insects or myriapods.

Order MONILIFORMIDA

Trunk usually pseudosegmented. Proboscis cylindrical, with long, approximately straight rows of hooks; sensory papillae absent or present. Proboscis receptacle double-walled; outer wall with muscle fibers arranged spirally. Proboscis retractor muscles pierce posterior end of receptacle or somewhat ventral. Brain near posterior end or near middle of receptacle. Protonephridial organs absent.

Family: Moniliformidae

Order GIGANTORHYNCHIDA

Trunk occasionally pseudosegmented. Proboscis a truncate cone, with approximately longitudinal rows of rooted hooks on the anterior portion, and rootless spines on the basal portion. Sensory pits present on apex of proboscis and each side of neck. Proboscis receptacle single-walled with numerous accessory muscles, complex, thickest dorsally. Proboscis retractor muscles pierce ventral wall of receptacle. Brain near ventral, middle surface of receptacle. Protonephridial organs absent.

Family: Gigantorhynchidae

Order OLIGACANTHORHYNCHIDA

Trunk may be wrinkled but not pseudosegmented. Proboscis subspherical, with short, approximately longitudinal rows of few hooks each. Sensory papillae present on apex of proboscis and each side of neck. Proboscis receptacle single-walled, complex, thickest dorsally. Proboscis retractor muscle pierces dorsal wall of receptacle. Brain near ventral, middle surface of receptacle. Protonephridial organs present.

Family: Oligacanthorhynchidae

Class APORORHYNCHIDA

Main longitudinal lacunar canals dorsal and ventral. Hypodermal nuclei giant, ameboid, few. Giant nuclei present in lemnisci and cement glands. Ligament sacs persist in females. Proboscis receptacle absent. Brain near middle of proboscis. Cement glands separate, piriform. Eggs oval, thick-shelled. Parasites of birds.

Order APORORHYNCHIDEA

Trunk short, conical, may be curved ventrally. Proboscis large, globular, with tiny spine-like hooks (which may not pierce the surface of the proboscis) arranged in several spiral rows. Proboscis not retractable. Neck absent or reduced. Protonephridial organs absent.

Family: Apororhynchidae

Class PALAEACANTHOCEPHALA

Main longitudinal lacunar canals lateral. Hypodermal nuclei fragmented, numerous, occasionally restricted to anterior half of trunk. Nuclei of lemnisci and cement glands fragmented. Spines present on trunk of some species. Single ligament sac of female not persistent throughout life. Protonephridia absent. Cement glands separate, tubular to spheroid. Eggs oval to elongate, sometimes with polar thickenings of second membrane. Parasites of fishes, amphibians, reptiles, birds, and mammals.

Order ECHINORHYNCHIDA

Trunk never pseudosegmented. Proboscis cylindrical to spheroid, with longitudinal, regularly alternating rows of hooks; sensory papillae present or absent. Proboscis receptacle double-walled. Proboscis retractor muscles pierce posterior end of receptacle. Brain near middle or posterior end of receptacle. Parasites of fishes and amphibians.

Families: Illiosentidae, Rhadinorhynchidae, Heterosentidae, Fessisentidae, Diplosentidae, Heteracanthocephalidae, Hypoechinorhynchidae, Echinorhynchidae, Pomphorhynchidae

Order POLYMORPHIDA

Proboscis spheroid to cylindrical, armed with numerous hooks in alternating longitudinal rows. Proboscis receptacle double-walled, with brain near center. Parasite of reptiles, birds, and mammals.

Families: Polymorphidae, Plagiorhynchidae, Centrorhynchidae

Class EOACANTHOCEPHALA

Main longitudinal lacunar canals dorsal and ventral, often no larger in diameter than irregular transverse commissures. Hypodermal nuclei few, giant, sometimes ameboid. Proboscis receptacle single-walled. Proboscis retractor muscle pierces posterior end of receptacle. Brain near anterior or middle of receptacle. Nuclei of lemnisci few, giant. Two persistent ligament sacs in female. Protonephridia absent. Cement gland single, syncytial, with several nuclei, with cement reservoir appended. Eggs variously shaped. Parasites of fishes, amphibians, and reptiles.

Order GYRACANTHOCEPHALIDA

Trunk small or medium-sized, spined. Proboscis small, spheroid, with a few spiral rows of hooks.

Family: Quadrigyridae

Order NEOACANTHOCEPHALA

Trunk small to large, unarmed. Proboscis spheroid to elongate, with hooks arranged variously.

Families: Neoechinorhynchidae, Dendronucleatidae, Tenuisentidae

REFERENCES

1. Bryant, C. 1970. Electron transport in parasitic helminths and protozoa. In Dawes, B., editor. Advances in parasitology, vol. 8. Academic Press, Inc., New York. pp. 139-172.
2. Bryant, C., and W. L. Nicholas. 1966. Studies on the oxidative metabolism of *Moniliformis dubius* (Acanthocephala). Comp. Biochem. Physiol. 17:825-840.
3. Butterworth, P. E. 1969. The development of the body wall of *Polymorphus minutus* (Acanthocephala) in the intermediate host, *Gammarus pulex.* Parasitology 59:373-388.
4. Byram, J. E., and F. M. Fisher, Jr. 1973. The absorptive surface of *Moniliformis dubius* (Acanthocephala). I. Fine structure. Tissue Cell 5:553-579.
5. Byram, J. E., and F. M. Fisher, Jr. 1974. The absorptive surface of *Moniliformis dubius* (Acanthocephala). II. Functional aspects. Tissue Cell 6:21-42.
6. Crompton, D. W. T. 1970. An ecological approach to acanthocephalan physiology. Cambridge University Press, Cambridge.
7. Goldsmid, J. M., M. E. Smith, and F. Fleming.

1974. Human infections with *Moniliformis* sp. in Rhodesia. Ann. Trop. Med. Parasitol. 68:363-364.
8. Golvan, Y. J. 1958. Le phylum des Acanthocephala. Premiére note. Sa place dans l'echelle zoologique. Ann. Parasitol. 33:539-602.
9. Grassi, B., and S. Calandruccio. 1888. Ueber einen *Echinorhynchus*, welcher auch in Menschen parasitiert und dessen Zwischenwirt ein *Blaps* ist. Zentr. Bakterior. Parasitenk. Orig. 3:521-525.
10. Hamann, O. 1892. Das system der Acanthocephalen. Zool. Anz 15:195-197.
11. Hammond, R. A. 1966. The proboscis mechanism of *Acanthocephalus ranae.* J. Exp. Biol. 45:203-213.
12. Hopp, W. B. 1954. Studies on the morphology and life cycle of *Neoechinorhynchus emydis* (Leidy), an acanthocephalan parasite of the map turtle, *Graptemys geographica* (La Sueur). J. Parasitol. 40:284-299.
13. Kates, K. C. 1943. Development of the swine thorn-headed worm, *Macracanthorhynchus hirudinaceus,* in its intermediate host. Am. J. Vet. Res. 4:173-181.
14. Körting, W., and D. Fairbairn. 1972. Anaerobic

energy metabolism in *Moniliformis dubius* (Acanthocephala). J. Parasitol. 58:45-50.

15. Lankester, R. 1900. A Treatise on Zoology. Adam and Charles Black, London.

16. Lumsden, R. D. 1975. Surface ultrastructure and cytochemistry of parasitic helminths. Exp. Parasitol. 37:267-339.

17. Nicholas, W. L. 1967. The biology of the Acanthocephala. In Dawes, B., editor. Advances in parasitology, vol. 5. Academic Press, Inc., New York. pp. 205-246.

18. Nicholas, W. L. 1973. The biology of the Acanthocephala. In Dawes, B., editor. Advances in parasitology, vol. 11. Academic Press, Inc., New York. pp. 671-706.

19. Nickol, B. B., and D. F. Oetinger. 1968. *Prosthorhynchus formosus* from the short-tailed shrew *(Blarina brevicauda)* in New York State. J. Parasitol. 54:456.

20. Schmidt, G. D. 1969. Acanthocephala as agents of disease in wild mammals. Wildl. Dis. 53:1-10.

21. Schmidt, G. D. 1971. Acanthocephalan infections of man, with two new records. J. Parasitol. 57:582-584.

22. Schmidt, G. D. 1972. Acanthocephala of captive primates. In Fiennes, R. N. T. W., editor. Pathology of simian primates, vol. 2. S. Karger AG, Basel.

23. Schmidt, G. D. 1973. Early embryology of the acanthocephalan *Mediorhynchus grandis* Van Cleave, 1916. Trans. Am. Microsc. Soc. 92:512-516.

24. Schmidt, G. D., and O. W. Olsen. 1964. Life cycle and development of *Prosthorhynchus formosus* (Van Cleave 1918) Travassos, 1926, an acanthocephalan parasite of birds. J. Parasitol. 50:721-730.

25. Starling, J. A. 1975. Tegumental carbohydrate transport in intestinal helminths: correlation between mechanisms of membrane transport and the biochemical environment of absorptive surfaces. Trans. Am. Microsc. Soc. 94:508-523.

26. Starling, J. A., and F. M. Fisher, Jr. 1975. Carbohydrate transport in *Moniliformis dubius* (Acanthocephala). I. The kinetics and specificity of hexose absorption. J. Parasitol. 61:977-990.

27. Uglem, G. L., and O. R. Larson. 1969. The life history and larval development of *Neoechinorhynchus saginatus* Van Cleave and Bangham, 1949 (Acanthocephala: Neoechinorhynchidae). J. Parasitol. 55:1212-1217.

28. Uglem, G. L., and C. P. Read. 1973. *Moniliformis dubius*: uptake of leucine and alanine by adults. Exp. Parasitol. 34:148-153.

29. Uglem. G. L., P. W. Pappas, and C. P. Read. 1973. Surface amino peptidase in *Moniliformis dubius* and its relation to amino acid uptake. Parasitology 67:185-195.

30. Van Cleave, H. J. 1941. Relationships of the Acanthocephala. Am. Natur. 75:31-47.

31. Van Cleave, H. J. 1948. Expanding horizons in the recognition of a phylum. J. Parasitol. 34:1-20.

32. Wallcott, C. D. 1911. Middle Cambrian annelids. Cambrian Geology and Paleontology II. Smithsonian Misc. Coll. 57:109-144.

33. Ward, H. L. 1940. Studies on the life-history of *Neoechinorhynchus cylindratus* (Van Cleave 1913) (Acanthocephala). Trans. Am. Microsc. Soc. 59:327-347.

34. Whitfield, P. J. 1970. The egg sorting function of the uterine bell of *Polymorphus minutus* (Acanthocephala). Parasitology 61:111-126.

35. Whitfield, P. J. 1971a. Spermiogenesis and spermatozoan ultrastructure in *Polymorphus minutus* (Acanthocephala). Parasitology 62:415-430.

36. Whitfield, P. J. 1971b. The locomotion of the acanthor of *Moniliformis dubius* (Archiacanthocephala). Parasitology 62:35-47.

37. Wright, R. D., and R. D. Lumsden. 1968. Ultrastructural and histochemical properties of the acanthocephalan epicuticle. J. Parasitol. 54:1111-1123.

SUGGESTED READING

Baer, J. C. 1961. Embranchement des Acanthocéphales. In Grassé, P. editor. Traité de zoologie, anatomie, systématique, biologie, vol. 4. Plathelminthes, Mésozoaires, Acanthocéphales, Nemertiens. Masson et cie., Paris. pp. 733-782. (A well-illustrated synthésis of the group.)

Bullock, W. L. 1969. Morphological features as tools and as pitfalls in acanthocephalan systematics. In Schmidt, G. D., editor. Problems in systematics of parasites. University Park Press, Baltimore. pp. 9-43. (A useful, philosophical discussion of the subject, with recommended techniques for study.)

Crompton, D. W. T. 1970. An ecological approach to acanthocephalan physiology. Cambridge University Press, Cambridge. (An outstanding summation of the subject.)

Crompton, D. W. T. 1975. Relationships between Acanthocephala and their hosts. Symposium of the Society for Experimental Biology. XXIX. Symbiosis. Cambridge University Press, Cambridge. pp. 467-504.

Golvan, Y. J. 1959. Le phylum des Acanthocéphala, note 2. La classes des Eoacanthocephala (Van Cleave 1936). Ann. Parasitol. 34:5-52.

Golvan, Y. J. 1960-1961. Le Phylum des Acanthocephala, note 3. La classe de Palaeacanthocephala Meyer, 1931. Ann. Parasitol. 35:138-165, 350-386, 713-723 and 36:76-91.

Golvan, Y. J. 1962. Le phylum des Acanthocéphala, note 4. La classe des Archiacanthocephala (A. Meyer 1931). Ann. Parasitol. 37:1-72.

Golvan, Y. J. 1969. Systematiques des Acanthocéphales (Acanthocéphala Rudolphi 1801). L'order des Palaeacanthocephala Meyer 1931. La superfamille des Echinorhynchoidea (Cobbold 1876) Golvan et Houin 1963. Mem. Mus. Nat. Hist. Nat.

47:1-373.(This is a very up-to-date account of this important superfamily. Besides descriptions of each species, it contains a key to genera and a host list.

Meyer, A. 1932-1933. Acanthocephala. In Bronn's Klassen und Ordnungen des Tierreichs, vol. 4. Akad. Verlagsgesellschaft M. B. H., Leipzig. (This classical book is much out of date but is still a useful reference to some species.)

Pappas, P. W. and C. P. Read. 1975. Membrane transport in helminth parasites: a review. Exp. Parasitol. 37:469-530.

Petrochenko, V. I. 1956, 1958. Acanthocephala of domestic and wild animals, vols. 1 and 2. Akad. Nauk SSSR, Moscow. (English translations: Israel Program for Scientific Translations, 1971.) (An in-dispensable resource for students of the phylum. Descriptions are given for nearly every species known at the time of writing.)

Schmidt, G. D. 1972. Revision of the class Archi-acanthocephala Meyer, 1931 (Phylum Acanthoceph-ala), with emphasis on Oligacanthorhynchidae Southwell et MacFie, 1925. J. Parasitol. 58:290-297. (A modern classification of this difficult class.)

Whitfield, P. J. 1973. The egg envelopes of *Polymorphus minutus* (Acanthocephala). Parasitology 66: 387-403.

Yamaguti, S. 1963. Systema Helminthum, vol. 5. Acanthocephala. Interscience, New York. (In most regards, this is a practical key to all genera of Acanthocephala known to 1963. Lists of all species and their hosts are included.)

Chapter 32

PHYLUM PENTASTOMIDA: THE TONGUE-WORMS

The pentastomids, or tongue-worms, are worm-like parasites of the respiratory systems of vertebrates. About 60 species are known. As adults, most live in the lungs of reptiles, especially snakes, lizards, and crocodilians, but one species lives in the air sacs of gulls and terns, and another inhabits the nasopharynx of canines and felines. The latter species is occasionally found as transient nymphs in the nasopharynx of humans, while other species, in their nymphal stages, also parasitize people. Thus, pentastomids are certainly of zoological interest and are also of some medical importance.[16]

The evolutionary relationships of pentastomids are obscure. Certain similarities with the Annelida have been pointed out, but most modern taxonomists align them with the Arthropoda.[13] It is possible that the group reached its zenith in the Mesozoic age of reptiles and that today's few species are relicts derived from those ancestors.[1] Recently, Wingstrand proposed that the Pentastomida be regarded as an order of the crustacean subclass Branchiura[18] (see Chapter 33). His conclusion is based on a demonstration that the spermatozoa of the two groups are nearly identical with regard to structure and development and that this type of spermatozoan represents a type of its own, not encountered in other animals. His argument is strengthened by the fact that each major crustacean group is characterized by its own type of spermatozoa, and if the Pentastomida and the Branchiura were unrelated, their sperm structure and development would represent a most extraordinary example of convergence in detail. Certainly, the dissimilarity of the adult morphology is great, but this in itself would not preclude Wingstrand's thesis. The adult thoracican barnacle is equally as dissimilar to an adult *Sacculina* (Chapter 33), but they both belong to the subclass Cirripedia based on developmental evidence. For the present, however, we will adopt a conservative view and retain the Pentastomida in a phylum of its own.

MORPHOLOGY

The body of a pentastomid (Fig. 32-1) is elongate, usually tapering toward the posterior end, and often showing distinct segmentation, forming numerous **annuli.** It is indistinctly divided into an anterior **forebody** and a posterior **hindbody,** which is bifurcated at its tip is some species.

The exoskeleton contains chitin,[17] which is sclerotized around the mouth opening and accessory genitalia. A striking characteristic of all adult pentastomids is the presence of two pairs of sclerotized hooks in the mouth region (Fig. 32-2). These may be located at the ends of stumpy stalks or may be nearly flush with the surface of the cephlothorax; in either case, they can be withdrawn into cuticular pockets. The hooks are single in some species and double in others. The apparently double hooks are actually single, with an accessory hook-like protrusion of the cuticle. In some species, the hook articulates against a basal fulcrum. The hooks are manipulated by powerful muscles and are used to tear and embed the mouth region into host tissues.

The body cuticle in some species also has circular rows of simple spines; the annuli may overlap enough to make the abdomen look serrated. There usually are transverse rows of cuticular glands, with conspicuous pores, of unknown function.

The chitinous cuticle is similar to that of the arthropods.[17] The muscles, too, are arthropodan in nature, being striated and segmentally arranged. The only sensory

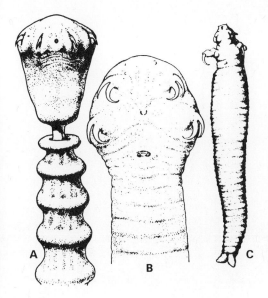

Fig. 32-1. Examples of pentastome body types. **A,** Anterior end of *Armillifer annulatus.* **B,** Head of *Leiperia gracilis.* **C,** Entire specimen of *Raillietiella mabuiae.* (From Baer, J. G. [adapted from Heymons]. 1952. Ecology of animal parasites. The University of Illinois Press, Urbana, Ill.)

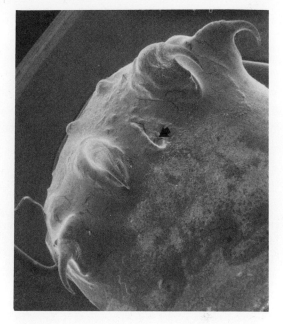

Fig. 32-2. Anterior end of a pentastome. Note both the mouth (arrow) between the middle hooks and the apical sensory papillae. (Photograph by John Ubelaker.)

structures so far recognized are papillae, especially on the exoskeleton of the cephalothorax. The digestive system is simple and complete, with the anus opening at the posterior end of the abdomen. The mouth is permanently held open by its sclerotized lining, the **cadre,** which may be circular, oval, or U-shaped, and is an important taxonomic character. The nervous system is arthropodan in nature and has been described by Doucet.[3]

Pentastomids are dioecious and usually show little sexual dimorphism except that males are usually smaller than females. The male has a single, tubular testis (two in *Linguatula*), which occupies one-third to one-half of the body cavity (Fig. 32-3, *A*). It is continuous with a seminal vesicle, which in turn connects to a pair of ejaculatory organs. These each have a duct that extends to a terminal cirrus, or penis, that fits into a **dilator organ.** The dilator organ, which is usually sclerotized, serves as an intromittent organ in some species and serves as a dilator and guide for the cirrus in others. The male genital pore is midventral on the anterior abdominal segment, near the mouth.

In the female, a single ovary extends nearly the length of the body cavity (Fig. 32-3, *B*). It may bifurcate at its distal end to become two oviducts. These unite to form the uterus. The oviducts and uterus usually are extensively coiled within the body. One or more diverticulae of the uterus serve as seminal receptacles. The uterus terminates as a short vagina that opens through the female gonopore, at the anterior end of the abdomen in the order Cephalobaenida or at the posterior end in the order Porocephalida.[6]

BIOLOGY (Fig. 32-4)

Adult pentastomids feed on tissue fluids and blood cells of their host. The female may produce several million fully embryonated eggs, which pass up the trachea of its host, are swallowed, then pass out with the feces. The intact egg appears to be surrounded by two shell membranes— an outer, thin membrane and an inner, thick one.[4] The inner layer, however, con-

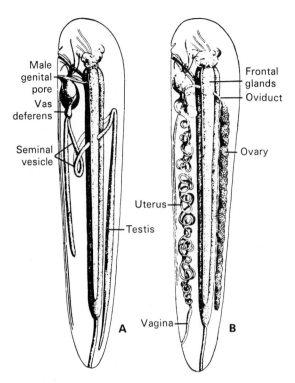

Male genital pore
Vas deferens
Seminal vesicle
Testis
Vagina
A

Frontal glands
Oviduct
Ovary
Uterus
B

Fig. 32-3. Reproductive systems of the pentastome *Waddycephalus teretiuscules*. **A,** Male; **B,** female. (From Baer, J. G. [adapted from Spomer in Heymons]. 1953. Ecology of animal parasites. The University of Illinois Press, Urbana, Ill.)

sists of three distinct layers. A characteristic of the pentastomid egg is the **facette,** a permanent, funnel-shaped opening through the inner membrane complex, with an inner opening extending toward the larva. In the embryo, a gland called the **dorsal organ** (to be described further) secretes a mucoid substance that pours through the facette and ruptures the original outer membrane, which is lost. The mucoid material then flows over the inner membrane to form a new outer membrane, which is sticky when wet.[10] The viscid eggs cling together, sometimes resulting in massive infections in the intermediate host. The eggs can withstand drying for at least 2 weeks, in the case of *Porocephalus crotali,* and they can remain viable in water at refrigerator temperatures for about 6 months.[4]

The larva that hatches from the egg is an oval, tailed creature with four stumpy legs, each with one or two retractable claws. The claws are manipulated by a combination of muscle fibers and an inner

hydraulic mechanism. A penetration apparatus is located at the anterior end of the body. This is composed of a median spear and two lateral, pointed forks; together with the clawed legs, these structures can tear through the tissues of the intermediate host. A pair of ducts open on either side of the median spear. Their origin and the nature of their secretion are unknown. Accessory spinelets are present around the penetration organ of many species.

Between the anterior legs is a simple mouth, surrounded by a U-shaped sclerotized cadre. An esophagus extends into the dorsal part of the body and expands into a blind sac; a thin hindgut is present in some species. Within the body cavity are a number of irregular giant cells, some with neutrophilic and others with eosinophilic granules. Their function is unknown. A consistent feature of the pentastomid embryo is the dorsal organ, referred to before. It consists of a number of gland cells surrounding a central hollow

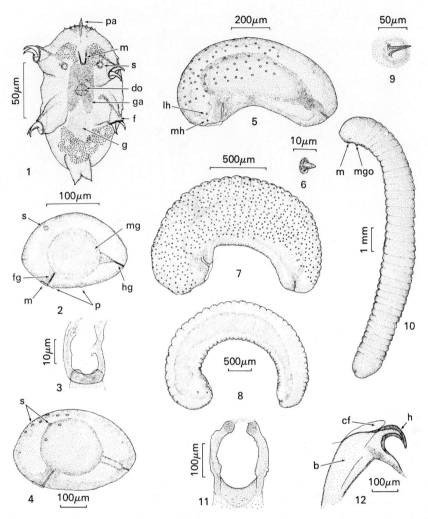

Fig. 32-4. Developmental stages of *Porocephalus crotali* in experimental intermediate hosts (camera lucida drawings made from living specimens). **1,** Primary larva (ventral view) after release from egg; **2,** first nymphal stage (nymph I) in left lateral view (all succeeding nymphs identically oriented); **3,** mouth ring of nymph I (en face view with anterior margin uppermost); **4,** nymph II; **5,** nymph III; **6,** lateral mouth hook of nymph III; **7,** nymph IV; **8,** nymph V (individual stigmata not shown); **9,** lateral mouth hook of nymph V; **10,** nymph VI (infective stage), male, removed from enveloping cuticle of nymph V; **11,** mouth ring of nymph VI; **12,** lateral mouth hook of nymph VI. *b,* Base of mouth hook; *cf,* cuticular fold or auxiliary hook; *do,* dorsal organ; *f,* foot or leg; *fg,* foregut; *g,* gut; *ga,* ganglion; *h,* external claw-like portion of mouth hook; *hg,* hindgut; *lh,* lateral mouth hook; *m,* mouth ring; *mg,* midgut; *mgo,* male genital opening; *mh,* medial mouth hook; *p,* papilla; *pa,* penetrating apparatus; *s,* stigma. (From Esslinger, J. H. 1962. J. Parasitol. 48:452-456.)

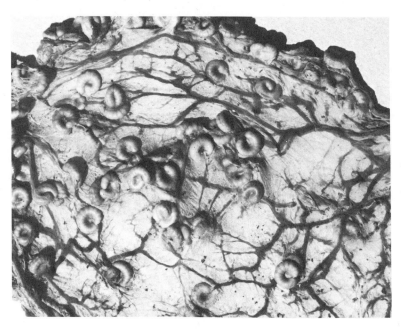

Fig. 32-5. Nymphs of *Porocephalus* sp. in the mesenteries of a vervet, *Cercopithicus aethops.* (Photograph by Robert E. Kuntz; from Self, J. T. 1972. Trans. Am. Microsc. Soc. 91:2-8.)

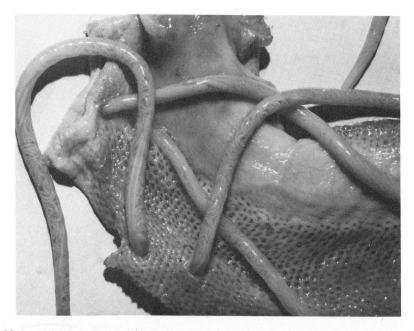

Fig. 32-6. *Kiricephalus pattoni* in the lung of an Oriental rat snake, *Ptyas mucosus.* (Photograph by Robert E. Kuntz.)

vesicle. The vesicle opens through the cuticle by a dorsal pore.

Complete life cycles are known for few species, but partial information is available for several. With the exceptions of *Reighardia sternae* in birds and a few species of *Linguatula* in mammals, all pentastomids mature in reptiles. The intermediate hosts are various fish, amphibians, reptiles, or, rarely, mammals. Typically, after ingestion by a poikilothermous vertebrate, the larva hatches and penetrates the intestine and migrates randomly in the body, finally becoming quiescent and metamorphosing into a nymph (Fig. 32-5). The nymph is infective to the definitive host; when eaten by the latter, it penetrates the host's intestine and bores into the lung, where it matures (Fig. 32-6). Each developmental stage undergoes one to several molts of the cuticle. The nymphal instars are difficult to differentiate. Some species even become sexually mature before completing the final ecdysis. A definitive host that eats the egg can also serve as intermediate host, similar to the case of *Trichinella*. The parasites, however, probably cannot migrate to the lung and mature.[1] While vertebrates are the intermediate hosts for the Porocephalida, cockroaches are used by at least one species of *Raillietiella*,[9] and *Reighardia sternae* in gulls has a direct life cycle.

The following two life cycles illustrate the biology of the group.

Porocephalus crotali

The life cycle of *Porocephalus crotali* of crotalid snakes was experimentally demonstrated by Esslinger (Fig. 32-4), using white mice as intermediate hosts.[5] On hatching, the larva penetrates the duodenal mucosa and works its way to the abdominal cavity. Complete penetration can be accomplished within an hour after the egg is swallowed. After wandering about for 7 or 8 days the larva molts and becomes lightly encapsulated in host tissue. The subsequent nymphal stages are devoid of the larval characteristics, having lost the legs, penetration apparatus, and tail. During the next 80 days or so, it molts five more times, gradually increasing in size and becoming segmented. The mouth hooks appear during the fourth nymphal instar and increase in size through subsequent ec-

dyses. The sexes can be differentiated after the fifth molt. After the sixth molt, the nymph becomes heavily encapsulated and dormant. When eaten by a snake, the nymph is activated; it quickly penetrates the snake's intestine and usually passes directly into the lung, as the lung and intestine are adjacent to each other. It buries its forebody into lung tissues, feeds on blood and tissue fluids, and matures.

Linguatula serrata

Linguatula serrata is unusual among the Pentastomida in that the adults live in the nasopharyngeal region of mammals. Cats, dogs, foxes, and other carnivores are the normal hosts of this cosmopolitan parasite. Apparently, nearly any mammal is a potential intermediate host.

Adult *L. serrata* embed their forebody into the nasopharyngeal mucosa, feeding on blood and fluids. Females live at least 2 years and produce millions of eggs.[7] The eggs are about 90 by 70 μm in size, with an outer shell that wrinkles when dry. Eggs exit the host in nasal secretions or, if swallowed, with the feces. When swallowed by an intermediate host, the four-legged larva hatches in the small intestine, penetrates the intestinal wall, and lodges in tissues, particularly in lungs, liver, and lymph nodes. There the nymphal instars develop, with the infective stage becoming surrounded by host tissues. When eaten by a definitive host, the infective nymph either attaches in the upper digestive tract or quickly travels there from the stomach, eventually reaching the nasopharynx. Females begin egg production in about 6 months.

PATHOGENESIS

There are two aspects of **pentastomiasis** in humans. **Visceral pentastomiasis** results when eggs are eaten and nymphs develop in various internal organs, and **nasopharyngeal pentastomiasis** results when nymphs that are eaten locate in the nasopharynx. Both types are rather commom in some parts of the world.

Visceral pentastomiasis

Several species of pentastomids have been found encysted in humans. Probably the most commonly involved species is *Ar-*

millifer armillatus, which has been reported from the liver, spleen, lungs, eyes, and mesenteries of people in, among other places, Africa, Malaysia, the Philippines, Java, and China.[2,15,16] Other reported species are *A. moniliformis, Pentastoma najae, Linguatula serrata,* and *Porocephalus* sp.

Most infections cause few if any symptoms and, therefore, go undetected. In fact, most recorded cases were found at autopsy, after death from other causes. However, infection of the spleen, liver, or other organs causes some tissue destruction. Ocular involvement may cause vision damage.[11] Prior visceral infection may sensitize a person, resulting in an allergy to subsequent infection.[8,14] The host response to nymphs is often highly inflammatory, although little pathological response is elicited in definitive reptilian hosts. Dead nymphs are often calcified and are sometimes detected in x-rays. Others begin a slow deterioration, causing a mononuclear cell response, with a subsequent abscess and granuloma formation. Experimentally produced heavy infections in rodents may kill them, indicating that visceral pentastomiasis possibly may be more important in human medicine than is usually thought.[14]

Nasopharyngeal pentastomiasis

When nymphs of *Linguatula serrata* invade the nasopharyngeal spaces of humans, they cause a condition usually called **halzoun,** also known as **marrara** or **nasopharyngeal linguatulosis.** According to Schacher and others, "Halzoun has been a clinically well recognized but aetiologically obscure disease in the Levant since its original description by Khouri (1905); in the Sudan it is known as the marrara syndrome. In Lebanon the disease is linked in the popular mind with the eating of raw or undercooked sheep or goat liver or lymph nodes; in the Sudan it is linked with the ingestion of various raw visceral organs of sheep, goats, cattle or camels. A few minutes to half an hour or more after eating, there is discomfort, and a prickling sensation deep in the throat; pain may later extend to the ears. Oedematous congestion of the fauces, tonsils, larynx, eustachian tubes, nasal passages, conjunctiva and lips is sometimes marked. Nasal and lachrymal discharges, episodic sneezing and coughing, dyspnoea, dysphagia, dysphonia and frontal headache are common. Complications may include abscesses in the auditory canals, facial swelling or paralysis and sometimes asphyxiation and death."[12]

At various times, this condition was suspected to be caused by the trematodes *Fasciola hepatica, Clinostomum complanatum,* or *Dicrocoelium dendriticum* or by leeches. However, the recovery of *L. serrata* nymphs from the nasal passages and throats of patients in India, Turkey, Greece, Morocco, and Lebanon indicate that this species is at least the main cause of the condition in those areas. It is possible that the parasites can become mature if not removed or lost initially.

"The epidemiology of this condition depends on cultural food patterns, in which nymphs are ingested when visceral organs, primarily liver or mesenteric lymph nodes of domestic herbivores, are consumed raw or undercooked."[12]

Classification of Phylum Pentastomida

Order CEPHALOBAENIDA

Mouth anterior to hooks. Hooks lacking fulcrum. Vulva at anterior end of abdomen.

Family RAILLIETIELLIDAE

Head somewhat pointed, may be quite prolonged. Hooks on ends of podial lobes, surrounded by small or conspicuous parapodial lobes. Parasites of reptiles.

Genera: *Cephalobaena, Raillietiella, Travassostulida, Mahafaliella, Gretillatia*

Family REIGHARDIIDAE

Podial and parapodial lobes absent, claws very small. Cuticle with small spines. Dorsal cephalic papillae absent. Posterior end rounded. Parasites of birds (Lariformes).

Genus: *Reighardia*

Order POROCEPHALIDA

Mouth between or below level of anterior hooks. Hooks with fulcrum. Vulva near posterior end of body.

Family SEBEKIDAE

Abdominal annuli short, serrated, numerous. Hooks form a trapezoid in ventral view. Mouth oval or U-shaped. Parasites of reptiles.

Genera: *Leiperia, Alofia, Sebekia, Diesingia*

Family SAMBONIDAE

Vulva separated from anus by 15 annuli. Buccal cadre heavily sclerotized, large, oval, or cordiform. Lateral lines poorly visible. Parasites of reptiles.

Genera: *Sambonia, Parasambonia, Waddycephalus, Elenia*

Family POROCEPHALIDAE

Vulva opens ventrally on last segment or in third annulus in front of anus. Buccal cadre heavily sclerotized, round, oval, or U-shaped. Parasites of reptiles.

Genera: *Porocephalus, Kiricephalus, Cubirea, Gigliolella, Armillifer, Ligamifer*

Family LINGUATULIDAE

Body flat, pointed toward posterior end. Testes double. Parasites of mammals (carnivores).

Genus: *Linguatula*

REFERENCES

1. Baer, J. G. 1952. Ecology of animal parasites. University of Illinois Press, Urbana, Ill.
2. Dönges, J. 1966. Parasitäre abdominalcysten bei Nigeriarern. Z. Trop. Parasitol. 17:252-256.
3. Doucet, J. 1965. Contribution à l'étude anatomique, histologique et histochimique des pentastomes (Pentastomida). Mem. Office Rech. Sci. Tech. Outre-Mer. Paris 14:1-150.
4. Esslinger, J. H. 1962a. Morphology of the egg and larva of *Porocephalus crotali* (Pentastomida). J. Parasitol. 48:457-462.
5. Esslinger, J. H. 1962b. Development of *Porocephalus crotali* (Humboldt, 1808) (Pentastomida) in experimental intermediate hosts. J. Parasitol 48:452-456.
6. Fain, A. 1961. Les pentastomides d'Afrique central. Mus. Roy. Afr. Cent. Ann. 92:1-115.
7. Hobmeier, A., and M. Hobmeier. 1940. On the life cycle of *Linguatula rhinaria*. Am. J. Trop. Med. 20:199-210.
8. Khalil, G. M., and J. F. Schacher. 1965. *Linguatula serrata* in relation to halzoun and the marrara syndrome. Am. J. Trop. Med. Hyg. 14:736-746.
9. Lavoippierre, M. M. J., and M. Lavoippierre. 1966. An arthropod intermediate host of a pentastomid. Nature 210:845-846.
10. Osche, G. 1963. Die systematische Stellung und Phylogenie der Pentastomids. Embryologische und vergleichend-anatomische Studien an *Reighardia sternae*. Z. Morph Ökol. Tiere 52:487-596.
11. Rendtdorff, R. C., M. W. Deiwesse, and W. Murrah. 1962. The occurrence of *Linguatula serrata*, a pentastomid, within the human eye. Am. J. Trop. Med. Hyg. 11:762-764.
12. Schacher, J. F., S. Saab, R. Germanos, and N. Boustany. 1969. The aetiology of halzoun in Lebanon: recovery of *Linguatula serrata* nymphs from two patients. Tr. R. Soc. Trop. Med. Hyg. 63:854-858.
13. Self, J. T. 1969. Biological relationships of the Pentastomida: a bibliography of the Pentastomida. Exp. Parasitol. 24:63-119.
14. Self, J. T. 1972. Pentastomiasis: host responses to larval and nymphal infections. Trans. Am. Micr. Soc. 91:2-8.
15. Self, J. T., H. C. Hopps, and A. O. Williams. 1972. Porocephaliasis in man and experimental mice. Exp. Parasitol. 32:117-126.
16. Self, J. T., H. C. Hopps, and A. O. Williams. 1975. Pentastomiasis in Africans, a review. Trop. Geogr. Pathol. 27:1-13.
17. Trainer, J. E. Jr., J. T. Self, and K. H. Richter. 1975. Ultrastructure of *Porocephalus crotali* (Pentastomida) cuticle with phylogenetic implications. J. Parasitol. 61:753-758.
18. Wingstrand, K. G. 1972. Comparative spermatology of a pentastomid, *Raillietiella hemidactyli*, and a branchiuran crustacean, *Argulus foliaceus*, with a discussion of pentastomid relationships. Det Kongelige Danske Videnskabernes Selskab Biologiske Skrifter 19(4):1-72.

Chapter 33

PARASITIC CRUSTACEA

A most interesting array of morphological adaptations for symbiosis can be found among the crustaceans. In addition to the academic interest they hold, numerous parasitic crustaceans are of substantial economic importance. In spite of this, they are often neglected in parasitology courses, as well as in courses in invertebrate zoology. Even though we are leaving coverage of other arthropod classes to entomology courses, the crustaceans are too good to pass up!

Some crustacean parasites have been known since antiquity, although the fact that they were crustaceans, or even arthropods, was not recognized until the early nineteenth century. Aristotle and Pliny recorded the affliction of tunny and swordfish by large parasites we would now recognize as lernaeocerid copepods. Rondelet (1554) figured a tunny with one of the copepods in place near the pectoral fin.[47] In 1746, Linnaeus first established the genus *Lernaea*,[35] and in his 1758 edition of *Systema Naturae,* he called the species (from European carp) *Lernaea cyprinacea*.[36] Various other highly modified copepods were described in the latter half of the eighteenth century and early nineteenth, but they had so few obvious arthropod features that they were variously classified as worms, gastropod molluscs, cephalopod molluscs, and annelids. Finally, Oken (1815-1816) associated these animals with other parasitic copepods that could be recognized as such.[40] Based on Surriray's important observation that their young, when hatched, resembled those of *Cyclops,* de Blainville (1822) firmly established these animals as crustaceans.[5]

Certain members of other crustacean subclasses have become so modified for parasitism as to be superficially unrecognizable as arthropods. Some of these will be discussed, but first it is necessary to consider some general characteristics of the class Crustacea of the phylum Arthropoda.

Crustaceans, along with other arthropods, are metameric Bilateria and have an exoskeleton containing chitin, jointed appendages, and a ventral nerve cord (primitively double). Their coelom is reduced, and the major body cavity is a hemocoel. The phylum is the most successful one on earth, judged both by the number of extant species and the number of individuals. The evolutionary plasticity of these animals has been fantastic, allowing them to exploit almost every type of niche capable of supporting metazoan life. The insects have radiated immensely in the terrestrial environment, and the crustaceans exhibit analogous radiation in marine habitats. Both classes are widely prevalent in freshwater niches. It is fair to ask what characteristics the arthropods possess that may have allowed such success. Certainly a major credit should go to the exoskeleton, or cuticle.

Cuticle. The cuticle is made up of several layers, all secreted by the hypodermis (Fig. 33-1), sometimes called the epidermis. The substance of the cuticle is nonliving, but the molting processes of the animal involve an active flux of resorption and depostion. The outermost layer, the **epicuticle,** is very thin and contains tanned protein, sometimes called **sclerotin,** but no chitin. Over the sclerotin, insects usually have a lipoidal layer that is of great value in preventing water loss through the cuticle, and over the lipid, they have a "varnish" layer that protects the wax from abrasion. The waxy and varnish layers are apparently missing in the Crustacea.

Beneath the epicuticle lies the thicker **procuticle,** the portion that lends the strength and weight to the exoskeleton. The procuticle contains protein, chitin, and in crustaceans, substantial deposits of calcium chloride along with some calcium

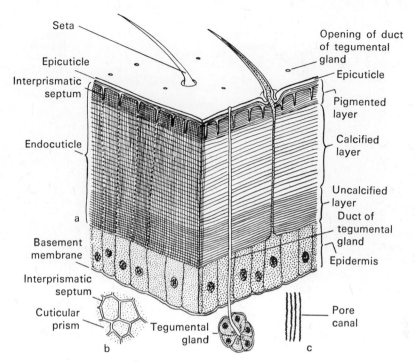

Fig. 33-1. *a*, Diagram showing structure of crustacean cuticle. All layers are secreted by the hypodermis (epidermis). The thin epicuticle is of tanned protein, and the procuticle (endocuticle) contains protein, chitin, and mineral salts. *b*, Horizontal section through pigmented layer of endocuticle. *c*, Pore canals as they appear in verticle sections. (From Dennell, R. 1960. In Waterman, T. H. editor. The physiology of crustacea, vol. 1. Academic Press, Inc., New York.)

phosphate and other inorganic salts. Chitin is a polysaccharide of N-acetyl glucosamine linked by 1,4-β-glycosidic bonds in long, unbranched molecules of high molecular weight. In contrast to the popular impression one may receive from the frequently used phrase "chitinous exoskeleton," chitin is flexible and contributes little to the rigidity of the cuticle. The hardness of the exoskeleton is conferred by tanned protein and especially by calcium carbonate in the **pigmented** and **unpigmented calcified layers.**

The **uncalcified layer** also contains chitin and protein, but here the protein is untanned, and the layer is membranous and flexible. Larger amounts of tanned protein (sclerotin), together with calcification, will lend more strength and mass to that area of the cuticle. (For discussion of quinonetanning, see Chapter 16.)

Obviously, all areas of the cuticle cannot be hard and massive, or the animal would be encased in an immovable box. The "joints" are thinner areas with little calcification or sclerotization and may be quite flexible. A sclerotized area limited by a suture line or flexible membrane is called a **sclerite.**

The protective function of the arthropodan exoskeleton is obvious, but therein does not lie the explanation for the success of its possessors. Metamerism conferred compartmentalization on the annelid ancestor, thus allowing the use of a "separate" hydrostatic skeleton compartment in each metamere, with localized movement of its appendages, greater efficiency, and development of greater complexity in muscle arrangement and nervous system. Nevertheless, by comparison with the arthropodan system of levers, the movements of worms are limited and inefficient. Such a system, in a genetically and

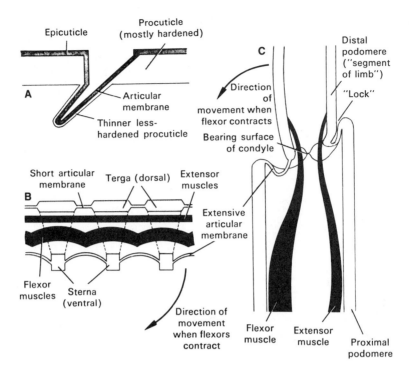

Fig. 33-2. Diagram of articulation and musculature of arthropod joints **A,** Flexibility is provided by thinner cuticle between sclerites. **B,** Greater flexion in one direction, such as in the abdomen of a crayfish, is made possible by the larger areas of thin cuticle on that side. **C,** Movement of a joint is by contraction of muscles inserted on opposite sides of the articulation, and the bearing surfaces of the joints are the condyles. (From Russell-Hunter, W. D. 1969. A biology of the higher invertebrates. The Macmillan Co., New York.)

evolutionarily plastic group, led to the enormous radiation we see today. Insertion of the muscles in the unyielding exoskeleton, coupled with fulcra provided by the condyles (Fig. 33-2) in the flexible joints, made very fine control of movements mechanically possible. Increase in complexity of movements meant corresponding evolution of nervous elements to coordinate the movements. Further, it should be noted that small changes in a given sclerite could substantially increase efficiency of a given body part for a given function, and arthropods have many sclerites. Thus, the arthropodan cuticle gave natural selection raw material on which to work.

Appendages have been modified to fulfill a variety of functions in a variety of ways. In addition to their primitive function of locomotion, appendages evidence a wide array of specialization for food-gathering, defense, reproduction, reception of stimuli, feeding, prehension, and respiration. A study of most arthropods requires a familiarity with the basic arrangement of appendages in that group, hence the introduction to crustacean appendages given later.

Molting. We must stress, however, that while the arthropod cuticle conferred many advantages of evolutionary potential, it conferred evolutionary problems as well. The chief of these was growth in an animal enclosed in a nonexpansible covering. The solution was a series of molts, or ecdyses, through which all arthropods go during their ontogeny. Much of the physiological activity of any arthropod is related to the molting cycle, and crustaceans are no exceptions. Growth in tissue mass occurs during an **interecdysis period,** and

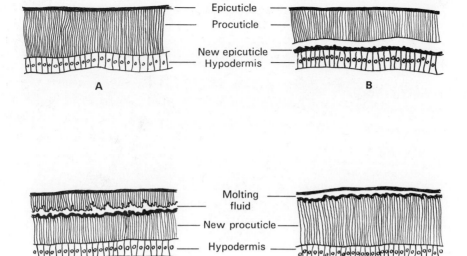

Epicuticle
Procuticle
New epicuticle
Hypodermis

A

B

Molting fluid
New procuticle
Hypodermis

C

D

Fig. 33-3. Cuticle secretion and resorption in preecdysis. **A,** Interecdysis condition. **B,** Old procuticle separates from hypodermis, which secretes new epicuticle. **C,** As new procuticle is secreted, molting fluid dissolves old procuticle, and the solution products are resorbed. **D,** At ecdysis, little more than the old epicuticle is left to discard. In postecdysis, new cuticle is stretched and unfolded, and more procuticle is secreted.

dimensional increase occurs immediately after molting, while the new cuticle is still soft. The immature stages of the animal between each molt are referred to as **instars.** Molting is controlled by hormones, at least in Malacostraca where the processes have been well investigated. A **molt-inhibiting hormone** (MIH) is produced in the **X-organ,** comprised of neurosecretory cells in the eyestalk, and **molting hormone** (MH) is produced in the **Y-organs,** a pair of glands near the mandibular adductor muscles (in decapods). The Y-organ is analogous, perhaps homologous, to the prothoracic glands that produce ecdysone in insects. Structures comparable to the X-organs are found within the head of sessile-eyed Malacostraca.[1]

As the titer of MIH decreases and that of MH increases, the organism undergoes a series of changes preparatory for a molt, the so-called **preecdysis** period. The hypodermis detaches from the old procuticle and starts secreting a new epicuticle (Fig. 33-3). At the same time, beneath the old procuticle, enzymes are liberated that begin to dissolve it. As the solution proceeds, the products of the reactions are re-

sorbed into the animal's body, including amino acids, N-acetylglucosamine, and calcium and other ions. These materials are thus salvaged and are incorporated into the new cuticle at the proper time.

Tanning in the new epicuticle occurs immediately, at least in species investigated,[32,53] thus protecting the newly secreted pigmented layer from the chitinase and other enzymes dissolving the old cuticle above. When almost all of the old cuticle has been absorbed, leaving essentially only old epicuticle, and much of the new cuticle has been secreted, the animal ecdyses. The old cuticle splits, normally along particular lines of weakness, or dehiscence, and the organism climbs out of its old clothes. The animal must expand to split the old cuticle, and before the new cuticle hardens, it must expand to larger body dimensions. Crustaceans accomplish this by rapid imbibition of water, a process aided by the fact that the osmotic pressure in the tissues and hemolymph has been increased prior to the molt by the calcium ions mobilized from the cuticle.[46] The increase in blood and tissue volume causes the small wrinkles in the still soft cuticle

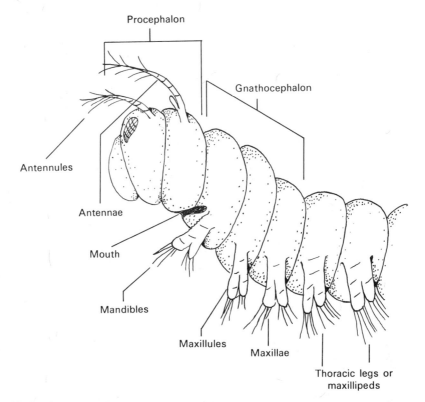

Fig. 33-4. Diagrammatic representation of a metameric precrustacean. The head of the modern crustacean has been derived from fusion of three ancestral preoral somites (procephalon) and three ancestral postoral somites (gnathocephalon). One or more additional thoracic somites may be incorporated into the head, and their appendages then are maxillipeds. Appendages of procephalon and gnathocephalon gave rise to head appendages as noted.

to smooth out, increasing the body dimensions, and the cuticle begins to harden again. In this **postecdysis period,** tanning of the protein and redeposition of calcium chloride in the procuticle occurs, and more procuticle is secreted. The animal is highly vulnerable to attack by predators while its cuticle is soft.

The length of the subsequent intermolt phase depends on the species involved, its age, and the season or annual cycle. Decapods that molt on an annual cycle and have a long intermolt period are said to be **anecdysic**.[30] Species in which one ecdysis cycle grades rapidly into another are **diecdysic**. Some crabs reach a maximum size and stop molting, undergoing "terminal anecdysis." These phases are controlled by MH and MIH produced by the Y- and X-organs described previously. In

crustaceans other than Malacostraca, little is known of the hormonal control of molting. At least one example, a barnacle, seems to be in a permanent diecdysis.[4] Furthermore, many copepod parasites of fish cease molting when they reach the adult stage, though they continue to grow actively. For example, a female *Lernaeocera* is about 2 mm long after her last molt, but she may attain an ultimate size of up to 60 mm *without molting*. What changes in the cuticle when the copepod reaches sexual maturity, and what causes the change? We do not know, but it is clear that the change permits continuous growth.

Appendages. The Insecta and Crustacea, along with the Myriapoda, have been considered by many zoologists to comprise the subphylum Mandibulata, as contrasted with the Chelicerata, which contains the

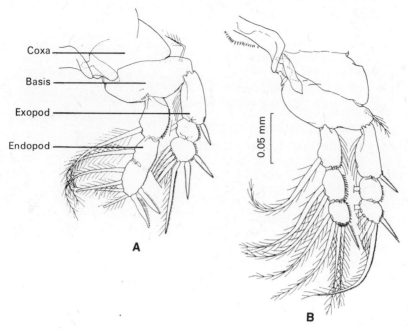

Coxa

Basis

Exopod

Endopod

0.05 mm

A

B

Fig. 33-5. First **(A)** and second **(B)** thoracic appendages of *Ergasilus megaceros* (Copepoda), a parasite of the sucker *Catastomus commersoni*. The terminal segments of the first endopod are fused, the ancestral condition being indicated by the presence of vestigial condyles. (From Roberts, L. S. 1970. Trans. Am. Microsc. Soc. 89:144.)

spiders, scorpions, ticks, and mites. Others maintain that the Mandibulata is an artificial grouping, advancing evidence that the mandibles of crustaceans and insects were evolved independently.[37] In any case, mandibles in the insects are homologous to those in crustaceans to the degree that they were derived from the appendage borne by the first (primitively) postoral somite of their ancestors.

The preoral portion of the head, the **procephalon** (Fig. 33-4), in both groups has been derived from at least three primitive somites, but the anteriormost of these bears no appendage, and an embryonic coelomic compartment in some forms points to its existence. The appendages of the second preoral somite gave rise to the **antennules** (first antennae) and those of the third to the **antennae** (second antennae) of crustaceans. The antennae of insects are homologous to the crustacean antennules, and the third preoral somite of insects bears no appendages (intercalary segment). The appendages of the three postoral somites, the **gnathocephalon,**

gave rise to the **mandibles, maxillules** (first maxillae), and **maxillae** (second maxillae) of crustaceans, respectively (mandibles, maxillae, and labium of insects). Therefore, the head of crustaceans is primitively made up of six fused somites, five of which bear appendages.

Crustacea is the only arthropodan class with two pairs of antennae, and except in rare cases, both pairs are present in all members of the class. In addition, one or more thoracic segments may be incorporated into the head, and the appendages of that segment may be modified as feeding appendages, or **maxillipeds.** They are followed by the other thoracic appendages, the **pereiopods,** and finally by the abdominal appendages, or **pleopods.** The pereiopods and pleopods may be variously modified for walking, swimming, or copulation.

The appendages of Crustacea were primitively **biramous** (having two main branches) (Fig. 33-5), and this condition prevails in at least some appendages of all living species during their lives. It should be pointed out that the terminology applied by various workers to crustacean appendages has not been blessed with uniformity. At least two systems are currently in wide use, and we have given the alternative term for each structure in parentheses. The lateral branch is the **exopod** (exopodite), and the medial one is the **endopod** (endopodite), each of which may contain several segments, varying by appendage and according to species. The endopod and exopod are borne on a **basis** (basipodite) and the basis, in turn, is attached to the **coxa** (coxopodite), together being referred to as the **protopod**. Processes from the protopod are termed **endites** and **exites,** and the exites may be called **epipods** (epipodite). Several of these terms are not usually applied to the appendages of the subclass Branchiopoda, but because no symbiotic branchopods are known, description of their limbs is not necessary. The two branches of the legs may not be homologous through all the crustacean subclasses.[48]

Development. Most crustaceans have **centrolecithal** eggs and superficial **cleavage,** that is, nuclei undergo several divisions without division of the cytoplasm, then migrate to the periphery to become the **blastoderm.** Yolk is concentrated in the interior of the embryo while differentiation proceeds in the superficial areas. Embryogenesis will not be detailed here, but the typical larva that hatches is called the **nauplius** (Fig. 33-6). The nauplius has only three pairs of appendages: antennules, antennae, and mandibles. These are quite different in form from the adult appendages and have locomotive function. The nauplius undergoes several ecdyses, usually adding somites and appendages at each molt. Nauplii typically have several instars, and the later ones may be referred to as **metanauplii.**

In the most primitive crustaceans, the body form becomes more like the adult with each molt, but most present-day species have larval instars and sometimes postlarval forms that are quite different from both the nauplius and the adult. The former type of development is known as **anamorphosis,** and the latter is **metamorphosis.** The suppression of larval instars, or passing stages comparable to them, be-

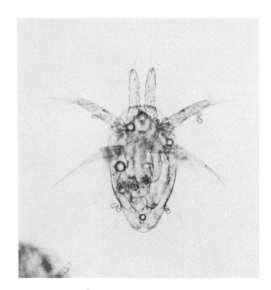

Fig. 33-6. Copepod nauplius. The anteriormost, uniramous appendages are the antennules, followed by the biramous antennae and mandibles. Note the internal, parasitic nematode and the external, phoretic Protozoa. (Photograph by Ralph Muller.)

fore hatching is common, and a juvenile, rather than a larva, may hatch with segmentation and appendages complete. Such development is called **epimorphosis.**

The adaptability and wide variation of the Crustacea are demonstrated by the variety of developmental patterns in the various groups. Thus, even within the same subclass, such as the Copepoda or the Malacostraca, some forms may be slightly metamorphic and some strongly metamorphic or even epimorphic. Some specific examples will be cited.

Other physiological functions. Other functions, such as feeding and digestion, osmoregulation, or excretion, of the Crustacea would have little applicability here because so many parasitic species are quite different from free-living forms in these respects. Some details of function that are known will be mentioned in relation to specific cases.

SUBCLASS COPEPODA

The copepods constitute one of the largest crustacean subclasses, second only to the Malacostraca. Their evolutionary versatility is displayed by the fact that several groups of copepods have been able to exploit symbiotic niches and by the spectrum of adaptations to symbiosis that they show, ranging from the slight to the extreme. It was pointed out that lernaeocerids are so bizarre that eighteenth century biologists did not recognize them as arthropods, but many parasitic and commensal copepods are comparatively much less highly modified. In fact, one can arrange examples of the various groups in an arbitrary series to demonstrate the progression from little to very high specialization.[27]

We shall cite but a few examples to illustrate the trends in adaptation to parasitism in copepods. Some of these trends are (1) reduction in locomotor appendages, (2) development of adaptations for adhesion, both by modification of appendages and by development of new structures, (3) increase in size and change in body proportions, caused by much greater growth of genital or reproductive regions, (4) fusion of body somites and loss of external evidence of segmentation, (5) reduction of sense organs, and (6) reduction in numbers of instars that are free-living, both by passing more stages before hatching and by larval instars becoming parasitic. "Typical" or primitive copepod development can be regarded as slightly metamorphic with a series of **copepodid** instars succeeding the nauplii. Copepodid larvae bear considerable similarity to the adults except in strongly metamorphic families like Lernaeopodidae and Lernaeoceridae.

Order Cyclopoida

The order Cyclopoida is a large group of copepods, important for its free-living members and symbiotic ones as well. Free-living cyclopoids occupy important niches as primary consumers in many aquatic habitats, and many species parasitize fish and invertebrates. Some of its members are among the economically most significant parasitic copepods. Fish farming, or intensive culture of fish in small ponds for food production, has been practiced in Europe and Asia for centuries. Under the crowded conditions of most fish ponds, chances for transmission of any infectious agent, including copepods, are greatly enhanced. Fish farming in North America is

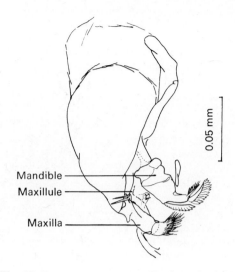

Fig. 33-7. Mouthparts (one side drawn) of *Ergasilus cerastes,* a parasite of catfishes *(Ictalurus* spp.). (From Roberts, L. S. 1969. J. Parasitol. 55:1268.)

growing, and with it the potential economic loss from such organisms as *Ergasilus* and *Lernaea.*

The diagnostic criteria for cyclopoids are, for the most part, based on characteristics of the free-living representatives, but some copepods now considered cyclopoids are highly modified as adults, such as *Lernaea.* It is clear that some criteria must be adduced that are applicable to the more specialized types, as well as to the unmodified forms. The buccal orifice of the cyclopoid does not project much above the body surface and is usually just a hole only partially covered anteriorly by the labrum (Fig. 33-7). The mandible consists of a short, stocky basal part and a sickle-shaped (falciform) distal part. The distal part is usually long, pointed, and with rows of denticles on its margins. In contrast, caligoids have tubular mouths, and the mandibles are long, flat blades, with a row of teeth on one margin on the distal end (Fig. 33-8). Another important characteristic is the fact that caligoids have a specialized, parasitic copepodid larva, the **chalimus,** which is attached to its host by means of a peculiar frontal filament, produced by a gland at the anterior of the cephalothorax (Fig. 33-9). No chalimus has been described in the life cycle of even the most specialized cyclopoids.

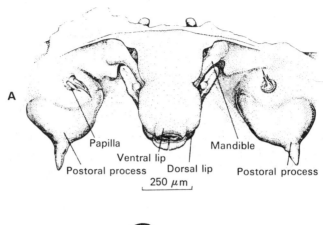

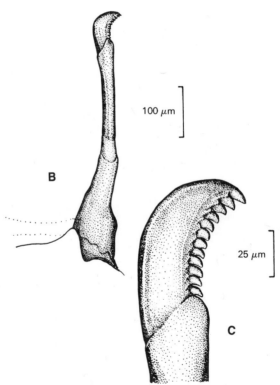

Fig. 33-8. Oral region of *Caligus curtus,* which parasitizes a variety of marine fishes. **A,** The base of the mandible can be seen as it extends into the tube formed by the dorsal and ventral lips. **B** and **C,** The mandible is a long, flat blade with teeth at its distal tip. (From Parker, R. R., Z. Kabata, L. Margolis, and M. D. Dean. 1968. J. Fish. Res. Bd. Can. 25:1960.)

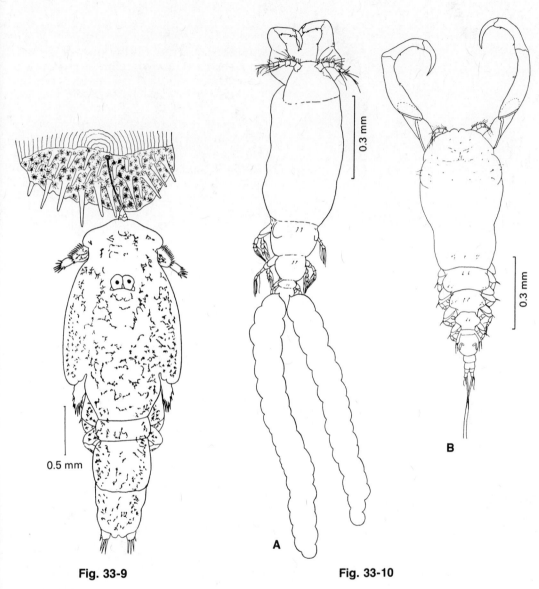

Fig. 33-9   **Fig. 33-10**

Fig. 33-9. Chalimus larva of *Caligus rapax*. (From Wilson, C. B. 1905. Proc. U.S. Nat. Mus. 28:549.)

Fig. 33-10. Examples of *Ergasilus,* a common parasite of freshwater and some marine fishes. **A**, *Ergasilus celestis,* from eels *(Anguilla rostrata)* and burbot *(Lota lota),* bearing egg sacs. **B,** *Ergasilus arthrosis,* reported from several species of freshwater hosts, nonovigerous. (From Roberts, L. S. 1969. J. Fish. Res. Bd. Can. 26:1000, 1008.)

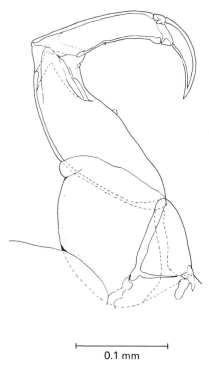

0.1 mm

Fig. 33-11. Antenna of *E. centrarchidarum*, a common parasite of members of the sunfish family (Centrarchidae). The antennae of *Ergasilus* are usually modified into a powerful organ used to grasp their host's gill filament, with the third and fourth joints opposable with the second. (From Roberts, L. S. 1970. Trans. Am. Microsc. Soc. 89:27.)

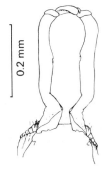

0.2 mm

Fig. 33-12. The tips of the antennae of *Ergasilus tenax* "lock" together, completely encircling the host's gill filament. (From Roberts, L. S. 1965. J. Parasitol. 51:989.)

Family Ergasilidae

Ergasilids are among the most common copepods parasitic of fish. They have been a "thorn in the flesh for many valuable fisheries in the Old World" for a long time,[27] and often frequent the gills of a variety of fishes in North America.[45,55] *Ergasilus* spp. are primarily parasites of freshwater hosts, but are common on several marine fishes, especially the more euryhaline ones such as sticklebacks, killifishes, and mullets.

Ergasilidae exhibit the primitive cyclopoid morphology with few, but effective, adaptations for parasitism. The antennules are sensory, but the antennae have become modified into powerful organs of prehension (Fig. 33-10). *Ergasilus* females usually are found clinging by their antennae to one of the fish's gill filaments. The primitive number of segments in the antenna has been reduced to four, and the terminal segment is characteristically in the form of a sharp claw. The third and fourth segments are opposable with the second (Fig. 33-11). Rather than depending on muscle and heavy sclerotization of the antennae, the antennal tips may be fused or locked so that the gill filament is completely encircled *(E. amplectens, E. tenax)* (Fig. 33-12). When removed from their position on the gill, most *Ergasilus* can swim reasonably well; their pereiopods retain the primitive flat copepod form, with setae and hairs well adapted for swimming. The first legs, however, show adaptation for their feeding habit. These appendages are supplied with heavy, blade-like spines, and in some species, the second and third endopodal segments are fused, presumably lending greater rigidity to the leg. Such modifications increase the ability of the animal to rasp off mucus and tissue from the gill to which it is clinging (Fig. 33-13). The first legs dislodge epithelial and underlying cells in this manner and sweep them forward to the mouth (Fig. 33-14).[14] It is easy to see that a heavy infestation with *Ergasilus* could severely damage gill tissue, interfere with respiration, open the way to secondary infection, and lead to death. Epizootics of *Ergasilus* on mullet *(Mugil)* were recorded in Israel; in one case up to 50% of the stock in some ponds was lost, and hundreds of dead mullet were found daily.[49]

Fig. 33-13. *Ergasilus labracis* in situ on gills of striped bass, *Morone saxatilis* (two specimens are indicated by arrows.) The gill operculum has been removed. Note also that the fish is infected by an isopod, *Lironeca ovalis*, partly hidden under gill.

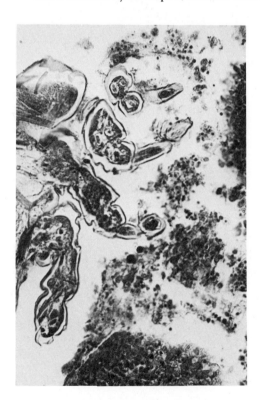

Fig. 33-14. Section of *Ergasilus sieboldi* in situ showing damage to gills inflicted by thoracic appendages. The tissue is rasped off, and the parasite feeds on detached epithelial, mucous, and blood cells. The first legs (at left) are particularly important in directing dislodged tissue anteriorly toward the mouth. (From Einszporn, T. 1965. Acta Parasitol. Polon. 13:380.)

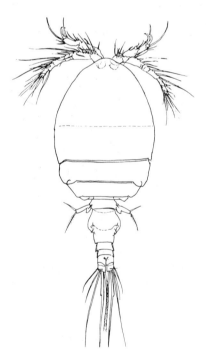

Fig. 33-15. A typical lichomolgid, *Ascidioxynus jamaicensis,* from the branchial sac of an ascidian, *Ascidia atra;* dorsal view of female. (From Humes, A. G., and J. H. Stock. 1973. Smithson. Contr. Zool., no. 127, p. 143.)

Ergasilus hatches as a typical nauplius; *E. sieboldi* has three naupliar and five copepodid stages, all free-living.[58] The adult males are planktonic as well, and the female is fertilized before attaching to the fish host. Only the female has been found as a parasite. The males of very few species are known, and in one of these, even the females are planktonic as adults (*E. chautauquaensis,* which may be the only nonparasitic species in the genus).

Family Lichomolgidae

Lichomolgids are symbionts with a wide variety of marine invertebrates, including serpulid polychaetes, alcyonarian and madreporarian corals, ascidians, sea anemones, nudibranchs, holothurians, starfishes, pelecypods, and sea urchins. Several other copepod families commonly are symbiotic with invertebrates, for example, the Notodelphyidae, in the body cavities of ascidians,[22] and the Mytilicolidae, in the intestines of pelecypods. It is evident that many species are involved, and many are yet to be discovered. The Lichomolgidae (family here broadly accepted) are divided by Humes and Stock into five families, embracing 76 genera and 324 species.[19,20]

Lichomolgids are generally cyclopoid in body form, with a retention of segmentation and swimming legs (Fig. 33-15). Segments of the antennae are reduced to three or four, and they often end in one to three terminal claws (Fig. 33-16). The antennae are apparently adapted for prehension in much the same manner as the ergasilids. Higher specialization is shown in some species, where one or more swimming legs may be reduced or vestigial. The copepodid larvae are often found parasitic on the same host as the adults, and relatively little time is apparently spent in the free-living naupliar stages. *Ostrincola koe* goes through six copepodid stages, all of which are parasitic in the mantle cavities of clams *(Tapes japonica),* where the adults also live.[31] *Lichomolgus canui,* a parasite of ascidians, goes through six free-living naupliar instars to the copepodid in 15 days.[12]

Family Lernaeidae

Lernaeidae is a relatively small family that parasitizes freshwater teleosts. Often quite large and conspicuous, some species,

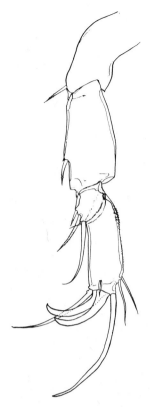

Fig. 33-16. Antenna of *Ascidioxynus jamaicensis* with three segments, ending in two claws. Ancestral form had four segments; third and fourth now fused, with a suture line marking former articulation. (From Humes, A. G., and J. H. Stock. 1973. Smithson. Contr. Zool., no. 127, p.143.)

especially *Lernaea cyprinacea,* are serious pests of economically important fishes. Therefore, they are among the best-known parasitic copepods. The genus *Lernaea* was established in 1746 by Linnaeus, and its ontogeny was clarified by Grabda in 1963.[16] *Lernaea cyprinacea* can infect a variety of fish hosts and even frog tadpoles.[51] The anterior of the parasite is embedded in the host's flesh and is anchored there by large processes that arise from the parasite's cephalothorax and thorax, hence the common name "anchor worm." They cause damage to the scales, skin, and underlying muscle tissue. There may be considerable inflammation, ulceration, and secondary bacterial and fungous infection. If the fish is small relative to the

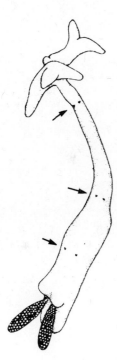

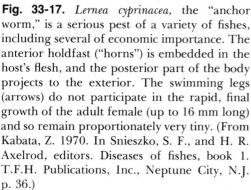

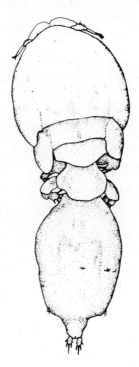

Fig. 33-17. *Lernea cyprinacea,* the "anchor worm," is a serious pest of a variety of fishes, including several of economic importance. The anterior holdfast ("horns") is embedded in the host's flesh, and the posterior part of the body projects to the exterior. The swimming legs (arrows) do not participate in the rapid, final growth of the adult female (up to 16 mm long) and so remain proportionately very tiny. (From Kabata, Z. 1970. In Snieszko, S. F., and H. R. Axelrod, editors. Diseases of fishes, book 1. T.F.H. Publications, Inc., Neptune City, N.J. p. 36.)

Fig. 33-18. *Dissonus nudiventris,* a more primitive caligid with three segments between the cephalothorax and genital somite. (From Kabata, Z. 1970. In Snieszko, S. F., and H. R. Axelrod, editors. Diseases of fishes, book 1. T.F.H. Publications, Inc., Neptune City, N.J. p. 21).

parasite, it can easily be killed by infection with several individuals. A fully developed *L. cyprinacea* may be over 12 mm long. Epizootics of this pest occur in wild fish populations, and it is a serious threat wherever fish are raised in hatcheries.[42]

Lernaeids are among the most highly specialized copepods. Once the sexually mature female is fertilized, she embeds her anterior end beneath a scale, near a fin base or in the buccal cavity. At that point, the parasite is less than 1.5 mm long and is superficially quite similar to *Cyclops* or other unspecialized cyclopoids. The female begins to grow rapidly, reaching "normal" size in little more than a week. The largest

specimen recorded was 15.9 mm (22 mm including the egg sacs).[16] Interestingly, the swimming legs and mouth parts do not take part in this growth, so that they quickly become very inconspicuous. At the same time, the large anchoring processes, two ventral and two dorsal (Fig. 33-17), grow into the fish's muscle. The body segmentation becomes blurred, the location of the somites being recognized only by finding the tiny legs. The result is an embryo-producing machine that bears practically no resemblance to an arthropod and that has its head permanently anchored in its food source. It is little wonder that the early taxonomists had such trouble correctly placing *Lernaea* in their system.

Nevertheless, as Blainville reported in 1882, the larvae can be clearly recognized as crustacean and are typical nauplii. The primitive series of naupliar instars has been shortened to three. When the nauplii

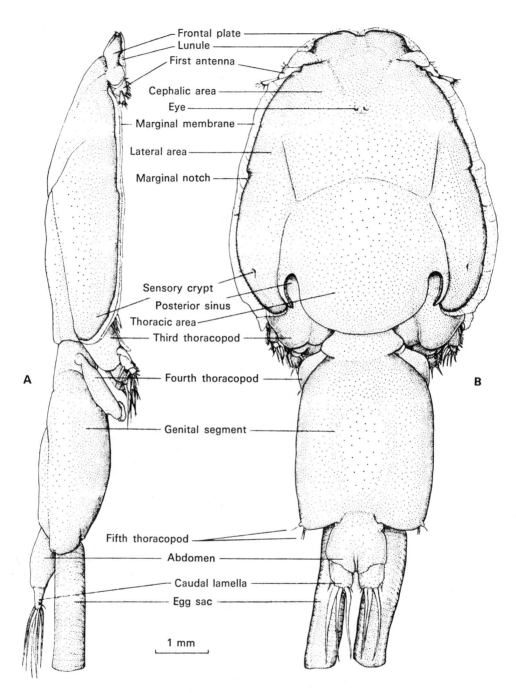

Frontal plate
Lunule
First antenna
Cephalic area
Eye
Marginal membrane
Lateral area
Marginal notch
Sensory crypt
Posterior sinus
Thoracic area
Third thoracopod
Fourth thoracopod
Genital segment
Fifth thoracopod
Abdomen
Caudal lamella
Egg sac

A
B

1 mm

Fig. 33-19. *Caligus curtus,* a more advanced caligid with only one segment between the cephalothorax and genital somite. **A** and **B,** female, lateral and dorsal views; **C,** female, ventral view; **D,** male, lateral view. (From Parker, R. R., Z. Kabata, L. Margolis, and M. D. Dean. 1968. J. Fish. Res. Bd. Can. 25:1951-1952.)

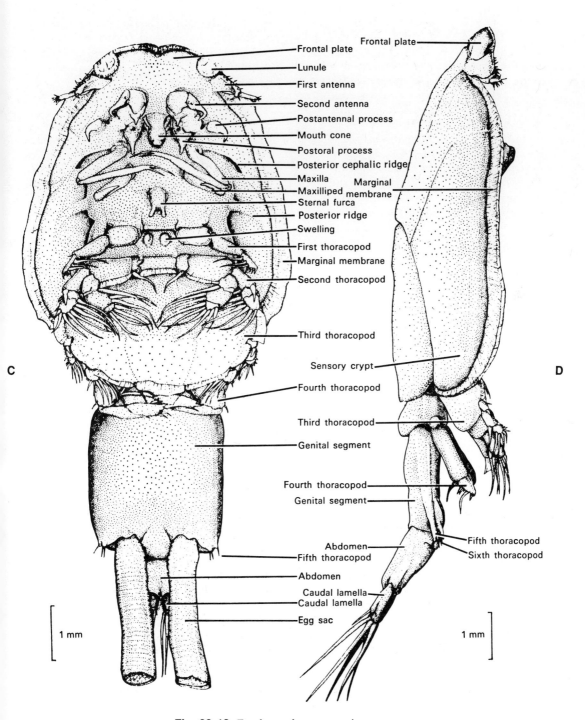

Frontal plate
Lunule
First antenna
Second antenna
Postantennal process
Mouth cone
Postoral process
Posterior cephalic ridge
Maxilla
Maxilliped
Sternal furca
Posterior ridge
Swelling
First thoracopod
Marginal membrane
Second thoracopod
Third thoracopod
Fourth thoracopod
Genital segment
Fifth thoracopod
Abdomen
Caudal lamella
Egg sac

Frontal plate
Marginal membrane
Sensory crypt
Third thoracopod
Fourth thoracopod
Genital segment
Abdomen
Fifth thoracopod
Sixth thoracopod
Caudal lamella

C

D

1 mm

1 mm

Fig. 33-19. For legend see opposite page.

hatch, they contain enough yolk material within their bodies to eliminate the need for feeding in any of the three naupliar stages. The third nauplius molts to give rise to the first copepodid, and this marks the end of the free-living life of a *Lernaea*. Thus, the length of time spent as a free-living organism has been markedly shortened, compared to the primitive condition, and the free-living instars do not even feed. The first copepodid has two pairs of swimming legs, is about 0.3 mm long, and is quite cyclopoid in form. If it does not find a host, further development apparently cannot proceed, but the first copepodid seems to be easy to satisfy. It can settle on the gills or skin of almost any available species of fish and proceed through four more copepodid instars. Each of these stages is marked by small increments in size and by additions of swimming legs or segments on the endopods and exopods of the legs.

The fifth copepodid finally molts to give rise to the sexually mature copepod, called the "cyclopoid" stage to distinguish it from the fully developed, sedentary female. The cyclopoid female is about 1.3 mm long, and the male is about 1.0 mm. The male attaches spermatophores to the genital segment of the female, giving her a lifetime supply of sperm. She then begins her anchored existence. All the larval and cyclopoid instars up to this point have been capable of free movement, even though they were parasites. However, a critical juncture is now reached. While the copepodids can develop on a wide variety of hosts, the inseminated females are more selective. They burrow beneath the scales and develop fully only on crucians *(Carassius carassius)*. If the copepodids develop on a crucian, they continue on the same fish, but if they develop on some other species, they detach and find the proper host. Thus, an intermediate host is not necessary in the strict sense, but occurrences of copepodids on other species probably contributes to the long confusion on this point in the literature.

It is curious that adult *L. cyprinacea* have been reported from such a variety of hosts belonging to remotely related families (Cyprinidae, Salmonidae, Centrarchidae, Catostomidae, and Ameiuridae) in several continents (Europe, North America, Asia and Africa). Some of these may represent mistaken identifications or strain or subspecies differences. Some 45 or so species of *Lernaea* have been described, but we consider only 28 valid[17]; these taxonomic problems of practical importance cannot be considered settled.

Order Caligoida

The order Caligoida is a large group, whose members are virtually all parasites of fish. Though even the most primitive caligoids show some adaptations to parasitism, like the cyclopoids, an array from generalized to extremely modified and bizarre can be demonstrated. The majority of caligoids are parasites of marine fishes, and since aquaculture of marine fishes has not yet been widely practiced, the actual or potential economic importance of the various caligoids is unknown. In western Japan, where culture of yellowtail *(Seriola)* is practiced intensively in small bays, *Caligus spinosus* has inflicted considerable damage.[23]

Family Caligidae

Adult caligids are quite evidently arthropods, although they have departed from the "typical" free-living copepod plan. They have, at least, some adaptations for prehension, tend to be larger than most free-living groups, and have some dorsoventral flattening for closer adhesion to the host surface. Some tend to be more sedentary, being mostly confined to the fish's branchial chamber, but the adults of many species can move rapidly over the host's surface (fins, gills, and mouth). They can swim and change hosts.

The usual caligid body form shows a fusion of the ancestral body somites: a large, flat cephalothorax, followed by one to three, free thoracic segments, a large genital segment, and a smaller unsegmented abdomen. The more primitive caligids, such as *Dissonus* (Fig. 33-18), have three segments between the cephalothorax and genital somite, while the more advanced forms, like *Caligus* (Fig. 33-19), have only one.[41] *Caligus curtus'* principal appendages for prehension are the antennae and maxillipeds. It has two **lunules** on the anterior margin of the cephalothorax,

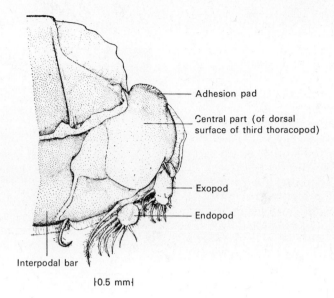

Fig. 33-20. Third thoracic leg of *Caligus curtus.* Note greatly enlarged, fused protopod with flexible marginal membrane. (From Parker, R. R., Z. Kabata, L. Margolis, and M. D. Dean. 1968. J. Fish Res. Bd. Can. 25:1966.)

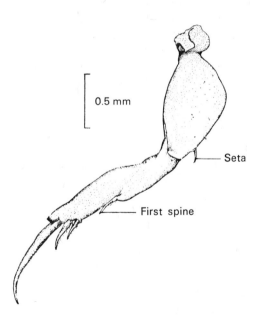

Fig. 33-21. Uniramous fourth thoracic leg of *Caligus curtus,* dorsal view. (From Parker, R. R., Z. Kabata, L. Margolis, and M. D. Dean. 1968. J. Fish Res. Bd. Can. 25:1966.)

which function as accessory organs of adhesion. The cephalothorax is roughly disc-shaped and bears a flexible, membranous margin. The posterior portion of the disc is not formed by the cephalothorax itself, but by the greatly enlarged, fused protopod of the third thoracic legs (Fig. 33-20). Membranes on the margin of the protopods match those on the cephalothorax, and the arrangement forms an efficient suction disc, when the cephalothorax is applied to the fish's surface and is arched.

Caligus on stationary or sluggishly moving fish perform settling movements at intervals to increase effectiveness of their "suction cup."[29] Settling involves slight, rapid, rotational movements on the longitudinal axis of the body, mediated by the maxillae, while the first and second thoracic legs pump water from beneath the parasite. Settling movements are rarely observed when caligids are attached to vigorously swimming fish, and water pressure against the parasite, always facing the current, is assumed sufficient to keep it pressed closely to its host. The first and second legs are the swimming structures, while the uniramous fourth legs (Fig. 33-21) are used least.

Interestingly, the third and fourth legs of the more primitive *Dissonus* are little modified; they are similar to the second legs, and all three pairs resemble the second legs of *C. curtus.*

The feeding apparatus of *C. curtus* is a good example of the tubular mouth type of the caligoids. The mouth tube is carried in a folded postion parallel to the body axis, but it can be erected so that its tip can be applied directly to the host surface. The tip of the tube bears flexible membranes analogous to those on the margin of the cephalothorax, again increasing the efficiency of the organ as a suction device. The bases of the mandibles are lateral to and outside the mouth tube. They enter the buccal cavity via longitudinal canals, so that their tips lie within the opening of the cone. The mandibular tips bear a sharp cutting blade on one side and a row of teeth on the other. Thus, the mandibles can work back and forth like little pistons in their canals, piercing and tearing off bits of host tissue to be sucked up by the muscular action of the mouth tube.

Although the developmental stages of caligids are mentioned in the older literature, reasonably complete descriptions for species of *Caligus* and *Lepeophtheirus* have only recently become available.[23,34] Both genera have only two naupliar stages, and these appear not to feed. The second nauplius molts to produce the first copepodid. The first copepodid must find a host or perish. If it finds its host, the copepodid clings to the fish with its prehensile antennae and molts to produce that specialized type of copepodid called the **chalimus.** Three more chalimus instars follow, all of them attached to the host by the frontal filament. The actual attachment process of a caligid is unknown, but it is probably similar to that of the lernaeopodid. The chalimus backs off from its point of attachment, thus pulling more filament out of the frontal organ, while stroking the filament with its maxillae.[34] The four chalimus instars are followed by two preadult stages in all caligids. The preadults are detached from the frontal filament, and they, as well as the adults, have the capacity for free movement over the host's body. Males are parasitic and not much smaller than females.

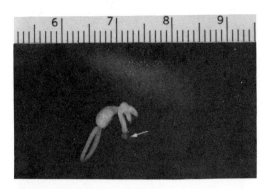

Fig. 33-22. *Salmincola inermis,* a lernaeopodid parasite of whitefish, *Coregonus* spp. The huge maxillae are fused to the bulla (arrow), which is embedded in the host's flesh, anchoring the female to that site. The powerful maxillipeds can be seen *anterior* to the maxillae, and the mouth is at the tip of the anteriormost cone-like projection.

Family Lernaeopodidae

The lernaeopodids are common, widespread parasites of fish, which frequently occur on freshwater hosts. *Salmincola,* a parasite of salmonids, has caused great damage to hatchery stocks in North America.[27] Lernaeopodids are substantially more modified away from the ancestral copepod form than are the Caligidae. Virtually all external signs of segmentation have disappeared in the adult (Fig. 33-22), as is the case with the Lernaeidae and Lernaeoceridae. Similarly, the adult females are permanently anchored in one place on the host. However, in contrast to these families, lernaeopodid females are attached almost completely outside the host; the anchor, or **bulla,** is nonliving and is formed from head and maxillary gland secretions. The maxillae themselves are fused to the bulla, and they are often huge. Occasionally, the maxillae are very short, as in *Clavella* (Fig. 33-23); but in these cases, a very long, mobile cephalothorax provides a "grazing range" similar in extent to that possible with longer maxillae.

Three types of bullae are recognized on the basis of morphology and are characteristically associated with lernaeopodids on freshwater teleosts, marine teleosts, and elasmobranchs.[27a] The first type, on fresh-

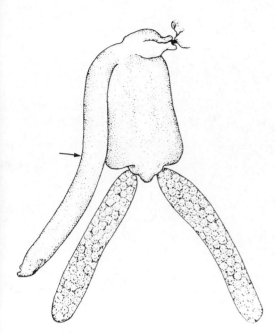

Fig. 33-23. The maxillae are very short in *Clavella,* but the cephalothorax (arrow) is long and mobile, providing an extended "grazing range." (From Kabata, Z. 1970. In Snieszko, S. F., and H. R. Axelrod, editors. Diseases of fishes, book 1. T.F.H. Publications, Inc., Neptune City, N.J. p. 39.)

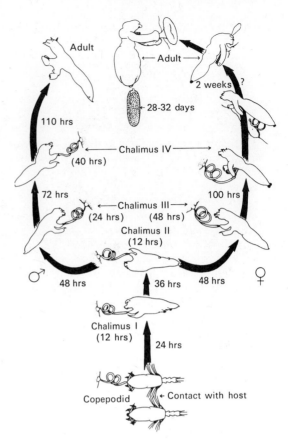

Fig. 33-24. Life cycle diagram of *Salmincola californiensis.* The times in parentheses refer to the duration of that particular stage, while those without parentheses denote time from first contact with host. (From Kabata, Z., and B. Cousens. 1973. J. Fish. Res. Bd. Can. 30:901.)

water hosts, is regarded as the most primitive. Bullae of this type tend to be large and to function mechanically as a "grappling iron" embedded in the host. The second and third types are relatively smaller and more intimately associated with host tissues, perhaps even secreting a cementing substance into them.

The maxillipeds are usually modified to form powerful grasping structures, and although they were primitively posterior to the maxillae, in most species they are now located and function more anteriorly. The bases of the maxillae mark the approximate posterior limit of the cephalothorax, and the rest of the body is the trunk, or fused thoracic and genital segments. The abdomen and swimming legs are absent or vestigial. There is extreme sexual dimorphism. The males are pygmies and are free to move around in search of females after the last chalimus stage. Both the maxillae and maxillipeds of the males are used as powerful grasping organs. The males,

however, do not use the bullae to anchor themselves.

The fascinating lernaeopodid development of *Salmincola californiensis* has been studied by Kabata and Cousens (Fig. 33-24).[28] As it hatches from the egg, the nauplius molts simultaneously to the copepodid. After the cuticle hardens, the copepodid must find a host within about 24 hours, or it dies. It attaches to the host with the prehensile hooks on the antennae and the powerful claws of the maxillae, and then it must find a suitable postion on the fish for placement of the frontal filament. It wanders over the host's skin until it finds a solid structure, such as a bone or fin ray, close to the surface. The maxillipeds exca-

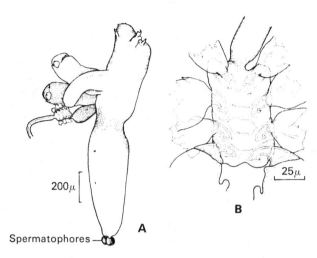

200μ

Spermatophores

A

25μ

B

Fig. 33-25. Early fourth chalimus of female *Salmincola californiensis* **A,** showing maxillae embedded in frontal filament. **B,** Enlarged end of frontal filament, showing tips of maxillae embedded in it (at bottom), along with the molted cuticle of maxillae tips from earlier chalimus stages. (From Kabata, Z., and B. Cousens. 1973. J. Fish. Res. Bd. Can. 30:888.)

vate a small cavity at that position and press the anterior end of the cephalothorax into the cavity. The terminal plug of the frontal filament detaches and is fixed to the underlying host structure by a rapidly hardening cement produced by the frontal gland. The copepodid moves backward, pulling the filament out of the frontal gland, and, if the attachment site is favorable and the copepodid has not been too much damaged by detachment of the frontal filament, it soon molts to the first chalimus stage.

These hazards destroy many copepodids. In the latter phases of the molt to the first chalimus, the maxillae work their way forward *within* the loosened copepodid cuticle and embed their sharp claws in the base of the frontal filament, which is now detached from the anterior margin of the body. With both of them firmly attached to the frontal filament, the maxillae must laboriously work back to their normal position over the obstacle posed by the mouth cone. One might think that the animal's problems are solved at this point, but, unfortunately, at next molt, the tips of the maxillae detach from the filament, as a hand is pulled from a glove. Therefore, at each molt the maxillae must renegotiate the mouth cone, reattach the end of the frontal filament to the anterior with a drop of cement from the frontal gland, pull free

and out of the maxillary sleeves, work their way forward again within the old cuticle to reembed in the filament, and then return to their normal position! After the cuticle is shed, it tears loose at the old maxillary tips, and the filament of the fourth chalimus still has the tips of the preceeding chalimi embedded in it (Fig. 33-25).

The fourth chalimus of the female is a long stage, during which the parasite frees herself from the frontal filament. To accomplish this, the maxillipeds, now large and well developed, step over the linked tips of the maxillae, and they, with their antennae, pull the parasite forward to the end of its filament tether. The parasite strains aginst the impeding filament until able to break loose, usually at the juncture of the last drop of cement, with the maxillae still joined. If one maxilla pulls loose, the animal is doomed because it cannot molt, being unable to withdraw the maxilla from its "sleeve" unless it can pull against the tip of the other maxilla.

After the fourth chalimus is free, it must find a suitable location for its permanent residence. With its antennae and mouth appendages, it rasps out a site for the bulla, now developing in the frontal organ. The maxillipeds cannot be used because they are the principal means of prehension. After molting, the bulla is everted, placed in the excavation, and detached from the anterior end. These processes are again dangerous for the parasite, which loses considerable body fluid, causing substantial mortality. Finally, the linked tips of the maxillae must find the opening in the implanted bulla, where they connect with small ducts and secrete cement from the maxillary glands. If and when this last maneuver is successful, the parasite is permanently attached to its host and can graze at will on the surface epithelium. Indeed it is no surprise that many copepods fail in this complicated series of developmental events; rather it is amazing that so many succeed.

Family Lernaeoceridae

The Lernaeoceridae are widespread and conspicuous parasites of marine fishes and mammals. They carry the evolutionary tendencies mentioned earlier to the extreme. Even the small ones are usually large by free-living copepod standards, and the large ones are the mammoths of the copepod world. *Pennella balaenopterae* from whales may be over a foot long! Their loss of external segmentation, obscuration of swimming appendages in the adult, and invasion of host tissue by their anterior ends are reminiscent of the cyclopoid family Lernaeidae. However, lernaeocerids tend to be more invasive of the circulatory system, sense organs, and viscera than are lernaeids. Each species usually has a characteristic site into which the anterior end grows and feeds. Several species, including all *Lernaeocera* spp., invade particular parts of the circulatory system, normally a large blood vessel. (It should be understood that the large trunk, bearing the reproductive organs and ovisacs, remains external.) Common sites are the heart, branchial vessels, and ventral aorta. On the Atlantic cod, *Gadus morhua*, *Lernaeocera branchialis* (Fig. 33-26) invades the bulbus arteriosus. The parasite generally attaches in the branchial area, and the cephalothorax may have to grow into and follow the ventral aorta for some distance. The associated pathogenesis is severe and is likely to have an impact on commercial fisheries. Two or more mature parasites on haddock *(Melanogrammus aeglefinus)* can cause the fish to be up to 29% underweight, have less than half the normal amount of liver fat, and have a decrease of one-half the hemoglobin content

Fig. 33-26. *Lernaeocera branchialis* from the Atlantic cod, *Gadus morhua*. The voluminous trunk of the organism, containing the reproductive organs, along with the coiled egg sacs, protrudes externally from the host in the region of the gills. The anterior end (at right) extends into the flesh of the host, and the antlers are embedded in the wall of the bulbus arteriosus, which is severely damaged. The antlers rarely penetrate the lumen of the bulbus, as this would lead to thrombus formation and death of both the parasite and host.

Fig. 33-27. *Phrixocephalus longicollum*, a lernaeocerid whose antlers proliferate into a luxuriant, intertwining growth. (From Kabata, Z. 1970. In Snieszko, S. F., and H. R. Axelrod, editors. Diseases of fishes, book 1. T.F.H. Publications., Inc., Neptune City, N.J. p. 33.)

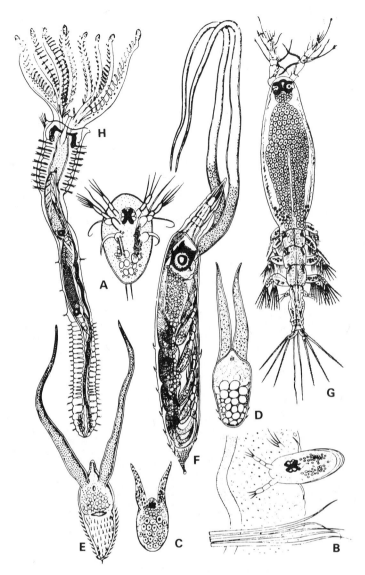

Fig. 33-28. *Haemocera danae,* a monstrilloid parasite of polychaete annelids. **A,** Nauplius **B,** Nauplius penetrating integument of host. **C-E,** Successive larval stages, showing development of absorptive appendages. **F,** Fully developed copepodid within spiny sheath. **G,** Adult female. **H,** Polychaete containing two copepodids in coelom. (From Baer, J. G. 1952. Ecology of animal parasites. University of Illinois Press, Urbana, Ill.)

of the blood.[26] At a 15% prevalence on haddock and 5% on cod, this parasite can be calculated to cause enormous losses.[25]

The form of adult females is grotesque. Anchoring processes, sometimes referred to as antlers, emanate from the anterior end. These are often more elaborate than those found in lernaeids. The greatest development of the antlers seems to be in *Phrixocephalus,* where many branches are

found (Fig. 33-27). *Lernaeolophus* and *Penella* have curious, branched outgrowths at the posterior part of the trunk, the function of which is unknown. As in the lernaeids, the appendages do not participate in the metamorphosis undergone by the rest of the female body; therefore, they are so small compared to the rest of the body as to be hardly discernible.

The life cycles of the lernaeocerids are

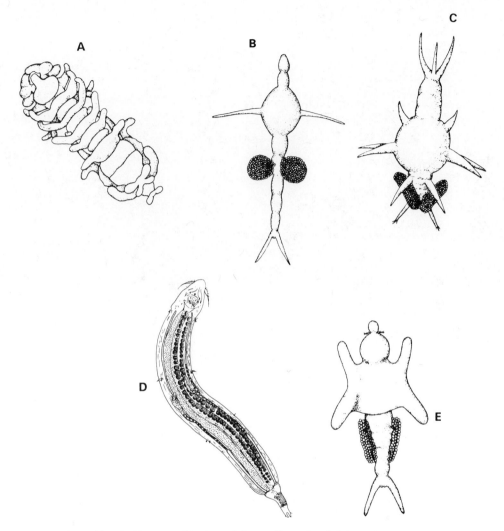

Fig. 33-29. Philichthyids, parasites in subdermal canals of fishes. **A,** *Philichthys xiphiae;* **B,** *Sphaerifer leydigi;* **C,** *Colobomatus sciaenae;* **D,** *Lerneascus nematoxys;* **E,** *Colobomatus muraenae.* (From Kabata, Z. 1970. In Snieszko, S. F., and H. R. Axelrod, editors. Diseases of fishes, book 1. T.F.H. Publications, Inc., Neptune City, N.J. p. 47.)

unique among the copepods in that they often require an intermediate host. Usually the intermediate host is another species of fish, but it may be an invertebrate (a cephalopod for *Pennella* and a pelagic gastropod for *Cardiodectes*). *Lernaeocera branchialis* apparently has only one naupliar stage, which leads a brief pelagic existence. The copepodid infects a flounder and undergoes several chalimus instars. The female is fertilized as a late chalimus while on the intermediate host and then detaches from the frontal filament. She undergoes another pelagic phase to search out the definitive host, a species of gadid

(cod family). The copepod attaches in the gill cavity, the anterior end burrows into the host tissue, aided by the strong antennae, and the dramatic metamorphosis begins. At the time she leaves the intermediate host, the female is only 2 to 3 mm long and is quite copepodan in appearance. In the metamorphosis, she loses all semblance of external segmentation and grows to 40 mm or more.

Orders Monstrilloida, Philichthyidea, Sarcotacidea

Little is known of the orders Monstrilloida, Philichthyidea, and Sarcotacidea, but they all deserve at least brief mention because of their great specialization for parasitism.

The Monstrilloida are parasites of polychaetes and gastropods. They have the distinction of being parasitic only during their larval stages. The adult is the free-living dispersal agent. They have the further distinction among the Crustacea of having only one pair of antennae. Neither monstrilloid larvae nor adults have a mouth or functional gut. The nauplius penetrates its host, which is either a polychaete or a prosobranch gastropod, depending on the species. It molts to become a rather undifferentiated larva with one to three pairs of apparently absorptive appendages (Fig. 33-28). Progressive differentiation and copepodid stages ensue, and the adult finally breaks out of the host to reproduce. Thus, only the adult and the nauplius are free-living, and the larval stages absorb food in a manner analogous to that of a tapeworm.

The general appearance of philichthyids is startling, to say the least, with unlikely looking processes emanating from their bodies (Fig. 33-29). This is a small group, completely endoparasitic in the subdermal canals of teleosts and elasmobranchs, that is, the frontal mucous passages and sinuses and the lateral line canal. Some species retain external evidence of segmentation, but in others it is less apparent. Because they have little use for organs of attachment, such appendages are reduced. The males are much smaller than the females and are less highly modified.

Sarcotacideans are also endoparasitic copepods and are probably the most highly specialized of any copepod parasite of a vertebrate (Fig. 33-30). They live in cysts in the muscle or abdominal cavity of their fish hosts. Their appendages are vestigial, and they appear to feed on blood from the vascular wall of the cyst. The adult female is little more than a reproductive bag within the cyst and may reach several centimeters in size. Males are much smaller, and one lives in each cyst, mashed between the wall of the cyst and the huge body of its mate.

Nothing is known of the development and many other aspects of the biology of sarcotacideans and philichthyideans. Their

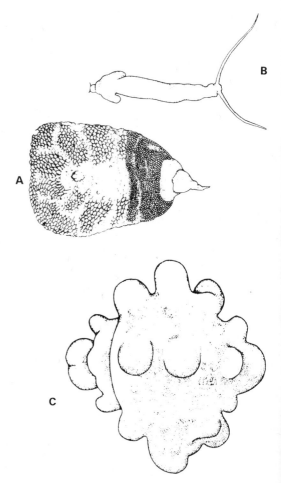

Fig. 33-30. Sarcotacideans may be the most highly specialized copepod parasites of vertebrates. **A,** *Sarcotaces* sp. female; **B,** *Sarcotaces* sp. male; **C,** *Ichthyotaces pteroisicola*. (From Kabata, Z. 1970. In Sniezko, S. F., and H. R. Axelrod, editors. Diseases of fishes, book 1. T.F.H. Publications, Inc. Neptune City, N.J. p. 49.)

sites on the hosts are so unobtrusive that these fascinating organisms probably occur much more widely than they have been reported; a diligent search would doubtless reveal a number of new species.

SUBCLASS BRANCHIURA

The subclass Branchiura is relatively small in numbers of species but great in its destructive potential in fish culture. All species are ectoparasites of fishes, although some can use frogs and tadpoles as hosts, too. They are dorsoventrally flattened,

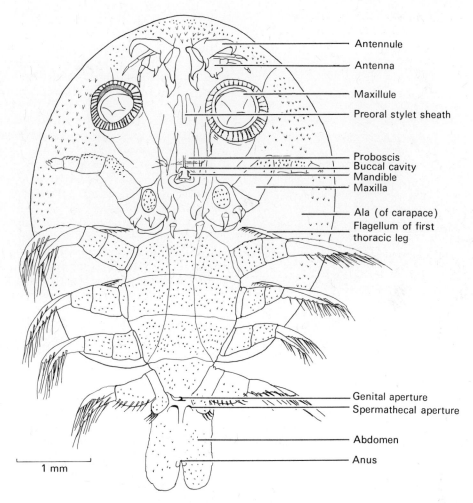

Antennule
Antenna
Maxillule
Preoral stylet sheath

Proboscis
Buccal cavity
Mandible
Maxilla

Ala (of carapace)
Flagellum of first
thoracic leg

Genital aperture
Spermathecal aperture

Abdomen
Anus

1 mm

Fig. 33-31. Ventral view of *Argulus viridis,* female. Note suctorial proboscis, modification of maxillules into sucking discs, and lateral expansion of carapace into alae. (Redrawn from Martin, M. F. 1932. Proc. Zool. Soc., p. 804.)

reminiscent of caligid copepods with which they are sometimes confused, and can adhere closely to the host's surface. Some species are moderately large, up to 12 mm or so. The most common, cosmopolitan genus is *Argulus* (Fig. 33-31). *Argulus* can swim well as an adult; the females must leave their hosts to deposit eggs on the substrate. Many *Argulus* spp. are not host specific and so have been recorded from a large number of fish species.

The branchiuran carapace is expanded laterally to form respiratory alae. They have two pairs of antennae. Homologies of the remaining head appendages have been disputed, but the best evidence suggests that the only appendages in the suctorial proboscis, or mouth tube, are the mandibles.[38] The large, prominent sucking discs are modified maxillules. Just posterior to the maxillular discs are the large maxillae, apparently used to maintain the animal's position on its host and to clean the other appendages. *Argulus* have four pairs of thoracic swimming legs of the typically crustacean biramous form. The exopods of the first two pairs often bear an odd, recurved process, the flagellum, thought by some to indicate affinities with the subclass Branchiopoda (Fig. 33-32). An unsegmented abdomen follows the four segments of the thorax.

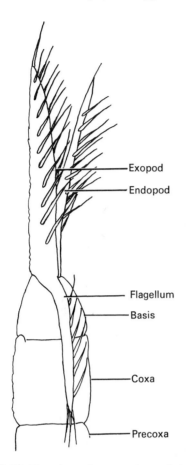

Exopod

Endopod

Flagellum

Basis

Coxa

Precoxa

Fig. 33-32. First thoracic appendage of *Argulus viridis,* showing flagellum. (Redrawn from Martin, M. F. 1932. Proc. Zool. Soc. p. 804.)

The Branchiura were long associated taxonomically with the Copepoda, but present knowledge does not justify this. Among other characteristics that differ from the copepods, branchiurans have a carapace, compound eyes, an unsegmented abdomen behind the genital apertures, and no thoracic segments are completely fused with the head. Another feature present in most branchiurans, but not in other Crustacea, is the piercing stylet or "sting." It is located on the midventral line, just posterior to the antennae (Fig. 33-33). The function of this curious organ is unknown. It has been observed to pierce the host's skin, and the gland cells with their ducts leading down the stylet are suggestive of toxic secretions. Many authors still believe that the stylet is used for feeding, and

the fact that its tip is too small to allow passage of host erythrocytes has led to the probably erroneous conclusion that the animals cannot feed on blood. Causey said that only "textbook species of *Argulus*" do not feed on blood.[8] In fact, Martin found that the stylet bore no direct relation to the proboscis. Her figures showed no connection to the digestive tract, and she considered it "possible that its [the stylet's] function is not very important."[38]

The development of *Argulus japonicus* and *Chonopeltis brevis* and various developmental stages of other species have been described.[15,52] Here again is a difference between the Branchiura and Copepoda. While the development of copepods is slightly too strongly metamorphic, that of most branchiurans can best be described as epimorphic. As noted, the eggs are laid on the substrate (no ovisacs, as in copepods), and the organism that hatches is not a larva, but a juvenile. In the first instar of *A. japonicus* even the sexes can be distinguished! *A. japonicus* has seven juvenile instars, but there may be fewer in other species.[39]

One of the most noteworthy developmental changes that occurs through the ecdysial series is that the primitive form of the maxillules is gradually lost, and the suckers develop in their place.

The development of *Chonopeltis* offers an intriguing puzzle.[15] The organism occurs on African fishes and seems to be one of the more modified branchiurans. Its thoracic legs have become reduced, and it has lost the ability to swim. Yet, in common with *Argulus*, the females must leave the host to deposit eggs, and the juveniles, which also cannot swim, are found on different host species than the adults! Their mode of dispersal remains unknown.

SUBCLASS CIRRIPEDIA

The most familiar cirripedes belong to the order Thoracica, the barnacles. They are important members of the littoral and sublittoral benthic fauna and are economically important as fouling organisms. Some members of the Thoracica are commonly found growing on other animals (see Fig. 2-1). Interestingly, *Conchoderma virgatum* is often found on the lernaeocerid, *Pennella,* a good example of hyperparasitism. (It

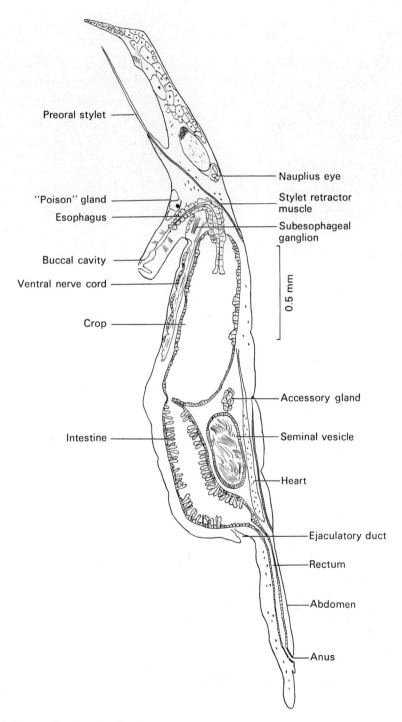

Fig. 33-33. Median longitudinal section of male *Argulus viridis,* semidiagrammatic, with preoral stylet extruded. (Redrawn from Martin, M. F. 1932. Proc. Zool. Soc., p. 804.)

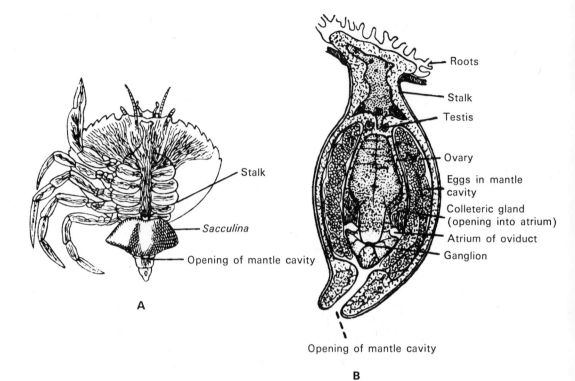

Fig. 33-34. A, A shore crab, *Carcinus,* infected with a mature rhizocephalan, *Sacculina.*
B, Longitudinal section of *Sacculina* at right angles to plane of greatest breadth. (Adapted from Calman, W. T. In Borradaile, L. A., F. A. Potts, L. E. S. Eastham, and J. T. Saunders, editors. 1956. The Invertebrata. A manual for the use of students, ed. 2. Cambridge University Press, Cambridge. p. 385.)

should be pointed out that numerous species of parasitic copepods frequently have epizooic suctorians, hydroids, algae, and so on growing on them, encouraged by the fact that the copepod is in terminal anecdysis, that is, it does not molt further.) However, other orders of cirripedes contain some fascinating organisms that are among the most highly specialized parasites known. These are parasites of other invertebrates, and space will permit consideration only of the most important order, the Rhizocephala.

Order Rhizocephala

Members of the order Rhizocephala are highly specialized parasites of decapod malacostracans. The decapods include the animals most of us know as crabs, crayfish, lobsters, and shrimps. The Sacculinidae are primarily parasites of a variety of brachyurans ("true" crabs), the Peltogastridae

are found on hermit crabs (anomurans), and the Lernaeodiscidae prefer the anomuran family Galtheidae as hosts.

As adults, the rhizocephalans resemble arthropods even less than lernaeids and lernaeocerids do. They have no gut or appendages, not even reduced ones, but get nutrients by means of root-like processes ramifying through the tissues of the crab host (Fig. 33-34). They start life much as many other crustaceans, with a nauplius larva, but the nauplius has no mouth or gut. The nauplius undergoes four molts, and the fifth larval instar is referred to as a **cypris** (Fig. 33-35) because of its resemblance to the free-living ostracod, *Cypris.*

Thus far its life cycle is not unlike that of a normal, thoracican barnacle. The cypris of a barnacle, however, would attach to a suitable spot on the substrate by its antennules and metamorphose to the adult

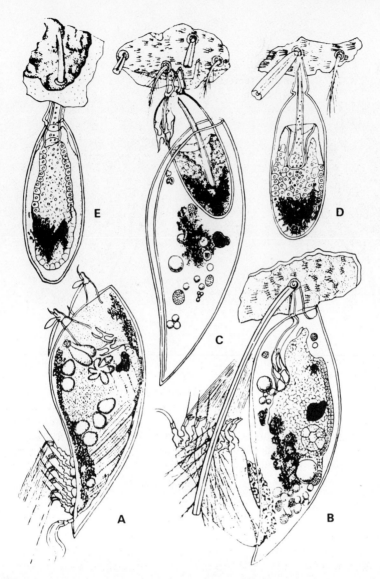

Fig. 33-35. Development of *Sacculina*. **A,** Cypris larva. **B,** Cypris attached to host by the parasite's antennae, shedding locomotory appendages. **C,** Kentrogon larva. **D** and **E,** Penetration of kentrogon into crab host. (From Baer, J. G. Ecology of animal parasites. 1952. University of Illinois Press, Urbana, Ill.)

form; the halves of the carapace become the mantle and secrete the calcareous covering plates. However, the cypris of *Sacculina carcini,* which is the best-known rhizocephalan, attaches to a brachyuran with its antennules.

Next, most of the differentiated structures, including swimming legs and their muscles, are shed from between the two valves of the carapace. The remaining, largely undifferentiated cell mass forms a peculiar larva called the **kentrogon.** The kentrogon may be likened to a living hypodermic syringe. The cell mass within is actually injected into the body of the crab at the base of a seta or other vulnerable spot where the cuticle is thin. The mass of cells migrates to a site just ventral to the host intestine and begins to grow (Fig. 33-36). As the absorptive processes grow

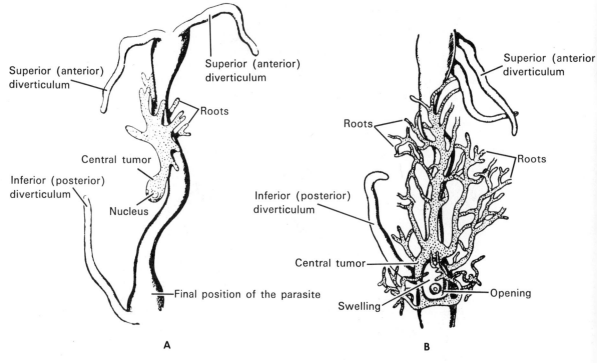

Fig. 33-36. Stages in the development of *Sacculina* on the midgut of a crab. **A,** Early stage. **B,** Later stage. Central "tumor" and absorptive "roots" of the parasite are indicated. There is a swelling caused by the body of the *Sacculina.* The opening indicated is that of a cavity in the central tumor, the "perisomatic cavity," from which the definitive body eventually protrudes (not the mantle opening). The superior and inferior diverticulae are parts of the host's intestine. (From Borradaile, L. A., F. A. Potts, L. E. S. Eastham, and J. T. Saunders, editors. 1956. The Invertebrata. A manual for the use of students, ed. 2. Cambridge University Press, Cambridge. p. 385.)

out into the crab's tissue, the central mass also begins to enlarge. This mass contains the developing gonads of both sexes; *Sacculina* is hermaphroditic, in common with most other Cirripedia. As it grows larger, it appears to press against the host hypodermis in the ventral cephalothorax and thereby prevents cuticle secretion. Finally, the weakened cuticle overlying the parasite breaks open and the gonadal mass of *Sacculina* becomes external.[13] After the parasite becomes externalized, the crab can no longer molt; therefore, further development of the host essentially ceases.

The life histories of peltogastrids and lernaeodiscids seem to be similar to sacculinids in many respects, although further ecdyses of the host are not hindered.[44] *Peltogasterella socialis* is not hermaphroditic; the female is fertilized by a cypris male after

the gonadal sac of the female becomes external.[21] The cypris male injects a cell mass into the mantle cavity of the female, and this mass migrates to the seminal receptacles, heretofore thought to be testes. It is possible that other rhizocephalans have not been examined carefully enough, and a similar situation may prevail in other species of Peltogastridae, as well as species in other families.

In light of the invasiveness of rhizocephalans, it is not surprising that there is a range of pathogenic effects of their hosts, including damage to the hepatic, blood, and connective tissues and to the thoracic nerve ganglion of infected crabs.[44] However, some of the most interesting effects are on the hormonal and reproductive processes of the host, the so-called parasitic castration. Crabs exhibit some degree of sexual dimorphism and morphological differences between the sexes are especially pronounced in the Brachyura. In the normal sequence of ecdyses, the secondary sexual characteristics of the respective sexes become increasingly apparent as the crab approaches maturity. The abdomen of the female is broad and apron-like with complete segmentation, but the male's abdomen is long and narrow, and the third, fourth, and fifth somites are usually fused. The female has four pairs of well-developed, biramous pleopods (abdominal appendages), well supplied with hairs on their margins, while the male has only the first and second pleopods, and these are long, slender, and adapted for copulation. When the young male crab is infected with *Sacculina,* various degrees of "feminization" are exhibited in the subsequent instars. The manifestations vary, depending on the species of host and its degree of development when infected, but they may include a broader, more completely segmented abdomen and alteration of the pleopods toward the female type. In female crabs, the effects seem to be more complex, involving some aspects of both hyperfeminization and hypofeminization. Somewhat similar effects of parasitic castration have been reported in the hosts of the Peltogastridae and Lernaeodiscidae.

Although many descriptions of the external changes occurring in the various decapod hosts infected with Rhizocephala have been made, remarkably little attention has been given to the mechanisms involved. The gonads of infected animals tend to be retarded in development, and in some cases, complete atrophy occurs. Interestingly, atrophy begins before the absorptive processes penetrate the gonad, and it is thought that damage occurs indirectly. Unfortunately, our knowledge of the mechanism of sex determination in crustaceans is rather meager. On the basis of some work on amphipods,[9,10] we can assume that decapod females are heterogametic (XY), that young individuals are ambipotent, and that development of the sexual characteristics of a genetic male depends not on its gonad, but on the presence of an **androgenic gland.** The males of all orders of higher Crustacea that have been examined have androgenic glands.

The androgenic gland is located near the vas deferens, and if it is removed from a young male, female characteristics will develop, including appearance of ovarian tissue in the testis and female appendages. It appears that parasitic castration of the male decapod involves destruction of its androgenic gland or antagonism of the gland's secretions—perhaps both. The crustacean testis apparently has no endocrine activity, but the ovary produces hormones responsible for female secondary sexual characteristics. It seems reasonable that the developing ovary of the rhizocephalan could produce hormones that would influence host development.

SUBCLASS MALACOSTRACA

Although the malacostracans constitute the largest subclass of Crustacea, with members widespread and abundant in marine and freshwater habitats, comparatively few are symbiotic. Those that are, by and large, are confined to the peracaridan orders Amphipoda and Isopoda. The isopods have been particularly successful in this regard, and some of them have become highly modified for parasitism. The eucaridan order Decapoda is the largest order of crustaceans, but few of its members are symbiotic.

Order Amphipoda

Free-living amphipods are widely prevalent aquatic organisms, often abundant along the seashore. Not many symbiotic species have been described, but some of the ones that have are quite common. Some Hyperiidae (Fig. 33-37) are frequent parasites of jellyfishes *(Aurelia, Cyanea)* and Phronimidae are found in the tunic of planktonic ascidians *(Salpa),* apparently killing the tunicate itself and taking over its gelatinous case. *Laphystius sturionis* (suborder Gammaridea) is a relatively unmodified amphipod, parasitizing a variety of marine fish.[27] The most interesting symbiotic amphipods are among the Caprellidea, and most unlikely looking amphipods they are. Caprellidae, or "skeleton shrimps," are predators that stalk around on hydroid or ectoproct colonies or algae and catch their prey with raptorial second legs, but the Cyamidae are curious ectoparasites of whales (Fig. 33-38). The

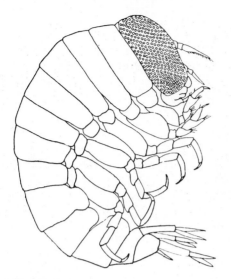

Fig. 33-37. *Hyperia galba,* an amphipod that is commonly found on jellyfish *(Aurelia)* in the North Atlantic. Hyperiidae typically have very large compound eyes that cover nearly the entire sides of the head. (From Kunkel, B. W. 1913. The Arthrostraca of Connecticut. State Geological and Natural History Survey. p. 46.)

abdomen is vestigial in both families. In contrast with most amphipods, cyamids are dorsoventrally flattened, a clearly adaptive characteristic in their ectoparasitic habitat. The second and the fifth through seventh legs are strongly modified adhesive organs.

Order Isopoda

Members of the order Isopoda have had more success in terrestrial environments than have other crustaceans, limited though that may be, and they are abundant in a variety of marine and freshwater habitats. Furthermore, they exploit the parasitic mode of existence more extensively than other malacostracans.

The gnathiidean and flabelliferan families parasitic on marine fish have relatively few modifications. *Gnathia* is a parasite only as a larva, the **praniza,** which was originally described as a separate genus before its true identity was recognized. The praniza stage (Fig. 33-39) attaches to a fish host and feeds on blood until its gut is hugely distended. It then leaves its host and molts to become an adult. The adults are benthic and do not feed. Because of con-

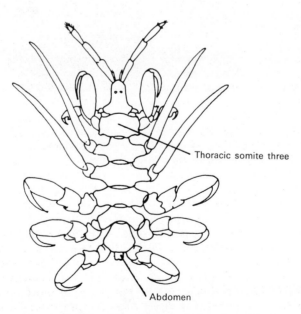

Thoracic somite three

Abdomen

Fig. 33-38. *Paracyamus,* an amphipod parasite of whales. Cyamids are extoparasites and are dorsoventrally flattened with several pairs of legs modified for clinging to their hosts. (Adapted from Sars, G. O. From Meglitsch, P. A. 1972. Invertebrate zoology, ed. 2. Oxford University Press, Oxford. p. 561.)

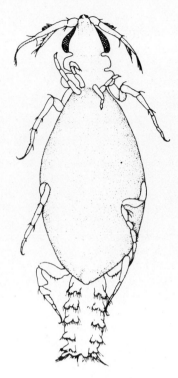

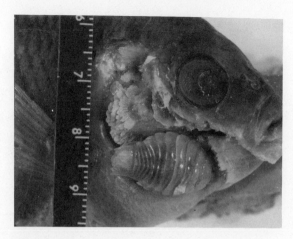

Fig. 33-40. *Livoneca ovalis* is a common parasite of fish along the U.S. Atlantic coast. Here it is found on a *Lepomis gibbosus* (operculum removed) from Chesapeake Bay.

Fig. 33-39. The praniza larva of the isopod *Gnathia.* The praniza is the only parasitic stage of this isopod. The gut becomes greatly distended with blood from its fish host. (From Kabata, Z. 1970. In Snieszko, S. F., and H. R. Axelrod, editors. Diseases of fishes, book 1. T.F.H. Publications, Inc., Neptune City, N.J. p. 49.)

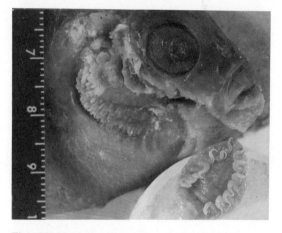

Fig. 33-41. *Lironeca ovalis,* same specimen as in Fig. 33-40, removed from its site on the gills. The gills show pressure atrophy and traumatic damage.

siderable sexual dimorphism, the male was also described in a separate genus, *Anceus.* Some of the Cymothoidae are of economic importance as fish parasites. The young of *Livoneca amurensis* and some other species burrow under a scale on their host. As the isopod grows, the underlying skin stretches to accommodate it; finally, the enveloped crustacean communicates to the exterior only by a small hole. *Livoneca ovalis* (Fig. 33-40) is a common parasite of a variety of teleosts in the Atlantic Ocean and has been reported along the U.S. coast from Mississippi to Massachusetts.[33,50] *L. ovalis* is usually found beneath the gill operculum, where, on small host individuals, it causes a marked pressure atrophy of the adjacent gills (Fig. 33-41).

The epicaridean isopods are highly specialized parasites of other Crustacea. The adult females of some species are compa-

rable in loss of external segmentation and appendages to the most advanced copepods and the Rhizocephala. Portions of the appendages that are not lost, however, are the oostegites forming the brood pouch. These may become enormously developed, while most of the other appendages disappear or become vestigial. A good example is *Pinnotherion vermiforme* (Fig. 33-42), an entoniscid whose structure was described by Atkins.[2] *Pinnotherion* is a parasite of a brachyuran crab, *Pinnotheres*

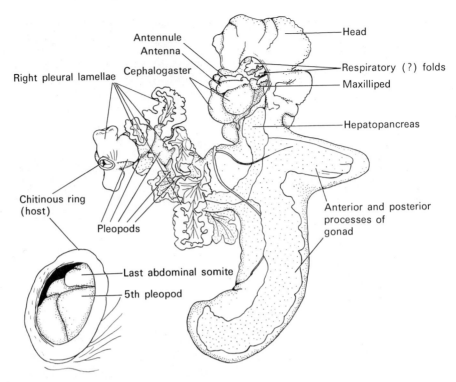

Fig. 33-42. Young female *Pinnotherion vermiforme,* an entoniscid isopod parasite of a crab, *Pinnotheres pisum.* The parasite develops in a closely investing, thin layer of the host's cuticle, which communicates with the branchial chamber by a small opening. The opening into the host's branchial chamber is surrounded by a somewhat thickened ring of cuticle (enlarged at left). The vascularized pleural lamellae extending from the abdomen are prominent. Note the vestigial nature of the appendages and presence of the peculiar contractile "cephalogaster." (Redrawn from Atkins, D. 1933. Proc. Zool. Soc. Lond., p. 331. The original author noted that this specimen probably had not yet spawned; the respiratory [?] folds and pleural lamellae had not yet reached their final stage of complexity; and the cephalogaster was abnormal in that the two lobes were unequal in size.)

pisum, itself a parasite of the mussel, *Mytilus edulis.* The juvenile isopod enters the branchial chamber of the crab and begins to grow there, pushing inward until it practically fills the hemocoel of the crab.

The parasite is still topologically outside its host, in the branchial chamber, since it does not break through into the hemocoel proper but is covered by a closely investing, very thin layer of crab cuticle. The enveloping pocket of cuticle in which the isopod lies communicates to the branchial chamber only by a small opening. The oostegites are produced into extensive, thin lamellae that form the brood chamber, and other parts of the pereiopods are rudimentary.

The huge brood chamber forms a hood over the anterior end of the organism and wraps around the large thoracic processes that contain the highly developed ovary. Extending from the sides of the abdomen are highly vascularized pleural lamellae, apparently of respiratory function, but the pleopods are vestigial. The short esophagus leads into a peculiar "cephalogaster," a contractile organ for sucking blood that apparently has absorptive function as well.

In the Cryptoniscidae, morphological modification appears even more extreme. After infection of the definitive host, the female *Ancyroniscus bonnieri* feeds heavily,

Fig. 33-43. *Pinnotheres ostreum* damages the gills of its host, the commercial oyster *(Crassostrea virginica).* The carapace is soft, and the eyes and chelae are reduced.

and the gorged isopod begins to produce eggs. During this process, most of the internal organs, including digestive and nervous system, disappear, and the animal becomes increasingly distended with eggs. Finally, all that is left is a large, pulsatile sac of eggs, and these are freed by rupture of the sac.[7]

The life histories of epicarideans are of great interest. The larva that hatches is an **epicaridium,** quite isopod-like in appearance. The epicaridium has blood-sucking mouth parts and attaches to a free-swimming copepod. There it feeds and molts to become a **microniscus** larva. The microniscus molts several times and develops into a **cryptoniscus,** which again is free-swimming and must find a suitable definitive host. The adult cryptoniscid is a protandrous hermaphrodite, each animal going through a male stage before developing into a female. In the Bopyridae, Entoniscidae, and Dajidae, the first isopod to infect the definitive host becomes a female and subsequent cryptoniscus larvae become small males, sometimes living as parasites within the female brood sac. In the bopyrid, *Stegophryxus hyptius,* the cryptoniscus has to actually enter the female's brood pouch to become a male; a masculinizing substance derived from feeding on the female may be responsible.[43] It does seem clear that sex determination in epicarideans is epigamic, depending on cir-

cumstances other than the chromosomal complement of the gametes. The cryptoniscids probably have a transitory androgenic gland that stimulates production of sperm, which are then stored, and development of the androgenic gland in the ambipotent individuals in other families depends on the presence of a female.

Effects of epicarideans on their hosts are similar to those described for rhizocephalans, including parasitic castration. Secondary sexual characteristics of the male host are lost, the host becomes feminized, and the gonads of both males and females are suppressed or atrophied. Feminization of brachyuran males by epicarideans is not as striking as that produced by rhizocephalans.[44] It is interesting that some rhizocephalans are themselves hyperparasitized by epicarideans, and these, in turn, induce castration of their rhizocephalan hosts!

Order Decapoda

The largest order of crustaceans, the decapods, contains a relatively small number of symbiotic species, although some of them are quite common. They are of interest because of the slight, but definite, modifications for parasitism that they illustrate. Pinnotherids are frequent commensals with polychaetes, in their tubes or burrows, or are parasites in the mantle cavity of pelecypods. *Pinnotheres pisum* has been mentioned, and *P. ostreum* is found in the commerically important oyster, *Crassostrea virginica* (Fig. 33-43). *P. ostreum* interferes with the feeding of its host and damages its gills sufficiently to cause female oysters to become males.[3] The same sex change can be produced experimentally by starving the oysters.

Pinnotherids are modified relatively little from the typical, free-living brachyuran. The adult females tend to be white or cream-colored with thin, soft cuticle in the carapace and with reduced eyes and chelae. Younger stages, including males, have hard carapaces and more well-developed eyes and chelae.[11]

A few species of symbiotic decapods belonging to other sections, such as Caridea and Anomura, are known. They live in such places as the mantle cavities of clams, the tubes of polychaetes, and on the stems of sea pens (Octocorallia) and are modified for symbiosis to about the same extent as pinnotherids. The supposed rarity of these forms is probably a result of the failure of collectors to look for them.[24] The decapods involved in cleaning symbiosis have already been alluded to in Chapter 2.

Classification of crustacean taxa with symbiotic members

Waterman and Chace remarked, "The diversity of crustaceans is reflected in their classification. In no other class of animals, except the mammals, has the generally accepted scheme of classification above the family level proved to be so inadequate. With new subclasses and orders of crustaceans still being discovered and with fresh evidence on the relationships of the various groups coming to light every year, the systematic arrangements of this heterogeneous assemblage is not likely to be stabilized in the near future."[54] The following system has been adapted from various sources; the part that refers to the several groups of parasitic copepods relies heavily on Yamaguti[57] and Kabata.[27]

Subclass OSTRACODA
Body entirely enclosed in bivalve carapace; body unsegmented or indistinctly segmented; no more than two pairs of trunk appendages. (Only a few species recorded, as gill parasites of marine teleosts and elasmobranchs.[18,56])

Order: MYODOCOPA

Family: Cypridinidae

Subclass COPEPODA
Typically with elongate, segmented body consisting of head, thorax, and abdomen. Thorax with seven somites, of which first and sometimes second fused with head to form cephalothorax. Thoracic appendages biramous except maxillipeds and often fifth swimming legs uniramous. No appendage on abdomen except uropods on anal somite.[6] No carapace. Compound eyes absent, but median nauplius eye often present. Gonopores on "genital segment," usually considered the last somite of the thorax. (Parasitic forms may not fit some or much of foregoing diagnosis and may be highly modified as adults and sometimes as juveniles.)

Order CALANOIDA
(No symbiotic members, but included for completeness.) With broad metasome and narrow urosome*; hinge joint between fifth and sixth free thoracic somites; females with long antennules (23 to 25 segments); antennae biramous; fifth legs little reduced and used in swimming; eggs carried in one or two ovisacs attached to genital segment. Large order of important marine and freshwater planktonic organisms, never symbiotic.

Order CYCLOPOIDA
In forms strongly modified for symbiosis: mouth not projecting much above surface of body, partly covered by labrum; mandible with basal segment short and stocky, distal segment falciform, long, pointed, with denticles often on one or both margins. Eggs in two ovisacs attached to genital somite, usually multiseriate. Parasitic forms with no chalimus larva attached to host by filament. In free-living forms and those little modified for symbiosis: broad metasome and narrow urosome; hinge joint between fourth and fifth free thoracic somites; antennules fairly short, not longer than metasome with six to seventeen segments; antennae uniramous; fifth legs vestigial and uniramous, with basal segment not enlarged. Large order, marine and freshwater, many benthic, symbiotic, and some planktonic species.

Families (representative) with symbiotic members: Ergasilidae, Bomolochidae, Chondracanthidae, Lernaeidae, Lichomolgidae, Stellicomitidae, Notodelphyidae, Pharodidae, Mytilicolidae; Taeniacanthidae, Clausidiidae, Xarifiidae; Asterocheridae

Order HARPACTICOIDA
Metasome and urosome approximately equal in width; hinge joint between fourth and fifth free thoracic somites; antennules very short, not longer than cephalothorax in females and with five to nine segments; fifth thoracic legs reduced and with inner expansion of basis; usually single ovisac carried medially on genital somite. Important marine and freshwater benthic order, rarely symbiotic.

Order CALIGOIDA
Urosome sometimes larger than metasome, and hinge between third and fourth free thoracic somites often fused so that metasome and urosome

---

*The anterior and posterior portions of many copepods are often designated metasome and urosome, respectively, though these do not correspond to the basic divisions cephalon, thorax, and abdomen. The location of a particular point of flexion between the metasome and urosome, the hinge joint, is of use in distinguishing the Calanoida, Cyclopoida, and Harpacticoida.

are indistinguishable; antennae very short, with two segments; antennules often prehensile; eggs carried in two long ovisacs; mouth tubular; mandibles flat, long blades with row of teeth on one margin; larval stages include chalimus, attached to host with frontal filament. Large and important order with many freshwater and marine species, all symbiotic.

Families (representative): Caligidae, Pandaridae, Dichelesthiidae, Lernaeoceridae, Lernaeopodidae, Sphyriidae, Eudactylinidae

Order MONSTRILLOIDA

Antennae and mouth parts absent as adult. Larvae parasitic in marine invertebrates; small group.

Order PHILICHTHYIDEA

Antennae and mouth parts rudimentary; no legs in adult females, male with 2 pairs antennae and 2 to 3 pairs legs; females elongate, with more or less fusion of somites that may be enlarged and with lateral or ventral processes. Small group living in canals, sinuses, and fossae of marine fishes; may be much more common than reports indicate.

Families: Lerneascidae, Philichthyidae

Order SARCOTACIDEA

Female ovoid or maggot-shaped; segmentation indistinct or absent; mouth parts much reduced; no legs, antennae, or ovisacs (virtually only copepod group without ovisacs). Male elongate, with or without distinct segmentation, but with mouth parts, antennae, and legs. Male lives with much larger female in cyst formed in viscera or musculature of their host, a marine fish. Another small group that may be more widespread than now thought.

Family: Sarcotacidae

Subclass BRANCHIURA

Body with head, thorax, and abdomen; head with flattened, bilobed, cephalic fold incompletely fused to first thoracic somite. Thorax with four pairs of appendages, biramous, and with proximal extension of exopod of first and second legs. Abdomen without appendages, unsegmented, bilobed. Eyes compound. Both pairs of antennae reduced; claws on antennules. Maxillules often forming pair of suctorial discs; maxillae uniramous. Gonopore at base of fourth legs. Ectoparasites of marine and freshwater fishes, occasionally of amphibians.

Order ARGULIDEA

Families: Argulidae, Dipteropeltidae

Subclass CIRRIPEDIA

Sessile or parasitic as adults, head reduced and abdomen rudimentary; paired, compound eyes absent. Body segmentation indistinct. Usually hermaphroditic. In nonsymbiotic and epizoic forms, carapace becomes mantle, which secretes calcareous plates; antennules become organs of attachment; antennae disappear. Young hatches as nauplius and develops to bivalved cypris larva. All marine.

Order THORACICA

With six pairs thoracic appendages, alimentary canal. Usually nonsymbiotic, though some epizoic and commensal on whales, fish, sea turtles, crabs. For example, *Xenobalanus, Coronula, Conchoderma, Chelonibia.*

Order ACROTHORACICA

Bores into mollusc shells or coral. Females usually with four pairs thoracic appendages, gut present, no abdomen. Dioecious, males very small, without gut and appendages except antennules, parasitic on outside of mantle of female.

Order ASCOTHORACICA

With segmented or unsegmented abdomen; usually six pairs thoracic appendages, gut present. Parasitic on echinoderms and soft corals. Example: *Trypetesa.*

Order APODA

Small, with body divided into rings; no mantle, thoracic appendages, or anus. Single genus described, *Proteolepas,* parasitic in mantle cavity of a stalked barnacle.

Order RHIZOCEPHALA

Adults with no segmentation, gut, or appendages; with root-like absorptive processes through tissue of host. Common parasite of decapod crustaceans.

Families: Sacculinidae, Peltogastridae, Lernaeodiscidae

Subclass MALACOSTRACA

Distinctly segmented bodies, typically with eight somites in the thorax and six somites plus the telson in the abdomen (except seven in Nebaliacea); all segments with appendages. Antennules often biramous; first one, two, or three thoracic appendages often maxillipeds. Primitively with carapace covering head and part or all of thorax, but lost in some orders. Gills usually thoracic epipods. Female gonopores on sixth thoracic segment, and male gonopores on eighth thoracic segment. Largest subclass, marine, freshwater, few terrestrial; many free-living, but parasitic members relatively few, found in only three of the 10 to 12 extant orders commonly recognized.

Superorder PERACARIDA

Without carapace or with carapace leaving at least four free thoracic somites; first thoracic somite fused with head. Brood pouch in female (typically formed from modified thoracic epipods, the oostegites). Several small, marine orders, and the two large orders with parasitic members.

Order AMPHIPODA

No carapace; ventral brood pouch of oostegites. Antennules often biramous; eyes usually sessile. Gills on thoracic coxae; first thoracic limbs maxillipeds, second and third pairs usually prehensile (gnathopods). Usually laterally compressed body form. Marine, freshwater, and terrestrial; free-living and symbiotic.

Suborder HYPERIIDEA

Head and eyes very large; only one thoracic somite fused with head. Pelagic or symbiotic in medusae or tunicates.

Families: Hyperiidae, Phronimidae

Suborder CAPRELLIDEA

Two thoracic somites fused with head; abdomen much reduced, with vestigial appendages. So-called skeleton shrimps and whale lice.

Families: Caprellidae, Cyamidae

Order ISOPODA

No carapace; ventral brood pouch of oostegites. Antennules usually uniramous, sometimes vestigial. Eyes sessile. Gills on abdominal appendages. Second and third thoracic appendages usually not prehensile. Body usually dorsoventrally flattened.

Suborder GNATHIIDEA

Thorax much wider than abdomen; first and seventh thoracic somites reduced, seventh without appendages. Larvae parasitic on marine fish.

Family: Gnathiidae

Suborder FLABELLIFERA

Flattened body; with ventral coxal plates sometimes joined to body; telson fused with next abdominal somite, and other abdominal somites may be fused. Uropods flattened, forming tail fan. Marine, free-living and ectoparasitic on fishes.

Families with parasitic members: Aegidae, Cymothoidae, Crallanidae

Suborder EPICARIDEA

Females greatly modified for parasitism, somites and appendages fused, reduced, or absent. Mouth parts modified for sucking: mandible for piercing and maxillae reduced or absent. Males, small, but less modified. Marine parasites of Crustacea.

Families: Bobyridae, Phryxidae, Entoniscidae, Dajidae, Cryptoniscidae

Superorder EUCARIDA

All thoracic segments fused with and covered by carapace. No oostegites or brood pouch. Eyes on stalks. Usually with zoea larval stage.

Order DECAPODA

First three pairs thoracic appendages modified to maxillipeds (therefore, appendages on remaining five thoracic somites equal 10). Crabs, lobsters, shrimps.

Suborder REPTANTIA

Dorsoventrally flattened, rostrum reduced or absent. Antennal scale reduced or absent. First pereiopods usually modified to form cheliped, pleopods reduced and not used for swimming. First abdominal somite reduced, shorter than ones following.

Section BRACHYURA

Abdomen reduced, folded under cephalothorax, only function of abdomen is reproductive.

Families with symbiotic members: Pinnotheridae, Parthenopidae

REFERENCES

1. Amar, R. 1948. Un organe endocrine chez *Idotea* (Crustacea Isopoda). Comp. R. Acad. Sci. Paris. 227:301-303.
2. Atkins, D. 1933. *Pinnotherion vermiforme* Giard and Bonnier, an entoniscid infecting *Pinnotheres pisum*. Proc. Zool. Soc. Lond. pp. 319-363.
3. Awati, P. R., and H. S. Rai. 1931. *Ostrea cucullata*. Indian Zool. Mem. 3:1-107.
4. Barnes, H., and J. J. Gonor. 1958. Neurosecretory cells in the cirripede, *Pollicipes polymerus*. J. Mar. Res. 17:81-102.
5. deBlainville, M. H. D. 1882. Mémoire sur les Lernées (Lernaea, Linn.). J. Phys. 95:372-380, 437-447.
6. Bowman, T. E. 1971. The case of the nonubiquitous telson and the fraudulent furca. Crustaceana 21:165-175.
7. Caullery, M., and F. Mesnil. 1920. *Ancyroniscus bonnieri* C. et M., epicaride parasite d'un sphéromide (*Dynamene bidentata* Mont.). Bull. Sci. Fr. Belg. 34:1-36.
8. Causey, D. 1959. "Ye crowlin' ferlie," or on the morphology of *Argulus*. Turtox News 37:214-217.
9. Charniaux-Cotton, H. 1960. Sex determination. In Waterman, T. H., editor. Physiology of Crus-

tacea, vol. I. Academic Press, Inc., New York, 411-447.

10. Charniaux-Cotton, H. 1962. Androgenic gland of crustaceans. Gen. Comp. Endocrinol. 1(Suppl.):241-247.

11. Christensen, A. M., and J. J. McDermott. 1958. Life-history and biology of the oyster crab, *Pinnotheres ostreum* Say. Biol. Bull. 114:146-179.

12. Costanzo. G. 1969. Stadi naupliari e primo copepodite di *Lichomolgus canui* G. O. Sars (Copepoda, Cyclopoida) del lago di Faro (Messina), allevata sperimentalments. Bol. Zool. 36:143-153.

13. Day, J. H. 1935. The life history of *Sacculina*. Q. J. Microsc. Sci. 77:549-583.

14. Einszporn, T. 1965. Nutrition of *Ergasilus sieboldi* Nordmann. II. The uptake of food and the food material. Acta Parasit. Polon. 13:373-380.

15. Fryer, G. 1961. Larval development in the genus *Chonopeltis* (Crustacea: Branchiura). Proc. Zool. Soc. Lond. 137:61-69.

16. Grabda, J. 1963. Life cycle and morphogenesis of *Lernaea cyprinacea* L. Acta Parasitol. Polon. 11:169-199.

17. Harding, J. P. 1950. On some species of *Lernaea*. Bull. Mus. (Nat. Hist.) Zool. 1:1-27.

18. Harding, J. P. 1966. Myodocopan ostracods from the gills and nostrils of fishes. In Barnes, H., editor. Some contemporary studies in marine science. George Allen & Unwin Ltd., London. pp. 369-374.

19. Humes, A. G., and J. H. Stock. 1972. Preliminary notes on a revision of the *Lichomolgidae,* cyclopoid copepods mainly associated with marine invertebrates. Bull. Zool. Museum, Amsterdam 2:121-133.

20. Humes, A. G., and J. H. Stock. 1973. A revision of the Family Lichomolgidae Kossmann, 1877, cyclopoid copepods mainly associated with marine invertebrates. Smithsonian Contrib. Zool., no. 127.

21. Ichikawa, A., and R. Yanagimachi. 1958. Studies on the sexual organization of the Rhizocephala I. The nature of the "testes" of *Peltogasterella socialis* Krüger. Annot. Zool. Jpn. 31:82-96.

22. Illg, P. L. 1958. North American copepods belonging to the family Notodelphydae. Proc. U.S. Nat. Mus. 107:463-649.

23. Izawa, K. 1969. Life history of *Caligus spinosus* Yamaguti, 1939 obtained from cultured yellow tail, *Seriola quinqueradiata* T. and S. (Crustacea: Caligoida). Report of Faculty of Fisheries, Perfectural University of Mie, 6:127-157.

24. Johnson, D. S. 1967. On some commensal decapod crustaceans from Singapore (Palaemonidae and Porcellanidae). J. Zool. Lond. 153:499-526.

25. Kabata, Z. 1955. The scientist, the fisherman, and the parasite. Scot. Fisheries Bull. 4:13-14.

26. Kabata, Z. 1958. *Lernaeocera obtusa* n. sp.; its biology and its effects on the haddock. Mar. Res. Scot. pp. 1-26.

27. Kabata, Z. 1970. Diseases of fishes, book I. T. F. H. Publishers, Inc., Neptune City, N.J.

27a.Kabata, Z. and B. Cousens. 1972. The structure of the attachment organ of Lernaeopodidae (Crustacea: Copepoda). J. Fish. Res. Bd. Can. 29:1015-1023.

28. Kabata, Z., and B. Cousens. 1973. Life cycle of *Salmincola californiensis* (Dana, 1852) (Copepoda: Lernaeopodidae). J. Fish. Res. Bd. Can. 30:881-903.

29. Kabata, Z., and G. C. Hewitt. 1971. Locomotion mechanisms in Caligidae (Crustacea: Copepoda). J. Fish. Res. Bd. Can. 28:1143-115.

30. Knowles, F. G. W. and D. B. Carlisle. 1956. Endocrine control in the Crustacea. Biol. Rev. 31:396-473.

31. Kô, Y. 1969. On the reproduction and metamorphosis of a commensal copepod, *Ostrincola koe,* in the Japanese clam, *Tapes japonica* (preliminary note). Bull. Fac. Fisheries, Nagasaki Univ., no. 27. (English summary)

32. Krishnan, G. 1951. Phenolic tanning and pigmentation of the cuticle in *Carcinus maenas*. Q. J. Microsc. Sci. 92:333-344.

33. Kunkel, B. W. 1918. The Arthrostraca of Connecticut. Conn. St. Geol. Nat. Hist. Surv. Bull., no. 26.

34. Lewis, A. G. 1963. Life history of the caligid copepod *Lepeophtheirus dissimulatus* Wilson, 1905 (Crustacea: Caligoida). Pacific Sci. 17:195-242.

35. Linnaeus, C. 1746. Fauna suecica sistems animalia suecica regni, ed. 1. Stockholm.

36. Linnaeus, C. 1758. Systema naturae, Ed. 10. Stockholm.

37. Manton, S. M. 1964. Mandibular mechanisms and the evolution of arthropods. Philos. Tr. R. Soc. Lond. [Biol.] 247:1-183.

38. Martin, M. F. 1932. On the morphology and classification of *Argulus* (Crustacea). Proc. Zool. Soc. Lond. pp. 771-806.

39. Meehean, O. L. 1940. A review of the parasitic Crustacea of the genus *Argulus* in the collection of the United States National Museum. Proc. U.S. Nat. Mus. 88:459-527.

40. Oken, L. 1816. Lehrbuch der Naturgeschichte, vols. 1 and 2. Dritter Theil, Zoologie, Jena.

41. Parker, R. R., Z. Kabata, L. Margolis, and M. D. Dean. 1968. A review and description of *Caligus curtus* Miller, 1785, type species of its genus. J. Fish. Res. Bd. Can. 25:1923-1969.

42. Putz, R. E., and J. T. Bowen. 1968. Parasites of freshwater fishes; IV. Miscellaneous, the anchor worm *(Lernaea cyprinacea)* and related species. U.S. Department of Interior, Bureau of Sport Fishing and Wildlife, Division of Fisheries Res. FDL-12.

43. Reinhard, E. G. 1949. Experiments on the determination and differentiation of sex in the bopyrid *Stegophryxus hyptius* Thompson. Biol. Bull. 96:17-31.

44. Reinhard, E. G. 1956. Parasitic castration of Crustacea. Exp. Parasitol. 5(1):79-107.

45. Roberts, L. S. 1970. *Ergasilus* (Copepoda: Cyclopodia): Revision and key to species in North America. Trans. Am. Microsc. Soc. 39:134-161.
46. Robertson, J. D. 1960. Ionic regulation in the crab, *Carcinus maenas* (L) in relation to the moulting cycle. Comp. Biochem. Physiol. 1:183-212.
47. Rondelet, G. 1554. Libri de piscibus marinus, vol. 1. Lyon.
48. Sanders, H. L. 1957. The Cephalocarida and crustacean phylogeny. Syst. Zool. 6:112-128, 148.
49. Sarig, S. 1971. Diseases of fishes, book 3. T. F. H. Publications, Inc., Neptune City, N.J.
50. Sindermann, C. J. 1970. Principal diseases of marine fish and shellfish. Academic Press, Inc., New York.
51. Tidd, W. M. 1962. Experimental infestations of frog tadpoles by *Lernaea cyprinacea*. J. Parasitol. 48:870.
52. Tokioka, T. 1936. Larval development and metamorphosis of *Argulus japonicus*. Mem. Coll. Sci. Kyoto. Ser. B, 12:93-114.
53. Travis, D. R. 1955. The moulting cycle of the spiny lobster, *Panulirus argus,* Latreille. II, Preecdysial histological and histochemical changes in the hepatopancreas and integumental tissues. Biol. Bull. 108:88-112.
54. Waterman, T. H., and F. A. Chace, Jr. 1960. In Waterman, T. H., editor. The physiology of Crustacea. Academic Press, Inc., New York. p. 23.
55. Wilson, C. B. 1911. North American parasitic copepods belonging to the Family Ergasilidae. Proc. U.S. Nat. Mus. 39:263-400.
56. Wilson, C. B. 1913. Crustacean parasites of West Indian fishes and land crabs, with descriptions of new genera and species. Proc. U.S. Nat. Mus. 44:189-277.
57. Yamaguti, S. 1963. Parasitic Copepoda and Branchiura of fishes. Interscience Publishers, New York.
58. Zmerzlaya, E. I. 1972. *Ergasilus sieboldii* Nordmann, 1832., its development biology and epizootic significance. (English summary) Izv. Gos-NIORKh, 80:132-177.

GLOSSARY

acanthella The developing acanthocephalan larva between an acanthor and a cystacanth, in which the definitive organ systems are developed.

acanthor The acanthocephalan larva that hatches from the egg.

accidental parasite A parasite found in other than its normal host. Also called an incidental parasite.

acetabulum A sucker; the ventral sucker of a fluke, a sucker on the scolex of a tapeworm.

aclid organ The spined introvert near the anterior end of an acanthor.

acquired immunity Immunity arising from a specific immune response, either humoral or cell mediated, stimulated by antigen in the host's own body (active) or in the body of another individual with the antibodies or lymphocytes transferred to the host (passive).

adoptive immunity An immune state conferred by inoculation of lymphocytes, not antibodies, from an immune animal rather than by exposure to the antigen itself.

agamete The germinative nucleus within an axial cell of a dicyemid mesozoan.

ala A term often applied to wing-like structures on plants or animals, for example, the lateral expansions of the branchiuran carapace to form respiratory alae, cuticular expansions of nematodes, and others.

amastigote A form of Trypanosomatidae that lacks a long flagellum. Also called a Leishman-Donovan (L-D) body; as in *Leishmania.*

ameboma The occasional result of a chronic amebic ulcer; a granuloma containing active trophozoites. Rare except in Central and South America.

amebula The daughter cell resulting from mitosis and cytokinesis of an encysted ameba.

amphid A sensory organ on each side of the "head" of nematodes.

amphistome A fluke with the ventral sucker transposed to the posterior end.

anamnestic response The immune response to a challenge or secondary antigen inoculation, marked by more rapid and stronger manifestation of the immune reaction (specifically, antibody titer) than after the primary immunizing dose.

anamorphosis The primitive type of arthropod development, in which the young gradually become more like the adult in body form after each ecdysis. Also called ametabolous development.

anapolysis The detachment of a senile proglottid after it has shed its eggs.

androgenic gland A gland located near the vas deferens in many Crustacea; its secretions are responsible for development of male secondary sexual characteristics.

anecdysis Ecdysis in which successive molts are separated by quite long intermolt phases, referred to as "terminal anecdysis," when maximum size is reached and no more ecdyses occur.

anisogametes Outwardly dissimilar male and female gametes.

antennae (second antennae of crustaceans) Second pair of appendages in Crustacea, with bases usually just posterior to antennules; primarily sensory but sometimes adapted for other functions; derived from appendages on primitive third preoral somite, no homologous appendage in insects.

antennules (first antennae) Anteriormost pair of appendages of Crustacea, primarily sensory but often adapted for additional or other functions in particular species, derived from appendages on primitive second peroral somite, homologous to antennae of insects.

anterior station The development of a protozoan in the middle or anterior intestinal portions of its insect host, such as the section Salivaria of Trypanosomatidae.

antibody An immunoglobulin protein, produced by B cells (or plasma cells derived from B cells), that binds with a specific antigen.

antibody titer A measure of the amount of antibody present, usually given in units per ml of serum.

antigen Any substance that will stimulate an immune response.

antigen challenge The dose or inoculation with an antigen given to an animal at some time after primary immunization with that antigen has been achieved.

antigenic determinant The area(s) on an antigen molecule that bind with antibody or specific receptor sites on the sensitized lymphocyte; they "determine" the specificity of the antibody or lymphocyte.

apical organ An organ of unknown function at the apex of a cestode's scolex.

apolysis The disintegration or detachment of a gravid tapeworm proglottid.

ascaridine A protein of unknown function in the sperm of *Ascaris.*

ascaroside A glycoside found in *Ascaris,* made of the sugar *ascarylose* and a series of secondary monol and diol alcohols.

579

ascites Edema, or accumulation of tissue fluid, in the mesenteries and abdominal cavity.

autoinfection Reinfection by a parasite larva without its leaving the host.

autotrophic nutrition Preformed organic molecules are not required as nutritive substances.

axial cells Central cells of a dicyemid mesozoan.

axoneme The core of a cilium or flagellum, comprised of microtubules.

axostyle A tube-like organelle in some flagellate protozoa, extending from the area of the kinetosomes to the posterior end, where it often protrudes.

B cell A type of lymphocyte that, on stimulation with an appropriate antigen, differentiates into a "blast" cell, this giving rise to plasma cells that liberate antibody to the antigen; so-called because in birds they are processed through a lymphoid organ called the bursa of Fabricius; of primary importance in humoral immune response.

bacillary bands Lateral zones in the body wall of some nematodes, consisting of glandular and nonglandular cells of unknown function.

Baer's disc The large, ventral sucker of an aspidogastrean trematode.

ballonets Four inflated areas within the "head" of nematodes of the family Gnathostomatidae. Each is connected to an internal cervical sac of unknown function.

basal body A centriole from which an axoneme arises. Also called a kinetosome or blepharoplast.

basis (basipodite) The joint of a crustacean appendage from which the exopod and the endopod originate, that is, the joint between the coxa and the exopod and endopod.

bilharziasis Schistosomiasis; disease caused by *Schistosoma* spp.

biotic potential The reproductive potential of a species.

biramous appendage An appendage with two main branches from a common basal joint, characteristic of Crustacea, though not all appendages of a crustacean may be biramous.

blackhead A disease of turkeys caused by the protozoan *Histomonas meleagridis*. Also called histomoniasis or infectious enterohepatitis.

blastoderm The "primary epithelium" formed in early embryonic development of many arthropods, when the nuclei migrate to the periphery and undergo superficial cleavage; usually encloses the central yolk mass.

blepharoplast A centriole from which arises an axoneme. Also called a basal body or kinetosome.

bothridium A muscular lappet on the dorsal or ventral side of the scolex of a tapeworm. Bothridia are often highly specialized, with many types of adaptations for adhesion.

bothrium A dorsal or ventral groove, which may be variously modified, on the scolex of a cestode.

bradyzoite A small stage in the Sarcocystidae that develops by endodyogeny in a zoitocyst. Similar to a merozoite in the Eimeriidae.

bubo A swollen lymph node.

bulla A nonliving structure, serves as an anchor and to which the maxillae are permanently attached, secreted by head and maxillary glands of female copepods in the family Lernaeopodidae.

cadre The sclerotized mouth lining of a pentastomid.

calabar swelling A transient subcutaneous nodule, provoked by the filarial nematode *Loa loa*.

calotte The "head" end of a dicyemid mesozoan.

carapace Structure formed by posterior and lateral extension of dorsal sclerites of the head in many Crustacea, usually covering and/or fusing with one or more thoracic somites; considered as arising from a fold of head exoskeleton.

cell-mediated immunity An immunity in which antigen is bound to receptor sites on the surface of sensitized T lymphocytes, such lymphocytes having been produced in response to prior immunizing experience with that antigen, and in which manifestation is via macrophage response with no intervention of antibody.

cellular immune response The binding of antigen with receptor sites on sensitized T lymphocytes to cause release of lymphokines that affect macrophages, a direct response with no intervention of antibody; also, the entire process by which the body responds to an antigen, resulting in a condition of cell-mediated immunity.

centrolecithal egg A type of egg, found in many arthropods, in which the nucleus is located centrally in a small amount of nonyolky cytoplasm, surrounded by a large mass of yolk; after fertilization and some nuclear divisions, the nuclei migrate to the periphery to proceed with superficial cleavage, the yolk remaining central.

cephalogaster A contractile organ in adult epicaridean isopods that functions in sucking blood and perhaps in respiration.

cercaria A larval digenetic trematode, produced by asexual reproduction within a sporocyst or redia.

cercomer A posterior, knob-like attachment on a procercoid or cysticercoid. It usually bears the hooks of the oncosphere.

Chagas' disease A disease of humans and other mammals caused by *Trypanosoma cruzi*.

chagoma A reddish nodule that forms at the site of entrance of *Trypanosoma cruzi* into the skin.

chalimus A specialized, parasitic copepodid, found in the copepod order Caligoida; attached to host by an anterior "frontal filament" that is secreted by the frontal gland.

chitin High molecular weight polymer of N-acetyl glucosamine linked by $1,4$-β-glycosidic bonds.

choanomastigote Like a promastigote but with the flagellum emerging from a collar-like process, as in *Crithidia*.

chromatoid bar Masses of RNA, visible with light microscopy, in young cysts of *Entamoeba* spp.

chyluria Lymph in the urine, characterized by a milky color.

ciliary organelles Organelles of specialized function formed by the fusion of cilia.

cirri Fused tufts of cilia in some protozoa, which function like tiny legs. Also, plural for cirrus.

cirrus A penis or copulatory organ of a flatworm.

clamp A complex set of sclerotized bars, forming a "pinching" organ on the opisthaptor of a monogenetic trematode.

coelozoic Living in the lumen of a hollow organ, such as the intestine.

coenurus A tapeworm larva in the family Taeniidae, in which several scolices bud from an internal germinative membrane, but none of which is enclosed in an internal secondary cyst.

commensalism A kind of symbiosis where one symbiont, the commensal, is benefited, while the other symbiont, the host, is neither harmed nor helped by the association.

complement A collective name for a series of proteins that bind in a complex series of reactions to antibody (either IgM or IgG) when the antibody is itself bound to an antigen; produces lysis of cells if the antibody is bound to antigens on the cell surface.

complement fixation test An immunological method used to detect presence of antibodies that bind (or fix) complement; a standard diagnostic test for many infections.

concomitant immunity Synonym of premunition.

condyles Bearing surfaces between arthropod joints, which provide the fulcra on which the joints move.

conoid A truncated cone of spiral fibrils located within the polar rings of the suborder Eimeriina.

contaminative antigen An antigen borne by the parasite that is common to both the host and the parasite, but which is genetically of host origin.

copepodid Juvenile stage(s) that succeed the naupliar stages in copepods, often quite similar in body form to the adult.

coracidium A larva with a ciliated epithelium, hatching from the egg of certain cestodes. A ciliated oncosphere.

costa The thickened base of an undulating membrane of some flagellate protozoa.

coxa (coxopodite) The most proximal joint of a crustacean appendage.

creeping eruption A skin condition caused by hookworm larvae not able to mature in a given host.

crura The branches of intestine of a flatworm.

cryptoniscus An intermediate, free-swimming larval stage of the isopod suborder Epicaridea, developing after microniscus; attaches to definitive host.

cryptozoite A preerythrocytic schizont of *Plasmodium*.

cypris A postnaupliar larva of barnacles (crustacean subclass Cirripedia) in which the carapace largely envelops the body; so-called because of its resemblance to the ostracod genus *Cypris*.

cystacanth A juvenile acanthocephalan that is infective to its definitive host.

cysticercoid A cestode larva developing from the oncosphere in most Cyclophyllidea. It usually has a "tail" and a well-formed scolex.

cysticercosis Infection with one or more cysticerci.

cystogenic cells Secretory cells in a cercaria that produce a metacercarial cyst.

cytophaneres Fibers radiating out from a zoitocyst into surrounding muscle. Found in some species of Sarcocystidae.

dauer larva A nematode larva with development arrested during unsuitable conditions, to resume development again when conditions improve.

decacanth The ten-hooked larva that hatches from the egg of a cestodarian tapeworm. Also called a lycophore.

definitive host The host in which a parasite achieves sexual maturity. If there is no sexual reproduction in the life of the parasite, the host most important to humans is the definitive host.

deirid A sensory papilla on each side near the anterior end of some nematodes.

delayed hypersensitivity (DH) A manifestation of cell-mediated immunity, distinguished from immediate hypersensitivity in that maximal response is reached about 24 hours or more after intradermal injection of the antigen; lesion site infiltrated primarily by monocytes and macrophages.

denticles (denticulate) Small, tooth-like projections.

deutomerite The posterior half of a cephaline gregarine protozoan.

diapolar cells Ciliated somatodermal cells located between the parapolar and uropolar cells of a mesozoan.

diecdysis Condition in which ecdysis processes are going on continuously, and one ecdysis cycle grades rapidly into another.

dioecious Separate sexes; males and females are different individuals.

diplokarya Having nuclei associated in pairs.

diplostomulum A strigeoid metacercaria in the family Diplostomatidae.

diporpa A larval stage in the life cycle of the monogean *Diplozoon*.

distome A fluke with two suckers: oral and ventral.

dourine A disease of horses and other equids caused by *Trypanosoma equiperdum*.

dyspnea Difficult or labored breathing.

ecdysis Molting or discard of inexpansible portions of cuticle, after which there is an increase in physical dimensions of the animal's body before newly secreted cuticle hardens.

eclipsed antigen An antigen borne by the parasite that is common to both the host and the parasite, but which is genetically of parasite origin.

ectocommensal A commensal symbiont that lives on the outer surface of its host.

ectoparasite A parasite that lives on the outer surface of its host.

ectopic An infection in a location other than normal or expected.

edema The accumulation of more than normal amounts of tissue fluid, or lymph, in the intercellular spaces, resulting in localized swelling of the area.

endite A medial process from the protopod.

endocommensal A commensal symbiont that lives inside its host.

endocytosis The ingestion of a parasite or other particle by a phagocytic host cell, the parasite thereby gaining entry into its host. Also called phagocytosis. Many cells normally feed by this process.

endodyogeny The same as endopolyogony except that only two daughter cells are formed.

endoparasite A parasite that lives inside its host.

endopod (endopodite) The medial branch of a biramous appendage.

endopolyogony The formation of daughter cells, each surrounded by its own membrane, while still in the mother cell.

endosome A nucleolus-like organelle that does not disappear during mitosis.

eosinophilia An elevated eosinophil count in the circulating blood. Commonly associated with chronic parasite infections.

epicaridium The first larval stage of the isopod suborder Epicaridea; attaches to free-living copepod.

epicuticle Thin, outermost layer of arthropod cuticle; contains sclerotin but not chitin.

epidemiology Study concerned with all ecological aspects of disease, in order to explain or anticipate an outbreak of disease, or its transmission, distribution, prevalence, and incidence.

epimastigote Like a promastigote but with a short undulating membrane, such as *Blastocrithidea*.

epimorphosis Arthropod development in which all specifically larval forms are suppressed or passed before hatching; the juvenile that hatches has the adult body form.

epipod (epipodite) A lateral process, from the protopod, usually with one or more joints; may be called an exite.

epizootic A massive infection rate among animals other than humans; identical to an epidemic in humans.

espundia A disease caused by *Leishmania braziliensis*. Also called chiclero ulcer, uta, pian bois, or mucocutaneous leishmaniasis.

eukaryote A cell containing membrane-bound nucleus.

eutely Cell or nuclear constancy; the adult has the same number of nuclei or cells as the larva. Eutely may exist in tissues, organs, or entire animals.

exflagellation The rapid formation of microgametes from a microgametocyte of *Plasmodium* and related genera.

exite A lateral process or joint from the protopod, sometimes referred to as an epipod.

exopod (exopodite) The lateral branch of a biramous appendage.

facette A funnel-shaped opening though the inner membrane complex of the egg of a pentastomid. It receives the product of the dorsal organ.

facultative symbiont When facultative, a symbiont is an opportunist, establishing a relationship with a host only if the opportunity presents itself; it is not physiologically dependent on doing so.

flagellum Applied to recurved process often found on the first two thoracic exopods of branchiuran crustaceans; however, such crustacean structures are not structurally or functionally flagella in the traditional sense.

genital atrium A cavity in the body wall of a flatworm into which male and female genital ducts open.

genitointestinal canal A duct connecting the oviduct and intestine of some polyopisthocotylean monogenea.

gid A disorientation caused by cysticerci in the brain; usually manifested by staggering or whirling.

glial cells Nonnerve cells in a brain or ganglion. Their function is obscure, but they may support the life processes of the neurons.

glycocalyx A finely filamentous layer containing carbohydrate, found on the outer surface of many cells, from 7.5 to 200 nm thick.

gnathocephalon That portion of the arthropod head (subphylum Mandibulata) derived from the first three primitively postoral somites, bearing the mandibles, maxillules, and maxillae in Crustacea.

gnathopod Term applied to prehensile appendages of some Crustacea, for example, the second and third thoracic legs of Amphipoda and the first thoracic legs of some Isopoda.

gonotyl A muscular sucker, or other perigenital specialization, surrounding or associated with the genital atrium of a digenetic trematode.

ground itch A skin rash caused by bacteria introduced by invasive hookworm larvae.

gynandry Maturation first of the female gonads within an individual, then of the male organs. Also called protogyny.

gynocophoral canal A longitudinal groove in the ventral surface of a male schistosome fluke.

halzoun The disease resulting from blockage of the nasopharynx by a parasite.

hamuli Large hooks on the opisthaptor of a monogenetic trematode, referred to as anchors by American authors.

haptens Molecules (usually) of small molecular weight, which are immunogenic only when attached to carrier molecules, usually proteins.

hemoglobinuria Bloody urine.

hepatosplenomegaly A swollen liver and spleen.

hermaphrodite An individual that is monoecious, having the gonads of both sexes in a single individual.

heterogonic life cycle A life cycle involving alterations of parasitic and free-living generations.

heterophile reaction An antigen-antibody reaction, in which the antibody was not specifically elicited by the antigen to which it binds.

heteroxenous Living within more than one host during a parasite's life cycle.

hexacanth An oncosphere. A six-hooked larva hatching from the egg of a eucestode.

histozoic Dwelling within the tissues of a host.

holophytic nutrition The formation of carbohydrates by chloroplasts.

holozoic nutrition Feeding by active ingestion of organisms or particles.

homogonic life cycle A life cycle in which all generations are parasitic or all are free living; there is no (or little) alternation of the two.

homothetogenic fission Mitotic fission across the rows of cilia of a protozoan.

host specificity The degree to which a parasite is able to mature in more than one host species.

humoral immune response The binding of antigen with soluble antibody in blood serum; also, the entire process by which the body responds to an antigen by producing antibody to that antigen.

hydatid cyst A larva of the cyclophyllidean cestode genus *Echinococcus,* with many protoscolices, some budding inside secondary brood cysts.

hydatid sand Free protoscolices forming sediment in a hydatid cyst.

hyperapolysis The detachment of a tapeworm proglottid while still juvenile, before eggs are formed.

hyperparasitism The condition when an organism is a parasite of another parasite.

ick A serious disease of freshwater fishes, caused by the ciliate protozoan *Ichthyophthirius multifiliis.*

icterus (jaundice) Yellowing of the skin and other organs because of bile pigments in the blood.

immediate hypersensitivity A biological manifestation of an antigen-antibody reaction in which the maximal response is reached in a few minutes or hours; intradermal injection of antigen produces local swelling and redness with heavy infiltration of polymorphonuclear leukocytes; intravenous injection may produce anaphylactic shock and death.

immune cross-reaction Binding of an antibody or cell receptor site with an antigen other than the one that would provide an exact "fit," that is, an antigen-antibody reaction in which the antigen is not precisely the same one that stimulated the production of that particular antibody.

immunity State in which a host is more or less resistant to an infective agent, preferably used in reference to resistance arising out of exposure to the agent and consequent stimulation of immune response (either humoral or cell-mediated).

immunogenic Refers to any substance that is antigenic, that is, stimulates production of antibody or cell-mediated immunity.

immunoglobulin Any one of five classes of proteins in blood serum that function as antibodies; abbreviated IgM, IgG, IgA, IgD, and IgE.

incidental parasite Same as an accidental parasite.

infraciliature All of the cilia basal bodies and their associated fibrils in a ciliate protozoan.

infusoriform larva A ciliated larva produced by an infusorigen within a dicyemid mesozoan.

infusorigen A mass of reproductive cells within a rhombogen.

innate immunity Anatomical or physiological features that make the host unsuitable for a certain range of parasites; characteristics with which the host is endowed as a species (according to Sprent [1963]), not evolved as defense mechanisms; nonspecific, as contrasted with responses to specific antigenic stimuli.

intermediate host A host in which a parasite develops to some extent but not to sexual maturity.

intermittent parasite A temporary parasite.

iodinophilous vacuole A vacuole within a protozoan that stains readily with iodine.

isogametes Outwardly similar male and female gametes.

jacket cells The ciliated somatoderm of an orthonectid mesozoan.

kala-azar A disease caused by *Leishmania donovani.* Also called Dum-dum fever or visceral leishmaniasis.

kentrogon A larva in the crustacean class Rhizocephala that is attached to its host crab; formed after the cypris larva molts and its appendages and carapace are discarded.

kinetid The axoneme of a cilium or flagellum together with its basal fibrils and organelles. Also called a mastigont.

kinetodesmose (kinetodesmata) A compound fiber joining cilia into rows.

kinetoplast A conspicuous part of a mitochondrion in a trypanosome, usually found near the blepharoplast.

kinetosome A centriole from which an axoneme arises. Also called a basal body or blepharoplast.

kinety A row of cilia basal bodies and their kinetodesmose. All of the kineties and kinetodesmata in the organism are its infraciliature.

Koch's blue bodies Schizonts of *Theileria parva* in circulating lymphocytes.

Kupffer cells Phagocytic epithelial cells lining the sinusoids of the liver.

labrum The sclerite forming the anterior closure of the mouth in arthropods, specifically, the free lobe overhanging the mouth.

lacunar system A system of canals in the body wall of an acanthocephalan, functioning as a circulatory system.

landscape epidemiology An approach to epidemiology that utilizes all ecological aspects of a nidus. By recognizing certain physical conditions, the epidemiologist can anticipate whether or not a disease can be expected to exist.

larval stem nematogen An early stage in the development of a dicyemid mesozoan.

Laurer's canal A (usually) blind canal extending from the base of the seminal receptacle of a digenetic trematode. It probably represents a vestigial vagina.

Leishman-Donovan body An amastigote in the Trypanosomatidae. Also known as a L-D body.

leishmaniasis An infection by a species of *Leishmania.*

lemniscus A structure occurring in pairs attached to the inner, posterior margin of the neck of an acanthocephalan, extending into the trunk cavity. Its function is unknown.

loculi Shallow, sucker-like depressions in an adhesive organ of a flatworm.

lumen The space within any hollow organ.

lunules Small, sucker-like discs on the anterior margin of some copepods in the family Caligidae, functioning as organs of adhesion.

lycophore The ten-hooked larva that hatches from the egg of a cestodarian tapeworm. Also called a decacanth.

lymph varices Dilated lymph ducts.

lymphadenitis An inflamed lymph node.

lymphokine Any one of several kinds of effector molecules released by T lymphocytes when antigen to which the lymphocyte is sensitized binds to the cell surface.

macrogamete The large, quiescent, "female" anisogamete.

macrogametocyte A cell giving rise to a macrogamete.

macrophage migration inhibitory factor (MIF) A lymphokine released by sensitized lymphocytes that tends to inhibit migration of macrophages in the immediate vicinity, thus contributing to accumulation of larger numbers of macrophages close to the site of MIF release.

mamelon A ventral, serrated projection on the ventral surface of a male nematode of the family Syphaciidae. Its function is unknown.

mandibles Third pair of appendages in Crustacea (second in Insecta), primarily feeding in function, derived from appendages on primitive fourth (first postoral) somite.

marginal bodies Sensory pits or short tentacles between the marginal loculi of the opisthaptor of an aspidogastrean trematode.

marrara Nasopharyngeal blockage by a parasite. Also called halzoun.

mastigont The axoneme of a cilium or flagellum together with its basal fibrils and organelles. Also called a kinetid.

Maurer's clefts Blotches on the surface of an erythrocyte infected with *Plasmodium falciparum*.

maxillae (second maxillae) Fifth pair of appendages in Crustacea, primarily feeding in function, derived from appendages on primitive sixth (third postoral) somite; homologous to labium in insects.

maxillipeds One or more pairs of head appendages originating posterior to maxillae in Crustacea; derived from appendages on somites that were primitively posterior to gnathocephalon; usually function in feeding but sometimes adapted for other functions, such as prehension, in parasitic forms.

maxillules (first maxillae) Fourth pair of appendages in Crustacea, primarily feeding in function; derived from appendages on primitive fifth (second postoral) somite; homologous to maxillae in insects.

megacolon A flabby, distended colon caused by chronic Chagas' disease.

megaesophagus A distended esophagus caused by chronic Chagas' disease.

Mehlis's glands Unicellular mucous and serous glands surrounding the ootype of a flatworm.

membranelle Short, trnasverse rows of cilia, fused at their bases, serving to move food particles toward the oral groove of a protozoan.

merozoite A daughter cell resulting from schizogony.

mesocercaria A larval stage of the digenetic trematode *Alaria*. It is an unencysted form between the cercaria and the metacercaria.

metacercaria A stage in the life cycle of most digenetic trematodes, between the cercaria and adult. Usually encysted and quiescent.

metacestode A larval cestode that is infective to its definitive host.

metacryptozoite A merozoite developed from a cryptozoite.

metacyclic The stage in the life cycle of a parasite that is infective to its definitive host.

metacyst The cystic stage of a parasite that is infective to a host.

metamere One of the segments in a metameric animal.

metamerism The division of the body along the anteroposterior axis into a serial succession of segments, each of which contains identical or similar representatives of all the organ-systems of the body; primitively in arthropods, including, externally, a pair of appendages and, internally, a pair of nerve ganglia, a pair of nephridia, a pair of gonads, paired blood vessels and nerves, and a portion of the digestive and muscular systems.

metamorphosis A type of development in which one or more juvenile types differ markedly in body form from the adult; occurs in numerous animal phyla. The term also is applied to the actual process of changing from larval to adult form.

metanauplius Term applied to later naupliar larvae of some crustaceans, that is, after several naupliar stages but before another larval type or preadult in developmental sequence.

metapolar cells The posterior tier of cells in the calotte of a dicyemid mesozoan.

metasome The portion of the body anterior to the major point of body flexion in many copepods; usually includes cephalothorax and several free thoracic segments.

metraterm The muscular, distended termination of the uterus of a digenetic trematode.

microfilaria The first-stage larva of any filariid nematode that is ovoviviparous; usually found in the blood or tissue fluids of the definitive host.

microgamete A slender, active, "male" anisogamete.

microgametocyte A cell giving rise to microgametes.

micronemes Slender, convoluted bodies that join a duct system with the rhoptries, opening at the tip of a sporozoite or merozoite.

microniscus Intermediate larval stages of the isopod suborder Epicaridea, parasitic on free-living copepods.

micropredator Temporary parasite.

micropyle A tiny pore in the membrane of some protozoans, such as *Plasmodium* and *Eimeria*. Also, a pore in the oocyst of some coccidians and in the egg of an insect.

microthrix (microtriches) Minute projections of the tegument of a cestode.

monoecious Hermaphroditic; one individual contains reproductive systems of both sexes.

monostome A fluke lacking a ventral sucker.

monoxenous Living within a single host during a parasite's life cycle.

monozoic A tapeworm whose "strobila" consists of a single unit.

mucron An apical anchoring device on an acephaline gregarine protozoan.

mutualism A type of symbiosis in which both host and symbiont benefit from the association.

myiasis Infection by fly maggots.

nagana A disease of ruminants caused by *Trypanosoma brucei brucei* or *T. congolense*.

natural immunity Physiological mechanisms evolved for the purpose of defense, possessed to some degree by all animals (according to Sprent, 1963). Though nonspecific in themselves, these mechanisms often depend on or are enhanced by specific immune responses.

nauplius Typically the earliest larval stage(s) of crustaceans; has only three pairs of appendages: antennules, antennae, and mandibles—all primarily of locomotive function.

neascus A strigeoid metacercaria with a spoon-shaped forebody.

nidus The specific locality where a given disease exists. A nidus is the result of a unique combination of ecological factors that favors the maintenance and transmission of the disease organism.

obligate symbiont An organism that is physiologically dependent on establishing a symbiotic relationship with another.

onchocercoma A subcutaneous nodule containing masses of the nematode *Onchocerca volvulus*.

oncomiracidium A ciliated larva of a monogenetic trematode.

oocyst The cystic form in the Apicomplexa, resulting from sporogony. The oocyst may be covered by a hard, resistant membrane (as in *Eimeria*) or it may not (as in *Plasmodium*).

oocyst residuum Cytoplasmic material not incorporated into the sporocyst within an oocyst. Seen as an amorphous mass with an oocyst.

oogenotop The female genital complex of a flatworm, including oviduct, ootype, Mehlis's glands, common vitelline duct, and upper uterus.

ookinete The motile, elongage zygote of a *Plasmodium* or related organism.

oostegites Modified thoracic epipods in females of the crustacean superorder Peracarida; they form a pouch for brooding embryos.

ootype An expansion of the flatworm female duct, surrounded by Mehlis's glands, where, in some flatworms, ducts from a seminal receptacle and vitelline reservoir join.

operculum A lid-like specialization of a parasite eggshell through which the larva escapes.

opisthaptor The posterior attachment organ of a monogenetic trematode.

opisthomastigote A form of Trypanosomatidae with the kinetoplast at the posterior end. The flagellum runs through a long reservoir to emerge at the anterior; no undulating membrane. For example, *Herpetomonas*.

opsonization Modification of the surface characteristics of an invading particle or organism by binding with antibody or a nonspecific molecule in such a manner as to facilitate phagocytosis by host cells.

orchitis Inflammation of the testis.

oriental sore A disease caused by *Leishmania tropica*. Also called Jericho boil, Delhi boil, Aleppo boil, or cutaneous leishmaniasis.

ovicapt A sphincter on the oviduct of a flatworm.

ovisac External sac attached to the somite that bears openings of gonoducts in females of many Copepoda. Fertilized eggs pass into the ovisacs for embryonation.

ovovitellarium A mixed mass of ova and vitelline cells. Found in the monogenean genus *Gyrodactylus* and in a few tapeworms.

pansporoblast A myxosporidean sporoblast that gives rise to more than one spore. Also called a sporoblast mother cell.

parabasal body A Golgi body located near the basal body of some flagellate protozoa, from which the parabasal filament runs to the basal body.

parabasal filament A fibril with periodicity visible in electron micrographs, that courses between the parabasal body and a kinetosome.

parapolar cells Cells making up the ciliated somatoderm immediately behind the calotte of a mesozoan.

parasite The raison d'etre for parasitologists.

parasitic castration A condition in which a parasite causes retardation in development or atrophy of host gonads, often accompanied by failure of secondary sexual characteristics to develop.

parasitism Symbiosis in which the symbiont benefits from the association, while the host is harmed in some way.

parasitologist A quaint person who seeks truth in strange places.

parasitophorous vacuole A vacuole within a host cell that contains a parasite.

paratenic host A host in which a parasite survives without undergoing further development. Also known as a transport host.

pars prostatica A dilation of the ejaculatory duct of

a flatworm, surrounded by unicellular prostate cells.

parthenogenesis The development of an unfertilized egg into a new individual.

paruterine organ A fibromuscular organ in some cestodes that replaces the uterus.

passive immunization An immune state in an animal created by inoculation with serum (containing antibodies) or lymphocytes from an immune animal, rather than by exposure to the antigen.

pathogenesis The production and development of disease.

pereiopods Thoracic appendages of Crustacea.

perikaryon (perikarya) That portion of the cell that contains the nucleus (karyon), sometimes called the cell body, used in reference to cells that have processes extending some distance away from the area of the nucleus, for example, nerve axons or tegumental cells of cestodes and trematodes.

peritrophic membrane A noncellular, delicate membrane lining an insect's midgut.

permanent parasite A parasite that lives its entire adult life within or on a host.

Peyer's patches Lymphoid tissue in the wall of the intestine; not circumscribed by a tissue capsule.

phagocytosis The active engulfment of a particle by a cell.

phasmid A sensory pit on each side near the end of the tail of nematodes of the class Phasmidea.

phoresis A form of symbiosis when the symbiont, the phoront, is mechanically carried about by its host. Neither is physiologically dependent on the other.

pipestem fibrosis Thickening of the walls of a bile duct as the result of the irritating presence of a parasite.

piroplasm Any of the class Piroplasmea, while in a circulating erythrocyte.

plasmotomy The division of a multinucleated cell into multinucleated daughter cells, without accompanying mitosis.

pleopods Abdominal appendages of Crustacea.

plerocercoid A cestode larva that develops from a procercoid. It usually shows little differentiation.

plerocercus A tapeworm larva in the order Trypanorhyncha in which the posterior forms a bladder, the blastocyst, into which the rest of the body withdraws.

polar granule A refractile granule within a coccidian oocyst.

polar ring Electron-dense organelles of unknown function, located under the cell membrane at the anterior tip of sporozoites and merozoites.

polaroplast An organelle, apparently a vacuole, near the polar filament of a microsporidean.

polyembryony The development of a single zygote into more than one offspring.

polyzoic A strobila, when consisting of more than one proglottid.

posterior station The development of a protozoan in the hindgut of its insect host, such as in the section Stercoraria of the Trypanosomatidae.

praesoma The proboscis, neck, and attached muscles and organs of an acanthocephalan.

praniza The parasitic larva of the isopod suborder Gnathiidea. It parasitizes fish and feeds on blood.

predation A short-term symbiosis in which the predator kills the prey outright; it does not subsist on the prey while it is alive.

premunition A resistance to reinfection or superinfection conferred by a still existing infection, but which does not destroy the organisms of the infection already present.

primite The anterior member of a pair of gregarines in syzygy.

procephalon The portion of the arthropod head (in the subphylum Mandibulata) that has been derived from the first three primitively preoral somites, bearing the antennules and antennae in Crustacea.

procercoid A cestode larva developing from a coracidium in some orders. It usually has a posterior cercomer.

procuticle The thicker layer beneath the epicuticle of arthropods, which lends mass and strength to the cuticle; it contains chitin, sclerotin, and also inorganic salts in Crustacea. Layers within procuticle vary in structure and composition.

proglottid One segment of a tapeworm strobila.

prohaptor The collective adhesive and feeding organs at the anterior end of a monogenetic trematode.

prokaryote A cell in which the chromosomes are not contained within a membrane-bound nucleus.

promastigote A form of Trypanosomatidae with the free flagellum anterior and the kinetoplast anterior to the nucleus, as in *Leptomonas*.

propolar cells The anterior tier of cells in the calotte of a dicyemid mesozoan.

protandry Maturation first of the male gonads, then of the female organs, within a hermaphroditic individual. Also called androgyny.

protomerite The anterior half of a cephaline gregarine protozoan.

protopod (protopodite) The coxa and basis taken together.

protoscolex The juvenile scolex budded within a coenurus or hydatid larva of a taeniid cestode.

pseudocyst Pocket of protozoans within a host cell but not surrounded by cyst wall of parasite origin.

pseudolabia Bilateral lips around the mouth of many nematodes of the order Spirurata. They are not homologous to the lips of most other nematodes, but develop from the inner wall of the buccal cavity.

quartan malaria Malaria with fevers recurring every 72 hours. Caused by *Plasmodium malariae*.

quotidian malaria Malaria with fevers recurring every 24 hours. Found in cases of overlapping infections.

rachis A central, longitudinal, supporting structure in the ovary of some nematodes.

redia A larval, digenetic trematode, produced by asexual reproduction within a miracidium, sporocyst, or mother redia.

reservoir A living or (rarely) nonliving source of infection. For instance, domestic cats are reservoirs of *Toxoplasma gondii.*

reticuloendothelial system The total complement of fixed macrophages in the body, especially reticular connective tissue and the lining epithelium of the blood vascular system. Some authorities also include the phagocytic white blood cells.

retrofection A process of reinfection, whereby larval nematodes hatch on the skin and reenter the body before molting to third-stage larvae.

rhombogen A stage in the life cycle of a dicyemid mesozoan.

rhoptries Elongate, electron-dense bodies extending within the polar rings of an apicomplexan.

Romaña's sign Symptoms of recent infection by *Trypanosoma cruzi,* consisting of edema of the orbit and swelling of the preauricular lymph node.

Romanovsky stain A complex stain, based on methylene blue and eosin, used to stain blood cells and hemoparasites. Wright's and Giemsa's stains are two common examples.

ruffles Slender projections of the exterior surface of a dicyemid mesozoan.

Saefftigen's pouch An internal, muscular sac near the posterior end of a male acanthocephalan. It contains fluid that aids in manipulating the copulatory bursa.

saprozoic nutrition The absorption of nutrients through a cell membrane.

sarcocystin A powerful toxin produced by zoitocysts of *Sarcocystis.*

satellite The posterior member of a pair of gregarines in syzygy.

schistosomule A juvenile stage of a blood fluke, between a cercaria and an adult. It is a migrating form taking the place of a metacercaria in the life cycle.

schizogony A form of asexual reproduction in which multiple mitoses take place, followed by simultaneous cytokineses, resulting in many daughter cells at once.

schizont A cell undergoing schizogony, in which nuclear divisions have occurred but cytokinesis is not completed. In its late phase, sometimes called a segmenter.

Schüffner's dots Stippling that appears on the membrane of an erythrocyte infected with *Plasmodium vivax.*

sclerite Any well-defined, sclerotized area of arthropod cuticle limited by suture lines or flexible, membranous portions of cuticle.

sclerotin A highly resistant and insoluble protein occurring in the cuticle of arthropods, also thought to occur in structures secreted by various other animals, such as in the eggshells of some trematodes, in which stabilization of the protein is achieved by orthoquinone cross links between free imino or amino groups of the protein molecules.

scolex The "head" or holdfast organ of a tapeworm.

scoliosis Lateral curvature of the spine.

slime ball A mass of mucus-covered cercariae of dicrocoeliid flukes, released from land snails.

somite A body segment or metamere, a term usually used in reference to arthropods.

sparganum A cestode plerocercoid of unknown identity.

spondylosis Degeneration of a vertebra.

sporadin A mature trophozoite of a gregarine protozoan.

sporoblast The cell mass that will differentiate into a sporocyst within an oocyst.

sporocyst A stage of development of a sporozoan protozoan, usually within an enclosing membrane, the oocyst. Also, an asexual stage of development in some trematodes.

sporocyst residuum Cytoplasmic material "left over" within a sporocyst after sporozoite formation. Seen as an amorphous mass.

sporogony Multiple fission of a zygote. Such a cell is also called a sporont.

sporont The undifferentiated cell mass within an unsporulated oocyst.

sporoplasm The ameba-like portion of a cnidosporan cyst that is infective to the next host.

sporozoite A daughter cell resulting from sporogony.

spring dwindling A disease of honeybees caused by the microsporan protozoan *Nosema apis.* Also called nosema disease, bee dysentery, bee sickness, and May sickness.

stichosome A column of large, rectangular cells called stichocytes, supporting and secreting into most of the esophagus of nematodes of the family Trichuridae.

Stieda body A plug in the inner wall of one end of a coccidian oocyst.

stigma An operculum-like area of an eggshell through which the miracidium of a schistosome fluke hatches.

strobilization The formation of a chain of zoids by budding, as in the strobila of a tapeworm.

strobilocercoid A cysticercoid that undergoes some strobilization; found only in *Schistotaenia.*

strobilocercus A simple cysticercus with some evident strobilization.

substiedal body Additional plug material underlying a Stieda body.

surra A disease of large mammals caused by *Trypanosoma evansi.*

swarmer Daughter trophozoites resulting from multiple transverse fissions of *Ichthyophthirius multifiliis* and a few other protozoans.

sylvatic disease A disease existing normally in the wild, not in the human environment.

symbiology The study of symbioses.

symbiont Any organism involved in a symbiotic relationship with another organism, the host.

symbiosis An interaction between two organisms, restricted in meaning by some scientists to interactions in which one organism lives in or on the body of another.

symmetrogenic fission Mitotic fission between the rows of flagella of protozoa.

syzygy A stage during sexual reproduction of some gregarines in which two or more sporadins connect end to end.

T cell A type of lymphocyte that, on stimulation with an appropriate antigen, gives rise to a population of sensitized lymphocytes with receptor sites for that antigen on their surface membranes; so-called because they are processed through the thymus; of primary importance in cell-mediated immunity.

tachyzoite Small, merozoite-like stages of *Toxoplasma*. They develop in the host cell's parasitophorous vacuole by endodyogeny.

temporary parasite A parasite that contacts its host only to feed and then leaves. Also called an intermittent parasite or micropredator.

tertian malaria Malaria with fevers recurring every 48 hours. Caused by *Plasmodium vivax, P. ovale,* and *P. falciparum.*

tetracotyle A strigeoid metacercaria in the family Strigeidae.

tetrathyridium The only postovic larval form known in the tapeworm cyclophyllidean genus *Mesocestoides.* A large, solid-bodied cysticercoid.

theileriosis A disease of cattle and other ruminants, caused by *Theileria parva.* Also called east coast fever.

thrombus A blood clot in a blood vessel or in one of the cavities of the heart.

trabecula In general anatomical usage, a septum extending from an envelope through enclosed substance, which, together with other trabeculae, forms part of the framework of various organs; here referring specifically to the cell processes connecting the perikarya of cestode and trematode tegumental cells with the distal cytoplasm.

transport host A paratenic host.

tribocytic organ A glandular, pad-like organ behind the acetabulum of a strigeoid trematode.

trophozoite The active, feeding stage of a protozoan, as contrasted by a cyst. Also called the vegetative stage.

trypomastigote A form of Trypanosomatidae with an undulating membrane and the kinetoplast located posterior to the nucleus; for example, *Trypanosoma.*

undulating membrane A name applied to two quite different structures in Protozoa. In some Mastigophora it is a fin-like ridge across the surface of a cell, with the axoneme of a flagellum near its surface. In some ciliates it is a line of cilia that are fused at their bases, usually beating to force food particles toward the gullet.

undulating ridges Undulatory waves in the surface of some protozoa, probably aided by subpellicular microtubules. The means of locomotion in some species.

uniramous appendage An arthropod appendage that is unbranched, characteristic of living arthropods other than Crustacea, though some crustacean appendages are uniramous.

urban disease A disease peculiar to the human environment, as contrasted with a disease normally in wild animals.

urn A region near the center of an infusoriform larva of a dicyemid mesozoan.

uropolar cells Somatoderm cells at the posterior end of the trunk of a dicyemid mesozoan.

urosome Portion of the body posterior to the major point of body flexion in many copepods; usually includes one or more free thoracic segments and abdomen.

vector Any agent such as water, wind, or insect, that transmits a disease organism.

vermicle The infective stage of *Babesia* in a tick.

verminous intoxication A variable condition of systemic poisoning caused by absorbed metabolites produced by parasites.

vesicular disease Any disease of the urinary bladder, such as vesicular schistosomiasis.

whirling disease A disease of fish, caused by the protozoan *Myxosoma cerebralis.*

Winterbottom's sign Swollen lymph nodes at the base of the skull, symptomatic of African sleeping sickness.

xenodiagnosis Diagnosing a disease by infecting a test animal.

xiphidiocercaria A cercaria with a stylet in the anterior rim of its oral sucker.

zoid An individual member of a colonial organism.

zoitocyst A tissue phase in the Sarcocystidae. They usually have internal septa and contain thousands of bradyzoites. Also called sarcocyst or Miescher's tubule.

zoonosis A disease of animals that is transmissible to humans. Some authors subdivide the concept into zooanthroponosis, infections humans can acquire from animals, and anthropozoonosis, a disease of humans transmissible to other animals.

INDEX

Italicized page numbers indicate illustrations.